PHYSICAL CONSTANTS

Quantity	SI	US Customary
acceleration of gravity (g)	9.80665 m/s^2	32.1740 ft/s^2
density, unit weight of water (at $4°C = 39.2°F$)	1000 kg/m^3	62.43 pcf
normal atmospheric pressure	101.325 kPa	14.6960 psi

GREEK ALPHABET

A	α	Alpha	N	ν	Nu
B	β	Beta	Ξ	ξ	Xi
Γ	γ	Gamma	O	o	Omicron
Δ	δ	Delta	Π	π	Pi
E	ϵ	Epsilon	P	ρ	Rho
Z	ζ	Zeta	Σ	σ	Sigma
H	η	Eta	T	τ	Tau
Θ	θ	Theta	Υ	υ	Upsilon
I	ι	Iota	Φ	ϕ	Phi
K	κ	Kappa	X	χ	Chi
Λ	λ	Lambda	Ψ	ψ	Psi
M	μ	Mu	Ω	ω	Omega

ADVANCED
MECHANICS
OF
MATERIALS

Fifth Edition

ADVANCED MECHANICS OF MATERIALS

ARTHUR P. BORESI

Professor and Head
Civil and Architectural Engineering
The University of Wyoming at Laramie

RICHARD J. SCHMIDT

Associate Professor
Civil and Architectural Engineering
The University of Wyoming at Laramie

OMAR M. SIDEBOTTOM

Professor Emeritus
Theoretical and Applied Mechanics
The University of Illinois at Champaign-Urbana

JOHN WILEY & SONS, INC.

New York   Chichester   Brisbane   Toronto   Singapore

Recognizing the importance of preserving what has been written,
it is a policy of John Wiley & Sons, Inc. to have books of enduring
value published in the United States printed on acid-free paper, and we
exert our best efforts to that end.

ACQUISITIONS EDITOR   Charity Robey
MARKETING MANAGER   Susan Elbe
PRODUCTION SUPERVISOR   Nancy Prinz
DESIGNER   Kevin Murphy
MANUFACTURING MANAGER   Andrea Price
COPY EDITING SUPERVISOR   Deborah Herbert

This book was set in Times Roman by Polyglot Compositors and
printed and bound by Hamilton Printing. The cover was printed
by Hamilton Printing.

Library of Congress Cataloging in Publication Data:

Boresi, Arthur P. (Arthur Peter), 1924–
    Advanced mechanics of materials/Arthur P. Boresi, Richard J.
Schmidt, Omar M. Sidebottom.
        p.   cm.
    Includes bibliographical references and index.
    ISBN 0-471-55157-0
    1. Strength of materials.   I. Schmidt, Richard J. (Richard Joseph),
1954– . II. Sidebottom, Omar M. (Omar Marion)   III. Title.
TA405.B66   1993
620.1′12—dc20   92-30349
                                                      CIP

Printed in the United States of America

10 9 8 7

To Our Students
and
To Our Wives
Jean, Pat, and Charlotte

PREFACE

This fifth edition represents a major revision of the fourth edition. However, as in previous editions, the blend of analysis, qualified approximations, and judgements based on practical experience is maintained. Each topic is developed from basic principles so that the applicability and limitations of the methods employed are clear. Introductory statements in each chapter serve as guidelines for the reader to the topics that are discussed. The topics are divided into three major parts: Part I—Fundamental Concepts; Part II—Classical Topics in Advanced Mechanics; and Part III—Selected Advanced Topics.

Part I, Chapters 1–5, includes topics from elasticity, plasticity, and energy methods that are important in the remainder of the book. In Chapter 1, the role and the limits of design are discussed. Basic concepts of one-dimensional load-stress, load-deflection, and stress-strain diagrams are introduced. A discussion of the tension test and associated material properties is presented, followed by an introduction to failure theories. These concepts are followed, in Chapter 2, by the theories of stress and strain, and by strain measurements (strain rosettes) and, in Chapter 3, by the theory of linear stress-strain-temperature relations. The discussion of anisotropic materials has been expanded, and example problems on orthotropic material behavior are given. Student problems for anisotropic materials are also included. Chapter 4 contains much new material related to inelastic (nonlinear) behavior and a broader treatment of yield criteria, including elastic-plastic behavior of beams, strain-hardening effects in bars, and residual stresses in elastic-plastic bars after unloading. The application of energy methods, Chapter 5, is expanded to include an in-depth discussion of the dummy-load method used by structural engineers and its relation to the Castigliano method. Additional worked examples and many new problems have been added. (In this edition, problems have been placed at the end of each Chapter, rather than at the end of each section.)

Part II, Chapters 6–12, treats some classical topics of advanced mechanics. Torsion is treated in Chapter 6, including new examples and problems. In addition, a finite difference solution of the rectangular cross section bar is presented. An example of limit analysis and residual stresses in a circular cross section shaft is also included. In Chapters 7 to 9, the three topics of unsymmetrical bending, shear center, and curved beams are examined on a rigorous basis, and limitations on existing analyses are indicated. A presentation of beams on elastic foundations, plus new problems and references, is given in Chapter 10. Some minor clarifications for the thick-wall cylinder and many new student problems are given in Chapter 11. In Chapter 12, the topic of stability of columns is expanded considerably, and a wide range of practical example problems and student exercises is included.

Part III, Chapters 13–19, presents the more advanced topics of flat plates, stress concentrations, fracture mechanics, fatigue, creep (time-dependent deformations), contact stresses, and the finite element method. The linear theory of flat plates is given in Chapter 13, including some illustrative problems and a collection of student exercises. The level is appropriate as an introduction for master-level students and for practicing engineers. Chapter 14 collects, in an integrated manner, material on stress concentrations previously presented in parts of Chapters 3, 12, and 13 of the fourth edition. New examples and exercise problems have been added, as well as some new charts of stress concentration factors for rectangular cross section beams. The topic of fracture mechanics is introduced in Chapter 15; it includes material previously given in Chapters 3 and 12 of the fourth edition and a brief discussion of other factors, such as elastic-plastic fracture, crack-growth analysis, load spectra and stress history, testing, and experimental data interpretation. A number of up-to-date books and papers are referenced. Progressive fracture (fatigue) is discussed in Chapter 16, including additional problems and references. An extended discussion of creep is presented in Chapter 17, including creep of metals and nonmetals (concrete, asphalt, and wood). Chapter 18, contact stresses, is essentially unchanged from Chapter 14 of the fourth edition. Chapter 19, the finite element method, is a completely rewritten treatment of Chapter 15 of the fourth edition. It includes discussions of the constant strain triangular element, the bilinear rectangular element, the linear isoparametric quadrilateral element, and the plane frame element. Example problems and exercise problems are included.

As a result of the new material and problems that have been added, this edition is larger than its predecessors. Consequently, it provides a greater choice of topics for study. It also has the advantage that the book can be used over a lifetime of practice, as a reference to topics of lasting importance in engineering. The book contains more material than can be covered in a one-quarter or a one-semester course. It is, however, with the proper selection of topics, suitable for a one-semester (one-quarter) course at either the senior level or the first-semester graduate level, for a two-semester (two- or three-quarter) course sequence, or as a reference work in several courses in mechanics.

The computer program listings in the fourth edition have been omitted from the current edition. However, revised versions of the programs from the fourth edition and new programs for applications in this edition are available on request from one of the authors (R. J. Schmidt, Department of Civil and Architectural Engineering, Box 3295, University of Wyoming, Laramie, WY 82071).

We thank Charity Robey, Wiley engineering editor, for her expert help and advice during the development of this edition. We also greatly appreciate the help of Suzanne Ingrao, with the difficult task of galley and page proof editing. We thank the reviewers of the preliminary format and content of the fifth edition for their constructive criticism and suggestions for improving the fourth edition. These reviewers are Stanley Chen, Arizona State University; Donald DaDeppo, University of Arizona; D. W. Haines, Manhattan College; Loren D. Lutes, Texas A. & M. University; Esmet M. Kamil, Pratt Institute; and Thomas A. Lenox, U.S. Military Academy, West Point, NY. We thank especially the reviewers of the draft manuscript for their helpful suggestions. These reviewers are J. A. M. Boulet, University of Tennessee-Knoxville; Ray W. James, Texas A. & M. University; A. P. Moser, Utah State University; William A. Nash, University of Massachusetts-Amherst; and Sam Y. Zamrik, Pennsylvania State University. We also acknowledge the contribution of Travis Finch for the artwork in Figs. 12.12 and 18.1.

Finally, we welcome comments, suggestions, questions, and corrections from the reader. They may be sent to Arthur P. Boresi, Department of Civil and Architectural Engineering, Box 3295, University of Wyoming, Laramie, WY 82071.

ARTHUR P. BORESI
RICHARD J. SCHMIDT
OMAR M. SIDEBOTTOM

October 1992

CONTENTS

PART III

SELECTED ADVANCED TOPICS   **509**

PART I

FUNDAMENTAL CONCEPTS

In Part I of this book, Chapters 1 to 5, we introduce and develop fundamental topics that are important in the remainder of the book. In Chapter 1, we emphasize basic material properties and their use in design. Theories of stress and strain are presented in Chapter 2, and linear stress-strain-temperature relations are introduced in Chapter 3. Inelastic material behavior is discussed in Chapter 4 and finally, energy methods are treated in Chapter 5.

1

INTRODUCTION

In this chapter, we present general concepts and definitions that are fundamental to many of the topics discussed in this book. The chapter serves also as a brief guide and introduction to the remainder of the book. The reader may find it fruitful to refer to this chapter, from time to time, in conjunction with the study of topics in other chapters.

1.1

THE ROLE OF DESIGN

This book emphasizes the methods of mechanics of materials and applications to the analysis and design of components of structural/machine systems. As such, it is directed to aeronautical, civil, mechanical, and nuclear engineers, as well as to specialists in the field of theoretical and applied mechanics. As engineers, we are problem solvers. The problems that we solve encompass practically all fields of human activity. We solve problems related to buildings, transportation (including automotive, rail, water, air and outer-space travel), water systems (e.g., dams and pipelines), manufacturing, specialized medical equipment, communication systems, computers, hazardous wastes, etc. These problems are generally encountered in the design, manufacture, and construction of engineering systems. Ordinarily, these systems are not built or manufactured before the design process is completed. The design process usually involves the development of many drawings and/or CAD files to describe the final system. One of the major purposes of the design process is to analyze or evaluate various design alternatives before a final design is selected. One of the simplest objectives of the analysis is to ensure that all components of the system will fit together and function properly. More complicated analysis involves the evaluation of forces in the proposed design to ensure that each component of the system functions properly (for instance, safely withstands loads or does not undergo excessive displacements). This analysis is essential in the process of refining the design to meet required conditions such as adequate strength, minimum weight, and minimum cost of production.

The process of refining the design can be very complicated and extremely time-consuming. For example, consider the design of a space vehicle, such as the shuttle. After the shuttle's mission or use has been established, the designer must decide on the shape of the vehicle and the materials to be used. The designer must analyze the vehicle's structure to determine if it is strong and stiff enough to withstand the aerodynamic and thermal loads to which it will be subjected. The designer must analyze the skin and individual component parts of the structure to determine how these loads will be carried and safely transmitted from part to part. This first analysis usually reveals evidence that a redesign of some members in the structure may provide a more efficient and safer distribution of load and perhaps a more cost-effective design. Unfortunately, the designer may also discover that improvements in one part of the system may require changes in another part and possible problems in still other parts. Thus, the designer may be faced with one or more iterations between analysis and design to ensure that the entire system will function properly. This type of iteration is a common feature of design (Cross, 1989; de Neufville, 1990).

Considerations other than resistance to and transfer of loads, such as those of form or appearance, cost, ease of manufacturing, time constraints, etc., may influence or even control the design. Indeed, these factors may not only govern the design of an individual component but also may have a strong influence on the design of a more general engineering system, such as an office building. However, considerations of this kind are secondary to the topics treated in this book.

The term *design* as used in this book is not limited to the detailed calculations required to determine the proper dimensions of a member; rather, this term is used in a broader sense that emphasizes the relation of the methods of mechanics of materials to the concepts and philosophy of a rational design code or specification. In particular, emphasis is placed on the development of equations, formulas, or methods by which detailed analyses can be performed. Thus, this text provides an analytical foundation that is fundamental to the design process. Readers interested in the general concepts and methods of design may refer to the books by Cross (1989) and de Neufville (1990).

1.2

TOPICS TREATED IN THIS BOOK

This book is intended for advanced undergraduate and graduate engineering students, as well as practicing engineers. The topics treated are separated into three groups: Part I, Fundamental Concepts; Part II, Classical Topics in Advanced Mechanics of Materials; Part III, Selected Advanced Topics. Part I treats general concepts that pertain to the entire book, theories of stress and strain, linear stress-strain-temperature relations, yield criteria for multiaxial stress states, and energy methods. These topics are intended to be read sequentially, more or less. However, depending on the background of the reader, some of these topics may be bypassed. Part II presents several chapters on classical applications of the methods of mechanics of materials, namely, torsion, nonsymmetrical bending of beams, shear center for thin-wall beam cross sections, curved beams, beams on elastic foundations, thick-wall cylinders, and buckling of columns. These chapters may be treated in any order, except that the chapter on shear centers should be studied after the chapters on torsion and nonsymmetrical bending of beams. Part III introduces chapters on selected advanced topics, namely, flat plates, stress concentration factors, contact stresses, fracture mechanics, high cycle fatigue, time-dependent deformation/creep, and finite element methods. Each of these chapters may be treated independently, more or less.

1.3

LOAD-STRESS AND LOAD-DEFLECTION RELATIONS

For most of the members considered in this book we derive relations, in terms of known loads and known dimensions of the member, for either the distributions of normal and shear stresses on a cross section of the member or for stress components that act at a point in the member. For a given member subjected to prescribed loads, the derivation of load-stress relations depends on satisfaction of the following requirements.

1. The equations of equilibrium (or equations of motion for bodies not in equilibrium)
2. The compatibility conditions (continuity conditions) that require deformed volume elements in the member to fit together without overlap or tearing
3. The constitutive relations

Two different methods are used to satisfy requirements 1 and 2: the method of mechanics of materials and the method of general continuum mechanics. Often, load-stress and load-deflection relations have not been derived in this book by general continuum mechanics methods, either because the beginning student does not have the necessary background or because of the complexity of the general solutions. Instead, the method of mechanics of materials is used to obtain either exact solutions or reliable approximate solutions. In the method of mechanics of materials, the load-stress relations are derived first. They are then used to obtain load-deflection relations for the member.

A simple member such as a circular shaft of uniform cross section may be subjected to complex loads that produce a multiaxial state of stress in the shaft. However, such complex loads can be reduced to several simple types of load, such as axial, bending, and torsion. Each type of load, when acting alone, produces mainly one stress component, which is distributed over the cross section of the shaft. The method of mechanics of materials can be used to obtain load-stress relations for each type of load. If the deformations of the shaft that result from one type of load do not influence the magnitudes of the other types of loads and if the material remains linearly elastic for the combined loads, the stress components due to each type of load can be added together (i.e., the method of superposition may be used).

In a complex member, each load may have a significant influence on each component of the state of stress. Then, the method of mechanics of materials becomes cumbersome, and the use of the method of continuum mechanics may be more appropriate.

Method of Mechanics of Materials

The method of mechanics of materials is based on simplified assumptions related to the geometry of deformation (requirement 2) so that strain distributions for a cross section of the member can be determined. A basic assumption is that plane sections before loading remain plane after loading. The assumption can be shown to be exact for axially loaded members of uniform cross sections, for slender straight torsion members having uniform circular cross sections, and for slender straight beams of uniform cross sections subjected to pure bending. The assumption is approximate for other beam problems. The method of mechanics of materials is used in this book to treat several advanced beam topics (Chapters 7 to 10). In a similar way, we assume that lines normal to the middle surface of an undeformed plate remain straight and normal to the middle surface after the load is applied. This assumption is used to simplify the plate problem in Chapter 13.

We review the steps used in the derivation of the flexure formula to illustrate the method of mechanics of materials and to show how the three requirements listed above are used. Consider a symmetrically loaded straight beam of uniform cross section subjected to a moment M that produces *pure bending* (Fig. 1.1a). (Note that

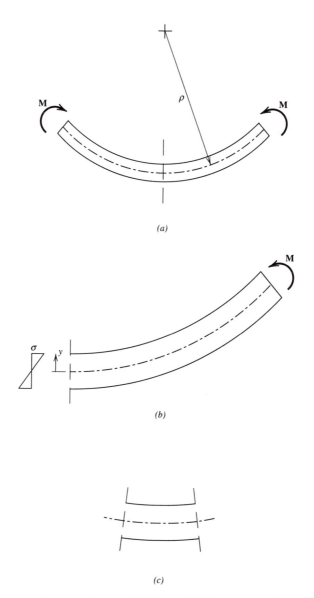

(a)

(b)

(c)

Figure 1.1   Pure bending of a long straight beam. (a) Circular curvature of beam in pure bending. (b) Free-body diagram of cut beam. (c) Infinitesimal segment of beam.

the plane of loads lies in a plane of symmetry of every cross section of the beam. See Sec. 6.1.) It is required that we determine the normal stress distribution σ for a specified cross section of the beam. We assume that σ is the major stress component and, hence, ignore other effects. Pass a section through the beam at the specified cross section so that the beam is cut into two parts. Consider a free-body diagram of one part (Fig. 1.1b). The applied moment M for this part of the beam is in equilibrium with internal forces represented by the sum of the forces that result from the normal stress σ that acts over the area of the cut section. Equations of equilibrium (requirement 1) relate the applied moment to internal forces. Since no axial external force acts, two integrals are obtained as follows: $\int \sigma \, dA = 0$ and $\int \sigma y \, dA = M$, where M is the applied external moment and y is the perpendicular distance from the neutral axis to the element of area dA.

Before the two integrals can be evaluated, we must know the distribution of σ over the cross section. Since the stress distribution is not known, it is determined indirectly through a strain distribution obtained by requirement 2. The continuity condition (requirement 2) is examined by consideration of two cross sections of the undeformed beam separated by an infinitesimal distance (Fig. 1.1c). Under the assumption that plane sections remain plane, the cross sections must rotate with respect to each other as the moment M is applied. There is a straight line in each cross section called the neutral axis along which the strains remain zero. Since plane sections remain plane, the strain distribution must vary linearly with the distance y as measured from this neutral axis.

Requirement 3 is now employed to obtain the relation between the assumed strain distribution and the stress distribution. Tension and compression stress-strain diagrams represent the response for the material in the beam. For sufficiently small strains, these diagrams indicate that the stresses and strains are linearly related. Their constant ratio, $\sigma/\epsilon = E$, is the modulus of elasticity for the material. In the linear range the modulus of elasticity is the same for tension and for compression for many engineering materials. Since other stress components in the beam are neglected, σ is the only stress component in the beam. Hence, the stress-strain relation for the beam is $\sigma = E\epsilon$. Therefore, both the stress σ and strain ϵ vary linearly with the distance y as measured from the neutral axis of the beam (Fig. 1.1). Hence, the equations of equilibrium can be integrated to obtain the flexure formula $\sigma = My/I$, where M is the applied moment at the given cross section of the beam and I is the moment of inertia of the beam cross section.

The method of mechanics of materials is used in Chapter 7 to treat nonsymmetrical bending, in Chapter 8 to treat shear center, in Chapter 9 to treat curved beams, and in Chapter 10 to treat beams on elastic foundations.

Method of Continuum Mechanics, Theory of Elasticity

Many of the problems treated in this book—noncircular torsion (Chapter 6), plates (Chapter 13), thick-wall cylinders (Chapter 11), contact stresses (Chapter 18), and stress concentrations (Chapter 14)—have multiaxial states of stress of such complexity that the method of mechanics of materials cannot be employed to derive load-stress and load-deflection relations as in the above example. Therefore, in such cases, the method of continuum mechanics is used. When we consider small displacements and linear elastic material behavior only, the general method of continuum mechanics reduces to the method of the theory of linear elasticity.

In the derivation of load-stress and load-deflection relations by the theory of linear elasticity, an infinitesimal volume element at a point in a body with faces normal to the coordinate axes is often employed. Requirement 1 is represented by the differential equations of equilibrium (Chapter 2). Requirement 2 is represented by the differential equations of compatibility (Chapter 2). The material response (requirement 3) for linearly elastic behavior is determined by one or more experimental tests that define the required elastic coefficients for the material. In this book we consider mainly isotropic materials for which only two elastic coefficients are needed (Chapter 3). These coefficients can be obtained from a tension specimen if both axial and lateral strains are measured for every load applied to the specimen. Requirement 3 is represented therefore by the isotropic stress-strain relations developed in Chapter 3. If the differential equations of equilibrium and the differential equations of compatibility can be solved subject to specified stress-strain relations and specified boundary conditions, the states of stress and displacements for every point in the member are obtained.

Deflections by Energy Methods

Certain structures are made up of members whose cross sections remain essentially plane during the deflection of the structures. The deflected position of a cross section of a member of the structure is defined by three orthogonal displacement components of the centroid of the cross section and by three orthogonal rotation components of the cross section. These six components of displacement and rotation of a cross section of a member are readily calculated by energy methods. For small displacements and small rotations and for linearly elastic material behavior, Castigliano's theorem is recommended as a method for the computation of the displacements and rotations. The method is employed in Chapter 5 for structures made up of axially loaded members, beams, and torsion members, and in Chapter 9 for curved beams.

1.4

STRESS-STRAIN RELATIONS

In Chapter 2, the state of stress at a point is defined by six stress components. The transformation of the stress components under a rotation of coordinate axes is developed, and equations of equilibrium (or equations of motion for accelerated bodies) are derived. The analogous theory of deformation, based on geometric concepts, is presented and strain-displacement relations, transformation of the strain components under a rotation of coordinate axes, and strain compatibility relations are derived.

To derive load-stress and load-deflection relations for specified structural members, the stress components must be related to the strain components. Consequently, in Chapter 3 we discuss linear stress-strain-temperature relations. These relations may be employed in the study of linearly elastic material behavior. In addition, they may be employed in plasticity theories to describe the linearly elastic part of the total response of materials. More generally, nonlinear (inelastic) stress-strain relations are required for the plastic part of material behavior. Unfortunately, these relations take on different forms, depending on the material behavior during plastic response. In this book, we consider only the limiting case of fully plastic loads for low-carbon structural steel. At the fully plastic load, stress components are assumed to be independent of the strain components and remain constant with increasing strain.

Since experimental studies are required to determine material properties (e.g. elastic coefficients for linearly elastic materials), the study of stress-strain relations is, in part, empirical. To obtain needed isotropic elastic material properties, we employ a tension specimen (Fig. 1.2). If lateral as well as longitudinal strains are measured for linearly elastic behavior of the tension specimen, the resulting stress-strain data represent the material response for obtaining the needed elastic constants for the material. The main structure of the stress-strain-temperature relations, however, is studied theoretically by means of the first law of thermodynamics (Chapter 3).

The stress-strain-temperature relations presented in Chapter 3 are limited mainly to small strains and small rotations. The reader interested in large strains and large rotations may refer to the works of Green and Adkins (1960).

Elastic and Inelastic Response of a Solid
Initially, we review the results of a simple tension test of a circular cylindrical bar that is subjected to an axially directed tensile load P (Fig. 1.2). It is assumed that the load is monotonically increased slowly (so-called static loading) from its initial value of zero to its final value, since the material response depends not only on the magnitude of the load, but also on other factors, such as the rate of loading, load cycling, etc. It is customary in engineering practice to plot the tensile stress σ in the bar as a function of the strain ϵ of the bar. In engineering practice, it is also

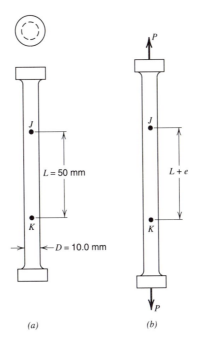

(a) (b)

Figure 1.2   Circular cross section tension specimen. (a) Undeformed specimen: Gage length L; diameter D. (b) Deformed specimen: Gage length elongation e.

customary to assume that the stress σ is uniformly distributed over the cross-sectional area of the bar and that it is equal in magnitude to P/A_0, where A_0 is the original cross-sectional area of the bar. Similarly, the strain ϵ is assumed to be constant over the gage length L and equal to $\Delta L/L = e/L$, where $\Delta L = e$ (Fig. 1.2b) is the change or elongation in the original gage length L (the distance JK in Fig. 1.2a). For these assumptions to be valid, the points J and K must be sufficiently far from the ends of the bar (a distance of one or more diameters D from the ends). However, according to the definition of stress (Sec. 2.1), the true stress is $\sigma_t = P/A_t$, where A_t is the true cross-sectional area of the bar when the load P acts. (The bar undergoes lateral contraction everywhere as it is loaded, with a corresponding change in cross-sectional area.) The difference between $\sigma = P/A_0$ and $\sigma_t = P/A_t$ is small, provided that the elongation e and, hence, the strain ϵ are sufficiently small (Sec. 2.8). If the elongation is large, A_t may differ significantly from A_0. In addition, the instantaneous or true gage length when load P acts is $L_t = L + e$ (Fig. 1.2b). Hence, like A_t, the true gage length L_t changes with the load P. Corresponding to the true stress σ_t, we may define the true strain ϵ_t as follows: In the tension test, assume that the load P is increased from zero (where $e = 0$ also) by successive infinitesimal increments dP. With each incremental increase dP in load P, there is a corresponding infinitesimal increase dL_t in the instantaneous gage length L_t. Hence, the infinitesimal increment $d\epsilon_t$ of the true strain ϵ_t due to dP is

$$d\epsilon_t = \frac{dL_t}{L_t} \tag{1.1}$$

Integration of Eq. (1.1) from L to L_t yields the true strain ϵ_t. Thus, we have

$$\epsilon_t = \int_L^{L_t} d\epsilon_t = \ln\left(\frac{L_t}{L}\right) = \ln\left(\frac{L + e}{L}\right) = \ln(1 + \epsilon) \tag{1.2}$$

In contrast to the engineering strain ϵ, the true strain ϵ_t is not linearly related to the elongation e of the original gage length L.

For many structural metals (e.g., alloy steels), the stress-strain relation of a tension specimen takes the form shown in Fig. 1.3. This figure is the *tensile stress-strain diagram* for the material. The graphical stress-strain relation (the curve $0ABCF$ in Fig. 1.3) was obtained by drawing a smooth curve through the tension test data for a certain alloy steel. Engineers use stress-strain diagrams to define certain properties of the material that are judged to be significant in the safe design of a statically loaded member. Some of these special properties are discussed briefly below. In addition, certain general material responses are addressed.

Material Properties

As noted above, Fig. 1.3 is the tensile stress-strain diagram that has been drawn from the test data of a specimen of alloy steel, Fig. 1.2. It is used by engineers to determine specific material properties used in the design of statically loaded members. There are also general characteristic behaviors that are somewhat common to all materials. To describe these properties and characteristics, it is convenient to expand the strain scale of Fig. 1.3 in the region $0AB$ (Fig. 1.4). It should be recalled that Figs. 1.3 and 1.4 are based on the following definitions of stress and strain: $\sigma = P/A_0$ and $\epsilon = e/L$, where A_0 and L are constants.

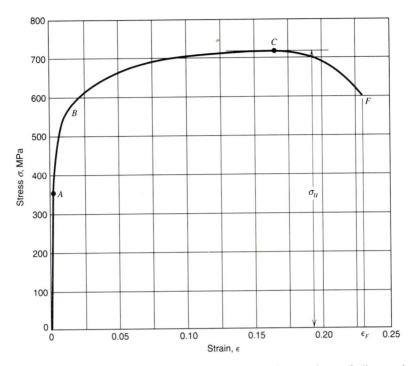

Figure 1.3   Engineering stress-strain diagram for tension specimen of alloy steel.

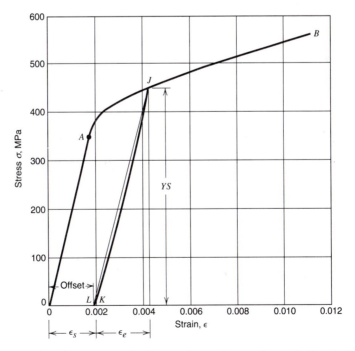

Figure 1.4   Engineering stress-strain diagram for tension specimen of alloy steel (expanded strain scale).

Consider a tensile specimen (bar) subjected to a strain ϵ under the action of a load P. If upon removal of the load P, the strain in the bar returns to zero as the load P goes to zero, the material in the bar is said to have been strained within the *elastic limit* or the material has remained *perfectly elastic*. If under loading the strain is linearly proportional to the load P (part $0A$ in Figs. 1.3 and 1.4), the material is said to be strained within the limit of *linear elasticity*. The maximum stress for which the material remains perfectly elastic is frequently referred to simply as the *elastic limit* σ_{EL}, whereas the stress at the limit of linear elasticity is referred to as the *proportional limit* σ_{PL} (point A in Figs. 1.3 and 1.4). Ordinarily, σ_{EL} is larger than σ_{PL}. The properties of elastic limit and proportional limit, although important from a theoretical viewpoint, are not of practical importance for materials like alloy steels. This is due to the fact that the transitions from elastic to inelastic behavior and from linear to nonlinear behavior are so gradual that these limits are very difficult to determine from the stress-strain diagram (part $0AB$ of the curves in Figs. 1.3 and 1.4).

When the load produces a stress σ that exceeds the elastic limit (e.g., the stress at point J in Fig. 1.4), the strain does not disappear upon unloading (curve JL in Fig. 1.4) A *permanent strain* $0L$ remains. For simplicity, it is assumed that the unloading occurs along the straight line JK, with a slope equal to that of the straight line $0A$. For this reason, the strain $0K$ is called the *offset strain* ϵ_s. The strain that is recovered when the load is removed is called the *elastic strain* ϵ_e. Hence, the total strain ϵ at point J is the sum of the offset strain and elastic strain, or $\epsilon = \epsilon_s + \epsilon_e$.

Yield Strength.   The value of stress associated with point J, Fig. 1.4, is called the *yield strength* and is denoted by σ_{YS} or simply by YS. In practice, a value of the offset strain is chosen and the yield strength is determined as the stress associated with the intersection of the curve $0AB$ and the straight line JK drawn from the offset strain value, with a slope equal to that of line $0A$ (Fig. 1.4). The value of the offset strain is arbitrary. However, a commonly agreed upon value of offset is 0.002 or 0.2% strain, as shown in Fig. 1.4. Typical values of yield strength for several structural materials are listed in Appendix A, for an offset of 0.2%. For materials with stress-strain curves like that of alloy steels (Figs. 1.3 and 1.4), the yield strength is used to predict the load that initiates inelastic behavior (yield) in a member.

Ultimate Tensile Strength.   Another important property determined from the stress-strain diagram is the *ultimate tensile strength* or *ultimate tensile stress* σ_u. It is defined as the maximum stress attained in the engineering stress-strain diagram, and in Fig. 1.3 it is the stress associated with point C. As seen from Fig. 1.3, the stress increases continuously beyond the elastic region $0A$, until point C is reached. This increase is because the material is hardening (gaining strength) because of the straining, at a faster rate than it is softening (losing strength) because of the reduction in cross-sectional area. At point C, the strain hardening effect is balanced by the effect of the area reduction. From point C to point F, the weakening effect of the area reduction controls, and the engineering stress decreases, until the specimen ruptures at point F.

Modulus of Elasticity.   In the straight-line region $0A$ of the stress-strain diagram, the stress is proportional to strain, that is, $\sigma = E\epsilon$. The constant of proportionality E is called the *modulus of elasticity*. It is also referred to as *Young's modulus*. Geometrically, it is equal in magnitude to the slope of the stress-strain relation in the region $0A$ (Fig. 1.4).

Percent Elongation.   The value of the elongation e_F of the gage length L at rupture (point F, Fig. 1.3) divided by the gage length L (in other words, the value of strain ϵ_F at rupture) multiplied by 100 is referred to as the *percent elongation* of the tensile specimen. The percent elongation is a measure of the *ductility* of the material. From Fig. 1.3, we see that the percent elongation of the alloy steel is approximately 23%.

An important structural metal, mild or structural steel, has a distinct stress-strain curve as shown in Fig. 1.5*a*. The portion 0*AB* of the stress-strain diagram is shown expanded in Fig. 1.5*b*. The stress-strain diagram for structural steel usually exhibits a so-called upper yield point, with stress σ_{YU}, and a lower yield point, with stress σ_{YL}. This is because the stress required to initiate yield in structural steel is larger than

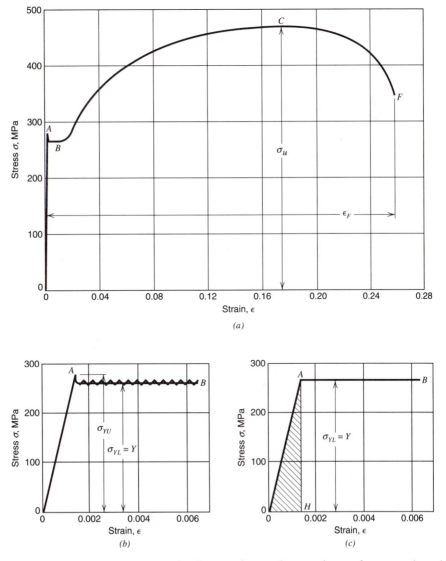

Figure 1.5   Engineering stress-strain diagram for tension specimen of structural steel. (*a*) Stress-strain diagram. (*b*) Diagram for small strain ($\epsilon < 0.007$). (*c*) Idealized diagram for small strain ($\epsilon < 0.007$).

the stress required to continue the yielding process. At the lower yield the stress remains essentially constant for increasing strain until strain hardening causes the curve to rise (Fig. 1.5a). The constant or flat portion of the stress-strain diagram may extend over a strain range of 10 to 40 times the strain at the yield point. Actual test data indicate that the curve from A to B bounces up and down as sketched in Fig. 1.5b. However, for simplicity, the data are represented by a horizontal straight line.

Yield Point for Structural Steel.   The upper yield point is usually ignored in design, and it is assumed that the stress initiating yield is the lower yield point stress, σ_{YL}. Consequently, for simplicity, the stress-strain diagram for the region $0AB$ is idealized as shown in Fig. 1.5c. Also for simplicity, we shall refer to the yield point stress as the *yield point* and denote it by the symbol Y. Recall that the yield strength (or yield stress) for alloy steel, and for materials such as aluminum alloys that have similar stress-strain diagrams, was denoted by YS (Fig. 1.4). However, for simplicity when there is no danger of confusion, we will also denote the yield strength by the symbol Y.

Modulus of Resilience.   Another property easily determined from the stress-strain diagram is the *modulus of resilience*. It is a measure of energy absorbed by a material up to the time it yields under load and is represented by the area under the stress-strain diagram to the yield point (the shaded area $0AH$ in Fig. 1.5c). In Fig. 1.5c, this area is given by $\frac{1}{2}\sigma_{YL}\epsilon_{YL}$. Since $\epsilon_{YL} = \sigma_{YL}/E$, and with the notation $Y = \sigma_{YL}$, we may express the modulus of resilience as follows:

$$\text{Modulus of resilience} = \frac{1}{2}\frac{Y^2}{E} \tag{1.3}$$

Necking of a Mild Steel Tension Specimen.   As noted above, the stress-strain curve for a mild steel tension specimen first reaches a local maximum called the upper yield or plastic limit σ_{YU}, after which it drops to a local minimum (the lower yield point Y) and runs approximately (in a wavy fashion) parallel to the strain axis for some range of strain. For mild steel, the lower yield point stress Y is assumed to be the stress at which yield is initiated. After some additional strain, the stress rises gradually; a relatively small change in load causes a significant change in strain. In this region (BC in Fig. 1.5a), substantial differences exist in the stress-strain diagrams, depending on whether area A_0 or A_t is used in the definition of stress. With area A_0, the curve first rises rapidly and then slowly, turning with its concave side down and attaining a maximum value σ_u, the ultimate strength, before turning down rapidly to fracture (point F, Fig. 1.5a). Physically, after σ_u is reached, the so-called *necking* of the bar occurs, Fig. 1.6. This necking is a drastic reduction of the cross-sectional area of the bar in the region where the fracture ultimately occurs. If the load P is referred to the true cross-sectional area A_t and, hence, $\sigma_t = P/A_t$, the true stress-strain curve differs considerably from the engineering stress-strain curve in the region BC (Figs. 1.5a and 1.7). In addition, the engineering stress-strain curves for tension and compression differ considerably in the plastic region (Fig. 1.7), because of the fact that in tension the cross-sectional area decreases with increasing load, whereas in compression it increases with increasing load. However, as can be seen from Fig. 1.7, little differences exist between the curves for small strains ($\epsilon_t < 0.01$).

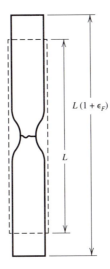

Figure 1.6   Necking of tension specimen.

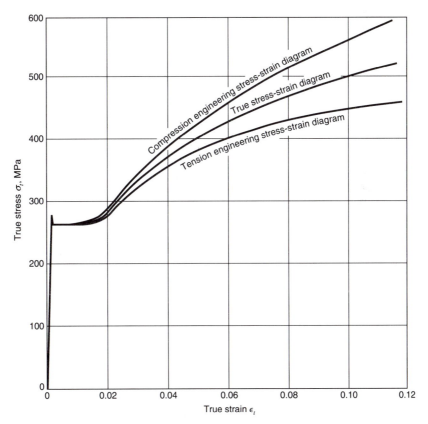

Figure 1.7   Comparison of tension and compression engineering stress-strain diagrams with the true stress-strain diagram for structural steel.

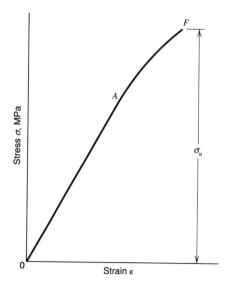

Figure 1.8   Stress-strain diagram for a brittle material.

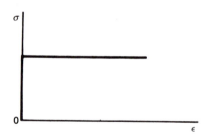

Figure 1.9   Stress-strain diagram for clay.

Other Materials. There are many materials whose tensile specimens do not undergo substantial plastic strain before fracture. These materials are called *brittle materials*. A stress-strain diagram typical of brittle materials is shown in Fig. 1.8. It exhibits little plastic range, and fracture occurs almost immediately at the end of the elastic range. On the other hand, there are materials that undergo extensive plastic deformation and little elastic deformation. Lead and clay are such materials. The idealized stress-strain diagram for clay is typical of such materials, Fig. 1.9. This response is referred to as *rigid-perfectly plastic*.

Load-Carrying Members

In the above discussion we have been concerned with the tension specimen used to obtain material properties. Stress-strain data obtained from such tension specimens are used to represent material responses for most members considered in this book. As indicated at the beginning of this section, the constitutive relations, the equations of equilibrium (equations of motion for accelerated members), and the compatibility conditions are used to derive load-stress and load-deflection relations for members. We restrict our study mainly to the behavior of solid members (i.e., members composed of materials that possess large cohesive forces, in contrast to fluids that can sustain only relatively small tensile forces) that have the ability to recover instantly their original size and shape when the forces producing the deformations are removed. This property of instant recovery of initial size and shape upon removal of load has been defined earlier in this section as perfect elasticity. In most of our discussion, we limit our consideration to linear perfect elasticity. We assume that the magnitudes of the stress components at any point P in a member depend at all times solely on the simultaneous deformation in the immediate neighborhood of the point P. In general, the state of stress at point P depends not only on the forces acting on the member at any instant, but also on the previous history of deformation of the member. For example, the state of stress at point P may depend on residual stresses due to previous history of cold work or cold forming of the member. The stress components at point P obtained from the load-stress relations derived later in this book must be added to the residual stresses at point P to obtain the actual state of stress at point P. However, in this book, we assume that residual stresses are negligibly small.

Generally, a structural member is acted on continuously by forces. For example, in the vicinity of the earth a member is acted on by the earth's gravitational force, even in the absence of other forces. Only in interstellar space does a member approach being free of the action of forces, although even there it is acted on by the gravitational attractions of the distant stars. Therefore, the *zero state* or *zero configuration* from which the deformations of the member are measured is arbitrary. However, once the zero configuration is specified, the strains of the member measured from the zero state determine the member's internal configuration.

Whenever a member exhibits the phenomenon of *hysteresis*—that is, of returning to its original size and shape only slowly or not at all after the load is removed—its behavior is not perfectly elastic. The study of members that recover their sizes and shapes only gradually after a load is removed is discussed in the theory of viscoelasticity (Brand, 1960; Flügge, 1967). The study of members that do not return to their original sizes after removal of load is generally considered in the theory of plasticity (Lubliner, 1990).

Finally, the complete description of the zero state of a member requires that the temperature at every point in the member, as well as its initial configuration, be specified. This is because, in general, a change in temperature produces a change in configuration. In turn, a change in configuration may or may not be accompanied by a change in temperature.

1.5

FAILURE AND LIMITS ON DESIGN

To design a structural system to perform a given function, the designer must have a clear understanding of the possible ways or modes by which the system may fail to perform its function. The designer must determine the possible *modes of failure* of the system and then establish suitable *failure criteria* that accurately predict the failure modes. In general, the determination of modes of failure requires extensive knowledge of the response of a structural system to loads. In particular, it requires a comprehensive stress analysis of the system. Since the response of a structural system depends strongly on the material used, so does the mode of failure. In turn, the mode of failure of a given material also depends on the manner or history of loading, such as the number of cycles of load applied at a particular temperature. Accordingly, suitable failure criteria must account for different materials, different loading histories, as well as factors that influence the stress distribution in the member.

A major part of this book is concerned with (1) stress analysis, (2) material behavior under load, and (3) the relationship between the mode of failure and a critical parameter associated with failure. The critical parameter that signals the onset of failure might be stress, strain, displacement, load, number of load cycles, or a combination of these. The discussion in this book is restricted to situations in which failure of a system is related to only a single critical parameter. In addition, we will examine the accuracy of the theories presented in the text with regard to their ability to predict system behavior. In particular, limits on design will be introduced utilizing factors of safety or reliability-based concepts that provide a measure of safety against failure.

Historically, limits on the design of a system have been established using a *factor of safety*. A factor of safety *SF* can be defined as

$$SF = \frac{R_n}{R_w} \tag{1.4}$$

where R_n is the nominal resistance (the critical parameter associated with failure) and R_w is the safe *working* magnitude of that same parameter. The letter R is used to represent the *resistance* of the system to failure. Generally, the magnitude of R_n is based on theory or experimental observation. The factor of safety is chosen on the basis of experiments or experience with similar systems made of the same material under similar loading conditions. Then the safe working parameter R_w is determined from Eq. (1.4). The factor of safety must account for unknowns, including variability of the loads, differences in material properties, deviations from the intended geometry, and our ability to predict the critical parameter.

Generally, a *design inequality* is employed to relate load effects to resistance. The design inequality is defined as

$$\sum_{i}^{N} Q_i \leq \frac{R_n}{SF} \tag{1.5}$$

where each Q_i represents the effect of a particular working (or service-level) load, such as internal pressure or temperature change, and N denotes the number of load types considered.

More recently, design philosophies based on *reliability concepts* (Salmon and Johnson, 1990) have been developed. It has been recognized that a single factor of safety is inadequate to account for all the unknowns mentioned above. Furthermore, each of the particular load types will exhibit its own statistical variability. Consequently, appropriate load and resistance factors are applied to both sides of the design inequality. So modified, the design inequality of Eq. (1.5) may be reformulated as

$$\sum_{i}^{N} \gamma_i Q_i \leq \phi R_n \tag{1.6}$$

where the γ_i are the load factors for load effects Q_i and ϕ is the resistance factor for the nominal capacity R_n. The statistical variation of the individual loads is accounted for in γ_i, whereas the variability in resistance (associated with material properties, geometry, and analysis procedures) is represented by ϕ. The use of this approach, known as *limit-states design*, is more rational than the factor-of-safety approach and produces a more uniform reliability throughout the system.

A *limit state* is a condition in which a system, or component, ceases to fulfill its intended function. This definition is essentially the same as the definition of *failure* used earlier in this text. However, some prefer the term limit state because the term failure tends to imply only some catastrophic event (brittle fracture), rather than an inability to function properly (excessive elastic deflections or brittle fracture). Nevertheless, the term failure will continue to be used in this book in the more general context.

EXAMPLE 1.1
Design of a Tension Rod

A steel rod is used as a tension brace in a structure. The structure is subjected to dead load, live load, and snow load. The effect of each of the individual loads on the tension brace is $D = 25$ kN, $L = 60$ kN, and $W = 30$ kN. Select a circular rod of appropriate size to carry these loads safely. Use steel with a yield strength of 250 MPa. Make the selection using (a) factor-of-safety design and (b) limit-states design.

SOLUTION

For simplicity in this example, the only limit state that will be considered is yielding of the cross section. Other limit states, including fracture of the member at the connections to the remainder of the structure, are ignored.

(a) In factor-of-safety design (also known as *allowable stress* or *working stress* design), the load effects are added without load factors. Thus, the total service-level load is

$$\sum Q_i = D + L + W = 115 \text{ kN} = 115{,}000 \text{ N} \tag{a}$$

The nominal resistance (capacity) of the tension rod is

$$R_n = YA_g = (250 \text{ MPa})A_g \tag{b}$$

where A_g is the gross area of the rod. In the design of tension members for steel structures, a factor of safety of 5/3 is used (AISC, 1989). Hence, the design inequality is

$$115{,}000 \leq \frac{250 A_g}{5/3} \tag{c}$$

which yields $A_g \geq 767$ mm². A rod of 32-mm diameter, with a cross-sectional area of 804 mm², is adequate.

(b)   In limit-states design, the critical load effect is determined by examination of several possible load combination equations. These equations represent the condition in which a single load quantity is at its maximum lifetime value, whereas the other quantities are taken at an arbitrary point in time. The relevant load combinations for this situation are specified (ASCE, 1990) as

$$1.4D \tag{d}$$
$$1.2D + 1.6L \tag{e}$$
$$1.2D + 0.5L + 1.3W \tag{f}$$

For the given load quantities, combination (e) is critical. The total load effect is

$$\sum \gamma_i Q_i = 126 \text{ kN} = 126{,}000 \text{ N} \tag{g}$$

In the design of tension members for steel structures, a resistance factor of $\phi = 0.9$ is used (AISC, 1986). Hence, the limit-states design inequality is

$$126{,}000 \le 0.9(250A_g) \tag{h}$$

which yields $A_g \ge 560 \text{ mm}^2$. A rod of 28-mm diameter, with a cross-sectional area of 616 mm^2, is adequate.

Discussion

The objective of this example has been to demonstrate the use of different design philosophies through their respective design inequalities, Eqs. (1.5) and (1.6). For the conditions posed, the limit-states approach produces a more economical design than the factor-of-safety approach. This can be attributed to the recognition in the load factor equations (d–f) that it is highly unlikely both live load and wind load would reach their maximum lifetime values at the same instant. Different combinations of dead load, live load, and wind load, which still give a total service-level load of 115 kN, could produce different factored loads and thus different area requirements for the rod under limit-states design.

Modes of Failure

When a structural member is subjected to loads, its response depends not only on the type of material from which it is made but also on the environmental conditions and the manner of loading. Depending on how the member is loaded, it may fail by *excessive deflection*, which results in the member being unable to perform its design function; it may fail by *plastic deformation* (*general yielding*), which may cause a permanent, undesirable change in shape; it may fail because of a *fracture* (break), which depending on the material and the nature of loading may be of a *ductile type* preceded by appreciable plastic deformation or of a *brittle type* with little or no prior plastic deformation. Materials such as glass, ceramics, rocks, plain concrete, and cast iron are examples of materials that fracture in a brittle manner under normal environmental conditions and the slow application of tension load. In uniaxial compression, they also fracture in a brittle manner, but the nature of the fracture is quite different from that in tension. Depending on a number of conditions such as environment, rate of load, nature of loading, and presence of cracks or flaws, structural metals may exhibit ductile or brittle fracture.

One type of loading that may result in brittle fracture of ductile metals is that of repeated loads. For example, consider a uniaxially loaded bar with a smooth surface that is subjected to repeated cycles of load. The bar may fail by fracture (usually, in a brittle manner). Fracture of a structural member under repeated loads is commonly called *fatigue fracture* or *fatigue failure*. Fatigue fracture may start by the initiation of one or more small cracks, usually in the neighborhood of the maximum critical stress in the member. Repeated cycling of the load causes the crack or cracks to propagate until the structural member is no longer able to carry the load across the cracked region, and the member ruptures.

Another manner in which a structural member may fail is that of elastic or plastic instability. In this failure mode, the structural member may undergo large displacements from its design configuration when the applied load reaches a critical value, the so-called *buckling load* (or *instability load*). This type of failure may result in excessive displacement or loss of ability (because of yielding or fracture) to carry the design load. In addition to the above failure modes, a structural member may fail because of environmental corrosion (chemical action).

To elaborate on the modes of failure of structural members, we discuss more fully the following categories of failure modes:

1. Failure by excessive deflection
    (a) Elastic deflection
    (b) Deflection due to creep
2. Failure by general yielding
3. Failure by fracture
    (a) Sudden fracture of brittle materials
    (b) Fracture of cracked or flawed members
    (c) Progressive fracture (fatigue)

These failure modes and their associated failure criteria are most meaningful for simple structural members (e.g., tension members, columns, beams, circular cross section torsion members). For more complicated two- and three-dimensional problems, the significance of such simple failure modes is open to question.

Many of these modes of failure for simple structural members are well-known to engineers. However, under unusual conditions of load or environment, other types of failure may occur. For example, in nuclear reactor systems, cracks in pipe loops have been attributed to stress-assisted corrosion cracking, with possible side effects attributable to residual welding stresses (Clarke and Gordon, 1973; Hakala et al., 1990; Scott and Tice, 1990).

The physical action in a structural member leading to failure is usually a complicated phenomenon, and in the following discussion the phenomena are necessarily oversimplified, but they nevertheless retain the essential features of the failures.

1. Failure by Excessive Elastic Deflection. The maximum load that may be applied to a member without causing it to cease to function properly may be limited by the permissible elastic strain or deflection of the member. Elastic deflection that may cause damage to a member can occur under these different conditions:

(a) Deflection under conditions of stable equilibrium, such as the stretch of a tension member, the angle of twist of a shaft, and the deflection of an end-loaded cantilever beam. Elastic deflections, under conditions of equilibrium, are computed in Chapter 5.

(b) Buckling, or the rather sudden deflection associated with unstable equilibrium and often resulting in total collapse of the member. This occurs, for example, when an axial load, applied gradually to a slender column, exceeds the Euler load. See Chapter 12.

(c) Elastic deflections that are the amplitudes of the vibration of a member sometimes are associated with failure of the member resulting from objectionable noise, shaking forces, collision of moving parts with stationary parts, etc., which result from the vibrations.

When a member fails by elastic deformation, the significant equations for design are those that relate loads and elastic deflection. For example, the equations, for the three members mentioned under (a) are $e = PL/AE$, $\theta = TL/GJ$, and $\delta = WL^3/3EI$. It is noted that these equations contain the significant property of the material involved in the elastic deflection, namely, the modulus of elasticity E (sometimes called the stiffness) or the shear modulus $G = E/[2(1 + v)]$, where v is Poisson's ratio. The stresses caused by the loads are not the significant quantities; that is, the stresses do not limit the loads that can be applied to the member. In other words, if a member of given dimensions fails to perform its load-resisting function because of excessive elastic deflection, its load-carrying capacity is not increased by making the member of stronger material. As a rule, the most effective method of decreasing the deflection of a member is by changing the shape or increasing the dimensions of its cross section, rather than by making the member of a stiffer material.

2. Failure by General Yielding. Another condition that may cause a member to fail is general yielding. General yielding is inelastic deformation of a considerable portion of the member, distinguishing it from localized yielding of a relatively small portion of the member. The following discussion of yielding addresses the behavior of metals at *ordinary* temperatures, that is, at temperatures that do not exceed the recrystallization temperature. Yielding at *elevated* temperatures (creep) is discussed in Chapter 17.

Polycrystalline metals are composed of extremely large numbers of very small units called crystals or grains. The crystals have slip planes on which the resistance to shear stress is relatively small. Under elastic loading, before slip occurs, the crystal itself is distorted due to stretching or compressing of the atomic bonds from their equilibrium state. If the load is removed, the crystal returns to its undistorted shape and no permanent deformation exists. When a load is applied that causes the yield strength to be reached, the crystals are again distorted but, in addition, defects in the crystal, known as dislocations (Eisenstadt, 1971), move in the slip planes by breaking and reforming atomic bonds. After removal of the load, only the distortion of the crystal (due to bond stretching) is recovered. The movement of the dislocations remains as permanent deformation.

After sufficient yielding has occurred in some crystals at a given load, these crystals will not yield further without an increase in load. This is due to the formation of dislocation entanglements that make motion of the dislocations more and more difficult. A higher and higher stress will be needed to push new dislocations through these entanglements. This increased resistance that develops after yielding is known as *strain hardening* or *work hardening*. Strain hardening is permanent. Hence, for strain-hardening metals, the plastic deformation and increase in yield strength are both retained after the load is removed.

When failure occurs by general yielding, stress concentrations usually are *not* significant because of the interaction and adjustments that take place between crystals in the regions of the stress concentrations. Slip in a few highly stressed crystals does not limit the general load-carrying capacity of the member, but merely causes readjustment of stresses that permit the more lightly stressed crystals to take higher stresses. The stress distribution approaches that which occurs in a member free from stress concentrations. Thus, the member as a whole acts substantially as an ideal homogeneous member, free from abrupt changes of section.

It is important to observe that if a member that fails by yielding is replaced by one with a material of a higher yield stress, the mode of failure may change to that of elastic deflection, buckling, or excessive mechanical vibrations. Hence, the entire basis of design may be changed when conditions are altered to prevent a given mode of failure.

3. Failure by Fracture.   Some members cease to function satisfactorily because they break (fracture) before either excessive elastic deflection or general yielding occurs. Three rather different modes or mechanisms of fracture that occur especially in metals are discussed briefly below.

(a) Sudden Fracture of Brittle Material.   Some materials—so-called brittle materials—function satisfactorily in resisting loads under static conditions until the material breaks rather suddenly with little or no evidence of plastic deformation. Ordinarily, the tensile stress in members made of such materials is considered to be the significant quantity associated with the failure, and the ultimate strength σ_u is taken as the measure of the maximum utilizable strength of the material (Fig. 1.8).

(b) Fracture of Flawed Members.   A member made of a ductile metal and subjected to static tensile loads will not fracture in a brittle manner so long as the member is free of flaws (cracks, notches, or other stress concentrations) and the temperature is not unusually low. However, in the presence of flaws, ductile materials may experience brittle fracture at normal temperatures. The flaw often contributes to development of a high hydrostatic tension stress (hydrostatic stress is discussed in Chapter 2). Yielding of ductile metals is not influenced significantly by hydrostatic stress so plastic deformation may be small or nonexistent even though fracture is impending. Thus, yield strength is not the critical material parameter when failure occurs by brittle fracture. Instead, *notch toughness*, the ability of a material to absorb energy in the presence of a notch (or other flaw), is the parameter that governs the failure mode. Dynamic loading and low temperatures also increase the tendency of a material to fracture in a brittle manner. Failure by brittle fracture is discussed in Chapter 15.

(c) Progressive Fracture (Fatigue).   If a metal that ordinarily fails by general yielding under a static load is subjected to repeated cycles of stress, it may fail by fracture without visual evidence of yielding, provided that the repeated stress is greater than a value called the *fatigue strength*. Under such conditions, minute cracks start at one or more points in the member, usually at points of high *localized* stress such as at abrupt changes in section, and gradually spread by fracture of the material at the edges of the cracks where the stress is highly concentrated. The *progressive fracture* continues until the member finally breaks. This mode of failure is usually called a *fatigue failure*, but it is better designated as *failure by progressive fracture* resulting from repeated loads. (See Chapter 16.)

The quantity usually considered most significant in failure by progressive fracture is *localized* tensile stress (although the fatigue crack sometimes occurs on the plane of maximum shear stress), and the maximum utilizable strength of the material is considered to be the stress (*fatigue strength*) corresponding to a given "life" (number of repetitions of stress). If the material has an *endurance limit* (fatigue strength for infinite life) and a design for so-called infinite life is desired, then the endurance limit is the limiting resistance value or maximum utilizable strength of the material (Chapter 16).

PROBLEMS

1.1. What requirements control the derivation of load-stress relations?

1.2. Describe the method of mechanics of materials.

1.3. How are stress-strain-temperature relations for a material established?

1.4. Explain the differences between elastic response and inelastic response of a solid.

1.5. What is a stress-strain diagram?

1.6. Explain the difference between elastic limit and proportional limit.

1.7. Explain the difference between the concepts of yield point and yield stress.

1.8. What is offset strain?

1.9. How does the engineering stress-strain diagram differ from the true stress-strain diagram?

1.10. What are modes of failure?

1.11. What are failure criteria? How are they related to modes of failure?

1.12. What is meant by the term factor of safety? How are factors of safety used in design?

1.13. What is a design inequality?

1.14. How is the usual design inequality modified to account for statistical variability?

1.15. What is a load factor? A load effect? A resistance factor?

1.16. What is a limit-states design?

1.17. What is meant by the phrase "failure by excessive deflection?"

1.18. What is meant by the phrase "failure by yielding?"

1.19. What is meant by the phrase "failure by fracture?"

1.20. Discuss the various ways that a structural member may fail.

1.21. Discuss the failure modes, critical parameters, and failure criteria that may apply to the design of a downhill snow ski.

1.22. For the steels whose stress-strain diagrams are represented by Figs. 1.3 to 1.5, determine the following properties as appropriate: the yield point, the yield strength, the upper yield point, the lower yield point, the modulus of resilience, the ultimate tensile strength, the strain at fracture, the percent elongation.

1.23. Use the mechanics of materials method to derive the load-stress and load-displacement relations for a solid circular rod of constant radius r and length L subjected to a torsional moment T as shown in Fig. P1.23.

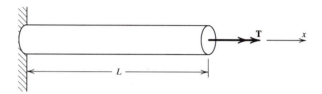

Figure P1.23   Solid circular rod in torsion.

1.24. Use the mechanics of materials method to derive the load-stress and load-displacement relations for a bar of constant width b, linearly varying depth d, and length L subjected to an axial tensile force P as shown in Fig. P1.24.

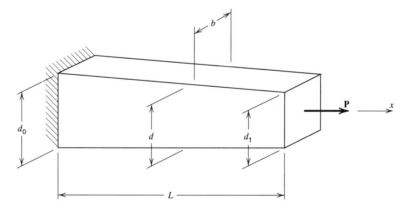

Figure P1.24   Tapered bar in tension.

1.25. Two flat plates are clamped to the ends of a pipe using four rods, each 15 mm in diameter, to form a cylinder that is to be subjected to internal pressure p, Fig. P1.25. The pipe has an outside diameter of 100 mm and an inside diameter of 90 mm. Steel is used throughout ($E = 200$ GPa). During assembly of the cylinder (before pressurization), the joints between the plates and ends of the pipe are sealed with a thin mastic and the rods are each pretensioned to 65 kN. Using the mechanics of materials method, determine the internal pressure that will cause leaking. Leaking is defined as a state of zero bearing pressure between the pipe ends and the plates. Also determine the change in stress in the rods. Ignore bending in the plates and radial deformation of the pipe.

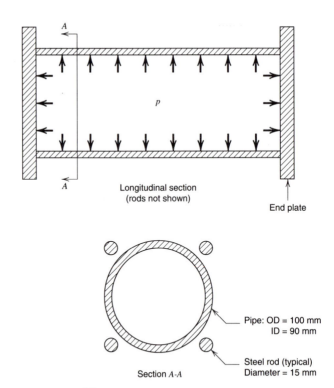

Longitudinal section
(rods not shown)

End plate

Pipe: OD = 100 mm
ID = 90 mm

Steel rod (typical)
Diameter = 15 mm

Section *A-A*

Figure P1.25   Pressurized cylinder.

REFERENCES

American Institute of Steel Construction (AISC) (1986). *Load and Resistance Factor Design Specification for Structural Steel Buildings.* Chicago, Ill.: September 1.

AISC (1989). *Specification for Structural Steel Buildings—Allowable Stress Design and Plastic Design.* Chicago, Ill.: June 1.

American Society of Civil Engineers (ASCE) (1990). *Minimum Design Loads for Buildings and Other Structures,* ASCE Std. 7–88 (formerly ANSI A58.1). New York: July.

Brand, D. R. (1960). *The Theory of Linear Viscoelasticity.* New York: Pergamon Press.

Clarke, W. L. and Gordon, G. M. (1973). Investigation of Stress Corrosion Cracking Susceptibility of Fe-Ni-Cr Alloys in Nuclear Reactor Water Environments. *Corrosion,* **29**(1): 1–12.

Cross, N. (1989). *Engineering Design Methods.* New York: Wiley.

de Neufville, R. (1990). *Applied Systems Analysis.* New York: McGraw-Hill.

Eisenstadt, M. M. (1971). *Introduction to Mechanical Properties of Metals.* New York: Macmillan.

Flügge, W. (1967). *Viscoelasticity.* London: Blaisdel.

Green A. E. and Adkins, J. E. (1960). *Large Elastic Deformations.* London: Oxford Univ. Press.

Hakala, J., Hanninen, H., and Asltonen, P. (1990). Stress Corrosion and Thermal Fatigue. *Nucl. Eng. Design*, **119**(2, 3): 389–398.

Lubliner, J. (1990). *Plasticity Theory*. New York: Macmillan.

Salmon, C. G. and Johnson, J. E. (1990). *Steel Structures—Design and Behavior*, 3rd ed. New York: Harper & Row.

Scott, P. M. and Tice, D. R. (1990). Stress Corrosion in Low-Alloy Steels. *Nucl. Eng. Design*, **119**(2, 3): 399–414.

2

THEORIES OF STRESS
AND STRAIN

In Chapter 1, we presented general concepts and definitions that are fundamental to many of the topics discussed in this book. In this chapter, we develop theories of stress and strain that are essential for the analysis of the behavior of a structural or mechanical system subjected to loads. The relations developed are used throughout the remainder of the book.

2.1

DEFINITION OF STRESS AT A POINT

Consider a general body subjected to forces acting on its surface (Fig. 2.1). Pass a fictitious plane Q through the body, cutting the body along surface A (Fig. 2.2). Designate one side of plane Q as positive and the other side as negative. The portion of the body on the positive side of Q exerts a force on the portion of the body on the negative side. This force is transmitted through the plane Q by direct contact of the parts of the body on the two sides of Q. Let the force that is transmitted through an incremental area ΔA of A by the part on the positive side Q be denoted by $\Delta \mathbf{F}$. In accordance with Newton's third law, the portion of the body on the negative side of Q transmits through area ΔA a force $-\Delta \mathbf{F}$.

The force $\Delta \mathbf{F}$ may be resolved into components $\Delta \mathbf{F}_N$ and $\Delta \mathbf{F}_S$, along unit normal $\mathbf{N}$ and unit tangent $\mathbf{S}$, respectively, to the plane Q. The force $\Delta \mathbf{F}_N$ is called the *normal (perpendicular) force* on area ΔA and $\Delta \mathbf{F}_S$ is called the *shear (tangential) force* on ΔA. The forces $\Delta \mathbf{F}$, $\Delta \mathbf{F}_N$, and $\Delta \mathbf{F}_S$ depend on the area ΔA and the orientation of plane Q. The magnitudes of the average forces per unit area are $\Delta F/\Delta A$, $\Delta F_N/\Delta A$, and $\Delta F_S/\Delta A$. These ratios are called the average stress, average normal stress, and average shear stress, respectively, acting on area ΔA. The concept of stress at a point is obtained by letting ΔA become an infinitesimal. Then the forces $\Delta \mathbf{F}$, $\Delta \mathbf{F}_N$, and $\Delta \mathbf{F}_S$ approach zero, but usually the ratios $\Delta F/\Delta A$, $\Delta F_N/\Delta A$, and $\Delta F_S/\Delta A$ approach limits different from zero. The limiting ratio of $\Delta F/\Delta A$ as ΔA goes to zero defines the stress vector $\boldsymbol{\sigma}$. Thus, the stress vector $\boldsymbol{\sigma}$ is given by

$$\boldsymbol{\sigma} = \lim_{\Delta A \to 0} \frac{\Delta \mathbf{F}}{\Delta A} \tag{2.1}$$

The stress vector $\boldsymbol{\sigma}$ (also called the traction vector) always lies along the limiting direction of the force vector $\Delta \mathbf{F}$, which in general is neither perpendicular nor tangent to the plane Q.

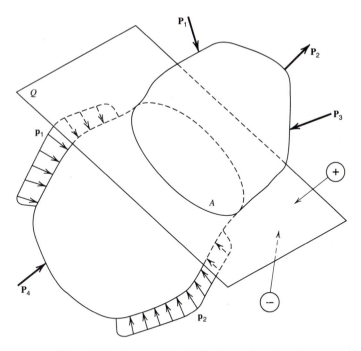

Figure 2.1   A general loaded body cut by plane Q.

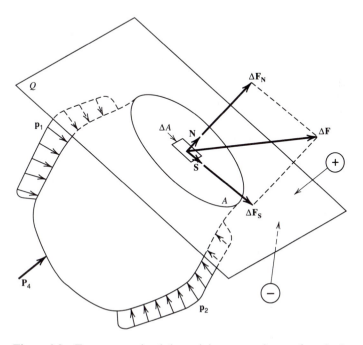

Figure 2.2   Force transmitted through incremental area of cut body.

Similarly, the limiting ratios of $\Delta \mathbf{F}_N / \Delta A$ and $\Delta \mathbf{F}_S / \Delta A$ define the *normal stress vector* $\boldsymbol{\sigma}_N$ and the *shear stress vector* $\boldsymbol{\sigma}_S$ that act at a point in the plane Q. These stress vectors are defined by the relations

$$\boldsymbol{\sigma}_N = \lim_{\Delta A \to 0} \frac{\Delta \mathbf{F}_N}{\Delta A}, \qquad \boldsymbol{\sigma}_S = \lim_{\Delta A \to 0} \frac{\Delta \mathbf{F}_S}{\Delta A} \tag{2.2}$$

The unit vectors associated with $\boldsymbol{\sigma}_N$ and $\boldsymbol{\sigma}_S$ are perpendicular and tangent, respectively, to the plane Q.

2.2

STRESS NOTATION

We use free-body diagrams to specify the state of stress at a point and to obtain relations between various stress components. In general a free-body diagram may be a diagram of a complete member, a portion of the member obtained by passing a cutting plane through the member, or a boxlike volume element of the member. The loads that act on any of these free bodies can be divided into two types as follows:

1.   Surface forces, which are forces that act on the surface of the free body
2.   Body forces, which are forces that act throughout the volume of that portion of the member considered in the free-body diagram

Examples of surface forces are contact forces and distributed loads. Concentrated loads and reactions at a point are considered contact forces. Distributed loads may be either line loads with dimensions of force per unit length or surface loads with dimensions of force per unit area (dimensions of pressure or stress). Distributed loads on beams are often indicated as loads per unit length. Examples of surface loads are pressure exerted by a fluid in contact with the body and normal and shear stresses that act on a cut section of the body.

Examples of body forces are gravitational forces, magnetic forces, and inertia forces. Since the body force is distributed throughout the volume of the free body, it is convenient to define body force per unit volume. We use the notation $\mathbf{B}$ or (B_x, B_y, B_z) for body force per unit volume, where $\mathbf{B}$ stands for body and subscripts (x, y, z) denote components in the (x, y, z) directions, respectively, of the rectangular coordinate system (x, y, z) (see Fig. 2.3).

Consider now a free-body diagram of a box-shaped volume element at a point 0 in a member, with sides parallel to the (x, y, z) axes (Fig. 2.4). For simplicity, we show the volume element with one corner at point 0 and assume that the stress components are uniform (constant) throughout the volume element. The surface forces are represented by the product of the stress components (Fig. 2.4) and the areas* on which they act. Body forces, represented by the product of the compo-

---

* The reader must multiply each stress component by an appropriate area before applying equations of force equilibrium. For example, σ_{xx} must be multiplied by the area $dy\,dz$.

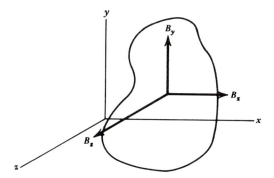

Figure 2.3   Body forces.

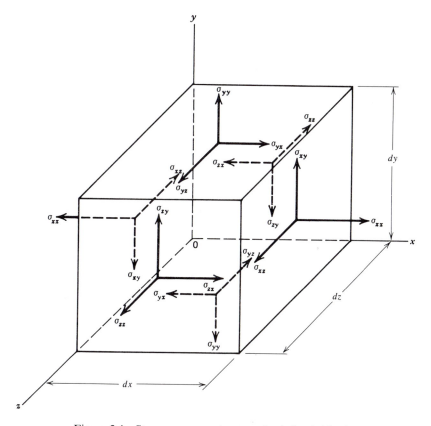

Figure 2.4   Stress components at a point in loaded body.

nents (B_x, B_y, B_z) and the volume of the element (product of the three infinitesimal
lengths of the sides of the element), are higher-order terms and are not shown on
the free-body diagram in Fig. 2.4. Consider the two faces perpendicular to the x
axis. The face from which the positive x axis is extended is taken to be the positive
face; the other face perpendicular to the x axis is taken to be the negative face. The
stress components σ_{xx}, σ_{xy}, and σ_{xz} acting on the positive face are taken to be in the
positive sense as shown when they are directed in the positive x, y, and z directions.

By Newton's third law, the positive stress components σ_{xx}, σ_{xy}, and σ_{xz} shown acting on the negative face in Fig. 2.4 are in the negative (x, y, z) directions, respectively. In effect, a positive stress component σ_{xx} exerts a tension (pull) parallel to the x axis. Equivalent sign conventions hold for the planes perpendicular to the y and z axes. Hence, associated with the concept of the state of stress at a point 0, nine components of stress exist.

$$(\sigma_{xx}, \sigma_{xy}, \sigma_{xz}), \quad (\sigma_{yy}, \sigma_{yx}, \sigma_{yz}), \quad (\sigma_{zz}, \sigma_{zx}, \sigma_{zy})$$

In the next section we show that the nine stress components may be reduced to six for most practical problems.

2.3

SYMMETRY OF THE STRESS ARRAY AND STRESS ON AN ARBITRARILY ORIENTED PLANE

Symmetry of Stress Components

The nine stress components relative to rectangular coordinate axes (x, y, z) may be tabulated in array form as follows:

$$\mathbf{T} = \begin{bmatrix} \sigma_{xx} & \sigma_{xy} & \sigma_{xz} \\ \sigma_{yx} & \sigma_{yy} & \sigma_{yz} \\ \sigma_{zx} & \sigma_{zy} & \sigma_{zz} \end{bmatrix} \tag{2.3}$$

where $\mathbf{T}$ symbolically represents the stress array called the stress tensor. In this array, the stress components in the first, second, and third rows act on planes perpendicular to the (x, y, z) axes, respectively. Seemingly, nine stress components are required to describe the state of stress at a point in a member. However, if the only forces that act on the free body in Fig. 2.4 are surface forces and body forces, we can demonstrate from the equilibrium of the volume element in Fig 2.4 that the three pairs of the shear stresses are equal. Summation of moments leads to the result

$$\sigma_{yz} = \sigma_{zy}, \qquad \sigma_{zx} = \sigma_{xz}, \qquad \sigma_{xy} = \sigma_{yx} \tag{2.4}$$

Thus, with Eq. (2.4), Eq. (2.3) may be written in the symmetric form

$$\mathbf{T} = \begin{bmatrix} \sigma_{xx} & \sigma_{xy} & \sigma_{xz} \\ \sigma_{xy} & \sigma_{yy} & \sigma_{yz} \\ \sigma_{xz} & \sigma_{yz} & \sigma_{zz} \end{bmatrix} \tag{2.5}$$

Hence, for this type of stress theory, only six components of stress are required to describe the state of stress at a point in a member.

Although we do not consider body couples or surface couples in this book (Boresi and Chong, 1987), it is possible for them to be acting on the free body in Fig. 2.4. This means that Eqs. (2.4) are no longer true and that nine stress components are required to represent the unsymmetrical state of stress.

TABLE 2.1
Stress Notations (Symmetric Stress Components)

I	σ_{xx}	σ_{yy}	σ_{zz}	$\sigma_{xy} = \sigma_{yx}$	$\sigma_{xz} = \sigma_{zx}$	$\sigma_{yz} = \sigma_{zy}$
II	σ_x	σ_y	σ_z	$\tau_{xy} = \tau_{yx}$	$\tau_{xz} = \tau_{zx}$	$\tau_{yz} = \tau_{zy}$
III	σ_{11}	σ_{22}	σ_{33}	$\sigma_{12} = \sigma_{21}$	$\sigma_{13} = \sigma_{31}$	$\sigma_{23} = \sigma_{32}$

The stress notation described above is widely used in engineering practice. It is the notation used in this book,* row I of Table 2.1. Two other frequently used symmetric stress notations are also listed in Table 2.1. The symbolism indicated in row III is employed where index notation is used (Boresi and Chong, 1987).

Stresses Acting on Arbitrary Planes

The stress vectors σ_x, σ_y, and σ_z on planes that are perpendicular, respectively, to the x, y, and z axes are

$$\sigma_x = \sigma_{xx}\mathbf{i} + \sigma_{xy}\mathbf{j} + \sigma_{xz}\mathbf{k}$$
$$\sigma_y = \sigma_{yx}\mathbf{i} + \sigma_{yy}\mathbf{j} + \sigma_{yz}\mathbf{k}$$
$$\sigma_z = \sigma_{zx}\mathbf{i} + \sigma_{zy}\mathbf{j} + \sigma_{zz}\mathbf{k} \tag{2.6}$$

where $\mathbf{i}$, $\mathbf{j}$, and $\mathbf{k}$ are unit vectors relative to the (x, y, z) axes (see Fig. 2.5 for σ_x). Now consider the stress vector σ_P on an arbitrary oblique plane P through point 0 of a member (Fig. 2.6). For clarity, the plane P is shown removed from point 0. The unit normal vector to plane P is

$$\mathbf{N} = l\mathbf{i} + m\mathbf{j} + n\mathbf{k} \tag{2.7}$$

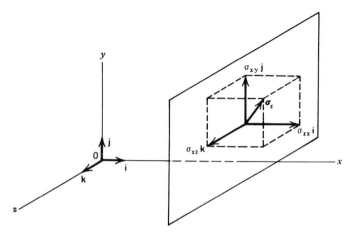

Figure 2.5   Stress vector and its components acting on a plane perpendicular to the x axis.

* Equivalent notations are used for other orthogonal coordinate systems (see Sec. 2.5.).

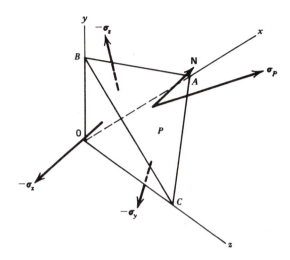

Figure 2.6   Stress vector on arbitrary plane having a normal **N**.

where (l, m, n) are the direction cosines of unit vector **N**. Therefore, *vectorial summation of forces* acting on the tetrahedral element $0ABC$ yields the following (note that the ratios of areas $0BC$, $0AC$, $0BA$ to area ABC are equal to l, m, and n, respectively):

$$\boldsymbol{\sigma}_P = l\boldsymbol{\sigma}_x + m\boldsymbol{\sigma}_y + n\boldsymbol{\sigma}_z \qquad (2.8)$$

Also, in terms of the projections $(\sigma_{Px}, \sigma_{Py}, \sigma_{Pz})$ of the stress vector $\boldsymbol{\sigma}_P$ along axes (x, y, z), we may write

$$\boldsymbol{\sigma}_P = \sigma_{Px}\mathbf{i} + \sigma_{Py}\mathbf{j} + \sigma_{Pz}\mathbf{k} \qquad (2.9)$$

Comparison of Eqs. (2.8) and (2.9) yields, with Eqs. (2.6)

$$\sigma_{Px} = l\sigma_{xx} + m\sigma_{yx} + n\sigma_{zx}$$
$$\sigma_{Py} = l\sigma_{xy} + m\sigma_{yy} + n\sigma_{zy}$$
$$\sigma_{Pz} = l\sigma_{xz} + m\sigma_{yz} + n\sigma_{zz} \qquad (2.10)$$

Equations (2.10) allow the computation of the components of stress on any oblique plane defined by unit normal $\mathbf{N}:(l, m, n)$, provided that the six components of stress

$$\sigma_{xx}, \sigma_{yy}, \sigma_{zz}, \sigma_{xy} = \sigma_{yx}, \sigma_{xz} = \sigma_{zx}, \sigma_{yz} = \sigma_{zy}$$

at point 0 are known. When point 0 lies on the surface of the member where the surface forces are represented by distributions of normal and shear stresses, Eqs. (2.10) represent the *stress boundary conditions at point 0*.

Normal Stress and Shear Stress on an Oblique Plane

The normal stress σ_{PN} on the plane P is the projection of the vector $\boldsymbol{\sigma}_P$ in the direction of **N**; that is, $\sigma_{PN} = \boldsymbol{\sigma}_P \cdot \mathbf{N}$. Hence, by Eqs. (2.7), (2.9), and (2.10)

$$\sigma_{PN} = l^2\sigma_{xx} + m^2\sigma_{yy} + n^2\sigma_{zz} + mn(\sigma_{yz} + \sigma_{zy}) + nl(\sigma_{xz} + \sigma_{zx})$$
$$+ lm(\sigma_{xy} + \sigma_{yx})$$
$$= l^2\sigma_{xx} + m^2\sigma_{yy} + n^2\sigma_{zz} + 2mn\sigma_{yz} + 2ln\sigma_{xz} + 2lm\sigma_{xy} \qquad (2.11)$$

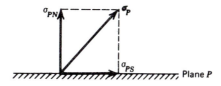

Figure 2.7   Normal and shear stress components of stress vector on an arbitrary plane.

By Eq. (2.11), the normal stress σ_{PN} on an oblique plane with unit normal $\mathbf{N}:(l, m, n)$ is expressed in terms of the six stress components $(\sigma_{xx}, \sigma_{yy}, \sigma_{zz}, \sigma_{xy}, \sigma_{xz}, \sigma_{zy})$. Often, the maximum value of σ_{PN} at a point is of importance in design (see Sec. 4.1). Of the infinite number of planes through point 0, σ_{PN} attains a maximum value called the maximum principal stress on one of these planes. The method of determining this stress and the orientation of the plane on which it acts is given in Sec. 2.4.

To compute the magnitude of the shear stress σ_{PS} on plane P, we note by geometry (Fig. 2.7) that

$$\sigma_{PS} = \sqrt{\sigma_P^2 - \sigma_{PN}^2} = \sqrt{\sigma_{Px}^2 + \sigma_{Py}^2 + \sigma_{Pz}^2 - \sigma_{PN}^2} \qquad (2.12)$$

Substitution of Eqs. (2.10) and (2.11) into Eq. (2.12) yields σ_{PS} in terms of $(\sigma_{xx}, \sigma_{yy}, \sigma_{zz}, \sigma_{xy}, \sigma_{xz}, \sigma_{yz})$ and (l, m, n). In certain criteria of failure, the maximum value of σ_{PS} at a point in the body plays an important role (see Sec. 4.4). The maximum value of σ_{PS} can be expressed in terms of the maximum and minimum principal stresses [see Eq. (2.39), Sec. 2.4].

2.4

TRANSFORMATION OF STRESS. PRINCIPAL STRESSES. OTHER PROPERTIES

Transformation of Stress

Let (x, y, z) and (X, Y, Z) denote two rectangular coordinate systems with a common origin (Fig. 2.8). The cosines of the angles between the coordinate axes (x, y, z) and the coordinate axes (X, Y, Z) are listed in Table 2.2. Each entry in Table 2.2 is the cosine of the angle between the two coordinate axes designated at the top of its column and to the left of its row. The angles are measured from the (x, y, z) axes to the (X, Y, Z) axes. For example, $l_1 = \cos \theta_{xX}$, $l_2 = \cos \theta_{xY}, \ldots$ (see Fig. 2.8.). Since the axes (x, y, z) and axes (X, Y, Z) are orthogonal, the direction cosines of Table 2.2 must satisfy the following relations:

For the Row Elements

$$l_i^2 + m_i^2 + n_i^2 = 1, \qquad i = 1, 2, 3, \qquad \text{and} \qquad l_1 l_2 + m_1 m_2 + n_1 n_2 = 0, \ldots, \ldots \quad (2.13)$$

For the Column Elements

$$l_1^2 + l_2^2 + l_3^2 = 1, \ldots, \ldots, \qquad \text{and} \qquad l_1 m_1 + l_2 m_2 + l_3 m_3 = 0, \ldots, \ldots \quad (2.14)$$

The stress components $\sigma_{XX}, \sigma_{XY}, \sigma_{XZ}, \ldots$ are defined with reference to (X, Y, Z) axes in the same manner as $\sigma_{xx}, \sigma_{xy}, \sigma_{xz}, \ldots$ are defined relative to the axes (x, y, z).

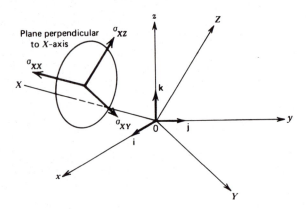

Figure 2.8 Stress components on plane perpendicular to transformed X axis.

TABLE 2.2
Direction Cosines

	x	y	z
X	l_1	m_1	n_1
Y	l_2	m_2	n_2
Z	l_3	m_3	n_3

Hence, σ_{XX} is the normal stress component on a plane perpendicular to axis X and σ_{XY}, σ_{XZ} are shear stress components on this same plane (Fig. 2.8), and so on. Hence, by Eq. (2.11),

$$\sigma_{XX} = l_1^2 \sigma_{xx} + m_1^2 \sigma_{yy} + n_1^2 \sigma_{zz} + 2m_1 n_1 \sigma_{yz} + 2n_1 l_1 \sigma_{zx} + 2l_1 m_1 \sigma_{xy}$$
$$\sigma_{YY} = l_2^2 \sigma_{xx} + m_2^2 \sigma_{yy} + n_2^2 \sigma_{zz} + 2m_2 n_2 \sigma_{yz} + 2n_2 l_2 \sigma_{zx} + 2l_2 m_2 \sigma_{xy}$$
$$\sigma_{ZZ} = l_3^2 \sigma_{xx} + m_3^2 \sigma_{yy} + n_3^2 \sigma_{zz} + 2m_3 n_3 \sigma_{yz} + 2n_3 l_3 \sigma_{zx} + 2l_3 m_3 \sigma_{xy} \quad (2.15)$$

The shear stress component σ_{XY} is the component of the stress vector in the Y direction on a plane perpendicular to the X axis; that is, it is the Y component of the stress vector σ_X acting on the plane perpendicular to the X axis. Thus, σ_{XY} may be evaluated by forming the scalar product of the vector σ_X [determined by Eqs. (2.9) and (2.10) with $l_1 = l$, $m_1 = m$, $n_1 = n$] with a unit vector parallel to the Y axis; that is, with the unit vector (Table 2.2)

$$\mathbf{N}_2 = l_2 \mathbf{i} + m_2 \mathbf{j} + n_2 \mathbf{k} \quad (2.16)$$

By Eqs. (2.9). (2.10), and (2.16), σ_{XY} is determined; similar procedures also determine σ_{XZ} and σ_{YZ}. Hence,

$$\sigma_{XY} = \sigma_X \cdot \mathbf{N}_2 = \sigma_Y \cdot \mathbf{N}_1$$
$$= l_1 l_2 \sigma_{xx} + m_1 m_2 \sigma_{yy} + n_1 n_2 \sigma_{zz} + (m_1 n_2 + m_2 n_1) \sigma_{yz}$$
$$+ (l_1 n_2 + l_2 n_1) \sigma_{zx} + (l_1 m_2 + l_2 m_1) \sigma_{xy} \quad (2.17a)$$

$$\sigma_{XZ} = \boldsymbol{\sigma}_X \cdot \mathbf{N}_3 = l_1 l_3 \sigma_{xx} + m_1 m_3 \sigma_{yy} + n_1 n_3 \sigma_{zz} + (m_1 n_3 + m_3 n_1)\sigma_{yz}$$
$$+ (l_1 n_3 + l_3 n_1)\sigma_{zx} + (l_1 m_3 + l_3 m_1)\sigma_{xy} \tag{2.17b}$$

$$\sigma_{YZ} = \boldsymbol{\sigma}_Y \cdot \mathbf{N}_3 = l_2 l_3 \sigma_{xx} + m_2 m_3 \sigma_{yy} + n_2 n_3 \sigma_{zz} + (m_2 n_3 + m_3 n_2)\sigma_{yz}$$
$$+ (l_2 n_3 + l_3 n_2)\sigma_{zx} + (l_2 m_3 + l_3 m_2)\sigma_{xy} \tag{2.17c}$$

Equations (2.15) and (2.17) determine the stress components relative to rectangular axes (X, Y, Z) in terms of the stress components relative to rectangular axes (x, y, z); that is, they determine how the stress components transform under a rotation of rectangular axes. A set of quantities that transform according to this rule is called a second-order symmetrical tensor. Later it will be shown that strain components (see Sec. 2.7) and moments and products of inertia (see Sec. B.3) also transform under rotation of axes by similar relationships; hence, they too are second-order symmetrical tensors.

Principal Stresses

It may be shown that for any general state of stress at any point 0 in a body, there exist three mutually perpendicular planes at point 0 on which the shear stresses vanish. The remaining normal stress components on these three planes are called *principal stresses*. Correspondingly, the three planes are called principal planes, and the three mutually perpendicular axes that are normal to the three planes (hence, that coincide with the three principal stress directions) are called principal axes. Thus, by definition, principal stresses are directed along principal axes that are perpendicular to the principal planes. A cubic element subjected to principal stresses is easily visualized, since the forces on the surface of the cube are normal to the faces of the cube. More complete discussions of principal stress theory are present elsewhere (Boresi and Chong, 1987). Here we merely sketch the main results.

Principal Values and Directions

Since the shear stresses vanish on principal planes, the stress vector on principal planes is given by $\boldsymbol{\sigma}_P = \sigma \mathbf{N}$, where σ is the magnitude of the stress vector $\boldsymbol{\sigma}_P$ and $\mathbf{N}$ the unit normal to a principal plane. Let $\mathbf{N} = l\mathbf{i} + m\mathbf{j} + n\mathbf{k}$ relative to rectangular axes (x, y, z) with associated unit vectors $\mathbf{i}$, $\mathbf{j}$, $\mathbf{k}$. Thus, (l, m, n) are the direction cosines of the unit normal $\mathbf{N}$. Projections of $\boldsymbol{\sigma}_P$ along (x, y, z) axes are $\sigma_{Px} = \sigma l$, $\sigma_{Py} = \sigma m$, $\sigma_{pz} = \sigma n$. Hence, by Eq. (2.10), we obtain

$$l(\sigma_{xx} - \sigma) + m\sigma_{xy} + n\sigma_{xz} = 0$$
$$l\sigma_{xy} + m(\sigma_{yy} - \sigma) + n\sigma_{yz} = 0$$
$$l\sigma_{xz} + m\sigma_{yz} + n(\sigma_{zz} - \sigma) = 0 \tag{2.18}$$

Since Eqs. (2.18) are linear homogeneous equations in (l, m, n) and the trivial solution $l = m = n = 0$ is impossible because $l^2 + m^2 + n^2 = 1$ [law of direction cosines, Eq. (2.13)], it follows from the theory of linear algebraic equations that Eqs. (2.18) are consistent if and only if the determinant of the coefficients of (l, m, n) vanishes identically. Thus, we have

$$\begin{vmatrix} \sigma_{xx} - \sigma & \sigma_{xy} & \sigma_{xz} \\ \sigma_{xy} & \sigma_{yy} - \sigma & \sigma_{yz} \\ \sigma_{xz} & \sigma_{yz} & \sigma_{zz} - \sigma \end{vmatrix} = 0 \tag{2.19}$$

or, expanding the determinant, we obtain

$$\sigma^3 - I_1\sigma^2 - I_2\sigma - I_3 = 0 \tag{2.20}$$

where

$$I_1 = \sigma_{xx} + \sigma_{yy} + \sigma_{zz}$$

$$I_2 = -\begin{vmatrix} \sigma_{xx} & \sigma_{xy} \\ \sigma_{xy} & \sigma_{yy} \end{vmatrix} - \begin{vmatrix} \sigma_{xx} & \sigma_{xz} \\ \sigma_{xz} & \sigma_{zz} \end{vmatrix} - \begin{vmatrix} \sigma_{yy} & \sigma_{yz} \\ \sigma_{yz} & \sigma_{zz} \end{vmatrix}$$

$$= \sigma_{xy}^2 + \sigma_{xz}^2 + \sigma_{yz}^2 - \sigma_{xx}\sigma_{yy} - \sigma_{xx}\sigma_{zz} - \sigma_{yy}\sigma_{zz}$$

$$I_3 = \begin{vmatrix} \sigma_{xx} & \sigma_{xy} & \sigma_{xz} \\ \sigma_{xy} & \sigma_{yy} & \sigma_{yz} \\ \sigma_{xz} & \sigma_{yz} & \sigma_{zz} \end{vmatrix} \tag{2.21}$$

The three roots $(\sigma_1, \sigma_2, \sigma_3)$ of Eq. (2.20) are the three principal stresses at point 0. The magnitudes and directions of σ_1, σ_2, and σ_3 for a given member depend only on the loads being applied to the member and cannot be influenced by the choice of coordinate axes (x, y, z) used to specify the state of stress at point 0. This means that I_1, I_2, and I_3 given by Eqs. (2.21) are *invariants of stress* and must have the same magnitudes for all choices of coordinate axes (x, y, z). Relative to principal axes, the stress invariants may be written in terms of the principal stresses as

$$I_1 = \sigma_1 + \sigma_2 + \sigma_3$$
$$I_2 = -\sigma_1\sigma_2 - \sigma_2\sigma_3 - \sigma_1\sigma_3$$
$$I_3 = \sigma_1\sigma_2\sigma_3$$

When $(\sigma_1, \sigma_2, \sigma_3)$ have been determined, the direction cosines of the three principal axes are obtained from Eqs. (2.18) by setting σ in turn equal to $(\sigma_1, \sigma_2, \sigma_3)$, respectively, and observing the direction cosine condition $l^2 + m^2 + n^2 = 1$ for each of the three values of σ. See Example 2.1.

In special cases, two principal stresses may be numerically equal. Then, Eqs. (2.18) show that the associated principal directions are not unique. In these cases, any two mutually perpendicular axes that are perpendicular to the unique third principal axis will serve as principal axes with corresponding principal planes. If all three principal stresses are equal, then $\sigma_1 = \sigma_2 = \sigma_3$ at point 0, and all planes passing through point 0 are principal planes. In this case, any set of three mutually perpendicular axes at point 0 will serve as principal axes. This stress condition is known as a state of hydrostatic stress, since it is the condition that exists in a fluid in static equilibrium.

EXAMPLE 2.1
Principal Stresses and Principal Directions

The state of stress at a point in a body is given by $\sigma_{xx} = -10$, $\sigma_{yy} = 30$, $\sigma_{xy} = 15$, and $\sigma_{zz} = \sigma_{xz} = \sigma_{yz} = 0$; see Fig. E2.1a. Determine the principal stresses and orientation of the principal axes at the point.

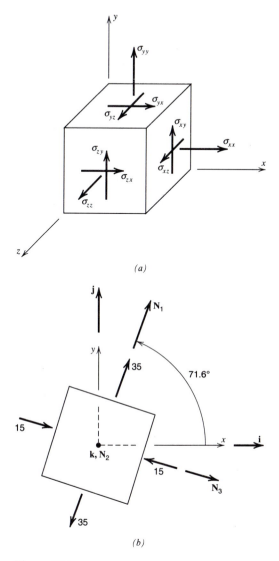

Figure E2.1

SOLUTION

By Eq. (2.21) the three stress invariants are

$$I_1 = 20, \qquad I_2 = 525, \qquad \text{and} \qquad I_3 = 0$$

Substituting the invariants into Eq. (2.20) and solving for the three roots of this equation, we obtain the principal stresses

$$\sigma_1 = 35, \qquad \sigma_2 = 0, \qquad \text{and} \qquad \sigma_3 = -15$$

To find the orientation of the first principal axis in terms of its direction cosines l_1, m_1, and n_1, we substitute $\sigma_1 = 35$ into Eq. (2.18) for σ. The direction cosines must also satisfy Eq. (2.13). Thus, we have

$$-45l_1 + 15m_1 = 0 \qquad \text{(a)}$$
$$15l_1 - 5m_1 = 0 \qquad \text{(b)}$$
$$-35n_1 = 0 \qquad \text{(c)}$$
$$l_1^2 + m_1^2 + n_1^2 = 1 \qquad \text{(d)}$$

Only two of the first three of these equations are independent. Equation (c) gives

$$n_1 = 0$$

Simultaneous solution of Eqs. (b) and (d) yields the result

$$l_1^2 = \tfrac{1}{10}$$

or

$$l_1 = \pm 0.3162$$

Substituting into Eq. (b) for l_1, we obtain

$$m_1 = \pm 0.9487$$

where the order of the + and − signs corresponds to those of l_1. Note also that Eq. (a) is satisfied with these values of l_1, m_1, and n_1. Thus, the first principal axis is directed along unit vector $\mathbf{N}_1$, where

$$\mathbf{N}_1 = 0.3162\mathbf{i} + 0.9487\mathbf{j} \qquad \text{or} \qquad \text{(e)}$$
$$\mathbf{N}_1 = -0.3162\mathbf{i} - 0.9487\mathbf{j} \qquad \text{(f)}$$

where $\mathbf{i}$ and $\mathbf{j}$ are unit vectors along the x and y axes, respectively.

The orientation of the second principal axis is found by substitution of $\sigma = \sigma_2 = 0$ into Eq. (2.18), which yields

$$l_2 = 0 \qquad \text{and} \qquad m_2 = 0$$

Proceeding as for σ_1, we then obtain

$$n_2 = \pm 1$$

from which

$$\mathbf{N}_2 = \pm\mathbf{k}$$

where $\mathbf{k}$ is a unit vector along the z axis.

The orientation of the third principal axis is found in a similar manner.

$$l_3 = \pm 0.9487$$
$$m_3 = \mp 0.3162$$
$$n_3 = 0$$

To establish a definite sign convention for the principal axes, we require them to form a right-handed triad. If $\mathbf{N}_1$ and $\mathbf{N}_2$ are unit vectors that define the directions of the first two principal axes, then the unit vector N_3 for the third principal axis is determined by the right-hand rule of vector multiplication. Thus, we have

$$N_3 = N_1 \times N_2$$

or

$$N_3 = (m_1 n_2 - m_2 n_1)i + (l_2 n_1 - l_1 n_2)j + (l_1 m_2 - l_2 m_1)k \qquad \text{(g)}$$

In our example, if we arbitrarily select N_1 from Eq. (e) and $N_2 = +k$, we obtain N_3 from Eq. (g) as:

$$N_3 = 0.9487i - 0.3162j$$

The principal stresses $\sigma_1 = 35$ and $\sigma_3 = -15$ and their orientations (the corresponding principal axes) are illustrated in Fig. E2.1b. The third principal axis is normal to the x-y plane shown and is directed toward the reader. The corresponding principal stress is $\sigma_2 = 0$. Since all the stress components associated with the z direction (σ_{zz}, σ_{xz}, and σ_{yz}) are zero, this stress state is said to be a state of plane stress in the x-y plane (see the discussion below on plane stress).

Octahedral Stress

Let (X, Y, Z) be principal axes. Consider the family of planes whose unit normals satisfy the relation $l^2 = m^2 = n^2 = \frac{1}{3}$ with respect to the principal axes (X, Y, Z). There are eight such planes (the octahedral planes, Fig. 2.9) that make equal angles with respect to the (X, Y, Z) directions. Therefore, the normal and shear stress components associated with these planes are called the *octahedral normal stress* σ_{oct} and *octahedral shear stress* τ_{oct}. By Eqs. (2.10), (2.11), and (2.12), we obtain

$$\sigma_{\text{oct}} = \tfrac{1}{3}(\sigma_1 + \sigma_2 + \sigma_3) = \tfrac{1}{3}I_1$$
$$9\tau_{\text{oct}}^2 = (\sigma_1 - \sigma_2)^2 + (\sigma_1 - \sigma_3)^2 + (\sigma_2 - \sigma_3)^2 = 2I_1^2 + 6I_2 \quad \text{(2.22)}$$

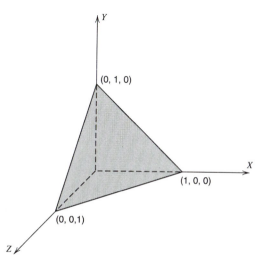

Figure 2.9   Octahedral plane for $l = m = n = 1/\sqrt{3}$, relative to principal axes (X, Y, Z).

since for the principal axes $\sigma_{XX} = \sigma_1$, $\sigma_{YY} = \sigma_2$, $\sigma_{ZZ} = \sigma_3$, and $\sigma_{XY} = \sigma_{YZ} = \sigma_{ZX} = 0$. [See Eqs. (2.21).] It follows that since (I_1, I_2, I_3) are invariants under rotation of axes, we may refer Eqs. (2.22) to arbitrary (x, y, z) axes by replacing I_1, I_2, I_3 by their general forms as given by Eqs. (2.21). Thus for arbitrary (x, y, z) axes,

$$\sigma_{oct} = \tfrac{1}{3}(\sigma_{xx} + \sigma_{yy} + \sigma_{zz})$$
$$9\tau_{oct}^2 = (\sigma_{xx} - \sigma_{yy})^2 + (\sigma_{xx} - \sigma_{zz})^2 + (\sigma_{yy} - \sigma_{zz})^2 + 6\sigma_{xy}^2 + 6\sigma_{xz}^2 + 6\sigma_{yz}^2 \quad (2.23)$$

The octahedral normal and shear stresses play a role in certain failure criteria (Sec. 4.4).

Mean and Deviator Stress

Experiments indicate that yielding and plastic deformation of many metals are essentially independent of the mean normal stress σ_m, where

$$\sigma_m = \frac{\sigma_{xx} + \sigma_{yy} + \sigma_{zz}}{3} = \frac{\sigma_1 + \sigma_2 + \sigma_3}{3} = \frac{1}{3}I_1 \quad (2.24)$$

Comparing Eqs. (2.22), (2.23), and (2.24), we note that the mean normal stresss σ_m is equal to σ_{oct}. Most plasticity theories postulate that plastic behavior of materials is related primarily to that part of the stress tensor that is independent of σ_m. Therefore, the stress array [Eq. 2.5] is rewritten in the following form:

$$\mathbf{T} = \mathbf{T}_m + \mathbf{T}_d \quad (2.25)$$

where $\mathbf{T}$ symbolically represents the stress array, Eq. (2.5), and

$$\mathbf{T}_m = \begin{bmatrix} \sigma_m & 0 & 0 \\ 0 & \sigma_m & 0 \\ 0 & 0 & \sigma_m \end{bmatrix} \quad (2.26a)$$

and

$$\mathbf{T}_d = \begin{bmatrix} \dfrac{2\sigma_{xx} - \sigma_{yy} - \sigma_{zz}}{3} & \sigma_{xy} & \sigma_{xz} \\[2mm] \sigma_{xy} & \dfrac{2\sigma_{yy} - \sigma_{xx} - \sigma_{zz}}{3} & \sigma_{yz} \\[2mm] \sigma_{xz} & \sigma_{yz} & \dfrac{2\sigma_{zz} - \sigma_{yy} - \sigma_{xx}}{3} \end{bmatrix} \quad (2.26b)$$

The array $\mathbf{T}_m$ is called the mean stress tensor. The array $\mathbf{T}_d$ is called the *deviator* stress tensor, since it is a measure of the deviation of the state of stress from a hydrostatic stress state, that is, from the state of stress that exists in an ideal (frictionless fluid).

Let (x, y, z) be the transformed axes that are in the principal stress directions. Then,

$$\sigma_{xx} = \sigma_1, \qquad \sigma_{yy} = \sigma_2, \qquad \sigma_{zz} = \sigma_3, \qquad \sigma_{xy} = \sigma_{xz} = \sigma_{yz} = 0$$

and Eq. (2.25) is simplified accordingly. Application of Eqs. (2.21) to Eq. (2.26b) yields the following stress invariants for T_d:

$$J_1 = 0$$
$$J_2 = I_2 + \tfrac{1}{3}I_1^2 = \tfrac{1}{6}[(\sigma_1 - \sigma_2)^2 + (\sigma_2 - \sigma_3)^2 + (\sigma_3 - \sigma_1)^2]$$
$$J_3 = I_3 + \tfrac{1}{3}I_1 I_2 + \tfrac{2}{27}I_1^3$$
$$= \tfrac{1}{27}(2\sigma_1 - \sigma_2 - \sigma_3)(2\sigma_2 - \sigma_3 - \sigma_1)(2\sigma_3 - \sigma_1 - \sigma_2) \qquad (2.27)$$

The principal directions for T_d are the same as those for T. It can be shown that since $J_1 = 0$, T_d represents a state of *pure shear*. The principal values of the deviator tensor T_d are

$$S_1 = \sigma_1 - \sigma_m = \frac{2\sigma_1 - \sigma_2 - \sigma_3}{3} = \frac{(\sigma_1 - \sigma_3) + (\sigma_1 - \sigma_2)}{3}$$

$$S_2 = \sigma_2 - \sigma_m = \frac{(\sigma_2 - \sigma_3) + (\sigma_2 - \sigma_1)}{3} = \frac{(\sigma_2 - \sigma_3) - (\sigma_1 - \sigma_2)}{3}$$

$$S_3 = \sigma_3 - \sigma_m = \frac{(\sigma_3 - \sigma_1) + (\sigma_3 - \sigma_2)}{3} = -\frac{(\sigma_1 - \sigma_3) + (\sigma_2 - \sigma_3)}{3} \qquad (2.28)$$

Since $S_1 + S_2 + S_3 = 0$, only two of the principal stresses (values) of T_d are independent. Many of the formulas of the mathematical theory of plasticity are often written in terms of the stress invariants of the deviator stress tensor T_d.

Plane Stress

In a large class of important problems, certain approximations may be applied to simplify the three-dimensional stress array [see Eq. (2.3)]. For example, simplifying approximations can be made in analyzing the deformations that occur in a thin flat plate subjected to in-plane forces. We define a thin plate to be a prismatic member (e.g., a cylinder) of a very small length or thickness h. Accordingly, the middle surface of the plate, located halfway between its ends (faces) and parallel to them, may be taken as the (x, y) plane. The thickness direction is then coincident with the direction of the z axis. If the plate is not loaded on its faces, $\sigma_{zz} = \sigma_{zx} = \sigma_{zy} = 0$ on its lateral surfaces $(z = \pm h/2)$. Consequently, since the plate is thin, as a first approximation, it may be assumed that

$$\sigma_{zz} = \sigma_{zx} = \sigma_{zy} = 0 \qquad (2.29)$$

throughout the plate thickness.

Furthermore, it is assumed that the remaining stress components σ_{xx}, σ_{yy}, and σ_{xy} are independent of z. With these approximations, the stress array reduces to a function of the two variables (x, y). Then it is called a *plane stress array* or the *tensor of plane stress*.

Consider a transformation from the (x, y, z) coordinate axes to the (X, Y, Z) coordinate axes for the condition that the z axis and Z axis remain coincident under the transformation. Then, for a state of plane stress in the (x, y) plane, Table 2.3 gives the direction cosines between the axes in a transformation from the (x, y) coordinate axes to the (X, Y) coordinate axes (Fig. 2.10). Hence, with

TABLE 2.3

	x	y	z
X	$l_1 = \cos\theta$	$m_1 = \sin\theta$	$n_1 = 0$
Y	$l_2 = -\sin\theta$	$m_2 = \cos\theta$	$n_2 = 0$
Z	$l_3 = 0$	$m_3 = 0$	$n_3 = 1$

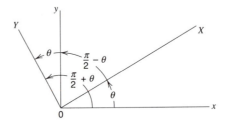

Figure 2.10   Location of transformed axes for plane stress.

Table (2.3) and Fig. (2.10), Eqs. (2.15) and (2.17) yield

$$\sigma_{XX} = \sigma_{xx}\cos^2\theta + \sigma_{yy}\sin^2\theta + 2\sigma_{xy}\sin\theta\cos\theta$$
$$\sigma_{YY} = \sigma_{xx}\sin^2\theta + \sigma_{yy}\cos^2\theta - 2\sigma_{xy}\sin\theta\cos\theta$$
$$\sigma_{XY} = -(\sigma_{xx} - \sigma_{yy})\sin\theta\cos\theta + \sigma_{xy}(\cos^2\theta - \sin^2\theta) \qquad (2.30)$$

By means of trigonometric double angle formulas, Eq. (2.30) may be written in the form

$$\sigma_{XX} = \tfrac{1}{2}(\sigma_{xx} + \sigma_{yy}) + \tfrac{1}{2}(\sigma_{xx} - \sigma_{yy})\cos 2\theta + \sigma_{xy}\sin 2\theta$$
$$\sigma_{YY} = \tfrac{1}{2}(\sigma_{xx} + \sigma_{yy}) - \tfrac{1}{2}(\sigma_{xx} - \sigma_{yy})\cos 2\theta - \sigma_{xy}\sin 2\theta$$
$$\sigma_{XY} = -\tfrac{1}{2}(\sigma_{xx} - \sigma_{yy})\sin 2\theta + \sigma_{xy}\cos 2\theta \qquad (2.31)$$

Equations (2.30) or (2.31) expresss the stress components σ_{XX}, σ_{YY}, and σ_{XY} in the (X, Y) coordinate system in terms of the corresponding stress components σ_{xx}, σ_{yy}, and σ_{xy} in the (x, y) coordinate system for the plane transformation defined by Fig. (2.10) and Table 2.3.

Mohr's Circle in Two Dimensions

In the form of Eq. (2.31), the plane transformation of stress components is particularly suited for graphical interpretation. Stress components σ_{XX} and σ_{XY} act on face BE in Fig. (2.11) that is located at a positive angle θ (counterclockwise) from face BC on which stress components σ_{xx} and σ_{xy} act. The variation of the stress components σ_{XX} and σ_{XY} with θ may be depicted graphically by constructing a diagram in which σ_{XX} and σ_{XY} are coordinates. For each plane BE, there is a point on the diagram whose coordinates correspond to values of σ_{XX} and σ_{XY}.

Rewriting the first of Eqs. (2.31) by moving the first term on the right side to the left side and squaring both sides of the resulting equation, squaring both sides of

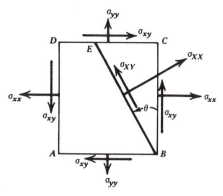

Figure 2.11   Stress components on plane perpendicular to transformed X axis for plane stress.

the last of Eq. (2.31), and adding, we obtain

$$[\sigma_{XX} - \tfrac{1}{2}(\sigma_{xx} + \sigma_{yy})]^2 + (\sigma_{XY} - 0)^2 = \tfrac{1}{4}(\sigma_{xx} - \sigma_{yy})^2 + \sigma_{xy}^2 \qquad (2.32)$$

Equation (2.32) is the equation of a circle in the σ_{XX}, σ_{XY} plane whose center C has coordinates

$$[\tfrac{1}{2}(\sigma_{xx} + \sigma_{yy}), 0] \qquad (2.33)$$

and whose radius R is given by the relation

$$R = \sqrt{\tfrac{1}{4}(\sigma_{xx} - \sigma_{yy})^2 + \sigma_{xy}^2} \qquad (2.34)$$

Consequently, the geometrical representation of the first and third of Eqs. (2.31) is a circle (Fig. 2.12). This stress circle is frequently called *Mohr's circle* in honor of Otto Mohr, who first employed it to study plane stress problems. It is necessary to take the positive direction of the σ_{XY} axis downward so that the positive direction of θ in both Figs. 2.11 and 2.12 is counterclockwise.

Since σ_{xx}, σ_{yy}, and σ_{xy} are known quantities, the circle in Fig. 2.12 can be constructed using Eqs. (2.33) and (2.34). The interpretation of Mohr's circle of stress requires that one known point be located on the circle. When $\theta = 0$ (Fig. 2.10), the first and third of Eqs. (2.31) give

$$\sigma_{XX} = \sigma_{xx} \qquad \text{and} \qquad \sigma_{XY} = \sigma_{xy} \qquad (2.35)$$

which are coordinates of point P in Fig. 2.12.

Principal stresses σ_1 and σ_2 are located at points Q and Q' in Fig. 2.12 and occur when $\theta = \theta_1$ and $\theta_1 + \pi/2$, measured counterclockwise from line CP. The two magnitudes of θ are given by the third of Eqs. (2.31) since $\sigma_{XY} = 0$ when $\theta = \theta_1$ and $\theta_1 + \pi/2$. Note that in Fig. 2.12, we must rotate through angle 2θ from line CP, which corresponds to a rotation of θ from plane BC in Fig. 2.10. [See also Eqs. (2.31)] Thus, by Eqs. (2.31), for $\sigma_{XY} = 0$, we obtain (see also Fig. 2.12)

$$\tan 2\theta = \frac{2\sigma_{xy}}{\sigma_{xx} - \sigma_{yy}} \qquad (2.36)$$

Solution of Eq. (2.36) yields the values $\theta = \theta_1$ and $\theta_1 + \pi/2$.

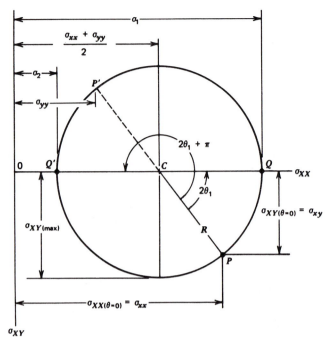

Figure 2.12   Mohr's circle for plane stress.

The magnitudes of the principal stresses from Mohr's circle of stress are

$$\sigma_1 = \frac{\sigma_{xx} + \sigma_{yy}}{2} + \sqrt{\tfrac{1}{4}(\sigma_{xx} - \sigma_{yy})^2 + \sigma_{xy}^2}$$

$$\sigma_2 = \frac{\sigma_{xx} + \sigma_{yy}}{2} - \sqrt{\tfrac{1}{4}(\sigma_{xx} - \sigma_{yy})^2 + \sigma_{xy}^2} \tag{2.37}$$

and are in agreement with the values predicted by the procedure outlined earlier in this section.

Another known point on Mohr's circle of stress can be located although it is not needed for the interpretation of the circle. When $\theta = \pi/2$, the first and third of Eqs. (2.31) give

$$\sigma_{XX} = \sigma_{yy} \quad \text{and} \quad \sigma_{XY} = -\sigma_{xy} \tag{2.38}$$

These coordinates locate point P' in Fig. 2.12, which is on the opposite end of the diameter from point P.

Note that Example 2.1 could also have been solved by means of Mohr's circle.

EXAMPLE 2.2
Mohr's Circle in Two Dimensions

A piece of chalk is subjected to combined loading consisting of a tensile load P and torque T (Fig. E2.2a). The chalk has an ultimate strength σ_u as determined in a simple tensile test. The load P remains constant at such a value that it produces a tensile stress $0.51 \sigma_u$ on any cross section. The torque T is increased gradually until fracture occurs on some inclined surface.

Assuming that fracture takes place when the maximum principal stress σ_1 reaches the ultimate strength σ_u, determine the magnitude of the torsional shear stress produced by torque T at fracture and determine the orientation of the fracture surface.

SOLUTION

Take the x and y axes with their origin at a point on the surface of the chalk as shown in Fig. E2.2a. Then a volume element taken from the chalk at the origin of the axes will be in plane stress (Fig. E2.2b) with $\sigma_{xx} = 0.51 \sigma_u$, $\sigma_{yy} = 0$, and σ_{xy}

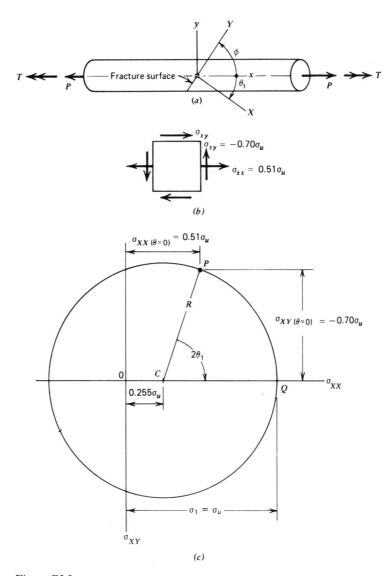

Figure E2.2

unknown. The magnitude of the shear stress σ_{xy} can be determined from the condition that the maximum principal stress σ_1 [given by Eq. (2.37)] is equal to σ_u; thus,

$$\sigma_u = 0.255\,\sigma_u + \sqrt{(0.255\,\sigma_u)^2 + \sigma_{xy}^2}$$
$$\sigma_{xy} = 0.700\,\sigma_u$$

Since the torque acting on the right end of the piece of chalk is counterclockwise, the shear stress σ_{xy} acts down on the front face of the volume element (Fig. E2.2b) and is therefore negative. Thus,

$$\sigma_{xy} = -0.700\,\sigma_u$$

In other words, σ_{xy} actually acts downward on the right face of Fig. E2.2b and upward on the left face. We determine the location of the fracture surface first using Mohr's circle of stress and then using Eq. (2.36). As indicated in Fig. E2.2c, the center C of Mohr's circle of stress lies on the σ_{XX} axis at distance $0.255\,\sigma_u$ from the origin 0 [see Eq. (2.33)]. The radius R of the circle is given by Eq. (2.34); $R = 0.745\,\sigma_u$. When $\theta = 0$, the stress components $\sigma_{XX(\theta=0)} = \sigma_{xx} = 0.51\,\sigma_u$ and $\sigma_{XY(\theta=0)} = \sigma_{xy} = -0.700\,\sigma_u$ locate point P on the circle. Point Q representing the maximum principal stress is located by rotating clockwise through angle $2\theta_1$ from point P; therefore, the fracture plane is perpendicular to the X axis, which is located at an angle θ_1 clockwise from the x axis. The angle θ_1 can also be obtained from Eq. (2.36), as the solution of

$$\tan\theta_1 = \frac{2\sigma_{xy}}{\sigma_{xx}} = -\frac{2(0.700\,\sigma_u)}{0.51\,\sigma_u} = -2.7452$$

Thus,

$$\theta_1 = -0.6107 \text{ rad}$$

Since θ_1 is negative, the X axis is located clockwise through angle θ_1 from the x axis. The fracture plane is at angle ϕ from the x axis. It is given as

$$\phi = \frac{\pi}{2} - |\theta_1| = 0.9601 \text{ rad}$$

The magnitude of ϕ depends on the magnitude of P. If $P = 0$, the chalk is subjected to pure torsion and $\phi = \pi/4$. If $P/A = \sigma_u$ (A is the cross-sectional area), the chalk is subjected to pure tension ($T = 0$) and $\phi = \pi/2$.

Mohr's Circles in Three Dimensions*

As discussed in Chapter 4, the failure of load-carrying members is often associated with either the maximum normal stress or the maximum shear stress at the point in the member where failure is initiated. The maximum normal stress is equal to the maximum of the three principal stresses σ_1, σ_2, and σ_3. In general, we will order the

---

* In the early history of stress analysis, Mohr's circles in three dimensions were used extensively. However, today, they are used principally as a heuristic device.

principal stresses so that $\sigma_1 > \sigma_2 > \sigma_3$. Then, σ_1 is the maximum (signed) principal stress and σ_3 the minimum principal stress (see Fig. 2.13.) Procedures have been presented for determining the values of the principal stresses for either the general state of stress or for plane stress. For plane stress states, two of the principal stresses are given by Eqs. (2.37) the third being $\sigma_{zz} = 0$.

Even though the construction of Mohr's circle of stress was presented for plane stress ($\sigma_{zz} = 0$), the transformation equations given by either Eqs. (2.30) or (2.31) are not influenced by the magnitude of σ_{zz} but require only that $\sigma_{zx} = \sigma_{zy} = 0$ (Problem 2.2). Therefore, in terms of the principal stresses, Mohr's circle of stress can be constructed by using any two of the principal stresses, thus giving three Mohr's circles for any given state of stress. Consider any point in a stressed body for which values of σ_1, σ_2, and σ_3 are known. For any plane through the point, let the N axis be normal to the plane and the S axis coincide with the shear component of the stress for the plane. If we choose σ_{NN} and σ_{NS} as coordinate axes in Fig. 2.13, three Mohr's circles of stress can be constructed. As will be shown later, the stress components σ_{NN} and σ_{NS} for any plane passing through the point locates a point either on one of the three circles in Fig. 2.13 or in one of the two shaded areas. The maximum shear stress $\tau_{\max}$ for the point is equal to the maximum value of σ_{NS} and is equal in magnitude to the radius of the largest of the three Mohr's circles of stress. Hence,

$$\tau_{\max} = \sigma_{NS(\max)} = \frac{\sigma_{\max} - \sigma_{\min}}{2} \tag{2.39}$$

where $\sigma_{\max} = \sigma_1$ and $\sigma_{\min} = \sigma_3$ (Fig. 2.13).

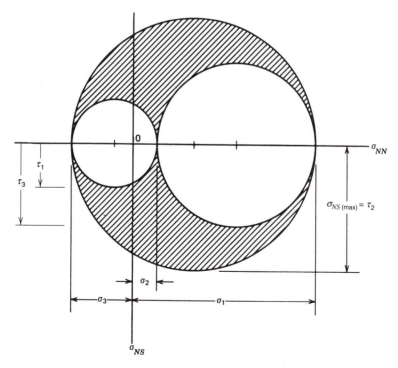

Figure 2.13   Mohr's circles in three dimensions.

Once the state of stress at a point is expressed in terms of the principal stresses, three Mohr's circles of stress can be constructed as indicated in Fig. 2.13. Consider plane P whose normal relative to the principal axes has direction cosines l, m, and n. The normal stress σ_{NN} on plane P is by Eq. (2.11)

$$\sigma_{NN} = l^2\sigma_1 + m^2\sigma_2 + n^2\sigma_3 \qquad (2.40)$$

Similarly, the square of the shear stress σ_{NS} on plane P is, by Eqs. (2.10) and (2.12),

$$\sigma_{NS}^2 = l^2\sigma_1^2 + m^2\sigma_2^2 + n^2\sigma_3^2 - \sigma_{NN}^2 \qquad (2.41)$$

For known values of the principal stresses σ_1, σ_2, and σ_3 and of the direction cosines l, m, and n for plane P, graphical techniques can be developed to locate the point in the shaded area of Fig. 2.13 whose coordinates $(\sigma_{NN}, \sigma_{NS})$ are the normal and shear stress components acting on plane P. However, we recommend the procedure in Sec. 2.3 to determine magnitudes for σ_{NN} and σ_{NS}. In the discussion to follow, we show that the coordinates $(\sigma_{NN}, \sigma_{NS})$ locate a point in the shaded area of Fig. 2.13.

Since

$$l^2 + m^2 + n^2 = 1 \qquad (2.42)$$

Eqs. (2.40), (2.41), and (2.42) are three simultaneous equations in l^2, m^2, and n^2. Solving for l^2, m^2, and n^2 and noting that $l^2 \geq 0$, $m^2 \geq 0$, and $n^2 \geq 0$, we obtain

$$l^2 = \frac{\sigma_{NS}^2 + (\sigma_{NN} - \sigma_2)(\sigma_{NN} - \sigma_3)}{(\sigma_1 - \sigma_2)(\sigma_1 - \sigma_3)} \geq 0$$

$$m^2 = \frac{\sigma_{NS}^2 + (\sigma_{NN} - \sigma_1)(\sigma_{NN} - \sigma_3)}{(\sigma_2 - \sigma_3)(\sigma_2 - \sigma_1)} \geq 0$$

$$n^2 = \frac{\sigma_{NS}^2 + (\sigma_{NN} - \sigma_1)(\sigma_{NN} - \sigma_2)}{(\sigma_3 - \sigma_1)(\sigma_3 - \sigma_2)} \geq 0 \qquad (2.43)$$

Ordering the principal stresses such that $\sigma_1 > \sigma_2 > \sigma_3$, we may write Eqs. (2.43) in the form

$$\sigma_{NS}^2 + (\sigma_{NN} - \sigma_2)(\sigma_{NN} - \sigma_3) \geq 0$$
$$\sigma_{NS}^2 + (\sigma_{NN} - \sigma_3)(\sigma_{NN} - \sigma_1) \leq 0$$
$$\sigma_{NS}^2 + (\sigma_{NN} - \sigma_1)(\sigma_{NN} - \sigma_2) \geq 0$$

These inequalities may be rewritten in the form

$$\sigma_{NS}^2 + \left(\sigma_{NN} - \frac{\sigma_2 + \sigma_3}{2}\right)^2 \geq \frac{1}{4}(\sigma_2 - \sigma_3)^2 = \tau_1^2$$

$$\sigma_{NS}^2 + \left(\sigma_{NN} - \frac{\sigma_1 + \sigma_3}{2}\right)^2 \leq \frac{1}{4}(\sigma_3 - \sigma_1)^2 = \tau_2^2$$

$$\sigma_{NS}^2 + \left(\sigma_{NN} - \frac{\sigma_1 + \sigma_2}{2}\right)^2 \geq \frac{1}{4}(\sigma_1 - \sigma_2)^2 = \tau_3^2 \qquad (2.44)$$

where $\tau_1 = \frac{1}{2}|\sigma_2 - \sigma_3|$, $\tau_2 = \frac{1}{2}|\sigma_3 - \sigma_1|$, $\tau_3 = \frac{1}{2}|\sigma_1 - \sigma_2|$ are the maximum (extreme) magnitudes of the shear stresses in three-dimensional principal stress space and $(\sigma_1, \sigma_2, \sigma_3)$ are the signed principal stresses (see Fig. 2.13). The inequalities of Eqs. (2.44) may be interpreted graphically as follows: Let $(\sigma_{NN}, \sigma_{NS})$ denote the abscissa and ordinate, respectively, on a graph (Fig. 2.13). Then, an admissible state of stress must lie within a region bounded by three circles obtained from Eqs. (2.44) where the equalities are taken (the shaded region in Fig. 2.13).

EXAMPLE 2.3
Three-Dimensional State of Stress

Let the state of stress at a point be given by $\sigma_{xx} = 120$ MPa, $\sigma_{yy} = 55$ MPa, $\sigma_{zz} = -85$ MPa, $\sigma_{xy} = -55$ MPa, $\sigma_{yz} = 33$ MPa, and $\sigma_{zx} = -75$ MPa. Determine the three principal stresses and the directions associated with them.

SOLUTION

Substituting the given stress components into Eq. (2.20), we obtain

$$\sigma^3 - 90\sigma^2 - 18{,}014\sigma + 471{,}680 = 0$$

The three principal stresses are the three roots of this equation. They are

$$\sigma_1 = 176.80 \text{ MPa}, \qquad \sigma_2 = -110.86 \text{ MPa}, \qquad \sigma_3 = 24.06 \text{ MPa}$$

The direction cosines for any one of the principal stress directions are given by substituting the given principal stress into Eqs. (2.18). Substitution of σ_1 into Eqs. (2.18) gives

$$
\begin{aligned}
(120 - 176.80)l_1 - 55m_1 - 75n_1 &= 0 \\
-55l_1 + (55 - 176.80)m_1 + 33n_1 &= 0 \\
-75l_1 + 33m_1 + (-85 - 176.80)n_1 &= 0
\end{aligned}
\tag{a}
$$

where $l_1, m_1,$ and n_1 are the direction cosines for the σ_1 direction. Only two of these equations are independent; in addition, the direction cosines must satisfy the equation

$$l_i^2 + m_i^2 + n_i^2 = 1, \qquad i = 1, 2, \text{ or } 3 \tag{b}$$

The simultaneous solution of any two of Eqs. (a) along with Eq. (b) gives

$$l_1 = 0.8372, \qquad m_1 = -0.4587, \qquad n_1 = -0.2977$$

In a similar manner, we obtain sets of direction cosines for σ_2 and σ_3.

$$
\begin{aligned}
l_2 &= 0.2872, & m_2 &= -0.0944, & n_2 &= 0.9532 \\
l_3 &= 0.4657, & m_3 &= 0.8834, & n_3 &= -0.0521
\end{aligned}
$$

2.5

DIFFERENTIAL EQUATIONS OF MOTION OF A DEFORMABLE BODY

In previous sections, we determined the stress components needed to specify the state of stress at a *point* 0 in a deformed body for a given set of orthogonal coordinate axes (x, y, z). We derived transformation equations that define the state of stress at point 0 for any other set of orthogonal axes (X, Y, Z) rotated with respect to (x, y, z). We derived relations that give at point 0 the principal stresses and their directions, the maximum shear stress, the octahedral normal and shear stresses, and the hydrostatic and deviatoric states of stress.

In this section, we derive differential equations of motion of a deformable solid body (differential equations of equilibrium if the deformed body has zero acceleration). These equations are needed when the theory of elasticity is used to derive load-stress and load-deflection relations for a member. We consider a general deformed body and choose a differential volume element at point 0 in the body as indicated in Fig. 2.14. The form of the differential equations of motion depends on the type of orthogonal coordinate axes employed. We choose rectangular coordinate axes (x, y, z) whose directions are parallel to the edges of the volume element. *In this book, we restrict our consideration mainly to small displacements and, therefore, do not distinguish between coordinate axes in the deformed state and in the undeformed state* (Boresi and Chong, 1987). Six cutting planes bound the volume element shown as a free-body diagram in Fig. 2.15. In general, the state of stress changes with the location of point 0. In particular, the stress components undergo changes from one face of the volume element to another face. Body forces (B_x, B_y, B_z) are included in the free-body diagram.

To write the differential equations of motion, each stress component must be multiplied by the area on which it acts and each body force must be multiplied by the volume of the element since (B_x, B_y, B_z) have dimensions of force per unit volume. The equations of motion for the volume element in Fig. 2.15 are then obtained by summation of these forces and summation of moments. In Sec. 2.3 we

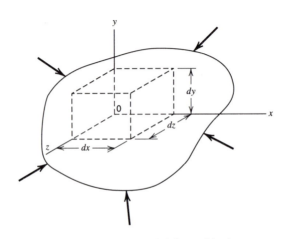

Figure 2.14 General deformed body.

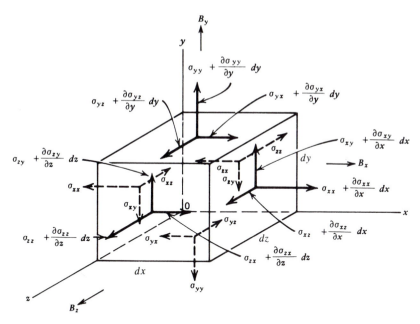

Figure 2.15   Stress components showing changes from face to face along with body force per unit volume including inertial forces.

have already used summation of moments to obtain the stress symmetry conditions [Eqs. (2.4)]. Summation of forces in the x direction gives*

$$\frac{\partial \sigma_{xx}}{\partial x} + \frac{\partial \sigma_{yx}}{\partial y} + \frac{\partial \sigma_{zx}}{\partial z} + B_x = 0$$

where σ_{xx}, $\sigma_{yx} = \sigma_{xy}$, and $\sigma_{xz} = \sigma_{zx}$ are stress components in the x direction and B_x is the body force per unit volume in the x direction including inertial (acceleration) forces. Summation of forces in the y and z directions yields similar results. The three equations of motion are thus

$$\frac{\partial \sigma_{xx}}{\partial x} + \frac{\partial \sigma_{yx}}{\partial y} + \frac{\partial \sigma_{zx}}{\partial z} + B_x = 0$$

$$\frac{\partial \sigma_{xy}}{\partial x} + \frac{\partial \sigma_{yy}}{\partial y} + \frac{\partial \sigma_{zy}}{\partial z} + B_y = 0$$

$$\frac{\partial \sigma_{xz}}{\partial x} + \frac{\partial \sigma_{yz}}{\partial y} + \frac{\partial \sigma_{zz}}{\partial z} + B_z = 0 \qquad (2.45)$$

We use Eqs. (2.45) in the treatment of torsion of noncircular sections (Chapter 6).

As noted earlier, the form of the differential equations of motion depends on the coordinate axes; Eqs. (2.45) were derived for rectangular coordinate axes. In this book we need differential equations of motion in terms of cylindrical coordinates and plane polar coordinates. These are not derived here; instead, we present the

---

* Note σ_{xx} on the left face of the element goes to $\sigma_{xx} + d\sigma_{xx} = \sigma_{xx} + (\partial\sigma_{xx}/\partial x)\,dx$ on the right face of the element, with similar changes for the other stress components (Fig. 2.15).

most general form from the literature (Boresi and Chong, 1987, pp. 218–222) and show how the general form can be reduced to desired forms. The equations of motion relative to orthogonal curvilinear coordinates (x, y, z) (see Fig. 2.16), are

$$\frac{\partial(\beta\gamma\sigma_{xx})}{\partial x} + \frac{\partial(\gamma\alpha\sigma_{yx})}{\partial y} + \frac{\partial(\alpha\beta\sigma_{zx})}{\partial z} + \gamma\sigma_{yx}\frac{\partial\alpha}{\partial y}$$

$$+ \beta\sigma_{zx}\frac{\partial\alpha}{\partial z} - \gamma\sigma_{yy}\frac{\partial\beta}{\partial x} - \beta\sigma_{zz}\frac{\partial\gamma}{\partial x} + \alpha\beta\gamma B_x = 0$$

$$\frac{\partial(\beta\gamma\sigma_{xy})}{\partial x} + \frac{\partial(\gamma\alpha\sigma_{yy})}{\partial y} + \frac{\partial(\alpha\beta\sigma_{zy})}{\partial z} + \alpha\sigma_{zy}\frac{\partial\beta}{\partial z}$$

$$+ \gamma\sigma_{xy}\frac{\partial\beta}{\partial x} - \alpha\sigma_{zz}\frac{\partial\gamma}{\partial y} - \gamma\sigma_{xx}\frac{\partial\alpha}{\partial y} + \alpha\beta\gamma B_y = 0$$

$$\frac{\partial(\beta\gamma\sigma_{xz})}{\partial x} + \frac{\partial(\gamma\alpha\sigma_{yz})}{\partial y} + \frac{\partial(\alpha\beta\sigma_{zz})}{\partial z} + \beta\sigma_{xz}\frac{\partial\gamma}{\partial x}$$

$$+ \alpha\sigma_{yz}\frac{\partial\gamma}{\partial y} - \beta\sigma_{xx}\frac{\partial\alpha}{\partial z} - \alpha\sigma_{yy}\frac{\partial\beta}{\partial z} + \alpha\beta\gamma B_z = 0 \tag{2.46}$$

where (α, β, γ) are metric coefficients that are functions of the coordinates (x, y, z). They are defined by

$$ds^2 = \alpha^2\,dx^2 + \beta^2\,dy^2 + \gamma^2\,dz^2 \tag{2.47}$$

where ds is the differential arc length representing the diagonal of a volume element (Fig. 2.16) with edge lengths $\alpha\,dx$, $\beta\,dy$, and $\gamma\,dz$, and where (B_x, B_y, B_z) are the components of body force per unit volume including inertial forces. For rectangular coordinates, $\alpha = \beta = \gamma = 1$ and Eqs. (2.46) reduce to Eqs. (2.45).

Specialization of Equations (2.46)

Commonly employed orthogonal curvilinear systems in three-dimensional problems are the cylindrical coordinate system (r, θ, z) and spherical coordinate system (r, θ, ϕ); in plane problems, the plane polar coordinate system (r, θ) is frequently used. We will now specialize Eqs. (2.46) for these systems.

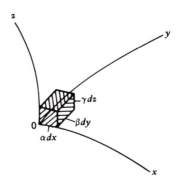

Figure 2.16   Orthogonal curvilinear coordinates.

(a) Cylindrical Coordinate System (r, θ, z). In Eqs. (2.46), we let $x = r$, $y = \theta$, $z = z$. Then the differential length ds is defined by the relation

$$ds^2 = dr^2 + r^2\, d\theta^2 + dz^2 \tag{2.48}$$

A comparison of Eqs. (2.47) and (2.48) yields

$$\alpha = 1, \qquad \beta = r, \qquad \gamma = 1 \tag{2.49}$$

Substituting Eq. (2.49) into Eqs. (2.46), we obtain the differential equations of motion

$$\frac{\partial \sigma_{rr}}{\partial r} + \frac{1}{r}\frac{\partial \sigma_{\theta r}}{\partial \theta} + \frac{\partial \sigma_{zr}}{\partial z} + \frac{\sigma_{rr} - \sigma_{\theta\theta}}{r} + B_r = 0$$

$$\frac{\partial \sigma_{r\theta}}{\partial r} + \frac{1}{r}\frac{\partial \sigma_{\theta\theta}}{\partial \theta} + \frac{\partial \sigma_{z\theta}}{\partial z} + \frac{2\sigma_{r\theta}}{r} + B_\theta = 0$$

$$\frac{\partial \sigma_{rz}}{\partial r} + \frac{1}{r}\frac{\partial \sigma_{\theta z}}{\partial \theta} + \frac{\partial \sigma_{zz}}{\partial z} + \frac{\sigma_{rz}}{r} + B_z = 0 \tag{2.50}$$

where $(\sigma_{rr}, \sigma_{\theta\theta}, \sigma_{zz}, \sigma_{r\theta}, \sigma_{rz}, \sigma_{\theta z})$ represent stress components defined relative to cylindrical coordinates (r, θ, z). We use Eqs. (2.50) in Chapter 11 to derive load-stress and load-deflection relations for thick-wall cylinders.

(b) Spherical Coordinate System (r, θ, ϕ). In Eqs. (2.46), we let $x = r, y = \theta, z = \phi$, where r is the radial coordinate, θ the colatitude, and ϕ the longitude. Since the differential length ds is defined by

$$ds^2 = dr^2 + r^2\, d\theta^2 + r^2 \sin^2\theta\, d\phi^2 \tag{2.51}$$

comparison of Eqs. (2.47) and (2.51) yields

$$\alpha = 1, \qquad \beta = r, \qquad \gamma = r\sin\theta \tag{2.52}$$

Substituting Eq. (2.52) into Eqs. (2.46), we obtain the differential equations of motion

$$\frac{\partial \sigma_{rr}}{\partial r} + \frac{1}{r}\frac{\partial \sigma_{\theta r}}{\partial \theta} + \frac{1}{r\sin\theta}\frac{\partial \sigma_{\phi r}}{\partial \phi} + \frac{1}{r}(2\sigma_{rr} - \sigma_{\theta\theta} - \sigma_{\phi\phi} + \sigma_{\theta r}\cot\theta) + B_r = 0$$

$$\frac{\partial \sigma_{r\theta}}{\partial r} + \frac{1}{r}\frac{\partial \sigma_{\theta\theta}}{\partial \theta} + \frac{1}{r\sin\theta}\frac{\partial \sigma_{\phi\theta}}{\partial \phi} + \frac{1}{r}[(\sigma_{\theta\theta} - \sigma_{\phi\phi})\cot\theta + 3\sigma_{r\theta}] + B_\theta = 0$$

$$\frac{\partial \sigma_{r\phi}}{\partial r} + \frac{1}{r}\frac{\partial \sigma_{\theta\phi}}{\partial \theta} + \frac{1}{r\sin\theta}\frac{\partial \sigma_{\phi\phi}}{\partial \phi} + \frac{1}{r}(3\sigma_{r\phi} + 2\sigma_{\theta\phi}\cot\theta) + B_\phi = 0 \tag{2.53}$$

where $(\sigma_{rr}, \sigma_{\theta\theta}, \sigma_{\phi\phi}, \sigma_{r\theta}, \sigma_{r\phi}, \sigma_{\theta\phi})$ are defined relative to spherical coordinates (r, θ, ϕ).

(c) Plane Polar Coordinate System (r, θ). In plane-stress problems relative to (x, y) coordinates, $\sigma_{zz} = \sigma_{xz} = \sigma_{yz} = 0$, and the remaining stress components are functions of (x, y) only (Sec. 2.4). Letting $x = r, y = \theta, z = z$ in Eqs. (2.50) and noting that $\sigma_{zz} = \sigma_{rz} = \sigma_{\theta z} = (\partial/\partial z) = 0$, we obtain from Eq. (2.50)

$$\frac{\partial \sigma_{rr}}{\partial r} + \frac{1}{r} \frac{\partial \sigma_{\theta r}}{\partial \theta} + \frac{\sigma_{rr} - \sigma_{\theta\theta}}{r} + B_r = 0$$

$$\frac{\partial \sigma_{r\theta}}{\partial r} + \frac{1}{r} \frac{\partial \sigma_{\theta\theta}}{\partial \theta} + 2\frac{\sigma_{r\theta}}{r} + B_\theta = 0 \tag{2.54}$$

2.6

DEFORMATION OF A DEFORMABLE BODY

In the first four sections of this chapter, we examined the six stress components that define the state of stress at a point in a loaded member, derived the transformation equations of stress, and derived expressions for the maximum principal stress, maximum shear stress, and maximum octahedral shear stress at a point. These relations are of interest throughout most of the book. Differential equations of equilibrium (differential equations of motion for members being accelerated) were derived in Sec. 2.5. These are needed in chapters in which the theory of elasticity is used to derive load-stress and load-deflection relations. Additionally, differential equations of compatibility, needed in the theory of elasticity, are derived in Sec. 2.8; the derivation employs small displacement approximations and the associated strain-displacement relations. Although small displacements are considered in most applications of this book, more general finite strain-displacement relations are derived in this chapter so that the reader may better understand the approximations that lead to the strain-displacement relations of small-displacement theory.

In the derivation of strain-displacement relations for a member, we consider the member first to be unloaded (undeformed and unstressed) and next to be loaded (stressed and deformed). We let R represent the closed region occupied by the undeformed member and R^* the closed region occupied by the deformed member. Asterisks are used to designate quantities associated with the deformed state of members throughout the book.

Let (x, y, z) be rectangular coordinates (Fig. 2.17). A particle P is located at the general coordinate point (x, y, z) in the undeformed body. Under a deformation, the particle moves to a point (x^*, y^*, z^*) in the deformed state defined by the equations

$$x^* = x^*(x, y, z)$$
$$y^* = y^*(x, y, z)$$
$$z^* = z^*(x, y, z) \tag{2.55}$$

where the values of (x, y, z) are restricted to region R and (x^*, y^*, z^*) are restricted to region R^*. Equations (2.55) define the final location of a particle P that lies at a given point (x, y, z) in the undeformed member. It is assumed that the functions (x^*, y^*, z^*) are continuous and differentiable in the independent variables (x, y, z), since a discontinuity of these functions would imply a rupture of the member. Mathematically, this means that Eqs. (2.55) may be solved for single-valued solu-

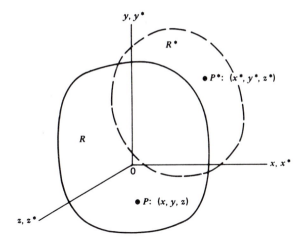

Figure 2.17 Location of general point P in undeformed and deformed body.

tions of (x, y, z); that is,

$$x = x(x^*, y^*, z^*)$$
$$y = y(x^*, y^*, z^*)$$
$$z = z(x^*, y^*, z^*) \tag{2.56}$$

Equations (2.56) define the initial location of a particle P that lies at point (x^*, y^*, z^*) in the deformed member. Functions (x, y, z) are continuous and differentiable in the independent variables (x^*, y^*, z^*).

When (x^*, y^*, z^*) are used as independent variables [Eq. (2.56)], the point of view is that of the *Eulerian or spatial coordinate method*. When (x, y, z) are used as independent variables, the point of view is that of the *Lagrangian or material coordinate method*. It may be shown that for classical, small-displacement theories of elasticity and plasticity, it is not necessary to distinguish between the variables (x^*, y^*, z^*) and (x, y, z). We employ material coordinates in this book.

2.7

STRAIN THEORY. TRANSFORMATION OF STRAIN. PRINCIPAL STRAINS*

The theory of stress of a continuous medium rests solely on Newton's laws. As will be shown in this section, the theory of strain rests solely on geometric concepts. Both the theories of stress and strain are, therefore, independent of material behavior and, as such, are applicable to the study of all materials. Furthermore, although the theories of stress and strain are based on different physical concepts,

---

* The theory presented in this article includes quadratic terms in the displacement components (u, v, w) and in the engineering strain ϵ_E. One may discard all quadratic terms in u, v, w, and ϵ_E and directly obtain the theory of strain for small deformations. (See Sec. 2.8.)

mathematically, they are equivalent, as will become evident in the following discussion.

Strain of a Line Element

When a body is deformed, the particle at point $P:(x, y, z)$ passes to the point $P^*:(x^*, y^*, z^*)$ (Fig. 2.18). Also, the particle at point $Q:(x + dx, y + dy, z + dz)$ passes to the point $Q^*:(x^* + dx^*, y^* + dy^*, z^* + dz^*)$, and the infinitesimal line element $PQ = ds$ passes into the line element $P^*Q^* = ds^*$. We define the *engineering strain* ϵ_E of the line element $PQ = ds$ as

$$\epsilon_E = \frac{ds^* - ds}{ds} \tag{2.57}$$

Therefore, by the definition, $\epsilon_E > -1$. Equation (2.57) is employed widely in engineering.

By Eqs. (2.55), we obtain the total differential

$$dx^* = \frac{\partial x^*}{\partial x} dx + \frac{\partial x^*}{\partial y} dy + \frac{\partial x^*}{\partial z} dz \tag{2.58}$$

with similar expressions for dy^*, dz^*. Noting that

$$x^* = x + u$$
$$y^* = y + v$$
$$z^* = z + w \tag{2.59}$$

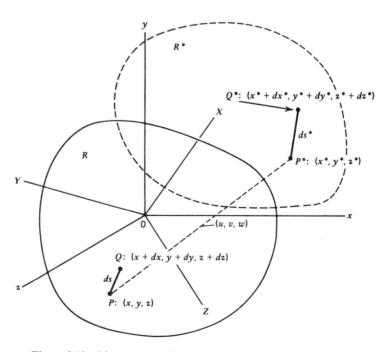

Figure 2.18   Line segment PQ in undeformed and deformed body.

where (u, v, w) denote the (x, y, z) components of the displacement of P to P^*, and also noting that

$$(ds)^2 = (dx)^2 + (dy)^2 + (dz)^2$$
$$(ds^*)^2 = (dx^*)^2 + (dy^*)^2 + (dz^*)^2 \tag{2.60}$$

we find[†] [retaining quadratic terms in derivatives of (u, v, w)]

$$M = \frac{1}{2}\left[\left(\frac{ds^*}{ds}\right)^2 - 1\right] = \epsilon_E + \tfrac{1}{2}\epsilon_E^2 = l^2\epsilon_{xx} + lm\epsilon_{xy} + ln\epsilon_{xz}$$

$$\quad + ml\epsilon_{yx} + m^2\epsilon_{yy} + mn\epsilon_{yz} + nl\epsilon_{zx} + nm\epsilon_{zy} + n^2\epsilon_{zz}$$

$$= l^2\epsilon_{xx} + m^2\epsilon_{yy} + n^2\epsilon_{zz} + 2lm\epsilon_{xy} + 2ln\epsilon_{xz} + 2mn\epsilon_{yz} \tag{2.61}$$

where M is called the *magnification factor* and

$$\epsilon_{xx} = \frac{\partial u}{\partial x} + \frac{1}{2}\left[\left(\frac{\partial u}{\partial x}\right)^2 + \left(\frac{\partial v}{\partial x}\right)^2 + \left(\frac{\partial w}{\partial x}\right)^2\right]$$

$$\epsilon_{yy} = \frac{\partial v}{\partial y} + \frac{1}{2}\left[\left(\frac{\partial u}{\partial y}\right)^2 + \left(\frac{\partial v}{\partial y}\right)^2 + \left(\frac{\partial w}{\partial y}\right)^2\right]$$

$$\epsilon_{zz} = \frac{\partial w}{\partial z} + \frac{1}{2}\left[\left(\frac{\partial u}{\partial z}\right)^2 + \left(\frac{\partial v}{\partial z}\right)^2 + \left(\frac{\partial w}{\partial z}\right)^2\right] \tag{2.62}$$

$$\epsilon_{xy} = \epsilon_{yx} = \frac{1}{2}\left(\frac{\partial v}{\partial x} + \frac{\partial u}{\partial y} + \frac{\partial u}{\partial x}\frac{\partial u}{\partial y} + \frac{\partial v}{\partial x}\frac{\partial v}{\partial y} + \frac{\partial w}{\partial x}\frac{\partial w}{\partial y}\right)$$

$$\epsilon_{xz} = \epsilon_{zx} = \frac{1}{2}\left(\frac{\partial w}{\partial x} + \frac{\partial u}{\partial z} + \frac{\partial u}{\partial x}\frac{\partial u}{\partial z} + \frac{\partial v}{\partial x}\frac{\partial v}{\partial z} + \frac{\partial w}{\partial x}\frac{\partial w}{\partial z}\right)$$

$$\epsilon_{yz} = \epsilon_{zy} = \frac{1}{2}\left(\frac{\partial w}{\partial y} + \frac{\partial v}{\partial z} + \frac{\partial u}{\partial y}\frac{\partial u}{\partial z} + \frac{\partial v}{\partial y}\frac{\partial v}{\partial z} + \frac{\partial w}{\partial y}\frac{\partial w}{\partial z}\right)$$

are the finite strain-displacement relations[‡] and where

$$l = \frac{dx}{ds}, \qquad m = \frac{dy}{ds}, \qquad n = \frac{dz}{ds} \tag{2.63}$$

are the direction cosines of line element ds.

We may interpret the quantities ϵ_{xx}, ϵ_{yy}, ϵ_{zz} physically, by considering line elements ds that lie parallel to the (x, y, z) axes, respectively. For example, let the line element ds (Fig. 2.18) lie parallel to the x axis. Then $l = 1$, $m = n = 0$, and

---

[†] Although one may compute ϵ_E directly from Eq. (2.57), it is mathematically simpler to form the quantity $M = \frac{1}{2}[(ds^*/ds)^2 - 1] = \frac{1}{2}[(1 + \epsilon_E)^2 - 1] = \epsilon_E + \frac{1}{2}\epsilon_E^2$. Then one may compute ϵ_E from Eq. (2.61). For small ϵ_E (Sec. 2.8), $\epsilon_E \cong M$. A more detailed derivation of Eq. (2.61) is given by Boresi and Chong (1987, Sec. 2-6).

[‡] In small displacement theory, the quadratic terms in Eqs. (2.62) are neglected. Then, Eqs. (2.62) reduce to Eqs. (2.81).

Eq. (2.61) yields

$$M_x = \epsilon_{Ex} + \tfrac{1}{2}\epsilon_{Ex}^2 = \epsilon_{xx} \tag{2.61a}$$

where M_x and ϵ_{Ex} denote the magnification factor and the engineering strain of the element ds (parallel to the x direction). Hence, ϵ_{xx}, physically, is the magnification factor of the line element at P that lies initially in the x direction. In particular, if the engineering strain is small ($\epsilon_{Ex} \ll 1$), we obtain the result $\epsilon_{xx} \approx \epsilon_{Ex}$: namely, that ϵ_{xx} is approximately equal to the engineering strain for small strains. Similarly, for the cases where initially ds lies parallel to the y axis and then the z axis, we obtain

$$M_y = \epsilon_{Ey} + \tfrac{1}{2}\epsilon_{Ey}^2 = \epsilon_{yy}$$
$$M_z = \epsilon_{Ez} + \tfrac{1}{2}\epsilon_{Ez}^2 = \epsilon_{zz} \tag{2.61b}$$

Thus, $(\epsilon_{xx}, \epsilon_{yy}, \epsilon_{zz})$ physically represent the magnification factors for line elements that initially lie parallel to the (x, y, z) axes, respectively.

To obtain a physical interpretation of the components $\epsilon_{xy}, \epsilon_{xz}, \epsilon_{yz}$, it is necessary to determine the rotation between two line elements initially parallel to the (x, y) axes, (x, z) axes, and (y, z) axes, respectively. To do this, we first determine the final direction of a single line element under the deformation. Then, we use this result to determine the rotation between two line elements.

Final Direction of Line Element

As a result of the deformation, the line element ds: (dx, dy, dz) deforms into the line element ds^*: (dx^*, dy^*, dz^*). By definition, the direction cosines of ds and ds^* are

$$l = \frac{dx}{ds}, \qquad m = \frac{dy}{ds}, \qquad n = \frac{dz}{ds}$$

$$l^* = \frac{dx^*}{ds^*}, \qquad m^* = \frac{dy^*}{ds^*}, \qquad n^* = \frac{dz^*}{ds^*} \tag{2.64}$$

Alternatively, we may write

$$l^* = \frac{dx^*}{ds}\frac{ds}{ds^*}, \qquad m^* = \frac{dy^*}{ds}\frac{ds}{ds^*}, \qquad n^* = \frac{dz^*}{ds}\frac{ds}{ds^*} \tag{2.65}$$

By Eqs. (2.58) and (2.59), we find

$$\frac{dx^*}{ds} = \left(1 + \frac{\partial u}{\partial x}\right)l + \frac{\partial u}{\partial y}m + \frac{\partial u}{\partial z}n$$

$$\frac{dy^*}{ds} = \frac{\partial v}{\partial x}l + \left(1 + \frac{\partial v}{\partial y}\right)m + \frac{\partial v}{\partial z}n$$

$$\frac{dz^*}{ds} = \frac{\partial w}{\partial x}l + \frac{\partial w}{\partial y}m + \left(1 + \frac{\partial w}{\partial z}\right)n \tag{2.66}$$

and by Eq. (2.57)

$$\frac{ds}{ds^*} = \frac{1}{1 + \epsilon_E} \tag{2.67}$$

Hence, Eqs. (2.65), (2.66), and (2.67) yield

$$(1 + \epsilon_E)l^* = \left(1 + \frac{\partial u}{\partial x}\right)l + \frac{\partial u}{\partial y}m + \frac{\partial u}{\partial z}n$$

$$(1 + \epsilon_E)m^* = \frac{\partial v}{\partial x}l + \left(1 + \frac{\partial v}{\partial y}\right)m + \frac{\partial v}{\partial z}n$$

$$(1 + \epsilon_E)n^* = \frac{\partial w}{\partial x}l + \frac{\partial w}{\partial y}m + \left(1 + \frac{\partial w}{\partial z}\right)n \qquad (2.68)$$

Equations (2.68) represent the final direction cosines of line element ds when it passes into the line element ds^* under the deformation.

Rotation Between Two Line Elements (Definition of Shear Strain)

Next, let us consider two infinitesimal line elements PA and PB of lengths ds_1 and ds_2 emanating from point P. For simplicity, let PA be perpendicular to PB[†] (Fig. 2.19). Let the direction cosines of lines PA and PB be (l_1, m_1, n_1) and (l_2, m_2, n_2), respectively. By the deformation, line elements PA, PB are transformed into line elements P^*A^*, P^*B^*, with direction cosines (l_1^*, m_1^*, n_1^*) and (l_2^*, m_2^*, n_2^*), respectively. Since PA is perpendicular to PB, by the definition of scalar product of vectors

$$\cos\frac{\pi}{2} = l_1 l_2 + m_1 m_2 + n_1 n_2 = 0 \qquad (2.69)$$

Similarly, the angle θ^* between P^*A^* and P^*B^* is defined by

$$\cos\theta^* = l_1^* l_2^* + m_1^* m_2^* + n_1^* n_2^* \qquad (2.70)$$

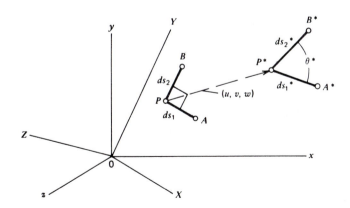

Figure 2.19   Line segments PA and PB before and after deformation.

[†] This restriction is not necessary but is used for simplicity. See Boresi and Chong (1987).

In turn, (l_1^*, m_1^*, n_1^*) and (l_2^*, m_2^*, n_2^*) are expressed in terms of (l_1, m_1, n_1) and (l_2, m_2, n_2), respectively, by means of Eq. (2.68). Hence, by Eqs. (2.68), (2.69), and (2.70), we may write with Eqs. (2.62)

$$
\begin{aligned}
\gamma_{12} &= (1 + \epsilon_{E1})(1 + \epsilon_{E2})\cos\theta^* \\
&= 2l_1 l_2 \epsilon_{xx} + 2m_1 m_2 \epsilon_{yy} + 2n_1 n_2 \epsilon_{zz} + 2(l_1 m_2 + l_2 m_1)\epsilon_{xy} \\
&\quad + 2(m_1 n_2 + m_2 n_1)\epsilon_{yz} + 2(l_1 n_2 + l_2 n_1)\epsilon_{xz}
\end{aligned}
\tag{2.71}
$$

where γ_{12} is defined to be the *engineering shear strain* between line elements PA and PB as they are deformed into P^*A^* and P^*B^* (Fig. 2.19).

To obtain a physical interpretation of ϵ_{xy}, we now let PA and PB be oriented initially parallel to axes (x, y), respectively. Hence, $l_1 = 1$, $m_1 = n_1 = 0$ and $l_2 = n_2 = 0$, $m_2 = 1$. Then Eq. (2.71) yields the result

$$
\gamma_{12} = \gamma_{xy} = 2\epsilon_{xy}
\tag{2.72}
$$

In other words, $2\epsilon_{xy}$ represents the engineering shear strain between two line elements initially parallel to the (x, y) axes, respectively. Similarly, we may consider PA and PB to be oriented initially parallel to the (y, z) axes and then to the (x, z) axes to obtain similar interpretations for $\epsilon_{yz}, \epsilon_{xz}$. Thus,

$$
\gamma_{xy} = 2\epsilon_{xy}, \qquad \gamma_{yz} = 2\epsilon_{yz}, \qquad \gamma_{xz} = 2\epsilon_{xz}
\tag{2.73}
$$

represent the engineering shear strains between two line elements initially parallel to the (x, y), (y, z) and (x, z) axes, respectively.

If the strains $\epsilon_{E1}, \epsilon_{E2}$ are small and the rotations are small (e.g., $\theta^* \approx \pi/2$), Eq. (2.71) yields the approximation

$$
\gamma_{12} = (1 + \epsilon_{E1})(1 + \epsilon_{E2})\cos\theta^* \approx \frac{\pi}{2} - \theta^*
\tag{2.74}
$$

and the engineering shear strain becomes approximately equal to the change in angle between line elements PA and PB.

Other results analogous to those of stress theory (Sec. 2.3 and 2.4) also hold. For example, the symmetric array

$$
\begin{bmatrix}
\epsilon_{xx} & \epsilon_{xy} & \epsilon_{xz} \\
\epsilon_{xy} & \epsilon_{yy} & \epsilon_{yz} \\
\epsilon_{xz} & \epsilon_{yz} & \epsilon_{zz}
\end{bmatrix}
\tag{2.75}
$$

is the *strain tensor*. Under a rotation of axes, the components of the strain tensor $(\epsilon_{xx}, \epsilon_{xy}, \epsilon_{xz}, \ldots)$ transform in exactly the same way as the stress tensor [Eqs. (2.15) and (2.17)]. [Compare Eqs. (2.5) and (2.75). Also compare Eqs. (2.11) and (2.61)]. To show this transformation, consider again axes (x, y, z) and (X, Y, Z), as in Sec. 2.4, Fig. 2.8 (also Fig. 2.18), and Table 2.2. The strain components $\epsilon_{XX}, \epsilon_{XY}, \epsilon_{XZ}, \ldots$, are defined with reference to axes (X, Y, Z) in the same manner as $\epsilon_{xx}, \epsilon_{xy}, \epsilon_{xz}, \ldots$, are defined relative to axes (x, y, z). Hence, ϵ_{XX} is the extensional strain of a line element at point P (Fig. 2.18) that lies in the direction of the X axis, and ϵ_{XY} and ϵ_{XZ} are shear components between line elements that are parallel to axes (X, Y) and (X, Z), respectively, and so on for $\epsilon_{YY}, \epsilon_{ZZ}, \epsilon_{YZ}$. Hence, if we let element ds lie parallel to the X axis, Eq. (2.61), with Table 2.2, yields

$$\epsilon_{XX} = l_1^2\epsilon_{xx} + m_1^2\epsilon_{yy} + n_1^2\epsilon_{zz} + 2l_1m_1\epsilon_{xy} + 2l_1n_1\epsilon_{xz} + 2m_1n_1\epsilon_{yz} \quad (2.76a)$$

Similarly for the line elements that lie parallel to axes Y and Z, respectively, we have

$$\epsilon_{YY} = l_2^2\epsilon_{xx} + m_2^2\epsilon_{yy} + n_2^2\epsilon_{zz} + 2l_2m_2\epsilon_{xy} + 2l_2n_2\epsilon_{xz} + 2m_2n_2\epsilon_{yz} \quad (2.76b)$$

$$\epsilon_{ZZ} = l_3^2\epsilon_{xx} + m_3^2\epsilon_{yy} + n_3^2\epsilon_{zz} + 2l_3m_3\epsilon_{xy} + 2l_3n_3\epsilon_{xz} + 2m_3n_3\epsilon_{yz} \quad (2.76c)$$

Similarly, if we take line elements PA and PB parallel, respectively to axes X and Y (Fig. 2.19), Eqs. (2.71) and (2.73) yield the result

$$\tfrac{1}{2}\gamma_{XY} = \epsilon_{XY} = l_1l_2\epsilon_{xx} + m_1m_2\epsilon_{yy} + n_1n_2\epsilon_{zz} + (l_1m_2 + l_2m_1)\epsilon_{xy}$$
$$+ (m_1n_2 + m_2n_1)\epsilon_{yz} + (l_1n_2 + l_2n_1)\epsilon_{xz} \quad (2.76d)$$

In a similar manner, we find

$$\tfrac{1}{2}\gamma_{YZ} = \epsilon_{YZ} = l_2l_3\epsilon_{xx} + m_2m_3\epsilon_{yy} + n_2n_3\epsilon_{zz} + (l_2m_3 + l_3m_2)\epsilon_{xy}$$
$$+ (m_2n_3 + m_3n_2)\epsilon_{yz} + (l_2n_3 + l_3n_2)\epsilon_{xz} \quad (2.76e)$$

$$\tfrac{1}{2}\gamma_{XZ} = \epsilon_{XZ} = l_1l_3\epsilon_{xx} + m_1m_3\epsilon_{yy} + n_1n_3\epsilon_{zz} + (l_1m_3 + l_3m_1)\epsilon_{xy}$$
$$+ (m_1n_3 + m_3n_1)\epsilon_{yz} + (l_1n_3 + l_3n_1)\epsilon_{xz} \quad (2.76f)$$

where (l_1, m_1, n_1), (l_2, m_2, n_2) and (l_3, m_3, n_3) are the direction cosines of axes X, Y, and Z, respectively.

Equations (2.76) represent the transformation of the strain tensor $(\epsilon_{xx}, \epsilon_{yy}, \ldots, \epsilon_{yz})$ under a rotation from axes (x, y, z) to axes (X, Y, Z). (See Figs. 2.18 and 2.19 and also Fig. 2.8.)

Principal Strains

Under a deformation of a body (Sec. 2.6), any infinitesimal sphere in the body is deformed into an ellipsoid, called the *strain ellipsoid*. The principal axes of the strain ellipsoid have the directions of the principal axes of strain (see below) at the center of the ellipsoid in the deformed member. The radii of the infinitesimal sphere that pass into the principal axes of the strain ellipsoid are initially perpendicular to each other, and they coincide with the principal axes of strain in the undeformed body. Hence, through any point in an undeformed member, there exist three mutually perpendicular line elements that remain perpendicular under the deformation. The strains of these three line elements are called the *principal strains* at the point. We denote them by $(\epsilon_{E1}, \epsilon_{E2}, \epsilon_{E3})$ and the corresponding principal values of the magnification factor $M = \epsilon_E + \tfrac{1}{2}\epsilon_E^2$ are denoted by (M_1, M_2, M_3). By analogy with stress theory (Sec. 2.4), the principal values of the magnification factor are the three roots of the determinantal equation

$$\begin{vmatrix} \epsilon_{xx} - M & \epsilon_{xy} & \epsilon_{xz} \\ \epsilon_{xy} & \epsilon_{yy} - M & \epsilon_{yz} \\ \epsilon_{xz} & \epsilon_{yz} & \epsilon_{zz} - M \end{vmatrix} = 0 \quad (2.77a)$$

or

$$M^3 - \bar{I}_1 M^2 - \bar{I}_2 M - \bar{I}_3 = 0$$
$$M = \epsilon_E + \tfrac{1}{2}\epsilon_E^2$$

(2.77b)

where

$$\bar{I}_1 = \epsilon_{xx} + \epsilon_{yy} + \epsilon_{zz}$$

$$\bar{I}_2 = - \begin{vmatrix} \epsilon_{xx} & \epsilon_{xy} \\ \epsilon_{xy} & \epsilon_{yy} \end{vmatrix} - \begin{vmatrix} \epsilon_{xx} & \epsilon_{xz} \\ \epsilon_{xz} & \epsilon_{zz} \end{vmatrix} - \begin{vmatrix} \epsilon_{yy} & \epsilon_{yz} \\ \epsilon_{yz} & \epsilon_{zz} \end{vmatrix}$$

$$= \epsilon_{xy}^2 + \epsilon_{xz}^2 + \epsilon_{yz}^2 - \epsilon_{xx}\epsilon_{yy} - \epsilon_{xx}\epsilon_{zz} - \epsilon_{yy}\epsilon_{zz}$$

$$\bar{I}_3 = \begin{vmatrix} \epsilon_{xx} & \epsilon_{xy} & \epsilon_{xz} \\ \epsilon_{xy} & \epsilon_{yy} & \epsilon_{yz} \\ \epsilon_{xz} & \epsilon_{yz} & \epsilon_{zz} \end{vmatrix}$$

(2.78)

are the *strain invariants* [see Eqs. (2.19), (2.20), (2.21)]. Because of the symmetry of the determinant of Eq. (2.77a), the roots $M_i : i = 1, 2, 3$ are always real. Also since $\epsilon_{Ei} > -1$, $M_i > -1$.

The three principal strain directions associated with the three principal strains $(\epsilon_{E1}, \epsilon_{E2}, \epsilon_{E3})$, Eq. (2.77b), are obtained as the solution for (l, m, n) of the equations

$$l(\epsilon_{xx} - M) + m\epsilon_{xy} + n\epsilon_{xz} = 0$$
$$l\epsilon_{xy} + m(\epsilon_{yy} - M) + n\epsilon_{yz} = 0$$
$$l\epsilon_{xz} + m\epsilon_{yz} + n(\epsilon_{zz} - M) = 0$$
$$l^2 + m^2 + n^2 = 1$$

(2.79)

Recall that only two of the first three of Eqs. (2.79) are independent. The solution $M = M_1$ yields the direction cosines for $\epsilon_E = \epsilon_{E1}$ and so on for $M = M_2(\epsilon_E = \epsilon_{E2})$, $M = M_3(\epsilon_E = \epsilon_{E3})$.

If (x, y, z) axes are principal strain axes, $\epsilon_{xx} = M_1$, $\epsilon_{yy} = M_2$, $\epsilon_{zz} = M_3$, $\epsilon_{xy} = \epsilon_{xz} = \epsilon_{yz} = 0$ and the expressions for the strain invariants $\bar{I}_1, \bar{I}_2, \bar{I}_3$ to reduce to

$$\bar{I}_1 = M_1 + M_2 + M_3$$
$$\bar{I}_2 = -M_1 M_2 - M_1 M_3 - M_2 M_3$$
$$\bar{I}_3 = M_1 M_2 M_3$$

(2.80)

2.8

SMALL-DISPLACEMENT THEORY

The deformation theory developed in Sec. 2.6 and 2.7 is purely geometrical and the associated equations are exact. In the small-displacement theory, the quadratic terms in Eqs. (2.62) are discarded. Then

$$\epsilon_{xx} \cong \frac{\partial u}{\partial x}, \qquad \epsilon_{yy} \cong \frac{\partial v}{\partial y}, \qquad \epsilon_{zz} \cong \frac{\partial w}{\partial z}$$

$$\epsilon_{xy} \cong \frac{1}{2}\left(\frac{\partial v}{\partial x} + \frac{\partial u}{\partial y}\right), \qquad \epsilon_{xz} \cong \frac{1}{2}\left(\frac{\partial w}{\partial x} + \frac{\partial u}{\partial z}\right), \qquad \epsilon_{yz} \cong \frac{1}{2}\left(\frac{\partial w}{\partial y} + \frac{\partial v}{\partial z}\right) \quad (2.81)$$

are the strain-displacement relations for small-displacement theory. Then, the magnification factor reduces to

$$M \cong \epsilon_E \tag{2.82}$$

The above approximations, which are the basis for small-displacement theory, imply that the strains and rotations (excluding rigid-body rotations) are small compared to unity. The latter condition is not necessarily satisfied in the deformation of thin flexible bodies, such as rods, plates, and shells. For these bodies the rotations may be large. Consequently, the small-displacement theory must be used with caution: It is usually applicable for massive (thick) bodies, but it may give results that are seriously in error when applied to thin flexible bodies.

Strain Compatibility Relations

The six strain components are determined by Eqs. (2.81) if the three displacement components (u, v, w) are known. However, the three displacement components (u, v, w) cannot be determined by the integration of Eqs. (2.81) if the six strain components are chosen arbitrarily. That is, certain relationships (the so-called *strain compatibility relations*) among the six strain components must exist in order that Eqs. (2.81) may be integrated to obtain the three displacement components. To illustrate this point, for simplicity, consider the case of *plane strain* relative to the (x, y) plane. This state of strain is defined by the condition that the displacement components (u, v) are functions of (x, y) only and $w = $ constant. Then Eqs. (2.81) yield

$$\epsilon_{xx} = \frac{\partial u}{\partial x}, \qquad \epsilon_{yy} = \frac{\partial v}{\partial y}, \qquad 2\epsilon_{xy} = \frac{\partial u}{\partial y} + \frac{\partial v}{\partial x}$$

$$\epsilon_{xz} = \epsilon_{yz} = \epsilon_{zz} = 0 \tag{a}$$

The strain compatibility condition is obtained by elimination of the two displacement components (u, v) from the three nonzero strain-displacement relations in Eqs. (a). This can be done by differentiation and addition as follows. Note that by differentiation, Eqs. (a) yield

$$\frac{\partial^2 \epsilon_{xx}}{\partial y^2} = \frac{\partial^3 u}{\partial x \, \partial y^2}, \qquad \frac{\partial^2 \epsilon_{yy}}{\partial x^2} = \frac{\partial^3 v}{\partial x^2 \, \partial y} \tag{b}$$

and

$$\frac{2 \partial^2 \epsilon_{xy}}{\partial x \, \partial y} = \frac{\partial^3 u}{\partial x \, \partial y^2} + \frac{\partial^3 v}{\partial x^2 \, \partial y} \tag{c}$$

Addition of the right-hand sides of Eqs. (b) shows that the right-hand side of Eq. (c) is obtained. Therefore, the relation

$$\frac{\partial^2 \epsilon_{xx}}{\partial y^2} + \frac{\partial^2 \epsilon_{yy}}{\partial x^2} = \frac{2 \partial^2 \epsilon_{xy}}{\partial x \, \partial y} \tag{d}$$

among the three strain components exists. This result, valid for small strains, is known as the *strain compatibility relation for plane strain*. In the general case, a similar elimination of (u, v, w) from Eqs. (2.81) yields the results (Boresi and Chong, 1987; Sec. 2-16)

$$\frac{\partial^2 \epsilon_{yy}}{\partial x^2} + \frac{\partial^2 \epsilon_{xx}}{\partial y^2} = 2 \frac{\partial^2 \epsilon_{xy}}{\partial x \, \partial y}$$

$$\frac{\partial^2 \epsilon_{zz}}{\partial x^2} + \frac{\partial^2 \epsilon_{xx}}{\partial z^2} = 2 \frac{\partial^2 \epsilon_{xz}}{\partial x \, \partial z}$$

$$\frac{\partial^2 \epsilon_{zz}}{\partial y^2} + \frac{\partial^2 \epsilon_{yy}}{\partial z^2} = 2 \frac{\partial^2 \epsilon_{yz}}{\partial y \, \partial z}$$

$$\frac{\partial^2 \epsilon_{zz}}{\partial x \, \partial y} + \frac{\partial^2 \epsilon_{xy}}{\partial z^2} = \frac{\partial^2 \epsilon_{yz}}{\partial z \, \partial x} + \frac{\partial^2 \epsilon_{zx}}{\partial y \, \partial z}$$

$$\frac{\partial^2 \epsilon_{yy}}{\partial x \, \partial z} + \frac{\partial^2 \epsilon_{xz}}{\partial y^2} = \frac{\partial^2 \epsilon_{xy}}{\partial y \, \partial z} + \frac{\partial^2 \epsilon_{yz}}{\partial x \, \partial y}$$

$$\frac{\partial^2 \epsilon_{xx}}{\partial y \, \partial z} + \frac{\partial^2 \epsilon_{yz}}{\partial x^2} = \frac{\partial^2 \epsilon_{xz}}{\partial x \, \partial y} + \frac{\partial^2 \epsilon_{xy}}{\partial x \, \partial z} \tag{2.83}$$

Equations (2.83) are known as the *strain compatibility equations of small-displacement theory*. It may be shown that if the strain components $(\epsilon_{xx}, \epsilon_{yy}, \epsilon_{zz}, \epsilon_{xy}, \epsilon_{xz}, \epsilon_{yz})$ satisfy Eqs. (2.83), there exist displacement components (u, v, w) that are solutions of Eqs. (2.81). More fully, in the small-displacement theory, the functions $(\epsilon_{xx}, \epsilon_{yy}, \epsilon_{zz}, \epsilon_{xy}, \epsilon_{xz}, \epsilon_{yz})$ are possible components of strain if, and only if, they satisfy Eqs. (2.83). For large displacement theory, the equivalent results are given by Murnahan (1951).

Strain-Displacement Relations for Orthogonal Curvilinear Coordinates

More generally, the strain-displacement relations [Eqs. (2.62)] may be written for orthogonal curvilinear coordinates (Fig. 2.16). The derivation of the expressions for $(\epsilon_{xx}, \epsilon_{yy}, \epsilon_{zz}, \epsilon_{xy}, \epsilon_{xz}, \epsilon_{yz})$ is a routine problem (Boresi and Chong, 1987). For small-displacement theory, the results are

$$\epsilon_{xx} = \frac{1}{\alpha} \left(\frac{\partial u}{\partial x} + \frac{v}{\beta} \frac{\partial \alpha}{\partial y} + \frac{w}{\gamma} \frac{\partial \alpha}{\partial z} \right)$$

$$\epsilon_{yy} = \frac{1}{\beta} \left(\frac{\partial v}{\partial y} + \frac{w}{\gamma} \frac{\partial \beta}{\partial z} + \frac{u}{\alpha} \frac{\partial \beta}{\partial x} \right)$$

$$\epsilon_{zz} = \frac{1}{\gamma} \left(\frac{\partial w}{\partial z} + \frac{u}{\alpha} \frac{\partial \gamma}{\partial x} + \frac{v}{\beta} \frac{\partial \gamma}{\partial y} \right)$$

$$\epsilon_{xy} = \frac{1}{2} \left(\frac{1}{\beta} \frac{\partial u}{\partial y} + \frac{1}{\alpha} \frac{\partial v}{\partial x} - \frac{v}{\alpha \beta} \frac{\partial \beta}{\partial x} - \frac{u}{\alpha \beta} \frac{\partial \alpha}{\partial y} \right)$$

$$\epsilon_{xz} = \frac{1}{2} \left(\frac{1}{\alpha} \frac{\partial w}{\partial x} + \frac{1}{\gamma} \frac{\partial u}{\partial z} - \frac{u}{\alpha \gamma} \frac{\partial \alpha}{\partial z} - \frac{w}{\alpha \gamma} \frac{\partial \gamma}{\partial x} \right)$$

$$\epsilon_{yz} = \frac{1}{2} \left(\frac{1}{\beta} \frac{\partial w}{\partial y} + \frac{1}{\gamma} \frac{\partial v}{\partial z} - \frac{w}{\beta \gamma} \frac{\partial \gamma}{\partial y} - \frac{v}{\beta \gamma} \frac{\partial \beta}{\partial z} \right) \tag{2.84}$$

where (u, v, w) are the projections of the displacement vector of point (x, y, z) on the tangents to the respective coordinate lines at that point and (α, β, γ) are the metric coefficients of the coordinate system [Eq. (2.47)]. Equations (2.84) are easily specialized for particular coordinates. For cylindrical coordinates $x = r$, $y = \theta$, $z = z$ and then $\alpha = 1$, $\beta = r$, $\gamma = 1$; for spherical coordinates, $x = r$, $y = \theta = $ colatitude, $z = \phi = $ longitude and then $\alpha = 1$, $\beta = r$, $\gamma = r \sin \theta$ (see Sec. 2.5), etc.

Thus, we obtain for

Cylindrical Coordinates

$$\epsilon_{rr} = \frac{\partial u}{\partial r}, \qquad \epsilon_{\theta\theta} = \frac{u}{r} + \frac{1}{r}\frac{\partial v}{\partial \theta}, \qquad \epsilon_{zz} = \frac{\partial w}{\partial z}$$

$$\gamma_{r\theta} = 2\epsilon_{r\theta} = \frac{1}{r}\frac{\partial u}{\partial \theta} + \frac{\partial v}{\partial r} - \frac{v}{r}, \qquad \gamma_{rz} = 2\epsilon_{rz} = \frac{\partial u}{\partial z} + \frac{\partial w}{\partial r}$$

$$\gamma_{\theta z} = 2\epsilon_{\theta z} = \frac{\partial v}{\partial z} + \frac{1}{r}\frac{\partial w}{\partial \theta} \tag{2.85}$$

Spherical Coordinates

$$\epsilon_{rr} = \frac{\partial u}{\partial r}, \qquad \epsilon_{\theta\theta} = \frac{u}{r} + \frac{1}{r}\frac{\partial v}{\partial \theta}, \qquad \epsilon_{\phi\phi} = \frac{u}{r} + \frac{v}{r}\cot \theta + \frac{1}{r \sin \theta}\frac{\partial w}{\partial \phi}$$

$$\gamma_{r\theta} = 2\epsilon_{r\theta} = \frac{1}{r}\frac{\partial u}{\partial \theta} + \frac{\partial v}{\partial r} - \frac{v}{r}, \qquad \gamma_{r\phi} = 2\epsilon_{r\phi} = \frac{1}{r \sin \theta}\frac{\partial u}{\partial \phi} + \frac{\partial w}{\partial r} - \frac{w}{r}$$

$$\gamma_{\theta\phi} = 2\epsilon_{\theta\phi} = \frac{1}{r}\left(\frac{\partial w}{\partial \theta} - w\cot \theta\right) + \frac{1}{r \sin \theta}\frac{\partial v}{\partial \phi} \tag{2.86}$$

Polar Coordinates

$$\epsilon_{rr} = \frac{\partial u}{\partial r}, \qquad \epsilon_{\theta\theta} = \frac{u}{r} + \frac{1}{r}\frac{\partial v}{\partial \theta}, \qquad \gamma_{r\theta} = 2\epsilon_{r\theta} = \frac{1}{r}\frac{\partial u}{\partial \theta} + \frac{\partial v}{\partial r} - \frac{v}{r} \tag{2.87}$$

EXAMPLE 2.4
Three-Dimensional State of Strain

The parallelopiped in Fig. E2.4 is deformed into the shape indicated by the dashed straight lines (small displacements). The displacements are given by the following relations: $u = C_1 xyz$, $v = C_2 xyz$, and $w = C_3 xyz$. (a) Determine the state of strain at point E when the coordinates of point E^* for the deformed body are $(1.504, 1.002, 1.996)$. (b) Determine the normal strain at E in the direction of line EA. (c) Determine the shear strain at E for the undeformed orthogonal lines EA and EF.

SOLUTION

The magnitudes of C_1, C_2, and C_3 are obtained from the fact that the displacements of point E are known as follows: $u_E = 0.004$ m, $v_E = 0.002$ m, and

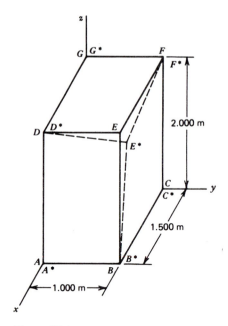

Figure E2.4

$w_E = -0.004$ m. Thus,

$$u = \frac{0.004}{3} xyz$$

$$v = \frac{0.002}{3} xyz$$

$$w = -\frac{0.004}{3} xyz$$

(a) The strain components for the state of strain at point E are given by Eqs. (2.81). At point E,

$$\epsilon_{xx} = \frac{\partial u}{\partial x} = \frac{0.004}{3} yz = 0.00267, \qquad \epsilon_{yy} = 0.00200, \qquad \epsilon_{zz} = -0.00200$$

$$\epsilon_{xy} = \frac{1}{2}\left(\frac{\partial v}{\partial x} + \frac{\partial u}{\partial y}\right) = \frac{1}{2}\left(\frac{0.002}{3} yz + \frac{0.004}{3} xz\right) = 0.00267$$

$$\gamma_{xy} = 2\epsilon_{xy} = 0.00583, \qquad \gamma_{xz} = 2\epsilon_{xz} = -0.00007$$

$$\gamma_{yz} = 2\epsilon_{yz} = -0.00300$$

(b) Let the X axis lie along the line from E to A. The direction cosines of EA are $l_1 = 0$, $m_1 = -1/\sqrt{5}$, and $n_1 = -2/\sqrt{5}$. Equations (2.61) and (2.82) give the magnitude for ϵ_{XX}. Thus,

$$\epsilon_{XX} = \epsilon_{yy}m_1^2 + \epsilon_{zz}n_1^2 + 2\epsilon_{yz}m_1 n_1$$

$$= \frac{0.00200}{5} - \frac{0.00200(4)}{5} - \frac{0.00300(2)}{5} = -0.00240$$

(c) Let the Y axis lie along the line from E to F. The direction cosines of EF are $l_2 = -1$, $m_2 = 0$, and $n_2 = 0$. The shear strain $\gamma_{XY} = 2\epsilon_{XY}$ is given by Eq. (2.76d). Thus,

$$\gamma_{XY} = 2\epsilon_{XY} = 2\epsilon_{xy}l_2m_1 + 2\epsilon_{xz}l_2n_1$$

$$= \frac{(0.00533)}{\sqrt{5}} + \frac{(-0.00007)(2)}{\sqrt{5}} = 0.00232$$

EXAMPLE 2.5
State of Strain in Torsion-Tension Member

A straight torsion-tension member with a solid circular cross section has a length $L = 6$ m and radius $R = 10$ mm. The member is subjected to tension and torsion loads that produce an elongation $\Delta L = 10$ mm and a rotation of one end of the member with respect to the other end of $\pi/3$ rad. Let the origin of the (r, θ, z) cylindrical coordinate axes lie at the centroid of one end of the member, with the z axis extending along the centroidal axis of the member. The deformations of the member are assumed to occur under conditions of constant volume. The end $z = 0$ is constrained so that only radial displacements are possible there. (a) Determine the displacements for any point in the member and the state of strain for a point on the outer surface. (b) Determine the principal strains for the point where the state of strain was determined.

SOLUTION

The change in radius ΔR for the member is obtained from the condition of constant volume. Thus,

$$\pi R^2 L = \pi (R + \Delta R)^2 (L + \Delta L)$$
$$10^2(6 \times 10^3) = (10 + \Delta R)^2(6010)$$
$$\Delta R = -0.00832 \text{ mm}$$

(a) The displacements components

$$u = -0.000832r \text{ (mm)}$$
$$v = 0.0001745rz \text{ (mm)}$$
$$w = 0.001667z \text{ (mm)}$$

satisfy the displacement boundary conditions at $z = 0$. The strain components at the outer radius are given by Eqs. (2.85). They are (rounded to six decimal places)

$$\epsilon_{rr} = \frac{\partial u}{\partial r} = -0.000832, \qquad \epsilon_{\theta\theta} = \frac{u}{r} + \frac{1}{r}\frac{\partial v}{\partial \theta} = -0.000832$$

$$\epsilon_{zz} = \frac{\partial w}{\partial z} = 0.001667, \qquad \gamma_{r\theta} = 2\epsilon_{r\theta} = \frac{1}{r}\frac{\partial u}{\partial \theta} + \frac{\partial v}{\partial r} - \frac{v}{r} = 0$$

$$\gamma_{rz} = 2\epsilon_{rz} = \frac{\partial u}{\partial z} + \frac{\partial w}{\partial r} = 0, \qquad \gamma_{\theta z} = 2\epsilon_{\theta z} = \frac{\partial v}{\partial z} + \frac{1}{r}\frac{\partial w}{\partial \theta} = 0.001745$$

(b)   The three principal strains are the three roots of a cubic equation, Eq. (2.77b), where the three invariants of strain are defined by Eqs. (2.78). Choose the (x, y, z) coordinate axes at the point on the outer surface of the member where the strain components have been determined in part (a). Let $x = r$, $y = \theta$, and $z = z$. From Eqs. (2.78),

$$\bar{I}_1 = \epsilon_{rr} + \epsilon_{\theta\theta} + \epsilon_{zz} = -0.000832 - 0.000832 + 0.001667 \simeq 0$$

$$\bar{I}_2 = -\epsilon_{rr}\epsilon_{\theta\theta} - \epsilon_{rr}\epsilon_{zz} - \epsilon_{\theta\theta}\epsilon_{zz} + \epsilon_{r\theta}^2 + \epsilon_{rz}^2 + \epsilon_{\theta z}^2$$

$$= -(-0.000832)(-0.000832) - (-0.000832)(0.001667)$$

$$-(-0.000832)(0.001667) + \left(\frac{0.001745}{2}\right)^2$$

$$= +2.838 \times 10^{-6}$$

$$\bar{I}_3 = \begin{vmatrix} \epsilon_{rr} & \epsilon_{r\theta} & \epsilon_{rz} \\ \epsilon_{\theta r} & \epsilon_{\theta\theta} & \epsilon_{\theta z} \\ \epsilon_{zr} & \epsilon_{z\theta} & \epsilon_{zz} \end{vmatrix} = \begin{vmatrix} -0.000832 & 0 & 0 \\ 0 & -0.000832 & \dfrac{0.001745}{2} \\ 0 & \dfrac{0.001745}{2} & 0.001667 \end{vmatrix}$$

$$= 1.785 \times 10^{-9}$$

Substitution of these results into Eq. (2.77b) gives the following cubic equation in $\epsilon(= M)$:

$$\epsilon^3 - 2.838 \times 10^{-6}\epsilon - 1.785 \times 10^{-9} = 0$$

One principal strain, $\epsilon_{rr} = -0.000832$, is known. Factoring out this root, we find

$$\epsilon^2 - 0.000832\epsilon - 2.146 \times 10^{-6} = 0$$

Solution of this quadratic equation yields the remaining two principal strains. Thus, the three principal strains are

$$\epsilon_1 = 0.001939$$
$$\epsilon_2 = -0.000832$$
$$\epsilon_3 = -0.001107$$

EXAMPLE 2.6
Mohr's Circle for Plane Strain

A state of plane strain at a point in a body is given, with respect to the (x, y) axes, as $\epsilon_{xx} = 0.00044$, $\epsilon_{yy} = 0.00016$, and $\epsilon_{xy} = -0.00008$. Determine the principal strains in the (x, y) plane, the orientation of the principal axes of strain, the maximum shear strain, and the strain state on a block rotated by an angle of $\theta' = 25°$ measured counterclockwise with respect to the reference axes.

SOLUTION

Since the components of strain form a symmetric second-order tensor, they are transformed in precisely the same way as stresses. Thus, plane strain states can be represented by Mohr's circle in the same way as plane stress states. By analogy to the development for plane stress, Mohr's circle for plane strain is defined by the equation, [see Eq. (2.32)]

$$[\epsilon_{XX} - \tfrac{1}{2}(\epsilon_{xx} + \epsilon_{yy})]^2 + (\epsilon_{XY} - 0)^2 = \tfrac{1}{4}(\epsilon_{xx} - \epsilon_{yy})^2 + \epsilon_{xy}^2 \qquad (a)$$

Equation (a) is the equation of a circle in the $(\epsilon_{XX}, \epsilon_{XY})$ plane with center coordinates

$$[\tfrac{1}{2}(\epsilon_{xx} + \epsilon_{yy}), 0] \qquad (b)$$

and radius

$$R = \sqrt{\tfrac{1}{4}(\epsilon_{xx} - \epsilon_{yy})^2 + \epsilon_{xy}^2} \qquad (c)$$

The orientation of the principal axes of strain is given by the angle θ, where

$$\tan 2\theta = \frac{2\epsilon_{xy}}{\epsilon_{xx} - \epsilon_{yy}} \qquad (d)$$

and θ is measured with respect to the reference x axis, positive in the counterclockwise sense. The principal strains are

$$\epsilon_1 = \frac{(\epsilon_{xx} + \epsilon_{yy})}{2} + \sqrt{\tfrac{1}{4}(\epsilon_{xx} - \epsilon_{yy})^2 + \epsilon_{xy}^2} \qquad (e)$$

$$\epsilon_2 = \frac{(\epsilon_{xx} + \epsilon_{yy})}{2} - \sqrt{\tfrac{1}{4}(\epsilon_{xx} - \epsilon_{yy})^2 + \epsilon_{xy}^2} \qquad (f)$$

The maximum shear strain is simply the radius of the circle as given by Eq. (c).

For the data given, the state of strain may be expressed as $\epsilon_{xx} = 440\,\mu$, $\epsilon_{yy} = 160\,\mu$, and $\epsilon_{xy} = -80\,\mu$, where $\mu = 10^{-6}$. This representation of strain is known as *microstrain*. The Mohr's circle for this data is shown in Fig. E2.6a. By Eq. (b) and (c), the center of the circle is located at point C with coordinates $(300\,\mu, 0)$ and its radius is $R = 161\,\mu$. By Eqs. (e) and (f), the principal strains are

$$\epsilon_1 = 300\,\mu + 161\,\mu = 461\,\mu \qquad (g)$$
$$\epsilon_2 = 300\,\mu - 161\,\mu = 139\,\mu \qquad (h)$$

On Mohr's circle, they correspond to points Q and Q', respectively. The reference strain state is plotted at points $P(\epsilon_{xx}, \epsilon_{xy})$ and $P'(\epsilon_{yy}, -\epsilon_{xy})$. Note that the positive ϵ_{XY} axis is directed downward, as is done with the plane stress case. By Eq. (d) the principal axis corresponding to ϵ_1 in the body is located at an angle of $\theta = -14.87°$ with respect to the x axis. On Mohr's circle, this corresponds to an angle $2\theta = -29.74°$ from line CP to line CQ.

The maximum value of shear strain is $\epsilon_{XY(\text{max})} = R = 161\,\mu$. It occurs at an orientation of $\pm 45°$ from the principal axis for ϵ_1 ($\pm 90°$ from line CQ on Mohr's circle). Note that this is the maximum shear strain using the tensor definition of strain. The maximum engineering shear strain is $\gamma_{XY(\text{max})} = 2\epsilon_{XY(\text{max})} = 322\,\mu$ [Eq. (2.73)]. The strain state on a block at $\theta' = 25°$ ($50°$ on Mohr's circle) is identified by points $S(\epsilon'_{xx}, \epsilon'_{xy})$ and $S'(\epsilon'_{yy}, -\epsilon'_{xy})$. By geometry of the circle, the strain

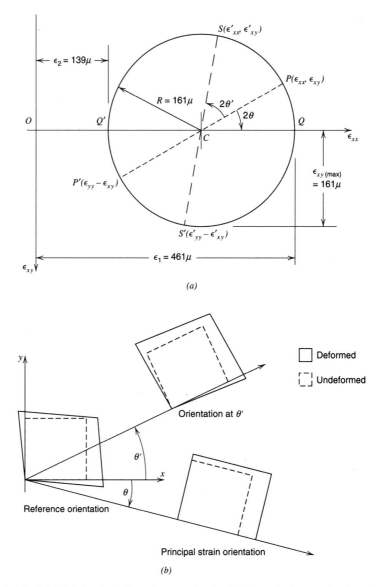

Figure E2.6 (a) Mohr's circle for plane strain. (b) Deformed element in three different orientations.

quantities are

$$\epsilon'_{xx} = OC + R\cos(2\theta' - 2\theta) = 329\,\mu$$
$$\epsilon'_{yy} = OC - R\cos(2\theta' - 2\theta) = 271\,\mu$$
$$\epsilon'_{xy} = -R\sin(2\theta' - 2\theta) = -295\,\mu$$

In Fig. E2.6b, the deformed shape of an element in the reference orientation is shown. Also illustrated is the deformed shape in the principal orientation, that is at an angle of $\theta = -14.87°$ with respect to the x axis. Notice that in this orientation, the deformed element is not distorted, since the shear strain is zero. Finally, the deformed shape at $\theta' = 25°$ with respect to the reference orientation is shown.

2.9

STRAIN MEASUREMENT. STRAIN ROSETTES

For members of complex shape subjected to loads, it may be mathematically impractical or impossible to derive analytical load-stress relations. Then, either numerical or experimental methods are used to obtain approximate results. Numerical methods (finite element methods) are treated in Chapter 19. Several experimental methods are used, the most common one being the use of strain gages. Strain gages are used to measure extensional strains on the free surface of a member or the axial extension/contraction of a bar. They cannot be used to measure the strain at an interior point of a member. To measure interior strains (or stresses), other techniques such as photoelasticity may be used, although this method has been largely superseded by modern numerical techniques. Nevertheless, photoelastic methods are still useful when augmented with modern computer data-acquisition techniques (Kobayashi, 1987). Additional experimental procedures are also available. They include holographic, Moiré, and laser speckle interferometry techniques. These specialized methods lie outside of the scope of this text (see Kobayashi, 1987). We shall discuss only the use of electrical resistance (bonded) strain gages. Electric strain gages are used to obtain average extensional strain over a given gage length. These gages are made of very fine wire or metal foil and are glued to the surface of the member being tested. When forces are applied to the member, the gage elongates or contracts with the member. The change in length of the gage alters its electrical resistance. The change in resistance can be measured and calibrated to indicate the average extensional strain that occurs over the gage length. To meet various requirements, gages are made in a variety of gage lengths, varying from 4 to 150 mm (approximately 0.15 to 6 in), and are designed for different environmental conditions.

A minimum of three extensional strain measurements in three different directions at a point on the surface of a member is required to determine the average state of strain at that point. Consequently, it is customary to cluster together three gages to form a *strain rosette* that may be cemented to the free surface of a member. Two common forms of rosettes are the *delta rosette* (with three gages spaced at 60° angles) and the *rectangular rosette* (with three gages spaced at 45° angles), Fig. 2.20. From the measurement of extensional strains along the gage arm directions (directions *a*, *b*, *c* in Fig. 2.20), one can determine the strain components $(\epsilon_{xx}, \epsilon_{yy}, \epsilon_{xy})$ at the point, relative to the (x, y) axes. Usually, one of the axes is

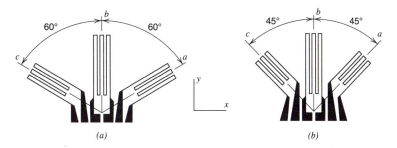

Figure 2.20   Rosette strain gages. (*a*) Delta rosette. (*b*) Rectangular rosette.

taken to be aligned with one arm of the rosette, say, the arm a. Hence, $\epsilon_{xx} = \epsilon_a$, the average extensional strain in the direction a. Then, the components $(\epsilon_{yy}, \epsilon_{xy})$ may be expressed in terms of the measured extensional strains ϵ_a, ϵ_b, and ϵ_c in the directions of the three rosette arms a, b, c, respectively. (See Example 2.7.)

EXAMPLE 2.7
Measurement of Strain on a Surface of a Member

A strain rosette with gages spaced at an angle θ is cemented to the free surface of a member (Fig. E2.7). Under a deformation of the member, the extensional strains measured by gages a, b, c are ϵ_a, ϵ_b, ϵ_c, respectively. (a) Derive equations that determine the strain components ϵ_{xx}, ϵ_{yy}, ϵ_{xy} in terms of ϵ_a, ϵ_b, ϵ_c and θ. (b) Specialize the results for the delta rosette ($\theta = 60°$) and rectangular rosette ($\theta = 45°$).

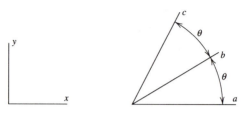

Figure E2.7

SOLUTION

(a)   The direction cosines of arms a, b, c, are respectively,

$$(l_a, m_a, n_a) = (1, 0, 0), (l_b, m_b, n_b) = (\cos\theta, \sin\theta, 0), (l_c, m_c, n_c) = (\cos 2\theta, \sin 2\theta, 0)$$

The extensional strain of a line element in the direction (l, m, n) is given by Eq. (2.61). Hence, by Eq. (2.61), the extensional strains in the directions of arms a, b, c are

$$\epsilon_a = \epsilon_{xx}$$
$$\epsilon_b = \epsilon_{xx}(\cos^2\theta) + \epsilon_{yy}(\sin^2\theta) + 2\epsilon_{xy}(\cos\theta)(\sin\theta)$$
$$\epsilon_c = \epsilon_{xx}(\cos^2 2\theta) + \epsilon_{yy}(\sin^2 2\theta) + 2\epsilon_{xy}(\cos 2\theta)(\sin 2\theta) \tag{a}$$

Equations (a) are three equations that may be solved for ϵ_{xx}, ϵ_{yy}, and ϵ_{xy} in terms of ϵ_a, ϵ_b, and ϵ_c for a given angle θ. The solution is

$$\epsilon_{xx} = \epsilon_a$$

$$\epsilon_{yy} = \frac{(\epsilon_a - 2\epsilon_b)\sin 4\theta + 2\epsilon_c \sin 2\theta}{4\sin^2\theta \sin 2\theta}$$

$$\epsilon_{xy} = \frac{2\epsilon_a(\sin^2\theta \cos^2 2\theta - \sin^2 2\theta \cos^2\theta) + 2(\epsilon_b \sin^2 2\theta - \epsilon_c \sin^2\theta)}{4\sin^2\theta \sin 2\theta} \tag{b}$$

(b) For $\theta = 60°$, $\cos\theta = 1/2$, $\sin\theta = \sqrt{3}/2$, $\cos 2\theta = -1/2$, and $\sin 2\theta = \sqrt{3}/2$. Therefore, for $\theta = 60°$, Eqs. (b) yield

$$\epsilon_{xx} = \epsilon_a$$

$$\epsilon_{yy} = \frac{2(\epsilon_b + \epsilon_c) - \epsilon_a}{3}$$

$$\epsilon_{xy} = \frac{\epsilon_b - \epsilon_c}{\sqrt{3}} \tag{c}$$

For $\theta = 45°$, $\cos\theta = 1/\sqrt{2}$, $\sin\theta = 1/\sqrt{2}$, $\cos 2\theta = 0$, and $\sin 2\theta = 1$. Therefore, for $\theta = 45°$, Eqs. (b) yield

$$\epsilon_{xx} = \epsilon_a, \epsilon_{yy} = \epsilon_c, \epsilon_{xy} = \epsilon_b - \tfrac{1}{2}(\epsilon_a + \epsilon_c) \tag{d}$$

PROBLEMS
Sections 2.1–2.4

2.1. Let the state of stress at a point be specified by the following stress components: $\sigma_{xx} = \sigma_{yy} = \sigma_{zz} = 0$, $\sigma_{xy} = -75$ MPa, $\sigma_{yz} = 65$ MPa, and $\sigma_{zx} = -55$ MPa. Determine the principal stresses, direction cosines for the three principal stress directions, and maximum shear stress.

2.2. Consider a state of stress in which the nonzero stress components are σ_{xx}, σ_{yy}, σ_{zz}, and σ_{xy}. Note that this is not a state of plane stress since $\sigma_{zz} \neq 0$. Consider another set of coordinate axes (X, Y, Z), with the Z axis coinciding with the z axis and the X axis located counterclockwise through angle θ from the x axis. Show that the transformation equations for this state of stress are identical to Eq. (2.30) or (2.31) for plane stress.

2.3. Let the state of stress at a point be specified by the following stress components: $\sigma_{xx} = 110$ MPa, $\sigma_{yy} = -86$ MPa, $\sigma_{zz} = 55$ MPa, $\sigma_{xy} = 60$ MPa, and $\sigma_{yz} = \sigma_{zx} = 0$. Determine the principal stresses, direction cosines of the principal stress directions, and maximum shear stress.

Ans. $\sigma_1 = 126.9$ MPa, $\sigma_2 = -102.9$ MPa, $\sigma_3 = 55.0$ MPa
$l_1 = 0.9625$, $m_1 = 0.2717$, $n_1 = 0$
$l_2 = 0.2717$, $m_2 = -0.9625$, $n_2 = 0$
$l_3 = 0$, $m_3 = 0$, $n_3 = -1$
$\sigma_{NS(max)} = 114.9$ MPa

2.4. Solve Problem 2.3 using the results of Problem 2.2

2.5. Let the state of plane stress be specified by the following stress components: $\sigma_{xx} = 90$ MPa, $\sigma_{yy} = -10$ MPa, $\sigma_{xy} = 40$ MPa. Let the X axis lie in the (x, y) plane and be located at $\theta = \pi/6$ clockwise from the x axis. The direction cosines for the X axis are $l = \cos(-\pi/6) = 0.8660$, $m = \sin(-\pi/6) = -0.5000$, $n = 0$. Determine the normal and shear stresses on a plane perpendicular to the X axis; use Eqs. (2.10), (2.11), and (2.12).

Ans. $\sigma_{XX} = 30.36$ MPa, $\sigma_{XY} = 63.30$ MPa

In Problems 2.6 through 2.9, the Z axis for the transformed axes coincides with the z axis for the volume element on which the known stress components act.

2.6. The nonzero stress components are $\sigma_{xx} = 200$ MPa, $\sigma_{yy} = 100$ MPa, and $\sigma_{xy} = -50$ MPa. Determine the principal stresses and maximum shear stress. Determine the angle between the X axis and the x axis when the X axis is in the direction of the principal stress with largest absolute magnitude.

2.7. The nonzero stress components are $\sigma_{xx} = -90$ MPa, $\sigma_{yy} = 50$ MPa, and $\sigma_{xy} = 60$ MPa. Determine the principal stresses and maximum shear stress. Determine the angle between the X axis and the x axis when the X axis is in the direction of the principal stress with largest absolute magnitude.

> *Ans.* $\sigma_1 = 72.2$ MPa, $\sigma_2 = -112.2$ MPa, $\sigma_3 = 0$, $\tau_{max} = 92.2$ MPa.
> X axis located 0.3543 rad clockwise from x axis.

2.8. The nonzero stress components are $\sigma_{xx} = 80$ MPa, $\sigma_{zz} = -60$ MPa, and $\sigma_{xy} = 30$ MPa. Determine the principal stresses and maximum shear stress. Determine the angle between the X axis and the x axis when the X axis is in the direction of the principal stress with largest absolute magnitude.

2.9. The nonzero stress components are $\sigma_{xx} = 150$ MPa, $\sigma_{yy} = 70$ MPa, $\sigma_{zz} = -80$ MPa, and $\sigma_{xy} = -45$ MPa. Determine the principal stresses and maximum shear stress. Determine the angle between the X axis and the x axis when the X axis is in the direction of the principal stress with largest absolute magnitude.

> *Ans.* $\sigma_1 = 170.2$ MPa, $\sigma_2 = 49.8$ MPa, $\sigma_3 = -80$ MPa,
> $\tau_{max} = 125.1$ MPa. X axis located 0.4221 rad clockwise from the x axis.

2.10. Using transformation equations of plane stress, determine σ_{XX} and σ_{XY} for the X axis located 0.5000 rad clockwise from the x axis. The nonzero stress components are given in Problem 2.6.

2.11. Using transformation equations of plane stress, determine σ_{XX} and σ_{XY} for the X axis located 0.1500 rad counterclockwise from the x axis. The nonzero stress components are given in Problem 2.7.

> *Ans.* $\sigma_{XX} = -69.1$ MPa, $\sigma_{XY} = 78.0$ MPa

2.12. Using transformation equations of stress (see Problem 2.2), determine σ_{XX} and σ_{XY} for the X axis located 1.00 rad clockwise from the x axis. The nonzero stress components are given in Problem 2.8.

2.13. Using transformation equations of stress (see Problem 2.2), determine σ_{XX} and σ_{XY} for the X axis located 0.70 rad counterclockwise from the x axis. The nonzero stress components are given in Problem 2.9.

> *Ans.* $\sigma_{XX} = 72.5$ MPa, $\sigma_{XY} = -47.1$ MPa

2.14. Solve Problem 2.10 using Mohr's circle of stress.

2.15. Solve Problem 2.11 using Mohr's circle of stress.

2.16. Solve Problem 2.12 using Mohr's circle of stress.

2.17. Solve Problem 2.13 using Mohr's circle of stress.

2.18. A volume element at the free surface is shown in Fig. P2.18. The state of stress is plane stress with $\sigma_{xx} = 100$ MPa. Determine the other stress components.

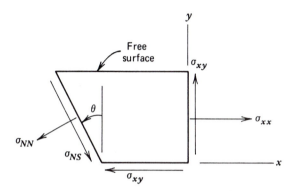

Figure P2.18

2.19. Determine the unknown stress components for the volume element in Fig. P2.19.

Ans. $\sigma_{xx} = 26.67$ MPa, $\sigma_{yy} = 172.50$ MPa

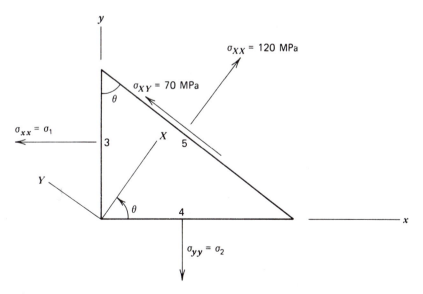

Figure P2.19

2.20. Determine the unknown stress components for the volume element in Fig. P2.20.

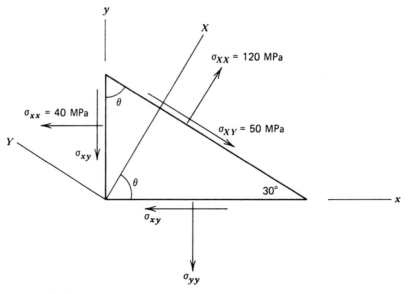

Figure P2.20

2.21. Determine the unknown stress components for the volume element in Fig. P2.21.

Ans. $\sigma_{xx} = -109.18$ MPa, $\sigma_{xy} = -10.01$ MPa

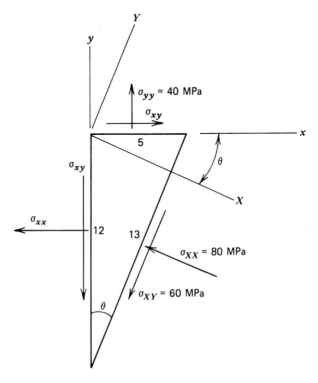

Figure P2.21

In Problems 2.22 through 2.26, determine the principal stresses, maximum shear stress, and octahedral shear stress.

2.22. The nonzero stress components are $\sigma_{xx} = -100$ MPa, $\sigma_{yy} = 60$ MPa, and $\sigma_{xy} = -50$ MPa.

2.23. The nonzero stress components are $\sigma_{xx} = 180$ MPa, $\sigma_{yy} = 90$ MPa, and $\sigma_{xy} = 50$ MPa.

Ans. $\sigma_1 = 202.3$ MPa, $\sigma_2 = 67.7$ MPa, $\sigma_3 = 0$, $\tau_{max} = 101.1$ MPa, $\tau_{oct} = 84.1$ MPa

2.24. The nonzero stress components are $\sigma_{xx} = -150$ MPa, $\sigma_{yy} = -70$ MPa, $\sigma_{zz} = 40$ MPa, and $\sigma_{xy} = -60$ MPa.

2.25. The nonzero stress components are $\sigma_{xx} = 80$ MPa, $\sigma_{yy} = -35$ MPa, $\sigma_{zz} = -50$ MPa, and $\sigma_{xy} = 45$ MPa.

Ans. $\sigma_1 = 95.5$ MPa, $\sigma_2 = -50.5$ MPa, $\sigma_3 = -50$ MPa, $\tau_{max} = 73.0$ MPa, $\tau_{oct} = 68.7$ MPa

2.26. The nonzero stress components are $\sigma_{xx} = 95$ MPa, $\sigma_{yy} = 0$, $\sigma_{zz} = 60$ MPa, and $\sigma_{xy} = -55$ MPa.

2.27. Let the state of stress at a point be given by $\sigma_{xx} = -120$ MPa, $\sigma_{yy} = 140$ MPa, $\sigma_{zz} = 66$ MPa, $\sigma_{xy} = 45$ MPa, $\sigma_{yz} = -65$ MPa, and $\sigma_{zx} = 25$ MPa. Determine the three principal stresses and directions associated with the three principal stresses.

Ans. $\sigma_1 = 180.2$ MPa, $\sigma_2 = 40.1$ MPa, $\sigma_3 = -134.3$ MPa
$l_1 = 0.0913$, $m_1 = 0.8740$, $n_1 = -0.4773$
$l_2 = 0.2584$, $m_2 = 0.4422$, $n_2 = 0.8589$
$l_3 = 0.9598$, $m_3 = -0.2062$, $n_3 = -0.1904$

2.28. Let the state of stress at a point be given by $\sigma_{xx} = 0$, $\sigma_{yy} = 100$ MPa, $\sigma_{zz} = 0$, $\sigma_{xy} = -60$ MPa, $\sigma_{yz} = 35$ MPa, and $\sigma_{zx} = 50$ MPa. Determine the three principal stresses.

2.29. Let the state of stress at a point be given by $\sigma_{xx} = 120$ MPa, $\sigma_{yy} = -55$ MPa, $\sigma_{zz} = -85$ MPa, $\sigma_{xy} = -55$ MPa, $\sigma_{yz} = 33$ MPa, and $\sigma_{zx} = -75$ MPa. Determine the three principal stresses and maximum shear stress.

Ans. $\sigma_1 = 162.5$ MPa, $\sigma_2 = -114.1$ MPa, $\sigma_3 = -68.4$ MPa, $\tau_{max} = 138.3$ MPa

2.30. Let the state of stress at a point be given by $\sigma_{xx} = -90$ MPa, $\sigma_{yy} = -60$ MPa, $\sigma_{zz} = 40$ MPa, $\sigma_{xy} = 70$ MPa, $\sigma_{yz} = -40$ MPa, and $\sigma_{zx} = -55$ MPa. Determine the three principal stresses and maximum shear stress.

2.31. Let the state of stress at a point be given by $\sigma_{xx} = -150$ MPa, $\sigma_{yy} = 0$, $\sigma_{zz} = 80$ MPa, $\sigma_{xy} = -40$ MPa, $\sigma_{yz} = 0$, and $\sigma_{zx} = 50$ MPa. Determine the three principal stresses and maximum shear stress.

Ans. $\sigma_1 = 91.2$ MPa, $\sigma_2 = 8.28$ MPa, $\sigma_3 = -169.5$ MPa, $\tau_{max} = 130.3$ MPa

2.32. **(a)** Solve Example 2.1 using Mohr's circle and show the orientation of the volume element on which the principal stresses act.

(b) Determine the maximum shear stress and show the orientation of the volume element on which it acts.

2.33. At a point on the flat surface of a member, load-stress relations give the following stress components relative to the (x, y, z) axes, where the z axis is perpendicular to the surface: $\sigma_{xx} = 240$ MPa, $\sigma_{yy} = 100$ MPa, $\sigma_{xy} = -80$ MPa, $\sigma_{zz} = \sigma_{xz} = \sigma_{yz} = 0$.

(a) Determine the principal stresses using Eq. (2.20), and then again using Eqs. (2.36) and (2.37).

(b) Determine the principal stresses using Mohr's circle and show the orientation of the volume element on which these principal stresses act.

(c) Determine the maximum shear stress and maximum octahedral shear stress.

Sections 2.5–2.8

2.34. The tension member in Fig. P2.34 has the following dimensions: $L = 5$ m, $b = 100$ mm, and $h = 200$ mm. The (x, y, z) coordinate axes are parallel to the edges of the member, with origin 0 located at the centroid of the left end. Under the deformation produced by load P, the origin 0 remains located at the centroid of the left end and the coordinate axes remain parallel to the edges of the deformed member. Under the action of load P, the bar elongates 20 mm. Assume that the volume of the bar remains constant with $\epsilon_{xx} = \epsilon_{yy}$.

(a) Determine the displacements for the member and the state of strain at point Q, assuming that the small-displacement theory holds.

(b) Determine ϵ_{zz} at point Q based on the assumption that displacements are not small.

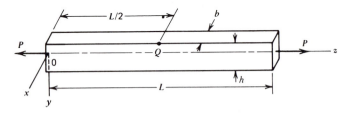

Figure P2.34

2.35. In many practical engineering problems, the state of strain is approximated by the condition that the normal and shear strains for some direction, say, the z direction, are zero, that is, $\epsilon_{zz} = \epsilon_{zx} = \epsilon_{zy} = 0$ (plane strain). In Chapter 3, it is shown that analogously, $\epsilon_{zx} = \epsilon_{zy} = 0$, but $\epsilon_{zz} \neq 0$ for members made of isotropic materials and loaded such that the state of stress may be approximated by the condition $\sigma_{zz} = \sigma_{zx} = \sigma_{zy} = 0$ (plane stress). Assume that ϵ_{xx}, ϵ_{yy}, and ϵ_{xy} for the (x, y) coordinate axes shown in Fig. P2.35 are known. Let the (X, Y) coordinate axes be defined by a counterclockwise rotation through angle θ as indicated in Fig. P2.35. Analogous to the transformation for plane stress, show that the transformation equations of plane strain are $\epsilon_{XX} = \epsilon_{xx} \cos^2 \theta + \epsilon_{yy} \sin^2 \theta + 2\epsilon_{xy} \sin \theta \cos \theta$ and $\epsilon_{XY} = -\epsilon_{xx} \sin \theta \cos \theta + \epsilon_{yy} \sin \theta \cos \theta + \epsilon_{xy}(\cos^2 \theta - \sin^2 \theta)$. [See Eq. (2.30).]

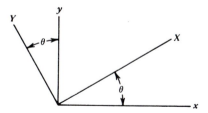

Figure P2.35

2.36. The square plate in Fig. P2.36 is loaded so that the plate is in a state of plane strain ($\epsilon_{zz} = \epsilon_{zx} = \epsilon_{zy} = 0$).

(a) Determine the displacements for the plate given the deformations shown and the strain components for the (x, y) coordinate axes.

(b) Determine the strain components for the (X, Y) axes.

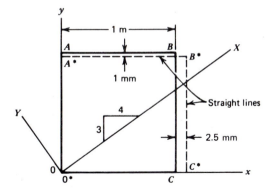

Figure P2.36

2.37. The square plate in Fig. P2.37 is loaded so that the plate is in a state of plane strain ($\epsilon_{zz} = \epsilon_{zx} = \epsilon_{zy} = 0$.)

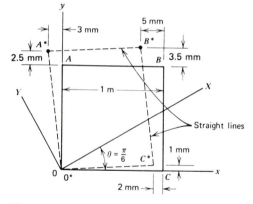

Figure P2.37

(a)   Determine the displacements for the plate for the deformations shown and the strain components for the (x, y) coordinate axes.

(b)   Determine the strain components for the (X, Y) axes.

Ans.   **(a)** $u = -0.0020x - 0.0030y$, $v = 0.0010x + 0.0025y$, $\epsilon_{xx} = -0.0020$,
$\epsilon_y = 0.0025$, $\gamma_{xy} = 2\epsilon_{xy} = -0.0020$;
(b) $\epsilon_{XX} = -0.00174$, $\epsilon_{YY} = 0.00224$, $\gamma_{XY} = 2\epsilon_{XY} = 0.00290$

2.38.   Determine the orientation of the (X, Y) coordinate axes for principal directions in Problem 2.37. What are the principal strains?

2.39.   The plate in Fig. P2.39 is loaded so that a state of plane strain $(\epsilon_{zz} = \epsilon_{zx} = \epsilon_{zy} = 0)$ exists.

(a)   Determine the displacements for the plate for the deformations shown and the strain components at point B.

(b)   Let the X axis extend from point 0 through point B. Determine ϵ_{XX} at point B.

Ans.   **(a)** (dimensions in m) $u = 0.000667xy$, $v = 0.001333xy$, $\epsilon_{xx} = 0.00200$,
$\epsilon_{yy} = 0.00200$, $\gamma_{xy} = 2\epsilon_{xy} = 0.00500$;
(b) $\epsilon_{XX} = 0.00400$

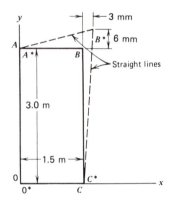

Figure P2.39

2.40.   The nonzero strain components at a point in a loaded member are $\epsilon_{xx} = 0.00180$, $\epsilon_{yy} = -0.00108$, and $\gamma_{xy} = 2\epsilon_{xy} = -0.00220$. Using the results of Problem 2.35, determine the principal strain directions and principal strains.

2.41.   Solve for the principal strains in Problem 2.40 by using Eqs. (2.77b) and (2.78).

Ans. $\epsilon_1 = 0.00217$, $\epsilon_2 = -0.00145$, $\epsilon_3 = 0$

2.42.   Determine the principal strains at point E for the deformed parallelepiped in Example 2.4.

2.43.   When solid circular torsion members are used to obtain material properties for finite strain applications, an expression for the engineering shear strain γ_{zx} is needed, where the (x, z) plane is a tangent plane and the z axis is parallel to

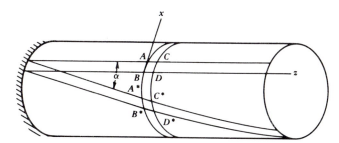

Figure P2.43

the axis of the member as indicated in Fig. P2.43. Consider an element *ABCD* in Fig. P2.43 for the undeformed member. Assume that the member deforms such that the volume remains constant and the diameter remains unchanged. (This is an approximation of the real behavior of many metals.) Thus, for the deformed element *A*B*C*D**, *A*B** = *AB*, *C*D** = *CD*, and the distance along the *z* axis of the member between the parallel curved lines *A*B** and *C*D** remains unchanged. Show that Eq. (2.71) gives the result $\gamma_{zx} = \tan \alpha$, where α is the angle between *AC* and *A*C**, where $\gamma_{zx} = 2\epsilon_{zx}$ is defined to be the engineering shear strain.

2.44. A state of plane strain exists at a point in a member, with the nonzero strain components $\epsilon_{xx} = -2000 \ \mu$, $\epsilon_{yy} = 400 \ \mu$, and $\epsilon_{xy} = -900 \ \mu$.

(a) Determine the principal strains in the (x, y) plane and the orientation of the rectangular element on which they act. (See Example 2.6.)

(b) Determine the maximum shear strain in the (x, y) plane and the orientation of the rectangular element on which it acts.

(c) Show schematically the deformed shape of a rectangular element in the reference orientation, along with the original undeformed element. (See Example 2.6.)

2.45. For the rectangular strain rosette, Fig. 2.20*b*, let arm *a* be directed along the positive *x* axis of axes (x, y).

(a) Show that the maximum principal strain is located at angle θ, counterclockwise to the *x* axis, where

$$\tan 2\theta = \frac{2\epsilon_b - \epsilon_a - \epsilon_c}{\epsilon_a - \epsilon_c}$$

(b) Show that the two principal surface strains ϵ_1 and ϵ_2 are given by

$$\epsilon_1 = \frac{\epsilon_a + \epsilon_c}{2} + R, \qquad \epsilon_2 = \frac{\epsilon_a + \epsilon_c}{2} - R$$

where

$$R = \tfrac{1}{2}[(\epsilon_a - \epsilon_c)^2 + (2\epsilon_b - \epsilon_a - \epsilon_c)^2]^{1/2}$$

(c) Construct the corresponding Mohr's circle for the rectangular rosette.

2.46. For the delta strain rosette, Fig. 2.20*a*, let arm *a* be directed along the positive *x* axis of axes (x, y).

(a) Show that the maximum principal strain is located at angle θ, counterclockwise to the x axis, where

$$\tan 2\theta = \frac{\sqrt{3}\,(\epsilon_b - \epsilon_c)}{2\epsilon_a - \epsilon_b - \epsilon_c}$$

(b) Show that the two principal surface strains ϵ_1 and ϵ_2 are given by

$$\epsilon_1 = \frac{\epsilon_a + \epsilon_b + \epsilon_c}{3} + R, \qquad \epsilon_2 = \frac{\epsilon_a + \epsilon_b + \epsilon_c}{3} - R$$

where

$$R = \tfrac{1}{3}[(2\epsilon_a - \epsilon_b - \epsilon_c)^2 + 3(\epsilon_b - \epsilon_c)^2]^{1/2}$$

(c) Construct the corresponding Mohr's circle for the delta rosette.

2.47. Let the arm a of a delta rosette, Fig. 2.20a, be directed along the positive x axis of axes (x, y). From measurements, $\epsilon_a = 2450\ \mu$, $\epsilon_b = 1360\ \mu$, and $\epsilon_c = -1310\ \mu$. Determine the two principal surface strains, the direction of the principal axes, and the associated maximum shear strain ϵ_{xy}.

2.48. Let the arm a of a rectangular rosette, Fig. 2.20b, be directed along the positive x axis of axes (x, y). Using Mohr's circle of strain, show that $2\epsilon_{XY} = \gamma_{XY} = 2\epsilon_b - \epsilon_a - \epsilon_c$.

REFERENCES

Boresi, A. P. and Chong, K. P. (1987). *Elasticity in Engineering Mechanics.* New York: Elsevier.

Kobayashi, A. S. (ed.) (1987). *Handbook on Experimental Mechanics.* Englewood Cliffs, N. J.: Prentice-Hall.

Murnahan, F. D. (1951). *Finite Deformation of an Elastic Solid.* New York: Wiley.

3

LINEAR STRESS-STRAIN-TEMPERATURE RELATIONS

In Chapter 2, we presented separate theories for stress and strain. These theories are based on the concept of a general continuum. Consequently, these theories are applicable to all continua. In particular, the theory of stress is based solely on the concept of force and the associated concept of force per unit area. Similarly, the theory of strain is based on geometrical concepts of infinitesimal line extensions and rotations between two infinitesimal lines. However, to relate the stress at a point in a material to the corresponding strain at that point, material properties are required. These properties enter into the stress-strain-temperature relations as material coefficients. The theoretical basis for these relations is the first law of thermodynamics.

In this chapter, we employ the first law of thermodynamics to derive linear stress-strain-temperature relations. In addition, certain concepts, such as complementary strain energy, that have application to nonlinear problems are introduced. These relations and concepts are utilized in many applications presented in subsequent chapters of this book.

3.1

FIRST LAW OF THERMODYNAMICS. INTERNAL-ENERGY DENSITY. COMPLEMENTARY INTERNAL-ENERGY DENSITY

The derivation of load-stress and load-deflection relations requires stress-strain relations that relate the components of the strain tensor to components of the stress tensor. The form of these relations depends on material behavior. In this book, we treat mainly materials that are isotropic, that is, at any point they have the same properties in all directions. Stress-strain relations for linearly elastic isotropic materials are well known and are presented in Sec. 3.4. Stress-strain relations may be treated theoretically by the use of the first law of thermodynamics, a precise statement of the law of conservation of energy. It is noted that the total amount of internal energy in a system is generally indeterminate. Hence, only changes of internal energy are measurable. These changes are determined by the first law of thermodynamics. If electromagnetic effects are disregarded, this law is described as follows:

The work that is performed on a mechanical system by external forces plus the heat that flows into the system from the outside equals the increase of internal energy plus the increase of kinetic energy.

Symbolically, the first law of thermodynamics is expressed by the equation

$$\delta W + \delta H = \delta U + \delta K \tag{3.1}$$

where δW is the worked performed on the system by *external forces,* δH the heat that flows into the system, δU the increase in internal energy, and δK the increase in kinetic energy.

To apply the first law of thermodynamics, we consider a loaded member in equilibrium. The deflections are assumed to be known. They are specified by known displacement components (u, v, w) for each point in the deflected member; positive u, v, and w are components of displacement of a point in the positive direction of rectangular orthogonal coordinate axes (x, y, z), respectively. We allow each point to undergo infinitesimal increments (variations) in the displacement components (u, v, w) indicated by $(\delta u, \delta v, \delta w)$. The stress components at every point of the member are considered to be unchanged under variations of the displacements. These displacement variations are arbitrary, except that two or more particles cannot occupy the same point in space, nor can a single particle occupy more than one position (the member does not tear). In addition, displacements of certain points in the member may be specified (e.g., at a fixed support); such specified displacements are referred to as forced boundary conditions (Langhaar, 1989). By Eq. (2.81), the variations of the strain components resulting from variations $(\delta u, \delta v, \delta w)$ are

$$\delta\epsilon_{xx} = \frac{\partial(\delta u)}{\partial x}, \qquad \delta\epsilon_{xy} = \frac{1}{2}\left[\frac{\partial(\delta v)}{\partial x} + \frac{\partial(\delta u)}{\partial y}\right]$$

$$\delta\epsilon_{yy} = \frac{\partial(\delta v)}{\partial y}, \qquad \delta\epsilon_{yz} = \frac{1}{2}\left[\frac{\partial(\delta w)}{\partial y} + \frac{\partial(\delta v)}{\partial z}\right]$$

$$\delta\epsilon_{zz} = \frac{\partial(\delta w)}{\partial z}, \qquad \delta\epsilon_{zx} = \frac{1}{2}\left[\frac{\partial(\delta w)}{\partial x} + \frac{\partial(\delta u)}{\partial z}\right] \tag{3.2}$$

These equations are used later in the analysis.

To introduce force quantities, consider an arbitrary volume V of the deformed member enclosed by a closed surface S. We assume that the member is in static equilibrium following the displacement variations $(\delta u, \delta v, \delta w)$. Therefore, the part of the member considered in volume V is in equilibrium under the action of surface forces (represented by stress distributions on surface S) and by body forces (represented by distributions of body forces per unit volume B_x, B_y, and B_z in volume V).

For adiabatic conditions (no net heat flow into V) and static equilibrium ($\delta K = 0$), the first law of thermodynamics states that, during the displacement variations $(\delta u, \delta v, \delta w)$, the variation in work of the external forces δW is equal to the variation of internal energy δU for each volume element. Hence, for V, we have

$$\delta W = \delta U \qquad (3.1a)$$

It is convenient to divide δW into two parts: the work of the surface forces δW_S and the work of the body forces δW_B. At point P of surface S, consider an increment of area dS. The stress vector σ_P acting on dS has components σ_{Px}, σ_{Py}, and σ_{Pz} defined by Eqs. (2.10). The surface force is equal to the product of these stress components and dS. The work δW_S is equal to the sum of the work of these forces over the surface S. Thus,

$$
\begin{aligned}
\delta W_S = & \int_S \sigma_{Px}\,\delta u\,dS + \int_S \sigma_{Py}\,\delta v\,dS + \int_S \sigma_{Pz}\,\delta w\,dS \\
= & \int_S \left[(\sigma_{xx}l + \sigma_{yx}m + \sigma_{zx}n)\,\delta u + (\sigma_{xy}l + \sigma_{yy}m + \sigma_{zy}n)\,\delta v \right. \\
& \left. + (\sigma_{xz}l + \sigma_{yz}m + \sigma_{zz}n)\,\delta w\right]dS
\end{aligned}
\qquad (3.3)
$$

For a volume element dV in volume V, the body forces are given by products of dV and the body force components per unit volume (B_x, B_y, B_z). The work δW_B of the body forces that act throughout V is

$$\delta W_B = \int_V (B_x\,\delta u + B_y\,\delta v + B_z\,\delta w)\,dV \qquad (3.4)$$

The variation of work δW of the external forces that act on volume V with surface S is equal to the sum of δW_S and δW_B. The surface integral in Eq. (3.3) may be converted into a volume integral by use of the divergence theorem (Boresi and Chong, 1987). Thus,

$$
\begin{aligned}
\delta W = \delta W_S + \delta W_B = & \int_V \left[\frac{\partial}{\partial x}(\sigma_{xx}\,\delta u + \sigma_{xy}\,\delta v + \sigma_{xz}\,\delta w)\right. \\
& + \frac{\partial}{\partial y}(\sigma_{yx}\,\delta u + \sigma_{yy}\,\delta v + \sigma_{yz}\,\delta w) \\
& + \frac{\partial}{\partial z}(\sigma_{zx}\,\delta u + \sigma_{zy}\,\delta v + \sigma_{zz}\,\delta w) \\
& \left. + B_x\,\delta u + B_y\,\delta v + B_z\,\delta w\right]dV
\end{aligned}
\qquad (3.5)
$$

With Eqs. (3.2) and (2.45), Eq. (3.5) reduces to

$$
\begin{aligned}
\delta W = & \int_V (\sigma_{xx}\,\delta\epsilon_{xx} + \sigma_{yy}\,\delta\epsilon_{yy} + \sigma_{zz}\,\delta\epsilon_{zz} + 2\sigma_{xy}\,\delta\epsilon_{xy} \\
& + 2\sigma_{yz}\,\delta\epsilon_{yz} + 2\sigma_{zx}\,\delta\epsilon_{zx})\,dV
\end{aligned}
\qquad (3.6)
$$

The internal energy U for volume V is expressed in terms of the internal energy per unit volume, that is, in terms of the *internal-energy density* U_0. Thus,

$$U = \int_V U_0 \, dV$$

and the variation of internal energy becomes

$$\delta U = \int_V \delta U_0 \, dV \tag{3.7}$$

Substitution of Eqs. (3.6) and (3.7) into Eq. (3.1a) gives the variation of the internal-energy density δU_0 in terms of the stress components and the variation in strain components. Thus,

$$\delta U_0 = \sigma_{xx} \, \delta\epsilon_{xx} + \sigma_{yy} \, \delta\epsilon_{yy} + \sigma_{zz} \, \delta\epsilon_{zz} + 2\sigma_{xy} \, \delta\epsilon_{xy} + 2\sigma_{xz} \, \delta\epsilon_{xz} + 2\sigma_{yz} \, \delta\epsilon_{yz} \tag{3.8}$$

This equation is used later in the derivation of expressions that relate the stress components to the strain energy density U_0 [see Eqs. (3.11)].

Elasticity and Internal-Energy Density

The strain-energy density U_0 is a function of certain variables; we need to determine these variables. For elastic material behavior, the total internal energy U in a loaded member is equal to the potential energy of the internal forces (called the *elastic strain energy*). Each stress component is related to the strain components; therefore, the internal-energy density U_0 at a given point in the member can be expressed in terms of the six components of the strain tensor. If the material is nonhomogeneous (has different properties at different points in the member), the function U_0 depends on location (x, y, z) in the member as well. The strain-energy density U_0 also depends on the temperature T (see Sec. 3.4). Generally, small elastic deformations do not cause significant changes in temperature (Boley and Weiner, 1960). Consequently, thermal stress problems may be treated approximately with the assumption that the time rate of change of temperature is sufficiently slow so that transient inertial effects may be ignored. Then, the stress distribution at any instant is the same as if the temperature distribution at that instant were maintained constant (Sec. 3.4).

Since the strain-energy density function U_0 generally depends on the strain components, the coordinates, and the temperature, we may express it as function of these variables. Thus,

$$U_0 = U_0(\epsilon_{xx}, \epsilon_{yy}, \epsilon_{zz}, \epsilon_{xy}, \epsilon_{xz}, \epsilon_{yz}, x, y, z, T) \tag{3.9}$$

Then, if the displacements (u, v, w) undergo a variation $(\delta u, \delta v, \delta w)$, the strain components take variations $\delta\epsilon_{xx}, \delta\epsilon_{yy}, \delta\epsilon_{zz}, \delta\epsilon_{xy}, \delta\epsilon_{xz}$, and $\delta\epsilon_{yz}$, and the function U_0 takes on the variation

$$\delta U_0 = \frac{\partial U_0}{\partial \epsilon_{xx}} \delta\epsilon_{xx} + \frac{\partial U_0}{\partial \epsilon_{yy}} \delta\epsilon_{yy} + \frac{\partial U_0}{\partial \epsilon_{zz}} \delta\epsilon_{zz} + \frac{\partial U_0}{\partial \epsilon_{xy}} \delta\epsilon_{xy} + \frac{\partial U_0}{\partial \epsilon_{xz}} \delta\epsilon_{xz} + \frac{\partial U_0}{\partial \epsilon_{yz}} \delta\epsilon_{yz} \tag{3.10}$$

Therefore, since Eqs. (3.8) and (3.10) are valid for arbitrary variations ($\delta u, \delta v, \delta w$), comparison yields for rectangular coordinate axes (x, y, z)

$$\sigma_{xx} = \frac{\partial U_0}{\partial \epsilon_{xx}}, \qquad \sigma_{yy} = \frac{\partial U_0}{\partial \epsilon_{yy}}, \qquad \sigma_{zz} = \frac{\partial U_0}{\partial \epsilon_{zz}}$$

$$\sigma_{xy} = \frac{1}{2} \frac{\partial U_0}{\partial \epsilon_{xy}}, \qquad \sigma_{xz} = \frac{1}{2} \frac{\partial U_0}{\partial \epsilon_{xz}}, \qquad \sigma_{yz} = \frac{1}{2} \frac{\partial U_0}{\partial \epsilon_{yz}} \qquad (3.11)$$

Elasticity and Complementary Internal-Energy Density

In many members of engineering structures, there may be one dominant component of the stress tensor; call it σ. This situation may arise in axially loaded members, simple columns, beams, or torsional members. Then the strain-energy density U_0 [Eq. (3.9)] depends mainly on the associated strain component ϵ; consequently, for a given temperature T, σ depends mainly on ϵ.

By Eq. (3.11), $\sigma = dU_0/d\epsilon$ and, therefore, $U_0 = \int \sigma \, d\epsilon$. It follows that U_0 is represented by the area under the stress-strain diagram (Fig. 3.1) The rectangular area $(0, 0), (0, \epsilon), (\sigma, \epsilon), (\sigma, 0)$ is represented by the product $\sigma\epsilon$. Hence, this area is given by

$$\sigma\epsilon = U_0 + C_0 \qquad (3.12)$$

where C_0 is called the *complementary internal-energy density* or *complementary strain energy density*. C_0 is represented by the area above the stress-strain curve and below the horizontal line from $(\sigma, 0)$ to (σ, ϵ). Hence by Fig. 3.1,

$$C_0 = \int \epsilon \, d\sigma \qquad (3.13)$$

or

$$\epsilon = \frac{dC_0}{d\sigma} \qquad (3.14)$$

The above graphical interpretation of the complementary strain energy is applicable only for the case of a single nonzero component of stress. However, an analytical generalization for several nonzero components of stress has been given by A. M. Legendre (Boresi and Chong, 1987). To achieve this generalization, we

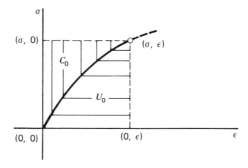

Figure 3.1

assume that Eqs. (3.11) may be integrated to obtain the strain components as functions of the stress components. Thus, we obtain

$$
\begin{aligned}
\epsilon_{xx} &= f_1(\sigma_{xx}, \sigma_{yy}, \sigma_{zz}, \sigma_{xy}, \sigma_{xz}, \sigma_{yz}) \\
\epsilon_{yy} &= f_2(\sigma_{xx}, \sigma_{yy}, \sigma_{zz}, \sigma_{xy}, \sigma_{xz}, \sigma_{yz}) \\
&\vdots \\
\epsilon_{yz} &= f_6(\sigma_{xx}, \sigma_{yy}, \sigma_{zz}, \sigma_{xy}, \sigma_{xz}, \sigma_{yz})
\end{aligned}
\tag{3.15}
$$

where $f_1, f_2, \ldots, f_6$ denote functions of the stress components. Substitution of Eqs. (3.15) into Eqs. (3.9) yields U_0 as a function of the six stress components. Then direct extension of Eq. (3.12) yields

$$
C_0 = -U_0 + \sigma_{xx}\epsilon_{xx} + \sigma_{yy}\epsilon_{yy} + \sigma_{zz}\epsilon_{zz} + 2\sigma_{xy}\epsilon_{xy} + 2\sigma_{xz}\epsilon_{xz} + 2\sigma_{yz}\epsilon_{yz} \tag{3.16}
$$

By Eqs. (3.15) and (3.16), the complementary energy density C_0 may be expressed in terms of the six stress components. Hence, differentiating Eq. (3.16) with respect to σ_{xx}, noting by the chain rule of differentiation that

$$
\frac{\partial U_0}{\partial \sigma_{xx}} = \frac{\partial U_0}{\partial \epsilon_{xx}} \frac{\partial \epsilon_{xx}}{\partial \sigma_{xx}} + \frac{\partial U_0}{\partial \epsilon_{yy}} \frac{\partial \epsilon_{yy}}{\partial \sigma_{xx}} + \frac{\partial U_0}{\partial \epsilon_{zz}} \frac{\partial \epsilon_{zz}}{\partial \sigma_{xx}}
$$
$$
+ \frac{\partial U_0}{\partial \epsilon_{xy}} \frac{\partial \epsilon_{xy}}{\partial \sigma_{xx}} + \frac{\partial U_0}{\partial \epsilon_{xz}} \frac{\partial \epsilon_{xz}}{\partial \sigma_{xx}} + \frac{\partial U_0}{\partial \epsilon_{yz}} \frac{\partial \epsilon_{yz}}{\partial \sigma_{xx}}
\tag{3.17}
$$

and employing Eq. (3.11), we find

$$
\epsilon_{xx} = \frac{\partial C_0}{\partial \sigma_{xx}} \tag{3.18}
$$

Similarly, taking derivatives of Eq. (3.16) with respect of the other stress components ($\sigma_{yy}, \sigma_{zz}, \sigma_{xy}, \sigma_{xz}, \sigma_{yz}$), we obtain the generalization of Eq. (3.14).

$$
\epsilon_{xx} = \frac{\partial C_0}{\partial \sigma_{xx}}, \qquad \epsilon_{yy} = \frac{\partial C_0}{\partial \sigma_{yy}}, \qquad \epsilon_{zz} = \frac{\partial C_0}{\partial \sigma_{zz}}
$$
$$
\epsilon_{xy} = \frac{1}{2}\frac{\partial C_0}{\partial \sigma_{xy}}, \qquad \epsilon_{xz} = \frac{1}{2}\frac{\partial C_0}{\partial \sigma_{xz}}, \qquad \epsilon_{yz} = \frac{1}{2}\frac{\partial C_0}{\partial \sigma_{yz}}
\tag{3.19}
$$

Because of their relationship to Eqs. (3.11), Eqs. (3.19) are said to be conjugate to Eqs. (3.11). Equations (3.19) are known also as the *Legendre transform* of Eqs. (3.11).

3.2

HOOKE'S LAW: ANISOTROPIC ELASTICITY

In the one-dimensional case, for a linear elastic material the stress σ is proportional to the strain ϵ; that is, $\sigma = E\epsilon$, where the proportionality factor E is called the *modulus of elasticity* of the material. The modulus of elasticity is a property of the material. Thus, for the one-dimensional case, only one material property is required to relate stress and strain for linear elastic behavior. The relation $\sigma = E\epsilon$ is known as *Hooke's law*. More generally in the three-dimensional case, Hooke's law asserts that each of the stress components is a linear function of the components of the strain tensor; that is, [with γ_{xy}, γ_{xz}, γ_{yz}, see Eq. (2.73)],

$$\sigma_{xx} = C_{11}\epsilon_{xx} + C_{12}\epsilon_{yy} + C_{13}\epsilon_{zz} + C_{14}\gamma_{xy} + C_{15}\gamma_{xz} + C_{16}\gamma_{yz}$$
$$\sigma_{yy} = C_{21}\epsilon_{xx} + C_{22}\epsilon_{yy} + C_{23}\epsilon_{zz} + C_{24}\gamma_{xy} + C_{25}\gamma_{xz} + C_{26}\gamma_y^z$$
$$\sigma_{zz} = C_{31}\epsilon_{xx} + C_{32}\epsilon_{yy} + C_{33}\epsilon_{zz} + C_{34}\gamma_{xy} + C_{35}\gamma_{xz} + C_{36}\gamma_{yz}$$
$$\sigma_{xy} = C_{41}\epsilon_{xx} + C_{42}\epsilon_{yy} + C_{43}\epsilon_{zz} + C_{44}\gamma_{xy} + C_{45}\gamma_{xz} + C_{46}\gamma_{yz}$$
$$\sigma_{xz} = C_{51}\epsilon_{xx} + C_{52}\epsilon_{yy} + C_{53}\epsilon_{zz} + C_{54}\gamma_{xy} + C_{55}\gamma_{xz} + C_{56}\gamma_{yz}$$
$$\sigma_{yz} = C_{61}\epsilon_{xx} + C_{62}\epsilon_{yy} + C_{63}\epsilon_{zz} + C_{64}\gamma_{xy} + C_{65}\gamma_{xz} + C_{66}\gamma_{yz} \qquad (3.20)$$

where the 36 coefficients, $C_{11}, \ldots, C_{66}$, are called elastic coefficients. Materials that exhibit such stress-strain relations involving a number of independent elastic coefficients are said to be *anisotropic*. (See also Sec. 3.5.)

In reality, Eq. (3.20) is no law, but merely an assumption that is reasonably accurate for many materials subjected to small strains. For a given temperature, time, and location in the body, the coefficients C_{ij} are constants that are characteristics of the material.

Equations (3.11) and (3.20) yield

$$\frac{\partial U_0}{\partial \epsilon_{xx}} = \sigma_{xx} = C_{11}\epsilon_{xx} + C_{12}\epsilon_{yy} + C_{13}\epsilon_{zz} + C_{14}\gamma_{xy} + C_{15}\gamma_{xz} + C_{16}\gamma_{yz}$$

$$\vdots$$

$$\frac{\partial U_0}{\partial \gamma_{yz}} = \sigma_{yz} = C_{61}\epsilon_{xx} + C_{62}\epsilon_{yy} + C_{63}\epsilon_{zz} + C_{64}\gamma_{xy} + C_{65}\gamma_{xz} + C_{66}\gamma_{yz} \qquad (3.21)$$

Hence, the appropriate differentiations of Eqs. (3.21) yield

$$\frac{\partial^2 U_0}{\partial \epsilon_{xx} \partial \epsilon_{yy}} = C_{12} = C_{21},$$

$$\frac{\partial^2 U_0}{\partial \epsilon_{xx} \partial \epsilon_{zz}} = C_{13} = C_{31}, \ldots, \quad \frac{\partial^2 U_0}{\partial \gamma_{yz} \partial \gamma_{xy}} = C_{46} = C_{64},$$

$$\frac{\partial^2 U_0}{\partial \gamma_{yz} \partial \gamma_{xz}} = C_{56} = C_{65} \qquad (3.22)$$

These equations show that $C_{12} = C_{21}$, $C_{13} = C_{31}, \ldots$, $C_{ik} = C_{ki}, \ldots$, $C_{56} = C_{65}$; that is, the elastic coefficients $C_{ij} = C_{ji}$ are symmetrical in the subscripts i, j. Therefore, there are only 21 *distinct* C's. In other words, the general anisotropic linear elastic material has 21 elastic coefficients. In view of the preceding relation, the strain energy density of a general anisotropic material is [by integration of

Eqs. (3.21); see Boresi and Chong, 1987]

$$
\begin{aligned}
U_0 = {} & \tfrac{1}{2}C_{11}\epsilon_{xx}^2 + \tfrac{1}{2}C_{12}\epsilon_{xx}\epsilon_{yy} + \cdots + \tfrac{1}{2}C_{16}\gamma_{xx}\gamma_{yz} \\
& + \tfrac{1}{2}C_{12}\epsilon_{xx}\epsilon_{yy} + \tfrac{1}{2}C_{22}\epsilon_{yy}^2 + \cdots + \tfrac{1}{2}C_{26}\gamma_{yy}\gamma_{yz} \\
& + \tfrac{1}{2}C_{13}\epsilon_{xx}\epsilon_{zz} + \tfrac{1}{2}C_{23}\epsilon_{yy}\epsilon_{zz} + \cdots + \tfrac{1}{2}C_{36}\gamma_{zz}\gamma_{yz} \\
& + \cdots\cdots\cdots\cdots\cdots\cdots\cdots\cdots\cdots\cdots\cdots\cdots\cdots \\
& + \tfrac{1}{2}C_{16}\gamma_{xx}\gamma_{yz} + \tfrac{1}{2}C_{26}\gamma_{yy}\gamma_{yz} + \cdots + \tfrac{1}{2}C_{66}\gamma_{yz}^2
\end{aligned}
\tag{3.23}
$$

In this form, Eq. (3.23) is important in the study of crystals (Nye, 1957).

3.3

HOOKE'S LAW: ISOTROPIC ELASTICITY

Isotropic Materials. Homogeneous Materials

If the constituents of the material of a solid member are distributed sufficiently randomly, any part of the member will display essentially the same material properties in all directions. If a solid member is composed of such randomly oriented constituents, it is said to be *isotropic*. Accordingly, if a material is isotropic, its physical properties at a point are invariant under a rotation of axes. A material is said to be *elastically isotropic* if its characteristic elastic constants are invariant under any rotation of coordinates.

If the material properties are identical for every point in a member, the member is said to be *homogeneous*. In other words, homogeneity implies that the physical properties of a member are invariant under a translation. Alternatively, a member whose material properties change from point to point is said to be nonhomogeneous. For example, since the elastic constants are functions of temperature, a member subjected to a nonuniform temperature distribution is nonhomogeneous. Accordingly, the property of nonhomogeneity is a scalar property; that is, it depends only on the location of a point in the member, not on any direction at the point. Consequently, the material in a member may be nonhomogeneous, but isotropic. For example, consider a flat sandwich plate formed by a layer of aluminum bounded by layers of steel. If the point considered is in a steel layer, the material properties have certain values that are generally independent of direction. That is, the steel is essentially isotropic. If the point considered is in the aluminum, the material properties differ from those of steel. Therefore, taken as a complete body, the sandwich plate exhibits nonhomogeneity. However, at any point in the sandwich plate, the properties are essentially independent of direction.*

---

* An exception occurs at the boundaries between the aluminum layer and steel layers. Here, the sandwich plate is anisotropic in nature.

Analogously, a member may be anisotropic, but homogeneous. For example, the physical properties of a crystal depend on direction in the crystal, but the properties vary little from one point to another (Nye, 1957).

If an elastic member is composed of isotropic materials, the strain energy density depends only on the principal strains (which are invariants), since for isotropic materials the elastic constants are invariants under arbitrary rotations [see Eq. (3.25)].

Strain Energy Density of Isotropic Elastic Materials

The strain energy density of an elastic isotropic material depends only on the principal strains $(\epsilon_1, \epsilon_2, \epsilon_3)$. Accordingly, if the elasticity is linear, Eq. (3.23) yields

$$U_0 = \tfrac{1}{2}C_{11}\epsilon_1^2 + \tfrac{1}{2}C_{12}\epsilon_1\epsilon_2 + \tfrac{1}{2}C_{13}\epsilon_1\epsilon_3 + \tfrac{1}{2}C_{12}\epsilon_1\epsilon_2 + \tfrac{1}{2}C_{22}\epsilon_2^2 + \tfrac{1}{2}C_{23}\epsilon_2\epsilon_3$$
$$+ \tfrac{1}{2}C_{13}\epsilon_1\epsilon_3 + \tfrac{1}{2}C_{23}\epsilon_2\epsilon_3 + \tfrac{1}{2}C_{33}\epsilon_3^2 \tag{3.24}$$

We note that a strain energy density function U_0 exists for either adiabatic or isothermal (constant temperature) deformations. However, the numerical values of the elastic coefficients C_{ij} differ in these two cases (Nye, 1957).

By symmetry, the naming of the principal axes is arbitrary. Hence, $C_{11} = C_{22} = C_{33} = C_1$, and $C_{12} = C_{23} = C_{13} = C_2$. Consequently, Eq. (3.24) contains only two distinct coefficients. For linear elastic isotropic materials, the strain energy density may be expressed in the form

$$U_0 = \tfrac{1}{2}\lambda(\epsilon_1 + \epsilon_2 + \epsilon_3)^2 + G(\epsilon_1^2 + \epsilon_2^2 + \epsilon_3^2) \tag{3.25}$$

where $\lambda = C_2$ and $G = (C_1 - C_2)/2$ are elastic coefficients called Lame's elastic coefficients. If the material is homogeneous and temperature is constant everywhere, λ and G are constants at all points. In terms of the strain invariants [see Eq. (2.78)], Eq. (3.25) may be written in the following form:

$$U_0 = (\tfrac{1}{2}\lambda + G)\bar{I}_1^2 + 2G\bar{I}_2 \tag{3.26}$$

Returning to orthogonal curvilinear coordinates (x, y, z) and introducing the general definitions of $\bar{I}_1$ and $\bar{I}_2$ from Eq. (2.78), we obtain

$$U_0 = \tfrac{1}{2}\lambda(\epsilon_{xx} + \epsilon_{yy} + \epsilon_{zz})^2 + G(\epsilon_{xx}^2 + \epsilon_{yy}^2 + \epsilon_{zz}^2 + 2\epsilon_{xy}^2 + 2\epsilon_{xz}^2 + 2\epsilon_{yz}^2) \tag{3.27}$$

where $(\epsilon_{xx}, \epsilon_{yy}, \epsilon_{zz}, \epsilon_{xy}, \epsilon_{xz}, \epsilon_{yz})$ are strain components relative to orthogonal coordinates (x, y, z); see Eqs. (2.84). Equations (3.11) and (3.27) now yield Hooke's law for a linear elastic isotropic material in the form (for orthogonal curvilinear coordinates x, y, z):

$$\sigma_{xx} = \lambda e + 2G\epsilon_{xx}, \qquad \sigma_{yy} = \lambda e + 2G\epsilon_{yy}, \qquad \sigma_{zz} = \lambda e + 2G\epsilon_{zz}$$
$$\sigma_{xy} = 2G\epsilon_{xy}, \qquad \sigma_{xz} = 2G\epsilon_{xz}, \qquad \sigma_{yz} = 2G\epsilon_{yz} \tag{3.28}$$

where $e \cong \epsilon_{xx} + \epsilon_{yy} + \epsilon_{zz} = \bar{I}_1$ is the classical small-displacement volumetric strain (also called cubical strain; see Boresi and Chong, 1987). Thus, we have shown that for isotropic linear elastic materials, the stress-strain relations involve only two elastic constants. An analytic proof of the fact that no further reduction is possible on a theoretical basis can be constructed (Jeffreys, 1957).

By means of Eqs. (3.28), we find [with Eqs. (2.21) and (2.78)]

$$
\begin{aligned}
I_1 &= (3\lambda + 2G)\bar{I}_1 \\
I_2 &= -\lambda(3\lambda + 4G)\bar{I}_1^2 + 4G^2\bar{I}_2 \\
I_3 &= \lambda^2(\lambda + 2G)\bar{I}_1^3 - 4\lambda G^2\bar{I}_1\bar{I}_2 + 8G^3\bar{I}_3
\end{aligned}
\tag{3.29}
$$

which relate the stress invariants, I_1, I_2, I_3 to the strain invariants $\bar{I}_1, \bar{I}_2, \bar{I}_3$.

Inverting Eqs. (3.28), we obtain

$$
\epsilon_{xx} = \frac{1}{E}(\sigma_{xx} - \nu\sigma_{yy} - \nu\sigma_{zz})
$$

$$
\epsilon_{yy} = \frac{1}{E}(\sigma_{yy} - \nu\sigma_{xx} - \nu\sigma_{zz})
$$

$$
\epsilon_{zz} = \frac{1}{E}(\sigma_{zz} - \nu\sigma_{xx} - \nu\sigma_{yy})
$$

$$
\epsilon_{xy} = \frac{1}{2G}\sigma_{xy} = \frac{1+\nu}{E}\sigma_{xy}
$$

$$
\epsilon_{xz} = \frac{1}{2G}\sigma_{xz} = \frac{1+\nu}{E}\sigma_{xz}
$$

$$
\epsilon_{yz} = \frac{1}{2G}\sigma_{yz} = \frac{1+\nu}{E}\sigma_{yz}
\tag{3.30}
$$

where

$$
E = \frac{G(3\lambda + 2G)}{\lambda + G}, \qquad \nu = \frac{\lambda}{2(\lambda + G)}
\tag{3.31}
$$

are elastic coefficients called Young's modulus and Poisson's ratio, respectively. Alternatively, Eqs. (3.28) may be written in terms of E and ν as follows:

$$
\sigma_{xx} = \frac{E}{(1+\nu)(1-2\nu)}[(1-\nu)\epsilon_{xx} + \nu(\epsilon_{yy} + \epsilon_{zz})]
$$

$$
\sigma_{yy} = \frac{E}{(1+\nu)(1-2\nu)}[(1-\nu)\epsilon_{yy} + \nu(\epsilon_{xx} + \epsilon_{zz})]
$$

$$
\sigma_{zz} = \frac{E}{(1+\nu)(1-2\nu)}[(1-\nu)\epsilon_{zz} + \nu(\epsilon_{xx} + \epsilon_{yy})]
$$

$$
\sigma_{xy} = \frac{E}{1+\nu}\epsilon_{xy}, \qquad \sigma_{xz} = \frac{E}{1+\nu}\epsilon_{xz}, \qquad \sigma_{yz} = \frac{E}{1+\nu}\epsilon_{yz}
\tag{3.32}
$$

Substitution of Eqs. (3.30) into Eq. (3.27) yields the strain-energy density U_0 in terms of stress quantities. Thus, we obtain

$$U_0 = \frac{1}{2E}[\sigma_{xx}^2 + \sigma_{yy}^2 + \sigma_{zz}^2 - 2v(\sigma_{xx}\sigma_{yy} + \sigma_{xx}\sigma_{zz} + \sigma_{yy}\sigma_{zz})$$

$$+ 2(1 + v)(\sigma_{xy}^2 + \sigma_{xz}^2 + \sigma_{yz}^2)]$$

$$= \frac{1}{2E}[I_1^2 + 2(1 + v)I_2] \tag{3.33}$$

If the axes (x, y, z) are directed along the principal axes of strain, then $\epsilon_{xy} = \epsilon_{xz} = \epsilon_{yz} = 0$. Hence, by Eq. (3.32), $\sigma_{xy} = \sigma_{xz} = \sigma_{yz} = 0$. Therefore, the axes (x, y, z) must also lie along the principal axes of stress. Consequently, for an isotropic material, the principal axes of stress are coincident with the principal axes of strain. *When we deal with isotropic materials, no distinction need be made between principal axes of stress and principal axes of strain. Such axes are called simply principal axes.*

EXAMPLE 3.1
Flat Plate Bent Around a Circular Cylinder

A flat rectangular plate lies in the (x, y) plane (Fig. E3.1a). The plate, of uniform thickness $h = 2.00$ mm, is bent around a circular cylinder (Fig. E3.1b) with the y axis parallel to the axis of the cylinder. The plate is made of an isotropic aluminum alloy ($E = 72.0$ GPa and $v = 0.33$). The radius of the cylinder is 600 mm. (a) Assuming that plane sections, $x = $ constant for the undeformed plate, remain plane after deformation, determine the maximum circumferential stress $\sigma_{\theta\theta(max)}$ in the plate for linearly elastic behavior. (b) The reciprocal of the radius of curvature R for a beam subject to pure bending is the curvature $\kappa = 1/R = M/EI$. For the plate, derive a formula for the curvature $\kappa = 1/R$ in terms of the applied moment M per unit width and the flexural rigidity $D = Eh^3/[12(1 - v^2)]$ of the plate.

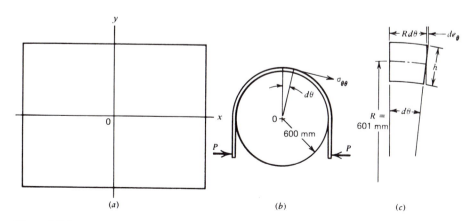

Figure E3.1

SOLUTION

(a) We assume that the middle surface of the plate remains unstressed and the stress through the thickness is negligible. Hence, the flexure formula is valid for the bending of the plate. Therefore, $\sigma_{\theta\theta} = \sigma_{yy} = 0$ for the middle surface and $\sigma_{rr} = 0$ throughout the plate thickness h. Equations (3.30) yield the results $\epsilon_{rr} = \epsilon_{\theta\theta} = \epsilon_{yy} = 0$ in the middle surface of the plate. Since the length of the plate in the y direction is large compared to the thickness h, the plate deforms approximately under conditions of plane strain; that is, $\epsilon_{yy} \approx 0$ throughout the plate thickness. Equations (3.30) give

$$\epsilon_{yy} = 0 = \frac{1}{E}\sigma_{yy} - \frac{v}{E}\sigma_{\theta\theta}$$

throughout the plate thickness. Thus, for plane strain relative to the (r,θ) plane

$$\sigma_{yy} = v\sigma_{\theta\theta} \tag{a}$$

With Eqs. (3.30), Eq. (a) yields

$$\epsilon_{\theta\theta} = \frac{1}{E}\sigma_{\theta\theta} - \frac{v}{E}\sigma_{yy} = \frac{(1 - v^2)}{E}\sigma_{\theta\theta} \tag{b}$$

The relation between the radius of curvature R of the deformed plate and $\epsilon_{\theta\theta}$ may be determined by the geometry of deformation of a plate segment (Fig. E3.1c). By similar triangles, we find from Fig. E3.1c that

$$\frac{R\,d\theta}{R} = \frac{2de_\theta}{h} = \frac{2\epsilon_{\theta\theta(max)}R\,d\theta}{h}$$

or

$$\epsilon_{\theta\theta(max)} = \frac{h}{2R} \tag{c}$$

Equations (b) and (c) yield the result

$$\sigma_{\theta\theta(max)} = \frac{Eh}{2(1 - v^2)R} = \frac{72.0 \times 10^3(2)}{2(1 - 0.33^2)(601)} = 134 \text{ MPa} \tag{d}$$

(b) In plate problems, it is convenient to consider a unit width of the plate (y direction) and let M be the moment per unit width. The moment of inertia for this unit width is $I = bh^3/12 = h^3/12$. Since $\sigma_{\theta\theta(max)} = M(h/2)/I$, this relation may be used with Eq. (d) to give

$$\frac{1}{R} = \frac{\sigma_{\theta\theta(max)}(2)(1 - v^2)}{Eh} = \frac{Mh(12)}{2h^3}\frac{2(1 - v^2)}{Eh} = \frac{M}{D} \tag{e}$$

where

$$D = \frac{Eh^3}{12(1 - v^2)} \tag{f}$$

is called the flexure rigidity of the plate. (See Chapter 13.)

EXAMPLE 3.2
The Simple Tension Test

In Sec. 1.4, the axial tension test and its role in the determination of material properties were discussed. The axial tension test in the linear elastic range of stress-strain may be used to interpret the Lamé coefficients λ and G. For example, consider a prismatic bar subjected to the following state of stress relative to the axes (x, y, z), with the z axis directed along the longitudinal axis of the bar:

$$\sigma_{xx} = \sigma_{yy} = \sigma_{xy} = \sigma_{xz} = \sigma_{yz} = 0, \qquad \sigma_{zz} = \sigma = \text{constant} \qquad \text{(a)}$$

For this state of stress to exist, the stresses on the lateral surface of the bar must be zero. On the ends of the bar, the normal stress is σ and the shear stress is zero. In other words, the state of stress in the bar is one of simple tension.

Equations (3.28) yield $\lambda e + 2G\epsilon_{xx} = \lambda e + 2G\epsilon_{yy} = \epsilon_{xy} = \epsilon_{xz} = \epsilon_{yz} = 0$. Solving these equations for the strain components, we obtain

$$\epsilon_{xx} = \epsilon_{yy} = -\frac{\lambda\sigma}{[2G(3\lambda + 2G)]}, \qquad \epsilon_{zz} = \frac{(\lambda + G)\sigma}{[G(3\lambda + 2G)]} \qquad \text{(b)}$$

It follows from Eqs. (b) that

$$-\frac{\epsilon_{xx}}{\epsilon_{zz}} = -\frac{\epsilon_{yy}}{\epsilon_{zz}} = \frac{\lambda}{[2(\lambda + G)]} = v, \qquad \epsilon_{zz} = \frac{\sigma}{E} \qquad \text{(c)}$$

where the quantities

$$E = \frac{G(3\lambda + 2G)}{(\lambda + G)}, \qquad v = \frac{\lambda}{[2(\lambda + G)]} \qquad \text{(d)}$$

are *Young's modulus of elasticity* and *Poisson's ratio*, respectively. See Sec. 1.4 for a further discussion of the modulus of elasticity. In terms of E and v, Eq. (b) becomes

$$\epsilon_{xx} = \epsilon_{yy} = -\frac{v\sigma}{E}, \qquad \epsilon_{zz} = \frac{\sigma}{E} \qquad \text{(e)}$$

Solving Eqs. (d) for the Lamé coefficients, λ and G, in terms of E and v, we obtain

$$\lambda = \frac{vE}{[(1 + v)(1 - 2v)]}, \qquad G = \frac{E}{[2(1 + v)]} \qquad \text{(f)}$$

The Lamé coefficient G is also called the *shear modulus of elasticity*. It may be given a direct physical interpretation (see Example 3.3). The Lamé coefficient λ has no direct physical interpretation. However, if the first of Eqs. (3.32) is written in the form of Eq. (g), the coefficient $E(1 - v)/[(1 + v)(1 - 2v)]$ can be called the *axial modulus*, since it relates the axial strain component ϵ_{xx} to its associated axial stress σ_{xx}.

$$\sigma_{xx} = \frac{E(1-v)}{(1+v)(1-2v)}\epsilon_{xx} + \frac{vE}{(1+v)(1-2v)}(\epsilon_{yy} + \epsilon_{zz}) \qquad \text{(g)}$$

Similarly, the Lamé coefficient $\lambda = vE/[(1+v)(1-2v)]$ may be called the *transverse modulus*, since it relates the strain components ϵ_{yy}, ϵ_{zz} (which act transversely to σ_{xx}) to the axial stress σ_{xx}. The second and third equations of Eqs. (3.32) may be written in a form similar to Eq. (g), with the same interpretation.

EXAMPLE 3.3
The Pure Shear Test

The pure shear test may be characterized by the stress state $\sigma_{xx} = \sigma_{yy} = \sigma_{zz} = \sigma_{xy} = \sigma_{xz} = 0$, and $\sigma_{yz} = \tau = $ constant. For this state of stress, Eqs. (3.28) yield the strain components

$$\epsilon_{xx} = \epsilon_{yy} = \epsilon_{zz} = \gamma_{xy} = \gamma_{xz} = 0, \qquad \gamma_{yz} = \frac{\tau}{G} \qquad \text{(a)}$$

where γ is used to represent engineering shear strain because of its convenient geometric interpretation [see Eq. (2.73)]. These formulas show that a rectangular parallelepiped $ABCD$ (Fig. E3.3) whose faces are parallel to the coordinate planes is sheared in the yz plane so that the right angle between the edges of the parallelepiped parallel to the y and z axes decreases by the amount γ_{yz}. For this reason, the coefficient G is called the *shear modulus of elasticity*. A pure shear state of stress can be obtained quite accurately by the torsion of a hollow circular cylinder with very thin walls (see Chapter 6).

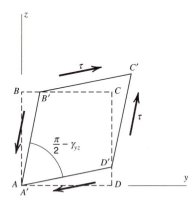

Figure E3.3

EXAMPLE 3.4
Elimination of Friction Effect in the Uniaxial
Compression Test

In a uniaxial compression test, the effect of friction between the test specimen and the testing machine platens restrains the ends of the specimen and prevents them from expanding freely in the lateral directions. This restraint may lead to erroneous measurement of the specimen strain. One way to eliminate this effect is to design the specimen and machine platens so that (1) the specimen and the end of the platens in contact with the specimen have the same cross sections, and (2) a certain relation exists between the material properties of the specimen and the platens.

To illustrate this point, let quantities associated with the specimen be denoted by subscript s and those associated with the platens be denoted by subscript p. Let P be the load applied to the specimen through the end platens. Because the cross-sectional shapes of the specimen and the platens are the same, we denote the areas by A. Let coordinate z be taken along the longitudinal axis of the specimen and coordinate x be perpendicular to axis z. Then, under a machine load P, the longitudinal strains in the specimen and platens are, respectively,

$$(\epsilon_{zz})_s = \frac{P}{(E_s A)}, \qquad (\epsilon_{zz})_p = \frac{P}{(E_p A)} \tag{a}$$

The associated lateral strains are

$$(\epsilon_{xx})_s = -v_s(\epsilon_{zz})_s = -\frac{v_s P}{(E_s A)}, \qquad (\epsilon_{xx})_p = -v_p(\epsilon_{zz})_p = -\frac{v_p P}{(E_p A)} \tag{b}$$

If the lateral strains in the specimen and platens are equal, they will expand laterally the same amount, thus eliminating friction that might be induced by the tendency of the specimen to move laterally relative to the platens. By Eq. (b), the requirement for friction to be nonexistent is that $(\epsilon_{xx})_s = (\epsilon_{xx})_p$, or

$$\frac{v_s}{E_s} = \frac{v_p}{E_p} \tag{c}$$

In addition to identical cross sections of specimen and platens, the moduli of elasticity and Poisson's ratios must satisfy Eq. (c). To reduce or eliminate the effect of friction on the tests results, it is essential to select the material properties of the platens to satisfy Eq. (c) as closely as possible.

3.4

EQUATIONS OF THERMOELASTICITY FOR ISOTROPIC MATERIALS

Consider an unconstrained member made of an isotropic elastic material in an arbitrary zero configuration. Let the uniform temperature of the member be increased by a small amount ΔT. Since the material is isotropic, all infinitesimal line elements in the volume undergo equal expansions. Furthermore, all line elements maintain their initial directions. Therefore, the strain components due to the temperature change ΔT are, with respect to rectangular Cartesian coordinates (x, y, z),

$$\epsilon'_{xx} = \epsilon'_{yy} = \epsilon'_{zz} = \alpha\,\Delta T, \qquad \epsilon'_{xy} = \epsilon'_{xz} = \epsilon'_{zy} = 0 \tag{3.34}$$

where α denotes the coefficient of thermal expansion of the material. For a non-homogeneous member, α may be a function of coordinates and of temperature; that is, $\alpha = \alpha(x, y, z, \Delta T)$.

Now let the member be subjected to forces that induce stresses $\sigma_{xx}, \sigma_{yy}, \ldots, \sigma_{yz}$ at point 0 in the member. Accordingly, if $\epsilon_{xx}, \epsilon_{yy}, \ldots, \epsilon_{yz}$ denote the strain components at point 0 after the application of the forces, the change in strain produced by the forces is represented by the equations

$$\epsilon''_{xx} = \epsilon_{xx} - \alpha\,\Delta T, \qquad \epsilon''_{yy} = \epsilon_{yy} - \alpha\,\Delta T, \qquad \epsilon''_{zz} = \epsilon_{zz} - \alpha\,\Delta T$$
$$\epsilon''_{xy} = \epsilon_{xy}, \qquad\qquad \epsilon''_{xz} = \epsilon_{xz}, \qquad\qquad \epsilon''_{yz} = \epsilon_{yz} \tag{3.35}$$

In general, ΔT may depend on the location of point 0 and time t. Hence $\Delta T = \Delta T(x, y, z, t)$. Substitution of Eq. (3.35) into Eqs. (3.28) yields

$$\sigma_{xx} = \lambda e + 2G\epsilon_{xx} - c\,\Delta T, \qquad \sigma_{yy} = \lambda e + 2G\epsilon_{yy} - c\,\Delta T$$
$$\sigma_{zz} = \lambda e + 2G\epsilon_{zz} - c\,\Delta T$$
$$\sigma_{xy} = 2G\epsilon_{xy}, \qquad \sigma_{xz} = 2G\epsilon_{xz}, \qquad \sigma_{yz} = 2G\epsilon_{yz} \tag{3.36}$$

where

$$c = (3\lambda + 2G)\alpha = \frac{E\alpha}{(1 - 2v)} \tag{3.37}$$

Similarly, substitution of Eqs. (3.36) into Eqs. (3.30) yields

$$\epsilon_{xx} = \frac{1}{E}[\sigma_{xx} - v(\sigma_{yy} + \sigma_{zz})] + \alpha\,\Delta T$$

$$\epsilon_{yy} = \frac{1}{E}[\sigma_{yy} - v(\sigma_{xx} + \sigma_{zz})] + \alpha\,\Delta T$$

$$\epsilon_{zz} = \frac{1}{E}[\sigma_{zz} - v(\sigma_{xx} + \sigma_{yy})] + \alpha\,\Delta T$$

$$\epsilon_{xy} = \frac{(1 + v)}{E}\sigma_{xy}, \qquad \epsilon_{xz} = \frac{(1 + v)}{E}\sigma_{xz}, \qquad \epsilon_{yz} = \frac{(1 + v)}{E}\sigma_{yz} \tag{3.38}$$

Finally, substituting Eqs. (3.38) into Eqs. (3.26) or (3.27), we find that

$$U_0 = (\tfrac{1}{2}\lambda + G)\bar{I}_1^2 + 2G\bar{I}_2 - c\bar{I}_1\,\Delta T + \tfrac{3}{2}c\alpha(\Delta T)^2 \tag{3.39}$$

In terms of the strain components [see Eqs. (2.78)], we obtain

$$U_0 = \tfrac{1}{2}\lambda(\epsilon_{xx} + \epsilon_{yy} + \epsilon_{zz})^2 + G(\epsilon_{xx}^2 + \epsilon_{yy}^2 + \epsilon_{zz}^2 + 2\epsilon_{xy}^2 + 2\epsilon_{xz}^2 + 2\epsilon_{yz}^2)$$
$$- c(\epsilon_{xx} + \epsilon_{yy} + \epsilon_{zz})\,\Delta T + \tfrac{3}{2}c\alpha(\Delta T)^2 \tag{3.40}$$

Equations (3.36) and (3.38) are the basic stress-strain relations of classical thermoelasticity for isotropic materials. For temperature changes ΔT, the strain energy density is modified by a temperature-dependent term that is proportional to the volumetric strain $e = \bar{I}_1 = \epsilon_{xx} + \epsilon_{yy} + \epsilon_{zz}$ and by a term proportional to $(\Delta T)^2$ [Eqs. (3.39) and (3.40)].

We find by Eqs. (3.38) and (3.40)

$$U_0 = \frac{1}{2E}[I_1^2 + 2(1+v)I_2] \tag{3.41}$$

and

$$U_0 = \frac{1}{2E}[\sigma_{xx}^2 + \sigma_{yy}^2 + \sigma_{zz}^2 - 2v(\sigma_{xx}\sigma_{yy} + \sigma_{xx}\sigma_{zz} + \sigma_{yy}\sigma_{zz})$$
$$+ 2(1+v)(\sigma_{xy}^2 + \sigma_{xz}^2 + \sigma_{yz}^2)] \tag{3.42}$$

in terms of stress components. Equation (3.42) does not contain ΔT explicitly. However, the temperature distribution may affect the stresses. Note Eqs. (3.41) and (3.42) are identical to the results in Eq. (3.33).

3.5

HOOKE'S LAW: ORTHOTROPIC MATERIALS

An important class of materials, called *orthotropic* materials, is discussed in this section. Materials such as wood, laminated plastics, cold rolled steels, reinforced concrete, various composite materials, and even forgings can be treated as orthotropic. Orthotropic materials possess three orthogonal planes of material symmetry and three corresponding orthogonal axes called the orthotropic axes. In some materials, for example, forged materials, these axes may vary from point to point. In other materials, for example, fiber-reinforced plastics and concrete reinforced with steel bars, the orthotropic directions remain constant as long as the fibers and steel reinforcing bars maintain constant directions. In any case, for an elastic orthotropic material, the elastic coefficients C_{ij} [Eq. (3.20)] remain unchanged at a point under a rotation of 180° about any of the orthotropic axes.

Let the (x, y, z) axes denote the orthotropic axes for an orthotropic material and let the (x, y) plane be a plane of material symmetry. Then, under the coordinate transformation $x \to x$, $y \to y$, and $z \to -z$, called a reflection with respect to the (x, y) plane, the elastic coefficients C_{ij} remain invariant. The direction cosines for this transformation (see Table 2.2) are defined by

$$l_1 = m_2 = 1, \qquad n_3 = -1, \qquad l_2 = l_3 = m_1 = m_3 = n_1 = n_2 = 0 \tag{3.43}$$

Substitution of Eqs. (3.43) into Eqs. (2.15), (2.17) and (2.76) reveals that for a reflection with respect to the (x, y) plane,

$$\sigma_{XX} = \sigma_{xx}, \sigma_{YY} = \sigma_{yy}, \sigma_{ZZ} = \sigma_{zz}, \sigma_{XY} = \sigma_{xy}, \sigma_{XZ} = -\sigma_{xz}, \sigma_{YZ} = -\sigma_{yz} \tag{3.44}$$

and

$$\epsilon_{XX} = \epsilon_{xx}, \epsilon_{YY} = \epsilon_{yy}, \epsilon_{ZZ} = \epsilon_{zz}, \gamma_{XY} = \gamma_{xy}, \gamma_{XZ} = -\gamma_{xz}, \gamma_{YZ} = -\gamma_{yz} \tag{3.45}$$

Since the C_{ij} are constant under the transformation of Eq. (3.43), the first of

Eqs. (3.20) yields

$$\sigma_{XX} = C_{11}\epsilon_{XX} + C_{12}\epsilon_{YY} + C_{13}\epsilon_{ZZ} + C_{14}\gamma_{XY} + C_{15}\gamma_{XZ} + C_{16}\gamma_{YZ} \quad (3.46)$$

Substitution of Eqs. (3.44) and (3.45) into Eq. (3.46) yields

$$\sigma_{xx} = \sigma_{XX} = C_{11}\epsilon_{xx} + C_{12}\epsilon_{yy} + C_{13}\epsilon_{zz} + C_{14}\gamma_{xy} - C_{15}\gamma_{xz} - C_{16}\gamma_{yz} \quad (3.47)$$

Comparison of the first of Eqs. (3.20) with Eq. (3.47) yields the conditions $C_{15} = -C_{15}$ and $C_{16} = -C_{16}$, or $C_{15} = C_{16} = 0$. Similarly, considering σ_{YY}, σ_{ZZ}, σ_{XY}, σ_{XZ}, σ_{YZ}, we find that $C_{25} = C_{26} = C_{35} = C_{36} = C_{45} = C_{46} = 0$. Thus, the coefficients for a material whose elastic properties are invariant under a reflection with respect to the (x, y) plane (i.e., for a material that possesses a plane of elasticity symmetry) are summarized by the matrix

$$\begin{bmatrix} C_{11} & C_{12} & C_{13} & C_{14} & 0 & 0 \\ C_{12} & C_{22} & C_{23} & C_{24} & 0 & 0 \\ C_{13} & C_{23} & C_{33} & C_{34} & 0 & 0 \\ C_{14} & C_{24} & C_{34} & C_{44} & 0 & 0 \\ 0 & 0 & 0 & 0 & C_{55} & C_{56} \\ 0 & 0 & 0 & 0 & C_{56} & C_{66} \end{bmatrix} \quad (3.48)$$

A general orthotropic material has two additional planes of elastic material symmetry, in this case, the (x, z) and (y, z) planes. Consider the (x, z) plane. Let $x \rightarrow x$, $y \rightarrow -y$, $z \rightarrow z$. Then, proceeding as above, noting that $l_1 = n_3 = 1$, $m_2 = -1$, $l_2 = l_3 = m_1 = m_3 = n_1 = n_2 = 0$, we find $C_{14} = C_{24} = C_{34} = C_{56} = 0$. Then, the matrix of Eq. (3.48) reduces to

$$\begin{bmatrix} C_{11} & C_{12} & C_{13} & 0 & 0 & 0 \\ C_{12} & C_{22} & C_{23} & 0 & 0 & 0 \\ C_{13} & C_{23} & C_{33} & 0 & 0 & 0 \\ 0 & 0 & 0 & C_{44} & 0 & 0 \\ 0 & 0 & 0 & 0 & C_{55} & 0 \\ 0 & 0 & 0 & 0 & 0 & C_{66} \end{bmatrix} \quad (3.49)$$

A reflection with respect to the (y, z) plane does not result in further reduction in the number of elastic coefficients C_{ij}. The matrix of coefficients in Eq. (3.49) contains nine elastic coefficients. Consequently, the stress-strain relations for the most general orthotropic material contain nine elastic coefficients relative to the orthotropic axes (x, y, z). Equations (3.20) are simplified accordingly. It should be noted, however, that this simplification occurs only when the orthotropic axes are used as the coordinate axes for which the C_{ij} are defined. The resulting equations are

$$\begin{aligned} \sigma_{xx} &= C_{11}\epsilon_{xx} + C_{12}\epsilon_{yy} + C_{13}\epsilon_{zz} \\ \sigma_{yy} &= C_{12}\epsilon_{xx} + C_{22}\epsilon_{yy} + C_{23}\epsilon_{zz} \\ \sigma_{zz} &= C_{13}\epsilon_{xx} + C_{23}\epsilon_{yy} + C_{33}\epsilon_{zz} \\ \sigma_{xy} &= C_{44}\gamma_{xy} \\ \sigma_{xz} &= C_{55}\gamma_{xz} \\ \sigma_{yz} &= C_{66}\gamma_{yz} \end{aligned} \quad (3.50)$$

The stress-strain relations for orthotropic materials in terms of orthotropic moduli of elasticity and orthotropic Poisson's ratios may be written in the form

$$\epsilon_{xx} = \frac{1}{E_x}\sigma_{xx} - \frac{v_{yx}}{E_y}\sigma_{yy} - \frac{v_{zx}}{E_z}\sigma_{zz}$$

$$\epsilon_{yy} = -\frac{v_{xy}}{E_x}\sigma_{xx} + \frac{1}{E_y}\sigma_{yy} - \frac{v_{zy}}{E_z}\sigma_{zz}$$

$$\epsilon_{zz} = -\frac{v_{xz}}{E_x}\sigma_{xx} - \frac{v_{yz}}{E_y}\sigma_{yy} + \frac{1}{E_z}\sigma_{zz}$$

$$\epsilon_{xy} = \frac{1}{2G_{xy}}\sigma_{xy} = \frac{1}{2}\gamma_{xy}$$

$$\epsilon_{xz} = \frac{1}{2G_{xz}}\sigma_{xz} = \frac{1}{2}\gamma_{xz}$$

$$\epsilon_{yz} = \frac{1}{2G_{yz}}\sigma_{yz} = \frac{1}{2}\gamma_{yz} \tag{3.51}$$

where E_x, E_y, E_z denote the orthotropic moduli of elasticity and G_{xy}, G_{xz}, G_{yz} denote the orthotropic shear moduli in the orthotropic coordinate system (x, y, z). The term v_{xy} is a Poisson ratio that characterizes the strain in the y direction produced by the stress in the x direction, with similar interpretations for the other Poisson ratios $v_{xz}, v_{yz}, \ldots, v_{zy}$. For example, by Eq. (3.51), for a tension specimen of orthotropic material subjected to a uniaxial stress $\sigma_{zz} = \sigma$ in the z direction, the axial strain in the z direction is $\epsilon_{zz} = \sigma/E_z$ and the lateral strains in the x and y directions are $\epsilon_{xx} = -v_{zx}\sigma/E_z$ and $\epsilon_{yy} = -v_{zy}\sigma/E_z$. (See Example 3.2 for the analogous isotropic tension test.)

Because of the symmetry of the coefficients in the stress-strain relations, we have by Eqs. (3.51) the identities

$$\frac{v_{xy}}{E_x} = \frac{v_{yx}}{E_y}, \qquad \frac{v_{xz}}{E_x} = \frac{v_{zx}}{E_z}, \qquad \frac{v_{yz}}{E_y} = \frac{v_{zy}}{E_z} \tag{3.52}$$

EXAMPLE 3.5
Stress-Strain Relations for Orthotropic Materials: The Plane Stress Case

Consider an orthotropic body with orthotropic axes (x, y, z) subjected to a plane stress state relative to the (x, y) plane. Let a rectangular region in the body be subjected to extensional stress σ_{xx}, Fig. E3.5a. By Eq. (3.51), the strain components are

$$\epsilon_{xx} = \frac{\sigma_{xx}}{E_x}$$

$$\epsilon_{yy} = -v_{xy}\epsilon_{xx} = -\frac{v_{xy}\sigma_{xx}}{E_x}$$

$$\epsilon_{zz} = -v_{xz}\epsilon_{xx} = -v_{xz}\frac{\sigma_{xx}}{E_x} \tag{a}$$

where v_{xy} and v_{xz} are orthotropic Poisson ratios.

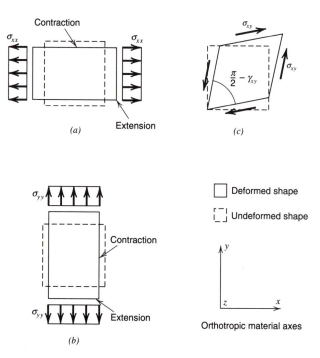

Figure E3.5   Orthotropic material. (*a*) Applied stress σ_{xx}. (*b*) Applied stress σ_{yy}. (*c*) Applied stress σ_{xy}.

Consider next the case where the rectangular region is subjected to an extensional stress σ_{yy}, Fig. E3.5b. By Eqs. (3.51), the strain components are

$$\epsilon_{yy} = \frac{\sigma_{yy}}{E_y}$$

$$\epsilon_{xx} = -v_{yx}\epsilon_{yy} = -\frac{v_{yx}\sigma_{yy}}{E_y}$$

$$\epsilon_{zz} = -v_{yz}\epsilon_{yy} = -\frac{v_{yz}\sigma_{yy}}{E_y} \qquad\qquad (b)$$

where v_{yx} and v_{yz} are orthotropic Poisson ratios.

For a combination of stresses (σ_{xx}, σ_{yy}), the addition of Eqs. (a) and (b) yields

$$\epsilon_{xx} = \frac{\sigma_{xx}}{E_x} - \frac{v_{yx}\sigma_{yy}}{E_y}$$

$$\epsilon_{yy} = -\frac{v_{xy}\sigma_{xx}}{E_x} + \frac{\sigma_{yy}}{E_y}$$

$$\epsilon_{zz} = -\frac{v_{xz}\sigma_{xx}}{E_x} - \frac{v_{yz}\sigma_{yy}}{E_y} \qquad\qquad (c)$$

Solving the first two of Eqs. (c) for $(\sigma_{xx}, \sigma_{yy})$ in terms of the in-plane strains $(\epsilon_{xx}, \epsilon_{yy})$. we obtain

$$\sigma_{xx} = \frac{E_x}{1 - \nu_{xy}\nu_{yx}}(\epsilon_{xx} + \nu_{yx}\epsilon_{yy})$$

$$\sigma_{yy} = \frac{E_y}{1 - \nu_{xy}\nu_{yx}}(\nu_{xy}\epsilon_{xx} + \epsilon_{yy}) \tag{d}$$

Finally, consider the element subjected to shear stress σ_{xy}, Fig. E3.5c. By Eqs. (3.51), we have

$$\sigma_{xy} = 2G_{xy}\epsilon_{xy} = G_{xy}\gamma_{xy} \tag{e}$$

where G_{xy} is the orthotropic shear modulus in the (x, y) plane and γ_{xy} is the engineering shear strain in the (x, y) plane. Thus, for the orthotropic material in a state of plane stress, we have the stress-strain relations [by Eqs. (d) and (e)]

$$\sigma_{xx} = \frac{E_x}{1 - \nu_{xy}\nu_{yx}}(\epsilon_{xx} + \nu_{yx}\epsilon_y)$$

$$\sigma_{yy} = \frac{E_x}{1 - \nu_{xy}\nu_{yx}}(\nu_{xy}\epsilon_{xx} + \epsilon_{yy})$$

$$\sigma_{xy} = 2G_{xy}\epsilon_{xy} = G_{xy}\gamma_{xy} \tag{f}$$

With these stress-strain relations, the theory for plane stress orthotropic problems follows in the same manner as for plane stress problems for isotropic materials.

EXAMPLE 3.6
Composite Thin-Wall Cylinder Subjected to Pressure and Temperature Increase

In Chapter 11, we consider the problem of thermal stresses in thick-wall cylinders. However, there are many situations in which the cylinder wall thickness t is small compared with the mean radius R_m of the cylinder. In these instances, certain approximations may be used to simplify the analysis. For example, consider a composite cylinder of length L formed from an inner cylinder of aluminum with outer radius R and thickness t_A, and an outer cylinder of steel with inner radius R and thickness t_S (Fig. E3.6a); $t_A \ll R$, $t_S \ll R$. The composite cylinder is supported snugly in an upright, unstressed state between rigid supports. An inner pressure p is applied to the cylinder (Fig. E3.6b), and the entire assembly is subjected to a uniform temperature change ΔT. Determine the stresses in both the aluminum and the steel cylinders for the case $t_A = t_S = t = 0.02\,R$. For aluminum, $E_A = 69$ GPa, $\nu_A = 0.333$, and $\alpha_A = 21.6 \times 10^{-6}$ per °C. For steel, $E_S = 207$ GPa, $\nu_S = 0.280$, and $\alpha_S = 10.8 \times 10^{-6}$ per °C. Subscripts A and S refer to aluminum and steel, respectively.

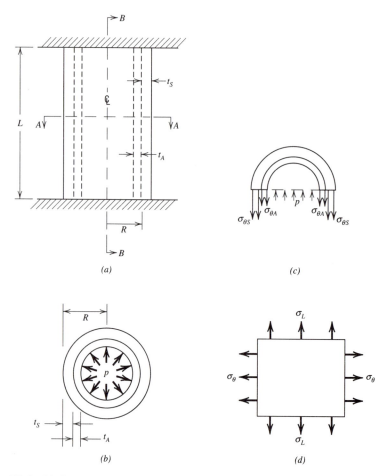

Figure E3.6   (a) Composite cylinder. (b) Cross section A-A. (c) Longitudinal section B-B. (d) Cylinder element.

SOLUTION

Since both cylinders are very thin, we may assume that the stresses in the tangential direction θ, $\sigma_{\theta A}$ and $\sigma_{\theta S}$ in the aluminum and steel, respectively, are constant through the thicknesses t_A and t_S, Fig. E3.6c. Also, it is sufficiently accurate to use the approximation $R - t = R$. From the free-body diagram of Fig. E3.6c, we have $\sum F = 2pRL - 2\sigma_{\theta S}tL - 2\sigma_{\theta A}tL = 0$. Hence,

$$\sigma_{\theta A} + \sigma_{\theta S} = \left(\frac{R}{t}\right)p = 50p \tag{a}$$

Since ordinarily the radial stress σ_r in the cylinder is very small (of the order p) compared with both the tangential stress σ_θ and the longitudinal stress σ_L, we assume that σ_r is negligible. Therefore, the cylinder is subjected approximately to a state of plane stress $(\sigma_L, \sigma_\theta)$, Fig. E3.6d. Hence, for plane stress, the strain-stress-temperature relations for each cylinder are

$$E\epsilon_L = \sigma_L - v\sigma_\theta + E\alpha(\Delta T)$$
$$E\epsilon_\theta = \sigma_\theta - v\sigma_L + E\alpha(\Delta T) \tag{b}$$

Equations (b) hold for all points in the cylinder, provided that the ends are free to expand radially. The cylinder is restrained from expanding longitudinally, since the end walls are rigid. Then, $\epsilon_L = 0$. Also at radial distance R (the interface between the aluminum and the steel sleeves), the radial displacement is u and the tangential strain is $\epsilon_\theta = [2\pi(R + u) - 2\pi R]/2\pi R = u/R$. Assuming that t is so small that this strain is the same throughout the aluminum and the steel sleeves, we have by Eqs. (b)

$$\epsilon_{LA} = \frac{1}{E_A}(\sigma_{LA} - v_A\sigma_{\theta A}) + \alpha_A(\Delta T) = 0$$

$$\epsilon_{LS} = \frac{1}{E_S}(\sigma_{LS} - v_S\sigma_{\theta S}) + \alpha_S(\Delta T) = 0$$

$$\epsilon_{\theta A} = \epsilon_{\theta S} = \frac{1}{E_A}(\sigma_{\theta A} - v_A\sigma_{LA}) + \alpha_A(\Delta T) = \frac{1}{E_S}(\sigma_{\theta S} - v_S\sigma_{LS}) + \alpha_S(\Delta T) \tag{c}$$

Also from the given data, $3E_A = E_S$ and $\alpha_A = 2\alpha_S$. Therefore, with Eqs. (a) and (c), we may write

$$\sigma_{LA} + \frac{1}{3}\sigma_{\theta S} - \frac{50}{3}p + \frac{2}{3}E_S\alpha_S(\Delta T) = 0$$

$$\sigma_{LS} - 0.28\sigma_{\theta S} + E_S\alpha_S(\Delta T) = 0$$

$$3(50p - \sigma_{\theta S}) - \sigma_{LA} + 2E_S\alpha_S(\Delta T) = \sigma_{\theta S} - 0.28\sigma_{LS} + E_S\alpha_S(\Delta T) = 0 \tag{d}$$

By the first two of Eqs. (d) and with $E_S\alpha_S = 2.236$ MPa per °C, we find that

$$\sigma_{LA} = \frac{50}{3}p - 1.491(\Delta T) - \tfrac{1}{3}\sigma_{\theta S}$$

$$\sigma_{LS} = 0.28\sigma_{\theta S} - 2.236(\Delta T) \tag{e}$$

Substitution of Eqs. (e) into the last of Eqs. (d) yields for the tangential stress in the steel cylinder

$$\sigma_{\theta S} = 37.16p + 0.8639(\Delta T) \tag{f}$$

By Eqs. (a) and (f), we find the tangential stress in the aluminum cylinder to be

$$\sigma_{\theta A} = 50p - 37.16p - 0.8639(\Delta T) = 12.84p - 0.8639(\Delta T)$$

and by Eqs. (e) and (f), we find the longitudinal stresses in the aluminum and steel cylinders, respectively,

$$\sigma_{LA} = 4.28p - 1.779(\Delta T)$$
$$\sigma_{LS} = 10.40p - 1.994(\Delta T)$$

Thus, for $p = 689.4$ kPa and $\Delta T = 100°C$

$$\sigma_{\theta A} = -77.4 \text{ MPa}$$
$$\sigma_{\theta S} = 112 \text{ MPa}$$
$$\sigma_{LA} = -175 \text{ MPa}$$
$$\sigma_{LS} = -192 \text{ MPa}$$

EXAMPLE 3.7
Douglas Fir Stress-Strain Relations

Wood is generally considered to be an orthotropic material. For example, a typical douglas fir strain-stress relationship [see Eqs. (3.51) and (3.52)] is, relative to material axes (x, y, z),

$$
\begin{aligned}
10^6 \times \epsilon_{xx} &= 87.0\, \sigma_{xx} - 34.8\, \sigma_{yy} - 43.5\, \sigma_{zz} \\
10^6 \times \epsilon_{yy} &= -34.8\, \sigma_{xx} + 1305.0\, \sigma_{yy} - 609.0\, \sigma_{zz} \\
10^6 \times \epsilon_{zz} &= -43.5\, \sigma_{xx} - 609.0\, \sigma_{yy} + 1740.0\, \sigma_{zz} \\
10^6 \times \epsilon_{xy} &= 696.0\, \sigma_{xy} \\
10^6 \times \epsilon_{xz} &= 290.0\, \sigma_{xz} \\
10^6 \times \epsilon_{yz} &= 3045.0\, \sigma_{yz}
\end{aligned}
\tag{a}
$$

where the x axis is longitudinal, the y axis radial in the tree, and the z axis tangent to the growth rings of the tree. The unit of stress is MPa.

At a point in a douglas fir log, the nonzero components of stress are

$$\sigma_{xx} = 7 \text{ MPa}, \quad \sigma_{yy} = 2.1 \text{ MPa}, \quad \sigma_{zz} = -2.8 \text{ MPa}, \quad \sigma_{xy} = 1.4 \text{ MPa} \tag{b}$$

(a) Determine the orientation of the principal axes of stress.
(b) Determine the strain components.
(c) Determine the orientation of the principal axes of strain.

SOLUTION

(a) Since $\sigma_{xz} = \sigma_{yz} = 0$, the z axis is a principal axis of stress and $\sigma_{zz} = -2.8$ MPa is a principal stress. Therefore, the orientation of the principal axes in the (x, y) plane is given by [Eq. (2.36)]

$$\tan 2\theta = \frac{2\sigma_{xy}}{(\sigma_{xx} - \sigma_{yy})} = \frac{2.8}{4.9} = 0.5714 \tag{c}$$

Equation (c) yields $2\theta = 29.74°$ or $209.74°$ and $\theta = 14.9°$ or $104.9°$. The maximum principal stress $\sigma_1 = 7.37$ MPa, in the (x, y) plane, occurs in the

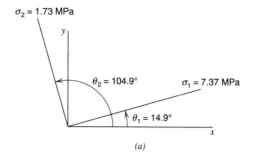

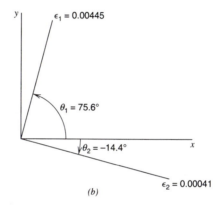

Figure E3.7 (a) Principal stress axes. (b) Principal strain axes.

direction $\theta = 14.9°$, and the minimum principal stress $\sigma_2 = 1.73$ MPa occurs in the direction $\theta = 104.9°$ (Fig. E3.7a).

(b) By Eqs. (a) and (b), we have

$$\epsilon_{xx} = 657.72 \times 10^{-6}, \quad \epsilon_{yy} = 4202.1 \times 10^{-6}, \quad \epsilon_{zz} = -6455 \times 10^{-6}$$
$$\epsilon_{xy} = 974.4 \times 10^{-6}, \quad \epsilon_{xz} = \epsilon_{yz} = 0. \tag{d}$$

(c) Since $\epsilon_{xz} = \epsilon_{yz} = 0$, the z axis is also a principal axis of strain. The orientation of the principal axes of strain in the (x, y) plane is given by

$$\tan 2\theta = \frac{2\epsilon_{xy}}{(\epsilon_{xx} - \epsilon_{yy})} = \frac{1948.8}{(-3544.38)} = -0.5498 \tag{e}$$

Hence, $2\theta = -28.80°$ or $151.20°$, and $\theta = -14.40°$ or $75.60°$. The maximum principal strain $\epsilon_1 = 0.00445$, in the (x, y) plane, occurs in the direction $\theta = 75.60°$, and the minimum principal strain $\epsilon_2 = 0.00041$ occurs in the direction $\theta = -14.40°$ (Fig. E3.7b). Thus, the principal axes of stress and strain do not coincide as they do for an isotropic material.

PROBLEMS
Sections 3.1–3.4

3.1. A square plate with 800-mm sides parallel to the x and y axes has a uniform thickness $h = 10$ mm and is made of an isotropic steel ($E = 200$ GPa and $v = 0.29$). The plate is subjected to a uniform state of stress. If $\sigma_{zz} = \sigma_{zx} = \sigma_{zy} = 0$ (plane stress), $\sigma_{xx} = \sigma_1 = 500$ MPa, and $\epsilon_{yy} = 0$ for the plate, determine $\sigma_{yy} = \sigma_2$ and the final dimensions of the plate, assuming linearly elastic conditions.

3.2. The plate in Problem 3.1 is subjected to plane strain ($\epsilon_{zz} = \epsilon_{zx} = \epsilon_{zy} = 0$). If $\sigma_{xx} = \sigma_1 = 500$ MPa and $\epsilon_{xx} = 2\epsilon_{yy}$, determine the magnitude of $\sigma_{yy} = \sigma_2$ and $\sigma_{zz} = \sigma_3$, assuming linearly elastic conditions.

Ans. $\sigma_{yy} = 377.2$ MPa, $\sigma_{zz} = 254.4$ MPa

3.3. A triaxial state of principal stress acts on the faces of a unit cube. Show that these stresses will not produce a volume change if $v = 1/2$. The material is a linearly elastic isotropic material. If $v \neq 1/2$, show that the condition necessary for the volume to remain unchanged is for $\sigma_1 + \sigma_2 + \sigma_3 = 0$.

3.4. A member is made of an isotropic linearly elastic aluminum alloy ($E = 72.0$ GPa and $v = 0.33$). Consider a point in the free surface that is tangent to the (x, y) plane. If $\sigma_{xx} = 250$ MPa, $\sigma_{yy} = -50$ MPa, and $\sigma_{xy} = -150$ MPa, determine the directions for strain gages at that point to measure two of the principal strains. What are the magnitudes of these principal strains?

Ans. $\epsilon_1 = 0.00485$ at 0.3927 rad clockwise from x axis,
$\epsilon_2 = -0.00299$

3.5. A member made of isotropic bronze ($E = 82.6$ GPa and $v = 0.35$) is subjected to a state of plane strain ($\epsilon_{zz} = \epsilon_{zx} = \epsilon_{zy} = 0$). Determine $\sigma_{zz}, \epsilon_{xx}, \epsilon_{yy}$, and $\gamma_{xy} = 2\epsilon_{xy}$, if $\sigma_{xx} = 90$ MPa, $\sigma_{yy} = -50$ MPa, and $\sigma_{xy} = 70$ MPa.

3.6. Solve Problem 3.1 for the condition that $\epsilon_{xx} = 2\epsilon_{yy}$

Ans. $\sigma_{yy} = 345.0$ MPa, $\epsilon_{xx} = 0.00200$, $\epsilon_{yy} = 0.00100$,
$\epsilon_{zz} = -0.00123$, $L_x = 801.60$ mm, $L_y = 800.80$ mm,
$L_z = 9.99$ mm

3.7. A rectangular rosette, Fig. 2.20b, is cemented to the free surface of a member made of an aluminum alloy 7075 T6 (see Appendix A). Under load, the strain readings are $\epsilon_a = \epsilon_{xx} = 0.00250$, $\epsilon_b = 0.00140$, $\epsilon_c = \epsilon_{yy} = -0.00125$.

(a) Determine the principal stresses. Note that the stress components on the free surface are zero.

(b) Show the orientation of the volume element on which the principal stresses in the plane of the rosette act.

(c) Determine the maximum shear stress τ_{max}.

(d) Show the orientation of the volume element on which τ_{max} acts.

3.8. The nonzero stress components at a point in a steel member ($E = 200$ GPa, $v = 0.29$) are $\sigma_{xx} = 80$ MPa, $\sigma_{yy} = 120$ MPa, and $\sigma_{xy} = 50$ MPa. Determine the principal strains.

3.9. Determine the extensional strain in Problem 3.8 in a direction 30° clockwise from the x axis.

3.10. A steel member ($E = 200$ GPa, $v = 0.29$) is subjected to a state of plane stress ($\sigma_{xx} = -80$ MPa, $\sigma_{yy} = 100$ MPa, $\sigma_{xy} = 50$ MPa). Determine the principal stresses and principal strains.

3.11. In Problem 3.10, determine the extensional strain in a direction 20° counterclockwise from the x axis.

3.12. The member in Fig. P3.12 is made of an aluminum alloy ($E = 72$ GPa, $v = 0.33$), and it has a square cross perpendicular to the plane of the figure. Stress components σ_{xx} and σ_{yy} are uniformly distributed as shown.

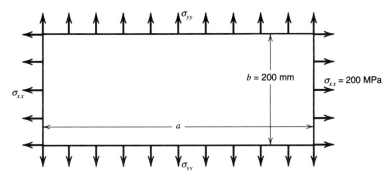

Figure P3.12

(a) If $\sigma_{xx} = 200$ MPa, determine the magnitude of σ_{yy} so that the dimension $b = 200$ mm does not change under the load.

(b) Determine the amount by which the dimension a changes.

(c) Determine the change in the cross-sectional area of the member.

3.13. Solve Example 3.6 for the case where $p = 689.4$ kPa is applied externally and $\Delta T = 100°$C is a decrease in temperature. Discuss the results.

Section 3.5

3.14. A member whose material properties remain unchanged (invariant) under rotations of 90° about axes (x, y, z) is called a *cubic material* relative to axes (x, y, z) and has three independent elastic coefficients (C_{11}, C_{12}, C_{13}). Its stress-strain relations relative to axes (x, y, z) are [a special case of Eq. (3.50)]

$$\sigma_{xx} = C_{11}\epsilon_{xx} + C_{12}\epsilon_{yy} + C_{12}\epsilon_{zz}$$
$$\sigma_{yy} = C_{12}\epsilon_{xx} + C_{11}\epsilon_{yy} + C_{12}\epsilon_{zz}$$
$$\sigma_{zz} = C_{12}\epsilon_{xx} + C_{12}\epsilon_{yy} + C_{11}\epsilon_{zz}$$
$$\sigma_{xy} = 2C_{13}\epsilon_{xy}$$
$$\sigma_{xz} = 2C_{13}\epsilon_{xz}$$
$$\sigma_{yz} = 2C_{13}\epsilon_{yz}$$

Although in practice aluminum is often assumed to be an isotropic material ($E = 69$ GPa, $v = 0.333$), it is actually a cubic material with $C_{11} = 103$ GPa, $C_{12} = 55$ GPa, $C_{13} = 27.6$ GPa. At a point in an aluminum body, the strain components are $\epsilon_{xx} = 0.0003$, $\epsilon_{yy} = 0.0002$, $\epsilon_{zz} = 0.0001$, $\epsilon_{xy} = 0.00005$, and $\epsilon_{xz} = \epsilon_{yz} = 0$.

(a) Determine the orientation of the principal axes of strain.

(b) Determine the stress components.

(c) Determine the orientation of the principal axes of stress.

(d) Calculate the stress components and determine the orientation of the principal axes of strain and stress under the assumption that the aluminum is isotropic.

3.15. A birch wood has the following strain-stress relations [see Eqs. (3.51) and (3.52)] relative to orthotropic axes (x, y, z):

$$10^6 \times \epsilon_{xx} = 72.50\,\sigma_{xx} - 36.25\,\sigma_{yy} - 36.25\,\sigma_{zz}$$
$$10^6 \times \epsilon_{yy} = -36.25\,\sigma_{xx} + 942.5\,\sigma_{yy} - 652.5\,\sigma_{zz}$$
$$10^6 \times \epsilon_{zz} = -36.25\,\sigma_{xx} - 652.5\,\sigma_{yy} - 1450.0\,\sigma_{zz}$$
$$10^6 \times \epsilon_{xy} = 507.5\,\sigma_{xy}$$
$$10^6 \times \epsilon_{xz} = 543.8\,\sigma_{xz}$$
$$10^6 \times \epsilon_{yz} = 2175.0\,\sigma_{yz}$$

where the x axis is longitudinal to the grain, the y axis radial in the tree, and the z axis tangent to the growth rings of the tree. The unit of stress is MPa. At a point in a birch log, the components of stress are $\sigma_{xx} = 7$ MPa, $\sigma_{yy} = 2.1$ MPa, $\sigma_{zz} = -2.8$ MPa, $\sigma_{xy} = 1.4$ MPa, and $\sigma_{xz} = \sigma_{yz} = 0$.

(a) Determine the orientation of the principal axes of stress.

(b) Determine the strain components.

(c) Determine the orientation of the principal axes of strain.

REFERENCES

Boley, B. A. and Weiner, J. H. (1960). *Theory of Thermal Stresses.* New York: Wiley.

Boresi, A. P. and Chong, K. P. (1987). *Elasticity in Engineering Mechanics.* New York: Elsevier.

Jeffreys, H. (1957). *Cartesian Tensors.* London: Cambridge Univ. Press.

Langhaar, H. L. (1989). *Energy Methods in Applied Mechanics.* Malabar, Florida: Krieger.

Nye, J. F. (1957), *Physical Properties of Crystals.* London: Oxford Univ. Press.

4

INELASTIC MATERIAL BEHAVIOR

In Chapter 1, we discussed the behavior of engineering materials subjected to uniaxial tension or compression. Various concepts related to elastic and inelastic response were introduced. The experimental stress-strain diagram was examined and various material properties for the uniaxial stress state were defined. The concept of failure was introduced as it applies to members subjected to a single (uniaxial) stress component. For a member subjected to a single, dominant stress component, appropriate uniaxial failure criteria may be applied. However, when a member is subjected to a multiaxial state of stress in which a single stress component does not dominate, failure criteria must account for the multiaxial nature of the stress state.

In this chapter, we examine certain criteria that are used to predict the initiation of the inelastic response of materials under multiaxial stress states. We use the term *inelastic* to denote material response characterized by a stress-strain diagram that is nonlinear and that retains a permanent strain or returns slowly to an unstrained state on complete unloading. The term *plastic* is used in certain descriptive expressions, such as fully plastic load. The term *plasticity* is used to describe the inelastic behavior of a material that retains a permanent set on complete unloading.

In this chapter, we concentrate on criteria for the initiation of yield of *ductile metals*, for example, structural steel. However, we will examine the inelastic behavior of some other materials as well. In Sec. 4.1, we discuss the limitations on the use of material properties obtained from the uniaxial stress-strain curve. In Sec. 4.2, we present a general discussion of nonlinear material behavior. This discussion establishes the perspective from which the remainder of the chapter is developed. The general concept of a yield criterion is treated in Sec. 4.3. The concept of a yield surface is introduced and presented in graphical form to facilitate interpretation of the yield criterion. In Sec. 4.4, specific yield criteria for ductile metals are developed and compared. The physical significance of these criteria is emphasized. In Sec. 4.5, yield criteria for other materials, including concrete, rock, and soil, are discussed. Various yield and failure criteria are compared and evaluated in Sec. 4.6.

4.1

LIMITATIONS ON THE USE OF UNIAXIAL STRESS-STRAIN DATA

As discussed in Chapter 1, tension or compression tests are used to obtain material properties under uniaxial conditions. These tests are usually run at room temperature in testing machines that have head speeds in the range of 0.20 to 10 mm/min.

The material properties determined under these conditions are often employed in a wide range of designs. However, in practice structural members are sometimes subjected to loads at lower or higher temperature than room temperature. In addition, loads may be applied at rates outside the range used in conventional tension or compression tests. Standard tension or compression tests employ a standard test specimen* to ensure a uniaxial stress state. In practice, the shape of the member, as well as the type of loading, may create a state of stress that is biaxial or triaxial. Such conditions place limitations on the use of material properties obtained from uniaxial tests. These limitations are discussed briefly below.

Rate of Loading

The rate at which load is applied can have a significant effect on the resulting stress-strain behavior of the material. Consider a material that responds in a ductile manner in a standard tension test. For such a material, the stress-strain curve will generally have an elastic range, followed by an inelastic region. Generally, the total strain ϵ can be separated into two parts: the elastic strain ϵ_e and inelastic (or plastic) strain ϵ_p, where $\epsilon_e = \sigma/E$ and $\epsilon_p = \epsilon - \epsilon_e$. If a tension test is run at a high rate of loading, the magnitude of inelastic strain that precedes fracture may be reduced considerably, relative to that under normal load rates. Then the material response is less ductile. Also, a high load rate increases both the apparent yield strength Y and apparent modulus of elasticity E. If the loading rate is very high, say, several orders of magnitude higher than 10 mm/min, the inelastic strain ϵ_p that precedes fracture may be eliminated almost entirely (see Sec. 1.4). Under these conditions, the material response may be characterized as *brittle* (see Fig. 1.8), although in the standard tension test the material responds in a ductile manner. Conversely, if the rate of loading is very low compared with that of the standard tension test, properties such as yield stress Y may be lowered (Morkovin and Sidebottom, 1947).

Temperature Lower than Room Temperature

If a metal tension specimen is tested at a temperature substantially below room temperature, it may fail in a brittle manner, even though it responds in a ductile manner in the standard tension test. If, in addition to the low temperature, the specimen is subjected to a very high rate of loading, the brittle response of the metal is amplified further. As a consequence, the strain that precedes failure is reduced more than when only one of these effects is present. Hence, if ductile behavior is required in a member, care must be employed in selecting an appropriate material when low temperature is combined with high load rates.

Temperature Higher than Room Temperature

If a metal is subjected to temperatures above the recrystallization temperature, the strain under a sustained, constant load will continue to increase until fracture occurs (the metal is said to be sensitive to load duration). This phenomenon is known as creep. The topic of creep is treated in Chapter 17.

---

* See ASTM Specification A370–77, *Annual Book of ASTM Standards*, Vol. 01.01, American Society for Testing and Materials, Philadelphia, PA 19103.

Effect of Unloading and Load Reversal

Consider the case of a tension specimen loaded into the inelastic range. Then, let the tension load be gradually removed and continue to load the specimen in compression as shown in Fig. 4.1. In an ideal model, it is assumed that the stress-strain relation for the material follows the path $OABCD$. However, actual test data for an annealed high-carbon steel indicate deviations from this path (Sidebottom and Chang, 1952). Two observations regarding these deviations are made. First, the *actual* unloading path does not follow the *ideal* linear elastic unloading path. Second, the subsequent yield strength in compression is reduced below the original value $-Y$ for the virgin material. The term *Bauschinger effect* is used to characterize this behavior.

The Bauschinger effect can also occur during simple cyclic loading in tension (or compression) without stress reversal (Lubahn and Felgar, 1961). Modern theories of plasticity attempt to account for the Bauschinger effect. For instance, the change in compressive yield strength due to tension hardening is often modeled by maintaining an elastic stress range of $2Y$ (see Fig. 4.1). However, the deviation from the ideal unload path BCD is often ignored.

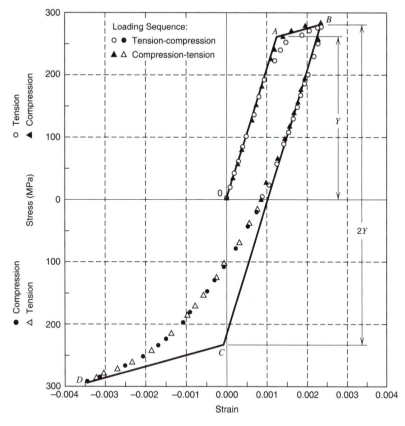

Figure 4.1   Tension and compression stress-strain diagrams for annealed high-carbon steel for initial and reversed loading. (From Sidebottom and Chang, 1952.)

Multiaxial States of Stress

Inelastic behavior can occur under multiaxial stress in a load-carrying member, even if none of the individual stress components exceeds the uniaxial yield stress for the material. The implication is that, under multiaxial stress states, the initiation of yielding is governed by some quantity other than the individual stress components themselves. Thus, it is necessary to combine the components of stress into an *effective uniaxial stress*. This effective stress is then compared with some material property, usually the uniaxial yield stress, by appropriate *yield criteria* to predict the beginning of inelastic response. The concept of yield criteria is discussed more completely in Sec. 4.3, 4.4, and 4.5.

4.2

NONLINEAR MATERIAL RESPONSE

As noted in Sec. 1.4 and 4.1, the shape of a tension stress-strain curve depends on the material itself and the test conditions. However, when the load is applied and removed slowly, certain features of the stress-strain curve are similar for all structural materials. For example, if the load is sufficiently small, the relation between stress and strain is linearly elastic; that is, the stress-strain curve is a straight line (line segment *OA* in Fig. 4.1). The stress associated with loading and unloading increases or decreases, respectively, along this straight-line path. As the load is increased to a sufficiently large value, the stress-strain curve becomes nonlinear. The material response may be classified as *elastic*, *plastic*, *viscoelastic*, *viscoplastic*, or *fracture*, depending on its response to the loading condition.

If the unloading path coincides with the loading path, the process is *reversible* and the material is said to be elastic (Fig. 4.2a). If the unloading path does not follow the loading path, the behavior is said to be *inelastic*. A material that behaves in a plastic manner does not return to an unstrained state after the load is released (Fig. 4.2b). If, after load removal, the material response continues to change with time, its response is said to be viscoelastic or viscoplastic. Upon removal of load, the stress-strain response of a viscoelastic material follows a path (*AB*, Fig. 4.2c) that is different from the loading path. But in time, after complete unloading, the material will return to an unstrained state (along path *BO*, Fig. 4.2c). Likewise, the initial unloading response of a viscoplastic material (*AB*, Fig. 4.2d) is different from its loading response, and after complete unloading, the response will also change with time. However, some permanent strain will remain (*OC*, Fig. 4.2d).

In practice, fracture may occur at various stress levels. The material response up to fracture may be almost linear and fracture may occur at relative small inelastic strains (Fig. 1.8). Conversely, the material response prior to fracture may be highly nonlinear, with large inelastic strains (Figs. 1.3 and 1.5). Fracture may occur due to slow crack growth caused by a large number of load repetitions at stress levels below the yield point, a process known as fatigue (Chapter 16). Fracture may also occur because of sufficiently high stress levels that cause microcracks to propagate rapidly (to increase in size rapidly until rupture occurs; see Chapter 15).

In the remainder of this chapter, we limit our discussion to failure of materials that undergo elastic, plastic, or fracture response.

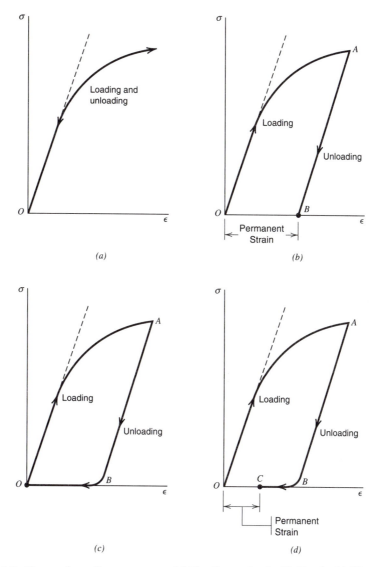

Figure 4.2   Types of nonlinear response. (a) Nonlinear elastic. (b) Plastic. (c) Viscoelastic. (d) Viscoplastic.

Models of Uniaxial Stress-Strain Curves

In a uniaxial tension test, the transition from linear elastic response to inelastic (nonlinear) response may be abrupt (Fig. 4.3a) or gradual (Fig. 4.3b). For an abrupt transition, the change is identified by the *kink* in the stress-strain curve. The stress level at this point is called the yield stress Y.* In the case of a gradual transition, the yield stress is arbitrarily defined as that stress which corresponds to a given permanent strain ϵ_s (usually, $\epsilon_s = 0.002$) that remains upon unloading along a straight-line path BB' parallel to AA' (Fig. 4.3b).

---

* In this chapter, the quantity Y represents *yield stress*, *yield point*, and *yield strength* without further distinction.

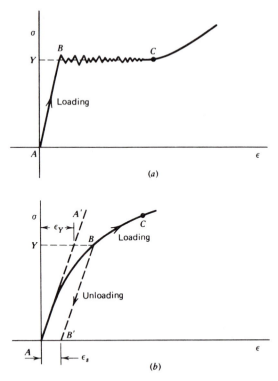

Figure 4.3   Experimental stress-strain curves. In (b), line BB' is parallel to line AA', which is tangent to curve ABC at A.

Actual stress strain curves, such as that in Fig. 4.3a, are difficult to use in mathematical solutions of complex problems. Therefore, idealized models of material response are used in analysis. For example, the uniaxial stress-strain curve shown in Fig. 4.3a may be modeled as shown in Fig. 4.4a, including unloading. Since part BC of the idealized curve in Fig. 4.4a is parallel to the strain axis (the stress remains constant with increasing strain), the material response is said to be *elastic-perfectly plastic*. For materials that strain harden in the initial nonlinear region, as does alloy steel, the stress continues to increase with increasing strain, (region AB in Fig. 4.1), although at a slower rate than in the elastic region (part OA in Fig. 4.1). The stress-strain curve for such a material might be idealized with a bilinear curve (region ABC, Fig. 4.4b). Such material response is referred to as *elastic-linear strain hardening*. For this idealized stress-strain diagram, the yield stress Y is the stress at point B (Fig. 4.4b), not the stress at a specified set ϵ_s. Finally, we recall that the real response of a material might not follow the assumed idealized stress-strain curve upon unloading. Some metals exhibit a Bauschinger effect, such as that exhibited by curves A'H in Figs. 4.4a and 4.4b and curve BD in Fig. 4.1.

Sometimes, the deformation imposed on a material may be so large that the elastic strain (point B in Fig. 4.4) is a small fraction of the total strain (say, the strain associated with point F, Fig. 4.4). In such cases, the elastic strain may have a negligible effect on the analysis. If the elastic strain is neglected, further idealizations (Fig. 4.5) of the stress-strain curves of Fig. 4.4 are possible. For the idealization of Fig. 4.5a, the material response is said to be *rigid-perfectly plastic*. The response shown in Fig. 4.5b is called *rigid-strain hardening*. In general, material response for which the elastic strain may be neglected is said to be simply *rigid-plastic*.

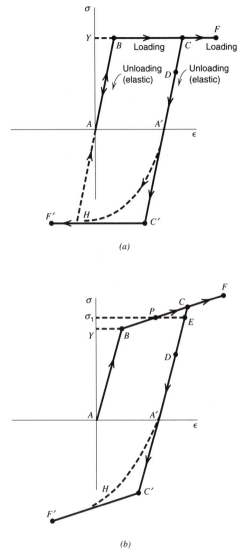

Figure 4.4   Idealized stress-strain curves. (a) Elastic-perfectly plastic response. (b) Elastic-strain-hardening response.

There exists an essential difference between nonlinear elastic response and plastic response of a material. In nonlinear elastic response there is always a unique relation between stress and strain throughout the load history; that is, for each value of stress, there is one and only one value of strain (Fig. 4.2a). However, for plastic response, there may be more than one value of strain for each value of stress. Specifically, if a plastic material is loaded into the inelastic region and then unloaded (Fig. 4.4b), a given value of stress may correspond to two values of strain, one for application of the load and one for removal of the load. This fact implies that the strain value for a given stress is *path-dependent*; that is, the strain can be determined uniquely, only if the stress and history of loading (the load path) are known. In contrast, elastic material response is *path-independent*, in that for a given stress the same value of strain is obtained during loading or unloading. To

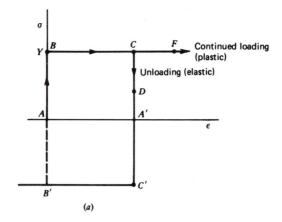

(a)

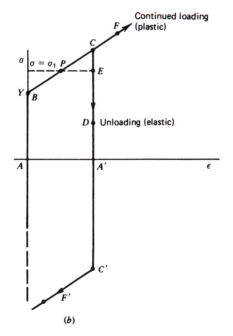

(b)

Figure 4.5   (a) Rigid elastic-perfectly plastic response. (b) Rigidly elastic-strain-hardening plastic response.

illustrate the path-dependent nature of plastic response, assume that the stress in a member is σ_1, and the stress-strain curve for the material is that of Fig. 4.4b. If the member is being loaded along BC, the strain is that associated with point P. If the member is being unloaded along CD, the strain is that associated with point E.

EXAMPLE 4.1
Strain-Hardening Axially Loaded Members

The stress-strain diagram for an isotropic metal at room temperature is approximated by two straight lines (Fig. 4.4b). Part AB has slope E and part BF has slope

βE, where E is the modulus of elasticity and β the strain-hardening factor for the metal. The intersection of the two lines defines the yield stress Y and yield strain $\epsilon_Y = Y/E$. The stress-strain relations in the region AB and BF are, respectively,

$$\sigma = E\epsilon, \qquad\qquad \epsilon \le \epsilon_Y \text{ (elastic stress-strain)} \qquad\qquad\text{(a)}$$
$$\sigma = (1 - \beta)Y + \beta E\epsilon, \qquad \epsilon > \epsilon_Y \text{ (inelastic stress-strain)} \qquad\text{(b)}$$

(a) Determine the constants β, Y, E, and ϵ_Y for the annealed high-carbon steel of Fig. 4.1.

(b) Consider the pin-joined structure in Fig. E4.1a. Each member has cross

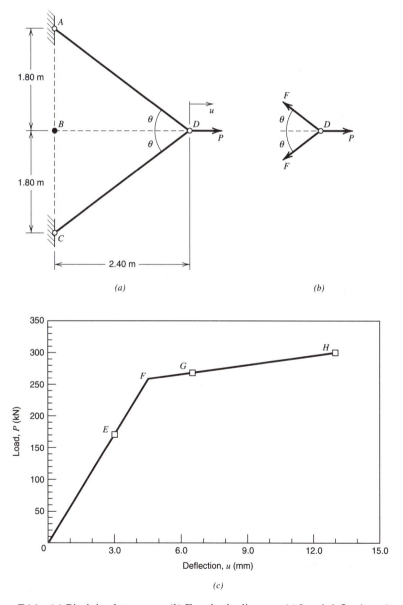

(a)   (b)

(c)

Figure E4.1   (a) Pin-joined structure. (b) Free-body diagram. (c) Load-deflection plot.

section area 645 mm^2 and is made of the steel of Fig. 4.1. A load $P = 170$ kN is applied. Compute the deflection u.

(c) Repeat part (b) for $P = 270$ kN and $P = 300$ kN.

(d) Use the results of parts (b) and (c) to plot a load-deflection graph for the structure.

SOLUTION

(a) By Fig. 4.1, $E = 211.4$ GPa, $Y = 252.6$ MPa, $\beta E = 16.9$ GPa (or $\beta = 0.0799$), and $\epsilon_Y = Y/E = 0.001195$. Equations (a) and (b) become

$$\sigma = 211{,}400\epsilon \text{ MPa}, \qquad \epsilon \leq 0.001195 \text{ (elastic stress-strain)} \qquad \text{(c)}$$
$$\sigma = 232.4 + 16{,}900\epsilon \text{ MPa}, \qquad \epsilon > 0.001195 \text{ (inelastic stress-strain)} \quad \text{(d)}$$

(b) By the geometry of the structure shown in Fig. E4.1a, $\cos \theta = 0.8$. By equilibrium of joint D (Fig. E4.1b), we find the force in members AD and CD to be $F = P/(2 \cos \theta) = 106.25$ kN. Therefore, the stress in members AD and CD is $\sigma = 164.73$ MPa < 252.6 MPa. Thus, for this load, the members are deformed elastically. By Eq. (c), the strain in members AD and CD is $\epsilon = 0.000779$. Since the length of members AD and CD is $L = 3.0$ m, their elongation is $e = \epsilon L = 2.338$ mm. Hence, $u = e/\cos \theta = 2.922$ mm.

(c) For $P = 270$ kN, $F = 168.75$ kN, $\sigma = 261.63$ MPa (> 252.6 MPa). Hence, bars AB and CD are strained inelastically. Therefore, by Eq. (d), $\epsilon = 0.001730$, and $e = \epsilon L = 5.189$ mm. Hence, $u = e/\cos \theta = 6.486$ mm. For $P = 300$ kN, $F = 187.5$ kN and $\sigma = 290.7$ MPa (> 252.6 MPa). By Eq. (d), the strain in the bars is $\epsilon = 0.003450$, the elongation of the bars is $e = \epsilon L = 10.349$ mm, and the deflection is $u = e/\cos \theta = 12.936$ mm.

(d) A summary of the load deflection data is given in Table E4.1, and the data are plotted in Fig. E4.1c. The yield load (point F) is located by the intersection of the extensions of lines $0E$ and GH. Note that the ratio of the slope of line FGH to the slope of line $0EF$ is $\beta = 0.0799$.

TABLE E4.1
Load-Deflection Data

Load P (kN)	Deflection u (mm)	Point (Fig. E4.1c)
0	0.0	0
170	2.922	E
270	6.486	G
300	12.936	H

EXAMPLE 4.2
Elastic-Perfectly Plastic Structure

The structure in Fig. E4.2a consists of a rigid beam AB and five rods placed symmetrically about line CD. A load P is applied to the beam as shown. The members are made of an elastic-perfectly plastic steel ($E = 200$ GPa), and they each

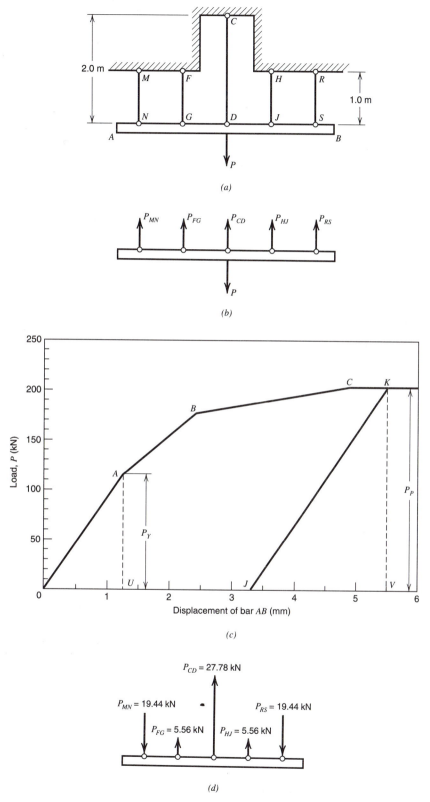

Figure E4.2 (*a*) Rod-supported beam. (*b*) Free-body diagram of beam. (*c*) Load-displacement diagram. (*d*) Residual forces in bars.

have a cross section area of 100 mm^2. Members CD, FG, and HJ have a yield point stress equal to $Y_1 = 500$ MPa, and members MN and RS have a yield point equal to $Y_2 = 250$MPa.

(a) Ignoring the weight of the beam, determine the magnitude of load P and the corresponding displacement of beam AB for $P = P_Y$, the load for which yield first occurs in the structure.

(b) Repeat part (a) for $P = P_P$, the fully plastic load, that is, the load for which all members are yielded.

(c) Construct the load-displacement diagram for beam AB.

(d) The fully plastic load P_P is gradually removed. Determine the residual forces that remain in the members of the structure.

SOLUTION

The load P produces elongations in the five members equal to the displacement of the beam. Since equilibrium can be maintained by the single member CD, the four other members may be considered redundant. However, since the members undergo equal elongations, the redundancy of the structure will not pose any problem.

(a) *Yield Load P_Y*. By inspection, yield is initiated first in members MN and RS. Yield is initiated in these members when the stress in them reaches the yield point stress Y_2. At this stress level, the load is $P = P_Y$. The corresponding load in members MN and RS is

$$P_1 = P_{MN} = P_{RS} = Y_2A = 25,000 \text{ N} = 25.0 \text{ kN} \qquad \text{(a)}$$

The elongation of members MN and RS at initiation of yield is $e_{MN} = e_{RS} = P_1L_{MN}/(EA) = 1.25$ mm. The displacement of the beam AB and the elongation of the other three members are also equal to 1.25 mm. The strain in members FG and HJ is 0.00125, and the stress is 250 MPa (<500 MPa). Therefore, bars FG and HJ are elastic. The strain in bar CD is 0.000625, and its stress is 125 MPa (<500 MPa). Therefore, it is also elastic. Hence, the axial forces in members FG, HJ, and CD are

$$P_2 = P_{FG} = P_{HJ} = \frac{EAe_{FG}}{L_{FG}} = 25.0 \text{ kN} \qquad \text{(b)}$$

$$P_3 = P_{CD} = \frac{EAe_{CD}}{L_{CD}} = 12.5 \text{ kN} \qquad \text{(c)}$$

The equilibrium force equation in the direction parallel to the members gives, with Eqs. (a–c),

$$P_Y = 2P_1 + 2P_2 + P_3 = 112.5 \text{ kN} \qquad \text{(d)}$$

and the corresponding beam displacement is 1.25 mm.

(b) *Fully Plastic Load P_P*. Since members FG and HJ are shorter than member CD, they yield next. Also, the axial forces in members MN and RS will remain constant as P is increased in magnitude beyond P_Y. Yield occurs in members FG and HJ when the stress in them reaches the yield stress $Y_1 = 500$ MPa. The load in each of these members is then

$$P_4 = P_{FG} = P_{HJ} = Y_1 A = 50.0 \text{ kN} \tag{e}$$

The elongation of these members at yield is $e_{FG} = e_{HJ} = P_4 L_{FG}/(EA) = 2.50$ mm. The displacement of beam AB and the elongation of all the other members are also 2.50 mm. The strain of member CD is 0.00125 and, therefore, the stress is 250 MPa (< 500 MPa). Hence, bar CD is elastic. Therefore, the force in member CD is

$$P_5 = P_{CD} = \frac{EAe_{CD}}{L_{CD}} = 25.0 \text{ kN} \tag{f}$$

The force equilibrium equation in the direction parallel to the members gives, with Eqs. (a), (e), and (f),

$$P = 2P_1 + 2P_4 + P_5 = 175.0 \text{ kN} \tag{g}$$

and the corresponding beam displacement is 2.50 mm.

The fully plastic load P_P occurs when yield is initiated in member CD. The loads in members MN, RS, FG, and HJ remain unchanged as P is increased beyond 175.0 kN. When member CD reaches yield, the axial force is

$$P_6 = P_{CD} = Y_2 A = 50.0 \text{ kN} \tag{h}$$

Also, when member CD reaches yield, its elongation is $e_{CD} = P_6 L_{CD}/(EA) = 5.00$ mm. The displacement of the beam, and that of all the other members, is 5.00 mm. The force equilibrium equation in the direction parallel to the members gives, with Eqs. (a), (e), and (h), the fully plastic load

$$P_P = 2P_1 + 2P_4 + P_6 = 200.0 \text{ kN} \tag{i}$$

and the beam displacement is 5.00 mm.

(c) *Load-Displacement Diagram.* The load-displacement diagram for beam AB is plotted in Fig. E4.2c. Point A corresponds to displacement 1.25 mm and the load P_Y. Point B occurs at the elastic-plastic load that initiates yield in members FG and HJ, with corresponding displacement 2.50 mm. Point C occurs at the fully plastic load P_P that initiates yield in member CD, with corresponding displacement 5.00 mm. The load-displacement curve is horizontal at the fully plastic load P_P.

(d) *Residual Forces.* Let the fully plastic load P_P be unloaded gradually from a displacement greater than 5.00 mm, say, 5.50 mm as indicated in Fig. E4.2c. Assume that the materials in the members respond elastically upon unloading. Then, the displacement of the beam will follow the path KJ, parallel to line $0A$. At the beginning of unloading, the stress in members MN and RS is equal to the yield stress $Y_2 = 250$ MPa. We assume that the material in these two members remains linearly elastic as the stress in the members goes to zero and also remains linearly elastic to -250 MPa in compression (i.e., there is no Bauschinger effect). Hence, the members will remain elastic for an increment of elongation Δe on unloading equal to twice the elongation at point A; that is, for an elongation $\Delta e = 2(1.25) = 2.50$ mm. The actual increment of elongation as the structure unloads from point K to point J is Δe_{VJ}, and it is given by the similarity of triangles $0AU$ and JKV in Fig. E4.2c. Thus, $\Delta e_{VJ}/200.0 = 1.25/112.5$, or $\Delta e_{VJ} = 2.222$ mm. Since Δe_{VJ} is less than 2.50 mm, all tension members unload elastically. The increment in load ΔP for each member for unloading is, therefore, $\Delta P_{MN} = \Delta P_{RS} = \Delta P_{FG} = \Delta P_{HJ} = EA(\Delta e_{VJ})/L_{MN} = 44.44$ kN and $\Delta P_{CD} = EA(\Delta e_{VJ})/L_{CD} = 22.22$ kN. At zero load, the residual forces in the members are equal to the axial forces in the members at the fully plastic load P_P minus the load increments ΔP. Thus,

$$P_{MN(\text{residual})} = P_{RS(\text{residual})} = 25.0 - 44.44 = -19.44 \text{ kN (compression)}$$
$$P_{FG(\text{residual})} = P_{HJ(\text{residual})} = 50.0 - 44.44 = 5.56 \text{ kN (tension)}$$
$$P_{CD(\text{residual})} = 50.0 - 22.22 = 27.78 \text{ kN (tension)}$$

The residual forces are shown in Fig. E4.2d.

4.3

YIELD CRITERIA: GENERAL CONCEPTS

The previous discussion of failure criteria has been limited to uniaxial stress states. However, more generally, we must apply failure criteria to multiaxial states of stress. We consider failure to occur at the initiation of inelastic material behavior through either yielding or fracture. In general, a complete plasticity theory has three components: a yield criterion (or yield function) that defines the initial inelastic response of the material, a flow rule that relates the plastic strain increments to the stress increments after initiation of the inelastic response, and a hardening rule that predicts changes in the yield surface (a geometrical representation of the yield criterion) due to the plastic strain (Mendelson, 1983). Since we limit ourselves to predicting initiation of inelastic behavior, we consider only the first component of plasticity theory: the yield criterion.

The primary objective of this section is to extend the concept of yield criteria to multiaxial stress states. The basis for this extension is the definition of an effective (or equivalent) uniaxial stress that is a particular combination of the components of the multiaxial stress state. It is postulated that yielding is initiated in a multiaxial stress state when this effective stress reaches a limiting value (assumed to be some function of the uniaxial yield stress). The same concept may be used to predict failure by fracture, provided that an appropriate failure criterion can be established. In the following, we examine various yield criteria and discuss their ability to predict the initiation of inelastic response of various materials subjected to multiaxial stress states. Unfortunately, no single yield criterion has been established that accurately predicts yielding (or fracture) for all materials. However, the initiation of yield in ductile metals can be predicted reasonably well by either the maximum shear-stress criterion or the maximum octahedral shear-stress criterion (the latter being equivalent to the distortional energy criterion).

A yield criterion can be any descriptive statement that defines conditions under which yielding will occur. It may be expressed in terms of specific quantities, such as the stress state, the strain state, a strain energy quantity, or others. A yield criterion is usually expressed in *mathematical* form by means of a yield function $f(\sigma_{ij}, Y)$, where σ_{ij} defines the state of stress and Y is the yield strength in uniaxial tension (or compression). The yield function is defined such that the yield criterion is satisfied when $f(\sigma_{ij}, Y) = 0$. When $f(\sigma_{ij}, Y) < 0$, the stress state is elastic. The condition $f(\sigma_{ij}, Y) > 0$ is undefined. To develop a yield function, the components of the multiaxial stress state are combined into a single quantity known as the *effective stress* σ_e. The effective stress is then compared with the yield stress Y, in some appropriate form, to determine if yield has occurred.

To help illustrate the nature of a yield criterion, the concept of a yield surface is used widely (Lubliner, 1990). A yield surface is a graphical representation of a yield function. For a three-dimensional stress state, the yield surface is plotted in *principal stress space*, also known as Haigh–Westergaard stress space. That is, the yield surface is plotted using principal stresses $\sigma_1, \sigma_2, \sigma_3$ as coordinates of three mutually perpendicular axes. In the following paragraphs, various yield criteria and their associated yield functions are discussed.

Maximum Principal Stress Criterion

To illustrate the concept of a yield criterion, consider the *maximum principal stress criterion*, often called Rankine's criterion. This criterion states that yielding begins at a point in a member when the maximum principal stress reaches a value equal to the tensile (or compressive) yield stress Y. For example, assume that a single nonzero principal stress σ_1 acts at a point in the member (Fig. 4.6a). According to Rankine's criterion, yielding will occur when σ_1 reaches the value Y. Next consider the case where principal stresses σ_1 and σ_2 ($|\sigma_1| > |\sigma_2|$) both act at the point as shown in Fig. 4.6b. Rankine's criterion again predicts that yielding will occur when $\sigma_1 = Y$, regardless of the fact that σ_2 also acts at the point. In other words, the maximum principal stress criterion ignores the effects of the other principal stresses.

If $\sigma_1 = -\sigma_2 = \sigma$, the shear stress τ is equal in magnitude to σ and occurs on 45° diagonal planes (Fig. 4.6c). Such a state of stress occurs in a cylindrical bar subjected to torsion. Thus, if the maximum principal stress criterion ($\sigma_1 = \sigma = Y$) is to be valid for a particular material under arbitrary loading, the shear yield stress τ_Y of the

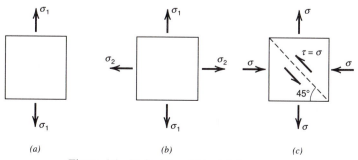

(a)   *(b)*   *(c)*

Figure 4.6   Uniaxial and biaxial stress states.

material must be equal to the tensile yield stress Y. For ductile metals, the shear yield stress τ_Y is much less than the tensile yield stress Y. It is evident that, for ductile metals, the maximum principal stress criterion is not applicable. However, for brittle materials that *fail by brittle fracture rather than yielding*, the maximum principal stress criterion may adequately predict tension fracture. In fact, the maximum principal stress criterion is often used in conjunction with other criteria to predict failure of brittle materials such as concrete (Chen and Han, 1988).

The maximum principal stress criterion can be expressed by the yield function

$$f = \max(|\sigma_1|, |\sigma_2|, |\sigma_3|) - Y \tag{4.1}$$

where the principal stresses *are not ordered*. From Eq. (4.1), we see that the effective stress is $\sigma_e = \max(|\sigma_1|, |\sigma_2|, |\sigma_3|)$.

The corresponding yield surface is defined by the locus of stress states that satisfy the yield criterion ($f = 0$). Hence, the yield surface for the maximum principal stress criterion is defined by the relations.

$$\sigma_1 = \pm Y, \qquad \sigma_2 = \pm Y, \qquad \sigma_3 = \pm Y \tag{4.2}$$

The yield surface consists of six planes, perpendicular to the principal stress coordinates axes (see Fig. 4.7).

Maximum Principal Strain Criterion

The *maximum principal strain criterion*, also known as St. Venant's criterion, states that yielding begins when the maximum principal strain at a point reaches a value equal to the yield strain $\epsilon_Y = Y/E$. For example, yielding in the block of Fig. 4.6a begins when $\epsilon_1 = \epsilon_Y$, which corresponds to $\sigma_1 = Y$. Under biaxial stress (Fig. 4.6b), the maximum principal strain in an isotropic material is $\epsilon_1 = (\sigma_1/E) - v(\sigma_2/E)$. For the stress state shown in Fig. 4.6b, where σ_2 is positive (tensile), yielding will begin for a value of $\sigma_1 > Y$. If σ_2 is negative (compressive), the maximum value of σ_1 that

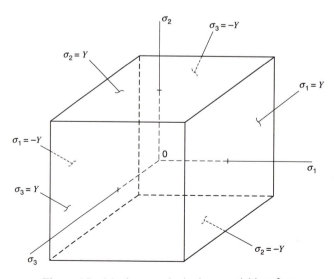

Figure 4.7   Maximum principal stress yield surface.

can be applied without causing yielding is less than Y. The maximum principal strain criterion does not accurately predict yielding of ductile metals but may show improved ability, relative to the maximum principal stress criterion, to predict fracture of brittle materials.

Even though the maximum principal strain criterion predicts yielding in terms of strain magnitudes, we still express its yield function in terms of stress. Writing the first of Eqs. (3.30) in terms of principal stresses, we have

$$\epsilon_1 = \frac{1}{E}(\sigma_1 - v\sigma_2 - v\sigma_3) \tag{4.3}$$

Assuming that ϵ_1 is the principal strain with the largest magnitude, we equate $|\epsilon_1|$ with ϵ_Y and obtain the yield function

$$f_1 = |\sigma_1 - v\sigma_2 - v\sigma_3| - Y = 0 \quad \text{or} \quad \sigma_1 - v\sigma_2 - v\sigma_3 = \pm Y \tag{4.4a}$$

If we assume that the principal strains are unordered, either ϵ_2 or ϵ_3 may have the largest magnitude. Then we obtain the additional possibilities

$$f_2 = |\sigma_2 - v\sigma_1 - v\sigma_3| - Y = 0 \quad \text{or} \quad \sigma_2 - v\sigma_1 - v\sigma_3 = \pm Y \tag{4.4b}$$

$$f_3 = |\sigma_3 - v\sigma_1 - v\sigma_2| - Y = 0 \quad \text{or} \quad \sigma_3 - v\sigma_1 - v\sigma_2 = \pm Y \tag{4.4c}$$

Hence, the effective stress σ_e may be defined as

$$\sigma_e = \max_{i \ne j \ne k} |\sigma_i - v\sigma_j - v\sigma_k| \tag{4.5}$$

and the yield function as

$$f = \sigma_e - Y \tag{4.6}$$

The yield surface for the maximum principal strain criterion for a biaxial stress state ($\sigma_3 = 0$) is shown in Fig. 4.8. The yield surface $ABCD$ illustrates that under biaxial tension (or biaxial compression), individual principal stresses greater then Y can occur without causing yielding.

Strain Energy Density Criterion

The *strain energy density criterion*, proposed by Beltrami (Mendelson, 1983), states that yielding at a point begins when the strain energy density at the point equals the strain energy density at yield in uniaxial tension (or compression). Written in terms of principal stresses, strain energy density is [Eq. (3.33)]

$$U_0 = \frac{1}{2E}[\sigma_1^2 + \sigma_2^2 + \sigma_3^2 - 2v(\sigma_1\sigma_2 + \sigma_1\sigma_3 + \sigma_2\sigma_3)] > 0 \tag{4.7}$$

By Eq. (4.7), the strain energy density at yield in a uniaxial tension test ($\sigma_1 = Y$, $\sigma_2 = \sigma_3 = 0$) is

$$U_{0Y} = \frac{Y^2}{2E} \tag{4.8}$$

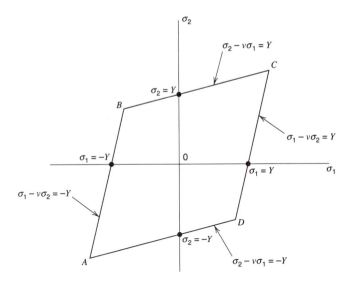

Figure 4.8   Maximum principal strain yield surface for biaxial stress state ($\sigma_3 = 0$); $f = 0$ along the perimeter $ABCD$.

Thus, the strain energy density criterion states that yield is initiated when the strain energy density U_0, Eq. (4.7), for any stress state equals U_{0Y}, Eq. (4.8).

We again consider the stress states depicted in Fig. 4.6, but apply the strain energy density criterion for yielding $U_0 = U_{0Y}$. For uniaxial tension (Fig. 4.6a), yielding is predicted to occur when $\sigma_1 = Y$, as it should. For a biaxial stress state in which $\sigma_1 = \sigma_2 = \sigma$ (Fig. 4.6b), yielding is predicted to occur when $2\sigma^2(1 - v) = Y^2$. If we assume that Poisson's ratio for the material is zero ($v = 0$), then yielding occurs when $\sigma = Y/\sqrt{2}$. If the biaxial stress state is $\sigma_1 = -\sigma_2 = \sigma$ (Fig 4.6c), a state of pure shear exists and yielding is again predicted to occur when $\sigma = Y/\sqrt{2}$, provided that $v = 0$.

The yield function for the strain energy density criterion is obtained by setting U_0, Eq. (4.7), equal to U_{0Y}, Eq. (4.8), to obtain

$$\sigma_1^2 + \sigma_2^2 + \sigma_3^2 - 2v(\sigma_1\sigma_2 + \sigma_1\sigma_3 + \sigma_2\sigma_3) - Y^2 = 0 \tag{4.9}$$

Hence, the yield function has the form

$$f = \sigma_e^2 - Y^2 \tag{4.10}$$

where the effective stress is

$$\sigma_e = \sqrt{\sigma_1^2 + \sigma_2^2 + \sigma_3^2 - 2v(\sigma_1\sigma_2 + \sigma_1\sigma_3 + \sigma_2\sigma_3)} \tag{4.11}$$

4.4

YIELDING OF DUCTILE METALS

Thus far, we have considered yield criteria that are based on point-wise values of stress or strain. However, in engineering, experimentally measured inelastic deformations (strains) are determined over gage lengths that are of finite length.

The engineering approach to defining inelastic deformation differs from that of the metallurgist who treats inelastic deformation as a microscopic quantity associated with the slip of crystals. Nevertheless, these engineering measurements and certain metallurgical concepts are employed to establish criteria to predict the inception of yielding. For example, it is well known in metallurgy that certain metal crystals have slip planes along which the resistance to shear force is relatively small. Thus, for such metals, yield criteria are based on limiting values of shear stress. Two such criteria are presented in the following discussion.

Maximum Shear-Stress (Tresca) Criterion

The *maximum shear-stress criterion*, also known as the *Tresca criterion*, states that yielding begins when the maximum shear stress at a point equals the maximum shear stress at yield in uniaxial tension (or compression). For a multiaxial stress state, the maximum shear stress is $\tau_{max} = (\sigma_{max} - \sigma_{min})/2$, where σ_{max} and σ_{min} denote the maximum and minimum ordered principal stress components, respectively. In uniaxial tension ($\sigma_1 = \sigma$, $\sigma_2 = \sigma_3 = 0$), the maximum shear stress is $\tau_{max} = \sigma/2$. Since yield in uniaxial tension must begin when $\sigma_1 = Y$, the shear stress associated with yielding is predicted to be $\tau_Y = Y/2$. Thus, the yield function for the maximum shear-stress criterion may be defined as

$$f = \sigma_e - \frac{Y}{2} \qquad (4.12)$$

where the effective stress is

$$\sigma_e = \tau_{max} \qquad (4.13)$$

The magnitudes of the extreme values of the shear stresses are [see Eq. (2.44) and Fig. 2.13]

$$\tau_1 = \frac{|\sigma_2 - \sigma_3|}{2}$$

$$\tau_2 = \frac{|\sigma_3 - \sigma_1|}{2}$$

$$\tau_3 = \frac{|\sigma_1 - \sigma_2|}{2} \qquad (4.14)$$

The maximum shear stress τ_{max} is the largest of (τ_1, τ_2, τ_3). If the principal stresses are unordered, yielding under a multiaxial stress state can occur for any one of the following conditions:

$$\sigma_2 - \sigma_3 = \pm Y$$
$$\sigma_3 - \sigma_1 = \pm Y$$
$$\sigma_1 - \sigma_2 = \pm Y \qquad (4.15)$$

By Eq. (4.15), the yield surface for the maximum shear-stress criterion is a regular hexagon in principal stress space (Fig. 4.9). For a biaxial stress state ($\sigma_3 = 0$), the yield surface takes the form of an elongated hexagon in the (σ_1, σ_2) plane (Fig. 4.10).

The Tresca criterion exhibits good agreement with experimental results for certain ductile metals. This can be anticipated by reexamination of the metallurgical basis for yielding. The movement of dislocations along slip planes, which is

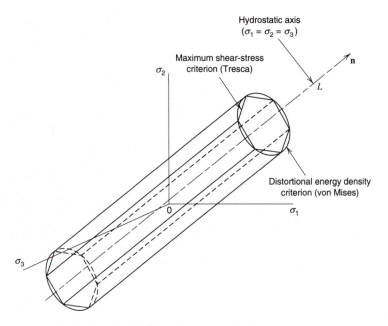

Figure 4.9   Yield surface in principal stress space.

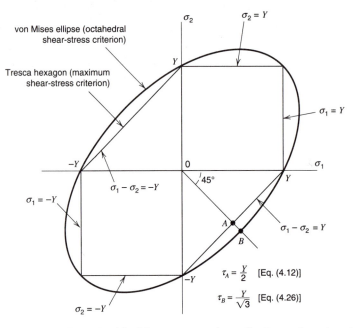

Figure 4.10   Yield surfaces for biaxial stress state ($\sigma_3 = 0$). For points A and B, $\sigma_1 = -\sigma_2 = \sigma$ (pure shear).

responsible for permanent deformation, is a shear-related phenomenon. Thus, the maximum shear-stress (Tresca) criterion has some physical basis. However, for a state of pure shear, such as occurs in a torsion test, the shear yield stress τ_Y of some ductile metals is found to be approximately 15% higher than the value predicted by the Tresca criterion. Thus, the Tresca criterion is conservative for these metals.

Nevertheless, since it is fairly accurate and computationally simple to use, the maximum shear-stress criterion is a reasonable choice for most ductile metals.

Distortional Energy Density (von Mises) Criterion

The *distortional energy density criterion*, often attributed to von Mises, states that yielding begins when the distortional strain energy density at a point equals the distortional strain energy density at yield in uniaxial tension (or compression). The distortional strain energy density is that energy associated with a change in the shape of a body. The total strain energy density U_0 (Eq. 4.7) can be broken into two parts: one part that causes volumetric change U_V and one that causes distortion U_D. The appropriate manipulations of Eq. (4.7) give

$$U_0 = \frac{(\sigma_1 + \sigma_2 + \sigma_3)^2}{18K} + \frac{(\sigma_1 - \sigma_2)^2 + (\sigma_2 - \sigma_3)^2 + (\sigma_3 - \sigma_1)^2}{12G} \tag{4.16}$$

where K is the bulk modulus ($K = E/[3(1 - 2\nu)]$) and G the shear modulus ($G = E/[2(1 + \nu)]$). The first term on the right side of Eq. (4.16) is U_V, the strain energy density associated with pure volume change. The second term is the distortional strain energy density

$$U_D = \frac{(\sigma_1 - \sigma_2)^2 + (\sigma_2 - \sigma_3)^2 + (\sigma_3 - \sigma_1)^2}{12G} \tag{4.17}$$

At yield under a uniaxial stress state ($\sigma_1 = \sigma$, $\sigma_2 = \sigma_3 = 0$), $U_D = U_{DY} = Y^2/6G$. Thus, for a multiaxial stress state, the distortional energy density criterion states that yielding is initiated when the distortional energy density U_D given by Eq. (4.17) equals $Y^2/6G$.

The distortional energy density criterion can be expressed in an alternate form as follows. By Eqs. (4.17) and (2.27), the distortional energy density U_D can be written in terms of the second deviator stress invariant J_2 as

$$U_D = \frac{1}{2G} J_2 \tag{4.18}$$

where

$$J_2 = \tfrac{1}{6}[(\sigma_1 - \sigma_2)^2 + (\sigma_2 - \sigma_3)^2 + (\sigma_3 - \sigma_1)^2] \tag{4.19a}$$

Relative to the general (x, y, z) axes, J_2 can be expressed in terms of the stress invariants I_1 and I_2 [see Eqs. (2.21) and (2.27)]

$$J_2 = I_2 + \tfrac{1}{3}I_1^2 \tag{4.19b}$$

At yield in uniaxial tension (or compression), $\sigma_1 = \pm Y$, $\sigma_2 = \sigma_3 = 0$. Then,

$$J_2 = \tfrac{1}{3}Y^2 \tag{4.20}$$

Therefore, by Eqs. (4.19a) and (4.20), we may write the yield function for the distortional energy density (von Mises) criterion as

$$f = \tfrac{1}{6}[(\sigma_1 - \sigma_2)^2 + (\sigma_2 - \sigma_3)^2 + (\sigma_3 - \sigma_1)^2] - \tfrac{1}{3}Y^2 \tag{4.21}$$

A more compact form for the yield function is

$$f = \sigma_e^2 - Y^2 \qquad (4.22)$$

where the effective stress is

$$\sigma_e = \sqrt{\tfrac{1}{2}[(\sigma_1 - \sigma_2)^2 + (\sigma_2 - \sigma_3)^2 + (\sigma_3 - \sigma_1)^2]} = \sqrt{3J_2} \qquad (4.23)$$

Also, noting the relationship between J_2 and the octahedral shear stress, Eqs. (2.27) and (2.22), we can rewrite the yield function, Eq. (4.21), for the von Mises yield criterion in the alternate form

$$f = \tau_{\text{oct}} - \frac{\sqrt{2}}{3} Y \qquad (4.24)$$

Thus, according to Eq. (4.24), when $f = 0$, the octahedral shear stress at a point reaches the value $(\sqrt{2/3})Y = 0.471\,Y$, and yielding occurs. This result agrees with that obtained by Eq. (4.21). For this reason, the distortional energy density (von Mises) criterion is also referred to as the *maximum octahedral shear-stress criterion*.

For a three-dimensional stress state, the yield surface for the von Mises criterion forms a cylinder that circumscribes the Tresca hexagon, (Fig. 4.9). For a biaxial stress state ($\sigma_3 = 0$), the von Mises yield surface reduces to an ellipse in the σ_1-σ_2 plane (Fig. 4.10).

As with the Tresca criterion, the von Mises criterion is fairly accurate in predicting initiation of yield for certain ductile metals. The von Mises criterion is more accurate for some materials than the Tresca criterion in predicting yield under pure shear. A state of pure shear exists for a principal stress state $\sigma_1 = -\sigma_2 = \sigma$, $\sigma_3 = 0$ (Fig. 4.6c). With this stress state, Eq. (4.22) predicts that yield occurs under pure shear when

$$3\sigma^2 = Y^2 \qquad (4.25)$$

For this stress state, the maximum shear stress is $\tau_{\max} = |\sigma_1 - \sigma_2|/2 = \sigma$. Therefore, at yield $\sigma = \tau_{\max} = \tau_Y$. Substitution of this value for σ into the von Mises criterion, Eq. (4.25), gives the shear stress at yield as

$$\tau_Y = \frac{Y}{\sqrt{3}} = 0.577\,Y \qquad (4.26)$$

as compared to $\tau_Y = Y/2$, which is predicted by the Tresca criterion, Eqs. (4.12) and (4.13). Thus, the von Mises criterion predicts that the pure-shear yield stress is approximately 15% greater than that predicted by the Tresca criterion (see also Fig. 4.10).

If the principal stresses are known, the Tresca criterion is easier to apply than the von Mises criterion. However, since the von Mises yield function is continuously differentiable, that is, its yield surface has a unique outward normal at all points (Figs. 4.9 and 4.10), it is preferred in plasticity studies in which plastic flow and strain hardening are considered. Materials that behave according to either the von Mises or Tresca criterion are often called J_2 materials, since the effective stress can be written solely in terms of this invariant. (Chen and Han, 1988; p. 75)

Effect of Hydrostatic Stress, The π-Plane

Examination of the yield functions for the Tresca and von Mises yield criteria indicates that the hydrostatic stress, $\sigma_m = (\sigma_1 + \sigma_2 + \sigma_3)/3$, has no influence on the initiation of yielding. This agrees with experimental evidence that indicates certain ductile metals do not yield under very high hydrostatic stress.* This independence of hydrostatic stress permits us to examine just a cross section of the Tresca and von Mises yield surfaces.

Consider an arbitrary stress state represented by point $B(\sigma_1, \sigma_2, \sigma_3)$ that lies on line M in principal stress space (Fig. 4.11). The vector $\mathbf{0B}$, which represents this stress state, can be decomposed into two components: $\mathbf{0A}$ that lies along the *hydrostatic axis* [line L with direction cosines: $(1/\sqrt{3}, 1/\sqrt{3}, 1/\sqrt{3})$] and $\mathbf{AB}$ that lies in a plane normal to line L. Vector $\mathbf{0A}$ represents the hydrostatic component of the stress state and vector $\mathbf{AB}$ represents its deviatoric component. Since hydrostatic stress has no influence on yielding, it is sufficient to discuss the yield surfaces only in terms of the deviatoric component (vector $\mathbf{AB}$). The plane that contains point B and is normal to the hydrostatic axis also contains all other points that represent stress states with the same hydrostatic stress. Such a plane is known as a *deviatoric plane*.

Now consider a second stress state represented by point D that also lies on line M and differs from point B only in its hydrostatic stress component. With respect to yielding, point D is identical to point B; they represent identical deviatoric stress states. Therefore, line M must be parallel to line L, the hydrostatic axis. The fact that the generator line M is parallel to the hydrostatic axis implies that all possible stress states may be viewed in terms of only their deviatoric components. Thus, only a

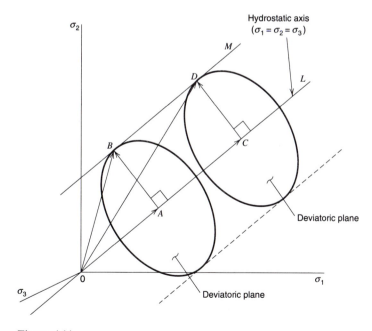

Figure 4.11

---

* However, hydrostatic tension can contribute to brittle fracture in otherwise ductile metals.

single deviatoric plane is needed to study the yield criteria. As our reference plane, we choose the deviatoric plane for which $\sigma_1 = \sigma_2 = \sigma_3 = 0$ (hence, $\sigma_m = 0$). This reference plane is known as the π-*plane* (Lubliner, 1990; Sec. 3.3).

Figure 4.12 shows the π-plane as the plane of the paper with the three, orthogonal, principal stress axes oblique to the plane. The Tresca and von Mises yield surfaces form a regular hexagon and circle, respectively, on the π-plane. Representation of the Tresca and von Mises yield surfaces on the π-plane is of fundamental importance in the theory of plasticity (which includes flow rules and hardening rules in addition to yield criteria). To construct these surfaces, we must determine the lengths of the line segments $0B$ and $0A$ (or $0C$), see Fig. 4.12. Consider the uniaxial stress state $(\sigma_1, \sigma_2 = \sigma_3 = 0)$ and unit vectors $\mathbf{i}$, $\mathbf{j}$, and $\mathbf{k}$ along axes σ_1, σ_2, and σ_3, respectively. The vector that represents this stress state is $\sigma_1 \mathbf{i}$. This vector may be decomposed into two perpendicular components: one component σ_L that lies along the hydrostatic axis L, with unit vector $\mathbf{n} = \frac{1}{\sqrt{3}}(\mathbf{i} + \mathbf{j} + \mathbf{k})$, and a second component σ_π that lies in the π-plane.

By geometry, $\sigma_L = \sigma_1 \mathbf{i} \cdot \mathbf{n} = \sigma_1/\sqrt{3}$. Since $\sigma_1^2 = \sigma_L^2 + \sigma_\pi^2$, we find $\sigma_\pi = \sqrt{\frac{2}{3}}\sigma_1$ to be the length of the projection of $\sigma_1 \mathbf{i}$ on the π-plane. Similar results are obtained for stress states $(\sigma_2, \sigma_1 = \sigma_3 = 0)$ and $(\sigma_3, \sigma_1 = \sigma_2 = 0)$. Thus, the components of a general stress state $(\sigma_1, \sigma_2, \sigma_3)$ have projections on the π-plane: $\sqrt{\frac{2}{3}}\sigma_1$, $\sqrt{\frac{2}{3}}\sigma_2$, $\sqrt{\frac{2}{3}}\sigma_3$.

The stress state $(\sigma_1 = Y, \sigma_2 = \sigma_3 = 0)$ represents the initiation of yield in a uniaxial tension test for both the Tresca and von Mises criteria. For this case, the length of the projection on the π-plane is $\sqrt{\frac{2}{3}}Y$ and the stress state lies at a common point on both yield surfaces, point A in Fig. 4.12. By inspection, the intersection of the von Mises cylinder (Fig. 4.9) with the π-plane is a circle of radius $\sqrt{\frac{2}{3}}Y$.

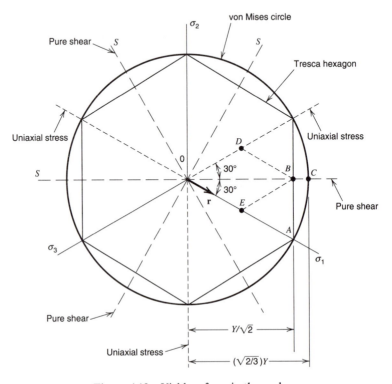

Figure 4.12   Yield surfaces in the π-plane.

Next consider a state of yield under pure shear according to the Tresca criterion ($\sigma_1 = -\sigma_3 = Y/2$, $\sigma_2 = 0$). The projections of the components of this stress state on the π-plane are $0E$ for σ_1 and $0D$ for σ_3. Line segments $0D$ and $0E$ are each of length $\sqrt{\frac{2}{3}}\sigma_1 = \sqrt{\frac{2}{3}}(Y/2) = \frac{1}{\sqrt{6}}Y$. We note that $0D = 0E = 0A/2$. The sum of the two vector projections gives the projection of the pure shear-stress state $0B$, where $0B = 2(0D)\cos 30 = Y/\sqrt{2}$. The point B lies on the Tresca yield surface. For pure shear, the von Mises criterion predicts that yield is initiated at point C, where $0C = \sqrt{\frac{2}{3}}Y$. The ratio $0C/0B$ is 1.15. This fact confirms the earlier statement that the von Mises criterion predicts yield at a shear stress that is 15% higher than that predicted by the Tresca criterion.

In general, uniaxial stress states exist along the projections of the $(\sigma_1, \sigma_2, \sigma_3)$ coordinate axes on the π-plane. Pure shear-stress states exist along the lines labeled S in Fig. 4.12, for which one principal stress is zero and the other two principal stresses are equal in magnitude but of opposite sign.

The values of the principal stresses that produce yielding in pure shear can be found as follows. A close-up view of a portion of the π-plane from Fig. 4.12 is shown in Fig. 4.13. The pure shear-stress state ($\sigma_1 = -\sigma_3$, $\sigma_2 = 0$) can be transformed back to principal stress coordinates using the law of sines. For the Tresca criterion, yield occurs at point B; lines $0D$ and $0E$ have length $Y/\sqrt{6}$. Hence, the principal stresses are

$$\left(\sigma_1 = \sqrt{\tfrac{3}{2}}\left(\frac{Y}{\sqrt{6}}\right) = \frac{Y}{2}, \quad \sigma_2 = 0, \quad \sigma_3 = -\sigma_1 = -\frac{Y}{2}\right)$$

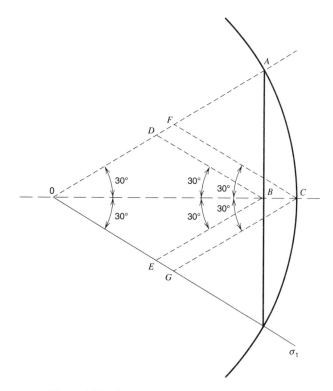

Figure 4.13 Geometry of yielding in pure shear.

and the shear yield stress is $\tau_Y = Y/2$. Similarly for the von Mises criterion, yield occurs at point C, lines $0F$ and $0G$ have length $(\sqrt{2}/3)Y$, the principal stresses are

$$\left(\sigma_1 = \frac{Y}{\sqrt{3}}, \quad \sigma_2 = 0, \quad \sigma_3 = -\frac{Y}{\sqrt{3}} \right),$$

and the shear yield stress is $\tau_Y = Y/\sqrt{3}$.

EXAMPLE 4.3
Comparison of Tresca and Von Mises Criteria

When the loads that act on a member reach their working values, the nonzero stress components at the critical point in the member where yield is initiated are $\sigma_{xx} = 100$ MPa, $\sigma_{yy} = -14.0$ MPa, and $\sigma_{xy} = 50.0$ MPa. The load-stress relations are linear so that the factor of safety SF [see Eq. (1.4)] can be applied to either the loads or stress components. The member material has a yield stress $Y = 300$ MPa.

(a) Assuming that the material is a Tresca material, determine the factor of safety SF against yield.

(b) Assuming that the material is a von Mises material, determine the factor of safety SF against yield.

(c) Determine which criterion, Tresca or von Mises, is more conservative.

(d) Illustrate the stress state and factors of safety in the π-plane for the material.

SOLUTION

(a) *Tresca (Maximum Shear-Stress) Criterion.* This criterion is defined by Eq. (4.12), with $\tau_{max} = Y/2 = (\sigma_{max} - \sigma_{min})/2$. To determine σ_{max} and σ_{min}, we compute the principal stresses $(\sigma_1, \sigma_2, \sigma_3)$ at the point. With the given data and Eqs. (2.20) and (2.21), the principal stresses are the roots of $\sigma^3 - 86\sigma^2 - 3900\sigma = 0$. The roots are $(118.8, 0, -32.8)$. Hence, $\sigma_{max} = 118.8$ MPa and $\sigma_{min} = -32.8$ MPa. When the loads are increased by the factor of safety SF, the principal stresses are also increased by the factor SF. Consequently, $\tau_{max} = Y/2 = 300/2 = SF(\sigma_{max} - \sigma_{min})/2 = SF(118.8 + 32.8)/2$. Therefore, $SF = 1.98$, if the material obeys the Tresca (maximum shear-stress) criterion.

(b) *von Mises (Maximum Octahedral Shear-Stress) Criterion.* By Eq. (4.21), including the factor of safety SF, we have

$$Y = \frac{SF}{\sqrt{2}} [(118.8)^2 + (32.8)^2 + (-32.8 - 118.8)^2]^{1/2} \qquad \text{(a)}$$

Hence, $SF = 2.17$ if the material obeys the von Mises criterion.

(c) The same design loads are applied to the member in parts (a) and (b). If the Tresca criterion is applicable, the design loads are increased by a factor of 1.98 to initiate yield. However, if the von Mises criterion is applicable, the design loads are increased by a factor 2.17 to initiate yield. Thus, the Tresca criterion is more conservative; it predicts yield initiation at smaller loads than the von Mises criterion.

(d) *Illustration in the π-Plane.* To illustrate this solution, we simply project each of the principal stress components onto the π-plane, sum the projected vectors, and determine the length of the resultant. The factors of safety are determined by comparing this length to the radial distances from the origin to the appropriate yield surfaces. As an alternative, we may work with the principal values of the deviatoric stress. Both approaches are considered.

In Fig. E4.3a, the projections of the components of principal stress are shown. Vector **0A** is the projection of σ_1 and has length $\sqrt{\frac{2}{3}}\sigma_1 = 97.0$ MPa. Likewise, vector **AB** is the projection of σ_3 and has length $\sqrt{\frac{2}{3}}\sigma_3 = -26.78$ MPa. The sum of these two projections is vector **0B** that has length 112.8 MPa. If the mean stress, $\sigma_m = 26.67$ MPa, is subtracted from each of the principal stress components, the principal values (S_1, S_2, S_3) of the deviator stress are obtained as (90.13 MPa, -28.67 MPa, -61.47 MPa) [see Eq. (2.28)]. The deviator stress components (S_1, S_2, S_3) are projected in the same way as the principal stress components. The projections are illustrated in Fig. E4.3b as vectors **0E**, **EF**, and **FB**, respectively. The sum of the three projections is vector **0B**, which is identical to that in Fig. E4.3a.

By inspection, we see that the stress state illustrated by vector **0B** is elastic, since it is within the boundary of the two yield surfaces under consideration. For the Tresca criterion, the factor of safety against yield is the ratio of the lengths of vectors **0C** and **0B**. The extension of vector **0B** to the Tresca hexagon defines point C, the point at which yielding would occur if the given stress

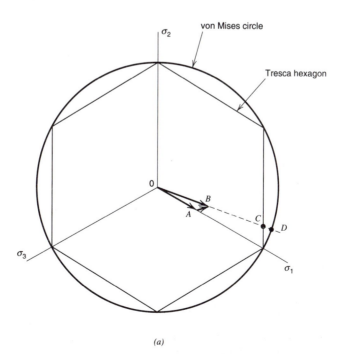

(a)

Figure E4.3   (a) Principal stress projections.

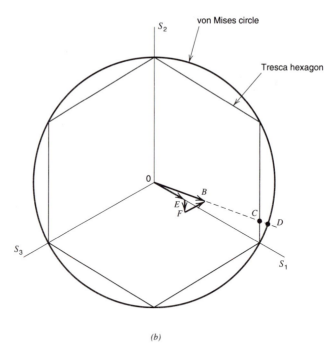

(b)

Figure E4.3   *(b)* Deviatoric stress projections.

state is increased proportionally. Vector **0C** has length 223.2 MPa. Hence, the factor of safety against yield for the Tresca criterion is $SF = 223.3/112.8 = 1.98$. For the von Mises criterion, the factor of safety is the ratio $0D/0B$. The length of vector **0D** is simply the radius of the von Mises circle that is $\sqrt{\frac{2}{3}}Y = 244.95$ MPa. Thus, for the von Mises criterion, $SF = 244.95/112.8 = 2.17$.

4.5

ALTERNATIVE YIELD CRITERIA

Interest in predicting the initiation of yielding is not limited to ductile metals. Many other materials used in engineering applications exhibit inelastic (yielding) behavior that is distinct from that of ductile metals. Thus, suitable yield criteria for these materials are needed. Some of these materials include soil, rock, concrete, and anisotropic composites. A brief discussion of yield criteria* for these materials follows in this section.

Mohr–Coulomb Yield Criterion
The yield behavior of many cohesive materials, including rock and concrete, has been observed to depend on hydrostatic stress. Specifically, an increase in hydro-

---

* The term *yield* is not entirely appropriate to describe the behavior of some of these materials in that their nonlinear response is often brittle unless relatively high confining pressure (hydrostatic compression) is applied.

static compressive stress produces an increased ability of some materials to resist yield. Also, these materials may exhibit different yield stresses in tension and compression. The Mohr–Coulomb yield criterion is a generalization of the Tresca criterion that accounts for the influence of hydrostatic stress. The yield function is written in terms of the stress state and two material properties: the cohesion c and angle of internal friction ϕ.

For principal stresses in the order $\sigma_1 > \sigma_2 > \sigma_3$, the Mohr–Coulomb yield function is (Lubliner, 1990)

$$f = \sigma_1 - \sigma_3 + (\sigma_1 + \sigma_3)\sin\phi - 2c\cos\phi \qquad (4.27)$$

For unordered principal stresses, the yield function becomes

$$f = \max_{i \neq j}\left[\sigma_i - \sigma_j + (\sigma_i + \sigma_j)\sin\phi\right] - 2c\cos\phi \qquad (4.28)$$

If we impose uniaxial tension until yield occurs ($\sigma_1 = Y_T$, $\sigma_2 = \sigma_3 = 0$), we can derive the yield stress in tension from Eq. (4.27) as

$$Y_T = \frac{2c\cos\phi}{1 + \sin\phi} \qquad (4.29)$$

Similarly, if we impose uniaxial compression with a stress state of ($\sigma_1 = \sigma_2 = 0$, $\sigma_3 = -Y_C$), the yield stress in compression is

$$Y_C = \frac{2c\cos\phi}{1 - \sin\phi} \qquad (4.30)$$

Equations (4.29) and (4.30) can be solved to obtain c and ϕ in terms of Y_T and Y_C as

$$c = \frac{Y_T}{2}\sqrt{\frac{Y_C}{Y_T}} \qquad (4.31)$$

$$\phi = \frac{\pi}{2} - 2\tan^{-1}\left(\sqrt{\frac{Y_T}{Y_C}}\right) \qquad (4.32)$$

If a material such as concrete is being studied and strength parameters Y_T and Y_C are known, Eqs. (4.31) and (4.32) can be used to find the properties c and ϕ needed by the Mohr–Coulomb yield function, Eq. (4.28).

The yield surface for the Mohr–Coulomb criterion has the form of an irregular hexagonal pyramid. The axis of the pyramid is the hydrostatic axis. The geometry of the pyramid depends on c and ϕ. The three-dimensional Mohr–Coulomb yield surface is shown in Fig. 4.14a and its intersection with the π-plane is shown in Fig. 4.14b. Compressive stress axes are used, as is common in studies of cohesive materials. Enforcing the yield criterion ($f = 0$) expressed in Eq. (4.27), we obtain a plane in the sextant $\sigma_1 > \sigma_2 > \sigma_3$. The other five planes are obtained by enforcing the yield criterion in Eq. (4.28) for the remaining principal stress ordering. Instead of writing equations for the six planes, the geometry of the irregular hexagon can be defined by two characteristic lengths, r_C and r_T, in the π-plane. These

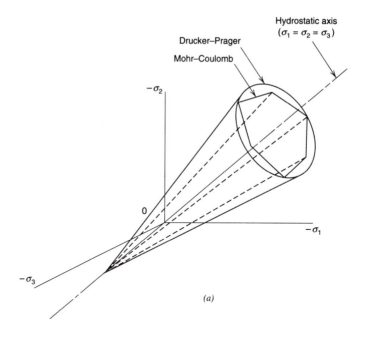

(a)

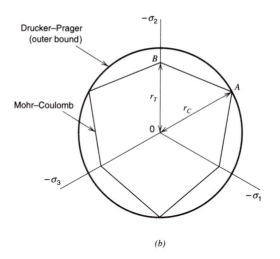

(b)

Figure 4.14   Mohr–Coulomb and Drucker–Prager yield surfaces. (*a*) In principal stress space. (*b*) Intersection with the π-plane.

lengths have the form (Chen and Han, 1988)

$$r_C = \frac{2\sqrt{6}\,c\cos\phi}{3 - \sin\phi} = \frac{\sqrt{6}\,Y_C(1 - \sin\phi)}{3 - \sin\phi}$$

$$r_T = \frac{2\sqrt{6}\,c\cos\phi}{3 + \sin\phi} = \frac{\sqrt{6}\,Y_T(1 + \sin\phi)}{3 + \sin\phi} \tag{4.33}$$

For a frictionless material ($\phi = 0$), the Mohr–Coulomb criterion reduces to the Tresca criterion, and $c = \tau_Y$, the yield stress in pure shear.

Drucker–Prager Yield Criterion

The Drucker–Prager yield criterion is a generalization of the von Mises criterion that includes the influence of hydrostatic stress. The yield function can be written in the form (Chen and Han, 1988)

$$f = \alpha I_1 + \sqrt{J_2} - K \tag{4.34}$$

where α and K are coefficients dependent on the cohesion c and the angle of internal friction ϕ. Viewed in principal stress space, the Drucker–Prager yield surface is a right circular cone. The size of the cone can be adjusted to match the Mohr–Coulomb pyramid by selecting appropriate values for α and K. The cone circumscribes the Mohr–Coulomb pyramid, as shown in Fig. 4.14, when α and K are determined from

$$\alpha = \frac{2 \sin \phi}{\sqrt{3}(3 - \sin \phi)}, \qquad K = \frac{6c \cos \phi}{\sqrt{3}(3 - \sin \phi)} \tag{4.35}$$

For these coefficients, the Drucker–Prager criterion provides an outer bound on the Mohr–Coulomb yield surface; the radius of the cone as it intersects the π-plane is r_C, see Fig. 4.14b. The Drucker–Prager yield surface can be linked more closely to the tension strength by using α and K from

$$\alpha = \frac{2 \sin \phi}{\sqrt{3}(3 + \sin \phi)}, \qquad K = \frac{6c \cos \phi}{\sqrt{3}(3 + \sin \phi)} \tag{4.36}$$

In this case, the cone has radius r_T in the π-plane, (Fig. 4.14b). For a frictionless material ($\phi = 0$), the Drucker–Prager criterion reduces to the von Mises criterion.

Hill's Criterion for Orthotropic Materials

The foregoing criteria for yielding are limited to isotropic materials. Yield behavior of anisotropic materials depends on the direction of the imposed stresses, relative to the material axes. Therefore, a criterion that predicts yielding of anisotropic materials must account for the different material properties in the various material directions. For orthotropic materials, materials that have three mutually orthogonal material directions, the yield criterion proposed by Hill (1950) is quite effective. Hill's criterion is a generalization of the von Mises criterion. The yield function has the form

$$\begin{aligned} f = {} & F(\sigma_{22} - \sigma_{33})^2 + G(\sigma_{33} - \sigma_{11})^2 + H(\sigma_{11} - \sigma_{22})^2 \\ & + L(\sigma_{23}^2 + \sigma_{32}^2) + M(\sigma_{13}^2 + \sigma_{31}^2) + N(\sigma_{12}^2 + \sigma_{21}^2) - 1 \end{aligned} \tag{4.37}$$

where subscripts 1, 2, and 3 are the material directions and the coefficients F, G, H, L, M, and N are obtained from uniaxial load tests. If we denote the three tensile yield stresses corresponding to the 1, 2, and 3 directions as X, Y, and Z, respectively, and the shear yield stresses as S_{12}, S_{13}, and S_{23}, the coefficients in Eq. (4.37) can be written as

$$2F = \frac{1}{Z^2} + \frac{1}{Y^2} - \frac{1}{X^2}$$

$$2G = \frac{1}{Z^2} + \frac{1}{X^2} - \frac{1}{Y^2}$$

$$2H = \frac{1}{X^2} + \frac{1}{Y^2} - \frac{1}{Z^2}$$

$$2L = \frac{1}{S_{23}^2}, \qquad 2M = \frac{1}{S_{13}^2}, \qquad 2N = \frac{1}{S_{12}^2} \qquad (4.38)$$

It is implicitly assumed in Hill's criterion that the yield stresses X, Y, and Z are the same in either tension or compression, and that yielding is insensitive to hydrostatic stress. For an isotropic material, the yield function coefficients reduce to

$$6F = 6G = 6H = L = M = N \qquad (4.39)$$

and the yield function in Eq. (4.37) reduces to the von Mises yield function, Eq. (4.21).

Since Hill's criterion is written for a specific reference coordinate system, that is, the principal material directions, we cannot change the reference coordinates without changing the function itself. Thus, Hill's criterion cannot be illustrated in the form of a yield surface in principal stress space for an arbitrary stress state.

4.6

COMPARISON OF FAILURE CRITERIA FOR GENERAL YIELDING

The yield criteria presented in the previous sections merely define methods for combining the individual components of a multiaxial stress state into an effective, uniaxial stress. The uniaxial stress is then compared to a material property, often the yield strength, to determine whether yielding has occurred. Depending on the type of member under consideration, the initiation of yielding might or might not provide an accurate measure for the limiting strength of the member. In this book, we define the failure load for the general yielding mode of failure as the load for which the load-deflection curve for the member becomes nonlinear. Since the effect of stress concentrations on the overall shape of the load-deflection curve for ductile materials is small (Smith and Sidebottom, 1969), we ignore this effect.

This definition of failure load leads to a lower-bound load for the general yielding mode of failure. For members made of an elastic-perfectly plastic material, the fully plastic load is an upper bound for the general yielding mode of failure. The fully plastic load is the load at which the entire cross section has yielded.

For example, consider the nondimensional load-strain curves for two simple structural members in Fig. 4.15. Member A is an axially loaded tension member and member B is a beam with a rectangular cross section subjected to pure bending. The members are made of the same material that, for simplicity, we take to be elastic-perfectly plastic. For member A, let P_Y denote the yield load, that is, the axial load that causes the stress at some point in the member to reach the yield stress Y. For member B, let M_Y denote the yield moment, that is, the moment for which a point on the outer fibers of the beam reaches the yield stress Y. Thus, for members of rectangular cross section of width b and depth h, loads P_Y and M_Y are

$$P_Y = Ybh, \qquad M_Y = Y\frac{bh^2}{6} \tag{4.40}$$

These loads are defined in this book as the failure loads for general yielding.

Now consider the fully plastic loads for these members. Let ϵ be the strain in member A (at any point) and in member B at a point on the outer fibers. Also, let ϵ_Y be the strain associated with the initiation of yield ($\epsilon_Y = Y/E$). Then the load-strain curve for member A is $0CD$ and for member B is $0CF$.

Since member A has a uniform distribution of stress over its cross section and the material is elastic-perfectly plastic, member A cannot support any load greater than P_Y. Thus, the load P_Y, which initiates yielding, is also the fully plastic load P_P. Hence, for member A, the fully plastic load is

$$P_P = Ybh = P_Y \tag{4.41}$$

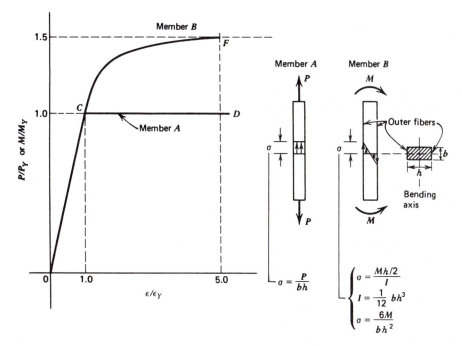

Figure 4.15   Nondimensional load-strain curves for members A and B.

The behavior of member B is substantially different from that of member A. Since $M = M_Y$ produces yielding only on the outer fibers of the beam while fibers in the interior of the beam have not yielded, the beam is capable of supporting substantially larger values of bending moment. Thus, the load-strain curve $0CF$ for member B continues to rise until all fibers through the beam cross section have yielded in either tension or compression. At that instant, the moment is $M = M_P$, where

$$M_P = Y\frac{bh^2}{4} = 1.5M_Y \qquad (4.42)$$

for a rectangular cross section.* Thus, we see that for member A, *initiation* of yielding and *extensive* yielding occur at the same load level $P = P_Y = P_P$. However, for member B, *initiation* of yielding occurs at $M = M_Y$, whereas extensive yielding occurs for $M = M_P = 1.5M_Y$.

In general, we assume that failure occurs when yielding is initiated at some point in a simple structural member. Thus, for the simple members of Fig. 4.15, failure occurs for loads corresponding to point C on the load-strain diagrams. However, current practice of plastic collapse design employs the concept of a fully plastic load. Therefore, we occasionally calculate fully plastic loads for simple structural members. The concept of fully plastic loads loses its meaning when the material in the member is capable of strain hardening (see Fig. 4.3b). For materials that strain-harden, a load corresponding to the fully plastic load (as defined above) is often calculated and serves as a lower bound on the actual member capacity when strain hardening is taken into account.

EXAMPLE 4.4
Partial Yielding due to Pure Bending

The rectangular beam identified as member B in Fig. 4.15 is subjected to pure bending such that $M = 1.25M_Y$. The material is elastic-perfectly plastic. Determine the extent to which the cross section has yielded.

SOLUTION

A short segment of the beam and the distribution of bending stress are shown in Fig. E4.4. First, consider only the portion of the cross section that is in compression. The internal force associated with the elastic stress distribution is shown as P_E and has magnitude

$$P_E = \tfrac{1}{2}Yby \qquad (a)$$

This force acts at a distance $e = \tfrac{2}{3}y$ from the neutral axis of the beam. Likewise, the internal force associated with the plastic stress distribution is shown as P_P and

---

* Note that $M_P = 1.5M_Y$ only for beams of rectangular cross section. In general, $M_P = fM_Y$, where f is known as the shape factor for the cross section, $f = Z/S$, Z is the plastic section modulus, and S is the elastic section modulus.

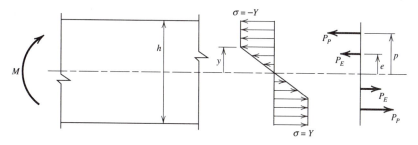

Figure E4.4

has magnitude

$$P_P = Yb\left(\frac{h}{2} - y\right) \tag{b}$$

It acts at a distance $p = y/2 + h/4$ from the neutral axis of the beam. Since the beam cross section is symmetric about the neutral axis, forces P_E and P_P act at the same relative locations on the tension side of the cross section but with opposite sense. Next, moment equilibrium is enforced to balance the applied moment against the internal force couples.

$$1.25M_Y = 2eP_E + 2pP_P \tag{c}$$

For a beam of rectangular cross section, $M_Y = bh^2Y/6$, Eq. (4.40). Therefore, Eqs (a–c) yield

$$y = \frac{h}{\sqrt{8}} = \frac{1}{\sqrt{2}}\left(\frac{h}{2}\right) \tag{d}$$

Hence, the elastic portion of the cross section has depth $2y = 0.707h$ and is centered on the neutral axis. The remaining 29.3% of the beam cross section has yielded.

Comparison of Failure Criteria

We noted in Sec. 4.4 that a large volume of experimental evidence indicates that the Tresca (maximum shear-stress) and von Mises (distortional energy density or maximum octahedral shear-stress) criteria are valid for the general yielding mode of failure for ductile materials. We now examine the critical values for these two criteria and the other criteria discussed in Sec. 4.3.

A typical stress-strain curve for a tensile specimen of ductile steel is shown in Fig. 4.16. When the specimen starts to yield, the following six quantities attain their critical values at the same load P_Y.

1. The maximum principal stress ($\sigma_{max} = P_Y/A$) reaches the tensile strength (yield stress) Y of the material.
2. The maximum principal strain ($\epsilon_{max} = \sigma_{max}/E$) reaches the value $\epsilon_Y = Y/E$.

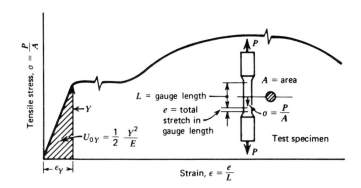

Figure 4.16   Typical stress-strain diagram for ductile steel.

3. The strain energy density U_0 absorbed by the material per unit volume reaches the value $U_{0Y} = Y^2/2E$ [from Eq. (4.8)].

4. The maximum shear stress $(\tau_{\max} = P_Y/2A)$ reaches the Tresca shear strength $\tau_Y = Y/2$.

5. The distortional energy density U_D absorbed by the material reaches a value $U_{DY} = Y^2/6G$ [from Eq. (4.17) with $\sigma_1 = Y$; all other stresses are zero].

6. The maximum octahedral shear stress τ_{oct} reaches the value $\tau_{\text{oct}Y} = \sqrt{2}\,Y/3 = 0.471\,Y$ [see Eq. (4.24)].

These six quantities, as determined from a simple tension test, are summarized in Table 4.1. Note that the maximum octahedral shear stress and distortional energy density, though different physical quantities, represent the same yield criterion, the von Mises criterion.

The six critical values in Table 4.1 occur simultaneously in a specimen that is loaded to yield in uniaxial tension. Hence, it is impossible to determine from a tension test which one of the six quantities is the best indicator of inelastic behavior for multiaxial stress states. However, in a biaxial or triaxial state of stress, these six quantities do not occur simultaneously.

Significance of Failure Criteria

A rational procedure for design requires that a general mode of failure be considered (e.g., failure by yielding or fracture). For a given mode of failure, each criterion of failure identifies a quantity that causes failure. That is, failure occurs when the value of the quantity reaches its critical value. Thus, we must choose a quantity, such as stress or strain, that can be associated with the initiation of failure. We must determine a critical value of the selected quantity that limits the loads; that is, a *suitable* test of the material must be made to determine the critical value. This critical value is frequently referred to as the maximum utilizable *strength* of the material. In most cases, a tension test is considered to be a suitable test for determining the critical or maximum value of this quantity. Ideally, we would like to test a material in the same state of stress that it will experience in the actual member. If this were feasible, there would be no need for failure criteria. If we were able to test the member under actual conditions, we would know its failure limit directly. Such an approach is not always necessary or economical.

TABLE 4.1

Quantity	Critical Value in Terms of Tension Test
1. Maximum principal stress	$Y = P_Y/A$

2. Maximum principal strain	$\epsilon_Y = Y/E$

3. Strain energy density	$U_{0Y} = Y^2/2E$

4. Maximum shear stress	$\tau_Y = P_Y/2A = Y/2$

5. Distortional energy density	$U_{DY} = \dfrac{Y^2}{6G}, \qquad G = \dfrac{E}{2(1+v)}$

6. Octahedral shear stress	$\tau_{oct} = (\sqrt{2}/3)Y = 0.471Y$

It is important to understand how criteria of failure fit into a rational design process. A single failure criterion that applies to all conditions in which load-resisting members are used is desirable. Unfortunately, this is too much to expect when we consider the radically different material types, modes of failure (ranging from ductile yielding to brittle fracture), and the simplifying conditions that are necessary to conduct a practical test. In general, we are limited to just two tests to obtain material properties: the tension test and torsion test.

Interpretation of Failure Criteria for General Yielding

Two criteria of failure, the maximum shear-stress (Tresca) and maximum octahedral shear-stress (von Mises) criteria, are represented graphically in Fig. 4.12. There it is shown that the greatest difference between the two criteria is exhibited when material properties obtained from a tension test are used to predict failure loads in pure shear, such as for a torsion member. In Table 4.2, we interpret five of the quantities discussed above as they apply to the tension test and torsion test (maximum strain energy density is omitted). First, we use each quantity to predict the tension yield stress Y (column 2). Then each quantity is used to obtain the shear yield stress τ_Y from a hollow torsion specimen (column 3). Finally, each of the five quantities is used to predict the relationship between Y and τ_Y (column 4). If all

TABLE 4.2
Comparison of Maximum Utilizable Values of a Material Quantity According to Various Yield Criteria for States of Stress in the Tension (a) and Torsion (b) Tests

(1)	(2)	(3)	(4)
			Relation between Values of Y and τ_Y if the Criterion is Correct for Both Stress States (col. 2 = col. 3)
Yield Criterion	**Predicted Maximum Utilizable Value as Obtained from a Tension Test (a)**	**Predicted Maximum Utilizable Value as Obtained from a Torsion Test (b)**	
Maximum principal stress	$\sigma_{max} = Y$	$\sigma_{max} = \tau_Y$	$\tau_Y = Y$
Maximum principal strain, $v = \frac{1}{4}$	$\epsilon_{max} = \dfrac{Y}{E}$	$\epsilon_{max} = \dfrac{5}{4}\dfrac{\tau_Y}{E}$	$\tau_Y = \dfrac{4}{5}Y$
Maximum shear stress	$\tau_{max} = \dfrac{1}{2}Y$	$\tau_{max} = \tau_Y$	$\tau_Y = \dfrac{1}{2}Y$
Maximum octahedral shear stress	$\tau_{oct\,Y} = \dfrac{\sqrt{2}}{3}Y$	$\tau_{oct\,Y} = \sqrt{\dfrac{2}{3}}\tau_Y$	$\tau_Y = \dfrac{1}{\sqrt{3}}Y$
Maximum distortional energy density	$U_{DY} = \dfrac{Y^2}{6G}$	$U_{DY} = \dfrac{\tau_Y^2}{2G}$	$\tau_Y = \dfrac{1}{\sqrt{3}}Y$

the quantities were accurate predictors of failure, they all would predict the same relationship between Y and τ_Y.

Experimental data indicate that $\tau_Y = 0.5Y$ for some metals (particularly some elastic-perfectly plastic metals), that $\tau_Y = Y/\sqrt{3}$ for some other metals, and that the value of τ_Y falls between $0.5Y$ and $Y/\sqrt{3}$ for most remaining metals. Thus, experimental evidence indicates that either the maximum shear-stress criterion or the maximum octahedral shear-stress criterion can be used to predict failure loads for metal members that fail by general yielding.

The states of stress in the tension and torsion tests represent about as wide a range as occurs in most members that fail by yielding under static loads. In the tension test, $\sigma_{max}/\tau_{max} = 2$, and in the torsion test, $\sigma_{max}/\tau_{max} = 1$. For some triaxial stress states, σ_{max}/τ_{max} is greater than 2, approaching infinity under pure hydrostatic stress. However in this case, failure occurs by brittle fracture rather than by yielding, if the hydrostatic stress is tensile.

Five different failure criteria are compared for a biaxial stress state (normal stress and shear) in Fig. 4.17 for stress states such that σ_{max}/τ_{max} lies between 1 and 2.

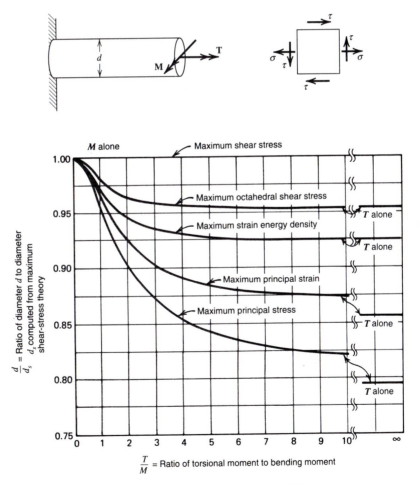

Figure 4.17 Comparison of criteria of failure.

Such a condition exists in a cylindrical shaft subjected to torque T and bending moment M. The shaft diameter necessary to prevent inelastic behavior according to the maximum shear-stress criterion is denoted by d_s. The minimum diameter to prevent inelastic behavior according to any of the other criteria is denoted by d. In Fig. 4.17, the ratio d/d_s is plotted against the ratio of the torque to the bending moment T/M. These ratios are obtained for combinations of T and M ranging from M acting alone ($T/M = 0$) to T acting alone ($T/M = \infty$). The case for which $T/M = \infty$ is shown by the horizontal asymptote (right side of the graph). The maximum shear stress criterion predicts the largest required diameter for all values of $T/M > 0$. For $T/M = 0$, all criteria predict equal diameters.

The application of the Tresca and von Mises criteria to the biaxial stress state (normal stress and shear) can be examined further by considering the intersection of the three-dimensional yield surfaces (Fig. 4.9) with the σ-τ plane. This intersection is shown in Fig. 4.18 in nondimensional form. The equations represented by these curves are found as follows. For any combination of σ and τ, yielding starts according to the maximum shear-stress criterion when

$$\sqrt{\left(\frac{\sigma}{2}\right)^2 + \tau^2} = \frac{Y}{2} \quad \text{or} \quad \left(\frac{\sigma}{Y}\right)^2 + 4\left(\frac{\tau}{Y}\right)^2 = 1 \tag{4.43}$$

Likewise according to the maximum octahedral shear-stress criterion, yielding starts when

$$\frac{\sqrt{2\sigma^2 + 6\tau^2}}{3} = \frac{\sqrt{2}\,Y}{3} \quad \text{or} \quad \left(\frac{\sigma}{Y}\right)^2 + 3\left(\frac{\tau}{Y}\right)^2 = 1 \tag{4.44}$$

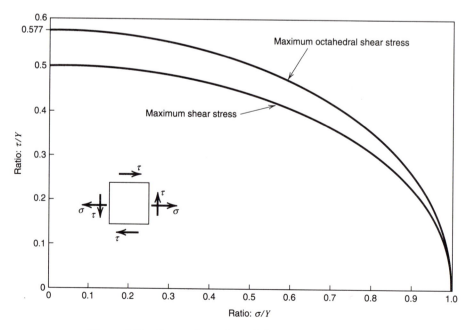

Figure 4.18 Comparison of von Mises and Tresca criteria.

EXAMPLE 4.5
Cylindrical Steel Shaft Subjected to Torsion and Bending

A circular cylindrical shaft is made of steel with a yield stress $Y = 700$ MPa. The shaft is subjected to a static bending moment $M = 13.0$ kN·m and a static torsional moment $T = 30.0$ kN·m (Fig. 4.17). Also, for the steel the modulus of elasticity is $E = 200$ GPa and Poisson's ratio is $v = 0.29$. Employing a factor of safety of $SF = 2.60$, determine the minimum required safe diameter for the shaft.

SOLUTION

Assuming that the failure is by yield initiation, we note that either the maximum octahedral shear-stress (or equivalently, the distortional energy) criterion or maximum shear-stress criterion is applicable.

For the octahedral shear-stress criterion we obtain by Eqs. (4.24) and (4.44)

$$\tau_{oct(max)} = \frac{\sqrt{2}}{3} Y = \frac{1}{3}\sqrt{2\sigma^2 + 6\tau^2} \tag{a}$$

which simplifies to

$$Y = \sqrt{\sigma^2 + 3\tau^2} \tag{b}$$

Yielding in the shaft will occur when the loads M and T are increased by the safety factor. Thus, $\sigma = (SF)Mc/I = 32(SF)M/\pi d^3$ and $\tau = (SF)Tc/J = 16(SF)T/\pi d^3$. Substitution into Eq. (b) yields

$$Y = \sqrt{\sigma^2 + 3\tau^2} = \frac{16SF}{\pi d^3}\sqrt{4M^2 + 3T^2} \tag{c}$$

or

$$d_{min} = \left(\frac{16SF}{\pi Y}\sqrt{4M^2 + 3T^2}\right)^{1/3}$$

Thus, with numerical values, we get

$$d_{min} = 103 \text{ mm}$$

Hence, by the octahedral shear-stress criterion a minimum shaft diameter of 103 mm is required to prevent inelastic action.

Alternatively, if the maximum shear-stress criterion is employed, the yield condition becomes [Eq. (4.43)]

$$\tau_{max} = \frac{Y}{2} = \frac{1}{2}\sqrt{\sigma^2 + 4\tau^2}$$

or

$$d_{min} = \left(\frac{32SF}{\pi Y}\sqrt{M^2 + T^2}\right)^{1/3}$$

Thus, with numerical values, we get

$$d_{min} = 107 \text{ mm}$$

A diameter not less than 107 mm would be required to prevent initiation of yielding of the shaft. We observe that the properties E and v of the steel do not enter into the computations. Note that the ratio of diameters $103/107 = 0.963$ agrees with Fig. 4.17 for $T/M = 30/13 = 2.31$.

Other Factors to Be Considered

Failure criteria do not account for all conditions that the engineer must consider in the problem of failure, even for failure by yielding of ductile materials subjected to static loads at ordinary temperatures. Recall that our definition of failure is limited to the *initiation* of yielding. It was shown that a beam has reserve capacity above the bending moment that causes first yield. In fact, for a rectangular cross section, the collapse moment is 50% higher than the yield moment. In other cases, some localized yielding may occur without destroying the usefulness of a member. This yielding causes a readjustment of stresses that may permit an appreciable increase in the loads on the member (Smith and Sidebottom, 1969).

In conclusion, we remark that although the capacity of a material to work-harden will permit higher applied loads to be incorporated into a design, present-day design specifications usually do not take work hardening into account. Hence, the so-called plastic design concept is based on the fully plastic load for an elastic-perfectly plastic material. As we have learned, it is relatively simple to calculate fully plastic loads for simple frame members, like beams. However, for more complicated massive parts or members, the calculation of fully plastic loads becomes extremely difficult because of the effect of the more general (triaxial) states of stress on general yielding. Also for thin-wall members, fully plastic loads might not be reached. Instead, instability (local buckling) may occur after the initiation of yielding, which complicates determination of ultimate load capacity of the member (see Chapter 12). Finally, in many instances, the uncertainty in the magnitudes of the loads applied to structures or mechanisms is quite large. Consequently, factors of safety (or load factors) are also large; see Eqs. (1.4) and (1.6). This leads to conservative, but relatively crude, approximations of behavior.

PROBLEMS
Sections 4.1–4.2

4.1. In Example 4.1, add a member *BD* of the same material and cross section as members *AD* and *CD*. Determine the loads P required to cause deflections $u = 2.00$, 4.00, 4.481, and 8 mm. Plot the load-deflection curve for the structure.

4.2. In Example 4.1, let $\beta = 0$. Determine the load P_{PC} for which the stress in members *AD* and *CD* of the truss first reaches the yield stress Y. This load P_{PC} is called the plastic collapse load of the truss. At this load, for an elastic-perfectly plastic material ($\beta = 0$), the truss becomes unable to sustain any load above P_{PC} and continues to deflect. This deflection will continue unabated, unless the material begins to strain harden after the initial region of a perfectly plastic response (see Fig. 1.5).

4.3. In Problem 4.1, assume that $\beta = 0$ and compute the plastic collapse load P_{PC} of the three bar truss (see Problem 4.2 for a discussion of plastic collapse load).

4.4. The members *AD* and *CF* in Fig. P4.4 are made of elastic-perfectly plastic structural steel, and member *BE* is made of 7075-T6 aluminum alloy (see Appendix A for properties). The members each have a cross-sectional area of 100 mm². Determine the load $P = P_Y$ that initiates yield of the structure and the fully plastic load P_P for which all the members yield.

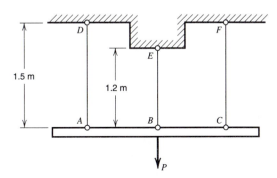

Figure P4.4

4.5. Two steel rods of equal length are supported and loaded as shown in Fig. P4.5. Rod *AB* is elastic-perfectly plastic and has yield strength $Y_{AB} = 250$ MPa. Rod *BC* is assumed to remain elastic. Both rods have cross-sectional area $A = 25$ mm² and $E = 200$ GPa. A horizontal load *P* is applied at *B*. Load *P* is initially zero and increases to 20 kN. From that point, the load is slowly cycled: 20 kN to −20 kN to 20 kN. Construct a load *P* vs displacement *u* curve for the system.

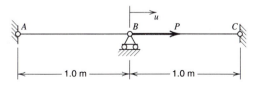

Figure P4.5

Section 4.3

4.6. Show that the maximum principal stress theory of failure predicts that $\tau_u = \sigma_u$, where σ_u is the fracture stress obtained in a tension test of a brittle material and τ_u is the fracture shear stress obtained in a torsion test of the same material.

4.7. The design loads of a member made of a brittle material produce the following nonzero stress components at the critical section in the member: $\sigma_{xx} = -60$ MPa, $\sigma_{yy} = 80$ MPa, and $\sigma_{xy} = 70$ MPa. The ultimate strength of the material is 460 MPa. Determine the factor of safety used in the design.

(a) Apply the maximum principal stress criterion.

(b) Apply the maximum principal strain criterion and use $v = 0.20$.

4.8. A brittle material has an ultimate strength $\sigma_u = 590$ MPa. A member made of this material is subjected to its design loads. At the critical point in the member, the nonzero stress components are $\sigma_{xx} = 160$ MPa and $\sigma_{xy} = 60$ MPa. Determine the factor of safety used in the design. Use the maximum principal stress criterion.

Section 4.4

4.9. A member is subjected to its design loads. The nonzero stress components at the point of maximum stress are $\sigma_{xx} = 150$ MPa and $\sigma_{xy} = 65$ MPa. The yield stress of the material is $Y = 450$ MPa.

(a) Determine the factor of safety used in the design, assuming that the material is a Tresca material.

(b) Repeat part (a), assuming that the material is a von Mises material.

4.10. A member made of steel ($E = 200$ GPa and $v = 0.29$) is subjected to a state of plane strain ($\epsilon_{zz} = \epsilon_{xz} = \epsilon_{yz} = 0$) when the design loads are applied. At the critical point in the member, three of the stress components are $\sigma_{xx} = 60$ MPa, $\sigma_{yy} = 240$ MPa, and $\sigma_{xy} = -80$ MPa. The material has a yield stress $Y = 490$ MPa. Based on the maximum shear-stress criterion, determine the factor of safety used in the design.

4.11. At the critical point in a member, the three principal stresses are nonzero. The yield point stress for the material is Y. If two of the principal stresses are equal, say, $\sigma_2 = \sigma_3$, show that the factor of safety based on the maximum shear-stress criterion is equal to the factor of safety based on the maximum octahedral shear-stress criterion.

Section 4.6

4.12. The rectangular beam considered in Example 4.4 is subjected to pure bending with a moment $M = \beta M_Y$, where β varies from 1.0 to 1.5 (note that at $\beta = 1.5$, $M = M_P$). The material is elastic-perfectly plastic. As β varies from 1.0 to 1.5, the dimension y (Fig. E4.4) varies from $h/2$ to 0, where y locates the boundary between the elastic and plastic regions of the beam. Develop a general relationship between β and the dimension y. Plot your relationship.

4.13. Compute the fully plastic moments for an elastic-perfectly plastic beam subjected to pure bending (see member B, Fig. 4.15) for the different cross sections shown in Fig. P4.13. Show that the ratio f of the fully plastic moment M_P to the yield moment M_Y ($f = M_P/M_Y$) is as given in Fig. P4.13. Note that the factor f is called the shape factor of the beam for the fully plastic state. The section at which $M = M_P$ occurs is called a plastic hinge, since the parts of the beam on either side of this section can rotate relative to one another, whereas the moment at the section remains unchanged. Thus, the so-called plastic hinge acts like an ordinary hinge with a constant amount of friction. A plastic hinge can also be developed in the presence of bending and axial loads, shear and bending loads, and axial and shear loads. Accordingly, more generally, plastic hinges can form in structures (e.g., in a frame) under different load combinations (see Chapter 5; Smith and Sidebottom, 1965; Hodge, 1959; Mendelson, 1983).

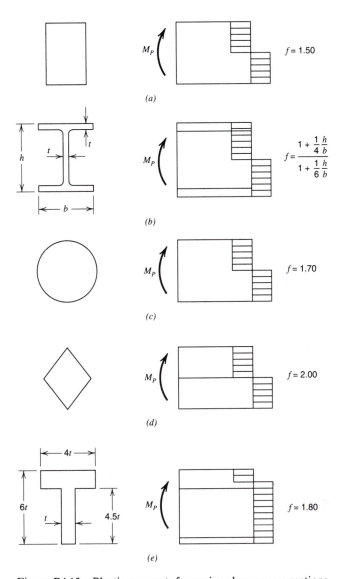

Figure P4.13   Plastic moments for various beam cross sections.

4.14. Consider a cantilever beam of rectangular cross section of width b and depth h. A lateral force P is applied at the free end of the beam (Fig. P4.14). The beam material is elastic-perfectly plastic (Fig. 4.4a).

Figure P4.14

(a) Determine the load $P = P_Y$ that causes initial yield in the beam.

(b) Determine the load $P = P_P$ that produces a plastic hinge.

(c) Compute the ratio P_P/P_Y.

Note: When the plastic hinge is produced at the wall section A of the beam, the beam is able to rotate freely about section A, with no increase in load P. In other words, the beam is said to collapse plastically. Also, it continues to rotate (kinematically) as a mechanism about point A. For this reason, in the study of structures the term *mechanism* or *kinematic mechanism* is used to describe this process.

4.15. Consider the indeterminate beam of Fig. P4.15. Let the beam material be elastic-perfectly plastic (Fig. 4.4a). Also let the cross section be rectangular with width b and depth h. Treat the internal moment at A as the redundant. Under sufficiently high load P, the moment at A will reach its limiting value M_P, the fully plastic moment. Thus, the redundant moment is known.

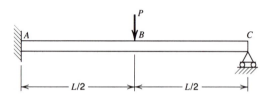

Figure P4.15

(a) Determine the magnitude of the force $P = P_{PC}$ that will cause a second hinge to form at B.

(b) Construct a moment diagram for the beam for load $P = P_{PC}$. The load P_{PC} is called the *plastic collapse load* for the member.

(c) Draw a sketch of the deformed shape of the beam for $P = P_{PC}$.

Note: The elastic segments of the beam rotate about a plastic hinge as rigid bodies. For this reason, the response of the beam at $P = P_{PC}$ is like a mechanism that rotates kinematically about hinges. Therefore, the term *mechanism* or *kinematic mechanism* is used in plastic collapse analysis (limit analysis) to describe this process.

4.16. A shaft has a diameter of 20 mm and is made of an aluminum alloy with yield stress $Y = 330$ MPa. The shaft is subjected to an axial load $P = 50.0$ kN.

(a) Determine the torque T that can be applied to the shaft to initiate yielding.

(b) Determine the torque T that can be applied to the shaft if the shaft is designed with a factor of safety $SF = 1.75$ for both P and T against initiation of yielding.
Use the maximum octahedral shear-stress criterion of failure.

4.17. A low-carbon steel shaft is designed with a diameter of 30 mm. It is subjected to an axial load $P = 30.0$ kN, a moment $M = 150$ N·m, and a torque

$T = 250$ N·m. If the yield point for the steel is $Y = 280$ MPa, determine the factor of safety used in the design of the shaft based on the maximum shear-stress criterion of failure, assuming that failure occurs at initiation of yielding.

Ans. $SF = 2.05$

4.18. A closed end thin-wall cylinder of titanium alloy Ti-6AL-4B ($Y = 800$ MPa) has an inside diameter of 38 mm and a wall of thickness of 2 mm. The cylinder is subjected to an internal pressure $p = 22.0$ MPa and axial load $P = 50.0$ kN. Determine the torque T that can be applied to the cylinder if the factor of safety for design is $SF = 1.90$. The design is based on the maximum shear-stress criterion of failure, assuming that failure occurs at initiation of yielding.

4.19. A load $P = 30.0$ kN is applied to the crank pin of the crank shaft in Fig. P4.19 to rotate the shaft at constant speed. The crank shaft is made of a ductile steel with a yield stress $Y = 276$ MPa. Determine the diameter of the crank shaft if it is designed using the maximum shear-stress criterion for initiation of yielding and the factor of safety is $SF = 2.00$.

Ans. $d = 89.2$ mm

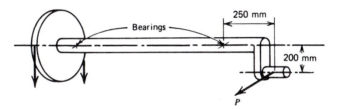

Figure P4.19

4.20. Solve Problem 4.19 using the maximum octahedral shear-stress criterion of failure.

4.21. The shaft in Fig. P4.21 is supported in flexible bearings at A and D, and two gears B and C are attached to the shaft at the locations shown. The gears are acted on by tangential forces as shown by the end view. The shaft is made of a ductile steel having a yield stress $Y = 290$ MPa. If the factor of safety for the design of the shaft is $SF = 1.85$, determine the diameter

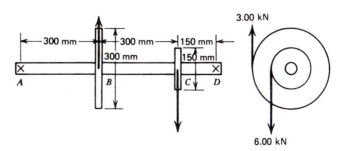

Figure P4.21

of the shaft using the maximum shear-stress criterion for the initiation of yielding failure.

Ans. $d = 35.7$ mm

4.22. Let the 6.00-kN load on the smaller gear of Problem 4.21 be horizontal instead of vertical. Determine the diameter of the shaft.

4.23. The 100-mm diameter bar shown in Fig. P4.23 is made of a ductile steel that has a yield stress $Y = 420$ MPa. The free end of the bar is subjected to a load P making equal angles with the positive directions of the three coordinate axes. Using the maximum octahedral shear-stress criterion of failure, determine the magnitude of P that will initiate yielding.

Ans. $P = 149.5$ kN

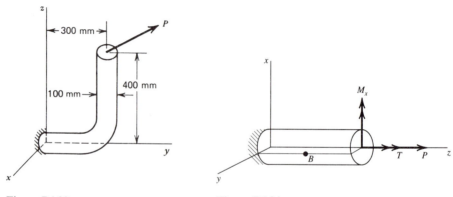

Figure P4.23   Figure P4.24

4.24. The shaft in Fig. P4.24 has a diameter of 20 mm and is made of a ductile steel ($Y = 400$ MPa). It is subjected to a combination of static loads as follows: axial load $P = 25.0$ kN, bending moment $M_x = 50.0$ N·m, and torque $T = 120$ N·m.

(a) Determine the factor of safety for a design based on the maximum octahedral shear-stress criterion of failure.

(b) Determine the factor of safety for a design based on the maximum shear-stress criterion of failure.

(c) Determine the maximum and minimum principal stresses and indicate the direction that they act at point B shown.

Ans. **(a)** $SF = 2.05$, **(b)** $SF = 1.91$,

   (c) $\sigma_{max} = 176.3$ MPa, $\sigma_{min} = -33.1$ MPa, $\theta = -0.4088$ rad

4.25. Let the material properties for the shaft in Problem 4.24 be obtained by using a hollow torsion specimen. The shear yield stress is found to be $\tau_Y = 200$ MPa. Re-solve parts (a) and (b).

4.26. The member in Fig. P4.26 has a diameter of 20 mm and is made of a ductile metal. Static loads P and Q are parallel to the y axis and z axis, respectively. Determine the magnitude of the yield stress Y of the material if yielding is impending.

Ans. $Y = 502.6$ MPa based on maximum shear-stress criterion of failure

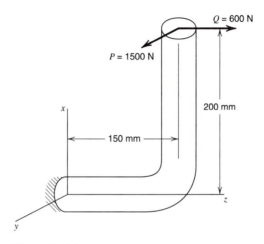

Figure P4.26

4.27. A 50-mm diameter structural steel shaft is subjected to a torque $T = 1.20$ kN·m and an axial load P. A hollow torsion specimen made of the same steel indicated a shear yield point $\tau_Y = 140$ MPa. If the shaft is designed for a factor of safety $SF = 2.00$, determine the magnitude of P based on

(a) the maximum shear-stress criterion of failure and

(b) the maximum octahedral shear-stress criterion of failure

4.28. A solid aluminum alloy ($Y = 320$ MPa) shaft extends 200 mm from a bearing support to the center of a 400-mm diameter pully (Fig. P4.28). The belt tensions T_1 and T_2 vary in magnitude with time. Their maximum values are $T_1 = 1800$ N and $T_2 = 180$ N. If the maximum values of the belt tensions are applied only a few times during the life of the shaft, determine the required diameter of the shaft if the factor of safety is $SF = 2.20$.

Ans. $d = 32.97$ mm based on the maximum shear-stress criterion of failure.

4.29. A closed thin-wall tube has a mean radius of 40.0 mm and wall thickness of 4.00 mm. It is subjected to an internal pressure of 11.0 MPa. The axis of the tube lies along the z axis. In addition to internal pressure, the tube is subjected to an axial load $P = 80.0$ kN, bending moments $M_x = 660$ N·m and $M_y = 480$ N·m, and torque $T = 3.60$ kN·m. If yielding is impending in the tube, determine the yield stress Y of the material based on the maximum shear-stress criterion of failure.

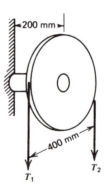

Figure P4.28

4.30. Solve Problem 4.29 by using the maximum octahedral shear-stress criterion of failure.

Ans. $Y = 216.6$ MPa

REFERENCES

Chen, W. F. and Han, D. J. (1988). *Plasticity for Structural Engineers.* New York: Springer-Verlag.

Hill, R. (1950). *The Mathematical Theory of Plasticity.* New York: Oxford Univ. Press.

Hodge, P. G. Jr., (1959). *Plastic Analysis of Structures.* New York: McGraw-Hill.

Lubahn, J. D. and Felgar, R. P. (1961). *Plasticity and Creep of Metals.* New York: Wiley.

Lubliner, J. (1990). *Plasticity Theory.* New York: Macmillan.

Mendelson, A. (1983). *PLASTICITY: Theory and Application.* Malabar, Florida: Krieger.

Morkovin, D. and Sidebottom, O. (1947). 'The effects of Non-Uniform Distribution of Stress on the Yield Strength of Steel.' Urbana Champaign, Ill.: Eng. Exp. Station, Univ. Ill., Bull. 373.

Sidebottom, O. M. and Chang, C. (1952). 'Influence of the Bauschinger Effect on Inelastic Bending of Beams.' In *Proceedings of the 1st U.S. National Congress of Applied Mechanics*, ASME, Chicago, Ill., June 11–16, pp. 631–639.

Smith, J. O. and Sidebottom, O. M. (1965). *Inelastic Behavior of Load-Carrying Members.* New York: Wiley.

Smith, J. O. and Sidebottom, O. M. (1969). *Elementary Mechanics of Deformable Bodies.* New York: Macmillan.

5

APPLICATIONS OF ENERGY METHODS

Energy methods are used widely to obtain solutions to elasticity problems and determine deflections of structures and machines. Since energy is a scalar quantity, energy methods are sometimes called scalar methods. In this chapter, energy methods are employed to obtain elastic deflections of statically determinate structures and to determine redundant reactions and deflections of statically indeterminate structures. The applications of energy methods in this book are limited mainly to linearly elastic material behavior and small displacements. However, in Sec. 5.1 and 5.2, energy methods are applied to two nonlinear problems to demonstrate their generality.

Castigliano's theorem on deflections is restricted to small displacements. It is used to obtain elastic deflections and determine redundant reactions. In applications to linearly elastic material behavior, the theorem is generally expressed in terms of the total strain energy of the structure. For the determination of the deflections of structures, two energy principles are presented: (1) the principle of stationary potential energy and (2) Castigliano's theorem on deflections. The general proofs of these principles are not presented in this book. Instead, the reader is referred to proofs given by H. L. Langhaar (1989).

In the application of the principle of stationary potential energy and of Castigliano's theorem on deflection to problems of structures it is assumed that every plane cross section of each member of a structure before deformation remains plane after deformation. Therefore, the displacement of a given cross section of a member is specified by three components of the displacement of the centroid of its cross section and by three angles that define the rotation of the plane of the cross section. Rectangular coordinate axes (x, y, z) are chosen for each member, with the z axis directed along the axis of the member and (x, y) axes taken as the principal axes of the cross section (see Appendix B). The principal axes (x, y) maintain the same orientation for every cross section of each straight member of the structure; that is, the member is prismatic.

5.1

PRINCIPLE OF STATIONARY POTENTIAL ENERGY

We employ the concept of generalized coordinates $(x_1, x_2, \ldots, x_n)$ to describe the shape of a structure in equilibrium (Langhaar, 1989; Sec. 1.2). Since plane cross sections of the members are assumed to remain plane, the changes of the generalized coordinates denote the translation and rotation of the cross section of the member. If a finite number of coordinates suffices to specify the configuration of the system, the system is said to possess a finite number of degrees of freedom. If a finite number of coordinates cannot specify the system configuration, the system is said to have infinitely many degrees of freedom. In this chapter we consider applications in which a finite number of degrees of freedom, equal to the number of generalized coordinates, specifies the configuration of the system.

Consider a system with a finite number of degrees of freedom that has the equilibrium configuration $(x_1, x_2, \ldots, x_n)$. A virtual (imagined) displacement is imposed such that the new configuration is $(x_1 + \delta x_1, x_2 + \delta x_2, \ldots, x_n + \delta x_n)$, where $(\delta x_1, \delta x_2, \ldots, \delta x_n)$ is the virtual displacement.* The virtual work δW corresponding to the virtual displacement is given by

$$\delta W = Q_1\, \delta x_1 + Q_2\, \delta x_2 + \cdots + Q_n\, \delta x_n \tag{5.1}$$

where $(Q_1, Q_2, \ldots, Q_n)$ are components of the generalized load. They are functions of the generalized coordinates. Let Q_i be defined for a given cross section of the structure; Q_i is a force if δx_i is a translation of the cross section, and Q_i is a moment (or torque) if δx_i is a rotation of the cross section.

The virtual work δW corresponding to virtual displacement of a mechanical system may be separated into the sum

$$\delta W = \delta W_e + \delta W_i \tag{5.2}$$

where δW_e is the virtual work of the external forces and δW_i the virtual work of the internal forces.

Analogous to the expression for δW in Eq. (5.1), under a virtual displacement $(\delta x_1, \delta x_2, \ldots, \delta x_n)$, we have

$$\delta W_e = P_1\, \delta x_1 + P_2\, \delta x_2 + \cdots + P_n\, \delta x_n \tag{5.3}$$

where $(P_1, P_2, \ldots, P_n)$ are functions of the generalized coordinates $(x_1, x_2, \ldots, x_n)$. By analogy to the Q_i in Eq. (5.1), the functions $(P_1, P_2, \ldots, P_n)$ are called the components of generalized *external* load. If the generalized coordinates $(x_1, x_2, \ldots, x_n)$ denote displacements and rotations that occur in a member or structure (system), the variables $(P_1, P_2, \ldots, P_n)$ may be identified as the components of the prescribed external forces and couples that act on the system.

Now imagine that the virtual displacement takes the system completely around any closed path. At the end of the closed path, we have $\delta x_1 = \delta x_2 = \cdots = \delta x_n = 0$. Hence, by Eq. (5.3), $\delta W_e = 0$. In our applications, we consider only systems that undergo elastic behavior. Then the virtual work δW_i of the internal forces is equal to the negative of the virtual change in the elastic strain energy δU, that is,

$$\delta W_i = -\delta U \tag{5.4}$$

---

* Note that the virtual displacement must not violate the essential boundary conditions (support conditions) for the structure.

where $U = U(x_1, x_2, \ldots, x_n)$ is the total strain energy of the system. Since the system travels around a closed path, it returns to its initial state and, hence, $\delta U = 0$. Consequently, by Eq. (5.4), $\delta W_i = 0$. Accordingly, the total virtual work δW [Eq. (5.2)] also vanishes around a closed path. The condition $\delta W = 0$ for virtual displacements that carry the system around a closed path indicates that the system is *conservative*. The condition $\delta W = 0$ is known as the *principle of stationary potential energy*.

For a conservative system (e.g., elastic structure loaded by conservative external forces), the virtual change in strain energy δU of the structure under the virtual displacement $(\delta x_1, \delta x_2, \ldots, \delta x_n)$ is

$$\delta U = \frac{\partial U}{\partial x_1} \delta x_1 + \frac{\partial U}{\partial x_2} \delta x_2 + \cdots + \frac{\partial U}{\partial x_n} \delta x_n \tag{5.5}$$

Then, Eqs. (5.1) through (5.5) yield the result

$$Q_1 \delta x_1 + Q_2 \delta x_2 + \cdots + Q_n \delta x_n = P_1 \delta x_1 + P_2 \delta x_2 + \cdots + P_n \delta x_n$$
$$- \frac{\partial U}{\partial x_1} \delta x_1 - \frac{\partial U}{\partial x_2} \delta x_2 - \cdots - \frac{\partial U}{\partial x_n} \delta x_n$$

or

$$Q_i = P_i - \frac{\partial U}{\partial x_i}, \qquad i = 1, 2, \ldots, n \tag{5.6}$$

For any system in static equilibrium with finite degrees of freedom, the vanishing of the components Q_i of the generalized force is sufficient for equilibrium. Therefore, by Eq. (5.6), an elastic system with n degrees of freedom is in equilibrium if (Langhaar, 1989; Sec. 1.9)

$$P_i = \frac{\partial U}{\partial x_i}, \qquad i = 1, 2, \ldots, n \tag{5.7}$$

The relation given in Eq. (5.7) is sometimes referred to as Castigliano's first theorem. For a structure, the strain energy U is obtained as the sum of the strain energies of the members of the structure. Note the similarity between Eqs. (5.7) and Eqs. (3.11).

As a simple example, consider a uniform bar loaded at its ends by an axial load P. Let the bar be made of a nonlinear elastic material with the load-elongation curve indicated in Fig. 5.1. The area below the curve represents the total strain energy U stored in the bar, that is, $U = \int P \, de$, then by Eq. (5.7), $P = \partial U / \partial e$, where P is the generalized external force and e the generalized coordinate. If the load-elongation data for the bar are plotted as a stress-strain curve (see Fig. 3.1), the area below the curve is the strain energy density U_0 stored in the bar. Then, $U_0 = \int \sigma \, d\epsilon$ and, by Eqs. (3.11), $\sigma = \partial U_0 / \partial \epsilon$.

Equation (5.7) is valid for nonlinear elastic (conservative) problems in which the nonlinearity is due either to finite geometry changes or material behavior, or both. The following example problem indicates the application of Eq. (5.7) for finite geometry changes.

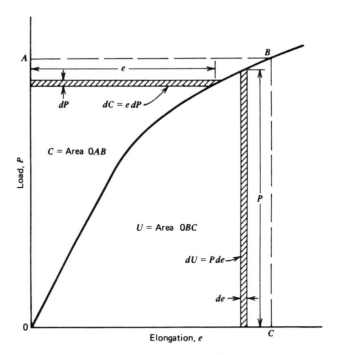

Figure 5.1   Nonlinear elastic load-elongation curve.

EXAMPLE 5.1
Equilibrium of Simple Linear Elastic Pin-Joined Truss

Two members AB and CB of lengths L_1 and L_2, respectively, of a pin-joined truss are attached to a rigid foundation at points A and C, as shown in Fig. E5.1a. The cross-sectional area of member AB is A_1 and that of member CB is A_2. The corresponding moduli of elasticity are E_1 and E_2. Under the action of horizontal and vertical forces P and Q, pin B undergoes finite horizontal and vertical displacement with components u and v, respectively (Fig. E5.1a). The bars AB and CB remain linearly elastic.

(a)   Derive formulas for P and Q in terms of u and v.

(b)   Let $E_1 A_1/L_1 = K_1 = 2.00$ N/mm and $E_2 A_2/L_2 = K_2 = 3.00$ N/mm, and let $b_1 = h = 400$ mm and $b_2 = 300$ mm. For $u = 30$ mm and $v = 40$ mm, determine the values of P and Q using the formulas derived in part (a).

(c)   Consider the equilibrium of the pin B in the displaced position B^* and verify the results of part (b).

(d)   For small displacement components u and v ($u, v \ll L_1, L_2$), linearize the formulas for P and Q derived in part (a).

SOLUTION

(a)   For this problem the generalized external forces are $P_1 = P$ and $P_2 = Q$ and the generalized coordinates are $x_1 = u$ and $x_2 = v$. For the geometry of

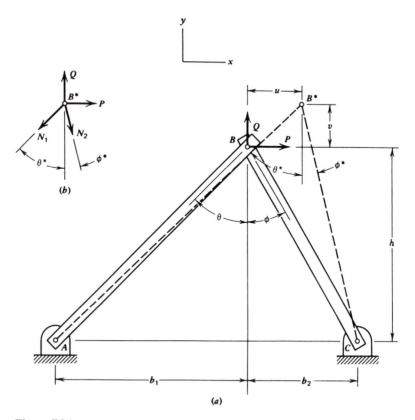

Figure E5.1

Fig. E5.1a, the elongations e_1 and e_2 of members 1 (members AB with length L_1) and 2 (member CB with length L_2) can be obtained in terms of u and v.

$$(L_1 + e_1)^2 = (b_1 + u)^2 + (h + v)^2, \qquad L_1^2 = b_1^2 + h^2$$
$$(L_2 + e_2)^2 = (b_2 - u)^2 + (h + v)^2, \qquad L_2^2 = b_2^2 + h^2 \qquad \text{(a)}$$

Solving for (e_1, e_2), we obtain

$$e_1 = \sqrt{(b_1 + u)^2 + (h + v)^2} - L_1$$
$$e_2 = \sqrt{(b_2 - u)^2 + (h + v)^2} - L_2 \qquad \text{(b)}$$

Since each member remains linearly elastic, the strain energies U_1 and U_2 of members AB and CB are

$$U_1 = \frac{1}{2} N_1 e_1 = \frac{E_1 A_1}{2 L_1} e_1^2$$

$$U_2 = \frac{1}{2} N_2 e_2 = \frac{E_2 A_2}{2 L_2} e_2^2 \qquad \text{(c)}$$

where N_1 and N_2 are the tension forces in the two members. The elongations of the two members are given by the relation $e_i = N_i L_i / E_i A_i$. The total strain energy U for the structure is equal to the sum $U_1 + U_2$ of the strain energies of the two members; therefore by Eqs. (c),

$$U = \frac{E_1 A_1}{2L_1} e_1^2 + \frac{E_2 A_2}{2L_2} e_2^2 \qquad \text{(d)}$$

The magnitudes of P and Q are obtained by differentiation of Eq. (d) with respect to u and v, respectively [see Eq. (5.7)]. Thus,

$$P = \frac{\partial U}{\partial u} = \frac{E_1 A_1 e_1}{L_1} \frac{\partial e_1}{\partial u} + \frac{E_2 A_2 e_2}{L_2} \frac{\partial e_2}{\partial u}$$

$$Q = \frac{\partial U}{\partial v} = \frac{E_1 A_1 e_1}{L_1} \frac{\partial e_1}{\partial v} + \frac{E_2 A_2 e_2}{L_2} \frac{\partial e_2}{\partial v} \qquad \text{(e)}$$

The partial derivatives of e_1 and e_2 with respect to u and v are obtained from Eqs. (b). Taking the derivatives and substituting in Eqs. (e), we find

$$P = \frac{E_1 A_1 (b_1 + u)}{L_1} \frac{\sqrt{(b_1 + u)^2 + (h + v)^2} - L_1}{\sqrt{(b_1 + u)^2 + (h + v)^2}}$$

$$- \frac{E_2 A_2 (b_2 - u)}{L_2} \frac{\sqrt{(b_2 - u)^2 + (h + v)^2} - L_2}{\sqrt{(b_2 - u)^2 + (h + v)^2}}$$

$$Q = \frac{E_1 A_1 (h + v)}{L_1} \frac{\sqrt{(b_1 + u)^2 + (h + v)^2} - L_1}{\sqrt{(b_1 + u)^2 + (h + v)^2}}$$

$$+ \frac{E_2 A_2 (h + v)}{L_2} \frac{\sqrt{(b_2 - u)^2 + (h + v)^2} - L_2}{\sqrt{(b_2 - u)^2 + (h + v)^2}} \qquad \text{(f)}$$

(b)   Substitution of the values $K_1, K_2, b_1, b_2, h, L_1, L_2, u,$ and v into Eqs. (f) gives

$$P = 43.8 \text{ N}$$
$$Q = 112.4 \text{ N} \qquad \text{(g)}$$

(c)   The values of P and Q may be verified by determining the tension forces N_1 and N_2 in the two members, determining directions of the axes of the two members for the deformed configuration, and applying equations of equilibrium to a free-body diagram of pin B^*. Elongations $e_1 = 49.54$ mm and $e_2 = 16.24$ mm are given by Eqs. (b). The tension forces N_1 and N_2 are

$$N_1 = e_1 K_1 = 99.08 \text{ N}$$
$$N_2 = e_2 K_2 = 48.72 \text{ N}$$

Angles θ^* and ϕ^* for the directions of the axes of the two members for the deformed configurations are found to be 0.7739 and 0.5504 rad, respectively. The free-body diagram of pin B^* is shown in Fig. E5.1b. The equations of equilibrium are

$$\sum F_x = 0 = P - N_1 \sin \theta^* + N_2 \sin \phi^*; \quad \text{hence, } P = 43.8 \text{ N}$$
$$\sum F_y = 0 = Q - N_1 \cos \theta^* - N_2 \cos \phi^*; \quad \text{hence, } Q = 112.4 \text{ N}$$

These values of P and Q agree with those of Eqs. (g).

(d) If displacements u and v are very small compared to b_1 and b_2, and, hence, with respect to L_1 and L_2, simple approximate expressions for P and Q may be obtained. For example, we find by the binomial expansion to linear terms in u and v that

$$\sqrt{(b_1 + u)^2 + (h + v)^2} = L_1 + \frac{b_1 u}{L_1} + \frac{hv}{L_1}$$

$$\sqrt{(b_2 - u)^2 + (h + v)^2} = L_2 - \frac{b_2 u}{L_2} + \frac{hv}{L_2}$$

With these approximations, Eqs. (f) yield the linear relations

$$P = \frac{E_1 A_1 b_1}{L_1^3}(b_1 u + hv) + \frac{E_2 A_2 b_2}{L_2^3}(b_2 u - hv)$$

$$Q = \frac{E_1 A_1 h}{L_1^3}(b_1 u + hv) + \frac{E_2 A_2 h}{L_2^3}(-b_2 u + hv)$$

If these equations are solved for the displacements u and v, the resulting relations are identical to those derived by means of Castigliano's theorem on deflections for linearly elastic materials (Sec. 5.3 and 5.4).

5.2

CASTIGLIANO'S THEOREM ON DEFLECTIONS

The derivation of Castigliano's theorem on deflections is based on the concept of complementary energy C of the system. Consequently, the theorem is sometimes called the "principle of complementary energy." The complementary energy C is equal to the strain energy U in the case of linear material response. However, for nonlinear material response, complementary energy and strain energy are not equal. For example, the complementary energy C of a nonlinear elastic tension member subject to an axial load is equal in magnitude to the area $0AB$ above the load-elongation curve $0B$ (see Fig. 5.1 and also Fig. 3.1 and Sec. 3.1), whereas the strain energy U is equal to the area $0BC$ below the curve $0B$. Hence, for this case, $C \neq U$.

In the derivation of Castigliano's theorem, the complementary energy C is regarded as a function of generalized forces $(F_1, F_2, \ldots, F_p)$ that act on a system

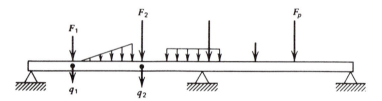

Figure 5.2

that is mounted on rigid supports (say the beam in Fig. 5.2). The complementary energy C depends also on distributed loads that act on the beam (Fig. 5.2), as well as the weight of the beam. However, these distributed forces do not enter explicitly into consideration in the derivation. In addition, the beam may be subjected to temperature effects (e.g., thermal strains; see Boresi and Chong, 1987; Chapter 4).

Castigliano's theorem may be stated generally as follows (Langhaar, 1989; Sec. 4.10):

> *If an elastic system is supported so that rigid-body displacements of the system are prevented, and if certain concentrated forces of magnitudes $F_1, F_2, \ldots, F_p$ act on the system, in addition to distributed loads and thermal strains, the displacement component q_i of the point of application of the force F_i, is determined by the equation*

$$q_i = \frac{\partial C}{\partial F_i}, \qquad i = 1, 2, \ldots, p \tag{5.8}$$

Note the similarity of Eqs. (5.8) and (3.19). The relation given by Eq. (5.8) is sometimes referred to as Castigliano's second theorem. With reference to Fig. 5.2, the displacement q_1 at the location of F_1 in the direction of F_1 is given by the relation $q_1 = \partial C/\partial F_1$.

The derivation of Eq. (5.8) is based on the assumption of small displacements; therefore, Castigliano's theorem is restricted to small displacements of the structure. The complementary energy C of a structure composed of m members may be expressed by the relation

$$C = \sum_{i=1}^{m} C_i \tag{5.9}$$

where C_i denotes the complementary energy of the ith member (Langhaar, 1989, Sec. 4.10; Charlton, 1959, Chapter V).

Castigliano's theorem on deflections may be extended to compute the rotation of line elements, in a system subjected to couples. For example, consider again a beam that is supported on rigid supports and subjected to external concentrated forces of magnitudes $F_1, F_2, \ldots, F_p$ (Fig. 5.3). Let two of the concentrated forces (F_1, F_2) be parallel, lie in a principal plane of the cross section, have opposite senses, and act perpendicular to the ends of a line element of length b in the beam (Fig. 5.3a). Then, Eq. (5.8) shows that the rotation θ (Fig. 5.3b) of the line segment due to the deformations is given by the relation

$$\theta = \frac{1}{b} \frac{\partial C}{\partial F_1} + \frac{1}{b} \frac{\partial C}{\partial F_2} \tag{a}$$

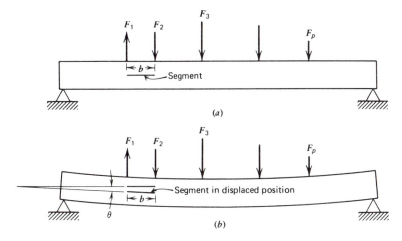

Figure 5.3   (*a*) Beam before deformation. (*b*) Beam after deformation.

where we have employed the condition of small displacements. To interpret this result, we employ the chain rule of partial differentiation of the complementary function C with respect to a scalar variable S. Considering the magnitudes of F_1 and F_2 to be functions of S, we have by the chain rule

$$\frac{\partial C}{\partial S} = \frac{\partial C}{\partial F_1}\frac{\partial F_1}{\partial S} + \frac{\partial C}{\partial F_2}\frac{\partial F_2}{\partial S} \tag{b}$$

In particular, we take the variable S equal to F_1 and F_2, that is, $S = F_1 = F_2 = F$, where F denotes the magnitudes of F_1 and F_2. Then, $\partial F_1/\partial S = \partial F_2/\partial S = 1$, and we obtain by Eq. (b)

$$\frac{\partial C}{\partial F} = \frac{\partial C}{\partial F_1} + \frac{\partial C}{\partial F_2} \tag{c}$$

Consequently, Eqs. (a) and (c) yield

$$\theta = \frac{1}{b}\frac{\partial C}{\partial F} \tag{d}$$

and since the equal and opposite forces F_1, F_2 constitute a couple of magnitude $M = bF$, Eq. (d) may be written in the form $\theta = \partial C/\partial M$. More generally, for couples M_i and rotations θ_i, we may write

$$\theta_i = \frac{\partial C}{\partial M_i}, \qquad i = 1, 2, \ldots, s \tag{5.10}$$

Hence, Eq. (5.10) determines the angular displacement θ_i of the arm of a couple of magnitude M_i that acts on an elastic structure. The sense of θ_i is the same as that of the couple M_i.

While Eqs. (5.8) and (5.10) are restricted to small displacements, they may be applied to structures that possess nonlinear elastic material behavior (Langhaar, 1989; Charlton, 1959). The following example problem indicates the application of Eq. (5.8) for nonlinear elastic material behavior.

EXAMPLE 5.2
Equilibrium of Simple Nonlinear Elastic Pin-Joined Truss

Let the two members of the pin-joined truss in Fig. E5.1 be made of a nonlinear elastic material whose stress-strain diagram is approximated by the relation $\epsilon = \epsilon_0 \sinh(\sigma/\sigma_0)$, where ϵ_0 and σ_0 are material constants. The truss is subjected to known loads P and Q. By means of Castigliano's theorem on deflections, determine the small displacement components u and v. Let $P = 10.0$ kN, $Q = 30.0$ kN, $\sigma_0 = 70.0$ MPa, $\epsilon_0 = 0.001$, $b_1 = h = 400$ mm, $b_2 = 300$ mm, and $A_1 = A_2 = 300$ mm^2. Show that the values for u and v so obtained agree with those obtained by a direct application of equations of equilibrium and the consideration of the geometry of the deformed truss.

SOLUTION

Let N_1 and N_2 be the tensions in members AB and CB. From the equilibrium conditions for pin B, we find

$$N_1 = \frac{L_1(Qb_2 + Ph)}{h(b_1 + b_2)}$$

$$N_2 = \frac{L_2(Qb_1 - Ph)}{h(b_1 + b_2)} \tag{a}$$

The complementary energy C for the truss is equal to the sum of the complementary energies for the two members. Thus,

$$C = C_1 + C_2 = \int_0^{N_1} e_1 \, dN_1 + \int_0^{N_2} e_2 \, dN_2 \tag{b}$$

With $e_1 = \epsilon_1 L_1$ and $e_2 = \epsilon_2 L_2$, Eq. (b) becomes

$$C = \int_0^{N_1} L_1 \epsilon_0 \sinh \frac{N_1}{A_1 \sigma_0} \, dN_1 + \int_0^{N_2} L_2 \epsilon_0 \sinh \frac{N_2}{A_2 \sigma_0} \, dN_2 \tag{c}$$

The displacement components u and v are obtained by substitution of Eq. (c) into Eq. (5.8). Thus, we find

$$u = q_P = \frac{\partial C}{\partial P} = L_1 \epsilon_0 \left(\sinh \frac{N_1}{A_1 \sigma_0} \right) \frac{\partial N_1}{\partial P} + L_2 \epsilon_0 \left(\sinh \frac{N_2}{A_2 \sigma_0} \right) \frac{\partial N_2}{\partial P}$$

$$v = q_Q = \frac{\partial C}{\partial Q} = L_1 \epsilon_0 \left(\sinh \frac{N_1}{A_1 \epsilon_0} \right) \frac{\partial N_1}{\partial Q} + L_2 \epsilon_0 \left(\sinh \frac{N_2}{A_2 \sigma_0} \right) \frac{\partial N_2}{\partial Q} \tag{d}$$

The partial derivatives of N_1 and N_2 with respect to P and Q are obtained by means of Eqs. (a). Taking derivatives and substituting into Eqs. (d), we obtain

$$u = \frac{L_1^2 \epsilon_0}{b_1 + b_2} \sinh \frac{L_1(Qb_2 + Ph)}{A_1 \sigma_0 h(b_1 + b_2)} - \frac{L_2^2 \epsilon_0}{b_1 + b_2} \sinh \frac{L_2(Qb_1 - Ph)}{A_2 \sigma_0 h(b_1 + b_2)}$$

$$v = \frac{L_1 b_2 \epsilon_0}{h(b_1 + b_2)} \sinh \frac{L_1(Qb_2 + Ph)}{A_1 \sigma_0 h(b_1 + b_2)} + \frac{L_2 b_1 \epsilon_0}{h(b_1 + b_2)} \sinh \frac{L_2(Qb_1 - Ph)}{A_2 \sigma_0 h(b_1 + b_2)} \tag{e}$$

Substitution of values for P, Q, σ_0, ϵ_0, b_1, b_2, h, A_1, and A_2 gives

$$u = 0.4709 \text{ mm}$$
$$v = 0.8119 \text{ mm} \qquad \text{(f)}$$

An alternate method of calculating u and v is as follows: Determine tensions N_1 and N_2 in the two members by Eqs. (a); next, determine elongations e_1 and e_2 for the two members and use these values of e_1 and e_2 along with geometric relations to calculate values for u and v. Equations (a) give $N_1 = 26.268$ kN and $N_2 = 14.286$ kN. Elongations e_1 and e_2 are given by the relations

$$e_1 = L_1\epsilon_0 \sinh\frac{N_1}{A_1\sigma_0} = 565.68(0.001)\sinh\frac{26{,}268}{300(70)} = 0.9071 \text{ mm}$$

$$e_2 = L_2\epsilon_0 \sinh\frac{N_2}{A_2\sigma_0} = 500.00(0.001)\sinh\frac{14{,}286}{300(70)} = 0.3670 \text{ mm}$$

With e_1 and e_2 known, values of u and v are given by the following geometric relations:

$$u = \frac{e_1\cos\phi - e_2\cos\theta}{\sin\theta\cos\phi + \cos\theta\sin\phi} = 0.4709 \text{ mm}$$

$$v = \frac{e_1\sin\phi + e_2\cos\theta}{\sin\theta\cos\phi + \cos\theta\sin\phi} = 0.8119 \text{ mm}$$

These values of u and v agree with those of Eqs. (f). Thus, Eq. (5.8) gives the correct values of u and v for this problem of nonlinear material behavior.

5.3

CASTIGLIANO'S THEOREM ON DEFLECTIONS FOR LINEAR LOAD-DEFLECTION RELATIONS

In the remainder of this chapter and in Chapter 9, we limit our consideration to linear elastic material behavior and small displacements. Consequently, the resulting load-deflection relation for either a member or structure is linear, the strain energy U is equal to the complementary energy C, and the principle of superposition applies. Then, Eqs. (5.8) and (5.10) may be written

$$q_i = \frac{\partial U}{\partial F_i}, \qquad i = 1, 2, \ldots, p \qquad \text{(5.11)}$$

$$\theta_i = \frac{\partial U}{\partial M_i}, \qquad i = 1, 2, \ldots, s \qquad \text{(5.12)}$$

where $U = U(F_1, F_2, \ldots, F_p, M_1, M_2, \ldots, M_s)$.

By Eq. (3.7), the strain energy U is

$$U = \int U_0 \, dV \tag{5.13}$$

where U_0 is the strain energy density. In this chapter we restrict ourselves to linear elastic, isotropic, homogeneous materials for which the strain energy density is [see Eq. (3.33)]

$$U_0 = \frac{1}{2E}(\sigma_{xx}^2 + \sigma_{yy}^2 + \sigma_{zz}^2) - \frac{v}{E}(\sigma_{xx}\sigma_{yy} + \sigma_{yy}\sigma_{zz} + \sigma_{zz}\sigma_{xx})$$

$$+ \frac{1}{2G}(\sigma_{yz}^2 + \sigma_{zx}^2 + \sigma_{xy}^2) \tag{5.14}$$

With load-stress formulas derived for the members of the structure, U_0 may be expressed in terms of the loads that act on the structure. Then, Eq. (5.13) gives U as a function of the loads. Equations (5.11) and (5.12) can then be used to obtain displacements at the points of applications of the concentrated forces or the rotations in the direction of the concentrated moments. Three types of loads are considered in this chapter for the various members of a structure as follows: (1) axial loading, (2) bending of beams, and (3) torsion. In practice, it is convenient to obtain the strain energy for each type of load acting alone and then add together these strain energies to obtain the total strain energy U, instead of using load-stress formulas and Eqs. (5.13) and (5.14) to obtain U.

Strain Energy U_N for Axial Loading

The equation for the total strain energy U_N due to axial loading is derived for the tension members shown in Figs. 5.4a and Fig. 5.4d. In general, the cross-sectional area A of the tension member may vary *slowly* with axial coordinate z. The line of action of the loads (the z axis) passes through the centroid of every cross section of the tension member. Consider two sections BC and DF of the tension member in Fig. 5.4a at distance dz apart. After the loads are applied, these sections are displaced to $B*C*$ and $D*F*$ (shown by the enlarged free-body diagram in Fig. 5.4b) and the original length dz has elongated an amount de_z. For linear elastic material behavior, de_z varies linearly with N as indicated in Fig. 5.4c. The shaded area below the straight line is equal to the strain energy dU_N for the segment dz of the tension member. The total strain energy U_N for the tension member becomes

$$U_N = \int dU_N = \int \tfrac{1}{2} N \, de_z \tag{5.15}$$

Noting that $de_z = \epsilon_{zz} \, dz$ and assuming that the cross-sectional area varies slowly, we have $\epsilon_{zz} = \sigma_{zz}/E$ [see Eq. (3.30)], and $\sigma_{zz} = N/A$, where A is the cross-sectional area of the member at section z. Then, we write Eq. (5.15) in the form

$$U_N = \int_0^L \frac{N^2}{2EA} \, dz \tag{5.16}$$

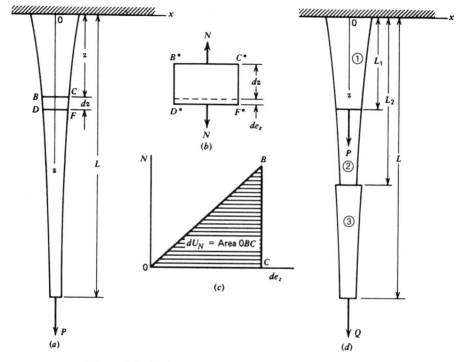

Figure 5.4   Strain energy due to axial loading of member.

At abrupt changes in material properties, load, and cross section, the values of E, N, and A change abruptly. Then, we must account approximately for these changes by writing U_N in the form (see Fig. 5.4d)

$$U_N = \int_0^{L_1} \frac{N_1^2}{2E_1 A_1} \, dz + \int_{L_1}^{L_2} \frac{N_2^2}{2E_2 A_2} \, dz + \int_{L_2}^{L} \frac{N_3^2}{2E_3 A_3} \, dz \qquad (5.16a)$$

where an abrupt change in load occurs at L_1 and an abrupt change in cross-sectional area occurs at L_2. The subscripts 1, 2, 3 refer to properties in parts 1, 2, 3 of the member (Fig. 5.4d).

Strain Energies U_M and U_S for Beams

Consider a beam of uniform cross section, as in Fig. 5.5a (or a beam with slowly varying cross section). We take (x, y, z) axes with origin at the centroid of the cross-section and with the z axis along the axis of the beam, the y axis down, and the x axis normal to the plane of the paper. The (x, y) axes are assumed to be principal axes for each cross section of the beam (see Appendix B). The loads P, Q, and R are assumed to lie in the (y, z) plane. The flexure formula

$$\sigma_{zz} = \frac{M_x y}{I_x} \qquad (5.17)$$

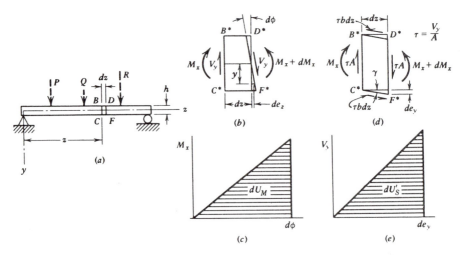

Figure 5.5   Strain energy due to bending and shear.

is assumed to hold, where M_x is the internal bending moment with respect to the principal x axis, I_x is the moment of inertia of the cross section at z about the x axis, and y is measured from the (x, z) plane. Consider two sections BC and DF of the unloaded beam at distance dz apart. After the loads are applied to the beam, plane sections BC and DF are displaced to B^*C^* and D^*F^* and are assumed to remain plane. An enlarged free-body diagram of the deformed beam segment is shown in Fig. 5.5b. Due to M_x, plane D^*F^* is rotated through angle $d\phi$ with respect to B^*C^*. For linear elastic material behavior, $d\phi$ varies linearly with M_x as indicated in Fig. 5.5c. The shaded area below the straight line is equal to the strain energy dU_M due to bending of the beam segment dz. An additional strain energy dU_S due to the shear V_y is considered later. The strain energy U_M for the beam due to M_x becomes

$$U_M = \int dU_M = \int \frac{1}{2} M_x \, d\phi \tag{5.18}$$

Noting that $d\phi = de_z/y$, $de_z = \epsilon_{zz} \, dz$, and assuming that $\epsilon_{zz} = \sigma_{zz}/E$ and $\sigma_{zz} = M_x y/I_x$, we may write Eq. (5.18) in the form

$$U_M = \int \frac{M_x^2}{2EI_x} \, dz \tag{5.19}$$

where in general M_x is a function of z. Equation (5.19) represents the strain energy due to bending about the x axis. A similar relation is valid for bending about the y axis for loads lying in the (x, z) plane. For abrupt changes in material E, moment M_x, or moment of inertia I_x, the value of U_M may be computed following the same procedure as for U_N [Eq. (5.16a)].

Equation (5.19) is exact for pure bending but is only approximate for shear loading as indicated in Fig. 5.5a. However, more exact solutions and experimental data indicate that Eq. (5.19) is fairly accurate, except for relatively short beams. An exact expression for the strain energy U_S due to shear loading of a beam is difficult to obtain. Consequently, an approximate expression for U_S is often used. When corrected by an appropriate coefficient, the use of this approximate expression often leads to fairly reliable results. The correction coefficients for various beam cross sections are discussed later.

Because of the shear V_y (Fig. 5.5b), shear stresses σ_{zy} are developed in each cross section; the magnitude of σ_{zy} is zero at both the top and bottom of the beam since the beam is not subjected to shear loads on the top or bottom surfaces. We define an average value of σ_{zy} as $\tau = V_y/A$. We assume that this average shear stress acts over the entire beam cross section (Fig. 5.5d) and, for convenience, assume that the beam cross section is rectangular with thickness b. Because of the shear, the displacement of face $D*F*$ with respect to face $B*C*$ is de_y. For linear elastic material behavior, de_y varies linearly with V_y, as indicated in Fig. 5.5e. The shaded area below the straight line is equal to the strain energy dU'_S for the beam segment dz. A correction coefficient k is now defined such that the exact expression for the shear strain energy dU_S of the element is equal to $k\,dU'_S$. Then, the shear strain energy U_S for the beam due to shear V_y is

$$U_S = \int k\,dU'_S = \int \frac{k}{2} V_y \, de_y \tag{5.20}$$

Noting that $de_y = \gamma\,dz$ and assuming that $\gamma = \tau/G$ and $\tau = V_y/A$, we may write Eq. (5.20) in the form

$$U_S = \int \frac{kV_y^2}{2GA}\,dz \tag{5.21}$$

Equation (5.21) represents the strain energy for shear loading of a beam. The value of V_y is generally a function of z. Also, the cross-sectional area A may vary slowly with z. For abrupt changes in material E, shear V_y or cross-sectional area A, the value of U_S may be computed following the same procedure as for U_N [Eq. (5.16a)].

An exact expression of U_S may be obtained, provided the exact shear stress distribution σ_{zy} is known. Then substitution into Eq. (5.14) to obtain U_0 (for σ_{zy}, the only nonzero stress component) and then substitution into Eq. (5.13) yields U_S. However, the exact distribution of σ_{zy} is often difficult to obtain, and approximate distributions are used. For example, consider a segment dy of thickness b of a beam cross section. In the engineering theory of beams, the stress component σ_{zy} is assumed to be uniform over thickness b. With this assumption (Popov, 1990)

$$\sigma_{zy} = \frac{V_y Q}{I_x b} \tag{5.22}$$

where Q is the first moment about the x axis of the area above the line of length b with ordinate y. Generally, σ_{zy} is not uniform over thickness b. Nevertheless, if for a beam of rectangular cross section, one assumes that σ_{zy} is uniform over b, it may be shown that $k = 1.20$.

Exact values of k are not generally available. Fortunately, in practical problems, the shear strain energy U_S is often small compared to U_M. Hence, for practical problems, the need for exact values of U_S is not critical. Consequently, as an expedient approximation, the correction coefficient k in Eq. (5.21) may be obtained as the ratio of the shear stress at the neutral surface of the beam calculated using Eq. (5.22) to the average shear stress V_y/A. For example, by this procedure, the magnitude of k for the rectangular cross section is

$$k = \frac{V_y Q}{I_x b} \frac{A}{V_y} = \frac{Qbh}{I_x b} = \frac{(bh/2)(h/4)h}{\frac{1}{12}bh^3} = 1.50 \tag{5.23}$$

This value is larger and hence more conservative than the more exact value 1.20. Nevertheless, since the more precise value is known, we recommend that $k = 1.20$ be used for rectangular cross sections. Approximate values of k, calculated by this method are listed in Table 5.1 for several beam cross sections. For I-sections, channels, and box-sections, $k = 1.00$, provided that the area A in Eq. (5.21) is taken as the area of the web for these cross sections.

Strain Energy U_T for Torsion

The strain energy U_T for a torsion member with circular cross section (Fig. 5.6a) may be derived as follows. Let the z axis lie along the centroidal axis of the torsion member. Before torsional loads T_1 and T_2 are applied, sections BC and DF are a distance dz apart. After the torsional loads are applied, these sections become sections $B*C*$ and $D*F*$, with section $D*F*$ rotated relative to section $B*C*$ through the angle $d\beta$, as shown in the enlarged free-body diagram of the element of length dz (Fig. 5.6b). For linear elastic material behavior, $d\beta$ varies linearly with T (Fig. 5.6c). The shaded area below the inclined straight line is equal to the torsional strain energy dU_T for the segment dz of the torsion member. Hence, the total torsional strain energy U_T for the torsional member becomes

$$U_T = \int dU_T = \int \tfrac{1}{2} T \, d\beta \tag{5.24}$$

Noting that $b \, d\beta = \gamma \, dz$ and assuming that $\gamma = \tau/G$ and $\tau = Tb/J$ (b is the radius and J the polar moment of inertia of the cross section), we may write Eq. (5.24) in

TABLE 5.1
Correction Coefficients for Strain Energy due to Shear

Beam Cross Section	k
Rectangle[a]	1.20
Solid circular[b]	1.33
Thin-wall circular[b]	2.00
I-section, channel, box-section[c]	1.00

[a] Exact value.

[b] Calculated by Eq. (5.23).

[c] The area A for the I-section, channel, or box-section is the area of the web hb, where h is the section depth and b the web thickness. The load is applied perpendicular to the axis of the beam and in the plane of the web.

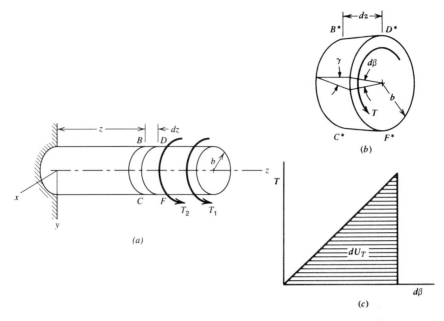

Figure 5.6 Strain energy due to torsion.

the form

$$U_T = \int \frac{T^2}{2GJ} dz \tag{5.25}$$

Equation (5.25) represents the strain energy for a torsion member with circular cross section. The unit angle of twist θ for a torsion member of circular cross section is given by $\theta = T/GJ$. Torsion of noncircular cross sections is treated in Chapter 6. Equation (5.25) is valid for other cross sections if the unit angle of twist θ for a given cross section replaces T/GJ in Eq. (5.25). For abrupt changes in material E, torsional load T or polar moment of inertia J, the value of U_T may be computed following the same procedure as for U_N [Eq. (5.16a)].

5.4

Deflections of Statically Determinate Structures

In the analysis of many engineering structures, the equations of static equilibrium are both necessary and sufficient to solve for unknown reactions and for internal actions in the members of the structure. For example, the simple structure shown in Fig. E5.1 is such a structure, since the equations of static equilibrium are sufficient to solve for the tensions N_1, and N_2 in members AB and CB, respectively. Structures for which the equations of static equilibrium are sufficient to determine the unknown tensions, shears, etc., uniquely are said to be *statically determinate structures*. Implied in the expression "statically determinate" is the condition that the

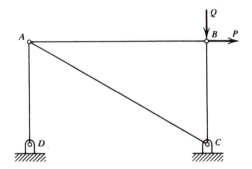

Figure 5.7 Statically determinate pin-joined truss.

deflections due to the loads are so small that the geometry of the initially unloaded structure remains essentially unchanged and the angles between members are essentially constant. If these conditions were not true, the internal tensions, etc., could not be determined without including the effects of the deformation and, hence, they could not be determined solely upon the basis of the equations of equilibrium.

The truss shown in Fig. 5.7 is a statically determinate truss. A physical characteristic of a statically determinate structure is that every member is essential for the proper functioning of the structure under the various loads to which it is subjected. For example, if member AC were to be removed from the truss of Fig. 5.7, the truss would be unable to support the loads; it would collapse.

Often, additional members are added to structures in order to stiffen the structure (reduce deflections), to strengthen the structure (increase its load-carrying capacity), or to provide alternate load paths (in the event of failure of one or more members). For example, for such purposes an additional diagonal member BD may be added to the truss of Fig. 5.7; see Fig. 5.8. Since the equations of static equilibrium are just sufficient for the analysis of the truss of Fig. 5.7, they are not adequate for the analysis of the truss of Fig. 5.8. Accordingly, the truss of Fig. 5.8 is said to be a *statically indeterminate structure*. The analysis of statically indeterminate structures requires additional information (additional equations) beyond that obtained from the equations of static equilibrium.

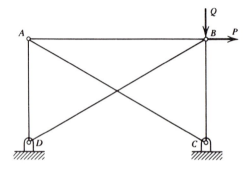

Figure 5.8 Statically indeterminate pin-joined truss.

In this section, the analysis of statically determinate structures is discussed. The analysis of statically indeterminate structures is presented in Sec. 5.5.

The strain energy U for a structure is equal to the sum of the strain energies of its members. The loading for the jth member of the structure is assumed to be such that the strain energy U_j for that member is

$$U_j = U_{Nj} + U_{Mj} + U_{Sj} + U_{Tj} \tag{5.26}$$

where U_{Nj}, U_{Mj}, U_{Sj}, and U_{Tj} are given by Eqs. (5.16), (5.19), (5.21), and (5.25), respectively. In the remainder of this chapter the limitations placed on the member cross section in the derivation of each of the components of the strain energy are assumed to apply. For instance, each beam is assumed to undergo bending about a principal axis of the beam cross section (see Appendix B and Chapter 7); Eqs. (5.19) and (5.21) are valid only for bending about a principal axis. For simplicity, we consider bending about the x axis (taken to be a principal axis) and let $M_x = M$ and $V_y = V$.

With the total strain energy U of the structure known, the deflection q_i of the structure at the location of a concentrated force F_i in the direction of F_i is [see Eq. (5.11)]

$$q_i = \frac{\partial U}{\partial F_i}$$

$$= \sum_{j=1}^{m} \left(\int \frac{N_j}{E_j A_j} \frac{\partial N_j}{\partial F_i} dz + \int \frac{k_j V_j}{G_j A_j} \frac{\partial V_j}{\partial F_i} dz \right.$$

$$\left. + \int \frac{M_j}{E_j I_j} \frac{\partial M_j}{\partial F_i} dz + \int \frac{T_j}{G_j J_j} \frac{\partial T_j}{\partial F_i} dz \right) \tag{5.27}$$

and the angle (slope) change θ_i of the structure at the location of a concentrated moment M_i in the direction of M_i is [see Eq. (5.12)]

$$\theta_i = \frac{\partial U}{\partial M_i}$$

$$= \sum_{j=1}^{m} \left(\int \frac{N_j}{E_j A_j} \frac{\partial N_j}{\partial M_i} dz + \int \frac{k_j V_j}{G_j A_j} \frac{\partial V_j}{\partial M_i} dz \right.$$

$$\left. + \int \frac{M_j}{E_j I_j} \frac{\partial M_j}{\partial M_i} dz + \int \frac{T_j}{G_j J_j} \frac{\partial T_j}{\partial M_i} dz \right) \tag{5.28}$$

where m is the number of members in the structure. Use of Castigliano's theorem on deflections, as expressed in Eqs. (5.27) and (5.28), to determine deflections or rotations at the location of a concentrated force or moment is outlined in the following procedure:

1. Write an expression for each of the internal actions (axial force, shear, moment, and torque) in each member of the structure in terms of the applied external loads.

2. To determine the deflection q_i of the structure at the location of a concentrated force F_i and in the directed sense of F_i, differentiate each of the internal action expressions with respect to F_i. Similarly, to determine the rotation θ_i of the structure at the location of a concentrated moment M_i and in the directed sense of M_i, differentiate each of the internal action expressions with respect to M_i.

3. Substitute the expressions for internal actions obtained in Step 1 and the derivatives obtained in Step 2 into Eq. (5.27) or (5.28) and perform the integration. The result is a relationship between the deflection q_i (or rotation θ_i) and the externally applied loads.

4. Substitute the magnitudes of the external loads into the result obtained in Step 3 to obtain a numerical value for the displacement q_i or rotation θ_i.

Curved Beams Treated as Straight Beams

The strain energy due to bending [see Eq. (5.19)] was derived by assuming that the beam is straight. The magnitude of U_M for curved beams is derived in Chapter 9, where it is shown that the error in using Eq. (5.19) to determine U_M is negligible as long as the radius of curvature of the beam is more than twice its depth. Consider the curved beam in Fig. 5.9 whose strain energy is the sum of U_N, U_S, and U_M, each of which is caused by the same load P. If the radius of curvature R of the curved beam is large compared to the beam depth, the magnitudes of U_N and U_S will be small compared to U_M and can be neglected. We assume that U_N and U_S can be neglected when the ratio of length to depth is greater than 10. The resulting error is often less than 1% and will seldom exceed 5%. A numerical result is obtained in Example 5.7.

Figure 5.9

EXAMPLE 5.3
End Load on a Cantilever Beam

Determine the deflection under load P of the cantilever beam shown in Fig. E5.3. Assume that the beam length L is more than five times the beam depth h.

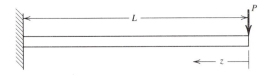

Figure E5.3   End-loaded cantilever beam.

SOLUTION

Since $L > 5h$, the strain energy U_S is small and will be neglected. Therefore, the total strain energy is [Eq. (5.19)]

$$U = U_M = \int_0^L \frac{M_x^2}{2EI_x} dz \tag{a}$$

By Castigliano's theorem, the deflection q_P is [Eq. (5.8)]

$$q_P = \frac{\partial U}{\partial P} = \int_0^L \frac{M_x}{EI_x} \frac{\partial M_x}{\partial P} dz \tag{b}$$

By Fig. E5.3, $M_x = Pz$. Therefore, $\partial M_x/\partial P = z$, and by Eq. (b), we find

$$q_P = \int_0^L \frac{Pz^2}{EI_x} dz = \frac{PL^3}{3EI_x} \tag{c}$$

This result agrees with elementary beam theory.

EXAMPLE 5.4
Cantilever Beam Loaded in Its Plane

The cantilever beam in Fig. E5.4 has a rectangular cross section and is subjected to equal loads P at the free end and at the center as shown.

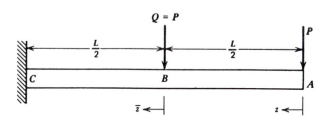

Figure E5.4

(a) Determine the deflection of the free end of the beam.

(b) What is the error in neglecting the strain energy due to shear if the beam length L is five times the beam depth h? Assume that the beam is made of steel ($E = 200$ GPa and $G = 77.5$ GPa).

SOLUTION

(a) To determine the dependencies of the shear V and moment M on the end load P, it is necessary to distinguish between the loads at A and B. Let the load at B be designated by Q. The moment and shear functions are continuous from

A to *B* and from *B* to *C*. From *A* to *B*, we have

$$V = P, \qquad \therefore \frac{\partial V}{\partial P} = 1$$

$$M = Pz, \qquad \therefore \frac{\partial M}{\partial P} = z$$

From *B* to *C*, we have

$$V = P + Q, \qquad \therefore \frac{\partial V}{\partial P} = 1$$

$$M = P\left(\bar{z} + \frac{L}{2}\right) + Q\bar{z}, \qquad \therefore \frac{\partial M}{\partial P} = \bar{z} + \frac{L}{2}$$

where we have chosen point *B* as the origin of coordinate $\bar{z}$ for the length from *B* to *C*. Equation (5.27) gives (with $Q = P$)

$$q_P = \int_0^{L/2} \frac{1.2P}{GA}(1)\,dz + \int_0^{L/2} \frac{Pz}{EI}(z)\,dz + \int_0^{L/2} \frac{2.4P}{GA}(1)\,d\bar{z}$$

$$+ \int_0^{L/2} \frac{P(2\bar{z} + L/2)}{EI}\left(\bar{z} + \frac{L}{2}\right)d\bar{z} = \frac{1.8PL}{GA} + \frac{7PL^3}{16EI} \qquad \text{(a)}$$

(b) Since the beam has a rectangular section, $A = bh$ and $I = bh^3/12$. Equation (a) can be rewritten as follows:

$$\frac{Ebq_P}{P} = \frac{1.8LE}{Gh} + \frac{7(12)L^3}{16h^3}$$

$$= \frac{1.8(5)(200)}{77.5} + \frac{7(12)(5^3)}{16}$$

$$= 23.23 + 656.25$$

$$= 679.48$$

$$\text{Error in neglecting shear term} = \frac{23.23(100)}{679.48} = 3.42\%$$

Alternatively, one could have used the approximate value $k = 1.50$ [Eq. (5.23)]. Then the estimate of shear contribution would have been increased by the ratio $1.50/1.20 = 1.25$. Overall the shear contribution would still remain small.

EXAMPLE 5.5
A Shaft-Beam Mechanism

A shaft *AB* is attached to the beam *CDFH*; see Fig. E5.5. A torque of 2.00 kN·m is applied to the end *B* of the shaft. Determine the rotation of section *B*. The shaft and beam are made of an aluminum alloy for which $E = 72.0$ GPa and $G = 27.0$ GPa. Neglect the strain energy of the hub *DF*.

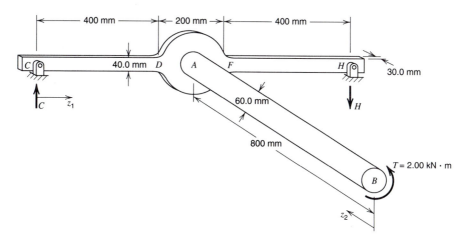

Figure E5.5   Shaft-beam mechanism.

SOLUTION

Since the beam is slender, the strain energy due to shear is neglected. Hence, the total strain energy of the mechanism is

$$U = U_M + U_T = 2 \int_0^{400} \frac{M_x^2}{2EI_x} dz + \int_0^{800} \frac{T^2}{2GJ} dz \tag{a}$$

The pin reactions at C and H have the same magnitude, but opposite sense. Therefore, moment equilibrium yields the result

$$H = C = 0.001 \ T \tag{b}$$

By Castigliano's theorem, the angular rotation at B is

$$\theta_B = \frac{\partial U}{\partial T} = 2 \int_0^{400} \frac{M_x}{EI_x} \frac{\partial M_x}{\partial T} dz_1 + \int_0^{800} \frac{T}{GJ} dz_2 \tag{c}$$

By Fig. E5.5 and Eq. (b), we have

$$T = 2,000,000 \ \text{N·mm}, \qquad M_x = Cz_1 = 2000z_1 \tag{d}$$

Therefore, Eqs. (c) and (d) yield, with $I_x = (30)(40)^3/12 = 160,000 \ \text{mm}^4$ and $J = \pi(60)^4/64 = 636,000 \ \text{mm}^4$, $\theta_B = 0.101$ rad.

EXAMPLE 5.6
Uniformly Loaded Cantilever Beam

The cantilever beam in Fig. E5.6a is subjected to a uniformly distributed load w. Determine the deflection of the free end. Neglect the shear strain energy U_S.

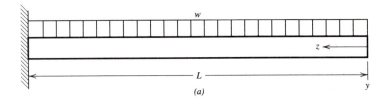

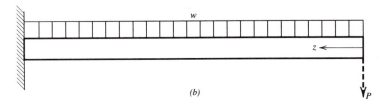

Figure E5.6 Uniformly loaded cantilever beam.

SOLUTION

Since there is no force acting at the free end, we introduce the fictitious force P, (Fig. E5.6b). See the discussion that follows Example 5.8. Hence, the deflection q_P at the free end is given by [see Eq. (5.8)]

$$q_P = \int_0^L \frac{M_x}{EI_x} \frac{\partial M_x}{\partial P}\bigg|_{P=0} dz \tag{a}$$

The bending moment M_x due to P and w is

$$M_x = Pz + \tfrac{1}{2}wz^2 \tag{b}$$

and

$$\frac{\partial M_x}{\partial P} = z \tag{c}$$

Substitution of Eqs. (b) and (c) into Eq. (a) yields

$$q_P = \frac{1}{EI_x} \int_0^L (Pz + \tfrac{1}{2}wz^2)\big|_{P=0}\, z\, dz = \frac{wL^3}{8EI} \tag{d}$$

This result agrees with elementary beam theory.

EXAMPLE 5.7
Curved Beam Loaded in Its Plane

The curved beam in Fig. E5.7 has a 30-mm square cross section and radius of curvature $R = 65$ mm. The beam is made of a steel for which $E = 200$ GPa and $v = 0.29$.

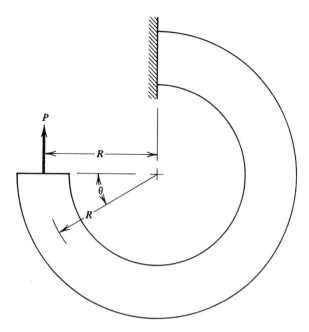

Figure E5.7

(a) If $P = 6.00$ kN, determine the deflection of the free end of the curved beam in the direction of P.

(b) What is the error in the deflection if U_N and U_S are neglected?

SOLUTION

The shear modulus for the steel is $G = E/[2(1 + v)] = 77.5$ GPa.

(a) It is convenient to use polar coordinates. For a cross section of the curved beam located at angle θ from the section on which P is applied (Fig. E5.7),

$$N = P\cos\theta, \qquad \frac{\partial N}{\partial P} = \cos\theta$$

$$V = P\sin\theta, \qquad \frac{\partial V}{\partial P} = \sin\theta$$

$$M = PR(1 - \cos\theta), \qquad \frac{\partial M}{\partial P} = R(1 - \cos\theta) \tag{a}$$

Substitution into Eq. (5.27) gives

$$q_P = \int_0^{3\pi/2} \frac{P\cos\theta}{EA}(\cos\theta)R\,d\theta + \int_0^{3\pi/2} \frac{kP\sin\theta}{GA}(\sin\theta)R\,d\theta$$

$$+ \int_0^{3\pi/2} \frac{PR(1 - \cos\theta)}{EI}R(1 - \cos\theta)R\,d\theta \tag{b}$$

Using the trigonometric identities $\cos^2 \theta = \frac{1}{2} + \frac{1}{2}\cos 2\theta$ and $\sin^2 \theta = \frac{1}{2} - \frac{1}{2}\cos 2\theta$, we find that

$$
q_P = \frac{PR}{EA} \int_0^{3\pi/2} (\tfrac{1}{2} + \tfrac{1}{2}\cos 2\theta) \, d\theta + \frac{1.2PR}{GA} \int_0^{3\pi/2} (\tfrac{1}{2} - \tfrac{1}{2}\cos 2\theta) \, d\theta
$$

$$
+ \frac{PR^3}{EI} \int_0^{3\pi/2} (1 - 2\cos \theta + \tfrac{1}{2} + \tfrac{1}{2}\cos 2\theta) \, d\theta
$$

$$
q_P = \frac{3\pi PR}{4EA} + \frac{1.2(3\pi)PR}{4GA} + \left(\frac{9\pi}{4} + 2\right)\frac{PR^3}{EI}
$$

$$
= \frac{3\pi(65)(6000)}{4(200 \times 10^3)(30)^2} + \frac{1.2(3\pi)(65)(6000)}{4(77,500)(30)^2}
$$

$$
+ \left(\frac{9\pi}{4} + 2\right)\frac{(65)^3(6000)(12)}{(200 \times 10^3)(30)^4}
$$

$$
= 0.0051 + 0.0158 + 1.1069 = 1.1278
$$

Again, as in Example 5.4, we could have used the value $k = 1.50$, with a resulting slight overall change in the shear contribution.

(b) In case U_N and U_S are neglected,

$$
q_P = 1.1069 \text{ mm}
$$

and the percentage error in the deflection calculation is

$$
\text{error} = \frac{(1.1278 - 1.1069)100}{1.1278} = 1.85\%
$$

This error is small enough to be neglected for most engineering applications. The ratio of length to depth for this beam is $3\pi(65)/[2(30)] = 10.2$.

EXAMPLE 5.8
Semicircular Cantilever Beam

The semicircular cantilever beam in Fig. E5.8a has a radius of curvature R and a circular cross section of diameter d. It is subjected to loads of magnitude P at points B and C.

(a) Determine the vertical deflection at C in terms of P, modulus of elasticity E, shear modulus G, radius of curvature R, area of the cross section A, and moment of inertia of the cross section I.

(b) Let $P = 150$ N, $R = 200$ mm, $d = 20.0$ mm, $E = 200$ GPa, and $G = 77.5$ GPa. Determine the effect of neglecting the strain energy U_s due to shear.

SOLUTION

(a) Since we wish to determine the vertical deflection at C, the contribution to the total strain energy of the load at C must be distinguished from the contribu-

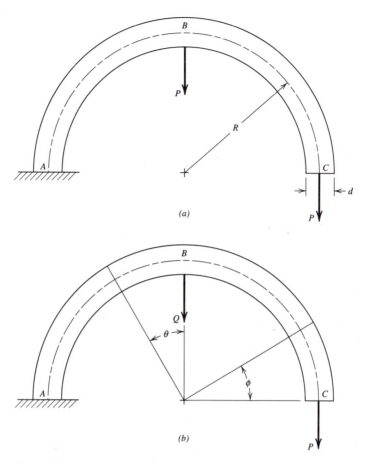

Figure E5.8

tion of the load at B. Therefore, as in Example 5.4, we denote the load P at B by Q ($= P$ in magnitude) (Fig. E5.8b). For section BC (Fig. E5.8b), we have

$$N = P\cos\phi, \qquad \frac{\partial N}{\partial P} = \cos\phi$$

$$V = P\sin\phi, \qquad \frac{\partial V}{\partial P} = \sin\phi$$

$$M = PR(1 - \cos\phi), \qquad \frac{\partial M}{\partial P} = R(1 - \cos\phi) \qquad \text{(a)}$$

For section AB, we have

$$N = (P + Q)\sin\theta, \qquad \frac{\partial N}{\partial P} = \sin\theta$$

$$V = (P + Q)\cos\theta, \qquad \frac{\partial V}{\partial P} = \cos\theta$$

$$M = PR(1 + \sin\theta) + QR\sin\theta, \qquad \frac{\partial M}{\partial P} = R(1 + \sin\theta) \qquad \text{(b)}$$

Substitution of Eqs. (a) and (b) into Eq. (5.27) yields, with $k = 1.33$,

$$q_c = \int_0^{\pi/2} \frac{P \cos^2 \phi}{EA} R \, d\phi + \int_0^{\pi/2} \frac{1.33 P \sin^2 \phi}{GA} R \, d\phi$$

$$+ \int_0^{\pi/2} \frac{PR^2 (1 - \cos \phi)^2}{EI} R \, d\phi + \int_0^{\pi/2} \frac{(P + Q) \sin^2 \theta}{EA} R \, d\theta$$

$$+ \int_0^{\pi/2} \frac{1.33(P + Q) \cos^2 \theta}{GA} R \, d\theta$$

$$+ \int_0^{\pi/2} \frac{PR(1 + \sin \theta) + QR \sin \theta}{EI} R(1 + \sin \theta) R \, d\theta \qquad (c)$$

Integration yields, if we note that $Q = P$,

$$q_c = \frac{3\pi}{4} \frac{PR}{EA} + \frac{3\pi}{4} \frac{1.33PR}{GA} + \left(\frac{7\pi}{4} + 1 \right) \frac{PR^3}{EI} \qquad (d)$$

The three terms on the right-hand side of Eq. (d) are the contributions of the axial force, shear, and moment, respectively, to the displacement q_c.

(b) For $P = 150$ N, $R = 200$ mm, $d = 20.0$ mm (hence, $A = 314$ mm^2 and $I = 7850$ mm^4), $E = 200$ GPa, and $G = 77.5$ GPa, Eq. (d) yields

$$q_c = 0.0011 + 0.0039 + 4.9666 = 4.97 \text{ mm}$$

where 0.0011 is due to the axial force, 0.0039 is due to shear, and 4.9666 is due to moment. The contributions of axial load and shear are very small. Since $R/d = 200/20 = 10$, this result confirms the statement at the end of Sec. 5.4.

Dummy Load Method and Dummy Unit Load Method

As illustrated in the preceding examples, Castigliano's theorem on deflections, as expressed in Eqs. (5.27) and (5.28), is useful for the determination of deflections and rotations at the locations of concentrated forces and moments. Frequently, it is necessary to determine the deflection or rotation at a location that has no corresponding external load (see Example 5.6). For example, we might want to know the rotation at the free end of a cantilever beam that is subjected to concentrated loads at midspan and at the free end, but no concentrated moment at the free end (as is the case in Example 5.4). Castigliano's theorem on deflections can be applied in these situations as well. The modified procedure is as follows:

1. Apply a fictitious force F_i (or fictitious moment M_i) at the location and in the direction of the displacement q_i (or rotation θ_i) to be determined.

2. Write an expression for each of the internal actions (axial force, shear, moment, and torque) in each member of the structure in terms of the applied external forces and moments, including the fictitious force (or moment).

3. To determine the deflection q_i of the structure at the location of a fictitious force F_i and in the sense of F_i, differentiate each of the internal action expres-

sions with respect to F_i. Similarly, to determine the rotation θ_i of the structure at the location of a fictitious moment M_i and in the sense of M_i, differentiate each of the internal action expressions with respect to M_i.

4. Substitute the expressions for the internal actions obtained in Step 2 and the derivatives obtained in Step 3 into Eq. (5.27) or (5.28) and perform the integration. The result is a relationship between the deflection q_i (or rotation θ_i) and the externally applied loads, including the fictitious force F_i (or moment M_i).

5. Since, in fact, the fictitious force (or moment) does not act on the structure, set its value to zero in the relation obtained in Step 4. Then substitute the numerical values of the external loads into this result to obtain numerical value for the displacement q_i (or rotation θ_i).

The above procedure is known as the *dummy load method*. The name derives from the procedure. A fictitious (or *dummy*) load is applied, its effect on internal actions is determined, and then it is removed. If the procedure is limited to small deflections of linear elastic structures (consisting of tension members, compression members, beams, and torsion bars), then the derivatives of the internal actions with respect to the fictitious loads are equivalent to the internal actions that result from a *unit force* (or *unit moment*) applied at the point of interest. When the method is used in this manner, it is referred to as the *dummy unit load method*. The net effect of this procedure is that it eliminates the differentiation in Eqs. (5.27) and (5.28). The internal actions (axial, shear, moment, and torque, respectively) in member j due to a *unit force* at location i may be represented as

$$n_{ji}^F = \frac{\partial N_j}{\partial F_i}, \quad v_{ji}^F = \frac{\partial V_j}{\partial F_i}, \quad m_{ji}^F = \frac{\partial M_j}{\partial F_i}, \quad t_{ji}^F = \frac{\partial T_j}{\partial F_i} \qquad (5.29a)$$

Similarly, the internal actions in member j due to a *unit moment* at location i may be represented as

$$n_{ji}^M = \frac{\partial N_j}{\partial M_i}, \quad v_{ji}^M = \frac{\partial V_j}{\partial M_i}, \quad m_{ji}^M = \frac{\partial M_j}{\partial M_i}, \quad t_{ji}^M = \frac{\partial T_j}{\partial M_i} \qquad (5.29b)$$

In the dummy unit load approach, Eq. (5.27) and (5.28) take the form

$$q_i = \sum_{j=1}^{m} \left(\int \frac{N_j n_{ji}^F}{E_j A_j} dz + \int \frac{k_j V_j v_{ji}^F}{G_j A_j} dz + \int \frac{M_j m_{ji}^F}{E_j I_j} dz + \int \frac{T_j t_{ji}^F}{G_j J_j} dz \right) \qquad (5.30a)$$

$$\theta_i = \sum_{j=1}^{m} \left(\int \frac{N_j n_{ji}^M}{E_j A_j} dz + \int \frac{k_j V_j v_{ji}^M}{G_j A_j} dz + \int \frac{M_j m_{ji}^M}{E_j I_j} dz + \int \frac{T_j t_{ji}^M}{G_j J_j} dz \right) \qquad (5.30b)$$

The use of this technique is illustrated in the following examples.

EXAMPLE 5.9
Cantilever Beam Deflections and Rotations

The cantilever beam in Fig. E5.9 has a rectangular cross section and is subjected to a midspan load P as shown.

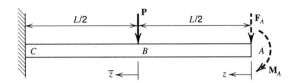

Figure E5.9

(a) Determine the vertical deflection and rotation of the free end of the beam by the dummy load method.

(b) Show that the same results are obtained by the dummy unit load method. Neglect strain energy due to shear.

SOLUTION

(a) The first step in the dummy load method is to apply a fictitious load F_A and a fictitious moment M_A at point A as shown in Fig. E5.9. Next, we write the moment expressions for the two intervals of the beam.

Interval $A-B$

$$M_{AB} = M_A + F_A z \tag{a}$$

Interval $B-C$

$$M_{BC} = M_A + F_A\left(\bar{z} + \frac{L}{2}\right) + P\bar{z} \tag{b}$$

Differentiation of Eqs. (a) and (b) with respect to the fictitious force and moment yields

Interval $A-B$

$$\frac{\partial M_{AB}}{\partial F_A} = z \tag{c}$$

$$\frac{\partial M_{AB}}{\partial M_A} = 1 \tag{d}$$

Interval $B-C$

$$\frac{\partial M_{BC}}{\partial F_A} = \bar{z} + \frac{L}{2} \tag{e}$$

$$\frac{\partial M_{BC}}{\partial M_A} = 1 \tag{f}$$

To find the vertical deflection at point A, we substitute Eqs. (a), (b), (c), and (e) into Eq. (5.27) and perform the integration.

$$q_A = \int_0^{L/2} \frac{M_A + F_A(z)}{EI}(z)\,dz + \int_0^{L/2} \frac{M_A + F_A(\bar{z} + L/2) + P(\bar{z})}{EI}\left(\bar{z} + \frac{L}{2}\right)d\bar{z} \quad \text{(g)}$$

$$= M_A\left(\frac{L^2}{2EI}\right) + F_A\left(\frac{L^3}{3EI}\right) + P\left(\frac{5L^3}{48EI}\right) \tag{h}$$

Since, in fact, the fictitious loads F_A and M_A do not exist, they are set to zero. Then Eq. (h) yields the deflection of point A as

$$q_A = \frac{5PL^3}{48EI} \tag{i}$$

To find the rotation of the section at A, we substitute Eqs. (a), (b), (d), and (f) into Eq. (5.28) and perform the integration.

$$\theta_A = \int_0^{L/2} \frac{M_A + F_A(z)}{EI}(1)\,dz + \int_0^{L/2} \frac{M_A + F_A(\bar{z} + L/2) + P(\bar{z})}{EI}(1)\,dz \quad \text{(j)}$$

$$= M_A\left(\frac{L}{EI}\right) + F_A\left(\frac{L^2}{2EI}\right) + P\left(\frac{L^2}{8EI}\right) \tag{k}$$

Again, the fictitious loads F_A and M_A are set to zero. Then Eq. (k) yields the rotation of the section at A as

$$\theta_A = \frac{PL^2}{8EI} \tag{l}$$

(b)   In the dummy unit load method, F_A and M_A are set to unity. Then the internal moment M due to the real force at B and the internal moments m^F and m^M [see Eqs. (5.29a) and (5.29b)] due to the unit force and unit moment at A are

Interval $A-B$

$$M_{AB} = 0 \tag{m}$$
$$m_{AB}^F = 1.0(z) = z \tag{n}$$
$$m_{AB}^M = 1.0 \tag{o}$$

Interval $B-C$

$$M_{BC} = P\bar{z} \tag{p}$$

$$m_{BC}^F = 1.0\left(\bar{z} + \frac{L}{2}\right) = \bar{z} + \frac{L}{2} \tag{q}$$

$$m_{BC}^M = 1.0 \tag{r}$$

The deflection at point A is obtained by the substitution of Eqs. (m), (n), (p), and (q) into Eq. (5.30a). The result is [see Eq. (i)]

$$q_A = \int_0^{L/2} \frac{P\bar{z}}{EI}\left(\bar{z} + \frac{L}{2}\right)d\bar{z} \tag{s}$$

$$= \frac{5PL^3}{48EI} \tag{t}$$

The rotation of the section at A is obtained by the substitution of Eqs. (m), (o), (p), and (r) into Eq. (5.30b). The result is [see Eq. (l)]

$$\theta_A = \int_0^{L/2} \frac{P\bar{z}}{EI}(1)\,d\bar{z} \tag{u}$$

$$= \frac{PL^2}{8EI} \tag{v}$$

Thus, the equivalence of the dummy load approach with the dummy unit load approach is demonstrated for this example.

EXAMPLE 5.10
Pin-Connected Structure

The pin-connected structure in Fig. E5.10 is made of an aluminum alloy for which $E = 72.0$ GPa. The magnitudes of the loads are $P = 10$ kN and $Q = 5$ kN. Members BC, CD, and DE each have cross-sectional areas of 900 mm². The remaining members have cross-sectional areas of 150 mm². Determine the rotation of member BE caused by the loads P and Q.

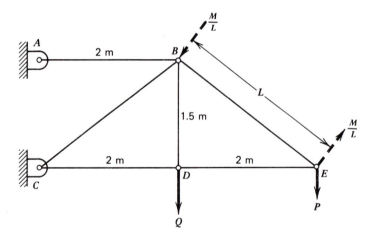

Figure E5.10

SOLUTION

To determine the rotation of member BE by energy methods, a moment M must be acting on member BE. Let M be an imaginary counterclockwise moment represented by a couple with equal and opposite forces $M/L(L = BE = 2.5 \times 10^3$ mm) applied perpendicular to BE at points B and E as indicated in Fig. E5.10. Equations of equilibrium give the following values for the axial forces in the members of the structure.

$$N_{AB} = \frac{4}{3}(Q + 2P) - \frac{5M}{3L}, \qquad \frac{\partial N_{AB}}{\partial M} = -\frac{5}{3L}$$

$$N_{BC} = -\frac{5}{3}(Q + P), \qquad\qquad \frac{\partial N_{BC}}{\partial M} = 0$$

$$N_{BD} = Q, \qquad\qquad\qquad \frac{\partial N_{BD}}{\partial M} = 0$$

$$N_{BE} = \frac{5P}{3} - \frac{4M}{3L}, \qquad\qquad \frac{\partial N_{BE}}{\partial M} = -\frac{4}{3L}$$

$$N_{CD} = N_{DE} = -\frac{4P}{E} + \frac{5M}{3L}, \qquad \frac{\partial N_{CD}}{\partial M} = \frac{5}{3L}$$

After the partial derivatives $\partial N_j/\partial M$ have been taken, the magnitude of M in the N_j is set to zero. The values of N_j and $\partial N_j/\partial M$ are then substituted into Eq. (5.28) to give

$$\theta_{BE} = \sum_{j=1}^{6} \frac{N_j L_j}{E_j A_j} \frac{\partial N_j}{\partial M} = \frac{N_{AB} L_{AB}}{EA_{AB}} \frac{\partial N_{AB}}{\partial M} + \frac{N_{BC} L_{BC}}{EA_{BC}} \frac{\partial N_{BC}}{\partial M}$$

$$+ \frac{N_{BD} L_{BD}}{EA_{BD}} \frac{\partial N_{BD}}{\partial M} + \frac{N_{BE} L_{BE}}{EA_{BE}} \frac{\partial N_{BE}}{\partial M}$$

$$+ 2\frac{N_{CD} L_{CD}}{EA_{CD}} \frac{\partial N_{CD}}{\partial M}$$

$$\theta_{BE} = \frac{4(25,000)(2000)}{3(72,000)(150)}\left[-\frac{5}{3(2500)} \right] + \frac{5(10,000)(2500)}{3(72,000)(150)}\left[-\frac{4}{3(2500)} \right]$$

$$- \frac{2(4)(10,000)(2000)}{3(72,000)(900)}\left[\frac{5}{3(2500)} \right] = -0.004115 - 0.002058 - 0.000549$$

$$= -0.00672 \text{ rad}$$

The negative sign for θ_{BE} indicates that the angle change is clockwise; that is, the angle change has a sign opposite to that assumed for M.

EXAMPLE 5.11
Curved Beam Loaded Perpendicular to Its Plane

The semicircular curved beam of radius R in Fig. E5.11 has a circular cross section of radius r. The curved beam is fixed at 0 and lies in the (x, y) plane with center of curvature at C on the x axis. Load P parallel to the z axis acts at a section $\pi/2$ from the fixed end. Determine the z component of the deflection of the free end. Assume that R/r is sufficiently large for U_s to be negligible.

SOLUTION

To find the z component of the deflection of the free end of the curved beam, a dummy unit load parallel to the z axis is applied at B as indicated in Fig. E5.11; the curved beam is indicated by its centroidal axis in order to simplify the figure. Consider a section D of the curved beam at an angle θ measured from section A at the load P. The internal moment and torque at section D due to forces at A and

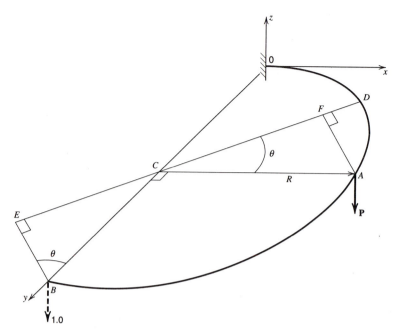

Figure E5.11

B are

$$M_D = P(AF) = PR \sin \theta$$
$$T_D = P(DF) = PR(1 - \cos \theta)$$
$$m_D^F = 1.0(BE) = R \cos \theta$$
$$t_D^F = 1.0(DC + CE) = R(1 + \sin \theta)$$

These values are substituted into Eqs. (5.30a) and (5.30b) to give

$$q_B = \int_0^{\pi/2} \left[\frac{PR \sin \theta (R \cos \theta)}{EI} + \frac{PR(1 - \cos \theta)[R(1 + \sin \theta)]}{GJ} \right] R \, d\theta$$

$$= \frac{1}{2} \frac{PR^3}{EI} + \frac{PR^3}{GJ} \left(\frac{\pi}{2} - \frac{1}{2} \right)$$

$$= \frac{2PR^3}{\pi E r^4} [1 + (1 + v)(\pi - 1)]$$

5.5

STATICALLY INDETERMINATE STRUCTURES

As we observed in Sec. 5.4, a statically determinate structure (Fig. 5.7) may be made statically indeterminate by the addition of a member (member *BD* in Fig. 5.8). Alternatively, a statically indeterminate structure is rendered statically determinate if certain members, internal actions, or supports are removed. For example, the truss in Fig. 5.8 is rendered statically determinate if member BD (or equally well member AC) is removed. Such a member in a statically indeterminate structure is said to be *redundant*, since after its removal the structure will remain in static equilibrium under arbitrary loads. In general, statically indeterminate structures contain one or more redundant members or supports.

Generally in the analysis of structures, internal actions in each member of the structure must be determined. For statically indeterminate structures, the equations of static equilibrium are not sufficient to determine these internal actions. For example, in Fig. 5.10a, the propped cantilever beam has four unknown support reactions, whereas there are only three equations of equilibrium for a planar structure. If the support at B were removed, the beam would function as a simple cantilever beam. Hence, we may consider the support at B to be redundant and, if it is removed, the beam is rendered statically determinate. Since the choice of the redundant is arbitrary, the moment at the wall (point A) could be considered the redundant.

If we consider the support at B to be redundant, additional information is required to determine the magnitude of the reaction R (see Fig. 5.10c). As we shall see, the fact that the support at B prevents the tip of the beam from displacing vertically may be used, in conjunction with Castigliano's theorem on deflections, to obtain the additional equation needed to determine the redundant reaction R.

Likewise, the three support reactions at A (or E) for member $ABCDE$ in Fig. 5.10b can be chosen as the redundants. Hence, either the support at A or E (but not both) may be removed to render the structure statically determinate. Let us assume that the support reactions at E are chosen as the redundants (Fig. 5.10d). The three redundant reactions are a vertical force V_E, which prevents vertical deflection at E; a bending moment M_E, which prevents bending rotation of the section at E; and a torque T_E, which prevents torsional rotation of the section at E. The

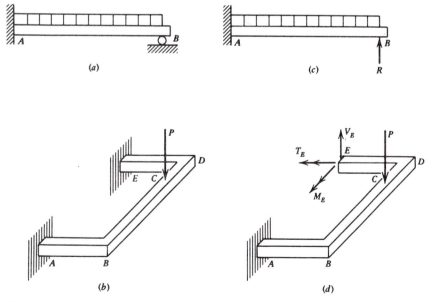

Figure 5.10

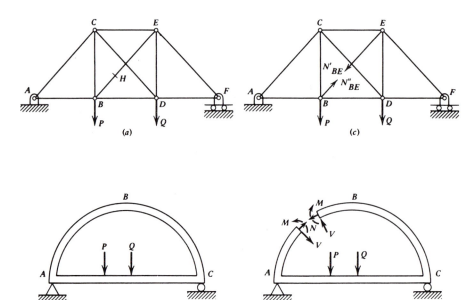

Figure 5.11

fact that vertical deflection, bending rotation, and torsional rotation are prevented at section E may be used, in conjunction with Castigliano's theorem on deflections, to obtain the additional equations needed to determine the support reactions at E.

The structures in Fig. 5.11 do not contain redundant reactions but do contain redundant members. In Fig. 5.11a, the member BE (or CD) of the truss is redundant. Hence, the truss is statically indeterminate. If either member BE or member CD is removed, the truss is rendered statically determinate. Likewise, the member ABC of the statically indeterminate structure in Fig. 5.11b is redundant. It may be removed to render the structure statically determinate.

Since the truss of Fig. 5.11a is pin-joined, the redundant member BE is subject to an internal axial force. Hence, the only redundant internal force for the truss is the tension in member BE (Fig. 5.11c). However, the redundant member ABC of the structure in Fig. 5.11d may support three internal reactions: the axial force N, shear V, and moment M. The additional equations (in addition to the equations of static equilibrium) required to determine the additional unknowns (the redundant internal actions caused by redundant members) in statically indeterminant structures may be obtained by the application of Castigliano's theorem on deflections.

In particular, we can show that

$$\frac{\partial U}{\partial F_i} = 0 \tag{5.31}$$

for every internal redundant force $(F_1, F_2, \ldots,)$ in the structure, and

$$\frac{\partial U}{\partial M_i} = 0 \tag{5.32}$$

for every internal redundant moment $(M_1, M_2, \ldots,)$ in the structure. Equations (5.31) and (5.32) are readily verified for the structures in Fig. 5.10. The beam in Fig. 5.10a has a redundant external reaction R at B that produces an internal shear at section B equal to R. Since the deflection at point B is zero, Eq. (5.27) gives $q_R = \partial U/\partial R = 0$, which agrees with Eq. (5.31). The structure in Fig. 5.10b has three internal redundant reactions (V_E, M_E, T_E) at section E, as indicated in Fig. 5.10d. Since the deflection and rotations at E remain zero as the structure is loaded, Eqs. (5.27) and (5.28) yield the results $\partial U/\partial V_E = \partial U/\partial M_E = \partial U/\partial T_E = 0$, which agree with Eqs. (5.31) and (5.32).

It is not directly apparent that Eqs. (5.31) and (5.32) are valid for the internal redundant reactions in the structures in Fig. 5.11. To show that they are valid, let N_{BE} be the redundant internal action for the pin-joined truss (Fig. 5.11a). Pass a section through some point H of member BE and apply equal and opposite tensions N'_{BE} and N''_{BE}, as indicated in Fig. 5.11c. Since the component of the deflection of point H along member BE is not zero, it is not obvious that

$$\frac{\partial U}{\partial N_{BE}} = 0 \tag{5.33}$$

In order to prove that Eq. (5.33) is valid, it is necessary to distinguish between tensions N'_{BE} and N''_{BE}. The displacement of point H in the direction of N'_{BE} is given by [see Eq. (5.27)]

$$q_{N'_{BE}} = \frac{\partial U}{\partial N'_{BE}} \tag{5.34}$$

and in the direction of N''_{BE}, the displacement is given by

$$q_{N''_{BE}} = \frac{\partial U}{\partial N''_{BE}} \tag{5.35}$$

These displacements $q_{N'_{BE}}$ and $q_{N''_{BE}}$ are collinear, have equal magnitudes, but have opposite senses. Hence, by Eqs. (5.34) and (5.35) we have

$$\frac{\partial U}{\partial N'_{BE}} + \frac{\partial U}{\partial N''_{BE}} = 0 \tag{5.36}$$

The reduction of Eq. (5.36) to Eq. (5.33) then follows by the same technique employed in the reduction of Eq. (a) of Sec. 5.2 to Eq. (d) of Sec. 5.2, since $N'_{BE} = N''_{BE} = N_{BE}$. In a similar manner, it may be shown for the structure in Fig. 5.11b that

$$\frac{\partial U}{\partial N} = \frac{\partial U}{\partial V} = \frac{\partial U}{\partial M} = 0 \tag{5.37}$$

where N, V, and M are the internal reactions for any given section of member ABC.

Note: In the application of Eqs. (5.31) and (5.32) to the system with redundant supports or redundant members, it is assumed that the unloaded system is stress-free. Consequently, redundant supports exert no force on the structure initially.

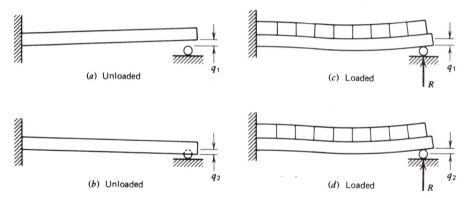

Figure 5.12   Effect of support settlement or thermal expansion or contraction on redundant supports of loaded beams. (*a*) Unloaded. (*b*) Unloaded. (*c*) Loaded. (*d*) Loaded.

However, in certain applications, these conditions do not hold. For example, consider the beam in Fig. 5.10. Initially, the right end of the beam may be lifted off the support, or the end support may exert a force on the beam because of either support settlement or thermal expansion or contraction. As a result, the end of the beam (in the absence of the redundant support) may be raised a distance q_1 above the location of the support before the beam is loaded (Fig. 5.12*a*) or it may be a distance q_2 below the support location (Fig. 5.12*b*).

If the displacement magnitudes q_1 or q_2 of the end of the beam (in the absence of the support) are known, we may compute the reaction R for the loaded beam (Figs. 5.12*c* and *d*) by the relations

$$q_1 = -\frac{\partial U}{\partial R} \quad \text{or} \quad q_2 = \frac{\partial U}{\partial R} \tag{5.38}$$

where the minus sign indicates that displacement q_1 and force R have opposite senses.

If known residual stresses (say, due to fabrication processes) are present in a structure before it is loaded, the total stresses in the structure after it is loaded may be computed in two steps. First, we assume that the structure is stress-free when unloaded, and we calculate the redundant reactions by means of Eqs. (5.31) and (5.32). Stresses are calculated for the known loads and known reactions. Next, we superimpose these calculated stresses on the residual stresses to determine the total stresses in the structure.

EXAMPLE 5.12
Statically Indeterminate Cantilever Beam

The beam in Fig. E5.12*a* is fixed at the left end, simply supported at the right end, and subjected to a concentrated load P at the center.

(a)   Determine the magnitude of the reaction R (Fig. E5.12*b*) at the right end.

(b)   Determine the deflection of the beam under load P.

(c)   If the simple support at the right end settles a vertical distance $PL^3/32EI$, determine the new magnitude of the reaction R.

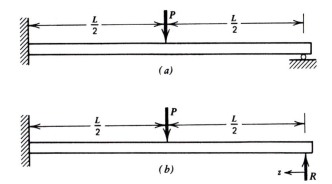

Figure E5.12

SOLUTION

(a) Since the vertical displacement of the beam at the simple support is zero, Eq. (5.31) gives (with z measured to the left from R)

$$\frac{\partial U}{\partial R} = 0 = \int_0^{L/2} \frac{Rz}{EI}(z)\,dz + \int_{L/2}^{L} \frac{Rz - P(z - L/2)}{EI}(z)\,dz$$

$$0 = \frac{R}{3}(L/2)^3 + \frac{R}{3}[L^3 - (L/2)^3] - \frac{P}{3}[L^3 - (L/2)^3]$$

$$+ \frac{PL}{4}[L^2 - (L/2)^2]$$

$$R = \frac{5P}{16}$$

(b) Reaction R is treated as independent of P in determining the deflection of the beam under load P. Therefore, the strain energy for the right half of the beam is independent of P. With z measured to the left from P, we have

$$q_P = \int_0^{L/2} \frac{Pz - R(z + L/2)}{EI}(z)\,dz = \frac{PL^3}{24EI} - \frac{5RL^3}{48EI} = \frac{7PL^3}{768EI}$$

(c) The vertical displacement at the simple support has a sense opposite to the sense of R; therefore, Eq. (5.31) gives (with z measured to the left from R)

$$\frac{\partial U}{\partial R} = -\frac{PL^3}{32EI} = \int_0^{L/2} \frac{Rz}{EI}(z)\,dz + \int_{L/2}^{L} \frac{Rz - P(z - L/2)}{EI}(z)\,dz$$

Hence,

$$-\frac{PL^3}{32} = \frac{RL^3}{3} - \frac{5PL^3}{48} \qquad \text{or} \qquad R = \frac{7P}{32}$$

EXAMPLE 5.13
Statically Indeterminate System

Determine the reactions at C for member ABC in Fig. E5.13a and the deflection of point B in the direction of P. Assume U_N and U_S are so small that they can be neglected.

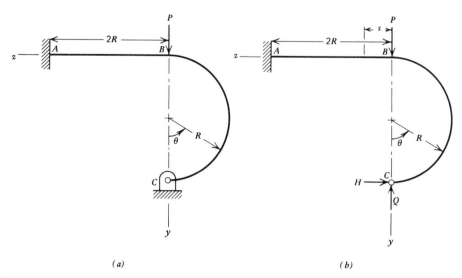

(a)   (b)

Figure E5.13

SOLUTION

The support at C allows rotation but prevents displacements. Our first problem is to determine the redundant reactions Q and H (Fig. E5.13b) at C. Since the y displacement at C is zero, Eq. (5.31) gives

$$\frac{\partial U}{\partial Q} = 0 = \int_0^\pi \frac{[QR\sin\theta - HR(1-\cos\theta)]}{EI} R(\sin\theta)R\,d\theta$$

$$+ \int_0^{2R} \frac{[(Q-P)s + 2HR]}{EI} s\,ds$$

or

$$Q\left(\frac{\pi}{2} + \frac{8}{3}\right) + 2H - \frac{8P}{3} = 0$$

or

$$4.2375Q + 2H - 2.6667P = 0 \qquad\qquad (a)$$

Since the z displacement at C is zero, Eq. (5.31) gives

$$\frac{\partial U}{\partial H} = 0 = \int_0^\pi \frac{[QR\sin\theta - HR(1-\cos\theta)]}{EI}[-R(1-\cos\theta)]R\,d\theta$$

$$+ \int_0^{2R} \frac{[(Q-P)s + 2HR]}{EI} 2R\,ds$$

or

$$2Q + H\left(\frac{3\pi}{2} + 8\right) - 4P = 0$$

or

$$2Q + 12.7124H - 4P = 0 \qquad (b)$$

The simultaneous solution of Eqs. (a) and (b) gives

$$Q = 0.5193P$$

$$H = 0.2329P$$

In the application of Castigliano's theorem to determine the deflection of the point B, we must differentiate the strain energy in member ABC with respect to the load at B; that is, with respect to P. However, since H and Q, the reactions at C, are expressed in terms of the load P, the chain rule must be used. That is,

$$q_P = \frac{\partial U}{\partial P} + \frac{\partial U}{\partial H}\frac{\partial H}{\partial P} + \frac{\partial U}{\partial Q}\frac{\partial Q}{\partial P} \qquad (c)$$

Boundary conditions at C require that $\partial U/\partial H = 0$ and $\partial U/\partial Q = 0$ and Eq. (c) is simplified accordingly. Also, for the released structure [Fig. 5.13(b)], the load at B does not cause bending moment in the curved part BC so that we need consider only the strain energy of the part AB. Thus,

$$q_P = \frac{\partial U}{\partial P} = \int_0^{2R} \frac{[(Q - P)s + 2HR]}{EI}(-s)\,ds$$

$$= \frac{1}{EI}\left(\frac{8}{3}PR^3 - \frac{8}{3}QR^3 - 4HR^3\right)$$

Following substitution for Q and H in terms of P,

$$q_P = 0.3503\frac{PR^3}{EI}$$

Notice that the above argument is applicable to indeterminate structures in general. That is, the boundary condition requirement that $\partial U/\partial H = 0$ and $\partial U/\partial Q = 0$ can be used to simplify the expression for strain energy in the structure. Only the strain energy in the released structure due to the applied load needs to be considered. (See also Example 5.12.)

EXAMPLE 5.14
Statically Indeterminate Truss

The inverted king post truss in Fig. E5.14 is constructed of a 160-mm deep by 60-mm wide rectangular steel beam ABC ($E_{AC} = 200$ GPa and $Y_{AC} = 240$ MPa), a 15-mm diameter steel rod ADC ($E_{DC} = 200$ GPa and $Y_{DC} = 500$ MPa), and a 40 mm by 40 mm white oak compression member BD ($E_{BD} = 12.4$ GPa and $Y_{BD} = 29.6$ MPa). Determine the magnitude of the load P that can be applied to the king post truss if all parts are designed using a factor of safety $SF = 2.00$. Neglect stress concentrations.

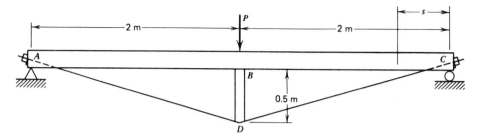

Figure E5.14

SOLUTION

Let member BD be the redundant member of the king post truss. We will include strain energy U_N for both member BD and member ADC; however, U_N and U_S for the beam are so small compared to U_M that they can be neglected. Let the compression load in member BD be N_{BD}. Equations of equilibrium at joint D give

$$N_{DC} = \sqrt{4.25}\, N_{BD}$$

The bending moment in the beam at distance s from either C or A is

$$M = \left(\frac{P}{2} - \frac{N_{BD}}{2}\right)s$$

Equation (5.31) gives

$$\frac{\partial U}{\partial N_{BD}} = 0 = \frac{N_{BD}L_{BD}}{E_{BD}A_{BD}} + 2\frac{N_{DC}L_{DC}}{E_{DC}A_{DC}}\frac{\partial N_{DC}}{\partial N_{BD}}$$

$$+ 2\int_0^{L_{BC}}\frac{M}{E_{AC}I_{AC}}\frac{\partial M}{\partial N_{BD}}ds$$

$$= \frac{500N_{BD}}{E_{BD}A_{BD}} + \frac{2(\sqrt{4.25}\times 10^3)(4.25)N_{BD}}{E_{DC}A_{DC}}$$

$$+ 2\int_0^{2000}\frac{\left(\dfrac{P}{2} - \dfrac{N_{BD}}{2}\right)s}{E_{AC}I_{AC}}\left(-\frac{s}{2}\right)ds$$

which can be simplified to give

$$P = N_{BD}\left[1 + \frac{500}{E_{BD}A_{BD}}\frac{3E_{AC}I_{AC}}{4\times 10^9} + \frac{\sqrt{4.25}\,(8.5\times 10^3)}{E_{DC}A_{DC}}\frac{3E_{AC}I_{AC}}{4\times 10^9}\right] \qquad \text{(a)}$$

But $A_{BD} = 40(40) = 1600 \text{ mm}^2$, $A_{DC} = \pi(15)^2/4 = 176.7 \text{ mm}^2$, and

$$I_{AC} = 60(160)^3/12 = 20.48\times 10^6 \text{ mm}^4.$$

These along with other given values when substituted in Eq. (a) give

$$P = 2.601 N_{BD}$$

The axial loads in members BD and ADC and the maximum moment in member ABC can now be written as functions of P.

$$N_{BD} = 0.384P(\text{N})$$
$$N_{DC} = 0.793P(\text{N})$$
$$M_{max} = 616P(\text{N} \cdot \text{mm})$$

Since the working stress for each member is half the yield stress for the member, a limiting value of P is obtained for each member. For compression member BD

$$\frac{Y_{BD}}{2} = \frac{29.6}{2} = \frac{N_{BD}}{A_{BD}} = \frac{0.384P}{1600}$$

$$P = 61,700 \text{ N}$$

For tension member ADC

$$\frac{Y_{DC}}{2} = \frac{500}{2} = \frac{N_{DC}}{A_{DC}} = \frac{0.797P}{176.7}$$

$$P = 55,700 \text{ N}$$

For beam ABC

$$\frac{Y_{AC}}{2} = \frac{240}{2} = \frac{M_{max}c}{I_{AC}} = \frac{616P(80)}{20.48 \times 10^6}$$

$$P = 49,900 \text{ N}$$

Thus, the design load for the king post truss is 49.9 kN.

EXAMPLE 5.15
Spring-Supported I-Beam

An aluminum alloy I-beam (depth = 100 mm, $I = 2.45 \times 10^6$ mm^4, $E = 72.0$ GPa) has a length of 6.8 m and is supported by seven springs ($K = 110$ N/mm) spaced at distance $l = 1.10$ m center to center along the beam (Fig. E5.15a). A load $P = 12.0$ kN is applied at the center of the beam over the center spring. Determine the load carried by each spring, the deflection of the beam under the load, the maximum bending moment, and the maximum bending stress in the beam.

SOLUTION

It is assumed that the springs are attached to the beam so that the springs can develop tensile as well as compressive forces. Because of symmetry, there are only four unknown spring forces: A, B, C, D. A free-body diagram of the beam with springs attached is shown in Fig. E5.15b. Let the loads B, C, and D carried by the springs be redundant reactions. The magnitudes of these redundants are obtained

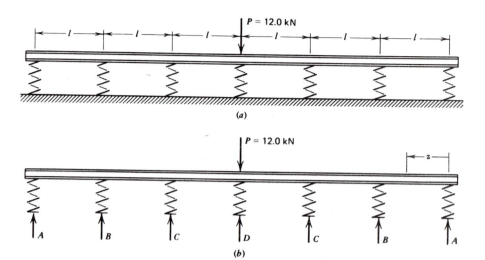

Figure E5.15

using Eq. (5.31)

$$\frac{\partial U}{\partial B} = \frac{\partial U}{\partial C} = \frac{\partial U}{\partial D} = 0 \tag{a}$$

The strain energy U for the beam and springs (if we neglect U_S for the beam) is given by the relation

$$U = 2\int_0^l \frac{M^2}{2EI}\,dz + 2\int_l^{2l} \frac{M^2}{2EI}\,dz + 2\int_{2l}^{3l} \frac{M^2}{2EI}\,dz$$

$$+ 2\left(\frac{A^2}{2K} + \frac{B^2}{2K} + \frac{C^2}{2K}\right) + \frac{D^2}{2K} \tag{b}$$

The moments in the three integrals are functions of the reaction A, which can be eliminated from Eq. (b) by the equilibrium force equation for the y direction.

$$A = \frac{P}{2} - B - C - \frac{D}{2} \tag{c}$$

The moments for the three segments of the beam are

$0 \le z \le l$

$$M = Az \qquad\qquad = \frac{P}{2}z - Bz - Cz - \frac{D}{2}z$$

$l \le z \le 2l$

$$M = Az + B(z - l) \qquad\qquad = \frac{P}{2}z - Bl - Cz - \frac{D}{2}z$$

$2l \le z \le 3l$

$$M = Az + B(z - l) + C(z - 2l) = \frac{P}{2}z - Bl - 2Cl - \frac{D}{2}z \tag{d}$$

Substitution of Eqs. (b–d) into the first of Eqs. (a) gives

$$\frac{\partial U}{\partial B} = 0 = \frac{2}{EI}\int_0^l \left(\frac{P}{2}z - Bz - Cz - \frac{D}{2}z\right)(-z)\,dz$$

$$+ \frac{2}{EI}\int_l^{2l} \left(\frac{P}{2}z - Bl - Cz - \frac{D}{2}z\right)(-l)\,dz$$

$$+ \frac{2}{EI}\int_{2l}^{3l} \left(\frac{P}{2}z - Bl - 2Cl - \frac{D}{2}z\right)(-l)\,dz$$

$$+ \frac{2}{K}\left(\frac{P}{2} - B - C - \frac{D}{2}\right)(-1) + \frac{2B}{K}$$

which can be simplified to give

$$0 = 12BEI + 6CEI + 3DEI - 3PEI - 13PKl^3 + 14BKl^3$$
$$+ 23CKl^3 + 13DKl^3 \tag{e}$$

Substitution of Eqs. (b–d) into the second and third of Eqs. (a) gives, after simplification,

$$0 = 6BEI + 12CEI + 3DEI - 3PEI - 23PKl^3 + 23BKl^3$$
$$+ 40CKl^3 + 23DKl^3 \tag{f}$$

$$0 = 6BEI + 6CEI + 9DEI - 3PEI - 27PKl^3 + 26BKl^3$$
$$+ 46CKl^3 + 27DKl^3 \tag{g}$$

Equations (e–g) are three simultaneous equations in the three unknowns, B, C, and D. Their magnitudes depend on the magnitudes of E, I, and K. Using the values specified in the problem, we have

$$0 = B + 1.0622C + 0.5838D - 0.5838P$$
$$0 = B + 1.8015C + 0.8804D - 0.8804P$$
$$0 = B + 1.6019C + 1.1389D - 0.9213P \tag{h}$$

The solutions of Eqs. (h) and (c) are

$$A = -0.0379P = -455 \text{ N}$$
$$B = 0.1014P = 1217 \text{ N}$$
$$C = 0.2578P = 3094 \text{ N}$$
$$D = 0.3573P = 4288 \text{ N}$$

The maximum deflection of the beam is the deflection under the load P, which is equal to the deflection of the spring at D.

$$q_P = \frac{D}{K} = \frac{4288}{110} = 38.98 \text{ mm}$$

$$M_{max} = 3lA + 2lB + lC = 4.58 \times 10^6 \text{ N} \cdot \text{mm}$$

$$\sigma_{max} = \frac{M_{max}c}{I} = 93.5 \text{ MPa}$$

Except for the simplifying assumptions that the shear introduced negligible error in the flexure formula and contributed negligible strain energy, the above solution is exact. A simple approximate solution of the same problem is presented in Chapter 10.

PROBLEMS
Sections 5.1–5.4

5.1. Determine the horizontal component of deflection of the free end of the curved beam described in Example 5.7. Assume that U_N and U_S are so small that they can be neglected.

5.2. For the pin-connected structure in Fig. E5.10, determine the component of the deflection of point E in the direction of force P.

Ans. $q_P = 25.60$ mm

5.3. Find the vertical deflection of point C in the truss shown in Fig. P5.3. All members have the same cross section and are made of the same material.

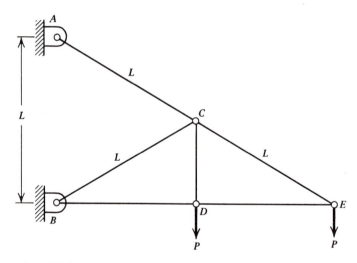

Figure P5.3

5.4. The beam in Fig. P5.4 has the central half of the beam enlarged so that the moment of inertia I is twice the value for each end section. Determine the deflection at the center of the beam.

Ans. $q_{max} = 65wL^4/(6144EI)$

5.5. Member ABC in Fig. P5.5 has a uniform symmetrical cross section and depth that is small compared to L and R. Determine the component of the deflection of point C in the direction of load P.

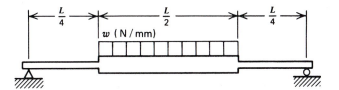

Figure P5.4

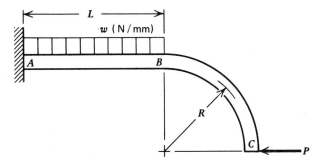

Figure P5.5

5.6. Member $0AB$ in Fig. P5.6 lies in one plane and has the shape of two quadrants of a circle. Assuming that U_S and U_N can be neglected, determine the vertical component of the deflection of point B.

Ans. Vertical $q_B = \dfrac{\pi R^2 M_B}{EI}$ (down)

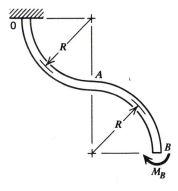

Figure P5.6

5.7. Determine the horizontal deflection of point B for the member in Fig. P5.6.

5.8. Determine the change in slope of the cross section at point B for the member in Fig. P5.6.

Ans. $\theta_B = \pi R M_B / EI$

5.9.   Determine the x and y components of the deflection of point B of the semicircular beam in Fig. P5.9. The depth of the beam is small compared with R.

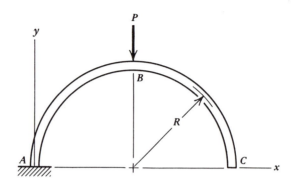

Figure P5.9

5.10.   Determine the vertical component of the deflection of point C for the semicircular beam in Problem 5.9.

Ans.   Vertical $q_c = PR^3\left(1 + \dfrac{\pi}{4}\right)/EI$ (down)

5.11.   The structure in Fig. P5.11 is made up of a cantilever beam $AB(E_1, I_1, A_1)$ and two identical rods BC and $CD(E_2, A_2)$. Let A_1 be large compared with A_2 and L_1 be large compared with the beam depth.

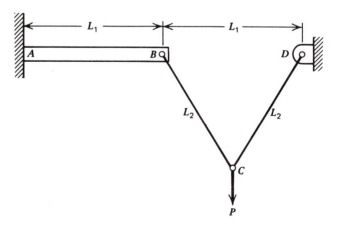

Figure P5.11

(a)   Determine the component of the deflection of point C in the direction of load P.

(b)   If $E_1 = E_2 = E$, the beam and rods have solid circular cross sections with radii r_1 and r_2, respectively, and $L_1 = L_2 = 25r_1$, determine the ratio of r_1 to r_2 such that the beam and rods contribute equally to q_P.

5.12. Beam *ABC* in Fig. P5.12 is simply supported and subjected to a linearly varying distributed load as shown. Determine the deflection of the center of the beam.

Ans. $q_B = wL^4/120EI$

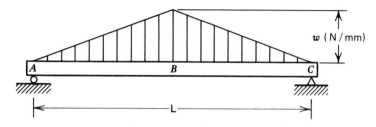

Figure P5.12

5.13. Member *ABC* in Fig. P5.13 has a circular cross section with radius *r*. It has a right angle bend at *B* and is loaded by a load *P* perpendicular to the plane of *ABC*. Determine the component of deflection of point *C* in the direction of *P*. Assume that L_1 and L_2 are each large compared to *r*.

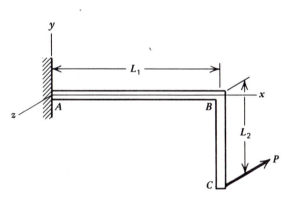

Figure P5.13

5.14. Member *ABC* in Fig. P5.14 lies in the plane of the paper, has a uniform circular cross section, and is subjected to torque T_0, also in the plane of the

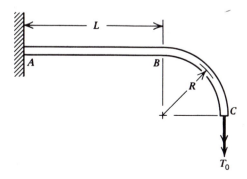

Figure P5.14

paper, as shown. Determine the displacement of point C perpendicular to the plane of ABC. $G = E/2(1 + v)$.

Ans. $q_C = T_0(4R^2 + 4R^2v - \pi R^2v + 4LR + 2L^2)/4EI$

5.15. For the member in Problem 5.14, determine the rotation of the section at C in the direction of T_0.

5.16. Member ABC in Fig. P5.16. lies in the plane of the paper, has a uniform circular cross section, and is subjected to a uniform load w (N/mm) that acts perpendicular to the plane of ABC. Determine the deflection of point C perpendicular to ABC, if length L is large compared with the diameter of the member. $G = E/2(1 + v)$

Ans. $q_C = wL^4(13 + 6v)/12EI$

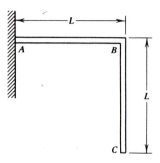

Figure P5.16

5.17. Member ABC in Fig. P5.17 lies in the (x, y) plane, has a uniform circular cross section, and is subjected to loads P perpendicular to the (x, y) plane. Determine the deflection of point C in the z direction, if R and L are large compared with the diameter of the member.

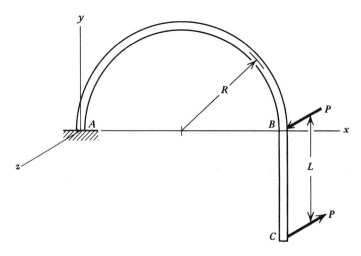

Figure P5.17

5.18. The semicircular member in Fig. P5.18 lies in the (x, y) plane and has a circular cross section with radius r. The member is fixed at A and is subjected to torque T_0 at the free end at B. Determine the angle of twist of the cross section at B. $G = E/2(1 + v)$.

Ans. $\theta_B = 2T_0R(2 + v)/Er^4$

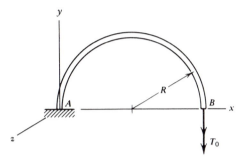

Figure P5.18

5.19. For the semicircular member in Problem 5.18, determine the x, y, and z components of the deflection of point B.

5.20. A bar having a circular cross section is fixed at the origin 0 as shown in Fig. P5.20 and has right angle bends at points A and B. Length $0A$ lies along the z axis; length AB is parallel to the x axis; length BC is parallel to the y axis. Determine the x, y, and z components of the deflection of point C. Moment M_C is a couple lying in a plane parallel to the (x, y) plane. $G = E/2(1 + v)$.

Ans. $u = 0$, $v = M_C L^2/2EI$, $w = -M_C L^2(5 + 2v)/2EI$

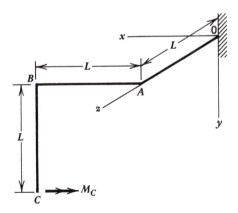

Figure P5.20

5.21. A stepped tension member has two sections of length 1.00 m, each section being circular in cross section with diameters of 120 mm and 80.0 mm, respectively; see Fig. P5.21. The member is made of an aluminum alloy that

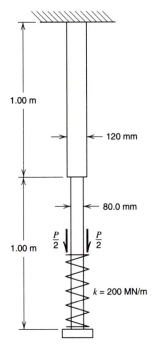

Figure P5.21

has a yield stress $Y = 330$ MPa and a modulus of elasticity $E = 72.0$ GPa. A spring slides freely over the bottom half of the member and bears on an end plate at the bottom end. The spring has a constant $k = 200$ MN/m. The member was designed using a safety factor of 1.80 for general yielding. Determine the deflection of the free end of the spring caused by the maximum allowable load P_{max}.

5.22. The beam ABC in Fig. P5.22 is made of steel ($E = 200$ GPa) and has a rectangular cross section, 70.0 mm by 50.0 mm. Member BD is made of an aluminum alloy ($E = 72.0$ GPa) and has a circular cross section of diameter 10.0 mm. Determine the vertical deflection under the load $Q = 8.50$ kN.

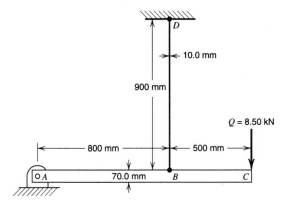

Figure P5.22

5.23. The beam ABC in Fig. P5.23 is made of steel ($E = 200$ GPa). It has a hollow circular section, with outer diameter 180 mm and inner diameter 150 mm. The spring has a constant $k = 2.00$ MN/m. A moment $M_0 = 40.0$ kN·m is applied at C. Determine the rotation of the cross section at C.

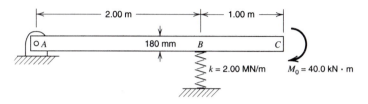

Figure P5.23

5.24. The beam in Fig. P5.24 is made of brass ($E = 83.0$ GPa) and has a square cross section with a dimension of 10.0 mm. The identical coil springs have constant $k = 30.0$ kN/m. A load $Q = 250$ N is applied at midspan ($a = 100$ mm). Determine the deflection at midspan.

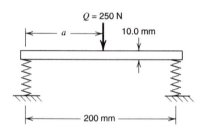

Figure P5.24

5.25. Let the location of the load $a = 150$ mm in Problem 5.24. Determine the deflection under Q.

5.26. The beam in Fig. P5.26 is made of an aluminum alloy ($E = 72.0$ GPa) and has a rectangular cross section with external dimensions 80.0 m by 100 mm and a wall thickness of 10.0 mm. The identical springs have constant $k = 300$ kN/m. A couple M_0 is applied at distance a from the left end. Let $M_0 = 15.0$ kN·m and $a = 1.50$ m. Determine the rotation of the section where M_0 is applied.

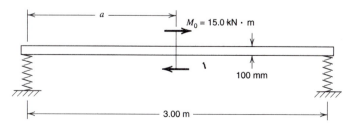

Figure P5.26

5.27. In Problem 5.26, let $a = 2.5$ m. Determine the rotation at the section where M_0 is applied.

5.28. A structure is fabricated by welding together three lengths of I-shape members ($Y = 250$ MPa, $E = 200$ GPa, and $G = 77.5$ GPa), as shown in Fig. P5.28. The members have cross-section properties $I_x = 695 \times 10^6$ mm^4, $I_y = 20.9 \times 10^6$ mm^4, $S_x = 2705 \times 10^3$ mm^3, $S_y = 228 \times 10^3$ mm^3, depth $= 515.6$ mm, and area $= 18190$ mm^2. The structure was designed with a safety factor of 2.00 for general yielding.

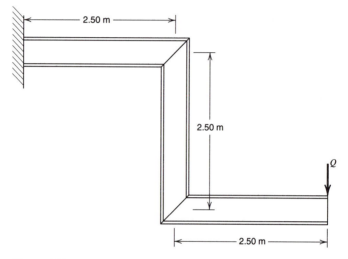

Figure P5.28

(a) Determine the maximum allowable load Q.

(b) For this load, what is the deflection at the point where Q is applied?

(c) Determine the error in neglecting the strain energy due to axial load, due to shear.

5.29. A structure (Fig. P5.29) is made by welding a circular cross-section steel shaft ($E = 200$ GPa, $G = 77.5$ GPa), of length 1.2 m and diameter 60.0 mm, to a rectangular cross-section steel beam of length 1.5 m and cross-section dimensions 70.0 mm by 30.0 mm. A torque $T_0 = 2.50$ kN·m is applied to the free end of the shaft as shown. Determine the rotation of the free end of the shaft.

5.30. A circular cross-section shaft AB, with diameter 80.0 mm and length 1.0 m, is made of an aluminum alloy ($G = 27.0$ GPa); see Fig. P5.30. It is attached at point A to a torsional spring ($\beta = 200$ kN·m per rad). A torque $T_0 = 4.00$ kN·m is applied at the free end B. Determine the rotation of the shaft at B.

5.31. A rectangular box-section beam is welded to a 180-mm diameter shaft (Fig. P5.31). The box-section has external dimensions 100 mm by 180 mm and a wall thickness of 20.0 mm. Both members are made of an aluminum alloy ($E = 72.0$ GPa and $G = 27.0$ GPa). For a load $Q = 16.0$ kN, determine the vertical deflections at the free ends of the beam and shaft.

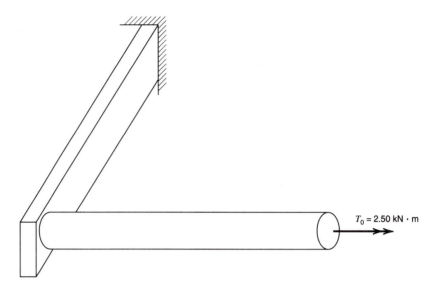

$T_0 = 2.50$ kN $\cdot$ m

Figure P5.29

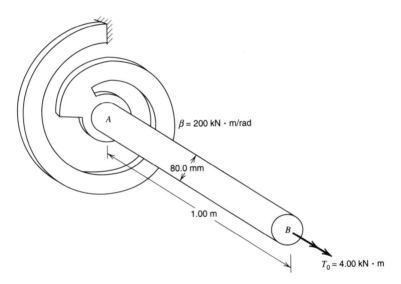

$\beta = 200$ kN $\cdot$ m/rad

A

80.0 mm

1.00 m

B

$T_0 = 4.00$ kN $\cdot$ m

Figure P5.30

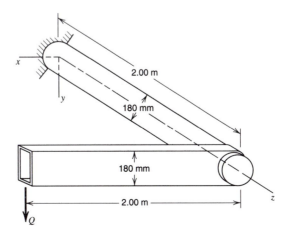

x

2.00 m

y

180 mm

180 mm

2.00 m

z

Q

Figure P5.31

5.32. For the beam shown in Fig. P5.32, determine the vertical deflection at midspan of the beam in terms of M_0, L, E, and I.

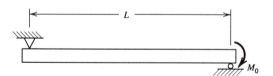

Figure P5.32

5.33. For the structure shown in Fig. P5.33, determine the horizontal and vertical displacement components of point C in terms of Q, w, L_1, L_2, E, and I.

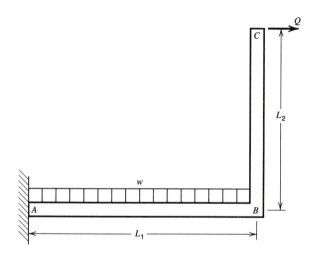

Figure P5.33

5.34. The circular curved beam AB in Fig. P5.34 has a radius of curvature R and circular cross section of diameter d. Determine the horizontal and vertical

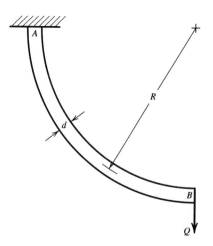

Figure P5.34

displacement components of point B in terms of E, R, d, and load Q. Neglect the strain energy due to axial load and shear.

5.35. Determine the rotation of the section at B in Problem 5.34.

Section 5.5

5.36. Let tension member EF be added to the structure in Fig. E5.10 as indicated in Fig. P5.36. Member EF is made of the same material and has the same cross-sectional area as member AB. The loads, material, and cross-sectional dimensions are indicated in Example 5.10. Determine the axial force in member EF and deflection of point E in the direction of force P.

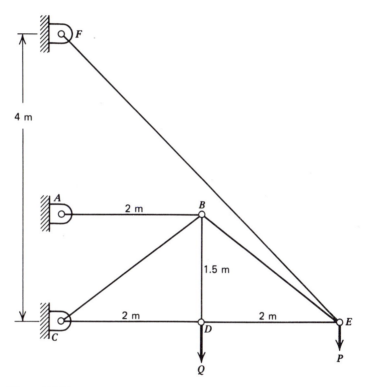

Figure P5.36

5.37. The beam in Fig. P5.37 is fixed at the right end and simply supported at the left end. Determine the reaction R at the left end, assuming that length L of the beam is large compared with its depth.

Ans. $R = 3wL/8$

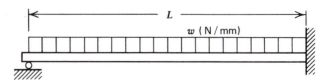

Figure P5.37

5.38. The beam in Problem 5.37 has a circular cross section with a diameter of 40 mm, has a length of 2.00 m, and is made of a steel ($E = 200$ GPa) having a working stress limit of 140 MPa.

(a) Determine the magnitude of w that will produce this limiting stress.

(b) How much would the stress in the beam be increased for the same value of w if the left end of the beam deflects 5.00 mm before making contact with the support?

5.39. The beam in Fig. P5.39 is subjected to two loads P and is supported at three locations A, B, and C as shown. Determine the rection at B, assuming that the beam length is large compared to its depth.

Ans. Reaction at $B = 11P/8$

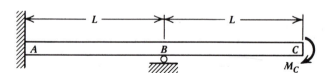

Figure P5.39

5.40. The beam in Fig. P5.40 is fixed at the left end and is supported on a roller at its center B. Assuming that the beam length is large compared to its depth, determine the reaction at B and slope of the beam over the support at B.

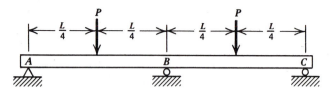

Figure P5.40

5.41. Member ABC in Fig. P5.41 has a constant cross section. Assuming that length R is large compared to the depth of the member, determine the horizontal H and vertical V components of the pin reaction at C.

Ans. $V = 0.673\,wR$, $H = 0.608\,wR$

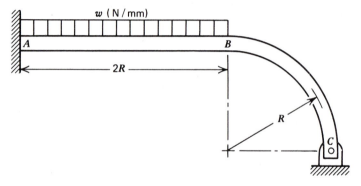

Figure P5.41

5.42. The beam in Fig. P5.42 is fixed at the right end and rests on a coil spring with spring constant K at the left end. Assuming that the beam length is large compared to its depth, determine the force R in the spring.

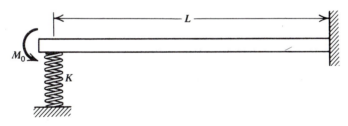

Figure P5.42

5.43. The structure in Fig. P5.43 is constructed of two steel columns AB and CD with moment of inertia I_1 and steel beam BC with moment of inertia I_2. Assume that lengths H and L are large compared with the depths of the members, determine the horizontal component of the pin reaction at D.

Ans. Horizontal reaction at $D = wL^3I_1/(8H^2I_2 + 12HLI_1)$

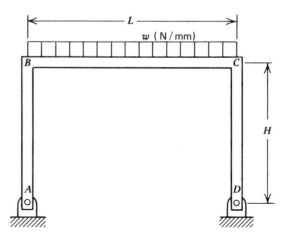

Figure P5.43

5.44. Assuming that dimensions R and L are large compared with the depth of the member, determine the maximum moment for the chain link shown in Fig. P5.44.

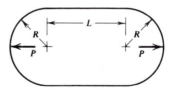

Figure P5.44

5.45. Member *ABCD* in Fig. P5.45 lies in the plane of the paper. If length *L* is large compared with the depth of the member, determine the pin reaction *V* at *D* and the horizontal displacement q_H of the pin at *D*.

Ans. $V = 81P/128, \quad q_H = 9PL^3/64EI$

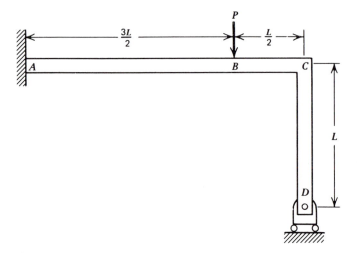

Figure P5.45

5.46. Let the pin at *D* for member *ABCD* in Problem 5.45 be prevented from displacing horizontally as load *P* is applied. Determine pin reactions *V* and *H* at *D*.

5.47. The structure in Fig. P5.47 is made up of a steel ($E = 200$ GPa) rectangular beam *ABC* with depth $h = 40.0$ mm and width $b = 30.0$ mm and two wood ($E = 10.0$ GPa) pin-connected members *BD* and *CD* with 25.0-mm square cross sections. If load $P = 9.00$ kN is applied to the beam at *C*, determine the reaction *V* at support *D* and the maximum stresses in the steel beam and wood members.

Ans. $V = 13.09$ kN, $\sigma_{\text{beam}} = 92.1$ MPa, $\sigma_{\text{compression member}} = 13.1$ MPa

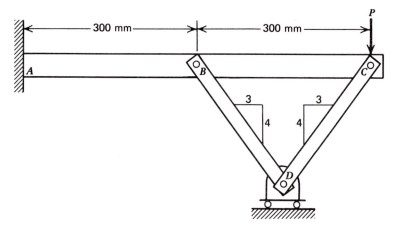

Figure P5.47

5.48. Member ABC in Fig. P5.48 has a uniform circular cross section with radius r that is small compared with R. Determine the pin reaction V at C and horizontal component of the displacement of point B.

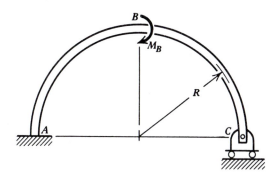

Figure P5.48

5.49. Member ABC in Fig. P5.49 has a right angle bend at B, lies in the (x, z) plane, and has a circular cross section with diameter d that is small compared with either length L_1 or length L_2. The reaction at C prevents deflection in the y direction only. Determine the reaction V at C when the moment M_0 is applied at C.

Ans. $V = M_0[3L_2^2 + 6L_1L_2(1 + v)]/[2L_1^3 + 2L_2^3 + 6L_1L_2^2(1 + v)]$

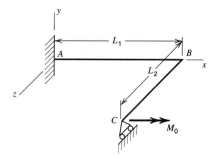

Figure P5.49

5.50. Member AB in Fig. P5.50 is a quadrant of a circle lying in the (x, z) plane, has a circular cross section of radius r, which is small compared with R, and is supported by a spring (spring constant K) at B, whose action line is parallel to the y axis. Determine the force in the spring when torque T_0 is applied at B with action line parallel to the negative z axis.

5.51. The structure in Fig. P5.51 has a uniform circular cross section with diameter d, which is small compared with either H or L. The structure is fixed at 0 and C and lies in the (x, z) plane. The load P is parallel to the y axis. Determine the magnitudes of the moment and torque at 0 and C.

Ans. $M_0 = M_C = PH/2$, $T_0 = T_C = PL^2/\{8[L + 2H(1 + v)]\}$

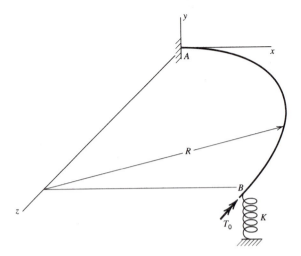

Figure P5.50

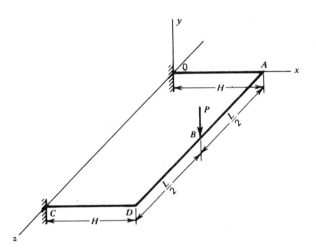

Figure P5.51

5.52. Each of the three members of the structure in Fig. P5.52 is made of a ductile steel ($E = 200$ GPa, $v = 0.29$) with yield stress $Y = 420$ MPa. Member $0A$ has a diameter of 100 mm, is fixed at 0, and is welded to beam AB, which has a rectangular cross section with a depth of 75.0 mm and width of 50.0 mm. Tension member BC has a circular cross section with a diameter of 7.50 mm. All the members are unstressed when $P = 0$. Determine the value of P based on a factor of safety of $SF = 2.00$ against initiation of yielding. Neglect stress concentrations.

5.53. Member BCD and tension member BD in Fig. P5.53 are made of materials having the same modulus of elasticity. Member BCD has a constant moment of inertia I, has a cross-sectional area that is large compared to area A of tension member BD, and has depth that is small compared to L. Determine the axial force N in member BD.

Ans. $N = 3PAL^2/(96I + 16\sqrt{2}\,AL^2)$

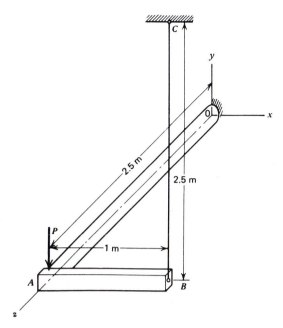

Figure P5.52

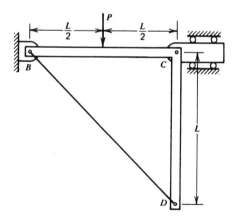

Figure P5.53

5.54. Member *BCDF* in Fig. P5.54 has the same moment of inertia *I* at every section. Determine the internal reactions N_D, V_D, and M_D at section *D*. Length *L* is large compared with the depth of the member.

5.55. Member *BCD* in Fig. P5.55 has the same moment of inertia *I* at every section. Determine the internal actions N_B, V_B, and M_B at section *B*. Radius *R* is large compared with the member's depth.

Ans. $N_B = 0$, $V_B = 8wR/(3\pi^2 + 6\pi - 24)$
$M_B = -2\pi R^2 w/(3\pi^2 + 6\pi - 24)$

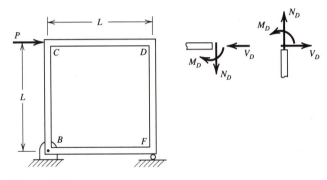

Figure P5.54

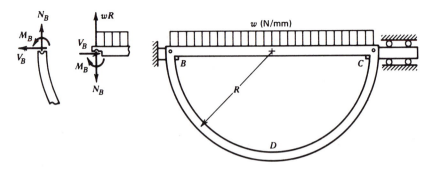

Figure P5.55

5.56. The beam in Fig. P5.56 is supported by three identical springs with spring constant k. It is subjected to a uniformly distributed load w. Determine the force in each spring in terms of w, k, L, E, and I.

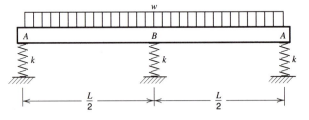

Figure P5.56

5.57. Show that the reaction at the roller support for the beam in Fig. P5.57 is equal to $5Q/2$.

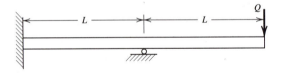

Figure P5.57

5.58. Determine the force in the spring (Fig. P5.58) in terms of w, L, k, E, and I.

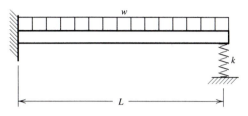

Figure P5.58

5.59. Show that the vertical reaction at the roller support for the curved member in Fig. P5.59 is equal to $2Q/(3\pi - 8)$.

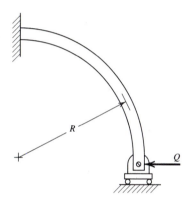

Figure P5.59

5.60. Determine the reaction at B for the beam in Fig. P5.60 in terms of w, L_1, L_2, E, and I.

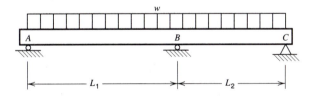

Figure P5.60

5.61. Determine the reaction at the roller support for the structure in Fig. P5.61, in terms of Q, R, E, and I.

5.62. In Problem 5.57, determine the vertical deflection at the point where Q is applied.

5.63. In Problem 5.59, determine the horizontal deflection at the point where load Q is applied.

5.64. In Problem 5.61, determine the vertical deflection at the point where load Q is applied.

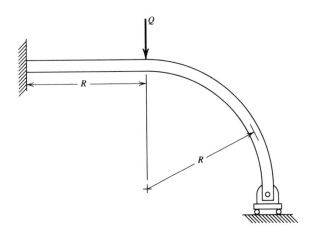

Figure P5.61

5.65. For the structure in Fig. P5.65, determine the force in the spring in terms of M_0, k, L, E, and I.

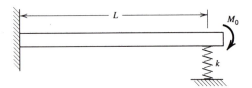

Figure P5.65

5.66. An I-beam is made of steel ($E = 200$ GPa) and is 5.0 m long (Fig. P5.66). It has cross section properties $I_x = 24.0 \times 10^6$ mm^4, $I_y = 1.55 \times 10^6$ mm^4, depth $= 203.2$ mm, flange width $= 101.6$ mm, and area $= 3490$ mm^2. The helical support spring has a constant $k = 1.00$ MN/m. For the case where $Q = 30.0$ kN, determine the force in the spring and the maximum bending stress in the beam.

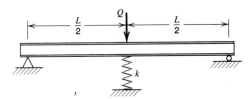

Figure P5.66

5.67. For the structure in Fig. P5.67, derive a formula for the force in the bar in terms of w, E_1, I, and L_1 for the beam and E_2, A_2, and L_2 for the bar.

5.68. The beam in Problem 5.67 is made of steel ($E = 200$ GPa). It has a rectangular cross section with dimensions 90.0 mm deep, 30.0 mm wide, and a length $L_1 = 2.00$ m. The rod is made of an aluminum alloy ($E = 72.0$ GPa). It has a diameter of 5.00 mm and a length $L_2 = 4.00$ m. Determine the tension in the bar in terms of load w.

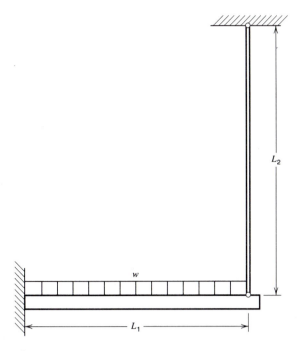

Figure P5.67

5.69. A shaft AB is attached to member $CDFH$ at A and fixed to a wall at B (Fig. P5.69). The shaft has a diameter of 60.0 mm and the parts CD and FH of member $CDFH$ have square cross sections 40 mm by 40 mm. The massive hub DF may be considered as rigid. A torque of magnitude 3.00 kN·m is applied to the midsection of the shaft as shown. All members are made of steel ($E = 200$ GPa and $G = 77.5$ GPa).

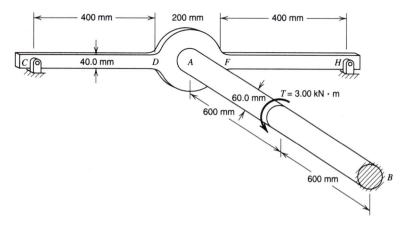

Figure P5.69

(a) Determine the maximum bending stress in members CD and FH.

(b) Determine the maximum shear stress in the shaft.

5.70. In Fig. P5.70, the shaft is attached to a torsional spring at one end and fixed to a rigid wall at the other end. The shaft has an 80.0-mm diameter and shear modulus $G = 77.5$ GPa. The torsional spring constant is $\beta = 200$ kN·m/rad. A torque of magnitude 5.00 kN·m is applied to the midsection of the shaft as shown. Determine the maximum shear stress in the shaft.

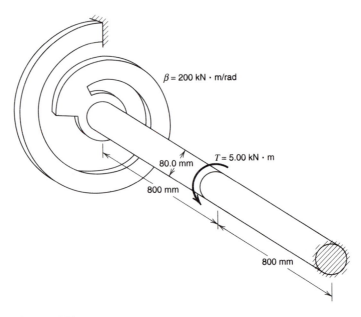

Figure P5.70

5.71. A rectangular box-beam, 100 mm by 200 mm and with a wall thickness of 10.0 mm, is welded to a shaft of diameter 180 mm, (Fig. P5.71). Determine the vertical reaction at the roller support. ($E = 200$ GPa and $G = 77.5$ GPa for all members.)

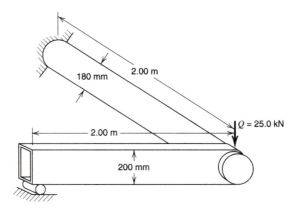

Figure P5.71

5.72. A steel torsion member has a length of 3.00 m and diameter of 120 mm (E = 200 GPa and G = 77.5 GPa). It is fixed to a rigid wall at one end. A steel beam of rectangular cross section 120 mm by 30.0 mm is welded perpendicularly to the torsion member at its midsection (Fig. P5.72). The beam is supported by a roller located 2.00 mm from the welded section. The free end of the member is subjected to a torque T = 16.0 kN·m. Determine the reaction at the roller.

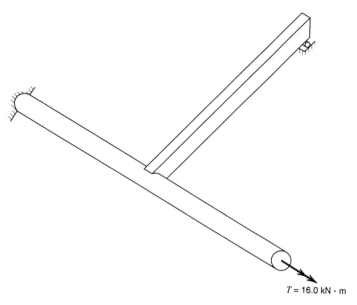

T = 16.0 kN · m

Figure P5.72

5.73. A beam ABC is fixed at its ends A and C (Fig. P5.73). Show that the shear V and moment M at C are given by the relations $V = Q(L^3 - 3a^2L + 2a^3)/L^3$ and $M = Q(aL^2 - 2a^2L + a^3)/L^2$.

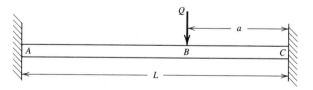

Figure P5.73

5.74. The beam in Problem 5.73 has a circular cross section of diameter d = 150 mm and length L = 2.00 m. For $a = L/3$, determine the magnitude of load Q that produces a maximum bending stress of 100 MPa.

5.75. The L-shaped beam $ABCD$ in Fig. P5.75 has a constant cross section along its length. Show that the horizontal and vertical reactions H and V, respectively, of the pin at D are given by the expressions $H = 9Q/22$ and $V = 4Q/11$.

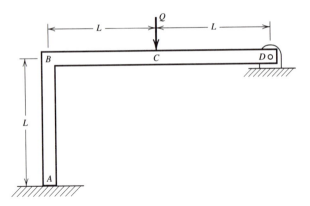

Figure P5.75

5.76. The L-shaped beam in Fig. P5.75 has a square cross section, 80.0 mm by 80.0 mm, and a length $3L = 2400$ mm.

   (a) Determine the magnitude of load Q that produces a maximum bending stress of 120 MPa in the beam.

   (b) The beam is made of steel ($E = 200$ GPa). Determine the vertical deflection of the beam at point C.

5.77. Consider the indeterminate beam of Example 5.12 (Fig. E5.12). Let the beam material be elastic-perfectly plastic (Fig. 4.4a). Let the cross section be rectangular with width b and depth h.

   (a) Determine the magnitude of the force $P = P_Y$ that causes yielding to initiate and locate the section at which it occurs.

   (b) Determine the magnitude of the force $P = P_P$ that causes a plastic hinge to form at the wall support.

   (c) Determine the magnitude of the force $P = P_{PC}$ that causes the beam to form a plastic hinge to occur at the section under load P. The load P_{PC} is called the *plastic collapse load* for the member.

   (d) Construct a moment diagram for the beam for load P_{PC}.

   (e) Draw a sketch of the deformed shape of the beam for $P = P_{PC}$.

Note: The elastic segments of the beam rotate about a plastic hinge as rigid bodies. For this reason, the response of the beam at $P = P_{PC}$ is like a mechanism that rotates kinematically about hinges. Therefore, the term *mechanism* or *kinematic mechanism* is used in plastic collapse analysis (limit analysis) to describe this process.

REFERENCES

Boresi, A. P. and Chong, K. P. (1987). *Elasticity in Engineering Mechanics*. New York: Elsevier.

Charlton, T. M. (1959). *Energy Principles in Applied Statics.* London: Blackie and Son, Ltd.

Langhaar, H. L. (1989). *Energy Methods in Applied Mechanics.* Malabar, Florida: Krieger.

Popov, E. P. (1990). *Engineering Mechanics of Solids.* Englewood Cliffs, N. J.: Prentice Hall.

PART II

CLASSICAL TOPICS IN ADVANCED MECHANICS

In Part II of this book, Chapters 6 to 12, we discuss topics that are considered classical subjects in advanced mechanics. In Chapter 6 the torsion of bars is treated. In Chapter 7 we present nonsymmetrical bending of straight beams. The concept of shear center is developed in Chapter 8. In Chapter 9 curved beam theory is discussed. Beams on elastic foundations are examined in Chapter 10, and thick-wall cylinders are studied in Chapter 11. Column buckling is explored in Chapter 12.

6

TORSION

In this chapter, we treat the problem of torsion of prismatic bars with noncircular cross sections. First, we treat the case in which each torsion member is made of a linearly elastic isotropic material. The last section of the chapter treats fully plastic torsion. For prismatic bars with circular cross sections, the torsion formulas are readily derived by the method of mechanics of materials. However, for noncircular cross sections, more general methods are required. In the following sections we treat noncircular cross sections by several methods, one of which is the semiinverse method of Saint-Venant (Boresi and Chong, 1987). General relations are derived that are applicable for both the linear elastic torsion problem and the fully plastic torsion problem. In order to aid in the solution of the resulting differential equation for some linear elastic torsion problems, the Saint-Venant solution is used in conjunction with the Prandtl elastic-membrane (soap-film) analogy.

In spirit, the semiinverse method of Saint-Venant is, in part, comparable to the mechanics of materials method in that certain assumptions, based on an understanding of the mechanics of the problem, are introduced initially. However, these assumptions are not so specific as to attempt to meet all the requirements of the problem. Rather, sufficient freedom is allowed so that the equations describing the torsion boundary value problem of solids may be employed to determine the solution more completely. For the case of circular cross sections, the method of Saint-Venant leads to an exact solution (subject to appropriate boundary conditions) for the torsion problem. Because of its importance in engineering, the torsion problem of circular cross sections is discussed first.

6.1

TORSION OF A PRISMATIC BAR OF CIRCULAR CROSS SECTION

Consider a solid cylinder with cross sectional area A and length L. Let the cylinder be subjected to a twisting couple $\mathbf{T}$ applied at the right end (Fig. 6.1). An equilibrating torque acts on the left end. The vectors that represent the torque are directed along the z axis, the centroidal axis of the shaft (Fig. 6.1). Under the action of the torque, an originally straight generator of the cylinder AB will deform into a helical curve AB^*. However, because of the radial symmetry of the circular cross section and because a deformed cross section must appear to be the same from both ends of the torsion member, plane cross sections of the torsion member normal to the z axis remain plane after deformation and all radii remain straight. Further-

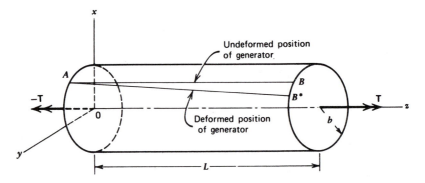

Figure 6.1 Circular cross-section torsion member.

more, for small displacements, each radius remains inextensible. In other words, the torque **T** causes each cross section to rotate as a rigid body about the z axis (axis of the couple); this axis is called the *axis of twist*. If we measure the rotation β of each section relative to the plane $z = 0$, the rotation β of a given section will depend on its distance from the plane $z = 0$. For small deformations, following Saint-Venant, we assume that the amount of rotation of a given section depends linearly on its distance z from the plane $z = 0$. Thus,

$$\beta = \theta z \tag{6.1}$$

where θ is the angle of twist per unit length of the shaft. Under the conditions that plane sections remain plane and Eq. (6.1) holds, we now seek to satisfy the equations of elasticity; that is, we employ the semiinverse method of seeking the elasticity solution.

Since cross sections remain plane and rotate as rigid bodies about the z axis, the displacement component w, parallel to the z axis, is zero. To calculate the (x, y) components of displacements u and v, consider a cross section at distance z from the plane $z = 0$. Consider a point in the circular cross section (Fig. 6.2) with radial distance $0P$. Under the deformation, radius $0P$ rotates into the radius $0P^*$ $(0P^* = 0P)$. In terms of the angular displacement β of the radius, the displacement components (u, v) are

$$u = x^* - x = 0P[\cos(\beta + \phi) - \cos \phi]$$
$$v = y^* - y = 0P[\sin(\beta + \phi) - \sin \phi] \tag{6.2}$$

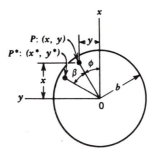

Figure 6.2

Expanding $\cos(\beta + \phi)$ and $\sin(\beta + \phi)$ and noting that $x = 0P \cos\phi$, $y = 0P \sin\phi$, we may write Eqs. (6.2) in the form

$$u = x(\cos\beta - 1) - y\sin\beta$$
$$v = x\sin\beta + y(\cos\beta - 1) \tag{6.3}$$

Restricting the displacement to be small, we obtain (since then $\sin\beta \approx \beta$, $\cos\beta \approx 1$), with the assumption that $w = 0$,

$$u = -y\beta, \qquad v = x\beta, \qquad w = 0 \tag{6.4}$$

to first degree terms in β. Substitution of Eq. (6.1) into Eq. (6.4) yields

$$u = -\theta yz, \qquad v = \theta xz, \qquad w = 0 \tag{6.5}$$

On the basis of the foregoing assumptions, Eqs. (6.5) represent the displacement components of a point in a circular shaft subjected to a torque **T**.

Substitution of Eqs. (6.5) into Eqs. (2.81) yields the strain components (if we ignore temperature effects)

$$\epsilon_{xx} = \epsilon_{yy} = \epsilon_{zz} = \epsilon_{xy} = 0, \qquad 2\epsilon_{zx} = \gamma_{zx} = -\theta y, \qquad 2\epsilon_{zy} = \gamma_{zy} = \theta x \tag{6.6}$$

Since the strain components are derived from admissible displacement components, compatibility is automatically satisfied (See Sec. 2.8. See also Boresi and Chong, 1987; Sec. 2.16). With Eqs. (6.6), Eqs. (3.32) yield the stress components for linear elasticity

$$\sigma_{xx} = \sigma_{yy} = \sigma_{zz} = \sigma_{xy} = 0, \qquad \sigma_{zx} = -\theta Gy, \qquad \sigma_{zy} = \theta Gx \tag{6.7}$$

Equations (6.7) satisfy the equations of equilibrium, provided the body forces are zero [Eqs. (2.45)].

To satisfy the boundary conditions, Eqs. (6.7) must yield no forces on the lateral surface of the bar; on the ends, they must yield stresses such that the net moment is equal to **T** and the resultant force vanishes. Since the direction cosines of the unit normal to the lateral surface are $(l, m, 0)$ (see Fig. 6.3), the first two of Eqs. (2.10) are satisfied identically. The last of Eqs. (2.10) yields

$$l\sigma_{zx} + m\sigma_{zy} = 0 \tag{6.8}$$

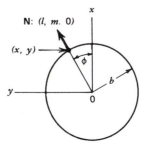

Figure 6.3

By Fig. 6.3,

$$l = \cos\phi = \frac{x}{b}, \qquad m = \sin\phi = \frac{y}{b} \tag{6.9}$$

Substitution of Eqs. (6.7) and (6.9) into Eqs. (6.8) yields

$$-\frac{xy}{b} + \frac{xy}{b} = 0$$

Therefore, the boundary conditions on the lateral surface are satisfied.

On the ends, the stresses must be distributed so that the net moment is **T**. Therefore, summation of moments on each end with respect to the z axis yields (Fig. 6.4)

$$\sum M_z = T = \int_A (x\sigma_{zy} - y\sigma_{zx})\,dA \tag{6.10}$$

Substitution of Eqs. (6.7) into Eq. (6.10) yields

$$T = G\theta \int_A (x^2 + y^2)\,dA = G\theta \int_A r^2\,dA \tag{6.11}$$

Since the last integral is the polar moment of inertia ($J = \pi b^4/2$) of the circular cross section, Eq. (6.11) yields

$$\theta = \frac{T}{GJ} \tag{6.12}$$

which relates the angular twist θ per unit length of the shaft to the magnitude T of the applied torque.

Since compatibility and equilibrium are satisfied, Eqs. (6.7) represent the solution of the elasticity problem. However, in applying torsional loads to most torsion members of circular cross section, the distributions of σ_{zx} and σ_{zy} on the ends probably do not satisfy Eqs. (6.7). In these cases, it is assumed that σ_{zx} and σ_{zy} undergo a redistribution with distance from the ends of the bar until, at a distance of a few bar diameters from the ends, the distributions are essentially given by Eqs. (6.7). This concept of redistribution of the applied end stresses with distance from the ends is known as the Saint-Venant principle (Boresi and Chong, 1987).

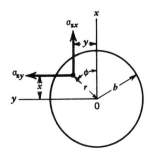

Figure 6.4

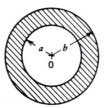

Figure 6.5

Since the solution of Eqs. (6.7) indicates that σ_{zx} and σ_{zy} are independent of z, the stress distribution is the same for all cross sections. Thus, the stress vector τ for any point P in a cross section is given by the relation

$$\tau = -\theta G y \mathbf{i} + \theta G x \mathbf{j} \tag{6.13}$$

The stress vector τ lies in the plane of the cross section, and it is perpendicular to the radius vector $\mathbf{r}$ joining point P to the origin 0. By Eq. (6.13), the magnitude of τ is

$$\tau = \theta G \sqrt{x^2 + y^2} = \theta G r \tag{6.14}$$

Hence, τ is a maximum for $r = b$; that is, τ attains a maximum value of $\theta G b$.
    Substitution of Eq. (6.12) into Eq. (6.14) yields the result

$$\tau = \frac{Tr}{J} \tag{6.15}$$

which relates the magnitude τ of the shear stress to the magnitude T of the torque. The above result holds also for cylindrical bars with hollow circular cross sections (Fig. 6.5), with inner radius a and outer radius b; for this cross section $J = \pi(b^4 - a^4)/2$ and $a \le r \le b$.
    The analysis for the torsion of noncircular cross sections proceeds in much the same fashion as for circular cross sections. However, in the case of noncircular cross sections, Saint-Venant assumed more generally that w is a function of (x, y), the cross-section coordinates. Then, the cross section does not remain plane but *warps*; that is, different points in the cross section, in general, undergo different displacements in the z direction.

EXAMPLE 6.1
Shaft with Hollow Circular Cross Section

A steel shaft has a hollow circular cross section (see Fig. 6.5), with radii $a = 22$ mm and $b = 25$ mm. It is subjected to a twisting moment $T = 500$ m·N.

(a) Determine the maximum shear stress in the shaft.

(b) Determine the angle of twist per unit length.

SOLUTION

(a)   The polar moment of inertia of the cross section is $J = \pi(b^4 - a^4)/2 = \pi(25^4 - 22^4)/2 = 245{,}600$ mm^4 = 24.56×10^{-8} m^4. Hence, by Eq. (6.15), $\tau_{max} = Tb/J = 500 \times 0.025/24.56 \times 10^{-8} = 50.9$ MPa.

(b)   By Eq. (6.12), with $G = 77$ GPa, $\theta = T/GJ = 500/(77 \times 10^9 \times 24.56 \times 10^{-8}) = 0.0264$ rad/m.

EXAMPLE 6.2
Circular Cross-Section Drive Shaft

Two pulleys, one at B and one at C, are driven by a motor through a stepped, steel drive shaft ($G = 77$ GPa) ABC, as shown in Fig. E6.2. Each pulley absorbs a torque of 113 m·N. The stepped shaft has two lengths $AB = L_1 = 1$ m long and $BC = L_2 = 1.27$ m long. The shafts are made of steel ($Y = 414$ MPa). Let the safety factor be $SF = 2.0$ for yield by the maximum shear-stress criterion.

(a)   Determine suitable diameter dimensions d_1 and d_2 for the two shaft lengths.

(b)   With the diameters selected in part (a), calculate the total angle of twist β_c of the shaft.

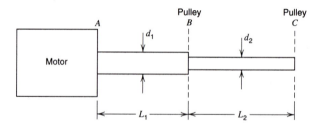

Figure E6.2   Circular cross-section shaft.

SOLUTION

Since each pulley removes 113 m·N, shaft AB must transmit a torque $T_1 = 226$ m·N, and shaft BC must transmit a torque $T_2 = 113$ m·N. Also, the maximum permissible shear stress in either shaft length is [by Eq. (4.12)] $\tau_{max} = \tau_Y/SF = 0.25Y = 103.5$ MPa.

(a)   By Eq. (6.15), we have $\tau_{max} = 2T/(\pi r_1^3)$. Consequently, we have $r_1 = [2T/(\pi\tau_{max})]^{(1/3)} = [2 \times 226/(\pi \times 103.5 \times 10^6)]^{1/3} = 0.0112$ m. Hence, the diameter $d_1 = 2r_1 = 0.0224$ m = 22.4 mm. Similarly, we find $d_2 = 2r_2 = 2 \times 0.00886$ m = 0.0177 m = 17.7 mm. Since these dimensions are not standard sizes, we choose $d_1 = 25.4$ mm and $d_2 = 19.05$ mm, since these sizes (1.0 and 0.75 in., respectively) are available in U.S. customary units.

(b)   By Eq. (6.12), the unit angle of twist in the shaft length AB is $\theta_1 = T_1/(GJ_1) = 2T_1/(G\pi r_1^4) = (2 \times 226)/(77 \times 10^9 \times \pi \times 0.0127^4) = 0.07183$ rad/m. Similarly, we obtain $\theta_2 = 0.1135$ rad/m. Therefore, the total angle of twist is $\beta_c = 1.0 \times 0.07183 + 1.27 \times 0.1135 = 0.216$ rad = 12.4°.

6.2

SAINT-VENANT'S SEMIINVERSE METHOD

Consider a torsion member with a uniform cross section of general shape as shown in Fig. 6.6. Axes (x, y, z) are taken as for the circular cross section (Fig. 6.1). The applied shear stress distribution on the ends $(\sigma_{zx}, \sigma_{zy})$ produces a torque **T**. In general, any number of stress distributions on the end sections may produce a torque **T**. According to Saint-Venant's principle, the stress distribution on sections sufficiently far removed from the ends depends principally on the magnitude of **T** and not on the stress distribution on the ends. Thus, for sufficiently long torsion members, the end stress distribution does not affect the stress distributions in a large part of the member.

Saint-Venant's semiinverse method starts by an approximation of the displacement components due to torque **T**. This approximation is based on observed geometric changes in the deformed torsion member.

Geometry of Deformation

As with circular cross sections, Saint-Venant assumed that every straight torsion member with constant cross section (relative to axis z) has an axis of twist, about which each cross section rotates approximately as a rigid body. Let the z axis in Fig. 6.6 be the axis of twist.

For the torsion member in Fig. 6.6, let $0A$ and $0B$ be line segments in the cross section for $z = 0$, which coincide with the x and y axes, respectively. After deformation, by rigid-body displacements, we may translate the new position of 0, that is, 0^* back to coincide with 0, align the axis of twist along the z axis, and rotate the deformed torsion member until the projection of 0^*A^* on the (x, y) plane coincides with the x axis. Because of the displacement (w displacement) of points in each cross section, 0^*A^* does not, in general, lie in the (x, y) plane. However, the amount of warping is small for small displacements; therefore, line $0A$ and curved line 0^*A^* are shown as coinciding in Fig. 6.6. Experimental evidence indicates also that the

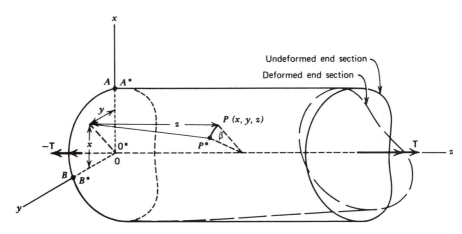

Figure 6.6 Torsion member.

distortion (warping) of each cross section is essentially the same. Furthermore, experimental evidence indicates that the cross-sectional dimensions of the torsion member are not changed significantly by the deformations, particularly for small displacements. In other words, deformation in the plane of the cross section is negligible. Hence, the projection of $0*B*$ on the (x, y) plane coincides approximately with the y axis, indicating that ϵ_{xy} $(\gamma_{xy} = 2\epsilon_{xy})$ is approximately zero [see Sec. 2.7; particularly, Eq. (2.74).].

Consider a point P with coordinates (x, y, z) in the undeformed torsion member (Fig. 6.6). Under deformation, P goes into $P*$. The point P, in general, is displaced by an amount w parallel to the z axis because of the warping of the cross section and by amounts u and v parallel to the x and y axes, respectively. The cross section in which P lies rotates through an angle β with respect to the cross at the origin. This rotation is the principal cause of the (u, v) displacements of point P. These observations led Saint-Venant to assume that $\beta = \theta z$, where θ is the angle of twist per unit length and therefore that the displacement components take the form

$$u = -\theta yz, \qquad v = \theta xz, \qquad w = \theta \psi(x, y) \qquad (6.16)$$

where ψ is the warping function [compare Eqs. (6.16) for the general cross section with Eqs. (6.5) for the circular cross section]. The function $\psi(x, y)$ may be determined such that the equations of elasticity are satisfied. Since we have assumed continuous displacement components (u, v, w), the small-displacement compatibility conditions [Eqs. (2.83)] are automatically satisfied.

The state of strain at a point in the torsion member is given by substitution of Eqs. (6.16) into Eqs. (2.81) to obtain

$$\epsilon_{xx} = \epsilon_{yy} = \epsilon_{zz} = \epsilon_{xy} = 0$$

$$2\epsilon_{zx} = \gamma_{zx} = \theta\left(\frac{\partial \psi}{\partial x} - y\right)$$

$$2\epsilon_{zy} = \gamma_{zy} = \theta\left(\frac{\partial \psi}{\partial y} + x\right) \qquad (6.17)$$

If the equation for γ_{zx} is differentiated with respect to y, the equation for γ_{zy} is differentiated with respect to x, and the second of these resulting equations is subtracted from the first, the warping function ψ may be eliminated to give the relation

$$\frac{\partial \gamma_{zx}}{\partial y} - \frac{\partial \gamma_{zy}}{\partial x} = -2\theta \qquad (6.18)$$

If the torsion problem is formulated in terms of $(\gamma_{zx}, \gamma_{zy})$, Eq. (6.18) is a geometrical condition (compatibility condition) to be satisfied for the torsion problem.

Stresses at a Point and Equations of Equilibrium

For torsion members made of isotropic materials, stress-strain relations for either elastic [the first of Eqs. (6.17) and Eqs. (3.32)] or inelastic conditions indicate that

$$\sigma_{xx} = \sigma_{yy} = \sigma_{zz} = \sigma_{xy} = 0 \qquad (6.19)$$

The stress components $(\sigma_{zx}, \sigma_{zy})$ are nonzero. If body forces and acceleration terms are neglected, these stress components may be substituted into Eqs. (2.45) to obtain equations of equilibrium for the torsion member.*

$$\frac{\partial \sigma_{zx}}{\partial z} = 0 \tag{6.20}$$

$$\frac{\partial \sigma_{zy}}{\partial z} = 0 \tag{6.21}$$

$$\frac{\partial \sigma_{yz}}{\partial y} + \frac{\partial \sigma_{xz}}{\partial x} = 0 \tag{6.22}$$

Equations (6.20) and (6.21) indicate that $\sigma_{zx} = \sigma_{xz}$ and $\sigma_{zy} = \sigma_{yz}$ are independent of z. These stress components must satisfy Eq. (6.22), which expresses a necessary and sufficient condition for the existence of a stress function $\phi(x, y)$ (the so-called Prandtl stress function) such that

$$\sigma_{zx} = \frac{\partial \phi}{\partial y}$$

$$\sigma_{zy} = -\frac{\partial \phi}{\partial x} \tag{6.23}$$

Thus, the torsion problem is transformed into the determination of the stress function ϕ. Boundary conditions put restrictions on ϕ.

Boundary Conditions

Since the lateral surface of a torsion member is free of applied stress, the resultant shear stress τ in the cross section of the torsion member, on the surface S of the cross section, must be directed tangent to the surface (Figs. 6.7a and b). The two shear stress components σ_{zx} and σ_{zy} that act on the cross-sectional element with sides dx, dy, and ds may be written in terms of τ (Fig. 6.7b) in the form

$$\sigma_{zx} = \tau \sin \alpha$$
$$\sigma_{zy} = \tau \cos \alpha \tag{6.24}$$

where according to Fig. 6.7a,

$$\sin \alpha = \frac{dx}{ds}, \qquad \cos \alpha = \frac{dy}{ds} \tag{6.25}$$

Since the component of τ in the direction of the normal $\mathbf{n}$ to the surface S is zero, projections of $\boldsymbol{\sigma}_{zx}$ and $\boldsymbol{\sigma}_{zy}$ in the normal direction (Fig. 6.7b) yield, with Eq. (6.25)

$$\sigma_{zx} \cos \alpha - \sigma_{zy} \sin \alpha = 0$$

$$\sigma_{zx} \frac{dy}{ds} - \sigma_{zy} \frac{dx}{ds} = 0 \tag{6.26}$$

---

* This approach was taken by Prandtl. See Sec. 7.3, Boresi and Chong (1987).

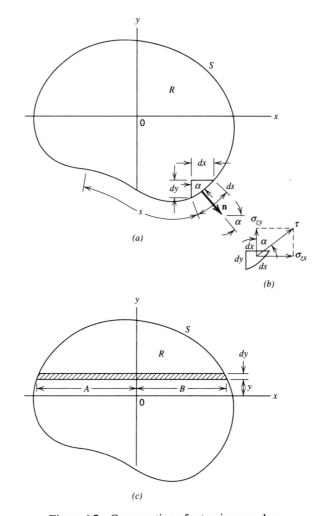

Figure 6.7 Cross section of a torsion member.

Substituting Eqs. (6.23) into Eq. (6.26), we find

$$\frac{\partial \phi}{\partial x}\frac{dx}{ds} + \frac{\partial \phi}{\partial y}\frac{dy}{ds} = \frac{d\phi}{ds} = 0$$

or

$$\phi = \text{constant on the boundary } S \qquad (6.27)$$

Since the stresses are given by partial derivatives of ϕ [see Eqs. (6.23)], it is permissible to take this constant to be zero; thus, we select

$$\phi = 0 \text{ on the boundary } S \qquad (6.28)$$

The preceding argument can be used to show that the shear stress

$$\tau = \sqrt{\sigma_{zx}^2 + \sigma_{zy}^2} \qquad (6.29)$$

at any point in the cross section is directed tangent to the contour ϕ = constant through the point.

The distributions of σ_{zx} and σ_{zy} on a given cross section must satisfy the following equations:

$$\sum F_x = 0 = \int \sigma_{zx}\, dx\, dy = \int \frac{\partial \phi}{\partial y}\, dx\, dy \tag{6.30}$$

$$\sum F_y = 0 = \int \sigma_{zy}\, dx\, dy = -\int \frac{\partial \phi}{\partial x}\, dx\, dy \tag{6.31}$$

$$\sum M_z = T = \int (x\sigma_{zy} - y\sigma_{zx})\, dx\, dy$$

$$= -\int \left(x\frac{\partial \phi}{\partial x} + y\frac{\partial \phi}{\partial y} \right) dx\, dy \tag{6.32}$$

In satisfying the second equilibrium equation, consider the strip across the cross section of thickness dy as indicated in Fig. 6.7c. Since the stress function does not vary in the y direction for this strip, the partial derivative can be replaced by the total derivative. For the strip, Eq. (6.31) becomes

$$dy \int \frac{\partial \phi}{\partial x}\, dx = dy \int \frac{d\phi}{dx}\, dx = dy \int_{\phi(A)}^{\phi(B)} d\phi$$

$$= dy[\phi(B) - \phi(A)] = 0 \tag{6.33}$$

since ϕ is equal to zero on the boundary. The same is true for every strip so that $\sum F_y = 0$ is satisfied. In a similar manner, Eq. (6.30) is verified. In Eq. (6.32), consider the term

$$-\int x\frac{\partial \phi}{\partial x}\, dx\, dy$$

which becomes for the strip in Fig. 6.7c

$$-dy \int x\frac{d\phi}{dx}\, dx = -dy \int_{\phi(A)}^{\phi(B)} x\, d\phi \tag{6.34}$$

Evaluating the latter integral by parts and noting that $\phi(B) = \phi(A) = 0$, we obtain

$$-dy \int_{\phi(A)}^{\phi(B)} x\, d\phi = -dy \left(x\phi\big|_A^B - \int_{x_A}^{x_B} \phi\, dx \right) = dy \int_{x_A}^{x_B} \phi\, dx \tag{6.35}$$

Summing for the other strips and repeating the process using strips of thickness dx for the other term in Eq. (6.32), we obtain the relation

$$T = 2 \iint \phi\, dx\, dy \tag{6.36}$$

The stress function ϕ can be considered to represent a surface over the cross section of the torsion member. This surface is in contact with the boundary of the cross section [see Eq. (6.28)]. Hence, Eq. (6.36) indicates that the torque is equal to twice the volume between the stress function and the plane of the cross section.

Note: Equations (6.18), (6.23), (6.28), and (6.36), as well as other equations in this section, have been derived for torsion members that have uniform cross sections that do not vary with z, that have simply connected cross sections, that are made of isotropic materials, and that are loaded so that deformations are small. These equations are used to obtain solutions for torsion members; they do not depend on any assumption as to material behavior except that the material is isotropic; therefore, they are valid for any specified material response (elastic or inelastic).

Two types of typical material response are considered in this chapter: linearly elastic response and elastic-perfectly plastic response (Fig. 4.4a). The linearly elastic response leads to the linearly elastic solution of torsion, whereas the elastic-perfectly plastic response leads to the fully plastic solution of torsion of a bar for which the entire cross section yields. The material properties associated with various material responses are determined by appropriate tests. Usually, as noted in Chapter 4, we assume that the material properties are determined by either a tension test or torsion test of a cylinder with thin-wall annular cross section.

6.3

LINEAR ELASTIC SOLUTION

Stress-strain relations for linear elastic behavior of an isotropic material are given by Hooke's law [see Eqs. (3.32)]. By Eqs. (3.32) and (6.23), we obtain

$$\sigma_{zx} = \frac{\partial \phi}{\partial y} = G\gamma_{zx}$$

$$\sigma_{zy} = -\frac{\partial \phi}{\partial x} = G\gamma_{zy} \tag{6.37}$$

Substitution of Eqs. (6.37) into Eq. (6.18) yields

$$\frac{\partial^2 \phi}{\partial x^2} + \frac{\partial^2 \phi}{\partial y^2} = -2G\theta \tag{6.38}$$

If the unit angle of twist θ is specified for a given torsion member and ϕ satisfies the boundary condition indicated by Eq. (6.28), then Eq. (6.38) uniquely determines the stress function $\phi(x, y)$. Once ϕ has been determined, the stresses are given by Eqs. (6.23) and the torque is given by Eq. (6.36). The elasticity solution of the torsion problem for many practical cross sections requires special methods (Boresi and Chong, 1987) for determining the function ϕ and is beyond the scope of this book. As indicated in the following paragraphs, an indirect method may be used to obtain solutions for certain types of cross sections, although it is not a general method.

Let the boundary of the cross section for a given torsion member be specified by the relation

$$F(x, y) = 0 \qquad (6.39)$$

Furthermore, let the torsion member be subjected to a specified unit angle of twist and define the stress function by the relation

$$\phi = BF(x, y) \qquad (6.40)$$

where B is a constant. This stress function is a solution of the torsion problem, provided $F(x, y) = 0$ on the lateral surface of the bar and $\partial^2 F/\partial x^2 + \partial^2 F/\partial y^2 =$ constant. Then, the constant B may be determined by substituting Eq. (6.40) into Eq. (6.38). With B determined, the stress function ϕ for the torsion member is uniquely defined by Eq. (6.40). This indirect approach may, for example, be used to obtain the solutions for torsion members whose cross sections are in the form of a circle, an ellipse, or an equilateral triangle.

Elliptical Cross Section

Let the cross section of a torsion member be bounded by an ellipse (Fig. 6.8). The stress function ϕ for the elliptical cross section may be written in the form

$$\phi = B\left(\frac{x^2}{h^2} + \frac{y^2}{b^2} - 1\right) \qquad (6.41)$$

since $F(x, y) = x^2/h^2 + y^2/b^2 - 1 = 0$ on the boundary [Eq. (6.39)]. Substituting Eq. (6.41) into Eq. (6.38), we obtain

$$B = -\frac{h^2 b^2 G\theta}{h^2 + b^2} \qquad (6.42)$$

in terms of the geometrical parameters (h, b), shear modulus G, and unit angle of twist θ. With ϕ determined, the shear stress components for the elliptical cross

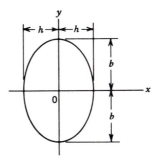

Figure 6.8 Ellipse.

section are, by Eqs. (6.23),

$$\sigma_{zx} = \frac{\partial \phi}{\partial y} = \frac{2By}{b^2} = -\frac{2h^2 G\theta y}{h^2 + b^2} \tag{6.43}$$

$$\sigma_{zy} = -\frac{\partial \phi}{\partial x} = -\frac{2Bx}{h^2} = \frac{2b^2 G\theta x}{h^2 + b^2} \tag{6.44}$$

The maximum shear stress τ_{max} occurs at the boundary nearest the centroid of the cross section. Its value is

$$\tau_{max} = \sigma_{zy(x=h)} = \frac{2b^2 hG\theta}{h^2 + b^2} \tag{6.45}$$

The torque T for the elliptical cross-section torsion member is obtained by substituting Eq. (6.41) into Eq. (6.36). Thus, we obtain

$$T = \frac{2B}{h^2} \int x^2 \, dA + \frac{2B}{b^2} \int y^2 \, dA - 2B \int dA = \frac{2B}{h^2} I_y + \frac{2B}{b^2} I_x - 2BA$$

Determination of I_x, I_y and A in terms of (b, h) allows us to write

$$T = -\pi Bhb \tag{6.46}$$

The torque may be expressed in terms of either τ_{max} or θ by means of Eqs. (6.42), (6.45), and (6.46). Thus,

$$\tau_{max} = \frac{2T}{\pi bh^2}, \qquad \theta = \frac{T(b^2 + h^2)}{G\pi b^3 h^3} \tag{6.47}$$

where $G\pi b^3 h^3/(b^2 + h^2) = GJ$ is called the torsional rigidity (stiffness) of the section and the torsional constant for the cross section is $J = \pi b^3 h^3/(b^2 + h^2)$.

Equilateral Triangle Cross Section

Let the boundary of a torsion member be an equilateral triangle (Fig. 6.9). The stress function is given by the relation

$$\phi = \frac{G\theta}{2h}\left(x - \sqrt{3}y - \frac{2h}{3}\right)\left(x + \sqrt{3}y - \frac{2h}{3}\right)\left(x + \frac{h}{3}\right) \tag{6.48}$$

Proceeding as for the elliptical cross section, we find

$$\tau_{max} = \frac{15\sqrt{3}\,T}{2h^3}, \qquad \theta = \frac{15\sqrt{3}\,T}{Gh^4} \tag{6.49}$$

where $Gh^4/15\sqrt{3} = GJ$ is called the torsional rigidity of the section. Hence, the torsional constant for the cross section is $J = h^4/(15\sqrt{3})$.

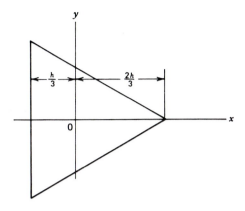

Figure 6.9   Equilateral triangle.

Rectangular Cross Sections

The indirect method outlined above fails for rectangular cross sections. Special methods (Boresi and Chong, 1987), which are beyond the scope of this book, are required to obtain the torsion solution for rectangular cross sections. We merely summarize some of the results here. Consider the rectangular cross section shown in Fig. 6.10. Relations between the cross-sectional dimensions, T, $\tau_{\max}$, and θ take the form

$$\theta = \frac{T}{k_1 G(2b)(2h)^3}, \qquad \tau_{\max} = \frac{T}{k_2(2b)(2h)^2} \tag{6.50}$$

where $\tau_{\max}$ is the maximum shear stress at the center of the long side at the boundary. Values of the parameters k_1 and k_2 are tabulated in Table 6.1 for several values of the ratio b/h. In Eq. (6.50), the factor $k_1 G(2b)(2h)^3 = GJ$ is the torsional rigidity of the section. The torsional constant for the cross section is $J = k_1(2b)(2h)^3$. Alternatively, by Eq. (6.50), $\tau_{\max}$ may be expressed in terms of θ as $\tau_{\max} = 2G\theta h k_3$, where $k_3 = k_1/k_2$.

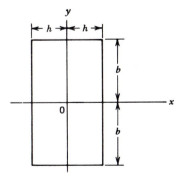

Figure 6.10   Rectangle.

TABLE 6.1
Torsional Parameters for Rectangular Cross Sections

b/h	1	1.5	2	2.5	3	4	6	10	∞
k_1	0.141	0.196	0.229	0.249	0.263	0.281	0.299	0.312	0.333
k_2	0.208	0.231	0.246	0.256	0.267	0.282	0.299	0.312	0.333

Other Cross Sections

There are many torsion members whose cross sections are so complex that exact analytical solutions are difficult to obtain. However, approximate solutions may be obtained by Prandtl's membrane analogy (see Sec. 6.4). An important class of torsion members are those with thin walls. Included in the class of thin-walled torsion members are open and box sections. Approximate solutions for these types of section are obtained in Sec. 6.5 and 6.6 by means of the Prandtl membrane analogy.

EXAMPLE 6.3
Rectangular Section Torsion Member

The rectangular section torsion member in Fig. E6.3 has a width of 40 mm. The first 3.00-m length of the torsion member has a depth of 60 mm, and the remaining 1.50-m length has a depth of 30 mm. The torsion member is made of steel for which $G = 77.5$ GPa. For $T_1 = 750$ N·m and $T_2 = 400$ N·m, determine the maximum shear stress in the torsion member. Determine the angle of twist of the free end. The support at the left end prevents rotation of this cross section but does not prevent warping.

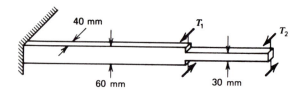

Figure E6.3

SOLUTION

For the left portion of the torsion member,

$$\frac{b}{h} = \frac{30}{20} = 1.5$$

From Table 6.1, we find $k_1 = 0.196$ and $k_2 = 0.231$. For the right portion of the torsion member,

$$\frac{b}{h} = \frac{20}{15} = 1.33$$

Linear interpolation between the values 1 and 1.5 in Table 6.1 gives $k_1 = 0.178$ and $k_2 = 0.223$. The torque in the left portion of the torsion member is $T = T_1 + T_2 = 1.15$ kN·m; the maximum shear stress in this portion of the torsion member is

$$\tau_{max} = \frac{T}{k_2(2b)(2h)^2} = \frac{1,150,000}{0.231(60)(40)^2} = 51.9 \text{ MPa}$$

The torque in the right portion of the torsion member is equal to $T_2 = 400$ N·m; the maximum shear stress in this portion of the torsion member is

$$\tau_{max} = \frac{400,000}{0.223(40)(30)^2} = 49.8 \text{ MPa}$$

Hence, the maximum shear stress occurs in the left portion of the torsion member and is equal to 51.9 MPa.

The angle of twist β is equal to the sum of the angles of twist for the left and right portions of the torsion member. Thus,

$$\beta = \frac{1,150,000(3000)}{0.196(77,500)(60)(40)^3} + \frac{400,000(1500)}{0.178(77,500)(40)(30)^3} = 0.0994 \text{ rad}$$

6.4

THE PRANDTL ELASTIC-MEMBRANE (SOAP-FILM) ANALOGY

In this section, we consider a solution of the torsion problem by means of an analogy proposed by Prandtl (1903). The method is based on the similarity of the equilibrium equation for a membrane subjected to lateral pressure and the torsion (stress function) equation [Eq. (6.38)]. Although this method is of historical interest, it is rarely used today to obtain quantitative results. It is discussed here primarily from a heuristic viewpoint, in that it is useful in the visualization of the distribution of shear-stress components in the cross section of a torsion member.

To set the stage for our discussion, consider an opening in the (x, y) plane that has the same shape as the cross section of the torsion bar to be investigated. Cover the opening with a homogeneous elastic membrane, such as a soap film, and apply pressure to one side of the membrane. The pressure causes the membrane to bulge out of the (x, y) plane, forming a curved surface. If the pressure is small, the slope of the membrane will also be small. Then, the lateral displacement $z(x, y)$ of the membrane and the Prandtl torsion stress function $\phi(x, y)$ satisfy the same equation in (x, y). Hence, the displacement $z(x, y)$ of the membrane is mathematically equivalent to the stress function $\phi(x, y)$, provided that $z(x, y)$ and $\phi(x, y)$ satisfy the same boundary conditions. This condition requires the boundary shape of the membrane to be identical to the boundary shape of the cross section of the torsion member. In the following discussion, we outline the physical and mathematical procedures that lead to a complete analogy between the membrane problem and the torsion problem.

As noted above, the Prandtl membrane analogy is based on the equivalence of the torsion equation, [Eq. (6.38), repeated here for convenience]

$$\frac{\partial^2 \phi}{\partial x^2} + \frac{\partial^2 \phi}{\partial y^2} = -2G\theta \tag{6.51}$$

and the membrane equation (to be derived in the next paragraph)

$$\frac{\partial^2 z}{\partial x^2} + \frac{\partial^2 z}{\partial y^2} = -\frac{p}{S} \tag{6.52}$$

where z denotes the lateral displacement of an elastic membrane subjected to a lateral pressure p in terms of force per unit area and an initial (large) tension S (Fig. 6.11) in terms of force per unit length.

For the derivation of Eq. (6.52), consider an element $ABCD$ of dimensions dx, dy of the elastic membrane shown in Fig. 6.11. The net vertical force due to the tension S acting along edge AD of the membrane is (if we assume small displacements so that $\sin \alpha \approx \tan \alpha$)

$$-S \, dy \sin \alpha \approx -S \, dy \tan \alpha = -S \, dy \frac{\partial z}{\partial x}$$

and, similarly, the net vertical force due to the tension S (assumed to remain constant for sufficiently small values of p) acting along edge BC is

$$S \, dy \tan\left(\alpha + \frac{\partial \alpha}{\partial x} dx\right) = S \, dy \frac{\partial}{\partial x}\left(z + \frac{\partial z}{\partial x} dx\right)$$

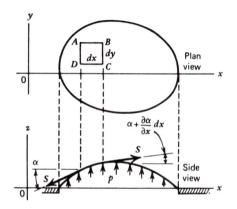

Figure 6.11

Similarly for edges AB and DC, we obtain

$$-S\,dx\frac{\partial z}{\partial y}, \qquad S\,dx\frac{\partial}{\partial y}\left(z + \frac{\partial z}{\partial y}dy\right)$$

Consequently, the summation of force in the vertical direction yields for the equilibrium of the membrane element $dx\,dy$

$$S\frac{\partial^2 z}{\partial x^2}dx\,dy + S\frac{\partial^2 z}{\partial y^2}dx\,dy + p\,dx\,dy = 0$$

or

$$\frac{\partial^2 z}{\partial x^2} + \frac{\partial^2 z}{\partial y^2} = -\frac{p}{S} \tag{6.52}$$

By comparison of Eqs. (6.51) and (6.52), we arrive at the following analogous quantities:

$$z = c\phi, \qquad \frac{p}{S} = c2G\theta \tag{6.53}$$

where c is a constant of proportionality. Hence,

$$\frac{z}{p/S} = \frac{\phi}{2G\theta}, \qquad \phi = \frac{2G\theta S}{p}z \tag{6.54}$$

Accordingly, the membrane displacement z is proportional to the Prandtl stress function ϕ, and since the shear-stress components σ_{zx}, σ_{zy} are equal to the appropriate derivatives of ϕ with respect to x and y [see Eqs. (6.23)], it follows that the stress components are proportional to the derivatives of the membrane displacement z with respect to the coordinates (x, y) in the flat plate to which the membrane is attached (Fig. 6.11). In other words, the stress components at a point (x, y) of the bar are proportional to the slopes of the membrane at the corresponding point (x, y) of the membrane. Consequently, the distribution of shear stress components in the cross section of the bar is easily visualized by forming a mental image of the slope of the corresponding membrane. Furthermore, for simply connected cross sections,* since z is proportional to ϕ, by Eqs. (6.36) and (6.54), we note that the twisting moment T is proportional to the volume enclosed by the membrane and the (x, y) plane (Fig. 6.11). For the multiply connected cross section, additional conditions arise (Sec. 6.6; see also Boresi and Chong, 1987).

Although rarely used today, recommended experimental techniques for the use of the membrane analogy are reported in a paper by Thoms and Masch (1965). The experimental technique requires that p/S be determined for the membrane.

---

*A region R is *simply connected* if every closed curve within it or on its boundary encloses only points in R. For example, the solid cross section in Fig. 6.7a (region R) is simply connected (as are all the cross sections in Sec. 6.3), since any closed curve in R or on its boundary contains only points in R. However, a region R between two concentric circles is not simply connected (see Fig. 6.5), since its inner boundary $r = a$ encloses points not in R. A region or cross section that is not simply connected is called *multiply connected*.

The procedure usually followed is to machine a circular hole in the plate in addition to the cross section of interest. If the same membrane is used for both openings, the values of p/S are the same for both openings, and the corresponding values of $G\theta$ are the same for the corresponding torsion members [see Eqs. (6.54)]. Since the relations among T, θ, and τ for a torsion member with a solid circular cross section are given by Eqs. (6.12) and (6.15), measurements obtained on the two elastic membranes may be used to compare torques and shear stresses for the two corresponding torsion members for *the same unit angle of twist*. In particular, then the ratio of the two volumes between the membrane surfaces and flat plate is equal to the ratio of the torques of the two corresponding torsion members. The ratio of the maximum slopes of the two membrane surfaces is equal to the ratio of the maximum shear stresses in the two corresponding torsion members.

Another important aspect of the elastic membrane analogy is that, without performing experiments, valuable deductions can be made by merely visualizing the shape that the membrane must take. For example, if a membrane covers holes machined in a flat plate, the corresponding torsion members have equal values of $G\theta$; therefore, the stiffnesses [see Eqs. (6.47), (6.49), and (6.50)] of torsion members made of materials having the same G are proportional to the volumes between the membranes and flat plate. For cross sections with equal area, one can deduce that a long narrow rectangular section has the least stiffness and the circular section has the greatest stiffness.

Important conclusions may also be drawn with regard to the magnitude of the shear stress and hence to the cross section for minimum shear stress. Consider the angle section shown in Fig. 6.12a. At the external corners A, B, C, E, and F, the membrane has zero slope and the shear stress is zero; therefore, external corners do not constitute a design problem. However, at the reentrant corner at D (shown as a right angle in Fig. 6.12a), the corresponding membrane would have an infinite slope, which indicates an infinite shear stress in the torsion member. In practical problems, the magnitude of the shear stress at D would be finite, but would be very large compared to that at other points in the cross section.

Remark on Reentrant Corners

If a torsion member with cross section shown in Fig. 6.12a is made of a ductile material and it is subjected to static loads, the material in the neighborhood of D yields and the load is redistributed to adjacent material, so that the stress concentration at point D is not particularly important. If, on the other hand, the material is brittle or the torsion member is subjected to fatigue loading, the shear stress at D limits the load-carrying capacity of the member. In such a case, the maximum shear stress in the torsion member may be reduced by removing some material as

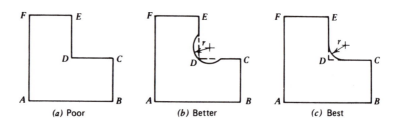

Figure 6.12   Angle section of a torsion member. (*a*) Poor. (*b*) Better. (*c*) Best.

shown in Fig. 6.12*b*. However, preferably, the member should be redesigned to alter the cross section (Fig. 6.12*c*). The maximum shear stress would then be about the same for the two cross sections shown in Figs. 6.12*b* and *c* for a given unit angle of twist; however, a torsion member with the cross section shown in Fig. 6.12*c* would be stiffer for a given unit angle of twist.

6.5

NARROW RECTANGULAR CROSS SECTION

The cross sections of many members of machines and structures are made up of narrow rectangular parts. These members are used mainly to carry tension, compression, and bending loads. However, they may be required also to carry secondary torsional loads. For simplicity, we use the elastic membrane analogy to obtain the solution of a torsion member whose cross section is in the shape of a narrow rectangle.

Consider a bar subjected to torsion. Let the cross section of the bar be a solid rectangle with width 2*h* and depth 2*b*, where $b \gg h$ (Fig. 6.13). The associated membrane is shown in Fig. 6.14.

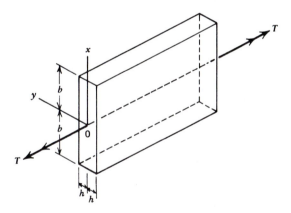

Figure 6.13

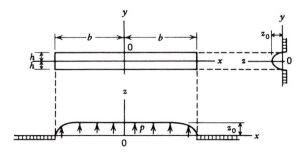

Figure 6.14

Except for the region near $x = \pm b$, the membrane deflection is approximately independent of x. Hence, if we assume that the membrane deflection is independent of x and parabolic with respect to y, the displacement equation of the membrane is

$$z = z_0\left[1 - \left(\frac{y}{h}\right)^2\right] \tag{6.55}$$

where z_0 is the maximum deflection of the membrane. Note that Eq. (6.55) satisfies the condition $z = 0$ on the boundaries $y = \pm h$. Also, if p/S is a constant in Eq. (6.52), the parameter z_0 may be selected so that Eq. (6.55) represents a solution of Eq. (6.52). Consequently, Eq. (6.55) is an approximate solution of the membrane displacement. By Eq. (6.55) we find

$$\frac{\partial^2 z}{\partial x^2} + \frac{\partial^2 z}{\partial y^2} = -\frac{2z_0}{h^2} \tag{6.56}$$

By Eqs. (6.56), (6.52), and (6.53), we may write $-2z_0/h^2 = -2cG\theta$ and Eqs. (6.55) becomes

$$\phi = G\theta h^2\left[1 - \left(\frac{y}{h}\right)^2\right] \tag{6.57}$$

Consequently, Eqs. (6.23) yield

$$\sigma_{zx} = \frac{\partial\phi}{\partial y} = -2G\theta y, \qquad \sigma_{zy} = -\frac{\partial\phi}{\partial x} = 0 \tag{6.58}$$

and we note that the maximum value of σ_{zx} is

$$\tau_{max} = 2G\theta h, \qquad \text{for } y = \pm h \tag{6.59}$$

Equations (6.36) and (6.57) yield

$$T = 2\int_{-b}^{b}\int_{-h}^{h} \phi\, dx\, dy = \frac{1}{3}G\theta(2b)(2h)^3 = GJ\theta \tag{6.60}$$

where

$$J = \tfrac{1}{3}(2b)(2h)^3 \tag{6.61}$$

is the torsional constant and GJ is the torsional rigidity.

In summary, we note that the solution is approximate and, in particular, the boundary condition for $x = \pm b$ is not satisfied. From Eqs. (6.59) and (6.60) we obtain

$$\tau_{max} = \frac{3T}{(2b)(2h)^2} = \frac{2Th}{J}, \qquad \theta = \frac{3T}{G(2b)(2h)^3} = \frac{T}{GJ} \tag{6.62}$$

Note that Eqs. (6.62) agree with Eqs. (6.50) since, from Table 6.1, $k_1 = k_2 = \tfrac{1}{3}$.

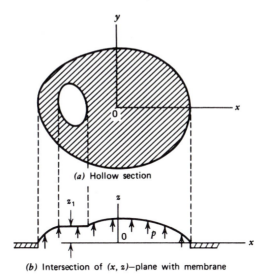

Figure 6.15   Membrane for hollow torsion member. (*a*) Hollow section. (*b*) Intersection of (x, z) plane with membrane.

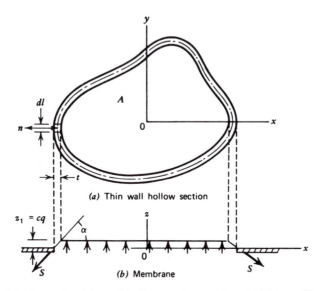

Figure 6.16   Membrane for thin-wall hollow torsion member. (*a*) Thin-wall hollow section. (*b*) Membrane.

the boundary is not constant, unless the thickness t is constant. This is apparent by Fig. 6.16*b* since $\tau = \partial\phi/\partial n$, where n is normal to a membrane contour curve $z = $ constant. Hence, by Eqs. (6.54) and Fig. 6.16*b*, $\tau = (2G\theta S/p)\,\partial z/\partial n = (2G\theta S/p)\tan\alpha$. Finally, by Eq. (6.53).

$$\tau = \frac{1}{c}\tan\alpha = \frac{1}{c}\sin\alpha \qquad \text{(since } \alpha \text{ is assumed to be small)} \qquad (6.65)$$

The quantity $q = \tau t$, with dimensions force/length, is commonly referred to in the literature as *shear flow*. As indicated in Fig. 6.16b, the shear flow is constant around the cross section of a thin-wall hollow torsion member and is equal to ϕ. Since ϕ is proportional to z [Eq. (6.53)], by Eq. (6.36), the torque is proportional to the volume under the membrane. Thus, we have approximately ($z_1 = c\phi_1$)

$$T = 2A\phi_1 = \frac{2Az_1}{c} = 2Aq = 2A\tau t \tag{6.66}$$

in which A is the area enclosed by the mean perimeter of the cross section (see the area enclosed by the dot-dashed line in Fig. 6.16a). A relation between τ, G, θ, and the dimensions of the cross section may be derived from the equilibrium conditions in the z direction. Thus,

$$\sum F_z = pA - \oint S \sin \alpha \, dl = 0,$$

and by Eqs. (6.65) and (6.53),

$$\frac{1}{A} \oint \tau \, dl = \frac{p}{cS} = 2G\theta \tag{6.67}$$

where l is the length of the mean perimeter of the cross section and S is the tensile force per unit length of the membrane.

Equations (6.66) and (6.67) are based on the simplifying assumption that the wall thickness is sufficiently small so that the shear stress may be assumed to be constant through the wall thickness. For the cross section considered in the following illustrative problem, the resulting error is negligibly small when the wall thickness is less than one-tenth of the minimum cross-sectional dimension.

EXAMPLE 6.5
Hollow Thin-Wall Circular Torsion Member

A hollow circular torsion member has an outside diameter of 22.0 mm and inside diameter of 18.0 mm, with mean diameter $D = 20.0$ mm and $t/D = 0.10$.

(a) Let the shear stress at the mean diameter be $\tau = 70.0$ MPa. Determine T and θ using Eqs. (6.66) and (6.67) and compare these values with values obtained using the elasticity theory. $G = 77.5$ GPa.

(b) Let a cut be made through the wall thickness along the entire length of the torsion member and let the maximum shear stress in the resulting torsion member be 70.0 MPa. Determine T and θ.

SOLUTION

(a) The area A enclosed by the mean perimeter is

$$A = \frac{\pi D^2}{4} = 100\pi \text{ mm}^2$$

The torque is given by Eq. (6.66)

$$T = 2A\tau t = 2(100\pi)(70)(2) = 87,960 \text{ N} \cdot \text{mm} = 87.96 \text{ N} \cdot \text{m}$$

Because the wall thickness is constant, Eq. (6.67) gives

$$\theta = \frac{\tau\pi D}{2GA} = \frac{70(\pi)(20)}{2(77,500)(100\pi)} = 0.0000903 \text{ rad/mm}$$

Elasticity values of T and θ are given by Eqs. (6.15) and (6.12). Thus, with

$$J = \frac{\pi}{32}(22^4 - 18^4) = 4040\pi \text{ mm}^4$$

we find that

$$T = \frac{\tau J}{r} = \frac{70(4040\pi)}{10} = 88,840 \text{ N} \cdot \text{mm} = 88.84 \text{ N} \cdot \text{m}$$

and

$$\theta = \frac{\tau}{Gr} = \frac{70}{77,500(10)} = 0.0000903 \text{ rad/mm}$$

The approximate solution agrees with the elasticity theory in the prediction of the unit angle of twist and yields torque that differs by only 1%. Note that the approximate solution assumes that the shear stress was uniformly distributed, whereas the elasticity solution indicates that the maximum shear stress is 10% greater than the value at the mean diameter, since the elasticity solution indicates that τ is proportional to r. Note that for a thin tube $J \approx 2\pi R^3 t = 4000\pi \text{ mm}^4$, where R is the mean radius and t the wall thickness (see Problem 6.17).

(b) When a cut is made through the wall thickness along the entire length of the torsion member, the torsion member becomes equivalent to a long narrow rectangle, for which the theory of Sec. 6.5 applies. Thus, with $h = 1$ and $b = 10\pi$

$$\theta = \frac{\tau_{\text{max}}}{2Gh} = \frac{70}{2(77,500)(1)} = 0.0004515 \text{ rad/mm}$$

$$T = \frac{8bh^2\tau_{\text{max}}}{3} = \frac{8(10\pi)(1)^2(70)}{3} = 5865 \text{ N} \cdot \text{mm} = 5.865 \text{ N} \cdot \text{m}$$

Hence, after the cut, the torque is 6.7% of the torque for part (a), whereas the unit angle of twist is 5 times greater than that for part (a). However, the maximum shear stress is essentially unchanged.

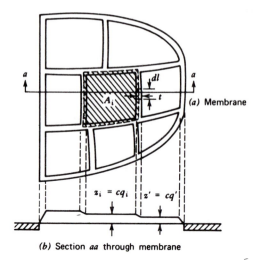

(b) Section *aa* through membrane

Figure 6.17   Multicompartment hollow thin-wall torsion member. *(a)* Membrane. *(b)* Section *a-a* through membrane.

Hollow Thin-Wall Torsion Member Having Several Compartments

Thin-wall hollow torsion members may have two or more compartments. Consider the torsion member whose cross section is shown in Fig. 6.17a. Section *a–a* through the membrane is shown in Fig. 6.17b. The plateau over each compartment is assumed to have a different elevation z_i. If there are N compartments, there are $N + 1$ unknowns to be determined. For a specified torque T, the unknowns are the N values for the q_i and the unit angle of twist θ, which is assumed to be the same for each compartment. The $N + 1$ equations are given by the equation

$$T = 2 \sum_{i=1}^{N} A_i \frac{z_i}{c}$$

$$= 2 \sum_{i=1}^{N} A_i q_i \qquad (6.68)$$

and N additional equations similar to Eq. (6.67)

$$\theta = \frac{1}{2GA_i} \oint_{l_i} \frac{q_i - q'}{t} \, dl, \qquad i = 1, 2, \ldots, N \qquad (6.69)$$

where A_i is the area bounded by the mean perimeter for the ith compartment, q' the shear flow for the compartment adjacent to the ith compartment where dl is located, t the thickness where dl is located, and l_i the length of the mean perimeter for the ith compartment. We note that q' is zero at the outer boundary. The maximum shear stress occurs where the membrane has the greatest slope, that is, where $(q_i - q')/t$ takes on its maximum value for the N compartments.

EXAMPLE 6.6
Two Compartment Hollow Thin-Wall Torsion Member

A hollow thin-wall torsion member has two compartments with cross-sectional dimensions as indicated in Fig. E6.6. The material is an aluminum alloy for which $G = 26.0\,\text{GPa}$. Determine the torque and unit angle of twist if the maximum shear stress, at locations away from stress concentrations, is 40.0 MPa.

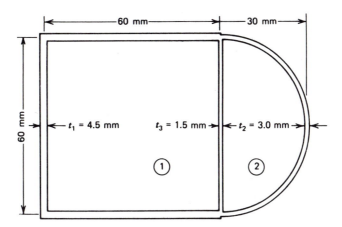

Figure E6.6

SOLUTION

Possible locations of the maximum shear stress are in the outer wall of compartment 1 where $t_1 = 4.5$ mm, in the outer wall of compartment 2 where $t_2 = 3.0$ mm, and the wall between the two compartments where $t_3 = 1.5$ mm. To determine the correct location, we must determine the ratio of q_1 to q_2. First, we write the three equations given by Eq. (6.68) and Eqs. (6.69).

$$T = 2(A_1 q_1 + A_2 q_2) = 7200\,q_1 + 2827q_2 \tag{a}$$

$$\theta = \frac{1}{2GA_1}\left[\frac{q_1 l_1}{t_1} + \frac{(q_1 - q_2)l_3}{t_3}\right] = \frac{1}{7200G}\left[\frac{180q_1}{4.5} + \frac{60(q_1 - q_2)}{1.5}\right] \tag{b}$$

$$\theta = \frac{1}{2GA_2}\left[\frac{q_2 l_2}{t_2} + \frac{(q_2 - q_1)l_3}{t_3}\right] = \frac{1}{900\pi G}\left[\frac{30\pi q_2}{3.0} + \frac{60(q_2 - q_1)}{1.5}\right] \tag{c}$$

Since the unit angle of twist given by Eq. (b) is equal to that given by Eq. (c), the ratio of q_1 to q_2 is found to be

$$\frac{q_1}{q_2} = 1.220$$

$q = \tau t$

This ratio is less than t_1/t_2; therefore, the maximum shear stress does not occur in the walls with thickness $t_1 = 4.5$ mm. Let us assume that it occurs in the wall with thickness t_2.

$$q_2 = \tau_{max}t_2 = 40.0(3) = 120.0 \text{ N/mm} \tag{d}$$

$$q_1 = 1.220(120.0) = 146.4 \text{ N/mm} \tag{e}$$

$$q_1 - q_2 = 26.4 \text{ N/mm}$$

$$\tau_1 = \frac{q_1}{t_1} = \frac{146.4}{4.5} = 32.5 \text{ MPa}, \quad \tau_2 = \frac{q_2}{t_2} = 40.0 \text{ MPa},$$

$$\tau_3 = \frac{q_1 - q_2}{t_3} = 17.6 \text{ MPa}$$

The magnitudes of q_1 and q_2 given by Eqs. (d) and (e) were based on the assumption that $\tau_2 = \tau_{max} = 40.0$ MPa; it is seen that the assumption is valid. These values for q_1 and q_2 may be substituted into Eqs. (a) and (b) to determine T and θ. Thus,

$$T = 7200(146.4) + 2827(120.0) = 1,393,000 \text{ N·mm} = 1.393 \text{ kN·m}$$

$$\theta = \frac{1}{7200(26,000)} \left[\frac{180(146.4)}{4.5} + \frac{60(26.4)}{1.5} \right]$$

$$= 0.0000369 \text{ rad/mm} = 0.0369 \text{ rad/m}$$

6.7

THIN-WALL TORSION MEMBERS WITH RESTRAINED ENDS

Torsion members with noncircular cross sections warp when subjected to torsional loads. However, if a torsion member is fully restrained by a heavy support at one end, warping at the end section is prevented. Hence, for a torsion member with a noncircular cross section, a normal stress distribution that prevents warping occurs at the restrained end. In addition, a shear stress distribution is developed at the restrained end to balance the torsional load.

Consider the I-section torsion member (Fig. 6.18) constrained against warping at the wall. At a section near the wall (Fig. 6.18a), the torsional load is transmitted mainly by lateral shear force V in each flange. This shear force produces lateral bending of each flange. As a result, on the basis of a linear bending theory, a linear normal stress distribution is produced at the wall. In addition, the shear stress distribution in each flange, at the wall, is similar to that for shear loading of a rectangular beam. At small distances away from the wall (Fig. 6.18b), partial warping occurs, and the torsional load is transmitted partly by the shear forces $V' < V$ induced by warping restraint and partly by Saint Venant's torsional shear. At greater distances from the wall (Fig. 6.18c), the effect of the restrained end diminishes rapidly, and the torque is transmitted mainly by torsional shear stresses.

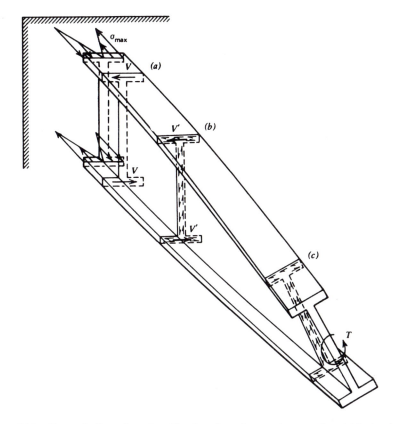

Figure 6.18 General effect of torsional load on I-section torsion member. (*a*) Lateral shear mainly. (*b*) Partly lateral shear and partly torsional shear. (*c*) Torsional shear mainly.

These remarks are illustrated further by a solution for an I-section torsion member presented later in this section.

Thin-wall hollow torsion members with restrained ends (see Fig. 6.19) may also warp under torsion. However, in contrast to torsion members with simply connected cross sections, noncircular thin-wall hollow torsion members may, under certain conditions, twist without warping. A solution presented by von Kármán and Chien (1946) for constant-thickness hollow torsion members indicates that a torsion member with equilateral polygon cross section does not warp. By contrast if $t_1 = t_2$, the rectangular section hollow torsion member ($a \neq b$) in Fig. 6.19 tends to warp when subjected to torsion loads and, hence, to develop a normal stress distribution at a restrained end. As a generalization of the von Kármán–Chien solution, a solution presented by Smith, Thomas, and Smith (1970) indicates that the torsion member in Fig. 6.19 does not warp if $bt_1 = at_2$. In addition, they presented a solution for the case when $a/b = 3/8$, $t_1 = b/32$, and $t_2 = b/16$. They found that the normal stress distribution at the end was nonlinear as indicated in Fig. 6.19 with $\sigma_{\max} = 0.0114T$, where $\sigma_{\max}$ has the units of MPa and T the units of N·m. For hollow torsion members with rectangular sections of constant thickness, similar normal stress distributions at a restrained end are predicted in the papers by von Kármán and Chien (1946) and Smith et al. (1970). Readers are referred to these two papers for thin-wall hollow torsion members with restrained ends.

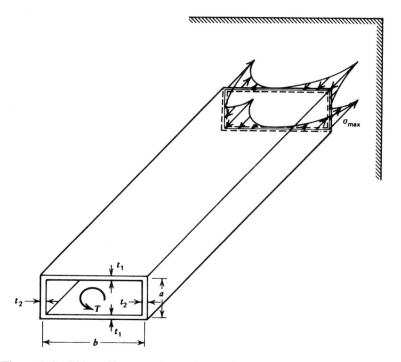

Figure 6.19 Thin-wall rectangular section torsion member with a restrained end.

I-Section Torsion Member Having One End Restrained from Warping

Consider an I-section torsion member subjected to a twisting moment T (Fig. 6.20a). Let the section at the wall be restrained from warping. A small distance from the wall, say, at section AB, partial warping takes place and the twisting moment T may be considered to be made up of two parts. One part is a twisting moment T_1 produced by the lateral shear forces since these forces constitute a couple with moment arm h. Hence,

$$T_1 = V'h \tag{6.70}$$

The second part is twisting moment T_2, which produces warping on the section. Hence, T_2 is given by Eq. (6.64) as

$$T_2 = JG\theta \tag{6.71}$$

The values of T_1 and T_2 are unknown since the values of V' and θ at any section are not known. Values of these quantities must be found before the lateral bending stresses in the flanges or the torsional shear stresses in the I-section can be computed. For this purpose, two equations are needed. From the condition of equilibrium, one of these equations is

$$T_1 + T_2 = T$$

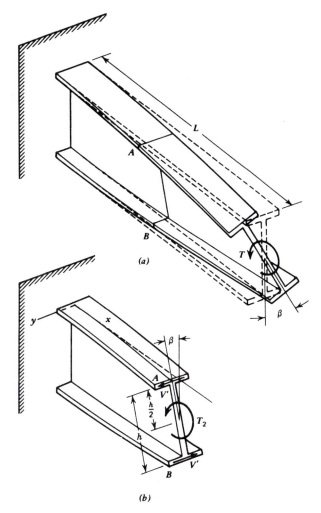

Figure 6.20 Effect of twisting moment applied to an I-section torsion member with one
end fixed.

which by Eqs. (6.70) and (6.71) may be written

$$V'h + JG\theta = T \tag{6.72}$$

For the additional equation, we may use the elastic curve equation for lateral bending of the upper flange in Fig. 6.20b, which is

$$\frac{EI}{2}\frac{d^2y}{dx^2} = -M \tag{6.73}$$

in which the x and y axes are chosen with positive directions as shown in Fig. 6.20; M is the lateral bending moment in the flange at any section, producing lateral bending in the flange; I is the moment of inertia of the entire cross section of the beam with respect to the axis of symmetry in the web, so that $\frac{1}{2}I$ closely approximates the value of the moment of inertia of a flange cross section. However, Eq. (6.73) does not contain either of the desired quantities V' and θ. These quantities are introduced into Eq. (6.73) as follows: In Fig. 6.20b the lateral deflection of the flange at section AB is

$$y = \frac{h}{2}\beta \tag{6.74}$$

Differentiation of Eq. (6.74) twice with respect to x gives

$$\frac{d^2y}{dx^2} = \frac{h}{2}\frac{d^2\beta}{dx^2} \tag{6.75}$$

and, since $d\beta/dx = \theta$, Eq. (6.75) may be written

$$\frac{d^2y}{dx^2} = \frac{h}{2}\frac{d\theta}{dx} \tag{6.76}$$

The substitution of this value of d^2y/dx^2 into Eq. (6.73) gives

$$\frac{EIh}{4}\frac{d\theta}{dx} = -M \tag{6.77}$$

In order to introduce V' into Eq. (6.77), use is made of the fact that $dM/dx = V'$. Thus, by differentiating both sides of Eq. (6.77) with respect to x, we obtain

$$\frac{EIh}{4}\frac{d^2\theta}{dx^2} = -V' \tag{6.78}$$

Equations (6.72) and (6.78) are simultaneous equations in V' and θ. The value of V' obtained from Eq. (6.78) is substituted into Eq. (6.72), which then becomes

$$-\frac{EIh^2}{4JG}\frac{d^2\theta}{dx^2} + \theta = \frac{T}{JG} \tag{6.79}$$

For convenience let

$$\alpha = \frac{h}{2}\sqrt{\frac{EI}{JG}} \tag{6.80}$$

so that Eq. (6.79) may be written

$$-\alpha^2\frac{d^2\theta}{dx^2} + \theta = \frac{T}{JG} \tag{6.81}$$

The solution of this equation is

$$\theta = Ae^{x/\alpha} + Be^{-x/\alpha} + \frac{T}{JG} \tag{6.82}$$

Appropriate boundary conditions in terms of θ or $d\theta/dx$ are needed to determine values of the constants A and B in Eq. (6.82). At the fixed end where $x = 0$, $\theta = d\beta/dx = (2/h)(dy/dx) = 0$ (since the slope is zero). At the free end where $x = L$, $d\theta/dx = 0$ [see Eq. (6.77)], since at the free end the bending moment M in the flange is zero. The values of A and B are determined from these two conditions and are substituted in Eq. (6.82), which gives the angle of twist per unit length

$$\theta = \frac{T}{JG}\left[1 - \frac{\cosh(L - x)/\alpha}{\cosh(L/\alpha)}\right] \tag{6.83}$$

The total angle of twist at the free end is

$$\beta = \int_0^L \theta \, dx = \frac{T}{JG}\left(L - \alpha \tanh\frac{L}{\alpha}\right) \tag{6.84}$$

The twisting moment T_2 at any section of the beam is obtained by substitution of the value of θ from Eq. (6.83) into Eq. (6.71). Thus,

$$T_2 = T\left[1 - \frac{\cosh(L - x)/\alpha}{\cosh(L/\alpha)}\right] \tag{6.85}$$

The maximum torsional shear stress at any section is computed by substituting this value of T_2 into Eq. (6.64). The lateral bending moment M in the flanges of the beam at any section is obtained by substituting $d\theta/dx$ from Eq. (6.83) into Eq. (6.77), which gives

$$M = -\frac{T}{h}\alpha\frac{\sinh(L - x)/\alpha}{\cosh(L/\alpha)} \tag{6.86}$$

Note that Eq. (6.86) shows that the maximum value of M occurs at the fixed end, for $x = 0$, and is

$$M_{\max} = \frac{T}{h}\alpha\tanh\frac{L}{\alpha} \tag{6.87}$$

Except for relatively short beams, the length L is large as compared with the value of α, and the value of $\tanh(L/\alpha)$ is approximately equal to 1 when $L/\alpha > 2.5$. In Eqs. (6.84) and (6.87) the substitution of $\tanh(L/\alpha) = 1$ gives

$$\beta = \frac{T(L - \alpha)}{JG} \tag{6.88}$$

$$M_{\max} = \frac{T\alpha}{h} \tag{6.89}$$

The approximate values of β and $M_{\max}$ obtained from Eqs. (6.88) and (6.89) lead to the following procedure for solving for the angle of twist and the maximum longitudinal stresses resulting from a twisting moment in an I-section torsion member with one section restrained from warping. Let Fig. 6.21b represent

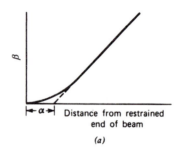

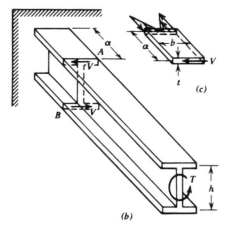

Figure 6.21 Lateral bending stresses at fixed section in a flange of an I-section torsion member.

an I-section torsion member that is fixed at one end and loaded at the free end by the twisting moment T. Figure 6.21a represents a typical curve showing the relation between the angle of twist β of the I-section torsion member and the distance from the fixed section of the beam. In Fig. 6.21a the distance from the fixed section to the section AB at which the straight-line portion of the curve intersects the horizontal axis is very nearly equal to the distance α as given by Eq. (6.80). Thus, from this fact and from Eq. (6.88), the length $L - \alpha$ of the beam between the free end and section AB may be considered as being twisted under pure torsion for the purpose of computing the angle of twist. From Eq. (6.89) the sections of the beam within the length α from the fixed section to section AB may be considered as transmitting the entire twisting moment T by means of the lateral shears V in the flanges. Therefore,

$$T = Vh \qquad \text{or} \qquad V = T/h \tag{6.90}$$

The force V causes each flange of length α to bend laterally, producing a normal stress in the flange, tensile stress at one edge, and compressive stress at the other (Fig. 6.21c). Assuming that each flange has a rectangular cross section, we have at the fixed end

$$\sigma = \frac{M\frac{1}{2}b}{I_f} = \frac{T}{h}\alpha\frac{\frac{1}{2}b}{\frac{1}{12}tb^3} = \frac{6T\alpha}{htb^2} \tag{6.91}$$

The value for α is given by Eq. (6.80), in which E and G are the tensile and shear moduli of elasticity, respectively, I is the moment of inertia of the entire section with respect to a centroidal axis parallel to the web, and J is an equivalent torsional constant of the section. Values of α calculated by this equation come close to values obtained from actual tests. For a section made up of slender, approximately rectangular elements, such as a rolled-steel channel, angle, or I-section, J is given by Eq. (6.63) if it is noted that b_i in this article replaces $2b_i$ in Eq. (6.63) and t_i replaces $2h_i$ in Eq. (6.63). All equations in this article have been derived for I-sections, but they apply as well to channels or Z-sections.

Various Loads and Supports for Beams in Torsion

The solution of Eq. (6.81) given by Eq. (6.82) is for the particular beam shown in Fig. 6.20. However, solutions of the equation have been obtained for beams loaded and supported as shown in Figs. A, B, C, and D in Table 6.2 by arranging the particular solution of the differential equation to suit the conditions of loading and support for each beam. The values of the maximum lateral bending moment M_{max} given in Table 6.2 may be used in Eq. (6.91) to compute the maximum

TABLE 6.2
Beams Subjected to Torsion

Type of Loading and Support	Maximum Lateral Bending Moment in Flange	Angle of Twist of Beam of Length L
	$M_{max} = \dfrac{T\alpha}{h}\tanh\dfrac{L}{2\alpha}$ $= \dfrac{T\alpha}{h}$, if $\dfrac{L}{2\alpha} > 2.5$	$\theta = \dfrac{T}{JG}\left(L - 2\alpha\tanh\dfrac{L}{2\alpha}\right)$ $= \dfrac{T}{JG}(L - 2\alpha)$ Error is small if $\dfrac{L}{2\alpha} > 2.5$
  $T = wLe$	$M_{max} = \dfrac{T\alpha}{2h}\left(\coth\dfrac{L}{2\alpha} - \dfrac{2\alpha}{L}\right)$ $= \dfrac{T\alpha}{2h}$, if $\dfrac{L}{2\alpha}$ is large	$\theta = \dfrac{T}{2JG}\left(\dfrac{L}{4} - \alpha\tanh\dfrac{L}{4\alpha}\right)$ $= \dfrac{T}{2JG}\left(\dfrac{L}{4} - \alpha\right)$ Error is small if $\dfrac{L}{4\alpha} > 2.5$
  w $T = wLe$ e	$M_{max} = \dfrac{T\alpha}{h}\left(\coth\dfrac{L}{\alpha} - \dfrac{\alpha}{L}\right)$ $= \dfrac{T\alpha}{h}$, if $\dfrac{L}{\alpha}$ is large	$\theta = \dfrac{T}{JG}\left(\dfrac{L}{2} - \alpha\tanh\dfrac{L}{2\alpha}\right)$ $= \dfrac{T}{JG}\left(\dfrac{L}{2} - \alpha\right)$ Error is small if $\dfrac{L}{2\alpha} > 2.5$

TABLE 6.2 (*Continued*)
Beams Subjected to Torsion

Type of Loading and Support	Maximum Lateral Bending Moment in Flange	Angle of Twist of Beam of Length L
		Approximate value
	$M_{max} = \dfrac{T\alpha}{h} \dfrac{\sinh\dfrac{L_1}{\alpha}\sinh\dfrac{L_2}{\alpha}}{\sinh\dfrac{L}{\alpha}}$	$\theta = \dfrac{1}{2}\dfrac{T}{JG}\left(\dfrac{L}{2} - \alpha\tanh\dfrac{L}{2\alpha}\right)$
	$= \dfrac{T\alpha}{2h}$, if	$= \dfrac{1}{2}\dfrac{T}{JG}\left(\dfrac{L}{2} - \alpha\right)$
	$\dfrac{L_1}{\alpha}$ and $\dfrac{L_2}{\alpha} > 2$	Error is small
	Error is small	if $\dfrac{L}{2\alpha} > 2.5$

lateral bending stress in the beam. The formulas in Table 6.2 where I-sections are shown may also be used for channels or Z-sections.

6.8

NUMERICAL SOLUTION OF THE TORSION PROBLEM

Considerable literature has been devoted to the Saint-Venant torsion problem of homogeneous, isotropic bars. However, studies of nonhomogeneous, anisotropic torsion members are less numerous. In modern aircraft structures, aerospace and industrial applications, laminated or anisotropic torsion members are used widely. For simple cross-sectional shapes, analytic solutions have been developed (Arutyunyan and Abramyan, 1963; Arutyunyan et al., 1988). For complicated, irregular cross sections, exact analytic solutions are difficult if not impossible to obtain. For such cases, both for homogeneous and isotropic, and for nonhomogeneous and anisotropic members, numerical or approximate solutions must be sought. The most commonly employed numerical techniques are finite element and finite difference methods (Noor and Anderson, 1975; Zienkiewicz and Taylor, 1989; Boresi and Chong, 1991). (See also Chapter 19 of this book.) Here for simplicity, we consider the finite difference solution of a homogeneous, isotropic torsion member with a square cross section R, boundary C, and dimension a (Fig. 6.22). We recall that Prandtl's formulation of the torsion problem for a homogeneous, isotropic material is [see Sec. 6.2 and 6.3; Eqs. (6.28) and (6.38)]

$$\nabla^2\phi = -2G\theta \qquad \text{on } R$$
$$\phi = 0 \qquad \text{on } C \qquad (6.92)$$

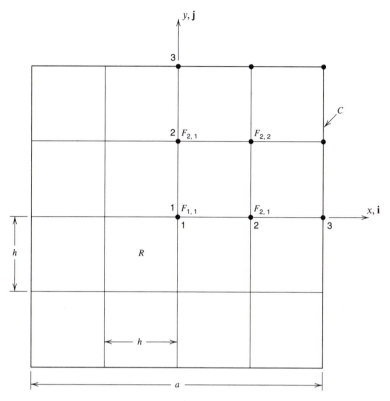

Figure 6.22   Square cross section

where ϕ is the Prandtl stress function, C the bounding curve of the bar cross section R, G the shear modulus of the material, and θ the angle of twist per unit length of the bar. To solve Eqs. (6.92) by the finite difference method, we must approximate $\nabla^2\phi$ by a finite difference formula. For this purpose, we employ the following second-order finite difference formula [Boresi and Chong, 1991, Eq. (3.48)],

$$(\nabla^2\phi)_{i,j} \approx \frac{1}{h^2}(\phi_{i-1,j} + \phi_{i+1,j} - 4\phi_{i,j} + \phi_{i,j-1} + \phi_{i,j+1}) \qquad (6.93)$$

where h is the mesh dimension and (i, j) denote the node point in R (Fig. 6.22). By symmetry, we need consider only a quarter of region R. Figure 6.22 shows a 4×4 square mesh subdivision with mesh dimension $h = a/4$. Numbering the node points as in Fig. 6.22, we need consider only three values of $\phi_{i,j}$ at the node points $(1, 1)$, $(2, 1)$, and $(2, 2)$, respectively. Then, by Eqs. (6.92) and (6.93), and with Fig. 6.22, we have

$$
\begin{aligned}
i = 1, j = 1; & \quad F_{2,1} + F_{2,1} - 4F_{1,1} + F_{2,1} + F_{2,1} = -2G\theta h^2 \\
i = 1, j = 2; & \quad F_{2,2} + F_{2,2} - 4F_{2,1} + F_{1,1} + 0 \quad = -2G\theta h^2 \\
i = 2, j = 2; & \quad F_{2,1} + 0 \quad - 4F_{2,2} + F_{2,1} + 0 \quad = -2G\theta h^2
\end{aligned} \qquad (6.94)
$$

Collecting terms in Eqs. (6.94), we have

$$-4F_{1,1} + 4F_{2,1} = -2G\theta h^2$$
$$F_{1,1} - 4F_{2,1} + 2F_{2,2} = -2G\theta h^2$$
$$2F_{2,1} - 4F_{2,2} = -2G\theta h^2 \qquad (6.95)$$

The solution of Eqs. (6.95) is

$$F_{1,1} = -2.250G\theta h^2$$
$$F_{2,1} = 1.750G\theta h^2$$
$$F_{2,2} = 1.375G\theta h^2 \qquad (6.96)$$

Next, we determine the maximum shear stress $\sigma_{zy} = -\partial\phi/\partial x$ that occurs at the point $x = a/2$, $y = 0$. Using a fourth-order backward-difference formula (Boresi and Chong, 1991; Table 3.3.2), we have the approximation

$$\left(\frac{\partial\phi}{\partial x}\right)_{i,j} = \frac{1}{24h}(6F_{i-4,j} - 32F_{i-3,j} + 72F_{i-2,j} - 96F_{i-1,j} + 50F_{i,j}) \qquad (6.97)$$

Thus, for $i = 3$, $j = 1$, that is, for $x = a/2$, $y = 0$, we have by Eq. (6.97)

$$(\sigma_{zy})_{\max} = -\left(\frac{\partial\phi}{\partial x}\right)_{3,1} \approx \frac{1}{24h}(0 - 32F_{2,1} + 72F_{1,1} - 96F_{2,1} + 0) \qquad (6.98)$$

Substitution of Eq. (6.96) into Eq. (6.98) yields

$$(\sigma_{zy})_{\max} \approx 2.583G\theta h = 0.646G\theta a \qquad (6.99)$$

The exact solution (Boresi and Chong, 1987; Sec. 7.10) yields $(\sigma_{zy})_{\max} = 0.675G\theta a$. If we take a finer mesh, we obtain a better approximation. For example, with $h = a/8$, we find $(\sigma_{zy})_{\max} = 0.666G\theta a$.

It can be shown (see Boresi and Chong, 1991; Chapter 3) that the error in the derivative of the stress function ϕ [and hence in $(\sigma_{zy})_{\max}$] is proportional to h^2. Thus, if δ denotes the error for a mesh size h, we may write $\delta = Ch^2$, where C is a constant. This fact can be used to extrapolate to a better approximation, given values for $h = a/4$ and $a/8$. For example, let $h = a/n$, where n is the number of segments into which a is divided. Let $h_1 = a/n_1$ and $h_2 = a/n_2$, $n_2 > n_1$, be two mesh sizes for which the approximate values S_1, S_2 of $(\sigma_{zy})_{\max}$ have been determined. Let δ_1 and δ_2 be the errors in S_1, S_2 compared with the exact value $(\sigma_{zy})_{\max}$. Then, we can obtain a better approximation by the equations

$$\delta_1 = Ch_1^2 = C\frac{a^2}{n_1^2} = (\sigma_{zy})_{\max} - S_1$$

and

$$\delta_2 = Ch_2^2 = C\frac{a^2}{n_2^2} = (\sigma_{zy})_{\max} - S_2$$

Solving for $(\sigma_{zy})_{max}$, we obtain the result

$$(\sigma_{zy})_{max} = \frac{n_2^2 S_2 - n_1^2 S_1}{n_2^2 - n_1^2} \tag{6.100}$$

For $n_1 = 4$, $S_1 = 0.646G\theta a$, and $n_2 = 8$, $S_2 = 0.666G\theta a$, Eq. (6.100) yields the approximate $(\sigma_{zy})_{max} = 0.673G\theta a$. This value is only 0.3% lower than the exact value $0.675G\theta a$.

Explicit closed-form analytical solutions of the torsion problem for complex irregular cross sections are not generally available. However, approximate solutions by the finite difference method are still feasible, although the generation of the approximating finite difference equations becomes more difficult.* The application of finite difference methods to irregular boundaries requires finite difference formulas for nonuniform mesh dimensions. The use of finite difference methods for irregular boundaries is discussed by Boresi and Chong (1991).

6.9

FULLY PLASTIC TORSION

Consider a torsion member made of an elastic-perfectly plastic material, that is, one whose shear stress-strain diagram is flat-topped at the shear yield stress τ_Y. As the torque is gradually increased, yielding starts at one or more places on the boundary of the cross section and spreads inward with increasing torque. Finally, the entire cross section becomes plastic at the limiting, fully plastic torque. The torsion analysis for the limiting torque or fully plastic torque is considered in this article.

Equations (6.23) and (6.29) are valid for both the elastic and plastic regions of each cross section of a torsion member. At the fully plastic torque, the resultant shear stress is $\tau = \tau_Y$ at every point in the cross section. Thus, Eqs. (6.23) and (6.29) give

$$\sigma_{zx}^2 + \sigma_{zy}^2 = \left(\frac{\partial \phi}{\partial y}\right)^2 + \left(\frac{\partial \phi}{\partial x}\right)^2 = \tau_Y^2 \tag{6.101}$$

Equation (6.101) uniquely determines the stress function $\phi(x, y)$ for a given torsion member for fully plastic conditions. Since the unit angle of twist does not appear in Eq. (6.101), the deformation (twist) of the torsion member is not specified at the fully plastic torque.

Now we consider a procedure by which Eq. (6.101) may be used to construct the stress function surface for the cross section of a given torsion member at fully plastic torque. Equation (6.28) indicates that $\phi = 0$ on the boundary, and Eq. (6.101) indicates that the absolute value of the maximum slope of ϕ everywhere in the

---

* It is shown in Chapter 19 that the inherent difficulties of the finite difference method in dealing with irregular boundaries or satisfying complicated boundary conditions can be circumvented by the use of finite element methods.

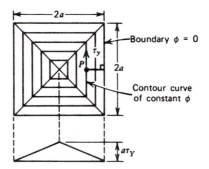

Figure 6.23   Stress function surface for a fully plastic square cross section.

cross section is a constant equal to τ_Y; therefore, the magnitude of ϕ at a point is equal to τ_Y times its distance from the nearest boundary, measured along the perpendicular from the point to the nearest boundary. The contour curves of constant ϕ are perpendicular to the direction of maximum slope and, hence, are parallel to the nearest boundary.

Consider the problem of constructing the stress function ϕ for a square cross section with sides $2a$, as indicated in Fig. 6.23. At a given point P, the resultant shear stress is τ_Y and is directed along a contour curve of constant ϕ; the elevation of the stress function at point P is equal to τ_Y times its perpendicular distance to the nearest boundary. The stress function ϕ for the square cross section is a pyramid of height $\tau_Y a$; this condition suggested to Nadai (1950) the so-called *sand-heap analogy* because sand poured on a flat plate with the same dimensions as the cross section of the torsion member tends to form a pyramid similar to that indicated in Fig. 6.23.

The fully plastic torque T_P for the square cross section can be obtained by means of Eq. (6.36), which indicates that the torque is equal to twice the volume under the stress function. Since the volume of a pyramid is equal to one-third of the area of the base times the height, we have

$$T_P = 2[\tfrac{1}{3}(2a)^2\tau_Y a] = \tfrac{8}{3}\tau_Y a^3 \tag{6.102}$$

The fully plastic torques for a few common cross sections are listed in Table 6.3 and are compared with the maximum elastic solutions for these cross sections. We see by Table 6.3 that the load-carrying capacity for the cross sections considered are greatly increased when we make it possible for yielding to spread throughout the cross sections.

Expressions for the fully plastic torques for a number of common structural sections have been derived (see Table 6.4, p. 280). We remark that the expressions in Table 6.4 are not exact. When the cross section has a reentrant corner as indicated in Fig. 6.24, the correct stress surface is as shown in Fig. 6.24a and not as in Fig. 6.24b. Since the expressions in Table 6.4 are based on the assumption that Fig. 6.4b is correct, these expressions are not exact for sections with reentrant corners. Finally, we note that the expressions for T_P in the second column of Table 6.4 hold for each of the two cross sections in the first column. Since the cross sections in Table 6.4 are made up of long narrow rectangular sections, the error in the expressions in Table 6.4 is small; furthermore, the error makes the expressions conservative. The maximum elastic torques are not given in Table 6.4 because of the

TABLE 6.3
θ_Y, T_Y, T_P, and T_P/T_Y for Five Common Cross Sections

Section	Maximum Elastic Torque T_Y and Unit Angle of Twist θ_Y	Fully Plastic Torque T_P	Ratio $\dfrac{T_P}{T_Y}$
Square 	$T_Y = 1.664\tau_Y a^3$ $\theta_Y = \dfrac{1.475\tau_Y}{2Ga}$	$\dfrac{8}{3}\tau_Y a^3$	1.605
Rectangle $\dfrac{b}{a} = 2$ 	$T_Y = 3.936\tau_Y a^3$ $\theta_Y = \dfrac{1.074\tau_Y}{2Ga}$ $T_Y = \dfrac{8}{3}\tau_Y b a^2$	$\dfrac{20}{3}\tau_Y a^3$	1.69
$\dfrac{b}{a} = \infty$	$\theta_Y = \dfrac{\tau_Y}{2Ga}$	$4\tau_Y b a^2$	1.50
Equilateral triangle 	$T_Y = \dfrac{2}{15\sqrt{3}}\tau_Y a^3$ $\theta_Y = \dfrac{2\tau_Y}{Ga}$	$\dfrac{2\sqrt{3}}{27}\tau_Y a^3$	1.67
Circle 	$T_Y = \dfrac{\pi}{2}\tau_Y a^3$ $\theta_Y = \dfrac{\tau_Y}{Ga}$	$\dfrac{2}{3}\pi\tau_Y a^3$	1.33

influence on initial yielding of the high stress-concentration factors at reentrant corners. In summary, Table 6.4 has limited applicability to the design of such sections in an actual structure for two reasons. First, failure by buckling, at least for thin sections, is likely to be the basis for design. Second, torsion of such sections is more often a secondary action that is usually accompanied by primary action, which produces bending or normal stresses.

In the calculation of the fully plastic torque for a hollow torsion member, the method of analysis is similar to that for elastic torsion of the hollow torsion member, since the stress function $\phi(x, y)$ is flat-topped (has zero slope) over the hollow region of the torsion member. Sadowsky (1941) has extended Nadai's sand-heap analogy to hollow torsion members. In order to simplify the analysis, only hollow torsion members of constant wall thickness are considered. For such torsion members, the fully plastic torque T_P is obtained by subtracting from the fully

TABLE 6.4
T_P **for Common Structural Sections**

Section	Fully Plastic Torque T_P
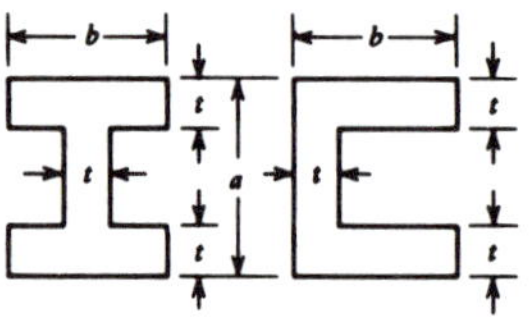	$\tau_Y t^2 \left(\dfrac{a}{2} + b - \dfrac{7}{6} t \right)$
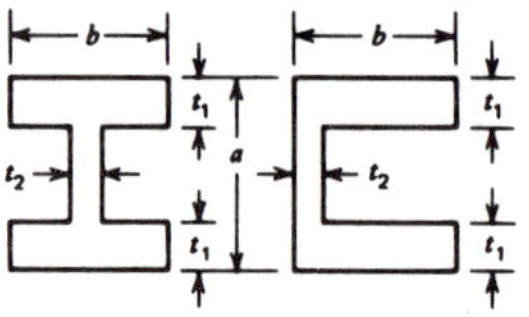	$\tau_Y \left[t_1^2 \left(b - \dfrac{t_1}{3} \right) + \dfrac{t_2^2}{2} \left(a + \dfrac{t_2}{3} \right) - t_1 t_2^2 \right]$
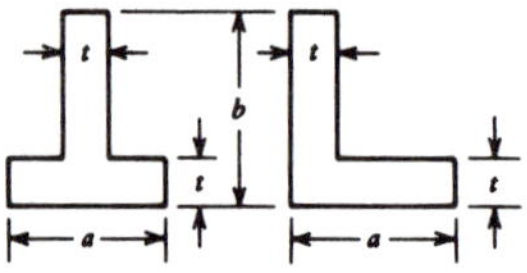	$\dfrac{\tau_Y t^2}{2} \left(a + b - \dfrac{4}{3} t \right)$
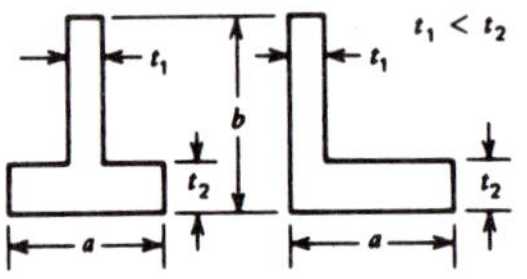	$\dfrac{\tau_Y}{2} \left(a t_2^2 + b t_1^2 - \dfrac{t_2^3}{3} - t_1^2 t_2 \right)$
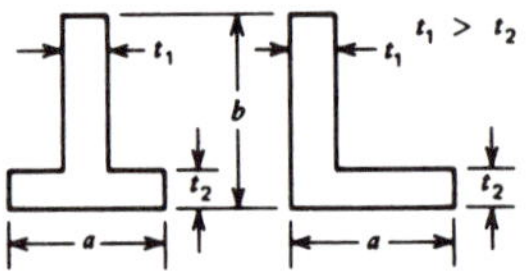	$\dfrac{\tau_Y}{2} \left(a t_2^2 + b t_1^2 - \dfrac{t_1^3}{3} - t_2^2 t_1 \right)$

Figure 6.24 Stress function surfaces for a fully plastic angle section. (a) Correct plastic-stress surface. (b) Incorrect plastic-stress surface.

plastic torque T_{PS} of a solid torsion member having the boundary of the outer cross section, the fully plastic torque T_{PH} of a solid torsion member having a cross section identical to the hollow region. That is, for such members

$$T_P = T_{PS} - T_{PH} \tag{6.103}$$

EXAMPLE 6.7
Limit Analysis and Residual Stress in a Circular
Cross-Section Shaft

A bar with circular cross section of radius b is subjected to a twisting moment T, which is increased until the bar is fully yielded.

(a) Determine the angle of twist θ_Y per unit length of the bar at initiation of yield.

(b) Determine the angle of twist $\theta > \theta_Y$, as the torque is increased beyond the torque T_Y that initiates yield to radius b_Y.

(c) Determine the elastic torque T_E, plastic torque T_P, and elastic-plastic torque T_{EP} (the sum of T_E and T_P) for part (b) above.

(d) Determine the limiting (fully plastic) torque T_L.

(e) After the limiting torque is reached, it is removed (released), and the torsion member springs back. As a result, residual stresses remain in the bar. Determine the residual stress distribution in the bar.

SOLUTION

As in Sec. 6.9, we consider a circular cross-section torsion member made of an elastic-perfectly plastic material, that is, one whose shear stress-strain diagram is flat-topped at the shear yield stress τ_Y (similar to Fig. 4.4a for tensile stress). As the torque is increased, yielding occurs when the shear stress τ on the boundary of the cross section reaches the value τ_Y, (Fig. E6.7a).

(a) When the shear stress reaches the value τ_Y at $r = b$, the corresponding torque is, by Eq. (6.15),

$$T_Y = \frac{\pi \tau_Y b^3}{2} \tag{a}$$

(see Table 6.3). As the torque is increased beyond the value T_Y, yielding spreads inward into the cross section of the member (Fig. E6.7b), so that the region from $r = b_Y$ to $r = b$ has yielded fully. The angle of twist per unit length of the bar at initiation of yield (Fig. E6.7a) is, by Eq. (6.14),

$$\theta_Y = \frac{\tau_Y}{Gb} \tag{b}$$

(b) As the torque increases beyond T_Y, the angle of twist θ per unit length increases, that is, $\theta > \theta_Y$, and the section yields to radius b_Y (Fig. E6.7b). Since the cross section is elastic to the radius b_Y, by Eq. (6.14),

$$b_Y = \frac{\tau_Y}{G\theta} \tag{c}$$

Dividing Eq. (c) by b, we find, with Eq. (b),

$$\theta = \frac{b\theta_Y}{b_Y} \tag{d}$$

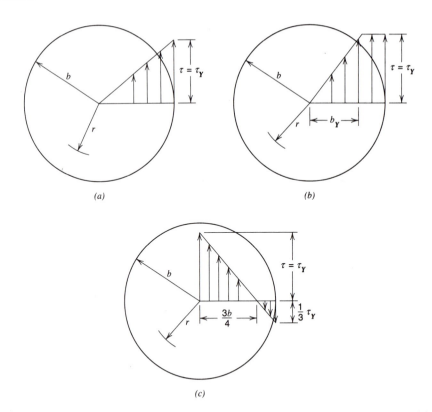

Figure E6.7   Solid circular cross section. (a) Yield at $r = b$. (b) Yield at $r = b_Y$. (c) Residual shear stress.

(c)   Likewise, the twisting moment T_E due to the shear stress up to $r = b_Y$, that is, in the elastic core, by Eq. (6.15) with $J = \pi b_Y^4/2$, is

$$T_E = \frac{\pi \tau_Y b_Y^3}{2} \tag{e}$$

The torque T_P due to the stress in the yielded (plastic) region of the bar $(r > b_Y)$ is

$$T_P = \int_{b_Y}^{b} r\tau_Y(2\pi r\, dr) = \frac{2}{3}\pi\tau_Y(b^3 - b_Y^3) \tag{f}$$

The total elastic-plastic moment T_{EP} is, by Eqs. (e) and (f),

$$T_{EP} = T_E + T_P = \frac{2}{3}\pi\tau_Y\left(b^3 - \frac{1}{4}b_Y^3\right) \tag{g}$$

(d)   The fully plastic or limiting moment T_L is given by Eq. (g), letting $b_Y \to 0$. Thus,

$$T_L = \frac{2}{3}\pi\tau_Y b^3 \tag{h}$$

This result agrees with that of Table 6.3. Dividing Eq. (h) by Eq. (a), we obtain

$$T_L = \frac{4}{3} T_Y \tag{i}$$

Thus, the limiting moment T_L is 4/3 times larger than the torque that initiates yield [recall the relation $M_P = 1.5 M_Y$ for bending of a rectangular beam, Eq. (4.42)].

(e) When the limiting torque T_L is released, the bar springs back (unwinds) elastically, and the shear stress-strain curve will follow a path similar to CA' in Fig. 4.4a. Since the limiting torque is $T_L = 4T_Y/3$, the elastic stress recovery at $r = b$ is $4\tau_Y/3$. Hence, since the shear stress is initially τ_Y at $r = b$, a residual shear stress $\tau_Y/3$ in the opposite sense of τ_Y remains at $r = b$. However, at the center of the cross section ($r = 0$), the residual stress remains τ_Y, in the original sense, as shown in Fig. E6.7c. Since the recovery is elastic, the shear stress varies linearly from $r = 0$ to $r = b$. Hence, the residual shear stress is zero at $r = 3b/4$.

PROBLEMS
Sections 6.1–6.3

6.1. Derive the relation for the shear stress distribution on the x axis for the equilateral triangle in Fig. 6.9.

6.2. Derive Eqs. (6.49) for the equilateral triangle.

6.3. A square shaft may be used to transmit power from a farm tractor to farm implements. A 25.0-mm square shaft is made of a steel having a yield stress of $Y = 380$ MPa. Determine the torque that can be applied to the shaft based on a factor of safety of $SF = 2.00$ by using the octahedral shear stress criterion of failure.

6.4. A square shaft has 42.0-mm sides and the same cross-sectional area as shafts having circular and equilateral triangular cross sections. If each shaft is subjected to a torque of 1.00 kN·m, determine the maximum shear stress for each of the three shafts.

Ans. $\tau_{square} = 64.89$ MPa, $\tau_{circle} = 47.82$ MPa, $\tau_{triangle} = 76.86$ MPa

6.5. The shafts in Problem 6.4 are made of a steel for which $G = 77.5$ GPa. Determine the unit angle of twist for each shaft.

6.6. The left-hand section of the torsion member in Fig. E6.3 is 2.00 m long, and the right-hand section is 1.00 m long. It is made of an aluminum alloy for which $G = 27.1$ GPa. Determine the magnitude of T_2 if $T_1 = 350$ N·m and the maximum shear stress is 45.0 MPa. Neglect stress concentrations at changes in section. Determine the angle of twist of the free end. The support at the left end prevents rotation of this cross section but does not prevent warping of the cross section.

Ans. $T_2 = 361.3$ N·m, angle of twist $= 0.1391$ rad

6.7. A torsion member has an elliptical cross section with major and minor dimensions of 50.0 mm and 30.0 mm, respectively. The yield stress of the material in the torsion member is $Y = 400$ MPa. Determine the maximum torque that can be applied to the torsion member based on a factor of safety of $SF = 1.85$ using the maximum shear stress criterion of failure.

6.8. A steel bar has a rectangular cross section 12.7 mm wide and 38.1 mm deep. The bar is subjected to a twisting moment $T = 135.7$ N·m. The shear yield stress of the material is 82.7 MPa.

(a) By Eq. (6.50), calculate the maximum shear stress in the bar and show in a diagram where it occurs.

(b) Calculate the shear stress in the bar at the center of the short side.

Ans. (a) $\tau_{max} = 82.7$ MPa, (b) $\tau = 27.6$ MPa

6.9. A bar of steel has a tensile yield stress $Y = 345$ MPa and shear yield stress $\tau_Y = 207$ MPa. The bar has a rectangular cross section, and it is subjected to a twisting moment $T = 565$ N·m. The working stress limit of the bar is two-thirds τ_Y. If the width of the cross section of the bar is $2h = 19$ mm, determine the length $2b$ of the cross section (Fig. 6.10).

6.10. A rectangular bar has a cross section such that $b/h = k$, and it is subjected to a twisting moment T. A cylindrical bar of diameter d is also subjected to T. Show that the maximum shear stresses in the two bars are equal, provided $d = 3.441h(kk_2)^{1/3}$ and the bars remain elastic.

6.11. The depth $2b$ of a rectangular cross section torsion bar is 38.1 mm (Fig. 6.10). Determine the required width $2h$ so that the maximum shear stress produced in it is the same as that in a cylindrical bar 51 mm in diameter, both bars being subjected to the same twisting moment.

6.12. Two bars, one with a square cross section and one with a circular cross section, have equal cross-sectional areas. The bars are subjected to equal twisting moments. Determine the ratio of the maximum shear stresses in the two bars, assuming that they remain elastic.

Ans. $\tau_{max(square\ bar)} = 1.36\ \tau_{max(circular\ bar)}$

6.13. A stepped steel shaft ABC has lengths $AB = L_1 = 1.0$ m and $BC = L_2 = 1.27$ m, with diameters $d_1 = 25.4$ mm and $d_2 = 19.05$ mm, respectively. The steel has a yield stress $Y = 450$ MPa and shear modulus $G = 77$ GPa. A twisting moment is applied at the stepped section B. Ends A and C are fixed.

(a) Determine the value of T that first causes yielding.

(b) For this value of T, determine the angle of rotation β_B at section B.

6.14. Consider a hollow elliptic cylinder with its outer elliptic surface defined by $(x/h)^2 + (y/b)^2 = 1$ and inner elliptic surface defined by $[x/(kh)]^2 + [y/(kb)]^2 = 1$. Show that

$$\theta = \frac{(h^2 + b^2)T}{\pi h^3 b^3 (1 - k^4)G} \qquad \tau_{max} = \frac{2T}{\pi bh^2 (1 - k^4)}$$

and

$$\sigma_{zx} = -\frac{2T}{\pi h b^3 (1 - k^4)} y, \qquad \sigma_{zy} = \frac{2T}{\pi b h^3 (1 - k^4)} x$$

Hint: By the theory of hollow torsional members (Boresi and Chong, 1987), the twisting moment T is related to ϕ by the relation

$$T = \int\!\!\int_R 2\phi \, dA + 2K_1 A_1, \text{ where } \phi = A\left(\frac{x^2}{h^2} + \frac{y^2}{b^2} - 1\right),$$

K_1 is the value of ϕ on the inner elliptic surface, A_1 is the area bounded by the inner ellipse and R is the solid region bounded by the inner and outer ellipses.

Sections 6.4–6.6

6.15. Find the maximum shear stress and unit angle of twist of the bar having the cross section shown in Fig. P6.15 when subjected to a torque at its ends of 600 N·m. The bar is made of a steel for which $G = 77.5$ GPa.

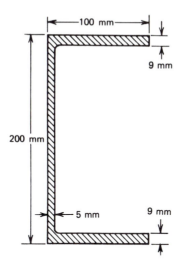

Figure P6.15

6.16. An aluminum alloy extruded section (Fig. P6.16) is subjected to a torsional load. Determine the maximum torque that can be applied to the member if the maximum shear stress is 75.0 MPa. Neglect stress concentrations at changes in section.

Ans. $T = 665.4$ N·m

6.17. For a thin-wall circular cross-section tube, show that the polar moment of inertia J of the cross section is approximately $J = 2\pi R^3 t$, where R is the mean radius of the tube and t is the wall thickness. Determine the percent error in J as t increases from 0.001 R to 0.2 R

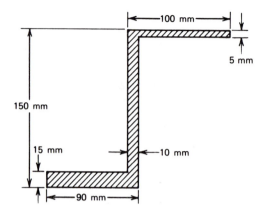

Figure P6.16

6.18. A thin-wall torsion member subjected to torque T has the cross section shown in Fig. P6.18. The wall thickness t is constant throughout the section. By the theory of Sec. 6.6, in terms of T, b, and t,

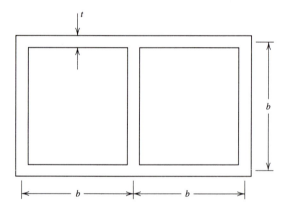

Figure P6.18

(a) derive formulas for the shear stress in the outer walls and the interior web and

(b) derive a formula for the unit angle of twist. Disregard the effects of stress concentrations at interior corners.

6.19. Remove the interior web in the cross section of Fig. P6.18. Derive formulas for the shear stress and unit angle of twist of the section. Ignore the effects of stress concentrations.

6.20. For the cross section in Problem 6.19, make a slit lengthwise along the member so as to form an open cross section. Derive formulas for the shear stress and unit angle of twist. Compare the results to those obtained in Problem 6.19.

6.21. A thin-wall brass tube ($G = 27.6$ GPa) has an equilateral triangular cross section. The mean length of one side of the triangle is 25.4 mm and the wall thickness 2.54 mm. The tube is subjected to a twisting moment $T = 20$ m·N. Determine the maximum shear stress and angle of twist per unit length of the tube.

Ans. $\tau_{max} = 14.1$ MPa, $\theta = 0.0696$ rad/m

6.22. An aluminum $(G = 26.7$ GPa) torsion bar has the cross section shown in Fig. P6.22. The bar is subjected to a twisting moment $T = 1356$ m·N.

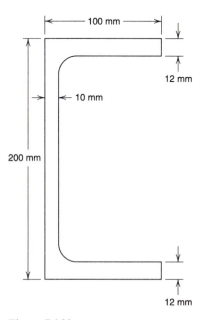

Figure P6.22

(a) Determine the maximum shear stress τ_{max} and angle of twist per unit length.

(b) At what location in the cross section does τ_{max} occur? Ignore stress concentrations.

6.23. Compare the shear stress and the unit angle of twist for three thin-wall sections: a circular tube, a square tube, and an equilateral triangle. The three sections have equal wall thicknesses and equal perimeters.

Ans. $\tau_{square} = 1.27\, \tau_{circle}$; $\tau_{triangle} = 1.65\, \tau_{circle}$; $\theta_{square} = 1.62\, \theta_{circle}$; $\theta_{triangle} = 2.74\, \theta_{circle}$

6.24. A steel $(G = 79$ GPa) torsion bar is subjected to twisting moments $T = 226$ m·N at its ends (Fig. P6.24). The bar is 1.2 m long.

(a) Determine the angle of twist of one end relative to the other end.

(b) Determine the maximum shear stress and its location in the section.

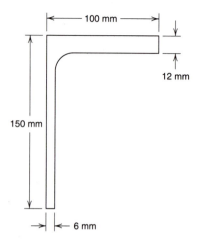

Figure P6.24

6.25. **(a)** Compare the twisting moments required to give the same angle of twist per unit length to the two sections shown in Fig. P6.25.

(b) Compare the twisting moments required to cause the same maximum shear stress. Ignore the effect of stress concentrations.

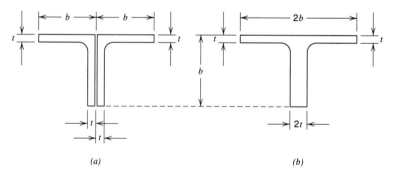

(a)   *(b)*

Figure P6.25

6.26. The hollow circular and square thin-wall torsion members in Fig. P6.26 have identical values for b and t. Neglecting the stress concentrations at

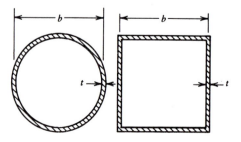

Figure P6.26

the corners of the square, determine the ratio of the torques and unit angle of twists for the two torsion members, for equal shear stresses in each.

6.27. A hollow thin-wall brass tube has an equilateral triangular cross section. The mean length of each side of the triangle is 40.0 mm. The wall thickness is 4.00 mm. Determine the torque and unit angle of twist for an average shear stress of 20.0 MPa, and for $G = 31.1$ GPa.

Ans. $T = 110.8$ N·m, $\theta = 0.0559$ rad/m

6.28. A hollow rectangular thin-wall steel torsion member has the cross section shown in Fig. P6.28. The steel has a yield stress $Y = 360$ MPa and shear modulus of elasticity of $G = 77.5$ GPa. Determine the maximum torque that may be applied to the torsion member, based on a factor of safety of $SF = 2.00$ for the octahedral shear stress criterion of failure. What is the unit angle of twist when the maximum torque is applied?

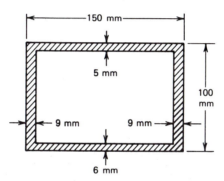

Figure P6.28

6.29. The hollow thin-wall torsion member of Fig. P6.29 has uniform thickness walls. Show that walls BC, CD, and CF are stress-free.

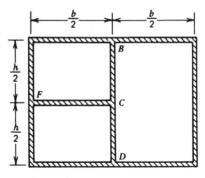

Figure P6.29

6.30. The aluminum ($G = 27.1$ GPa) hollow thin-wall torsion member in Fig. P6.30 has the dimensions shown. Its length is 3.00 m. If the member is subjected to a torque $T = 11.0$ kN·m, determine the maximum shear stress and angle of twist.

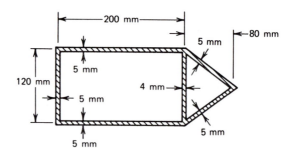

Figure P6.30

Section 6.7

6.31. A wide-flange steel ($E = 200$ GPa and $G = 77.5$ GPa) I-beam has a depth of 300 mm, web thickness of 15 mm, flange width of 270 mm, flange thickness of 20 mm, and length of 8.00 m. The I-beam is fixed at one end and free at the other end. A twisting moment $T = 7.00$ kN·m is applied at the free end. Determine the maximum normal stress and maximum shear stress in the I-beam and the angle of rotation β of the free end of the I-beam.

6.32. The I-beam in Fig. P6.32 is an aluminum alloy ($E = 72.0$ GPa and $G = 27.1$ GPa) extruded section. It is fixed at the wall and attached rigidly to the thick massive plate at the other end. Determine the magnitude of P for $\sigma_{max} = 160$ MPa.

Ans. $P = 1.095$ kN

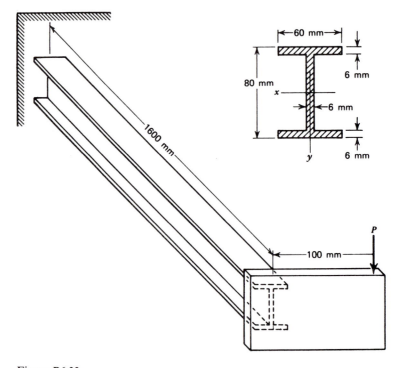

Figure P6.32

6.33. Let the thick plate in Problem 6.32 be subjected to a torque $T = 150$ N·m directed along the axis of the I-beam. Determine the maximum shear stress and angle of twist β of the plate.

Section 6.8

6.34. **(a)** By the finite difference method, determine the maximum shear stress for a solid rectangular torsion member subjected to a twisting moment T. The cross section of the member has dimensions $a \times 2a$. Use a mesh dimension of $h = a/4$.

 (b) Repeat part (a), using a mesh dimension $h = a/8$.

 (c) Compare the results of parts (a) and (b) with the results obtained in Sec. 6.8.

 (d) Use Eq. (6.100) to obtain an improved estimate of maximum shear stress.

Section 6.9

6.35. Derive the relation for the fully plastic torque for a rectangular cross section having dimensions $2a$ by $2b$.

6.36. Derive the relation for the fully plastic torque for the cross sections in the first row of Table 6.4.

6.37. A rectangular section torsion member has dimensions of 100 mm by 150 mm and is made of a steel for which the shear yield point is $\tau_Y = 100.0$ MPa. Determine T_P for the cross section and the ratio of T_P to T_Y, where T_Y is the maximum elastic torque.

6.38. A rectangular hollow torsion member has external dimensions of 200 mm by 400 mm. The cross section has a uniform thickness of 30 mm. For a material that has a shear stress $\tau_Y = 120$ MPa, determine the fully plastic torque.

 Ans. $T_P = 455.0$ kN·m

6.39. Repeat Example 6.7 for a hollow circular cross section with outer radius b and inner radius a.

REFERENCES

American Institute of Steel Construction, Inc. (AISC) (1983). *Torsion Analysis of Steel Members*. Chicago, Ill.

Arutyunyan, N. Kh. and Abramyan, B. L. (1963). *Torsion of Elastic Bodies*. Moscow: Fizmatgiz (in Russian).

Arutyunyan, N. Kh., Abramyan, B. L., and Naumov, V. E. (1988). 'Torsion of Inhomogeneous Shafts.' In *Mechanics of Deformable Solids* (A. Yu. Ishlinskii, ed.). New York: Allerton Press, Chapter 4.

Boresi, A. P. and Chong, K. P. (1987). *Elasticity in Engineering Mechanics*. New York: Elsevier.

Boresi, A. P. and Chong, K. P. (1991). *Approximate Solution Methods in Engineering Mechanics*. New York: Elsevier.

Nadai, A. (1950). *Theory of Flow and Fracture of Solids*, Vol. 1. New York: McGraw-Hill.

Noor, A. K. and Anderson, C. M. (1975). 'Mixed Isoparametric Elements for Saint-Venant Torsion'. In *Computer Methods Applied Mechanics and Engineering*, Vol. 6 (J. H. Argyris, W. Prager, and A. M. O. Smith, eds.) Amsterdam: North-Holland, pp. 195–218.

Prandtl, L. (1903). 'Zur Torsion von prisma atischen Stäben.' *Physik Ziet.*, **4**: 758–770.

Sadowsky, M. A. (1941). 'An Extension of the Sand Heap Analogy in Plastic Torsion Applicable to Cross-Sections Having One or More Holes.' *J. App. Mech.*, **8** (4): A166–A168.

Smith, F. A., Thomas, F. M., and Smith, J. O. (1970). 'Torsional Analysis of Heavy Box Beams in Structures.' *J. Structural Div.*, **ASCE 96** (ST3): 613–635 (Proc. Paper 7165).

Thoms, R. L. and Masch, F. D. (1965). 'Membrane Analogy Studies Employing Visible Contour Lines.' In *Developments in Theoretical and Applied Mechanics*, Vol. 2 (W. A. Shaw, ed.). New York: Pergamon Press, pp. 545–555.

von Kármán, T. and Chien, W. Z. (1946). 'Torsion with Variable Twist.' *J. Aero. Sci.*, **13**, (10): 503–510.

Zienkiewicz, O. C. and Taylor, R. C. (1989). *The Finite Element Method*, 4th ed., Vol. 1. New York: McGraw-Hill.

7

NONSYMMETRICAL BENDING OF STRAIGHT BEAMS

In this chapter we assume that there is a plane in which the forces that act on a beam lie. This plane is called the *plane of loads*. In addition, we assume that the plane of loads passes through a point (the *shear center*) in the beam cross section, so that there is no twisting (torsion) of the beam; that is, the resulting forces that act on any cross section of the beam consist only of bending moments and shear forces. The net torque is zero. For cross sections that have two or more axes of symmetry, the shear center is located at the intersection of the axes. For cross sections with one axis of symmetry, the shear center lies on the axis. Similarly, for a cross section with two axes of antisymmetry, the shear center is located at the point of intersection of the axes. For general cross sections, the theory of elasticity may be used to locate the shear center of a cross section (Boresi and Chong, 1987). However, in Chapter 8 the methods of mechanics of materials are used to locate the shear center approximately. We introduce the concepts of symmetrical and nonsymmetrical bending of straight beams and the plane of loads in Sec. 7.1. In Sec. 7.2, we develop formulas for stresses in beams subjected to nonsymmetrical bending. In Sec. 7.3, deflections of beams are computed. In Sec. 7.4, the effect of an inclined load relative to a principal plane is examined. Finally, in Sec. 7.5, a method is presented for computing fully plastic loads for cross sections in nonsymmetrical bending.

7.1

DEFINITION OF SHEAR CENTER IN BENDING. SYMMETRICAL AND NONSYMMETRICAL BENDING

The straight cantilever beam shown in Fig. 7.1 has a cross section of arbitrary shape. It is subjected to pure bending by the end couple M_0. Let the origin 0 of the coordinate system (x, y, z) be chosen at the centroid of the beam cross section at the left end of the beam, with the z axis directed along the centroidal axis of the beam, and the (x, y) axes taken in the plane of the cross section. Generally, the orientation of the (x, y) axis is arbitrary. However, we often choose the (x, y) axes so that the moments of inertia of the cross section I_x, I_y, and I_{xy} are easily calculated, or we may take them to be principal axes of the cross section (see Appendix B).

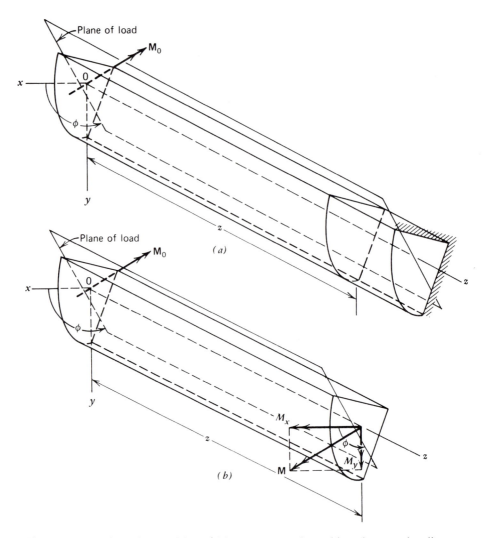

Figure 7.1 Cantilever beam with an arbitrary cross section subjected to pure bending.

The bending moment that acts at the left end of the beam (Fig. 7.1*a*) is repre-sented by the vector $\mathbf{M_0}$ directed perpendicular to a plane that forms an angle ϕ $(0 \leq \phi \leq \pi)$ taken positive when measured counterclockwise from the *x-z* plane as viewed from the positive *z* axis. This plane is called the *plane of load* or the *plane of loads*. A more complete description of the plane of loads is given later on in this section. Consider now a cross section of the beam at distance *z* from the left end. The free-body diagram of the part of the beam to the left of this section is shown in Fig. 7.1*b*. For equilibrium of this part of the beam, a moment $\mathbf{M}$, equal in magnitude but opposite in sense to $\mathbf{M_0}$, must act at section *z*. For the case shown $(\pi/2 \leq \phi \leq \pi)$, the (x, y) components (M_x, M_y) of $\mathbf{M}$ are related to the signed magnitude M of $\mathbf{M}$ by the relations $M_x = M \sin \phi$, $M_y = -M \cos \phi$. Since $\pi/2 \leq \phi \leq \pi$, $\sin \phi$ is positive and $\cos \phi$ negative. Since (M_x, M_y) are positive (Fig. 7.1*b*), the sign of M is positive. A more complete discussion of the sign con-vention for M is given in Sec. 7.2, following Eq. (7.13).

Shear Loading of a Beam. Shear Center Defined

Let the beam shown in Fig. 7.2*a* be subjected to a concentrated force **P** that lies in the end plane ($z = 0$) of the beam cross section. The vector representing **P** lies in a plane that forms angle ϕ $(0 \leq \phi \leq \pi)$, taken positive when measured counterclockwise from the z-x plane as viewed from the positive z axis. This plane is called the *plane of the load*. Consider a cross section of the beam at distance z from the left end. The free-body diagram of the part of the beam to the left of this section is shown in Fig. 7.2*b*. For equilibrium of this part of the beam, a moment **M**, with components M_x and M_y, shear components V_x and V_y, and in general, a twisting moment **T** (with vector directed along the positive z axis) must act on the section at z. However, if the line of action of force **P** passes through a certain point C (the shear center) in the cross section, $\mathbf{T} = \mathbf{0}$. In this discussion, we assume that

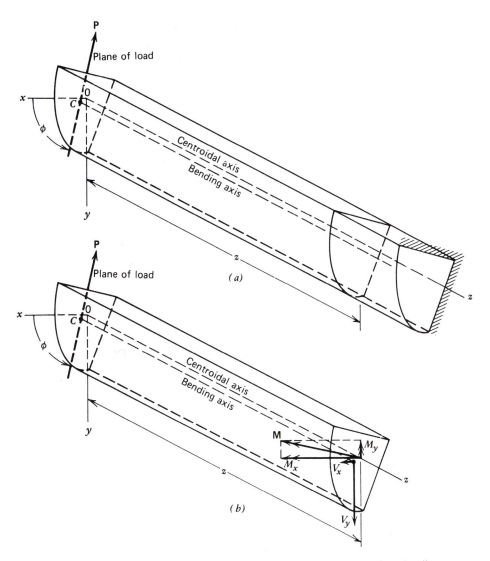

Figure 7.2   Cantilever beam with an arbitrary cross section subjected to shear loading.

the line of action of **P** passes through the shear center. Hence, **T** is not shown in Fig. 7.2b. Note that in Fig. 7.2b, the force **P** requires V_x, V_y to be positive [directed along positive (x, y) axes, respectively]. The component M_x is also directed along the positive x axis. However, since $\phi < \pi/2$, M_y is negative (directed along the negative y axis.

There is a particular axial line in the beam called the *bending axis of the beam*, which is parallel to the centroidal axis of the beam (the line that passes through the centroids of all of the cross sections of the beam). Except for special cases, the bending axis does not coincide with the centroidal axis (Fig. 7.2).

The intersection of the bending axis with any cross section of the beam locates a point C in that cross section called the *shear center* of the cross section (see Sec. 8.1). Thus, the bending axis passes through the shear centers of all the cross sections of the beam.

In Sec. 7.2, formulas are derived for the normal stress component σ_{zz} that acts on the cross section at z in terms of the bending moment components (M_x, M_y). Also, one may derive formulas for the shear stress components (τ_{zx}, τ_{zy}) due to the shear forces (V_x, V_y). However, if the length L of the beam is large compared to the maximum cross-section dimension D, such that $L/D > 5$, the maximum shear stress is small compared to the maximum normal stress. In this chapter we ignore the shear stresses due to (V_x, V_y); that is, we consider beams for which $L/D > 5$.

For bending of a beam by a concentrated force and for which the shear stresses are negligible, the line of action of the force must pass through the shear center of a cross section of the beam; otherwise, the beam will be subjected to both bending and torsion (twist). Thus, for the theory of pure bending of beams, we assume that the shear stresses due to concentrated loads are negligible and that the lines of action of concentrated forces that act on the beam pass through the shear center of a beam cross section. If the cross section of a beam has either an axis of symmetry or an axis of antisymmetry, the shear center C is located on that axis (Fig. 7.3). If the cross section has two or more axes of symmetry or antisymmetry, the shear center is located at the intersection of the axes (Figs. 7.3a and d). For a general cross section (Fig. 7.1) or for a relatively thick, solid cross section (Fig. 7.3c), the determination of the location of the shear center requires advanced computational methods (Boresi and Chong, 1987). For this reason, the location of the shear center is often determined in an approximate manner; the errors introduced by such approximations of the shear center location are discussed in the next paragraph.

Let the line of action of force **P** pass through an approximate location of the shear center of the beam, point B in Fig. 7.4. Let C be the location of the shear center. Since the line of action of force **P** does not pass through C, the force **P** is assumed to be replaceable by a couple (torque) that lies in the cross section and a force with an action line that passes through C. This representation or transformation of force **P** is assumed to be valid for the deformable beam cross section, although strictly speaking, it is applicable to rigid bodies only. The transformation is accomplished by adding self-equilibrating forces **P′** and **P″** at C that are parallel to **P** and have magnitudes equal to that of **P**. Thus, the force **P** is considered to be equivalent to a torque (couple) of magnitude $T = Pd$, due to forces **P** and **P″**, where d is the perpendicular distance between **P** and **P″** and a force **P′** acting at C.

Now pass a cutting plane through the member at distance z from the left end. The free-body diagram of the beam to the left of the cut is shown in Fig. 7.4b. For equilibrium, the forces at the cut include a bending moment **M** with compo-

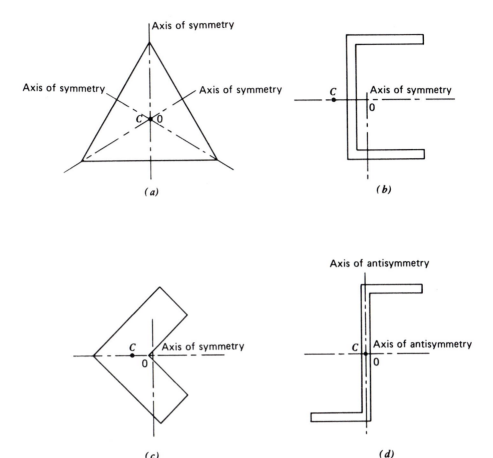

Figure 7.3 (*a*) Equilateral triangle section. (*b*) Open channel section. (*c*) Angle section. (*d*) Z-section.

nents M_x and M_y, torque of magnitude $T = Pd$, and shears V_x and V_y. The normal stress distribution σ_{zz} due to M_x and M_y can be calculated by the formulas derived in Sec. 7.2. The shear stresses due to V_x and V_y are considered to be negligible ($L/D > 5$). The shear stress due to torque T may be computed by the methods presented in Chapter 6. Cross sections with thick walls (Fig. 7.3*c*) require large torques if the maximum shear stress due to the torque is to be significant. For such cross sections, an approximate location of the shear center will suffice, since shear stresses due to T are small compared to the maximum value of σ_{zz}, provided Pz is large compared to Pd. However, caution must be used for cross sections made of connected narrow rectangular walls such as the open channel cross section shown in Fig. 7.3*b*, since, as noted in Chapter 6, such cross sections have little resistance to torsional loads. For these kinds of cross sections, an accurate estimate of the location of the shear center is necessary. Such problems are treated in Chapter 8.

In this chapter, unless the shear center is located by intersecting axes of symmetry or antisymmetry, its location is approximated. The reader should have a better understanding of such approximations after studying Chapter 8.

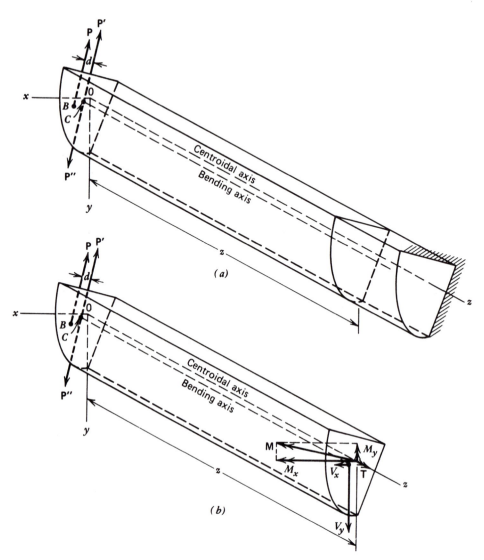

Figure 7.4 Cantilever beam with arbitrary cross section subjected to shear loading not at shear center.

Symmetrical Bending. Nonsymmetrical Bending

In Appendix B, it is shown that every beam cross section has principal axes (X, Y). With respect to principal axes (X, Y), the product of inertia of the cross section is zero; $I_{XY} = 0$. The principal axes (X, Y) for the cross section of the cantilever beam of Fig. 7.1 are shown in Fig. 7.5. For convenience, axes (X, Y) are also shown at a section of the beam at distance z from the left end of the beam. At the left end, let the beam be subjected to a couple $\mathbf{M_0}$ with sense in the negative X direction and a force $\mathbf{P}$ through the shear center C with sense in the negative Y direction (Fig. 7.5a). These loads are reacted by a bending moment $\mathbf{M} = \mathbf{M_X}$ at the cut section with sense in the positive X direction. By Bernoulli beam theory (Boresi and Chong, 1987), the stress σ_{zz} normal to the cross section is given by the flexure

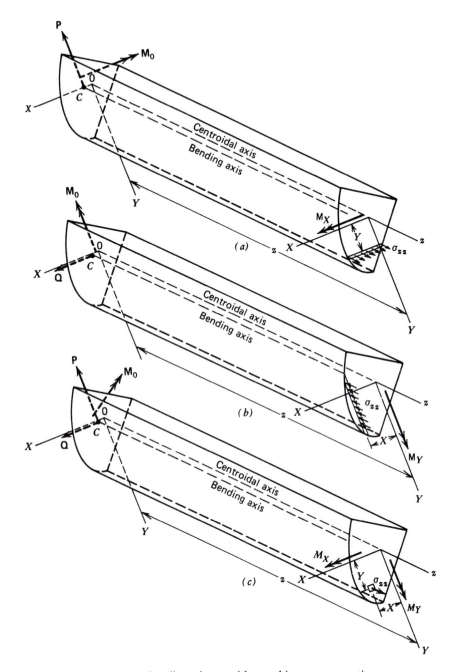

Figure 7.5   Cantilever beam with an arbitrary cross section.

formula

$$\sigma_{zz} = \frac{M_X Y}{I_X} \qquad (7.1)$$

where Y is the distance from the principal axis X to the point in the cross section
at which σ_{zz} acts, and I_X is the principal moment of inertia of the cross-sectional

area relative to the X axis. Equation (7.1) shows that σ_{zz} is zero for $Y = 0$ (the X axis). Consequently, the X axis is called the *neutral axis of bending of the cross section*, that is, the axis for which $\sigma_{zz} = 0$. We define the bending moment component M_X as positive when the sense of the vector representing $\mathbf{M}_X$ is in the positive X direction. Since M_X is related to σ_{zz} by Eq. (7.1), σ_{zz} is a tensile stress for positive values of Y and a compressive stress for negative values of Y. In addition to causing a bending moment component M_X, load P produces a positive shear V_Y at the cut section. It is assumed that the maximum shear stress τ_{ZY} resulting from V_Y is small compared to the maximum value of σ_{zz}. Hence, since this chapter treats bending effects only, we neglect shear stresses in this chapter.

Likewise, if a load $\mathbf{Q}$ (applied at the shear center C) directed along the positive X axis and a moment $\mathbf{M}_0$ directed along the negative Y axis are applied to the left end of the beam (Fig. 7.5b), they are reacted by a bending moment $\mathbf{M} = \mathbf{M}_Y$ directed along the positive Y axis. The normal stress distribution σ_{zz}, due to the component M_Y, is also given by the flexure formula. Thus,

$$\sigma_{zz} = -\frac{M_Y X}{I_Y} \tag{7.2}$$

where X is the distance from the principal axis Y to the point in the cross section at which σ_{zz} acts, and I_Y is the principal moment of inertia of the cross-sectional area relative to the Y axis. The negative sign arises from the fact that a positive M_Y produces compressive stresses on the positive side of the X axis. Now for $X = 0$ (the Y axis), $\sigma_{zz} = 0$. Hence, in this case, the Y axis is the *neutral axis of bending of the cross section*, that is, the axis for which $\sigma_{zz} = 0$. In either case [Eq. (7.1) or (7.2)], the beam is subjected to *symmetrical bending*. (Bending occurs about a neutral axis in the cross section that coincides with the corresponding principal axis.)

In Fig. 7.5c, the beam is subjected to moment $\mathbf{M}_0$ with components in the negative directions of both axis (X, Y), as well as concentrated forces $\mathbf{P}$ and $\mathbf{Q}$ acting through the shear center C. These loads result in a bending moment $\mathbf{M}$ at the cut section with positive components (M_X, M_Y). For this loading, the stress σ_{zz} normal to the cross section may be obtained by the superposition of Eqs. (7.1) and (7.2). Thus,

$$\sigma_{zz} = \frac{M_X Y}{I_X} - \frac{M_Y X}{I_Y} \tag{7.3}$$

In this case, the moment $\mathbf{M} = (M_X, M_Y)$ is not parallel to either of the principal axes (X, Y). Hence the bending of the beam occurs about an axis that is not parallel to either the X or Y axis. When the axis of bending does not coincide with a principal axis direction, the bending of the beam is said to be nonsymmetrical. The determination of the *neutral axis of bending of the cross section* for nonsymmetrical bending is discussed in Sec. 7.2.

Plane of Loads. Symmetrical and Nonsymmetrical Loading

Often, a beam is loaded by forces that lie in a plane which coincides with a plane of symmetry of the beam, (Fig. 7.6). In this figure, the y-axis is an axis of symmetry for the cross section; it is a principal axis. Hence, if axes (x, y) are principal axes for the cross section, the beams in Figs. 7.6a and b undergo symmetrical bending;

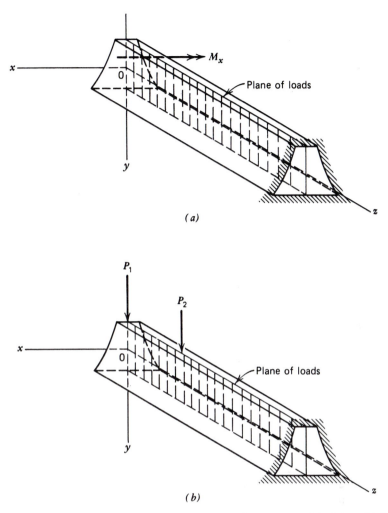

Figure 7.6   Plane of loads coincident with the plane of symmetry of the beam. (a) Couple loads. (b) Lateral loads.

that is, bending about a principal axis of a cross section, since the moment vector in Fig. 7.6a and the force vectors in Fig. 7.6b are parallel to principal axes. (See the discussion above in the section entitled "Symmetrical Bending. Nonsymmetrical Bending.") We further observe that since the shear center lies on the y axis, the plane of the load contains the axis of bending of the beam. It is shown later in this chapter that if the plane of loads does not coincide with a plane of symmetry of the beam, the beam may still deform symmetrically (bend about a principal axis), provided that the plane of loads contains the bending axis and is parallel to one of the principal planes [the (x, z) and (y, z) planes in Fig. 7.6].

Consider next two beams with cross sections shown in Fig. 7.7. Since a rectangular cross section (Fig. 7.7a) has two axes of symmetry that pass through its centroid 0, the shear center C coincides with the centroid 0. Let the intersection of the plane of the loads and the plane of the cross section be denoted by line L-L, which forms angle ϕ ($0 \leq \phi \leq \pi$) measured counterclockwise from the x-z plane

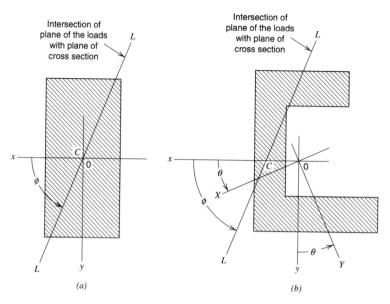

Figure 7.7   Unsymmetrically loaded beams. (a) Rectangular cross section. (b) Channel cross section.

and passes through the shear center C. Since the plane of loads contains point C, the bending axis of the rectangular beam lies in the plane of the loads. If the angle ϕ equals 0 or $\pi/2$, the rectangular beam will undergo symmetrical bending. For other values of ϕ, the beam undergoes nonsymmetrical bending; that is, bending for which the neutral axis of bending of the cross section does not coincide with either of the principal axes (X, Y).

In the case of a general channel section (Fig. 7.7b), the principal axes (X, Y) are located by a rotation through angle θ (positive θ is taken counterclockwise) from the (x, y) axes as shown. The value of θ is determined by Eq. (B.12) in Appendix B. Although the plane of loads contains the shear center C (and hence, the bending axis of the beam), it is not parallel to either of the principal planes (X, z), (Y, z). Hence, in general, the channel beam (Fig. 7.7b) undergoes nonsymmetrical bending. However, for the two special cases, $\phi = \theta$ or $\phi = \theta + \pi/2$, the channel beam does undergo symmetrical bending.

7.2

BENDING STRESSES IN BEAMS SUBJECTED TO NONSYMMETRICAL BENDING

Let a cutting plane be passed through a straight cantilever beam at section z. The free-body diagram of the beam to the left of the cut is shown in Fig. 7.8a. The beam has constant cross section of arbitrary shape. The origin 0 of the coordinate axes is chosen at the centroid of the beam cross section at the left end of the beam with the z axis taken parallel to the beam. The left end of the beam is subjected to a bending couple $\mathbf{M_0}$ that is equilibrated by bending moment $\mathbf{M}$ acting on the

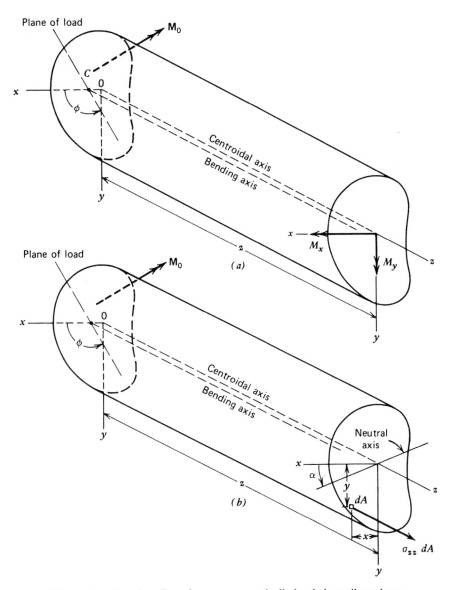

Figure 7.8   Pure bending of a nonsymmetrically loaded cantilever beam.

cross section at z, with positive components (M_x, M_y) as shown. The bending moment $\mathbf{M} = (M_x, M_y)$ is the resultant of the forces due to the normal stress σ_{zz} acting on the section (Fig. 7.8b). For convenience, we show (x, y) axes at the cross section z. It is assumed that the (x, y) axes are not principal axes for the cross section. In this article, we derive the load-stress formula that relates the normal stress σ_{zz} acting on the cross section to the components (M_x, M_y).

The derivation of load-stress and load-deformation relations for the beam requires that equations of equilibrium, compatibility conditions, and stress-strain relations be satisfied for the beam along with specified boundary conditions for the beam.

Equations of Equilibrium

Application of the equations of equilibrium to the free body in Fig. 7.8b yields (since there is no net resultant force in the z direction)

$$0 = \int \sigma_{zz} \, dA$$

$$M_x = \int y \sigma_{zz} \, dA$$

$$M_y = -\int x \sigma_{zz} \, dA \tag{7.4}$$

where dA denotes an element of area in the cross section and the integration is performed over the area A of the cross section. The other three equilibrium equations are satisfied identically, since σ_{zz} is the only nonzero stress component. To evaluate the integrals in Eq. (7.4), it is necessary that the functional relation between σ_{zz} and (x, y) be known. The determination of σ_{zz} as a function of (x, y) is achieved by considering the geometry of deformation and the stress-strain relations.

Geometry of Deformation

We assume that plane sections of an unloaded beam remain plane after the beam is subjected to pure bending. Consider two plane cross sections perpendicular to the bending axis of an unloaded beam such that the centroids of the two sections are separated by a distance Δz. These two planes are parallel since the beam is straight. These planes rotate with respect to each other when moments M_x and M_y are applied. Hence, the extension e_{zz} of longitudinal fibers of the beam between the two planes can be represented as a linear function of (x, y), namely

$$e_{zz} = a'' + b''x + c''y \tag{7.5}$$

where a'', b'', and c'' are constants. Since the beam is initially straight, all fibers have the same initial length Δz so that the strain ϵ_{zz} can be obtained by dividing Eq. (7.5) by Δz. Thus,

$$\epsilon_{zz} = a' + b'x + c'y \tag{7.6}$$

where $\epsilon_{zz} = e_{zz}/\Delta z$, $a' = a''/\Delta z$, $b' = b''/\Delta z$, and $c' = c''/\Delta z$.

Stress-Strain Relations

According to the theory of pure bending of straight beams, the only nonzero stress component in the beam is σ_{zz}. For linearly elastic conditions, Hooke's law states

$$\sigma_{zz} = E\epsilon_{zz} \tag{7.7}$$

Eliminating ϵ_{zz} between Eqs. (7.6) and (7.7), we obtain

$$\sigma_{zz} = a + bx + cy \tag{7.8}$$

where $a = Ea'$, $b = Eb'$, and $c = Ec'$.

Load-Stress Relation for Nonsymmetrical Bending

Substitution of Eq. (7.8) into Eqs. (7.4) yields

$$0 = \int (a + bx + cy)\, dA = a \int dA + b \int x\, dA + c \int y\, dA$$

$$M_x = \int (ay + bxy + cy^2)\, dA = a \int y\, dA + b \int xy\, dA + c \int y^2\, dA$$

$$M_y = -\int (ax + bx^2 + cxy)\, dA = -a \int x\, dA - b \int x^2\, dA - c \int xy\, dA \quad (7.9)$$

Since the z axis passes through the centroid of each cross section of the end beam, $\int x\, dA = \int y\, dA = 0$. The other integrals in Eqs. (7.9) are defined in Appendix B. Equations (7.9) simplify to

$$0 = aA$$
$$M_x = bI_{xy} + cI_x$$
$$M_y = -bI_y - cI_{xy} \quad (7.10)$$

where I_x and I_y are the centroidal moments of inertia of the beam cross section with respect to the x and y axes, respectively, and I_{xy} is the centroidal product of inertia of the beam cross section. Solving Eqs. (7.10) for the constants a, b, and c, we obtain

$$a = 0 \quad \text{(because } A \neq 0)$$

$$b = -\frac{M_y I_x + M_x I_{xy}}{I_x I_y - I_{xy}^2}$$

$$c = \frac{M_x I_y + M_y I_{xy}}{I_x I_y - I_{xy}^2} \quad (7.11)$$

The substitution of Eqs. (7.11) into Eq. (7.8) gives the normal stress distribution σ_{zz} on a given cross section of a beam subjected to unsymmetrical bending in the form

$$\sigma_{zz} = -\left(\frac{M_y I_x + M_x I_{xy}}{I_x I_y - I_{xy}^2}\right)x + \left(\frac{M_x I_y + M_y I_{xy}}{I_x I_y - I_{xy}^2}\right)y \quad (7.12)$$

Equation (7.12) is not the most convenient form for the determination of the maximum value of the flexural stress σ_{zz}. Also, Eq. (7.12) does not lend itself readily to visualization of the bending behavior of the beam. A more convenient, and a more visually meaningful, form follows.

Before the location of the points of maximum tensile and compressive stresses in the cross section are determined, it is useful to locate the neutral axis. For this purpose, it is desirable to express the neutral axis orientation in terms of the angle ϕ between the plane of the loads and the x axis; ϕ is measured positive counterclockwise (Fig. 7.7). The magnitude of ϕ is generally in the neighborhood of $\pi/2$ ($0 \leq \phi \leq \pi$). The bending moments M_x and M_y can be written in terms of ϕ as follows:

$$M_x = M \sin \phi$$
$$M_y = -M \cos \phi \quad (7.13)$$

in which M is the signed magnitude of moment **M** at the cut section. The sign of M is positive if the x projection of the vector **M** is positive; it is negative if the x projection of **M** is negative. Since the (x, y) axes are chosen for the convenience of the one making the calculations, they are chosen so that the magnitude of M_x is not zero. Therefore, by Eqs. (7.13),

$$\cot \phi = -\frac{M_y}{M_x} \tag{7.14}$$

Neutral Axis

The neutral axis of the cross section of a beam subjected to unsymmetrical bending is defined as the axis in the cross section for which $\sigma_{zz} = 0$. Thus, by Eq. (7.12), the equation of the neutral axis of the cross section is

$$y = \frac{M_x I_{xy} + M_y I_x}{M_x I_y + M_y I_{xy}} x = x \tan \alpha \tag{7.15}$$

where α is the angle between the neutral axis of bending and the x axis; α is measured positive counterclockwise (Fig. 7.8), and

$$\tan \alpha = \frac{M_x I_{xy} + M_y I_x}{M_x I_y + M_y I_{xy}} \tag{7.16}$$

Since $x = y = 0$ satisfies Eq. (7.15), the neutral axis passes through the centroid of the section. The right side of Eq. (7.16) can be expressed in terms of the angle ϕ by using Eq. (7.14). Thus,

$$\tan \alpha = \frac{I_{xy} - I_x \cot \phi}{I_y - I_{xy} \cot \phi} \tag{7.17}$$

More Convenient Form for the Flexure Stress σ_{zz}

Elimination of M_y between Eqs. (7.12) and (7.16) results in a more convenient form for the normal stress distribution σ_{zz} for beams subjected to nonsymmetrical bending; namely

$$\sigma_{zz} = \frac{M_x(y - x \tan \alpha)}{I_x - I_{xy} \tan \alpha} \tag{7.18}$$

where $\tan \alpha$ is given by Eq. (7.17). Once the neutral axis is located on the cross sections at angle α as indicated in Fig. 7.8b, points in the cross section where the tensile and compressive flexure stresses are maxima are easily determined. The coordinates of these points can be substituted into Eq. (7.18) to determine the magnitudes of these stresses. If M_x is zero, Eq. (7.12) may be used instead of Eq. (7.18) to determine magnitudes of these stresses, or axes (x, y) may be rotated by $\pi/2$ to obtain new reference axes (x', y').

Note: Equations (7.17) and (7.18) have been derived by assuming that the beam is subjected to pure bending. These equations are exact for pure bending. Although they are not exact for beams subjected to transverse shear loads, often the equations are assumed to be valid for such beams. The error in this assumption is usually small, particularly if the beam has a length of at least five times its maximum cross-sectional dimension.

In the derivation of Eqs. (7.17) and (7.18), the (x, y) axes are any convenient set of orthogonal axes that have an origin at the centroid of the cross-sectional area. The equations are valid if (x, y) are principal axes; in this case, $I_{xy} = 0$. If the axes are principal axes and $\phi = \pi/2$, Eq. (7.17) indicates that $\alpha = 0$ and Eq. (7.18) reduces to Eq. (7.1).

For convenience in deriving Eqs. (7.17) and (7.18), the origin for the x, y, z coordinate axes was chosen (see Fig. 7.8b) at the end of the free body opposite from the cut section with the positive z axis toward the cut section. The equations are equally valid if the origin is taken at the cut section with the positive z axis toward the opposite end of the free body. If ϕ_2 is the magnitude of ϕ for the second choice of axes and ϕ_1 is the magnitude of ϕ for the first choice of axes, then $\phi_2 = \pi - \phi_1$.

EXAMPLE 7.1
Channel Section Beam

The cantilever beam in Fig. E7.1a has a channel section as shown in Fig. E7.1b. A concentrated load $P = 12.0$ kN lies in the plane making an angle $\phi = \pi/3$ with the x axis. Load P lies in the plane of the cross section of the free end of the beam and passes through shear center C; in Chapter 8 we find that the shear center lies on the y axis as shown. Locate points of maximum tensile and compressive stresses in the beam and determine the stress magnitudes.

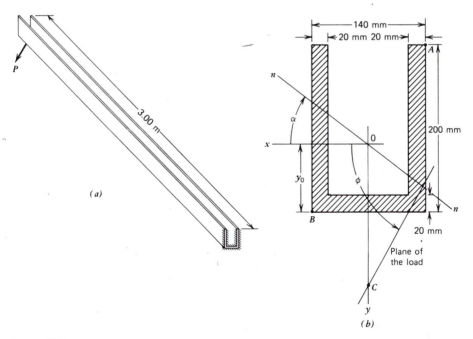

Figure E7.1

SOLUTION

Several properties of the cross-sectional area are needed (see Appendix B).

$$A = 10,000 \text{ mm}^2, \qquad I_x = 39.69 \times 10^6 \text{ mm}^4$$
$$y_0 = 82.0 \text{ mm}, \qquad I_y = 30.73 \times 10^6 \text{ mm}^4$$
$$I_{xy} = 0$$

The orientation of the neutral axis for the beam is given by Eq. (7.17).

$$\tan \alpha = -\frac{I_x}{I_y} \cot \phi = -\frac{39,690,000}{30,730,000}(0.5774) = -0.7457$$

$$\alpha = -0.6407 \text{ rad}$$

The negative sign indicates that the neutral axis n-n, which passes through the centroid ($x = y = 0$), is located clockwise 0.6407 rad from the x axis (Fig. 7.1b). The maximum tensile stress occurs at point A, whereas the maximum compressive stress occurs at point B. These stresses are given by Eq. (7.18) after M_x has been determined. From Fig. E7.1a

$$M = -3.00P = -36.0 \text{ kN·m}$$
$$M_x = M \sin \phi = -31.18 \text{ kN·m}$$
$$\sigma_A = \frac{M_x(y_A - x_A \tan \alpha)}{I_x}$$
$$= \frac{-31,180,000[-118 - (-70)(-0.7457)]}{39,690,000}$$
$$= 133.7 \text{ MPa}$$
$$\sigma_B = \frac{-31,180,000[82 - 70(-0.7457)]}{39,690,000} = -105.4 \text{ MPa}$$

EXAMPLE 7.2
Angle-Beam

Plates are welded together to form the 120 mm by 80 mm by 10 mm angle-section beam shown in Fig. E7.2a. The beam is subjected to a concentrated load $P = 4.00$ kN as shown. The load P lies in the plane making an angle $\phi = 2\pi/3$ with the x axis. Load P passes through shear center C; in Chapter 8 we find that the shear center is located at the intersection of the two legs of the angle section. Determine the maximum tensile and compressive bending stresses at the section of the beam where the load is applied.

(a) Solve the problem using the load-stress relations derived for nonsymmetrical bending.

(b) Solve the problem using Eq. (7.3).

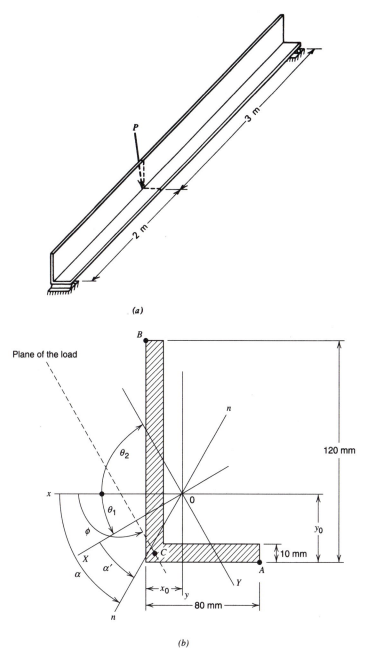

Figure E7.2

SOLUTION

(a) Several properties of the cross-sectional area are needed (see Appendix B).

$$A = 1900 \text{ mm}^2, \quad I_x = 2.783 \times 10^6 \text{ mm}^4$$
$$x_0 = 19.74 \text{ mm}, \quad I_y = 1.003 \times 10^6 \text{ mm}^4$$
$$y_0 = 39.74 \text{ mm}, \quad I_{xy} = -0.973 \times 10^6 \text{ mm}^4$$

The orientation of the neutral axis for the beam is given by Eq. (7.17). Thus,

$$\tan \alpha = \frac{I_{xy} - I_x \cot \phi}{I_y - I_{xy} \cot \phi}$$

$$= \frac{-0.973 \times 10^6 - 2.783 \times 10^6(-0.5774)}{1.003 \times 10^6 - (-0.973 \times 10^6)(-0.5774)} = 1.4363$$

$$\alpha = 0.9626 \text{ rad}$$

The positive sign indicates that the neutral axis n-n, which passes through the centroid ($x = y = 0$), is located counterclockwise 0.9628 rad from the x axis (Fig. E7.2b). The maximum tensile stress occurs at point A, whereas the maximum compressive stress occurs at point B. These stresses are given by Eq. (7.18) after M_x has been determined. From Fig. E7.2a

$$M = 1.2P = 4.80 \text{ kN·m}$$

$$M_x = M \sin \phi = 4.80 \times 10^3(0.8660) = 4.157 \text{ kN·m}$$

$$\sigma_A = \frac{M_x(y_A - x_A \tan \alpha)}{I_x - I_{xy} \tan \alpha}$$

$$= \frac{4.157 \times 10^6[39.74 - (-60.26)(1.4363)]}{2.783 \times 10^6 - (-0.973 \times 10^6)(1.4363)}$$

$$= 125.6 \text{ MPa}$$

$$\sigma_B = \frac{4.157 \times 10^6[-80.26 - 19.74(1.4363)]}{2.783 \times 10^6 - (-0.973 \times 10^6)(1.4363)}$$

$$= -108.0 \text{ MPa}$$

(b) To solve the problem using Eq. (7.3), it is necessary that the principal axes for the cross section be determined. The two values of the angle θ between the x axis and the principal axes are given by Eq. (B.12). Thus, we obtain

$$\tan 2\theta = -\frac{2I_{xy}}{I_x - I_y} = -\frac{2(-0.973 \times 10^6)}{2.783 \times 10^6 - 1.003 \times 10^6} = 1.0933$$

$$\theta_1 = 0.4150 \text{ rad} \qquad (\theta_2 = -1.156 \text{ rad})$$

The principal X and Y axes are shown in Fig. E7.2b. Thus [see Eq. (B.10) Appendix B]

$$I_X = I_x \cos^2 \theta_1 + I_y \sin^2 \theta_1 - 2I_{xy} \sin \theta_1 \cos \theta_1 = 3.212 \times 10^6 \text{ mm}^4$$

$$I_Y = I_x + I_y - I_X = 0.574 \times 10^6 \text{ mm}^4$$

Note that now angle ϕ is measured from the X axis and not from the x axis as for part (a). Hence,

$$\phi = \frac{2\pi}{3} - \theta_1 = 1.6794 \text{ rad}$$

Angle α', which determines the orientation of the neutral axis, is now measured from the X axis (Fig. E7.2b), and is given by Eq. (7.17). Hence, we find

$$\tan \alpha' = -\frac{I_X \cot \phi}{I_Y} = -\frac{3.212 \times 10^6(-0.1090)}{0.574 \times 10^6} = 0.6098$$

$$\alpha' = 0.5476 \text{ rad}$$

which gives the same orientation for the neutral axis as for part (a), that is,

$$\alpha = \alpha' + \theta_1$$
$$= 0.5476 + 0.4150$$
$$= 0.9626 \text{ rad}.$$

To use Eq. (7.3) relative to axes (X, Y), the X and Y coordinates of points A and B are needed. They are [Eq. (B.9)]

$$X_A = x_A \cos \theta_1 + y_A \sin \theta_1 = -60.26(0.9151) + 39.74(0.4032) = -39.12 \text{ mm}$$
$$Y_A = y_A \cos \theta_1 - x_A \sin \theta_1 = 39.74(0.9151) - (-60.26)(0.4032) = 60.66 \text{ mm}$$

and

$$X_B = 19.74(0.9151) - 80.26(0.4032) = -14.30 \text{ mm}$$
$$Y_B = -80.26(0.9151) - 19.74(0.4032) = -81.41 \text{ mm}$$

The moment components are

$$M_X = M \sin \phi = 4.80 \times 10^3(0.9941) = 4.772 \text{ kN·m}$$
$$M_Y = -M \cos \phi = -4.80 \times 10^3(-0.1084) = 520 \text{ N·m}$$

The stresses at A and B are calculated using Eq. (7.3).

$$\sigma = \frac{M_X Y_A}{I_X} - \frac{M_Y X_A}{I_Y}$$

$$= \frac{4.772 \times 10^6(60.66)}{3.212 \times 10^6} - \frac{0.520 \times 10^6(-39.12)}{0.574 \times 10^6}$$

$$= 125.6 \text{ MPa}$$

$$\sigma_B = \frac{M_X Y_B}{I_X} - \frac{M_Y X_B}{I_Y}$$

$$= \frac{4.772 \times 10^6(-81.41)}{3.212 \times 10^6} - \frac{0.520 \times 10^6(-14.30)}{0.574 \times 10^6}$$

$$= -108.0 \text{ MPa}$$

These values for σ_A and σ_B agree with the values calculated in part (a).

7.3

DEFLECTIONS OF STRAIGHT BEAMS SUBJECTED TO NONSYMMETRICAL BENDING

Consider a straight beam subjected to transverse shear loads and moments. The transverse shear loads lie in a plane and the moment vectors are normal to that plane. The neutral axes of all cross sections of the beam have the same orientation as long as the beam material remains linearly elastic. The deflections of the beam will be in a direction perpendicular to the neutral axis. It is relatively simple to determine the component of the deflection parallel to an axis, say, the y axis. The total deflection is easily determined once one component has been determined.

Consider the intersection of the (y, z) plane with the beam in Fig. 7.8. A side view of this section of the deformed beam is shown in Fig. 7.9. Before deformation, the lines FG and HJ were parallel and distance Δz apart. In the deformed beam, the two straight lines FG and HJ represent the intersection of the (y, z) plane with two planes perpendicular to the axis of the beam, a distance Δz apart at the neutral surface. Since plane sections remain plane and normal to the axis of the beam, the extensions of FG and HJ meet at the center of curvature $0'$. The distance from $0'$ to the neutral surface is the radius of curvature R_y of the beam in the (y, z) plane. Since the center of curvature lies on the negative side of the y axis, R_y is negative.

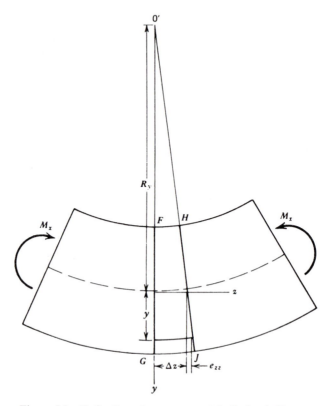

Figure 7.9   Deflection of a nonsymmetrically loaded beam.

We assume that the deflections are small so that $1/R_y \cong d^2v/dz^2$, where v is the y component of displacement. Under deformation of the beam, a fiber at distance y below the neutral surface elongates an amount $e_{zz} = (\Delta z)\epsilon_{zz}$. Initially, the length of the fiber is Δz. By the geometry of similar triangles,

$$-\frac{\Delta z}{R_y} = \frac{(\Delta z)\epsilon_{zz}}{y}$$

Dividing by Δz, we obtain

$$-\frac{1}{R_y} = \frac{\epsilon_{zz}}{y} \qquad \text{where} \qquad \frac{1}{R_y} \cong \frac{d^2v}{dz^2} \tag{7.19}$$

For linearly elastic behavior, Eqs. (7.18) and (7.12), with $x = 0$, and Eq. (7.7) yield

$$\frac{\epsilon_{zz}}{y} = \frac{M_x}{E(I_x - I_{xy}\tan\alpha)} = \frac{M_x I_y + M_y I_{xy}}{E(I_x I_y - I_{xy}^2)}$$

which, with Eq. (7.19), yields

$$\frac{d^2v}{dz^2} = -\frac{M_x}{E(I_x - I_{xy}\tan\alpha)} = -\frac{M_x I_y + M_y I_{xy}}{E(I_x I_y - I_{xy}^2)} \tag{7.20}$$

Note the similarity of Eq. (7.20) to the elastic curve equation for symmetrical bending. The only difference is that the term I has been replaced by $(I_x - I_{xy}\tan\alpha)$. The solution of the differential relation Eq. (7.20) gives the y component of the deflection v at any section of the beam. As is indicated in Fig. 7.10, the total deflection of the centroid at any section of the beam is perpendicular to the neutral axis. Therefore,

$$u = -v\tan\alpha \tag{7.21}$$

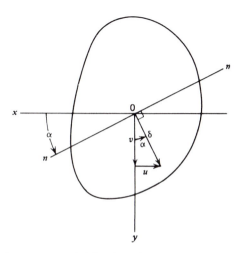

Figure 7.10   Components of deflection of a nonsymmetrically loaded beam.

and the total displacement is

$$\delta = \sqrt{u^2 + v^2} = \frac{v}{\cos \alpha} \tag{7.22}$$

EXAMPLE 7.3
Channel Section Simple Beam

Let the channel section beam in Fig. E7.1 be loaded as a simple beam with a concentrated load $P = 35.0$ kN acting at the center of the beam. Determine the maximum tensile and compressive stresses in the beam if $\phi = 5\pi/9$. If the beam is made of an aluminum alloy ($E = 72.0$ GPa), determine the maximum deflection of the beam.

SOLUTION

Amalogous to the solution of Example 7.1, we have

$$\tan \alpha = -\frac{I_x}{I_y} \cot \phi = -\frac{39,690,000}{30,730,000} \cot \frac{5\pi}{9} = 0.2277$$

$$\alpha = 0.2239 \text{ rad}$$

$$M = \frac{PL}{4} = \frac{35.0(3.00)}{4} = 26.25 \text{ kN·m}$$

$$M_x = M \sin \phi = 25.85 \text{ kN·m}$$

$$\sigma_{\text{tension}} = \frac{25,850,000[82 - (-70)(0.2277)]}{39,690,000} = 63.8 \text{ MPa}$$

$$\sigma_{\text{compression}} = \frac{25,850,000[-118 - 70(0.2277)]}{39,690,000} = -87.2 \text{ MPa}$$

Since the deflection of the center of a simple beam subjected to a concentrated load in the center is given by the relation $PL^3/48EI$, the y component of the deflection of the center of the beam is

$$v = \frac{PL^3 \sin \phi}{48EI_x} = \frac{35,000(3000)^3 \sin 5\pi/9}{48(72,000)(39,690,000)} = 6.78 \text{ mm}$$

The lateral deflection is

$$u = -v \tan \alpha = -6.78(0.2277) = -1.54 \text{ mm}$$

Finally, the total deflection is

$$\delta = \sqrt{u^2 + v^2} = 6.95 \text{ mm}$$

EXAMPLE 7.4
Cantilever I-Beam

A cantilever beam has a length of 3 m with cross section indicated in Fig. E7.4. The beam is constructed by welding two 40 mm by 40 mm steel ($E = 200$ GPa) bars longitudinally to the S-200 × 27 steel I-beam ($I_x = 24 \times 10^6$ mm^4 and $I_y = 1.55 \times 10^6$ mm^4). The bars and I-beam have the same yield stress, $Y = 300$ MPa. The beam is subjected to a concentrated load P at the free end at an angle $\phi = \pi/3$ with the x axis. Determine the magnitude of P necessary to initiate yielding in the beam and the resulting deflection of the free end of the beam.

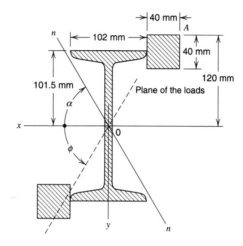

Figure E7.4

SOLUTION

Values of I_x, I_y, and I_{xy} for the composite cross section can be obtained using the procedure outlined in Appendix B.

$$I_x = 56.43 \times 10^6 \text{ mm}^4, \qquad I_y = 18.11 \times 10^6 \text{ mm}^4$$
$$I_{xy} = 22.72 \times 10^6 \text{ mm}^4$$

The orientation of the neutral axis for the beam is given by Eq. (7.17). We find

$$\tan \alpha = \frac{I_{xy} - I_x \cot \phi}{I_y - I_{xy} \cot \phi} = \frac{22.72 \times 10^6 - 56.43 \times 10^6 (0.5774)}{18.11 \times 10^6 - 22.72 \times 10^6 (0.5774)} = -1.9759$$
$$\alpha = -1.023 \text{ rad}$$

The orientation of the neutral axis $n - n$ is indicated in Fig. E7.4. The maximum tensile stress occurs at point A; the magnitude of the stress is obtained using Eq. (7.18).

$$M = -3P$$

$$M_x = M \sin \phi = -2.598P$$

$$\sigma_A = Y = \frac{M_x(y_A - x_A \tan \alpha)}{I_x - I_{xy} \tan \alpha}$$

$$P = \frac{Y(I_x - I_{xy} \tan \alpha)}{(-2.598 \times 10^3)(y_A - x_A \tan \alpha)}$$

$$= \frac{300[56.43 \times 10^6 - 22.72 \times 10^6(-1.9759)]}{-2.598 \times 10^3[-120 - (-91)(-1.9759)]} = 39.03 \text{ kN}$$

Since the deflection of the free end of a cantilever beam subjected to symmetrical bending is given by the relation $P_y L^3/3EI$, the y component of the deflection of the free end of the beam is

$$v = \frac{PL^3 \sin \phi}{3E(I_x - I_{xy} \tan \alpha)}$$

$$= \frac{39.03 \times 10^3(3 \times 10^3)^3(0.8660)}{3(200 \times 10^3)[56.43 \times 10^6 - 22.72 \times 10^6(-1.9759)]} = 17.33 \text{ mm}$$

Hence,

$$u = -v \tan \alpha = 34.25 \text{ mm}$$

and the total displacement of the free end of cantilever beam is

$$\delta = \sqrt{u^2 + v^2} = 38.39 \text{ mm}$$

7.4

EFFECT OF INCLINED LOADS

Some common rolled sections such as I-beams and channels are designed so that I_x is many times greater than I_y and $I_{xy} = 0$. Equation (7.17) indicates that the angle α may be large even though ϕ is nearly equal to $\pi/2$. Thus, the neutral axis of such I-beams and channels is steeply inclined to the horizontal axis (the x axis) of symmetry when the plane of the loads deviates slightly from the vertical plane of symmetry. As a consequence, the maximum flexure stress and maximum deflection may be quite large. These rolled sections should not be used as beams unless the lateral deflection is prevented. If lateral deflection of the beam is prevented, nonsymmetrical bending cannot occur.

In general, however, I-beams and channels make poor long-span cantilever beams. The following example illustrates this fact.

EXAMPLE 7.5
An Unsuitable Beam

An S-610 × 134 I-beam ($I_x = 937 \times 10^6$ mm⁴ and $I_y = 18.7 \times 10^6$ mm⁴) is subjected to a bending moment M in a plane with angle $\phi = 1.5533$ rad; the plane of the loads is 1° ($\pi/180$ rad) clockwise from the (y, z) plane of symmetry. Determine the neutral axis orientation and the ratio of the maximum tensile stress in the beam to the maximum tensile stress for symmetrical bending.

SOLUTION

The cross section of the I-beam with the plane of the loads is indicated in Fig. E7.5. The orientation of the neutral axis for the beam is given by Eq. (7.17).

$$\tan \alpha = \frac{-I_x \cot \phi}{I_y} = -\frac{937 \times 10^6 (0.01746)}{18.7 \times 10^6} = -0.8749$$

$$\alpha = -0.7188 \text{ rad}$$

The orientation of the neutral axis is indicated in Fig. E7.5. If the beam is subjected to a positive bending moment, the maximum tensile stress is located at point A. By Eqs. (7.13) and (7.18),

$$M_x = M \sin \phi = 0.9998M$$

$$\sigma_A = \frac{0.9998M[305 - 90.5(-0.8749)]}{937 \times 10^6} = 4.099 \times 10^{-7}M \qquad (a)$$

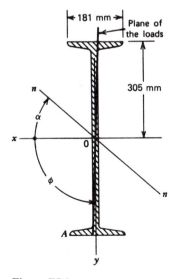

Figure E7.5

When the plane of the loads coincides with the y axis (Fig. E7.5), the beam is subjected to symmetrical bending and the maximum bending stress is

$$\sigma_A = \frac{My}{I_x} = \frac{305M}{937 \times 10^6} = 3.255 \times 10^{-7}M \tag{b}$$

The ratio of the stress σ_A given by Eq. (a) to that given by Eq. (b) is 1.259. Hence, the maximum stress in the I-beam is increased 25.9% when the plane of the loads is merely 1° from the symmetrical vertical plane.

7.5

FULLY PLASTIC LOAD FOR NONSYMMETRICAL BENDING

A beam of general cross section (Fig. 7.11) is subjected to pure bending. The material in the beam has a flat-top stress-strain diagram with yield point Y in both tension and compression (Fig. 4.4a). At the fully plastic load, the deformations of the beam are unchecked and continue (until possibly the material begins to strain harden). The fully plastic load is the upper limit for failure loads (Sec. 4.6) since the deformations of the beam at the outset of any strain hardening generally exceed design limits for the deformations.

In contrast to the direct calculation of fully plastic load in symmetrical bending (Sec. 4.6), an inverse method is required to determine the fully plastic load for a

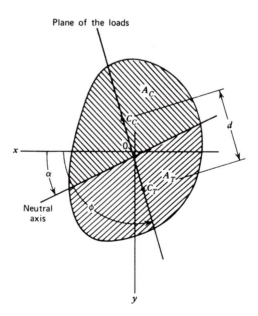

Figure 7.11   Location of a neutral axis for fully plastic bending of a nonsymmetrically loaded beam.

beam subjected to nonsymmetrical bending. Although the plane of the loads is generally specified for a given beam, the orientation and location of the neutral axis, when the fully plastic moment is developed at a given section of the beam, must be determined by trial and error. The analysis is begun by assuming a value for the angle α (Fig. 7.11). The neutral axis is inclined to the x axis by the angle α, but does not necessarily pass through the centroid as in the case of linearly elastic conditions. The location of the neutral axis is determined by the condition that it must divide the cross-sectional area into two equal areas. This follows from the fact that since the yield point stress is the same for tension and compression, the area A_T that has yielded in tension must be equal to the area A_C that has yielded in compression. In other words, the net resultant tension force on the section must be equal to the net resultant compression force.

The yield point stress Y is uniform over the area A_T that has yielded in tension; the resultant tensile force $P_T = YA_T$ is located at the centroid C_T of A_T. Similarly, the resultant compressive force $P_C = YA_C$ is located at the centroid C_C of A_C. The fully plastic moment M_P is given by

$$M_P = YA_T d = \frac{YAd}{2} \tag{7.23}$$

where A is the total cross-sectional area and d is the distance between the centroids C_T and C_C as indicated in Fig. 7.11. A plane through the centroids C_T and C_C is the plane of the loads for the beam. In case the calculated angle ϕ (Fig. 7.11) does not correspond to the plane of the applied loads, a new value is assumed for α and the calculations are repeated. Once the angle ϕ (Fig. 7.11) corresponds to the plane of the applied loads, the magnitude of the fully plastic load is calculated by setting the moment due to the applied loads equal to M_P given by Eq. (7.23).

EXAMPLE 7.6
Fully Plastic Moment for Unsymmetrical Bending

A steel beam has the cross section shown in Fig. E7.6. The beam is made of a steel having a yield point stress $Y = 280$ MPa. Determine the fully plastic moment for the condition that the neutral axis passes through point B. Determine the orientation of the neutral axis and the plane of the loads.

SOLUTION

The neutral axis must divide the cross section into two equal areas since the area that has yielded in tension A_T must equal the area that has yielded in compression A_C. The neutral axis bisects edge AC. Therefore,

$$\tan \alpha = \frac{30}{20} = 1.5$$

$$\alpha = 0.9828 \text{ rad}$$

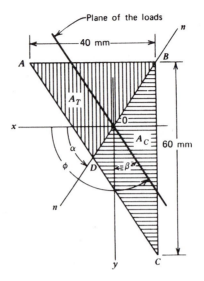

Figure E7.6

The plane of the loads passes through the centroids of area ABD and BCD. The centroids of these areas are located at $(\frac{20}{3}, -10)$ for ABD and $(-\frac{20}{3}, 10)$ for BCD.

$$\tan \beta = \frac{20/3 - (-20/3)}{10 - (-10)} = 0.6667$$

$$\beta = 0.5880 \text{ rad}$$

$$\phi = \frac{\pi}{2} + \beta = 2.1588 \text{ rad}$$

The fully plastic moment M_P is equal to the product of the force on either of the two areas (A_T or A_C) and the distance d between the two centroids.

$$d = \sqrt{\left(20 - \frac{20}{3}\right)^2 + (30 - 10)^2} = 24.04 \text{ mm}$$

$$M_P = A_T Yd = \tfrac{1}{2}(40)(30)(280)(24.04) = 4.039 \times 10^6 \text{ N·mm}$$

$$= 4.039 \text{ kN·m}$$

Since the orientation of the neutral axis is known *a priori*, iteration is not necessary in this example.

PROBLEMS
Section 7.1–7.2

7.1. A timber beam 250 mm wide by 300 mm deep by 4.2 m long is used as a simple beam on a span of 4 m. It is subjected to a concentrated load P at the midsection of the span. The plane of the loads makes an angle $\phi =$

$5\pi/9$ with the horizontal x axis. The beam is made of yellow pine with a yield stress $Y = 25.0$ MPa. If the beam has been designed with a factor of safety $SF = 2.50$ against initiation of yielding, determine the magnitude of P and orientation of the neutral axis.

7.2. The plane of the loads for the rectangular section beam in Fig. P7.2 coincides with a diagonal of the rectangle. Show that the neutral axis for the beam cross section coincides with the other diagonal.

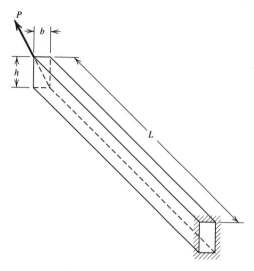

Figure P7.2

7.3. In Fig. P7.3 let $b = 300$ mm, $h = 300$ mm, $t = 25.0$ mm, $L = 2.50$ m, and $P = 16.0$ kN. Calculate the maximum tensile and compressive stresses in the beam, and determine the orientation of the neutral axis.

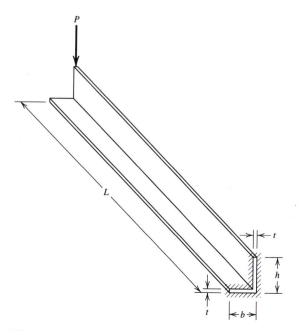

Figure P7.3

7.4. In Fig. P7.3 let $b = 200$ mm, $h = 300$ mm, $t = 25.0$ mm, $L = 2.50$ m, and $P = 16.0$ kN. Calculate the maximum tensile and compressive stresses in the beam and determine the orientation of the neutral axis.

Ans. $\sigma_{zz(\text{ten})} = 98.6$ MPa, $\sigma_{zz(\text{com})} = -81.9$ MPa

7.5. In Fig. P7.5 let $b = 150$ mm, $t = 50.0$ mm, $h = 150$ mm, and $L = 2.00$ m. The beam is made of a steel that has a yield point stress $Y = 240$ MPa. Using a factor of safety of $SF = 2.00$, determine the magnitude of P if $\phi = 2\pi/9$ from the horizontal x axis.

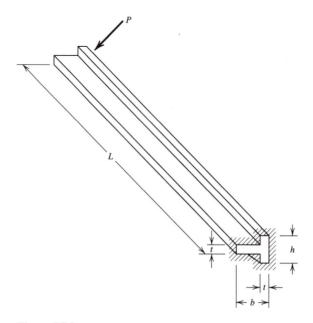

Figure P7.5

7.6. A simple beam is subjected to a concentrated load $P = 4.00$ kN at the mid-length of a span of 2.00 m. The beam cross section is formed by nailing together two 50.0 mm by 150 mm boards as indicated in Fig. P7.6. The plane of the loads passes through the centroid of the two boards as indicated. Determine the maximum flexure stress in the beam and orientation of the neutral axis.

Ans. $\sigma_{zz(\text{max})} = 4.17$ MPa, $\alpha = 1.3522$ rad

7.7. Solve Problem 7.6 if $\phi = 1.900$ rad.

7.8. A C-180 × 15 [mm] [kg/m] rolled steel channel ($I_x = 8.87 \times 10^6$ mm⁴, depth = 178 mm, width = 53 mm, $x_B = 13.7$ mm) is used as a simply supported beam as, for example, a purlin in a roof (Fig. P7.8). If the slope of the roof is 1/2 and the span of the purlin is 4 m, determine the maximum tensile and compressive stresses in the beam caused by a uniformly distributed vertical load of 1.00 kN/m.

Ans. $\sigma_{zz(\text{ten})} = 48.4$ MPa, $\sigma_{zz(\text{com})} = -105.2$ MPa

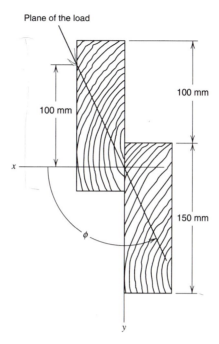

Figure P7.6

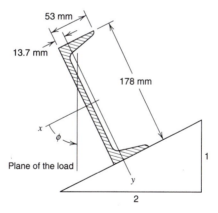

Figure P7.8

7.9. Two rolled steel angles ($I_{x_1} = 391 \times 10^3$ mm^4, $I_{y_1} = 912 \times 10^3$ mm^4, $I_{x_1y_1} = 349 \times 10^3$ mm^4, and $A = 1148$ mm^2) are welded to a 200 mm by 10 mm steel plate to form a composite Z-bar (Fig. P7.9). The Z-bar is a simply supported beam used as a purlin in a roof of slope $\frac{1}{2}$. The beam has a span of 4.00 m. The yield stress of the steel in the plate and angles is $Y = 300$ MPa. The beam has been designed using a factor of safety of $SF = 2.50$ against initiation of yielding. If the plane of the loads is vertical, determine the magnitude of the maximum distributed load that can be applied to the beam.

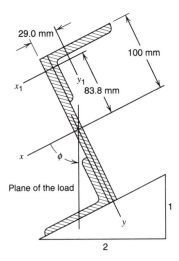

29.0 mm

100 mm

x_1

y_1

83.8 mm

x

ϕ

Plane of the load

y

1

2

Figure P7.9

7.10. A steel Z-bar is used as a cantilever beam having a length of 2.00 m. When viewed from the free end toward the fixed end of the beam, the cross section has the orientation and dimensions shown in Fig. P7.10. A concentrated load $P = 14.0$ kN acts at the free end of the beam at an angle $\phi = 1.25$ rad. Determine the maximum flexure stress in the beam.

Ans. $I_x = 39.36 \times 10^6$ mm^4, $I_y = 9.84 \times 10^6$ mm^4,
$I_{xy} = 14.40 \times 10^6$ mm^4, $\alpha = 0.2557$ rad, $\sigma_{zz(max)} = 76.6$ MPa

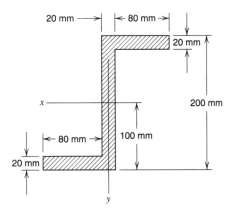

20 mm

80 mm

20 mm

x

200 mm

80 mm

100 mm

20 mm

y

Figure P7.10

7.11. An extruded bar of aluminum alloy has the cross section shown in Fig. P7.11. A 1.00-m length of this bar is used as a cantilever beam. A concentrated load $P = 1.25$ kN is applied at the free end and makes an angle of $\phi = 5\pi/9$ with the x axis. The view in Fig. P7.11 is from the free end toward the fixed end of the beam. Determine the maximum tensile and compressive stresses in the beam.

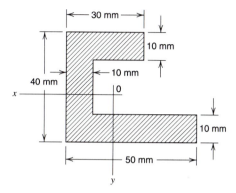

Figure P7.11

7.12. An extruded bar of aluminum alloy has the cross section shown in Fig. P7.12. A 2.10-m length of this bar is used as a simple beam on a span of 2.00 m. A concentrated load $P = 5.00$ kN is applied at midlength of the span and makes an angle of $\phi = 1.40$ rad with the x axis. Determine the maximum tensile and compressive stresses in the beam.

Ans. $x_0 = 28.0$ mm, $y_0 = 35.0$ mm, $I_x = 1.330 \times 10^6$ mm^4

$I_y = 917 \times 10^3$ mm^4, $I_{xy} = 30.0 \times 10^3$ mm^4

$\alpha = -0.2153$ rad, $\sigma_A = 81.5$ MPa, $\sigma_B = -75.8$ MPa

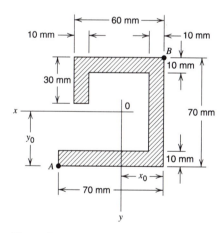

Figure P7.12

7.13. A cantilever beam has a right triangular cross section and is loaded by a concentrated load P at the free end (Fig. P7.13). Solve for the stresses at points A and C at the fixed end if $P = 4.00$ kN, $h = 120$ mm, $b = 75.0$ mm, and $L = 1.25$ m.

7.14. A girder that supports a brick wall is built up of an S-310 × 47 I-beam $(A_1 = 6030$ mm^2, $I_{x_1} = 90.7 \times 10^6$ mm^4, $I_{y_1} = 3.90 \times 10^6$ mm$^4)$, a C-310 × 31 channel $(A_2 = 3930$ mm^2, $I_{x_2} = 53.7 \times 10^6$ mm^4, $I_{y_2} = 1.61 \times 10^6$ mm$^4)$, and a cover plate 300 mm by 10 mm riveted together (Fig. P7.14). The girder is

6.00 m long and is simply supported at its ends. The load is uniformly distributed such that $w = 20.0$ kN/m. Determine the orientation of the neutral axis and the maximum tensile and compressive stresses.

Ans. $\alpha = -0.1653$ rad, $\sigma_A = 66.3$ MPa, $\sigma_B = -92.3$ MPa

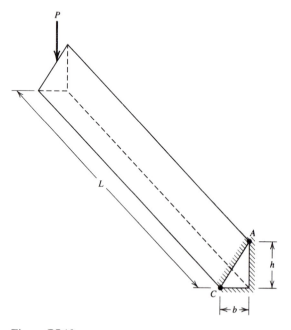

Figure P7.13

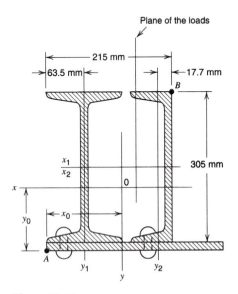

Figure P7.14

7.15. A load $P = 50$ kN is applied to a rolled steel angle ($I_x = I_y = 570 \times 10^3$ mm^4, $I_{xy} = -332.5 \times 10^3$ mm^4, $A = 1148$ mm^2) by means of a 76 mm by 6 mm plate riveted to the angle (Fig. P7.15). The action line of load P coincides with the centroidal axis of the plate. Determine the maximum stress at a section, such as AA, of the angle. *Hint*: Resolve the load P into a load (equal to P) at the centroid of the angle and a bending couple.

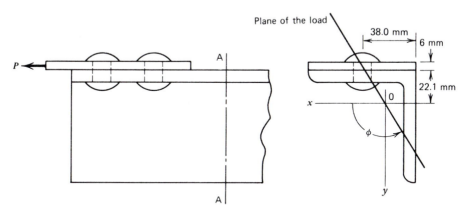

Figure P7.15

7.16. The beam shown in Fig. P7.16 has a cross section of depth 60 mm and width 30 mm. The load P and reactions R_1 and R_2 all lie in a plane that forms an angle of 20° counterclockwise from the y axis. Determine the point in the beam at which the maximum tensile flexural stress acts and the magnitude of that stress.

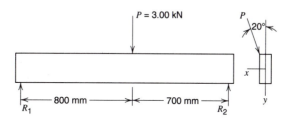

Figure P7.16

7.17. A beam has a square cross section (Fig. P7.17).

    (a) Determine an expression for σ_{max} in terms of M, h, and ψ.

    (b) Compare values of σ_{max} for $\psi = 0$, 15, and 45°.

7.18. Consider the beam shown in Fig. P7.18.

    (a) Derive an expression for σ_{max} in terms of M, h, and ψ.

    (b) Compare values of σ_{max} for $\psi = 0$, 30, 45, 60, and 90°.

7.19. An I-beam has the cross section shown in Fig. P7.19. The design flexural stress is limited to 120 MPa. Determine the allowable bending moment M.

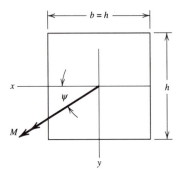

Figure P7.17

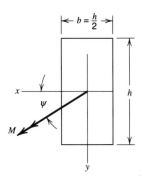

Figure P7.18

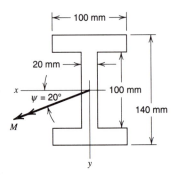

Figure P7.19

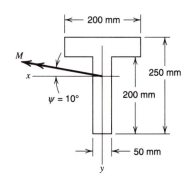

Figure P7.20

7.20. A T-beam has the cross section shown in Fig. P7.20. The design flexural stress is limited to 150 MPa. Determine the allowable bending moment M.

7.21. A beam has an isosceles triangular cross section (Fig. P7.21). The maximum flexural stress is limited to 90 MPa. Determine the magnitude of the allowable bending moment M.

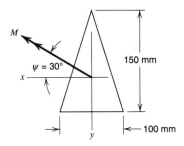

Figure P7.21

7.22. A circular cross section shaft is mounted in bearings that develop shear reactions only (Fig. P7.22). Determine the location and magnitude of the maximum flexural stress in the beam.

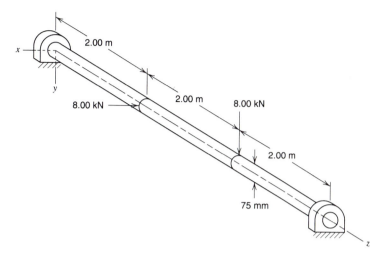

Figure P7.22

7.23. A wood beam of rectangular cross section 200 mm by 100 mm is simply supported at its ends (Fig. P7.23). Determine the location and magnitude of the maximum flexural stress in the beam.

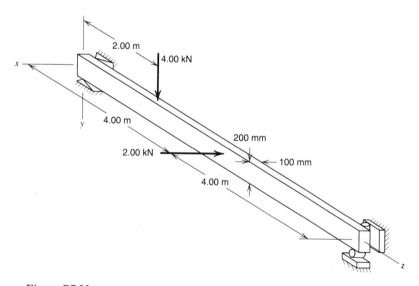

Figure P7.23

Section 7.3

7.24. Determine the deflection of the beam in Problem 7.1 if $E = 12.0$ GPa for the yellow pine.

7.25. The beam in Problem 7.3 is made of 7075-T6 aluminum alloy for which $E = 71.7$ GPa. Determine the deflection of the free end of the beam.

Ans. $v = 13.81$ mm, $u = -7.86$ mm, $\delta = 15.89$ mm

7.26. The beam in Problem 7.4 is made of 7075-T6 aluminum alloy for which $E = 71.7$ GPa. Determine the deflection of the free end of the beam.

7.27. The beam in Problem 7.6 is made of yellow pine for which $E = 12.0$ GPa. Determine the deflection at the center of the beam.

Ans. $v = 0.33$ mm, $u = -1.49$ mm, $\delta = 1.53$ mm

7.28. Determine the deflection of the center of the beam in Problem 7.8. $E = 200$ GPa.

7.29. If the beam in Problem 7.9 is subjected to a distributed load of $w = 6.5$ kN/m, determine the deflection at the center of the beam. $E = 200$ GPa.

Ans. $v = 1.58$ mm, $u = 8.25$ mm, $\delta = 8.40$ mm

7.30. Determine the deflection of the beam in Problem 7.10. $E = 200$ GPa.

7.31. Determine the deflection of the free end of the beam in Problem 7.11. $E = 72.0$ GPa.

Ans. $v = 33.16$ mm, $u = 6.25$ mm, $\delta = 33.74$ mm

7.32. Determine the deflection of the midspan of the beam in Problem 7.12. $E = 72.0$ GPa.

7.33. Determine the deflection of the free end of the beam in Problem 7.13. $E = 200$ GPa.

Ans. $v = 4.82$ mm, $u = -3.86$ mm, $\delta = 6.18$ mm

Section 7.5

7.34. The cantilever beam in Problem 7.11 is made of a low-carbon steel that has a yield stress $Y = 200$ MPa.

    (a) Determine the fully plastic load P_P for the beam for the condition that $\alpha = 0$.

    (b) Determine the fully plastic load P_P for the beam for the condition that $\alpha = \pi/6$.

7.35. The cantilever beam in Problem 7.13 is made of a mild steel that has a yield point stress $Y = 240$ MPa. Determine the fully plastic load P_P for the condition that $\alpha = 0$.

Ans. $P_P = 21.21$ kN at $\phi = 1.2679$ rad

REFERENCES

Boresi, A. P. and Chong, K. P. (1987). *Elasticity in Engineering Mechanics.* New York: Elsevier.

8

SHEAR CENTER FOR THIN-WALL BEAM CROSS SECTIONS

8.1

APPROXIMATIONS FOR SHEAR IN THIN-WALL BEAM CROSS SECTIONS

The definition of the bending axis of a straight beam with constant cross section was given in Chapter 7; it is the axis through the shear centers of the cross sections of the beam. To bend a beam without twisting, the plane of the loads must contain the axis of bending; that is, the plane of the loads must pass through the shear center of every cross section of the beam.

For a beam with a cross section that possess two or more axes of symmetry or antisymmetry, the bending axis is the same as the longitudinal centroidal axis, because for each cross section the shear center and centroid coincide. However, for cross sections with only one axis of symmetry, the shear center and centroid do not coincide. For example, consider the equal-leg angle section shown in Fig. 8.1. Let the beam cross section be oriented so that the principal axes of inertia (X, Y) are directed horizontally and vertically. The beam bends without twist (Chapter 7), if it is loaded by a force P that passes through the shear center C (Fig. 8.1b). As

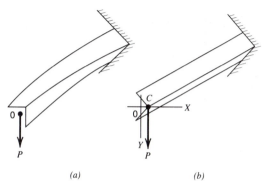

| (a) | (b) |

Figure 8.1 Effect of applying load through shear center. (a) Load P applied at point 0 produces twist and bending. (b) Load P applied at point C produces bending only.

is shown later, the shear center C coincides approximately with the intersection of the center lines of the two legs of the angle section. When the load P is applied at the centroid 0 of the cross section, the beam bends and twists (Fig. 8.1a).

The determination of the exact location of the shear center for an arbitrary cross section is beyond the scope of this book (see Boresi and Chong, 1987). However, an approximate solution is presented in this chapter, which gives reasonably accurate results for "thin-wall" cross sections. The simplifying assumptions on which the approximate solution is based may be illustrated by reference to Fig. 8.2. In Fig. 8.2, the cross section shown is that of the beam in Fig. 8.1b and is obtained by passing a cutting plane perpendicular to the bending axis through the beam. The view shown is obtained by looking from the support toward the end of the beam at which P is applied.

'For equilibrium of the beam element so obtained, the shear stresses on the cut cross section must balance the load P. However, the shear stresses in the cross section are difficult to compute exactly. Hence, simplifying approximations are employed. Accordingly, consider a portion of the legs of the cross section, shown enlarged in Fig. 8.2b. Let axes x, y, z be chosen so that the (x, y) axes are tangent and normal, respectively, to the upper leg, and let the z axis be taken perpendicular to the cross section (the plane of Fig. 8.2b) and directed positively along the axis of the beam from the load P to the support. Then, the shear stress components in the cross section of the beam are σ_{zx} and σ_{zy} as shown. Since the shear stresses on the lateral surfaces of the beam are zero, $\sigma_{yz} = 0$. Hence, σ_{zy} vanishes at BD and EF [since $\sigma_{yz} = \sigma_{zy}$; see Eq. (2.4)]. Since $\sigma_{zy} = 0$ at BD and EF and the wall thickness between BD and EF is small (thin-wall) with respect to the length of the legs of the cross section, we assume that σ_{zy} does not change greatly (remains approximately zero) through the wall. The effect of σ_{zy} (the shear stress in the thickness direction) is ignored in the following discussion. In addition, it is as-

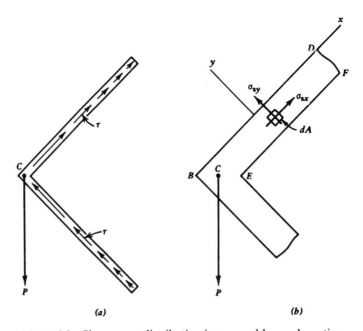

(a) *(b)*

Figure 8.2   Shear stress distribution in an equal-leg angle section.

sumed that the shear stress component σ_{zx} (along the legs) is approximately constant through the wall thickness and is equal to the average tangential shear stress τ in the wall (Fig. 8.2a). With these approximations for σ_{zy} and σ_{zx}, we find that a reasonably accurate and simple estimate of the shear center location may be obtained.

Since the above approximations are based on the concept of "thin walls," the question as to what constitutes a thin wall arises. It so happens that the answer to this question is not of great importance in locating the shear center, as the following example illustrates.

Consider the cross section, shown in Fig. 8.3, of a "thick-wall" cantilever beam. The view is taken looking from the support to the loaded (unrestrained) end. The approximate solution presented later in this chapter indicates that the shear center is located at point C', a distance $0.707t$ from the corner as shown. The applied load P is shown acting at point C'. By the theory of elasticity, the shear center is located (Kelber, 1948) at point C, $0.237t$ to the right of point C'. To examine the effects of this discrepancy, let us apply at point C two forces P' and P'' equal in magnitude and opposite in sense. (The forces P' and P'' do not disturb the equilibrium of the beam.) Hence, the beam may be considered to be loaded by force P acting at point C' or, equivalently, by a force P' (equal in magnitude to force P) and twisting moment due to P and P''. For example at the support, the bending moment is equal to PL (L is the length of the beam) and the torque T is equal to $0.237tP$. In practice, the ratio of the bending moment to the twisting moment is equal to 100 or more. In addition, the torsion theory presented in Chapter 6 shows that a thick-wall section, such as the one shown in Fig. 8.3, has considerable torsional resistance. Therefore, in applications in which length L of the beam

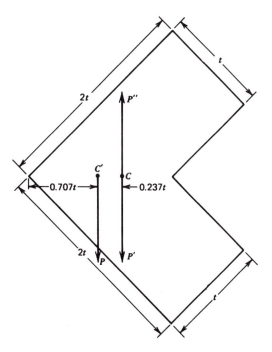

Figure 8.3   Shear center location for a thick-wall angle section.

is large compared to t, the shear stresses due to the torque ($T = 0.237tP$) are so small compared to the maximum bending stress that they can be neglected. Thus, for these beams the shear stresses due to the torque are no greater than the shear stresses due to shear load P'; both are often neglected.

8.2

SHEAR FLOW IN THIN-WALL BEAM CROSS SECTIONS

The average shear stress τ at each point in the walls of the beam cross section is assumed to have a direction tangent to the wall. The product of this shear stress and the wall thickness t defines the *shear flow* q (Sec. 6.6); thus,

$$q = \tau t \tag{8.1}$$

In the equation that will be derived for determining the shear flow q, we assume that the beam material remains linearly elastic and that the flexure formula is valid. Hence, we assume that the plane of the loads contains the bending axis of the beam and is parallel to one of the principal axes of inertia. It is convenient to consider a beam cross section that has one axis of symmetry (the x axis in Fig. 8.4). If the load P is parallel to the y axis and passes through the shear center C'', the x axis is the neutral axis for linearly elastic behavior and the flexure formula is valid. The derivation of the formula for q requires that both the bending moment M_x and total shear V_y be defined; load P is taken in the negative y direction so that both M_x and V_y are positive.

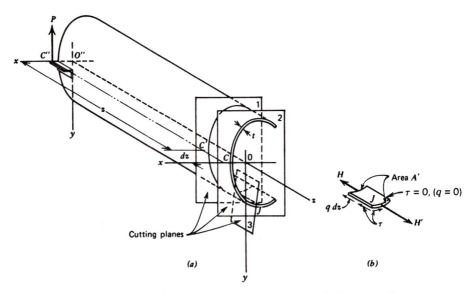

Figure 8.4   Shear flow in a beam having a symmetrical cross section.

We wish to determine the shear flow q at point J in the cross section of the beam in Fig. 8.4a at a distance $z + dz$ from load P. The free-body diagram necessary to determine q is obtained by three cutting planes. Cutting planes 1 and 2 are perpendicular to the z axis at distances z and $z + dz$ from the load P. Cutting plane 3 is parallel to the z axis and perpendicular to the lateral surface of the beam at J. The free body removed by the three cutting planes is indicated in Fig. 8.4b. The normal stress distributions σ_{zz} as given by the flexure formula act on the faces made by cutting planes 1 and 2. The resulting forces on these faces of area A' are parallel to the z axis and are indicated in Fig. 8.4b as H and H', respectively. Since the forces H and H' are unequal in magnitude, equilibrium of forces in the z-direction is maintained by the force $q\,dz$ on the face made by cutting plane 3. Therefore,

$$q\,dz = H' - H \tag{8.2}$$

Now, integrations of σ_{zz} over the faces with area A' at sections 1 and 2 yield (with the flexure formula)

$$H = \int_{A'} \sigma_{zz}\,dA = \int_{A'} \frac{M_x y}{I_x}\,dA$$

and

$$H' = \int_{A'} (\sigma_{zz} + d\sigma_{zz})\,dA = \int_{A'} \frac{(M_x + dM_x)y}{I_x}\,dA$$

Substitution of these two relations into Eq. (8.2) and solution for q yields

$$q = \frac{dM_x}{dz}\frac{1}{I_x}\int_{A'} y\,dA$$

According to beam theory, the total shear V_y in the cross section of a beam is given by $V_y = dM_x/dz$. Also, since $\int_{A'} y\,dA = A'\bar{y}'$ where $\bar{y}'$ is the distance from the x axis to the centroid of A', we may express q as

$$q = \frac{V_y A' \bar{y}'}{I_x}$$

Furthermore, since the value of the shear stress τ in the longitudinal section cut by plane 3 (Fig. 8.4) is the same as the shear stress in the cross section cut by plane 2, the shear flow in the cross section at point J is

$$q = \tau t = \frac{V_y A' \bar{y}'}{I_x} = \frac{V_y Q}{I_x} \tag{8.3}$$

where t is the wall thickness at point J. The first moment of area A', that is, $A'\bar{y}'$, is commonly denoted by Q.

Equation (8.3) is used to locate the shear center of thin-wall beam cross sections for both symmetrical and unsymmetrical bending. The method is demonstrated in Sec. 8.3 for beam cross sections made up of moderately thin walls. Local buckling is not considered (see Chapter 12).

In many applications (e.g., girders), the beam cross sections are built up by joining stiff longitudinal stringers by thin webs. The webs are generally stiffened at several locations along the length of the beam. The shear center location for beams of this type is considered in Sec. 8.4.

8.3

SHEAR CENTER FOR A CHANNEL SECTION

A cantilever beam subjected to a bending load V at C' in a plane perpendicular to the axis x of symmetry of the beam is shown in Fig. 8.5. We wish to locate the plane of the load so that the channel bends without twisting. In other words, we wish to locate the bending axis CC' of the beam, or the shear center C of any cross section AB.

In Fig. 8.5a let V be transformed into a force and couple at section AB by introducing, at the shear center C whose location is as yet unknown, two equal and opposite forces V' and V'', each equal in magnitude to V. The forces V and V'' constitute the external bending couple at section AB, which is held in equilibrium by the internal resisting moment at section AB in accordance with the flexure formula, Eq. (7.1); the distribution of the normal stress σ_{zz} on section AB is shown in Fig. 8.5a. The force V' is located at a distance e from the center of the web of the channel, as indicated in Figs. 8.5a and 8.5b. Force V' is resisted by shear stress τ or shear flow q [Eq. (8.3)], in cross section AB. Since the shear flow is directed along the straight sides of the channel, it produces forces F_1, F_2, and F_3, which lie in the cross section as indicated in Fig. 8.5b. Accordingly, by equilibrium

$$\sum F_x = F_2 - F_1 = 0 \tag{8.4}$$
$$\sum F_y = V' - F_3 = 0 \tag{8.5}$$
$$\sum M_z = V'e - F_1 h = 0 \tag{8.6}$$

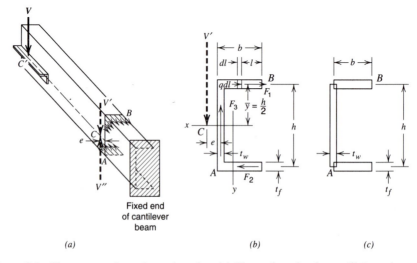

Fixed end
of cantilever
beam

(a)   (b)   (c)

Figure 8.5   Shear center for a channel section. (a) Channel section beam. (b) Location of C. (c) Idealized areas.

TABLE 8.1
Locations of Shear Centers for Sections Having One Axis of Symmetry

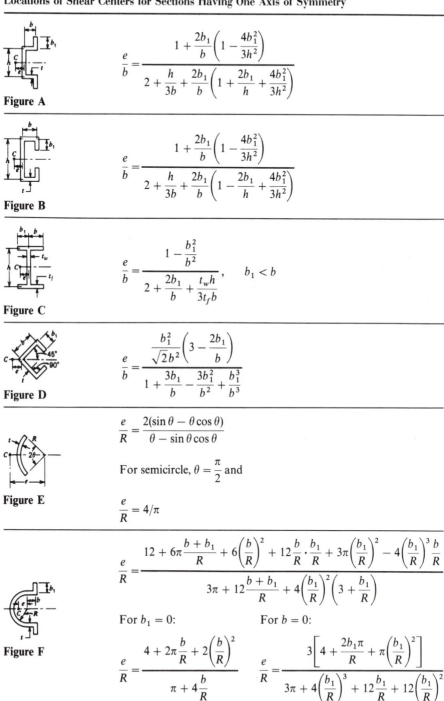

Figure A

$$\frac{e}{b} = \frac{1 + \dfrac{2b_1}{b}\left(1 - \dfrac{4b_1^2}{3h^2}\right)}{2 + \dfrac{h}{3b} + \dfrac{2b_1}{b}\left(1 + \dfrac{2b_1}{h} + \dfrac{4b_1^2}{3h^2}\right)}$$

Figure B

$$\frac{e}{b} = \frac{1 + \dfrac{2b_1}{b}\left(1 - \dfrac{4b_1^2}{3h^2}\right)}{2 + \dfrac{h}{3b} + \dfrac{2b_1}{b}\left(1 - \dfrac{2b_1}{h} + \dfrac{4b_1^2}{3h^2}\right)}$$

Figure C

$$\frac{e}{b} = \frac{1 - \dfrac{b_1^2}{b^2}}{2 + \dfrac{2b_1}{b} + \dfrac{t_w h}{3t_f b}}, \qquad b_1 < b$$

Figure D

$$\frac{e}{b} = \frac{\dfrac{b_1^2}{\sqrt{2}b^2}\left(3 - \dfrac{2b_1}{b}\right)}{1 + \dfrac{3b_1}{b} - \dfrac{3b_1^2}{b^2} + \dfrac{b_1^3}{b^3}}$$

Figure E

$$\frac{e}{R} = \frac{2(\sin\theta - \theta\cos\theta)}{\theta - \sin\theta\cos\theta}$$

For semicircle, $\theta = \dfrac{\pi}{2}$ and

$$\frac{e}{R} = 4/\pi$$

Figure F

$$\frac{e}{R} = \frac{12 + 6\pi\dfrac{b + b_1}{R} + 6\left(\dfrac{b}{R}\right)^2 + 12\dfrac{b}{R}\cdot\dfrac{b_1}{R} + 3\pi\left(\dfrac{b_1}{R}\right)^2 - 4\left(\dfrac{b_1}{R}\right)^3\dfrac{b}{R}}{3\pi + 12\dfrac{b + b_1}{R} + 4\left(\dfrac{b_1}{R}\right)^2\left(3 + \dfrac{b_1}{R}\right)}$$

For $b_1 = 0$:
$$\frac{e}{R} = \frac{4 + 2\pi\dfrac{b}{R} + 2\left(\dfrac{b}{R}\right)^2}{\pi + 4\dfrac{b}{R}}$$

For $b = 0$:
$$\frac{e}{R} = \frac{3\left[4 + \dfrac{2b_1\pi}{R} + \pi\left(\dfrac{b_1}{R}\right)^2\right]}{3\pi + 4\left(\dfrac{b_1}{R}\right)^3 + 12\dfrac{b_1}{R} + 12\left(\dfrac{b_1}{R}\right)^2}$$

The magnitude of the load V' is assumed to be known. Therefore, the determination of the distance e from the center line of the web to the shear center requires only that the force F_1 $(=F_2)$ be determined.

To determine F_1, it is convenient to think of the beam cross section as made up of line segments (Fig. 8.5a) with specified thicknesses. Since the forces F_1, F_2, F_3 are assumed to lie along the center line of the walls, the cross section is idealized as three narrow rectangles of lengths b, h, and b as indicated in Fig. 8.5c; note that the actual and idealized cross-sectional areas are equal since the three areas overlap. However, the moments of inertia of the actual and idealized cross sections differ from each other slightly. The moment of inertia of the idealized area is

$$I_x = \frac{1}{12}t_w h^3 + 2bt_f\left(\frac{h}{2}\right)^2 + 2\frac{1}{12}bt_f^3$$

This result may be simplified further by neglecting the third term, since for the usual channel section t_f is small compared to b or h. Thus, we write

$$I_x = \tfrac{1}{12}t_w h^3 + \tfrac{1}{2}t_f bh^2 \tag{8.7}$$

The force F_1 may be found from the shear flow equation

$$F_1 = \int_0^b q\,dl = \frac{V_y}{I_x}\int_0^b A'\bar{y}'\,dl = \frac{V_y t_f h}{2I_x}\int_0^b l\,dl = \frac{V_y t_f b^2 h}{4I_x} \tag{8.8}$$

where q is given by Eq. (8.3). The distance e to the shear center of the channel section is determined by substituting Eqs. (8.7) and (8.8) into Eq. (8.6) with the magnitude of V' set equal to that of V_y. Thus, we find

$$e = \frac{b}{2 + \dfrac{1}{3}\dfrac{t_w h}{t_f b}} \tag{8.9}$$

Because of the assumptions employed and approximations used, Eq. (8.9) gives an approximate location of the shear center for channel sections. The error is small for thin-wall sections. The approximate locations of the shear center for several other thin-wall sections with an axis of symmetry are given in Table 8.1.

EXAMPLE 8.1
Shear Center for Channel with Sloping Flanges

A 4-mm thickness plate of steel is formed into the cross section shown in Fig. E8.1a. Locate the shear center for the cross section.

SOLUTION

For simplicity in finding the moment of inertia, we approximate the actual cross section (Fig. E8.1a) by the cross section shown in Fig. E8.1b. The moment of inertia

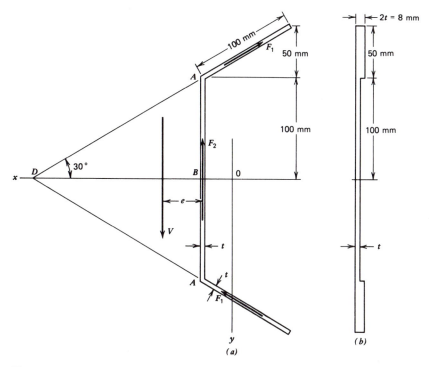

Figure E8.1

about the x axis for the cross section in Fig. E8.1b closely approximates that for the actual cross section in Fig. E8.1a.

$$I_x = \frac{8(300)^3}{12} - \frac{4(200)^3}{12} = 15{,}330{,}000 \text{ mm}^4$$

Because of the shear flow, forces F_1 and F_2 are developed in the three legs of the cross section. The magnitude of force F_1 requires integration; therefore, it is convenient to take moments about point D so that the magnitude of F_1 is not required. Since the shear flow from A to B to A varies parabolically, the average shear flow is equal to the shear flow at A plus 2/3 of the difference between the shear flow at B and shear flow at A.

$$q_A = \frac{V}{I_x} A' \bar{y} = \frac{V}{I_x}(100)(4)(125) = 50{,}000 \frac{V}{I_x}$$

$$q_B = q_A + \frac{V}{I_x}(100)(4)(50) = 70{,}000 \frac{V}{I_x}$$

$$q_{ave} = q_A + \frac{2}{3}(q_B - q_A) = 63{,}330 \frac{V}{I_x}$$

$$F_2 = 200 q_{ave} = 63{,}330 \frac{V}{I_x}(200) = 12{,}670{,}000 \frac{V}{I_x}$$

With point D as the moment center, the clockwise moment of V must equal the counterclockwise moment of F_2. Thus, we have $(173.2 - e) V = 173.2 \, F_2$, and hence, $e = 30.1$ mm.

EXAMPLE 8.2
Shear Center for Unequal-Leg Channel

A beam has an unsymmetrical section whose shape and dimensions are as shown in Fig. E8.2a. Locate the shear center.

SOLUTION

Centroidal x and y axes are chosen that are parallel to the sides of the thin-wall legs of the cross section. The origin 0 of the coordinates axes is located at $x_D = 25.0$ mm and $y_D = 40.0$ mm. To apply the theory to unsymmetrical sections, we use principal X and Y axes. As indicated in Appendix B, the principal axes may be described in terms of I_x, I_y, and I_{xy}. These values are $I_x = 1.734 \times 10^6$ mm^4, $I_y = 0.876 \times 10^6$ mm^4, and $I_{xy} = -0.500 \times 10^6$ mm^4. The angle θ between the x axis and X axis is obtained by the relation [Eq. (B.12)]

$$\tan 2\theta = -\frac{2I_{xy}}{I_x - I_y} = \frac{-2(-0.500 \times 10^6)}{1.734 \times 10^6 - 0.876 \times 10^6} = 1.166$$

from which $\theta = 0.4308$ rad. Since θ is positive, the X axis is located counterclockwise from the x axis. By using the equations in Appendix B, we find the principal moments of inertia to be $I_X = 1.964 \times 10^6$ mm^4, and $I_Y = 0.646 \times 10^6$ mm^4. The principal axes are shown in Figs. E8.2b and c.

The shear center C is located by considering two separate cases of loading (without twisting) in two orthogonal planes of the loads. The intersection of these two planes of loads determines the shear center C. Thus, assume that the resultant V'_Y of unbalanced loads on one side of the section in Fig. E8.2b is parallel to the Y axis. Since V'_Y is assumed to pass through the shear center, the beam bends without twisting and the X axis is the neutral axis; hence, the flexure formula and

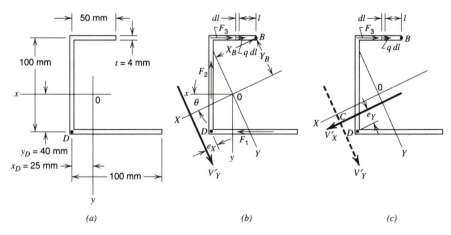

(a)   (b)   (c)

Figure E8.2

Eq. (8.3) apply. Because of the shear flow, forces F_1, F_2, and F_3 are developed in the three legs of the cross section (Fig. E8.2b). Only the magnitude of F_3 is required if point D is chosen as the moment center. In order to determine F_3, it is necessary that the shear flow q be determined as a function of l, the distance from point B. The coordinates of point B, the shear flow q, and force F_3 are determined as follows:

$$X_B = x_B \cos \theta + y_B \sin \theta = -25(0.9086) - 60(0.4176) = -47.77 \text{ mm}$$
$$Y_B = y_B \cos \theta - x_B \sin \theta = -60(0.9086) + 25(0.4176) = -44.08 \text{ mm}$$

$$q = \frac{V_Y}{I_X} A'\bar{Y}' = \frac{V_Y}{I_X} tl\left(|Y_B| + \frac{1}{2}l\sin\theta\right)$$

$$F_3 = \int_0^{50} q \, dl = \frac{V_Y t}{I_X} \int_0^{50} l\left(44.08 + \frac{0.4176}{2}l\right) dl = 0.1299 V_Y$$

Using the fact that $V'_Y = V_Y$ (the total shear at the section), we obtain the distance e_X from point D to force V'_Y, which passes through the shear center, from the equilibrium moment equation. Therefore,

$$V_Y e_X = 100 F_3$$

or

$$e_X = 12.99 \text{ mm}$$

Next assume that the resultant of the unbalanced loads on one side of the section in Fig. E8.2c is V'_X and it is parallel to the X axis. Since V'_X is assumed to pass through the shear center, the beam bends without twisting and the Y axis is the neutral axis. The shear flow q and force F_3 are given by

$$q = \frac{V_X}{I_Y} A'\bar{X}' = \frac{V_X}{I_Y} tl\left(|X_B| - \frac{1}{2}l\cos\theta\right)$$

$$F_3 = \int_0^{50} q \, dl = \frac{V_X t}{I_Y} \int_0^{50} l\left(47.77 - \frac{0.9086}{2}l\right) dl = 0.2525 V_X$$

Set $V'_X = V_X$ (the total shear at the section) and take moments about point D. Therefore,

$$V_X e_Y = 100 F_3$$
$$e_Y = 25.25 \text{ mm}$$

In terms of principal coordinates, the shear center C is located at

$$X_C = x_D \cos \theta + y_D \sin \theta + e_X = 52.41 \text{ mm}$$
$$Y_C = y_D \cos \theta - x_D \sin \theta - e_Y = 0.66 \text{ mm}$$

The x and y coordinates of the shear center C are

$$x_C = X_C \cos \theta - Y_C \sin \theta = 47.35 \text{ mm}$$
$$y_C = Y_C \cos \theta + X_C \sin \theta = 22.49 \text{ mm}$$

8.4

SHEAR CENTER OF COMPOSITE BEAMS FORMED FROM STRINGERS AND THIN WEBS

Often, particularly in the aircraft industry, beams are built up by welding or riveting longitudinal stiffeners, called stringers, to thin webs. Such beams are often designed to carry large bending loads and small shear loads. Two examples of cross sections of such beams are shown in Fig. 8.6. A beam whose cross section consists of two T-section stringers joined to a semicircular web is shown in Fig. 8.6a, and a beam whose cross section consists of a vertical web joined to two angle section stringers that, in turn, are joined to two horizontal webs that support two T-section stringers is shown in Fig. 8.6b.

Caution: In practice, beams with cross sections similar to those shown in Fig. 8.6b have webs so thin that they may buckle before they fail due to yielding (Chapter 12), particularly in aircraft applications. Consider, for example, a cantilever beam subject to end load (Fig. 8.7). Before buckling, the state of stress at the neutral axis is pure shear, as indicated on the volume element A in Fig. 8.7. After buckling, the state of stress is as indicated on volume element B in Fig. 8.7. A photo (Langhaar, 1942) of a similar beam with a buckled web is indicated in Fig. 8.8. After buckling of the web, the shear in the beam is carried by diagonal tension (block B, Fig. 8.7) (Bleich, 1952; Timoshenko and Gere, 1961; Kuhn, 1956). To strengthen such beams, transverse stiffeners are placed at each end and along the beam, as indicated in Fig. 8.8. These stiffeners restrain relative motion between the longitudinal stiffeners so that the beam may develop resistance to the diagonal tension. In addition, transverse stiffeners are located at sections where loads are applied to the beam. In this chapter, we assume that the web thickness is sufficiently thick so that the shear flow does not cause web buckling.

The calculation of the shear center location for beam cross sections similar to those shown in Fig. 8.6 is based on two simplifying assumptions: (1) that the web

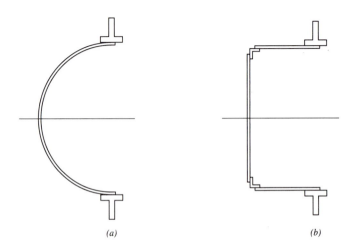

(a)   (b)

Figure 8.6   Beam cross sections built up of stringers and thin webs.

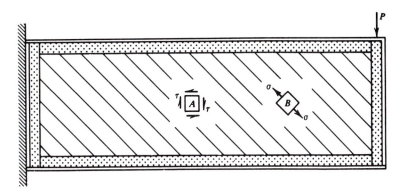

Figure 8.7   Web buckling in a cantilever beam.

Figure 8.8   Diagonal-tension beam.

does not support tensile or compressive stresses due to bending loads and (2) that the shear flow is constant in a web between pairs of transverse stiffners. The actual webs of these composite beams are often so thin that they may buckle under small compressive stresses. Therefore, the webs should not be expected to carry compressive flexure stresses. In general, the webs can carry tensile flexure stresses. However, this capability is sometimes ignored in their design.

Since the web walls are usually very thin, the moment of inertia for symmetrical cross sections of composite beams is approximated by the relation

$$I_x = 2 \sum_{i=1}^{n} A_i \bar{y}_i^2 \tag{8.10}$$

where $2n$ is the number of stringers, A_i are the cross-sectional areas of the stringers on one side of the neutral axis (x axis), and $\bar{y}_i$ are the distances from the neutral axis to the centroids of the areas A_i. Equation (8.10) discards the effect of the web. Hence I_x is underestimated. With this value of I_x, the computed flexure stresses are overestimated (higher than the true stresses).

Note: Transverse shear stresses are developed in the areas A_i of the stringers so that the stringers carry part of the total shear load V_y applied to the beam. However, the part of V_y carried by each stringer is usually ignored. This error is corrected in part by assuming that each web is extended to the centroid of the area of each stringer, thus increasing the contribution of the web. The procedure is demonstrated in the following example.

EXAMPLE 8.3
Shear Center for Composite Beam

A composite beam has a symmetrical cross section as shown in Fig. E8.3. A vertical web with a thickness of 2 mm is riveted to two square stringers. Two horizontal webs, with a thickness of 1 mm, are riveted to the square stringers and the T-section stringers. Locate the shear center of the cross section.

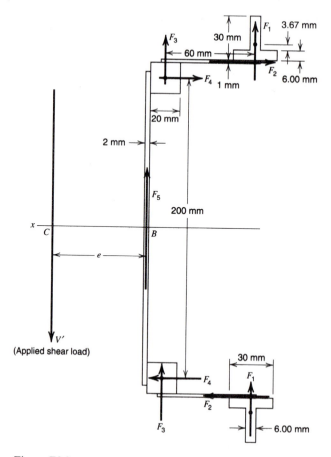

Figure E8.3

SOLUTION

The centroid of each T-section is located 9.67 mm from its base. The distance from the x axis to the centroid of each T-section is

$$\bar{y}_2 = 100 + 10 + 1 + 9.67 = 120.67 \text{ mm}$$

The approximate value of I_x [Eq. (8.10)] is

$$I_x = 2A_1\bar{y}_1^2 + 2A_2\bar{y}_2^2 = 2(400)(100)^2 + 2(324)(120.67)^2$$
$$= 17.44 \times 10^6 \text{ mm}^4$$

In these calculations, the shear flow q_1 is assumed to be constant from the centroid of the T-section to the centroid of the square stringers. The magnitude of q_1 is [Eq. (8.3)]

$$q_1 = \frac{V_y}{I_x}A'\bar{y}' = \frac{V_y}{I_x}(324)(120.67) = 39.10 \times 10^3 \frac{V_y}{I_x}$$

where V_y is the total shear at the section. The forces F_1, F_2, and F_3 are given by the relations

$$F_1 = (9.67 + 0.5)q_1 = 397.6 \times 10^3 \frac{V_y}{I_x}$$

$$F_2 = 60q_1 = 2.346 \times 10^6 \frac{V_y}{I_x}$$

$$F_3 = (10 + 0.5)q_1 = 410.5 \times 10^3 \frac{V_y}{I_x}$$

The shear flow q_2 is also assumed to be constant between centroids of the square stringers. Hence,

$$q_2 = q_1 + \frac{V_y}{I_x}(400)(100) = 79.10 \times 10^3 \frac{V_y}{I_x}$$

The forces F_4 and F_5 are given by the relations

$$F_4 = (10 + 1)q_2 = 870.1 \times 10^3 \frac{V_y}{I_x}$$

$$F_5 = 200q_2 = 15.82 \times 10^6 \frac{V_y}{I_x}$$

These forces with V' (Fig. E8.3) must satisfy equilibrium in the y direction, that is,

$$\sum F_y = V' - 2F_1 - 2F_3 - F_5 = 0$$

Hence,

$$V' = \frac{2(397.6 \times 10^3 V_y) + 2(410.5 \times 10^3 V_y) + 15.82 \times 10^6 V_y}{17.44 \times 10^6} = V_y$$

Thus, the applied shear load V' is equal to the total internal shear V_y in the section. The moment equilibrium equation for moments about point B determines the shear center location. Thus,

$$\sum M_B = V'e + 2F_1(71) + 2F_3(11) - F_2(221) - F_4(200) = 0$$
$$e = [2.346 \times 10^6(221) + 870.1 \times 10^3(200) - 2(397.6 \times 10^3)(71)$$
$$-2(410.5 \times 10^3)(11)]/17.44 \times 10^6$$
$$e = 35.95 \text{ mm}$$

This estimate of the location of the shear center C (Fig. E8.3) may be in error by several percent because of the simplifying assumptions. Hence, if the transverse bending loads are placed at C, they may introduce a small torque load in addition to bending loads. In most applications, the shear stresses resulting from this small torque are relatively insignificant (Chapter 6). In addition, it is questionable that the beam can be manufactured to precise dimensions and that the loads can be placed with great accuracy. Thus, the need for greater accuracy in our computations is also questionable.

8.5

SHEAR CENTER OF BOX BEAMS

Another class of practical beams is the box beam (with boxlike cross section) (Fig. 8.9). Box beams ordinarily have thin walls. However, they usually have walls sufficiently thick so that the walls will not buckle when subjected to elastic compressive stresses developed by bending. Box beams may be composed of several legs of different thickness (Fig. 8.9) or they may be a composite of longitudinal stringers and very thin webs (Fig. 8.10). The beams in Figs. 8.9 and 8.10 are one-compartment, box beams. In general, box beams may contain two or more compartments.

For convenience, let the x axis be an axis of symmetry in Figs. 8.9 and 8.10. Let the beams be subjected to symmetrical bending. Hence, let the plane of the loads be parallel to the y axis and let it contain the shear center C. The determination of the location of the shear center requires that the shear stress distribution in the cross section be known. However, the shear stress distribution cannot be obtained using Eq. (8.3) alone, since area A' is not known. (A' is the area of the wall from a point of interest in the wall to a point in the wall where $q = 0$.) Consequently, an additional equation, Eq. (6.67), along with Eq. (8.3), is required to obtain the shear stress distribution for a cross section of a box beam. Since there is no twist-

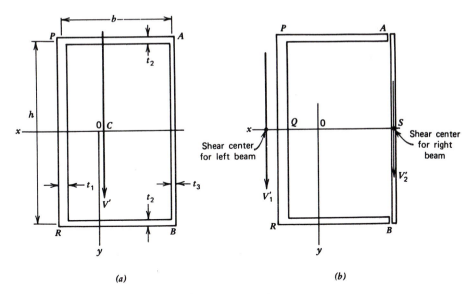

Figure 8.9 Box beam.

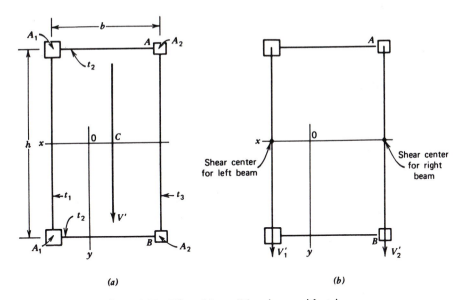

Figure 8.10 Ultra thin-wall box beam with stringers.

ing, the unit angle of twist in the beam is zero and, hence, Eq. (6.67) yields

$$\int_0^l \frac{q}{t}\,dl = 0 \tag{8.11}$$

where dl is an infinitesimal length of the wall of the box beam cross section at a point where the thickness is t and the shear flow is q. The length l of the perimeter of the box beam cross section is measured counterclockwise from any convenient point in the wall.

The shear flow q_A at any point, say, point A in Figs. 8.9 or 8.10, is an unknown. If this shear flow is subtracted from the actual shear flow at every point of the box

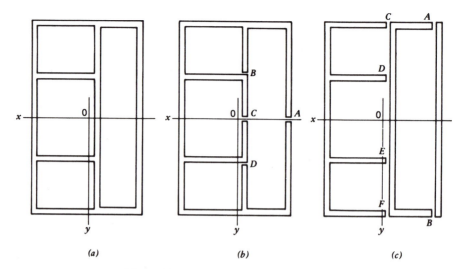

Figure 8.11 Multicompartment box beam.

beam wall, the resulting shear flow at A (and in this case, at B because of symmetry) is zero. We refer to such a point (of zero shear flow) as a *cut*. Then the resulting shear flow is the same as if the two beams (Figs. 8.9*b* and 8.10*b*) have no shear resistance at points A and B, but still have continuity of displacement at points A and B. Since the subtraction of q_A results in a subtraction of a zero force resultant, the subtraction produces no additional horizontal or vertical components of load on the cross section. The portions V'_1, V'_2 of the shear load V' acting on each of the two parts AB and BA (Figs. 8.9*b* and 8.10*b*) are proportional to the moments of inertia of the two parts of the beam since the curvature of the two parts must be continuous at points A and B. For convenience, let $V' = I$ (in magnitude) so that $V'_1 = I_1$ and $V'_2 = I_2$. Then, the shear flow at any point in the wall of either of the two parts of the beams (Fig. 8.9*b*) can be obtained using Eq. (8.3). The shear flow q_A is then added to the resulting shear flows for the two parts of the beam. The magnitude of q_A is obtained by satisfying Eq. (8.11). The force in each wall of the cross section can then be determined. The location of the shear center is obtained from the fact that the moment of these forces about any point in the plane of the cross section must be equal to the moment of the applied shear load V' about the same point.

For beams whose cross sections contain more than one compartment (Fig. 8.11), the above procedure must be repeated for a point in the wall of each compartment, such as at A, B, C, and D in Fig. 8.11*b* or at A, B, C, D, E, and F in Fig. 8.11*c*. The magnitudes of the shear flows that must be subtracted for each compartment are obtained by satisfying Eq. (8.11) for each compartment.

Unsymmetrical box beam cross sections can also be treated by the above procedure. In this case, it is desirable to refer the calculation to principal axes, say, X and Y. The method proceeds as follows: First, locate the plane of the loads for bending about the X axis; second, locate the plane of the loads for bending about the Y axis. The shear center of the cross section is given by the intersection of these two planes. The bending axis intersects each cross section of the box beam at the shear center.

EXAMPLE 8.4
Shear Center for Box Beam

For the box beam in Fig. 8.9, let $b = 300$ mm, $h = 500$ mm, $t_1 = 20$ mm, and $t_2 = t_3 = 10$ mm. Determine the location of the shear center for the cross section.

SOLUTION

The moment of inertia for the x axis is $I_x = 687.9 \times 10^6$ mm^4. Cuts are taken at points A and B to divide the beam into two parts (Fig. 8.9b). For convenience, let the magnitude of the shear load V' for the box beam be equal to the magnitude of I_x so that $V'_1 = I_{x1}$ and $V'_2 = I_{x2}$. The shear flow q is determined at points P, Q, and S for the two parts of the cut beam cross section (Fig. 8.9b) as follows (with $V'_1 = V_1$, $V'_2 = V_2$):

$$q_P = \frac{V_1 A' \bar{y}'}{I_{x1}} = (bt_2)\frac{h}{2} = 300(10)(250) = 750.0 \text{ kN/mm}$$

$$q_Q = q_P + \left(\frac{h}{2}t_1\right)\frac{h}{4} = 1{,}375.0 \text{ kN/mm}$$

$$q_S = \left(\frac{h}{2}t_3\right)\frac{h}{4} = 312.5 \text{ kN/mm}$$

The senses of the shear flows oppose those of V'_1 and V'_2. For the left part of the beam (Fig. E8.4a), the shear flow increases linearly from zero at B to q_P at R and decreases linearly from q_P at P to zero at A. The shear flow changes parabolically from q_P at R to q_Q at Q and back to q_P at P. For the right of the beam, the shear flow changes parabolically from zero at B to q_S at S and back to zero at A. Now, we add q_A (assumed positive in a counterclockwise direction) to the value of q at

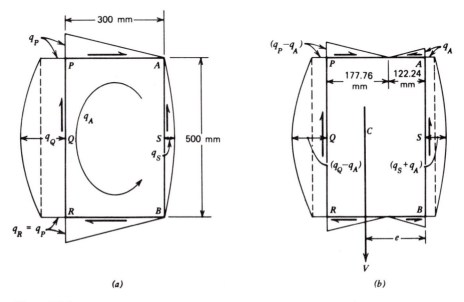

(a)   (b)

Figure E8.4

every point in the cross section (Fig. E8.4b), and we require that Eq. (8.11) be satisfied. Starting at P, we find that

$$0 = \left[q_A - q_P - \frac{2}{3}(q_Q - q_P) \right]\frac{h}{t_1}$$

$$+ \left(q_A - \frac{q_P}{2} \right)\frac{b}{t_2} + \left(q_A + \frac{2}{3}q_S \right)\frac{h}{t_3} + \left(q_A - \frac{q_P}{2} \right)\frac{b}{t_2}$$

$$0 = 135.0q_A - (750.0 + \tfrac{2}{3} \times 625.0)25 - (375.0)30$$
$$+ (\tfrac{2}{3} \times 312.5)50 - (375.0)30$$
$$q_A = 305.6 \text{ kN/mm}$$

This value of q_A must be added to the values computed for the cross section with the cuts to give the shear flow (Fig. E8.4b). The equilibrium moment equation for moments about point B gives

$$0 = V'e - \left(444.4 + \frac{2}{3}625 \right)(500)(300) - \frac{444.4}{2}(177.76)(500)$$

$$+ \frac{305.6}{2}(122.24)(500)$$

$$e = \frac{139.57 \times 10^9 \text{ N·mm}}{687.9 \times 10^6 \text{ N}} = 202.9 \text{ mm}$$

The shear center C lies on the x axis at a point 202.9 mm to the left of the center line of the right leg of the box section.

PROBLEMS
Section 8.3

8.1. Locate the shear center for the hat section beam shown in Fig. *A* of Table 8.1 by deriving the expression for *e*.

8.2. Verify the relation for *e* for the cross section shown in Fig. *B* of Table 8.1.

8.3. Locate the shear center for an unsymmetrical I-beam shown in Fig. *C* of Table 8.1 by deriving the expression for *e*.

8.4. Show that the shear center for the cross section in Fig. *D* of Table 8.1 is located at distance *e* as shown.

8.5. Derive the relation for *e* for the circular arc cross section shown in Fig. *E* of Table 8.1.

8.6. Derive the relation for *e* for the helmet cross section shown in Fig. *F* of Table 8.1

8.7. An extruded bar of aluminum alloy has the cross section shown in Fig. P8.7. Locate the shear center for the cross section.

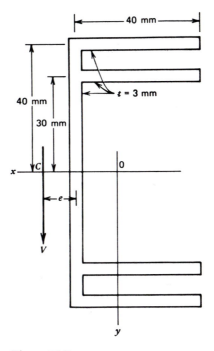

Figure P8.7

Note: Small differences in the value of e may occur because of differences in the approximations of I_x.

8.8. A 2.50-mm thick plate of steel is formed into the cross section shown in Fig. P8.8. Locate the shear center for the cross section.

Ans. $e = 37.14$ mm

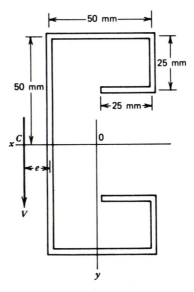

Figure P8.8

8.9. A rolled steel channel has the dimensions shown in Fig. P8.9. Locate the shear center for the cross section.

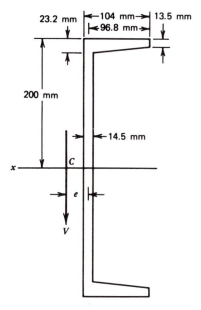

Figure P8.9

8.10. A beam has the cross section shown in Fig. P8.10. Locate the shear center for the cross section.

Ans. $\theta = -0.4215$ rad, $e_X = 45.50$ mm, $e_Y = 126.88$

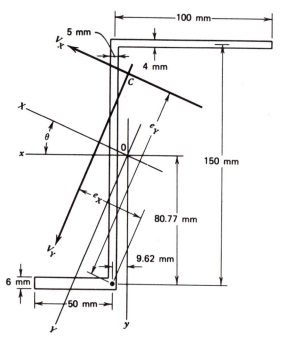

Figure P8.10

8.11. An extruded bar of aluminum alloy has the cross section shown in Fig. P8.11. Locate the shear center for the cross section.

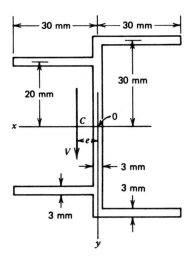

Figure P8.11

8.12. A 4-mm thick plate of steel is formed into the cross section shown in Fig. P8.12. Locate the shear center for the cross section.

Ans. $e = 28.50$ mm

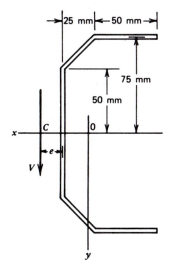

Figure P8.12

8.13. A 5-mm thick plate of steel is formed into the cross section shown in Fig. P8.13. Locate the shear center for the cross section.

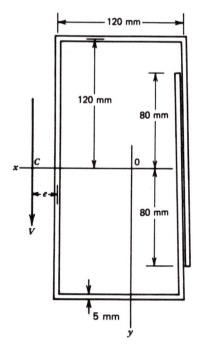

Figure P8.13

8.14. A 5-mm thick plate of steel is formed into the semicircular shape shown in Fig. P8.14. Locate the shear center for the cross section.

Ans. $e = 318.31$ mm

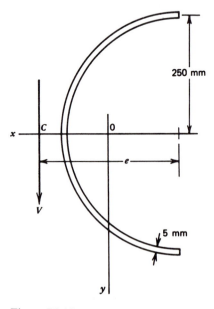

Figure P8.14

8.15. The horizontal top-most and bottom-most arms of the extruded bar of Fig. P8.7 are removed. Locate the shear center for the modified section.

8.16. An aluminum alloy extrusion has the cross section shown in Fig. P8.16. The member is to be used as a beam with the x axis as the neutral axis. Locate the shear center for the cross section.

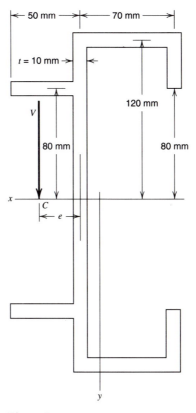

Figure P8.16

8.17. Locate the shear center for the beam cross section shown in Fig. P8.17. Both flanges and the web have thickness $t = 3.00$ mm.

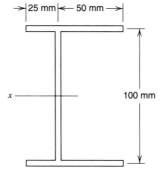

Figure P8.17

8.18. Locate the shear center for the beam cross section shown in Fig. P8.18. The walls of the cross section have constant thickness $t = 2.50$ mm.

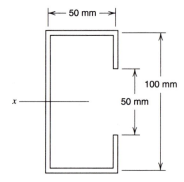

Figure P8.18

8.19. Locate the shear center for the beam cross section shown in Fig. P8.19. The walls of the cross section have constant thickness $t = 2.00$ mm.

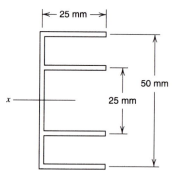

Figure P8.19

8.20. Locate the shear center for the beam cross section shown in Fig. P8.20. The walls of the cross section have constant thickness $t = 2.00$ mm.

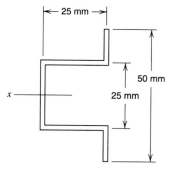

Figure P8.20

8.21. Locate the shear center for the beam cross section shown in Fig. P8.21. The walls of the cross section have constant thickness $t = 2.00$ mm.

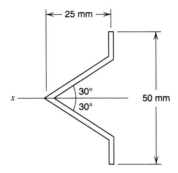

Figure P8.21

8.22. Locate the shear center for the beam cross section shown in Fig. P8.22. The walls of the cross section have constant thickness $t = 2.00$ mm.

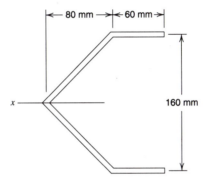

Figure P8.22

8.23. Locate the shear center for the beam cross section shown in Fig. P8.23. The walls of the cross section have constant thickness $t = 2.50$ mm.

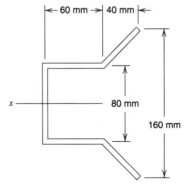

Figure P8.23

8.24. For the beam cross section shown in Fig. P8.24, $b \gg t$. Show that the moment of inertia $I_x = 5.609b^3t$ and locate the shear center for the cross section.

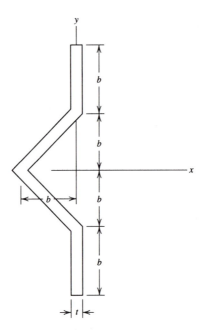

Figure P8.24

8.25. The channel shown in Fig. P8.25 is subjected to nonsymmetric bending. The associated shear forces, which act through the shear center, are $V_x = -2400$ N and $V_y = 1800$ N. Determine the distribution of the shear stress throughout the cross section. Make a sketch, to scale, of the shear-stress distribution in the channel walls.

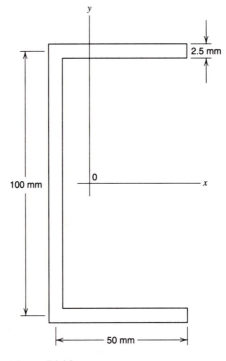

Figure P8.25

Section 8.4

8.26. A beam is built up of a thin steel sheet of thickness $t = 0.60$ mm bent into a semicircle as shown in Fig. P8.26. Two 25-mm square stringers are welded to the thin web as shown. Locate the shear center for the cross section.

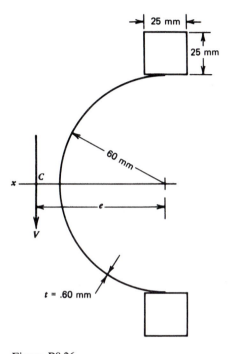

25 mm

25 mm

60 mm

x — C

e

V

$t = .60$ mm

Figure P8.26

8.27. A beam has a symmetrical cross section (Fig. P8.27). A vertical web with a thickness of 0.60 mm is welded to two 20 mm by 20 mm by 4 mm angle section ($A = 146$ mm^2 and centroid location 6.4 mm) stringers. The two horizontal webs have a thickness of 0.60 mm and are welded to the angle sections and 20 mm by 20 mm by 4 mm T-section stringers. Locate the shear center for the cross section.

Ans. $e = 28.20$ mm

8.28. A composite beam has a symmetrical cross section as shown in Fig. P8.28. A vertical web with a thickness of 2 mm is welded to the center of the flange of two 50 mm by 60 mm by 10 mm T-section stringers. Two horizontal webs, with a thickness of 1 mm, are welded to these stringers and to two additional T-section stringers. Locate the shear center of the cross section.

8.29. A composite beam has a symmetrical cross section, as shown in Fig. P8.29. A vertical web with a thickness of 2 mm is riveted to four rolled 30 mm by 30 mm by 5 mm angle sections ($A = 278$ mm^2 and centroid location 7.7 mm). Two horizontal webs, with thickness of 1 mm, are riveted to the angles and to areas A_1 (25 mm by 25 mm) and A_2 (40 mm by 40 mm). Locate the shear center of the cross section.

Ans. $e = 28.44$ mm

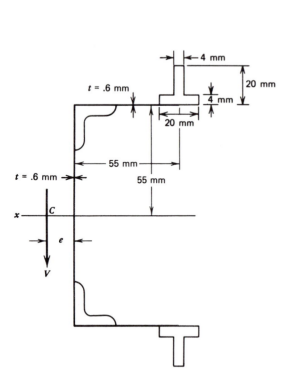

Figure P8.27

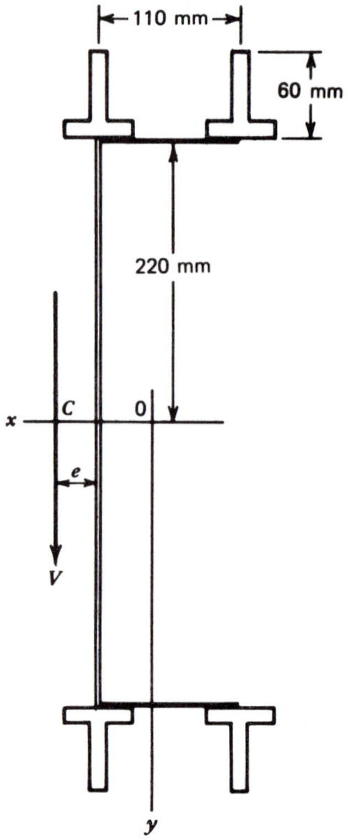

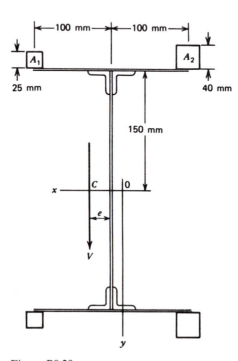

Figure P8.28

Figure P8.29

Section 8.5

8.30. For the box beam in Fig. 8.9, let $b = 100$ mm, $h = 200$ mm, $t_1 = 20$ mm, $t_2 = 10$ mm, and $t_3 = 5$ mm. Determine the location of the shear center for the cross section.

8.31. For the box beam in Fig. 8.10, let $b = 200$ mm, $h = 400$ mm, $t_1 = t_2 = t_3 = 1$ mm, and $A_1 = 3A_2 = 900$ mm^2. Determine the location of the shear center for the cross section.

Ans. $e = 83.33$ mm

8.32. Let $t_1 = 2$ mm with other dimensions from Problem 8.31 remaining unchanged. Determine the location of the shear center.

REFERENCES

Bleich, F. (1952). *Buckling Strength of Metal Structures*. New York: McGraw-Hill, pp. 337–343.

Boresi, A. P. and Chong, K. P. (1987) *Elasticity in Engineering Mechanics*. New York: Elsevier, Chap. 7.

Kelber, C. C. (1948). Numerical Determination of the Shear Center Coordinates of Heavy L-Shaped Sections. Champaign/Urbana, Ill: Univ. Ill., Dept. of Theoretical and Appl. Mech., M. S. Thesis.

Kuhn P. (1956). *Stresses in Aircraft and Shell Structures*. New York: McGraw-Hill.

Langhaar, H. L. (1942). Theoretical and Experimental Investigations of Thin-Webbed Plate-Girder Beams. San Diego, Calif.: Consolidate Aircraft Corp., Rept. SG-895, Sept. 1.

Timoshenko, S. and Gere, J. M. (1961). *Theory of Elastic Stability*: New York: McGraw-Hill, pp. 360–439.

9

CURVED BEAMS

9.1

INTRODUCTION

The flexure formula [Eq. (7.1)] is accurate for symmetrically loaded straight beams subjected to pure bending. It is also generally used to obtain approximate results for the design of straight beams subjected to shear loads, when the plane of loads contains the shear center and is parallel to a principal axis of the beam; the resulting errors in the computed stresses are small enough to be negligible as long as the beam length is at least five times the maximum cross-sectional dimension. In addition, the flexure formula is reasonably accurate in the analysis of curved beams for which the radius of curvature is more than five times the beam depth. However, for curved beams the error in the computed stress predicted by the flexure formula increases as the ratio of the radius of curvature of the beam to the depth of the beam decreases in magnitude. Hence, as this ratio decreases, one needs a more accurate solution for curved beams.

Timoshenko and Goodier (1970) have presented a solution based on the theory of elasticity for the linear elastic behavior of curved beams of rectangular cross sections for the loading shown in Fig. 9.1a. They used polar coordinates and obtained relations for the radial stress σ_{rr}, the circumferential stress $\sigma_{\theta\theta}$, and the shear stress $\sigma_{r\theta}$ (Fig. 9.1b). However, most curved beams do not have rectang-

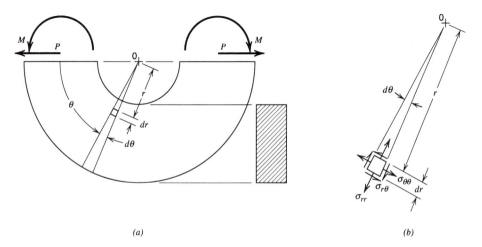

(a)   *(b)*

Figure 9.1   Rectangular section curved beam. (*a*) Curved beam loading. (*b*) Stress components.

ular cross sections. Therefore, in the following we present an approximate curved beam solution that is generally applicable to all symmetrical cross sections. This solution is based on two simplifying assumptions: (1) plane sections before loading remain plane after loading and (2) the radial stress σ_{rr} and shear stress $\sigma_{r\theta}$ are sufficiently small so that the state of stress is essentially one-dimensional. The resulting formula for the circumferential stress $\sigma_{\theta\theta}$ is the curved beam formula.

9.2

CIRCUMFERENTIAL STRESSES IN A CURVED BEAM

Consider the curved beam shown in Fig. 9.2a. The cross section of the beam has a plane of symmetry and the polar coordinates (r, θ) lie in the plane of symmetry, with origin at 0, the center of curvature of the beam. We also assume that the applied loads lie in the plane of symmetry. A positive moment is defined as one that causes the radius of curvature at each section of the beam to increase in magnitude. Thus, the applied loads on the curved beams in Figs. 9.1 and 9.2a cause positive moments. We wish to determine an approximate formula for the circumferential stress distribution $\sigma_{\theta\theta}$ on section BC. A free-body diagram of an element FBCH of the beam is shown in Fig. 9.2b (see Fig. 9.2a). The normal traction N, at the centroid of the cross section, the shear V, and moment M_x acting on face FH are shown in their positive directions. These forces must be balanced by the resultants due to the normal stress $\sigma_{\theta\theta}$ and shear stress $\sigma_{r\theta}$ that act on face BC. The effect of the shear stress $\sigma_{r\theta}$ on the computation of $\sigma_{\theta\theta}$ is usually small, except for curved beams with very thin webs. However, since ordinarily, practical curved beams are not designed with thin webs because of the possibility of failure by excessive radial stresses (see Sec. 9.3), neglecting the effect of $\sigma_{r\theta}$ on the computation of $\sigma_{\theta\theta}$ is reasonable.

Let the z axis be normal fo face BC (Fig. 9.2b). By equilibrium of forces in the z direction and of moments about the centroidal x axis, we find

$$\sum F_z = \int \sigma_{\theta\theta} \, dA - N = 0$$

$$\sum M_x = \int \sigma_{\theta\theta}(R - r) \, dA - M_x = 0$$

or

$$N = \int \sigma_{\theta\theta} \, dA \tag{9.1}$$

$$M_x = \int \sigma_{\theta\theta}(R - r) \, dA \tag{9.2}$$

where R is the distance from the center of curvature of the curved beam to the centroid of the beam cross section and r locates the element of area dA from the center of curvature. The integrals of Eqs. (9.1) and (9.2) cannot be evaluated until $\sigma_{\theta\theta}$ is expressed in terms of r. The functional relationship between $\sigma_{\theta\theta}$ and r is obtained from the assumed geometry of deformation and stress-strain relations for the material.

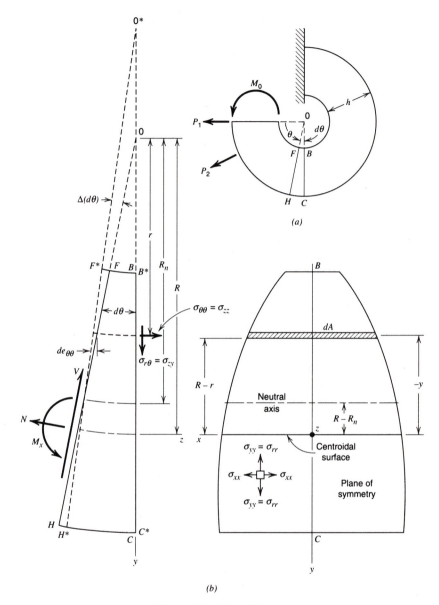

Figure 9.2 Curved beam.

The curved beam element $FBCH$ in Fig. 9.2b represents the element in the undeformed state. The element $F^*B^*C^*H^*$ represents the element after it is deformed
by the loads. For convenience, we have positioned the deformed element so that
face B^*C^* coincides with face BC. As in the case of straight beams, we assume that
planes B^*C^* and F^*H^* remain plane under the deformation. Face F^*H^* of the
deformed curved beam element forms an angle $\Delta(d\theta)$ with respect to FH. Line
F^*H^* intersects line FH at the neutral axis of the cross section (axis for which
$\sigma_{\theta\theta} = 0$) at distance R_n from the center of curvature. The movement of the center
of curvature from point 0 to point 0* is exaggerated in Fig. 9.2b in order to illustrate the geometry changes. For infinitesimally small displacements, the movement

of the center of curvature is infinitesimal. The elongation $de_{\theta\theta}$ of a typical element in the θ direction is equal to the distance between faces FH and F^*H^* and varies linearly with the distance $(R_n - r)$. However, the corresponding strain $\epsilon_{\theta\theta}$ is a nonlinear function of r, since the element length $r\,d\theta$ also varies with r. This fact distinguishes a curved beam from a straight beam. Thus, by Fig. 9.2b, we obtain for the strain

$$\epsilon_{\theta\theta} = \frac{de_{\theta\theta}}{r\,d\theta} = \frac{(R_n - r)\Delta(d\theta)}{r\,d\theta} = \left(\frac{R_n}{r} - 1\right)\omega \tag{9.3}$$

where

$$\omega = \frac{\Delta(d\theta)}{d\theta} \tag{9.4}$$

It is assumed that σ_{xx} is sufficiently small so that it may be neglected. Hence, the curved beam is considered to be a problem in plane stress. Although radial stress σ_{rr} may, in certain cases, be of importance (see Sec. 9.3), here we neglect its effect on $\epsilon_{\theta\theta}$. Then, by Hooke's law, we find

$$\sigma_{\theta\theta} = E\epsilon_{\theta\theta} = \frac{R_n - r}{r}E\omega = \frac{E\omega R_n}{r} - E\omega \tag{9.5}$$

Substituting Eq. (9.5) into Eqs. (9.1) and (9.2), we obtain

$$N = R_n E\omega \int \frac{dA}{r} - E\omega \int dA = R_n E\omega A_m - E\omega A \tag{9.6}$$

$$M_x = R_n R E\omega \int \frac{dA}{r} - (R + R_n)E\omega \int dA + E\omega \int r\,dA$$

$$= R_n R E\omega A_m - (R + R_n)E\omega A + E\omega R A = R_n E\omega(RA_m - A) \tag{9.7}$$

where A is the cross-sectional area of the curved beam and A_m has the dimensions of length and is defined by the relation

$$A_m = \int \frac{dA}{r} \tag{9.8}$$

Equation (9.7) can be rewritten in the form

$$R_n E\omega = \frac{M_x}{RA_m - A} \tag{9.9}$$

Then substitution into Eq. (9.6) gives

$$E\omega = \frac{A_m M_x}{A(RA_m - A)} - \frac{N}{A} \tag{9.10}$$

The circumferential stress distribution for the curved beam is obtained by substituting Eq. (9.9) and (9.10) into Eq. (9.5) to obtain the curved beam formula

$$\sigma_{\theta\theta} = \frac{N}{A} + \frac{M_x(A - rA_m)}{Ar(RA_m - A)} \tag{9.11}$$

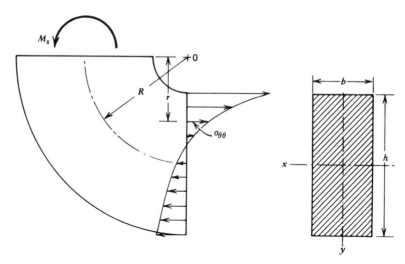

Figure 9.3 Circumferential stress distribution in a rectangular section curved beam ($R/h = 0.75$).

The normal stress distribution given by Eq. (9.11) is hyperbolic in form; that is, it varies as $1/r$. For the case of a curved beam with rectangular cross section ($R/h = 0.75$) subjected to pure bending, the normal stress distribution is shown in Fig. 9.3.

Since Eq. (9.11) has been based on several simplifying assumptions, it is essential that its validity be verified. Results predicted by the curved beam formula can be compared with those obtained from the elasticity solution for curved beams with rectangular sections and with those obtained from experiments on, or finite element analysis of, curved beams with other kinds of cross sections. The maximum value of circumferential stress $\sigma_{\theta\theta(CB)}$ as given by the curved beam formula may be computed from Eq. (9.11) for curved beams of rectangular cross sections subjected to pure bending and shear (Fig. 9.4). For rectangular cross sections, the ratios of $\sigma_{\theta\theta(CB)}$ to the elasticity solution $\sigma_{\theta\theta(\text{elast})}$ are listed in Table 9.1 for pure bending (Fig. 9.4a) and for shear loading (Fig. 9.4b), for several values of the ratio R/h, where h denotes the beam depth (Fig. 9.2a). The nearer this ratio is to 1, the less error in Eq. (9.11). The curved beam formula is more accurate for pure bending than shear loading. Most curved beams are subjected to a combination of bending and shear. The value of R/h is usually greater than 1.0 for curved beams, so that the error in the curved beam formula is not particularly significant. However, possible

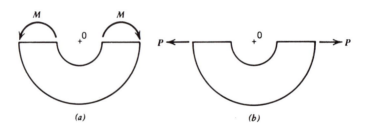

Figure 9.4 Types of curved beam loadings. (a) Pure bending. (b) Shear loading.

TABLE 9.1
Ratios of the Maximum Circumferential Stress in Rectangular Section Curved Beams as Computed by Elasticity Theory, the Curved Beam Formula and the Flexure Formula

$\dfrac{R}{h}$	Pure Bending		Shear Loading	
	$\dfrac{\sigma_{\theta\theta(CB)}}{\sigma_{\theta\theta(\text{elast})}}$	$\dfrac{\sigma_{\theta\theta(st)}}{\sigma_{\theta\theta(\text{elast})}}$	$\dfrac{\sigma_{\theta\theta(CB)}}{\sigma_{\theta\theta(\text{elast})}}$	$\dfrac{\sigma_{\theta\theta(st)}}{\sigma_{\theta\theta(\text{elast})}}$
0.65	1.046	0.439	0.855	0.407
0.75	1.012	0.526	0.898	0.511
1.0	0.997	0.654	0.946	0.653
1.5	0.996	0.774	0.977	0.776
2.0	0.997	0.831	0.987	0.834
3.0	0.999	0.888	0.994	0.890
5.0	0.999	0.933	0.998	0.934

errors occur in the curved beam formula for I- and T-section curved beams. These errors are discussed in Sec. 9.4. Also listed in Table 9.1 are the ratios of the maximum circumferential stress $\sigma_{\theta\theta(st)}$ given by the straight beam flexure formula [Eq. (7.1)] to the value $\sigma_{\theta\theta(\text{elast})}$. The straight beam solution is appreciably in error for small values of R/h and is in error by 7% for $R/h = 5.0$; the error is nonconservative. Generally, for curved beams with R/h greater than 5.0, the straight beam formula may be used.

As R becomes large compared to h, the right-hand term in Eq. (9.11) reduces to $-M_x y/I_x$. The negative sign results because the sign convention for positive moments for curved beams is opposite to that for straight beams [see Eq. (7.1)]. To prove this reduction, note that $r = R + y$. Then the term RA_m in Eq. (9.11) may be written as

$$RA_m = \int \left(\frac{R}{R + y} + 1 - 1 \right) dA = A - \int \frac{y}{R + y} dA \tag{a}$$

Hence, the denominator of the right-hand term in Eq. (9.11) becomes, for $R/h \to \infty$,

$$Ar(RA_m - A) = -A \int \left(\frac{Ry}{R + y} + y - y \right) dA - Ay \int \frac{y}{R + y} dA$$

$$= A \int \frac{y^2}{R + y} dA - A \int y \, dA - Ay \int \frac{y}{R + y} dA$$

$$= \frac{A}{R} \int \frac{y^2}{1 + (y/R)} dA - A \int y \, dA - \frac{Ay}{R} \int \frac{y}{1 + (y/R)} dA$$

$$= \frac{AI_x}{R} \tag{b}$$

since as $R/h \to \infty$, then $y/R \to 0$, $1 + y/R \to 1$, $\int [y^2\, dA/(1 + y/R)] \to I_x$, and $\int [y\, dA/(1 + y/R)] \to 0$. The right-hand term in Eq. (9.11) then simplifies to

$$\frac{M_x R}{AI_x}(A - RA_m - yA_m) = \frac{M_x R}{AI_x}\left(\int \frac{y/R}{1 + (y/R)}\, dA - \frac{y}{R}\int \frac{dA}{1 + (y/R)}\right)$$

$$= -\frac{M_x y}{I_x} \tag{c}$$

The curved beam formula [Eq. (9.11)] requires that A_m, defined by Eq. (9.8), be calculated for cross sections of various shapes. The number of significant digits retained in calculating A_m must be greater than that required for $\sigma_{\theta\theta}$ since RA_m approaches the value of A as R/h becomes large [see Eq. (a) above]. Explicit formulas for A, A_m, and R for several curved beam cross-sectional areas are listed in Table 9.2. Often, the cross section of a curved beam is composed of two or more of the fundamental areas listed in Table 9.2. The values of A, A_m, and R for

TABLE 9.2

Analytical Expressions for A, R, and $A_m = \displaystyle\int \frac{dA}{r}$

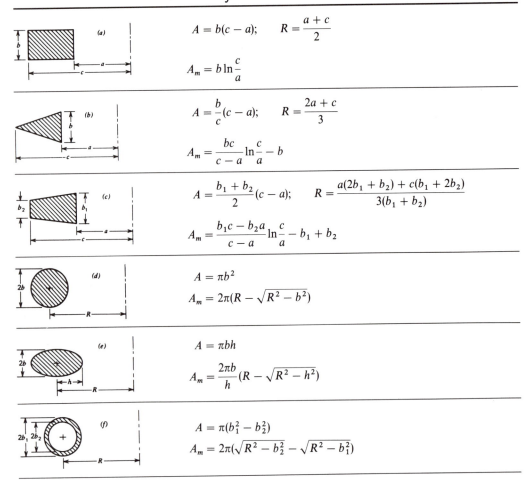

(a) $\quad A = b(c - a); \quad R = \dfrac{a + c}{2}$

$\quad A_m = b\ln\dfrac{c}{a}$

(b) $\quad A = \dfrac{b}{c}(c - a); \quad R = \dfrac{2a + c}{3}$

$\quad A_m = \dfrac{bc}{c - a}\ln\dfrac{c}{a} - b$

(c) $\quad A = \dfrac{b_1 + b_2}{2}(c - a); \quad R = \dfrac{a(2b_1 + b_2) + c(b_1 + 2b_2)}{3(b_1 + b_2)}$

$\quad A_m = \dfrac{b_1 c - b_2 a}{c - a}\ln\dfrac{c}{a} - b_1 + b_2$

(d) $\quad A = \pi b^2$

$\quad A_m = 2\pi(R - \sqrt{R^2 - b^2})$

(e) $\quad A = \pi bh$

$\quad A_m = \dfrac{2\pi b}{h}(R - \sqrt{R^2 - h^2})$

(f) $\quad A = \pi(b_1^2 - b_2^2)$

$\quad A_m = 2\pi(\sqrt{R^2 - b_2^2} - \sqrt{R^2 - b_1^2})$

TABLE 9.2 (*Continued*)

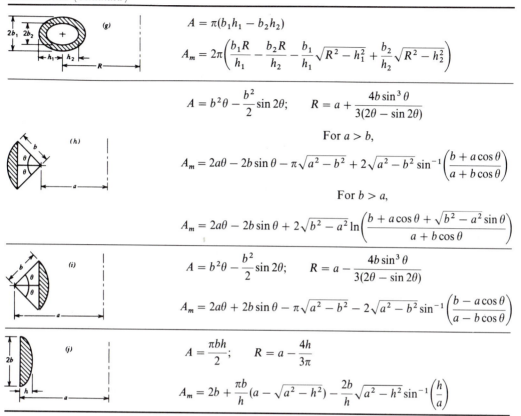

$$A = \pi(b_1 h_1 - b_2 h_2)$$

$$A_m = 2\pi \left(\frac{b_1 R}{h_1} - \frac{b_2 R}{h_2} - \frac{b_1}{h_1}\sqrt{R^2 - h_1^2} + \frac{b_2}{h_2}\sqrt{R^2 - h_2^2} \right)$$

$$A = b^2\theta - \frac{b^2}{2}\sin 2\theta; \qquad R = a + \frac{4b\sin^3\theta}{3(2\theta - \sin 2\theta)}$$

For $a > b$,

$$A_m = 2a\theta - 2b\sin\theta - \pi\sqrt{a^2 - b^2} + 2\sqrt{a^2 - b^2}\,\sin^{-1}\left(\frac{b + a\cos\theta}{a + b\cos\theta}\right)$$

For $b > a$,

$$A_m = 2a\theta - 2b\sin\theta + 2\sqrt{b^2 - a^2}\ln\left(\frac{b + a\cos\theta + \sqrt{b^2 - a^2}\sin\theta}{a + b\cos\theta}\right)$$

$$A = b^2\theta - \frac{b^2}{2}\sin 2\theta; \qquad R = a - \frac{4b\sin^3\theta}{3(2\theta - \sin 2\theta)}$$

$$A_m = 2a\theta + 2b\sin\theta - \pi\sqrt{a^2 - b^2} - 2\sqrt{a^2 - b^2}\,\sin^{-1}\left(\frac{b - a\cos\theta}{a - b\cos\theta}\right)$$

$$A = \frac{\pi b h}{2}; \qquad R = a - \frac{4h}{3\pi}$$

$$A_m = 2b + \frac{\pi b}{h}(a - \sqrt{a^2 - h^2}) - \frac{2b}{h}\sqrt{a^2 - h^2}\,\sin^{-1}\left(\frac{h}{a}\right)$$

the composite area are given by summation. Thus, for composite cross sections,

$$A = \sum_{i=1}^{n} A_i \tag{9.12}$$

$$A_m = \sum_{i=1}^{n} A_{m_i} \tag{9.13}$$

$$R = \frac{\displaystyle\sum_{i=1}^{n} R_i A_i}{\displaystyle\sum_{i=1}^{n} A_i} \tag{9.14}$$

where n is the number of fundamental areas that form the composite area.

Location of Neutral Axis of Cross Section

The neutral axis of bending of the cross section is defined by the condition $\sigma_{\theta\theta} = 0$. The neutral axis is located at distance R_n from the center of curvature. The distance R_n is obtained from Eq. (9.11) with the condition that $\sigma_{\theta\theta} = 0$ on the neutral surface $r = R_n$. Thus, Eq. (9.11) yields

$$R_n = \frac{A M_x}{A_m M_x + N(A - R A_m)} \tag{9.15}$$

For pure bending, $N = 0$, and then Eq. (9.15) yields

$$R_n = \frac{A}{A_m} \tag{9.16}$$

EXAMPLE 9.1
Stress in Curved Beam Portion of a Frame

The frame shown in Fig. E9.1 has a square cross section 50.0 mm by 50.0 mm. The load P is located 100 mm from the center of curvature of the curved beam portion of the frame. The radius of curvature of the inner surface of the curved beam is $a = 30$ mm. For $P = 9.50$ kN, determine the values for the maximum tensile and compressive stresses in the frame.

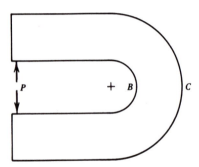

Figure E9.1

SOLUTION

The circumferential stresses $\sigma_{\theta\theta}$ are calculated using Eq. (9.11). Required values for A, A_m, and R for the curved beam are calculated using the equations in row (a) of Table 9.2. For the curved beam $a = 30$ mm and $c = 80$ mm.

$$A = b(c - a) = 50(80 - 30) = 2500 \text{ mm}^2$$

$$A_m = b \ln \frac{c}{d} = 50 \ln \frac{80}{30} = 49.04 \text{ mm}$$

$$R = \frac{a + c}{2} = \frac{80 + 30}{2} = 55 \text{ mm}$$

Hence, the maximum tensile stress is (at point B)

$$\sigma_{\theta\theta B} = \frac{P}{A} + \frac{M_x(A - r A_m)}{A r (R A_m - A)} = \frac{9500}{2500} + \frac{155(9500)[2500 - 30(49.04)]}{2500(30)[55(49.04) - 2500]}$$

$$= 106.2 \text{ MPa}$$

The maximum compressive stress is (at point C)

$$\sigma_{\theta\theta C} = \frac{9500}{2500} + \frac{155(9500)[2500 - 80(49.04)]}{2500(80)[55(49.04) - 2500]} = -49.3 \text{ MPa}$$

EXAMPLE 9.2
Stresses in a Crane Hook

Section BC is the critically stressed section of a crane hook (Fig. E9.2a). For a large number of manufactured crane hooks, the critical section BC can be closely approximated by a trapezoidal area with half of an ellipse at the inner radius and an arc of a circle at the outer radius. Such a section is shown in Fig. E9.2b, including dimensions for the critical cross section. The crane hook is made of a ductile

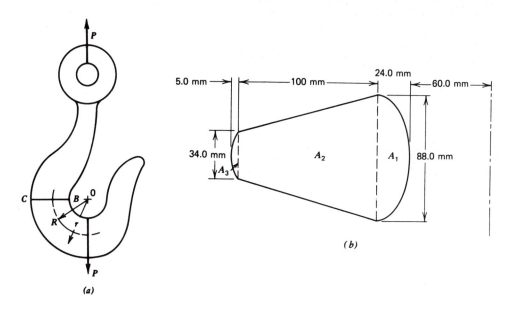

(b)

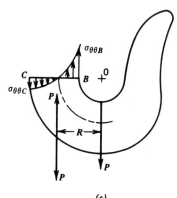

(a)

(c)

Figure E9.2   Crane hook.

steel that has a yield stress of $Y = 500$ MPa. Assuming that the crane hook is designed with a factor of safety of $SF = 2.00$ against initiation of yielding, determine the maximum load P that can be carried by the crane hook.

Note: An efficient algorithm to analyze crane hooks has been developed by Wang (1985).

SOLUTION

The circumferential stresses $\sigma_{\theta\theta}$ are calculated using Eq. (9.11). To calculate values of A, R, and A_m for the curved beam cross section, we divide the cross section into basic areas A_1, A_2, and A_3 (Fig. E9.2b).

For area A_1, $a = 84$ mm. Substituting this dimension along with other given dimensions into Table 9.2, row (j), we find

$$A_1 = 1658.76 \text{ mm}^2, \qquad R_1 = 73.81 \text{ mm}, \qquad A_{m_1} = 22.64 \text{ mm} \qquad \text{(a)}$$

For the trapezoidal area A_2, $a = 60 + 24 = 84$ mm and $c = a + 100 = 184$ mm. Substituting these dimensions along with other given dimensions into Table 9.2, row (c), we find

$$A_2 = 6100.00 \text{ mm}^2, \qquad R_2 = 126.62 \text{ mm}, \qquad A_{m_2} = 50.57 \text{ mm} \qquad \text{(b)}$$

For area A_3, $\theta = 0.5721$ rad, $b = 31.40$ mm, and $a = 157.60$ mm. When these values are substituted into Table 9.2, row (h), we obtain

$$A_3 = 115.27 \text{ mm}^2, \qquad R_3 = 186.01 \text{ mm}, \qquad A_{m_3} = 0.62 \text{ mm} \qquad \text{(c)}$$

Substituting values of A_i, R_i, and A_{mi} from Eqs. (a), (b), and (c) into Eqs. (9.12), (9.13), and (9.14), we calculate

$$A = 6100.00 + 115.27 + 1658.76 = 7874.03 \text{ mm}^2$$
$$A_m = 50.57 + 0.62 + 22.64 = 73.83 \text{ mm}$$
$$R = \frac{6100.00(126.62) + 115.27(186.01) + 1658.76(73.81)}{7874.03}$$
$$= 116.37 \text{ mm}$$

As indicated in Fig. E9.2c, the circumferential stress distribution $\sigma_{\theta\theta}$ is due to the normal load $N = P$ and moment $M_x = PR$. The maximum tension and compression values of $\sigma_{\theta\theta}$ occur at points B and C, respectively. For points B and C, by Fig. E9.2b, we find

$$r_B = 60 \text{ mm}$$
$$r_C = 60 + 24 + 100 + 5 = 189 \text{ mm}$$

Substituting the required values into Eq. (9.11), we find

$$\sigma_{\theta\theta B} = \frac{P}{7874.03} + \frac{116.37P[7874.03 - 60(73.83)]}{7874.03(60)[116.37(73.83) - 7874.03]}$$
$$= 0.000127P + 0.001182P$$
$$= 0.001309P \quad \text{(tension)}$$

$$\sigma_{\theta\theta C} = \frac{P}{7874.03} + \frac{116.37P[7874.03 - 189(73.83)]}{7874.03(189)[116.37(73.83) - 7874.03]}$$

$$= 0.000127P - 0.000662P$$

$$= -0.000535P \quad \text{(compression)}$$

Since the absolute magnitude of $\sigma_{\theta\theta B}$ is greater than $\sigma_{\theta\theta C}$, initiation of yield occurs when $\sigma_{\theta\theta B}$ equals the yield stress Y. The corresponding value of the failure load (P_f) is the load at which yield occurs. Dividing the failure load $P_f = Y/(0.001309)$ by the factor of safety $SF = 2.00$, we obtain the design load P; namely,

$$P = \frac{500}{2.00(0.001309)} = 190{,}900 \text{ N}$$

9.3

RADIAL STRESSES IN CURVED BEAMS

The curved beam formula for circumferential stress $\sigma_{\theta\theta}$ [Eq. (9.11)] is based on the assumption that the effect of radial stress is small. This assumption is quite accurate for curved beams with circular, rectangular, or trapezoidal cross sections; that is, cross sections that do not possess thin webs. However, in curved beams with cross sections in the form of an H, T, or I, the webs may be so thin that deformation of the cross section may produce a maximum radial stress in the web that may exceed the maximum circumferential stress. Also, although the radial stress is usually small, it may be significant relative to radial strength, for example, when anisotropic materials such as wood are formed into curved beams. The beam should be designed to take such conditions into account.

To illustrate the above remarks, we consider the tensile radial stress, due to a positive moment, that occurs in a curved beam at radius r from the center of curvature 0 of the beam (Fig. 9.5a). Consider equilibrium of the element $BDGF$ of the

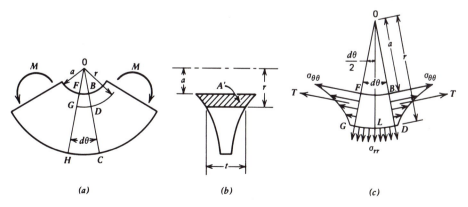

Figure 9.5   Radial stress in a curved beam. (a) Side view. (b) Cross-sectional shape. (c) Element $BDGF$.

beam shown enlarged in the free-body diagram in Fig. 9.5c. The faces *BD* and *GF*, which subtend the infinitesimal angle $d\theta$, have the area A' shown shaded in Fig. 9.5b. The distribution of $\sigma_{\theta\theta}$ on each of these areas produces a resultant circumferential force T (Fig. 9.5c) given by the expression

$$T = \int_a^r \sigma_{\theta\theta}\, dA \tag{9.17}$$

The components of the circumferential forces along line 0*L* are balanced by the radial stress σ_{rr} acting on the area $tr\, d\theta$, where t is the thickness of the cross section at the distance r from the center of curvature 0 (Fig. 9.5b). Thus for equilibrium in the radial direction along 0*L*, $\sum F_r = 0 = \sigma_{rr} tr\, d\theta - 2T\sin(d\theta/2) = (\sigma_{rr} tr - T)\, d\theta$, since for infinitesimal angle $d\theta/2$, $\sin(d\theta/2) = d\theta/2$. Thus, the tensile stress due to the positive moment is

$$\sigma_{rr} = \frac{T}{tr} \tag{9.18}$$

The force T is obtained by substitution of Eq. (9.11) into Eq. (9.17). Thus,

$$T = \frac{N}{A}\int_a^r dA + \frac{M_x}{RA_m - A}\int_a^r \frac{dA}{r} - \frac{M_x A_m}{A(RA_m - A)}\int_a^r dA$$

$$T = \frac{A'}{A}N + \frac{AA'_m - A'A_m}{A(RA_m - A)}M_x \tag{9.19}$$

where

$$A'_m = \int_a^r \frac{dA}{r} \quad \text{and} \quad A' = \int_a^r dA \tag{9.20}$$

Substitution of Eq. (9.19) into Eq. (9.18) yields the relation for the radial stress. For .rectangular cross-section curved beams subjected to shear loading (Fig. 9.4b), a comparison of the resulting approximate solution with the elasticity solution indicates that the approximate solution is conservative. Furthermore, for such beams it remains conservative to within 6% for values of $R/h > 1.0$ even if the term involving N in Eq. (9.19) is discarded. Consequently, if we retain only the moment term in Eq. (9.19), the expression for the radial stress may be approximated by the formula,

$$\sigma_{rr} = \frac{AA'_m - A'A_m}{trA(RA_m - A)}M_x \tag{9.21}$$

to within 6% of the elasticity solution for rectangular cross-section curved beams subjected to shear loading (Fig. 9.4b).

EXAMPLE 9.3
Radial Stress in T-Section

The curved beam in Fig. E9.3 is subjected to a load $P = 120$ kN. The dimensions of section *BC* are also shown. Determine the circumferential stress at *B* and radial stress at the junction of the flange and web at section *BC*.

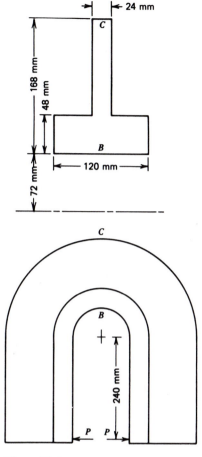

Figure E9.3

SOLUTION

The magnitudes of A, A_m, and R are given by Eqs. (9.12); (9.13), and (9.14), respectively. They are

$$A = 48(120) + 120(24) = 8640 \text{ mm}^2$$

$$R = \frac{48(120)(96) + 120(24)(180)}{8640} = 124.0 \text{ mm}$$

$$A_m = 120 \ln \frac{120}{72} + 24 \ln \frac{240}{120} = 77.93 \text{ mm}$$

The circumferential stress is given by Eq. (9.11). It is

$$\sigma_{\theta\theta B} = \frac{120,000}{8640} + \frac{364.0(120,000)[8640 - 72(77.93)]}{8640(72)[124.0(77.93) - 8640]}$$

$$= 13.9 + 207.8 = 221.7 \text{ MPa}$$

The radial stress at the junction of the flange and web is given by Eq. (9.21), with $r = 120$ mm and $t = 24$ mm. Magnitudes of A' and A'_m are

$$A' = 48(120) = 5760 \text{ m}^2$$

$$A'_m = 120 \ln \frac{120}{72} = 61.30 \text{ mm}$$

Substitution of these values in Eq. (9.21) gives

$$\sigma_{rr} = \frac{364.0(120{,}000)[8640(61.30) - 5760(77.93)]}{24(120)(8640)[124.0(77.93) - 8640]} = 138.5 \text{ MPa}$$

Hence, the magnitude of this radial stress is appreciably less than the maximum circumferential stress ($|\sigma_{\theta\theta B}| > |\sigma_{\theta\theta C}|$) and may not be of concern for the design engineer. However, in the solution of this problem, the effect of the stress concentration at the fillet joining the flange to the web has not be considered. This stress concentration increases the magnitude of the radial stress at the junction. However, the increase in stress is localized. Hence, it is not significant for curved beams made of ductile metal and subjected to static loads. However, for curved beams made of brittle materials or for curved beams of ductile material subjected to repeated loads, the localized stresses are significant. The effects of stress concentrations at fillets are considered in Chapter 14.

Curved Beams Made from Anisotropic Materials

Typically, the radial stresses developed in curved beams of *stocky* (rectangular, circular, etc.) cross sections are small enough that they can be neglected in analysis and design. However, some anisotropic materials may have low strength in the radial direction. Such materials include fiber-reinforced composites (fiberglass) and wood. For these materials, the relatively small radial stress developed in a curved beam may control the design of the beam due to the corresponding relatively low strength of the material in the radial direction. Hence, it may be important to properly account for radial stresses in curved beams of certain materials. When radial stresses exceed allowable stress limits, design changes are necessary. Possible changes to the beam include increasing the radius of curvature, modifying the cross section proportions, or adding mechanical reinforcement (deformed bars or lag bolts in the case of curved wood beams).

EXAMPLE 9.4
Radial Stress in Glulam Beam

A glued laminated timber (glulam) beam is used in a roof system. The beam has a simple span of 15 m and the middle half of the beam is curved with a mean radius of 10 m. The beam depth and width are both constant: $d = 0.800$ m and $b = 0.130$ m. Dead load is 2400 N/m and snow load is 4800 N/m. The geometry of the beam and assumed loading are shown in Fig. E9.4.

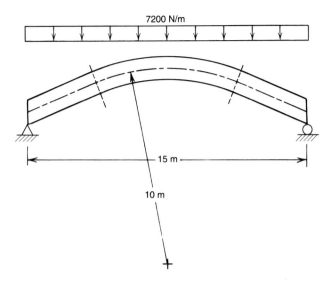

Figure E9.4

(a) Determine the maximum circumferential and radial stresses in the beam.

(b) Compare the maximum circumferential stress to that obtained from the straight beam flexure formula.

(c) Compare the maximum circumferential and radial stresses to the allowable stress limits for Douglas fir: $\sigma_{\theta\theta(allow)} = 15.8$ MPa, $\sigma_{rr(allow)} = 0.119$ MPa (AITC, 1985)

SOLUTION

(a) The maximum bending moment occurs at midspan and has magnitude $M_x = wl^2/8 = 202{,}500$ N·m. Circumferential stress $\sigma_{\theta\theta}$ is calculated using Eq. (9.11). For the curved beam described

$$a = R - \frac{d}{2} = 9.6 \text{ m}$$

$$c = r + \frac{d}{2} = 10.4 \text{ m}$$

$$A = 0.13 \times 0.80 = 0.104 \text{ m}^2$$

$$A_m = 0.13 \ln \frac{10.4}{9.6} = 0.0104056$$

The maximum circumferential stress occurs at the inner edge of the beam $r = a$. It is

$$\sigma_{\theta\theta(max)} = \frac{M_x(A - aA_m)}{Aa(RA_m - A)} = \frac{202{,}500[0.104 - 9.6(0.0104056)]}{0.104(9.6)[10.0(0.0104056) - 0.104]} = 15.0 \text{ MPa}$$

The maximum radial stress $\sigma_{rr(max)}$ is calculated using Eq. (9.21). However, the location at which $\sigma_{rr(max)}$ occurs is unknown. Thus, we must maximize σ_{rr} with respect to r. For a rectangular cross section, the quantities in Eq. (9.21) are

$$t = b = \text{width of cross section}$$
$$d = c - a = \text{depth of cross section}$$
$$A = bd$$
$$A' = b(r - a)$$
$$A_m = b \ln \frac{c}{a}$$
$$A'_m = b \ln \frac{r}{a}$$

Substitution of these expressions into Eq. (9.21) gives

$$\sigma_{rr} = \frac{M_x}{b} \left[\frac{d \ln (r/a) - (r - a) \ln (c/a)}{rd[R \ln (c/a) - d]} \right] \tag{a}$$

Maximizing σ_{rr} with respect to r, we find that $\sigma_{rr(max)}$ occurs at

$$r = a \exp \left(1 - \frac{a}{d} \ln \frac{c}{a} \right) \tag{b}$$

We evaluate Eq. (b) for the particular cross section of this example to obtain $r = 9.987$ m. At that location, the radial stress is, by Eq. (a),

$$\sigma_{rr(max)} = \frac{202,500}{0.13} \left[\frac{0.80 \ln (9.987/9.6) - [(9.987 - 9.6) \ln (10.4/9.6)]}{9.987(0.80)[10.0 \ln (10.4/9.6) - 0.80]} \right]$$
$$= 0.292 \text{ MPa} \tag{c}$$

An approximate formula for computing radial stress in curved beams of rectangular cross section is (AITC, 1985; p. 218)

$$\sigma_{rr} = \frac{3M}{2Rbd} \tag{d}$$

Using this expression, we determine the radial stress to be $\sigma_{rr} = 0.292$ MPa. The approximation of Eq. (d) is quite accurate in this case! In fact, for rectangular curved beams with $R/d > 3$, the error in Eq. (d) is less than 3%. However, as R/d becomes small, the error grows substantially and Eq. (d) is unconservative.

(b)   Using the curved beam formula, Eq. (9.11), we obtain the maximum circumferential stress as $\sigma_{\theta\theta(max)} = 15.0$ MPa. Using the straight beam flexure formula, Eq. (7.1), with $I_x = bd^3/12 = 0.005547$ m^4, we obtain $\sigma_{\theta\theta} = 202,500(0.40)/0.005547 = 14.6$ MPa. Thus, the straight beam flexure formula is within 3% of the curved beam formula. One would generally consider the flexure formula adequate for this case, in which $R/d = 12.5$.

(c) The maximum circumferential stress is just within its limiting value; the beam is understressed just 5%. However, the maximum radial stress is 245% over its limit. It would be necessary to modify beam geometry or add mechanical reinforcement to make this design acceptable.

9.4

CORRECTION OF CIRCUMFERENTIAL STRESSES IN CURVED BEAMS HAVING I-, T-, OR SIMILAR CROSS SECTIONS

If the curved beam formula is used to calculate circumferential stresses in curved beams having thin flanges, the computed stresses are considerably in error and the error is nonconservative. The error arises because the radial forces developed in the curved beam causes the tips of the flanges to deflect radially, thereby distorting the cross section of the curved beam. The resulting effect is to decrease the stiffness of the curved beam, to decrease the circumferential stresses in the tips of the flanges, and to increase the circumferential stresses in the flanges near the web.

Consider a short length of a thin-flanged I-section curved beam included between faces BC and FH that form an infinitesimal angle $d\theta$ as indicated in Fig. 9.6a. If the curved beam is subjected to a positive moment M_x, the circumferential stress distribution results in a tensile force T acting on the inner flange and a compressive force C acting on the outer flange, as shown. The components of these forces in the radial direction are $Td\theta$ and $C\,d\theta$. If the cross section of the curved beam did not distort, these forces would be uniformly distributed along each flange, as indicated in Fig. 9.6b. However, the two portions of the tension and compression flanges act as cantilever beams fixed at the web. The resulting bending due to cantilever beam action causes the flanges to distort, as indicated in Fig. 9.6c.

The effect of the distortion of the cross section on the circumferential stresses in the curved beam can be determined by examining the portion of the curved beam $ABCD$ in Fig. 9.6d. Sections AC and BD are separated by angle θ in the unloaded beam. When the curved beam is subjected to a positive moment, the center of curvature moves from 0 to 0*, section AC moves to $A*C*$, section BD moves to $B*D*$, and the included angle becomes $\theta*$. If the cross section does not distort, the inner tension flange AB elongates to length $A*B*$. Since the tips of the inner flange move radially inward relative to the undistorted position (Fig. 9.6c), the circumferential elongation of the tips of the inner flange is less than that indicated in Fig. 9.6d. *Therefore, $\sigma_{\theta\theta}$ in the tips of the inner flange is less than that calculated using the curved beam formula.* In order to satisfy equilibrium, it is necessary that $\sigma_{\theta\theta}$ for the portion of the flange near the web be greater than that calculated using the curved beam formula. Now consider the outer compression flange. As indicated in Fig. 9.6d, the outer flange shortens from CD to $C*D*$ if the cross section does not distort. Because of the distortion (Fig. 9.6c), the tips of the compressive flange move radially outward, requiring less compressive contraction. *Therefore, the magnitude of $\sigma_{\theta\theta}$ in the tips of the compression outer flange is less than that calculated by*

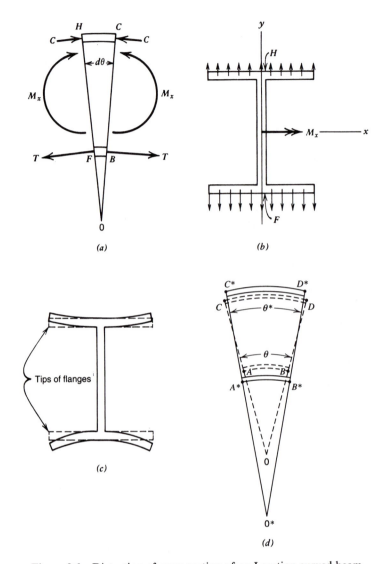

Figure 9.6   Distortion of cross section of an I-section curved beam.

the curved beam formula, and the magnitude of $\sigma_{\theta\theta}$ in the portion of the compression flange near the web is larger than that calculated by the curved beam formula.

The resulting circumferential stress distribution is indicated in Fig. 9.7. Since in developing the curved beam formula we assume that the circumferential stress is independent of x (Fig. 9.2), corrections are required if the formula is to be used in the design of curved beams having I- or T-cross sections and similar cross sections. There are two approaches that can be employed in the design of these curved beams. One approach is to prevent the radial distortion of the cross section by welding radial stiffeners to the curved beams. If distortion of the cross section is prevented, the use of the curved beam formula is appropriate. A second approach, suggested by H. Bleich (1933), is discussed below.

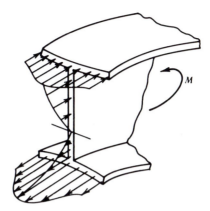

Figure 9.7   Stresses in I-section of curved beam.

Bleich's Correction Factors

Bleich reasoned that the actual maximum circumferential stresses in the tension and compression flanges for the I-section curved beam (Fig. 9.8a) may be calculated by the curved beam formula applied to an I-section curved beam with *reduced flange widths*, as indicated in Fig. 9.8b. By Bleich's method, if the same bending moment is applied to the two cross sections in Fig. 9.8, the computed maximum circumferential tension and compression stresses for the cross section shown in Fig. 9.8b, with no distortion, are equal to the actual maximum circumferential tension and compression stresses for the cross section in Fig. 9.8a, with distortion.

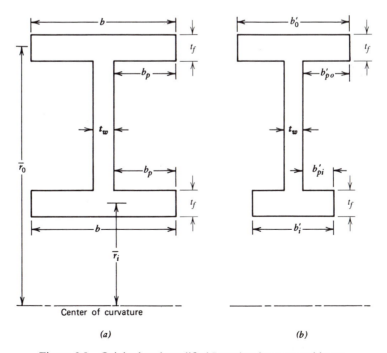

Figure 9.8   Original and modified I-section for a curved beam.

The approximate solution proposed by Bleich gives the results presented in tabular form in Table 9.3. In order to use the table, the ratio $b_p^2/\bar{r}t_f$ must be calculated, where

$$b_p = \text{projecting width of flange (see Fig. 9.8}a\text{)}$$
$$\bar{r} = \text{radius of curvature to the center of flange}$$
$$t_f = \text{thickness of flange}$$

The reduced width b_p' of the projecting part of each flange (Fig. 9.8b) is given by the relation

$$b_p' = \alpha b_p \tag{9.22}$$

where α is obtained from Table 9.3 for the computed value of the ratio $b_p^2/\bar{r}t_f$. The reduced width of each flange (Fig. 9.8b) is given by

$$b' = 2b_p' + t_w \tag{9.23}$$

where t_w is the thickness of the web. The curved beam formula [Eq. 9.11)] when applied to an undistorted cross section corrected by Eq. (9.23) predicts the maximum circumferential stress in the actual (distorted) cross section. This maximum stress occurs at the center of the inner flange. It should be noted that the state of stress at this point in the curved beam is not uniaxial. Because of the bending of the flanges (Fig. 9.6c), an x component of stress σ_{xx} (Fig. 9.2) is developed; the sign of σ_{xx} is opposite to that of $\sigma_{\theta\theta(\max)}$. Bleich obtained an approximate solution for σ_{xx} for the inner flange. It is given by the relation

$$\sigma_{xx} = -\beta\bar{\sigma}_{\theta\theta} \tag{9.24}$$

where β is obtained from Table 9.3 for the computed value of the ratio $b_p^2/\bar{r}t_f$, and where $\bar{\sigma}_{\theta\theta}$ is the magnitude of the circumferential stress at midthickness of the inner flange; the value of $\bar{\sigma}_{\theta\theta}$ is calculated based on the corrected cross section.

Although Bleich's analysis was developed for curved beams with relatively thin flanges, the results agree closely with a similar solution obtained by C. G. Anderson (1950) for I-beams and box beams, in which the analysis was not restricted to thin-flanged sections. Similar analyses of tubular curved beams with circular

TABLE 9.3
Table for Calculating the Effective Width and Lateral Bending Stress of Curved I- or T-Beams

$b_p^2/\bar{r}t$	0.2	0.3	0.4	0.5	0.6	0.7	0.8	0.9	1.0
α	0.977	0.950	0.917	0.878	0.838	0.800	0.762	0.726	0.693
β	0.580	0.836	1.056	1.238	1.382	1.495	1.577	1.636	1.677

$b_p^2/\bar{r}t$	1.1	1.2	1.3	1.4	1.5	2.0	3.0	4.0	5.0
α	0.663	0.636	0.611	0.589	0.569	0.495	0.414	0.367	0.334
β	1.703	1.721	1.728	1.732	1.732	1.707	1.671	1.680	1.700

and rectangular cross sections have been made by T. von Kármán (1911) and by S. Timoshenko (1923). An experimental investigation by D. C. Broughton, M. E. Clark, and H. T. Corten (1950) showed that another type of correction is needed if the curved beam has extremely thick flanges and thin webs. For such beams each flange tends to rotate about a neutral axis of its own in addition to the rotation about the neutral axis of the curved beam cross section as a whole. Curved beams for which the circumferential stresses are appreciably increased by this action probably fail by excessive radial stresses.

Note: The radial stress can be calculated using either the original or the modified cross section.

EXAMPLE 9.5
Bleich Correction Factors for T-Section

A T-section curved beam has the dimensions indicated in Fig. E9.5a and is subjected to pure bending. The curved beam is made of a steel having a yield stress $Y = 280$ MPa.

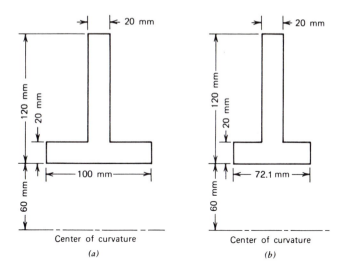

Figure E9.5 (a) Original section. (b) Modified section.

(a) Determine the magnitude of the moment that indicates yielding in the curved beam if Bleich's correction factors are not used.

(b) Use Bleich's correction factors to obtain a modified cross section. Determine the magnitude of the moment that initiates yielding for the modified cross section and compare with the result of part (a)

SOLUTION

(a) The magnitudes of A, A_m, and R for the original cross section are given by Eqs. (9.12), (9.13), and (9.14), respectively, as follows: $A = 4000$ mm^2, $A_m = 44.99$ mm, and $R = 100.0$ mm. By comparison of the stresses at the locations

$r = 180$ mm and $r = 60$ mm, we find that the maximum magnitude of $\sigma_{\theta\theta}$ occurs at the outer radius ($r = 180$ mm). See Eq. (9.11). Thus,

$$\sigma_{\theta\theta(\text{max})} = \left| \frac{M_x[4000 - 180(44.99)]}{4000(180)[100.0(44.99) - 4000]} \right|$$

$$= |-1.141 \times 10^{-5} M_x|$$

where M_x has the units of N·mm. Since the state of stress is assumed to be uniaxial, the magnitude of M_x to initiate yielding is obtained by setting $\sigma_{\theta\theta} = -Y$. Thus,

$$M_x = \frac{280}{1.141 \times 10^{-5}} = 24,540,000 \text{ N·mm} = 24.54 \text{ kN·m}$$

(b) The dimensions of the modified cross section are computed by Bleich's method; hence $b_p^2 / \bar{r} t_f$ must be calculated. It is

$$\frac{b_p^2}{\bar{r} t_f} = \frac{40(40)}{70(20)} = 1.143$$

Linear interpolation in Table 9.3 yields $\alpha = 0.651$ and $\beta = 1.711$. Hence, by Eqs. (9.22) and (9.23), the modified flange width is $b_p' = \alpha b_p = 0.651(40) = 26.04$ mm and $b' = 2b_p' + t_w = 2(26.04) + 20 = 72.1$ mm (Fig. E9.5b). For this cross section, by means of Eqs. (9.12), (9.13), and (9.14), we find

$$A = 72.1(20) + 20(100) = 3442 \text{ mm}^2$$

$$R = \frac{72.1(20)(70) + 20(100)(130)}{3442} = 104.9 \text{ mm}$$

$$A_m = 72.1 \ln \frac{80}{60} + 20 \ln \frac{180}{80} = 36.96 \text{ mm}$$

Now by means of Eq. (9.11), we find that the maximum magnitude of $\sigma_{\theta\theta}$ occurs at the inner radius of the modified cross section. Thus, with $r = 60$ mm, Eq. (9.11) yields

$$\sigma_{\theta\theta(\text{max})} = \frac{M_x[3442 - 60(36.96)]}{3442(60)[104.9(36.96) - 3442]} = 1.363 \times 10^{-5} M_x$$

The magnitude of M_x that causes yielding can be calculated by means of either the maximum shear-stress criterion of failure or the maximum octahedral shear-stress criterion of failure. If the maximum shear-stress criterion is used, the maximum principal stress must be computed. The minimum principal stress is σ_{xx}. Hence, by Eqs. (9.11) and (9.24), we find

$$\bar{\sigma}_{\theta\theta} = \frac{M_x[3442 - 70(36.96)]}{3442(70)[104.9(36.96) - 3442]} = 8.15 \times 10^{-6} M_x$$

$$\sigma_{xx} = -\beta\bar{\sigma}_{\theta\theta} = -1.711(8.15 \times 10^{-6} M_x) = -1.394 \times 10^{-5} M_x$$

and

$$\tau_{max} = \frac{\sigma_{max} - \sigma_{min}}{2} = \frac{Y}{2} = \frac{\sigma_{\theta\theta(max)} - \sigma_{xx}}{2}$$

$$M_x = 10{,}140{,}000 \text{ N·mm} = 10.14 \text{ kN·m}$$

A comparison of the moment M_x as determined in parts (a) and (b) above indicates that the computed M_x required to initiate yielding is reduced by 58.8% because of the distortion of the cross section. Since the yielding is highly localized, its effect is not of concern unless the curved beam is subjected to fatigue loading. If the second principal stress σ_{xx} is neglected, the moment M_x is reduced by 16.5% because of the distortion of the cross section. The distortion is reduced if the flange thickness is increased.

9.5

DEFLECTIONS OF CURVED BEAMS

A convenient method for determining the deflections of a linearly elastic curved beam is by the use of Castigliano's theorem (Chapter 5). For example, the deflections of the free end of the curved beam in Fig. 9.2a are given by the relations

$$\delta_{P_1} = \frac{\partial U}{\partial P_1} \tag{9.25}$$

$$\phi = \frac{\partial U}{\partial M_0} \tag{9.26}$$

where δ_{P_1} is the component of the deflection of the free end of the curved beam in the direction of load P_1, ϕ the angle of rotation of the free end of the curved beam in the direction of M_0, and U the total elastic strain energy in the curved beam. The total strain energy U [see Eq. (5.13)] is equal to the integral of the strain energy density U_0 over the volume of the curved beam [see Eqs. (3.33) and (5.14)].

Consider the strain energy density U_0 for a curved beam (Fig. 9.2) Because of the symmetry of loading relative to the (y, z) plane, $\sigma_{xy} = \sigma_{xz} = 0$, and since the effect of the transverse normal stress σ_{xx} (Fig. 9.2b) is ordinarily neglected, the formula for the strain energy density U_0 reduces to the form

$$U_0 = \frac{1}{2E}\sigma_{\theta\theta}^2 + \frac{1}{2E}\sigma_{rr}^2 - \frac{\nu}{E}\sigma_{rr}\sigma_{\theta\theta} + \frac{1}{2G}\sigma_{r\theta}^2$$

where the radial normal stress σ_{rr}, the circumferential normal stress $\sigma_{\theta\theta}$, and the shear stress $\sigma_{r\theta}$ are, relative to the (x, y, z) axes of Fig. 9.2b, $\sigma_{rr} = \sigma_{yy}$, $\sigma_{\theta\theta} = \sigma_{zz}$, and $\sigma_{r\theta} = \sigma_{yz}$. In addition, the effect of σ_{rr} is often small for curved beams of practical dimensions. Hence, the effect of σ_{rr} is often discarded from the expression for U_0. Then,

$$U_0 = \frac{1}{2E}\sigma_{\theta\theta}^2 + \frac{1}{2G}\sigma_{r\theta}^2$$

The stress components $\sigma_{\theta\theta}$ and $\sigma_{r\theta}$, respectively, contribute to the strain energies U_N and U_S because of the normal traction N and shear V (Fig. 9.2b). In addition, $\sigma_{\theta\theta}$ contributes to the bending strain energy U_M, as well as to the strain energy U_{MN} because of a coupling effect between the moment M and traction N, as we shall see in the derivation below.

Ordinarily, it is sufficiently accurate to approximate the strain energies U_S and U_N that are due to shear V and traction N, respectively, by the formulas for straight beams (see Sec. 5.3). However, the strain energy U_M due to bending must be modified. To compute the strain energy due to bending, consider the curved beam shown in Fig. 9.2b. Since the strain energy increment dU for a linearly elastic material undergoing small displacement is independent of the order in which loads are applied, let the shear load V and normal load N be applied first. Next, let the moment be increased from zero to M_x. The strain energy increment due to bending is

$$dU_M = \tfrac{1}{2}M_x \Delta(d\theta) = \tfrac{1}{2}M_x\omega \, d\theta \tag{9.27}$$

where $\Delta(d\theta)$, the change in $d\theta$, and $\omega = \Delta(d\theta)/d\theta$ are due to M_x alone. Hence, ω is determined from Eq. (9.10) with $N = 0$. Consequently, Eqs. (9.27) and Eq. (9.10) yield (with $N = 0$)

$$dU_M = \frac{A_m M_x^2}{2A(RA_m - A)E} \, d\theta \tag{9.28}$$

During the application of M_x, additional work is done by N because the centroidal (middle) surface (Fig. 9.2b) is stretched an amount $d\bar{e}_{\theta\theta}$. Let the corresponding strain energy increment due to the stretching of the middle surface be denoted by dU_{MN}. This strain energy increment dU_{MN} is equal to the work done by N as it moves through the distance $d\bar{e}_{\theta\theta}$. Thus,

$$dU_{MN} = N \, d\bar{e}_{\theta\theta} = N\bar{\epsilon}_{\theta\theta}R \, d\theta \tag{9.29}$$

where $d\bar{e}_{\theta\theta}$ and $\bar{\epsilon}_{\theta\theta}$ refer to the elongation and strain of the centroidal axis, respectively. The strain $\bar{\epsilon}_{\theta\theta}$ is given by Eq. (9.3) with $r = R$. Thus, Eq. (9.3) (with $r = R$) and Eqs. (9.29), (9.9), and (9.10) (with $N = 0$) yield the strain energy increment dU_{MN} due to coupling of the moment M_x and traction N.

$$dU_{MN} = \frac{N}{E}\left[\frac{M_x}{RA_m - A} - R\frac{A_m M_x}{A(RA_m - A)}\right]d\theta = -\frac{M_x N}{EA} \, d\theta \tag{9.30}$$

By Eqs. (5.16), (5.21), (9.28), and (9.30), the total strain energy U for the curved beam is obtained in the form

$$U = \int \frac{kV^2R}{2AG} \, d\theta + \int \frac{N^2R}{2AE} \, d\theta + \int \frac{A_m M_x^2}{2A(RA_m - A)E} \, d\theta - \int \frac{M_x N}{EA} \, d\theta \tag{9.31}$$

Equation (9.31) is an approximation, since it is based on the assumptions that plane sections remain plane and that the effect of the radial stress σ_{rr} on U is negligible. It might be expected that the radial stress increases the strain energy. Hence,

TABLE 9.4
Ratios of Deflections in Rectangular Section Curved Beams as Computed by
Elasticity Theory and by Approximate Strain-Energy Solution

	Neglecting U_{MN}		Including U_{MN}	
	Pure Bending	Shear Loading	Pure Bending	Shear Loading
$\left(\dfrac{R}{h}\right)$	$\left(\dfrac{\delta_U}{\delta_{elast}}\right)$	$\left(\dfrac{\delta_U}{\delta_{elast}}\right)$	$\left(\dfrac{\delta_U}{\delta_{elast}}\right)$	$\left(\dfrac{\delta_U}{\delta_{elast}}\right)$
0.65	0.923	1.563	0.697	1.215
0.75	0.974	1.381	0.807	1.123
1.0	1.004	1.197	0.914	1.048
1.5	1.006	1.085	0.968	1.016
2.0	1.004	1.048	0.983	1.008
3.0	1.002	1.021	0.993	1.003
5.0	1.000	1.007	0.997	1.001

Eq. (9.31) yields a low estimate of the actual strain energy. However, if M_x and N
have the same sign, the coupling U_{MN}, the last term in Eq. (9.31), is negative. Ordinarily, U_{MN} is small and, in many cases, it is negative. Hence, we recommend that
U_{MN}, the coupling strain energy be discarded from Eq. (9.31) when it is negative.
The discarding of U_{MN} from Eq. (9.31) raises the estimate of the actual strain energy
when U_{MN} is negative, and compensates to some degree for the lower estimate due
to discarding σ_{rr}.

The deflection δ_{elast} of rectangular cross section curved beams has been given by
Timoshenko and Goodier (1970) for the two types of loading shown in Fig. 9.4. The
ratio of the deflection δ_U given by Castigliano's theorem and the deflection δ_{elast} is
presented in Table 9.4 for several values of R/h. The shear coefficient k [see Eqs.
(5.21) and (5.23)] was taken to be 1.5 for the rectangular section, and Poisson's ratio
v was assumed to be 0.30.

Note: The deflection of curved beams is much less influenced by the curvature of
the curved beam than is the circumferential stress $\sigma_{\theta\theta}$. If R/h is greater than 2.0,
the strain energy due to bending can be approximated by that for a straight beam.
Thus, for $R/h > 2.0$, for computing deflections the third and fourth terms on the
right-hand side of Eq. (9.31) may be replaced by

$$U_M = \int \frac{M_x^2}{2EI_x} R \, d\theta \tag{9.32}$$

In particular, we note that the deflection of a rectangular cross section curved beam
with $R/h = 2.0$ is 7.7% greater when the curved beam is assumed to be straight than
when it is assumed to be curved.

Cross Sections in the Form of an I, T, etc.

As discussed in Sec. 9.4, the cross sections of curved beams in the form of an I, T,
etc. undergo distortion when loaded. One effect of the distortion is to decrease the

stiffness of the curved beam. As a result, deflections calculated on the basis of the undistorted cross section are less than the actual deflections. Therefore, the deflection calculations should be based on modified cross sections determined by Bleich's correction factors (Table 9.3). The strain energy terms U_N and U_M for the curved beams should also be calculated using the modified cross section. We recommend that the strain energy U_S be calculated with $k = 1.0$, and with the cross-sectional area A replaced by the area of the web $A_w = th$, where t is the thickness of the web and h the curved beam depth. Also, as a working rule, we recommend that the coupling energy U_{MN} be neglected if it is negative, and that it be doubled if it is positive.

EXAMPLE 9.6
Deformations in a Curved Beam Subjected to Pure Bending

The curved beam in Fig. E9.6 is made of an aluminum alloy ($E = 72.0\,\text{GPa}$), has a rectangular cross section with a thickness of 60 mm, and is subjected to a pure bending moment $M = 24.0\,\text{kN·m}$.

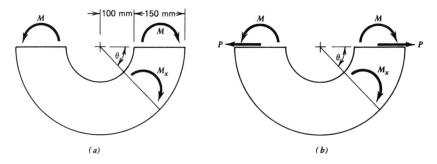

(a)   (b)

Figure E9.6

(a) Determine the angle change between the two horizontal faces where M is applied.

(b) Determine the relative displacement of the centroids of the horizontal faces of the curved beam.

SOLUTION

Required values for A, A_m, and R for the curved beam are calculated using equations in row (a) of Table 9.2.

$$A = 60(150) = 9000 \text{ mm}^2$$

$$A_m = 60 \ln \frac{250}{100} = 54.98 \text{ mm}$$

$$R = 100 + 75 = 175 \text{ mm}$$

(a) The angle change between the two faces where M is applied is given by Eq. (9.26). As indicated in Fig. E9.6a, the magnitude of M_x at any angle θ is $M_x = M$. Thus, by Eq. (9.26), we obtain

$$\phi = \frac{\partial U}{\partial M} = \int_0^\pi \frac{A_m M_x}{A(RA_m - A)E}(1)\, d\theta$$

$$= \frac{54.98(24,000,000)\pi}{9000[175(54.98) - 9000](72,000)}$$

$$= 0.01029 \text{ rad}$$

(b) In order to determine the deflection of the curved beam, a load P must be applied as indicated in Fig. E9.6b. In this case, $M_x = M + PR \sin \theta$ and $\partial U/\partial P = R \sin \theta$. Then the deflection is given by Eq. (9.25), in which the integral is evaluated with $P = 0$. Thus, the relative displacement is given by the relation

$$\delta_P = \frac{\partial U}{\partial P} = \int_0^\pi \frac{A_m M_x}{A(RA_m - A)E}\bigg|_{P=0} (R \sin \theta)\, d\theta$$

or

$$\delta_P = \frac{54.98(24,000,000)(175)(2)}{9000[175(54.98) - 9000](72,000)} = 1.147 \text{ mm}$$

EXAMPLE 9.7
Deflections in a Press

A press (Fig. E9.7a) has the cross section shown in Fig. E9.7b. It is subjected to a load $P = 11.2$ kN. The press is made of steel with $E = 200$ GPa, and $v = 0.30$. Determine the separation of the jaws of the press due to the load.

SOLUTION

The press is made up of two straight members and a curved member. We compute the strain energies due to bending and shear in the straight beams, without modification of the cross sections. The moment of inertia of the cross section is $I_x = 181.7 \times 10^3$ mm⁴. We choose the origin of the coordinate axes at load P, with z measured from P toward the curved beam. Then the applied shear V and moment M_x at a section in the straight beam are

$$V = P$$
$$M_x = Pz$$

In the curved beam portion of the press, we employ Bleich's correction factor to obtain a modified cross section. With the dimensions in Fig. E9.7b, we find

$$\frac{b_p^2}{\bar{r}t_f} = \frac{15^2}{35(10)} = 0.643$$

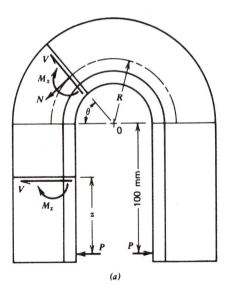

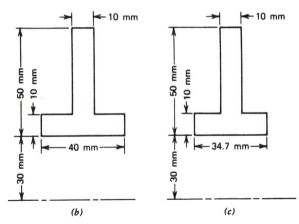

Figure E9.7   (a) Curved beam. (b) Original section. (c) Modified section.

A linear interpolation in Table 9.3 yields the result $\alpha = 0.822$. The modified cross section is shown in Fig. E9.7c. Equations (9.12), (9.13), and (9.14) give

$$A = 34.7(10) + 10(40) = 747 \text{ mm}^2$$

$$R = \frac{34.7(10)(35) + 10(40)(60)}{747} = 48.4 \text{ mm}$$

$$A_m = 10 \ln\frac{80}{40} + 34.7 \ln\frac{40}{30} = 16.9 \text{ mm}$$

With θ defined as indicated in Fig. E9.7a, the applied shear V, normal load N, and moment M_x for the curved beam are

$$V = P\cos\theta$$

$$N = P\sin\theta$$

$$M_x = P(100 + R\sin\theta)$$

Summing the strain energy terms for the two straight beams and the curved beam and taking the derivative with respect to P [Eq. (9.25)], we compute the increase in distance δ_P between the load points as

$$\delta_P = 2\int_0^{100} \frac{P}{A_w G}\,dz + 2\int_0^{100} \frac{Pz^2}{EI_x}\,dz + \int_0^{\pi} \frac{P\cos^2\theta}{A_w G}\,R\,d\theta + \int_0^{\pi} \frac{P\sin^2\theta}{AE}\,R\,d\theta$$

$$+ \int_0^{\pi} \frac{P(100 + R\sin\theta)^2 A_m}{A(RA_m - A)E}\,d\theta$$

The shear modulus is $G = E/[2(1 + v)] = 76{,}900$ MPa. Hence,

$$\delta_P = \frac{2(11{,}200)(100)}{76{,}900(500)} + \frac{2(11{,}200)(100)^3}{3(200{,}000)(181{,}700)}$$

$$+ \frac{11{,}200(48.4)\pi}{500(76{,}900)(2)} + \frac{11{,}200(48.4)\pi}{747(200{,}000)(2)}$$

$$+ \frac{16.9(11{,}200)}{747[48.4(16.9) - 747](200{,}000)}\left[(100)^2\pi + (48.4)^2\frac{\pi}{2} + 2(100)(48.4)(2)\right]$$

or

$$\delta_P = 0.058 + 0.205 + 0.022 + 0.006 + 0.972 = 1.263 \text{ mm}$$

9.6

STATICALLY INDETERMINATE CURVED BEAMS. CLOSED RING SUBJECTED TO A CONCENTRATED LOAD

Many loaded curved members, such as closed rings and chain links, are statically indeterminate (see Sec. 5.5). For such members, equations of equilibrium are not sufficient to determine all the internal resultants (V, N, M_x) at a section of the member. The additional relations needed to solve for the loads are obtained using Castigliano's theorem with appropriate boundary conditions. Since closed rings are commonly used in engineering, we present the computational procedure for a closed ring.

Consider a closed ring subjected to a central load P (Fig. 9.9a). From the condition of symmetry, the deformation of each quadrant of the ring is identical. Hence, we need consider only one quadrant. The quadrant (Fig. 9.9b) may be considered fixed at section FH with a load $P/2$ and moment M_0 at section BC. Because of the symmetry of the ring, as the ring deforms, section BC remains perpendicular to section FH. Therefore, by Castigliano's theorem, we have for the rotation of face BC

$$\phi_{BC} = \frac{\partial U}{\partial M_0} = 0 \qquad (9.33)$$

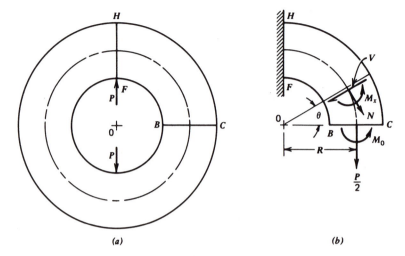

Figure 9.9   Closed ring.

The applied loads V, N, and M_x at a section forming angle θ with the face BC are

$$V = \frac{P}{2}\sin\theta$$

$$N = \frac{P}{2}\cos\theta$$

$$M_x = M_0 - \frac{PR}{2}(1 - \cos\theta) \tag{9.34}$$

Substituting Eqs. (9.31) and (9.34) into Eq. (9.33), we find

$$0 = \int_0^{\pi/2} \frac{[M_0 - (PR/2)(1 - \cos\theta)]A_m}{A(RA_m - A)E}\, d\theta - \int_0^{\pi/2} \frac{(P/2)\cos\theta}{AE}\, d\theta \tag{9.35}$$

where U_{MN} has been included. The solution of Eq. (9.35) is

$$M_0 = \frac{PR}{2}\left(1 - \frac{2A}{RA_m\pi}\right) \tag{9.36}$$

If R/h is greater than 2.0, we take the bending energy U_M as given by Eq. (9.32) and ignore the coupling energy U_{MN}. Then, M_0 is given by the relation

$$M_0 = \frac{PR}{2}\left(1 - \frac{2}{\pi}\right) \tag{9.37}$$

With M_0 known, the loads at every section of the closed ring [Eqs. (9.34)] are known. The stresses and deformations of the closed ring may be calculated by the methods of Sec. 9.2 to 9.5.

FULLY PLASTIC LOADS FOR CURVED BEAMS

In this section we consider curved beams made of elastic-perfectly plastic materials with yield point stress Y (Fig. 1.5b). For a curved beam made of elastic-perfectly plastic material, the fully plastic moment M_P under pure bending is the same as that for a straight beam with identical cross section and material. However, because of the nonlinear distribution of the circumferential stress $\sigma_{\theta\theta}$ in a curved beam, the ratio of the fully plastic moment M_P under pure bending to maximum elastic moment M_Y is much greater for a curved beam than for a straight beam with the same cross section.

Most curved beams are subjected to complex loading other than pure bending. The stress distribution for a curved beam at the fully plastic load P_P for a typical loading condition is indicated in Fig. 9.10. Since the tension stresses must balance the compression stresses and load P_P, the part A_T of the cross-sectional area A that has yielded in tension is larger than the part A_C of area A that has yielded in compression. In addition to the unknowns A_T and A_C, a third unknown is P_P, the load at the fully plastic condition. This follows from the fact that R (the distance from the center of curvature 0 to the centroid $\bar{0}$) can be calculated, and D is generally specified rather than P_P. The three equations necessary to determine the

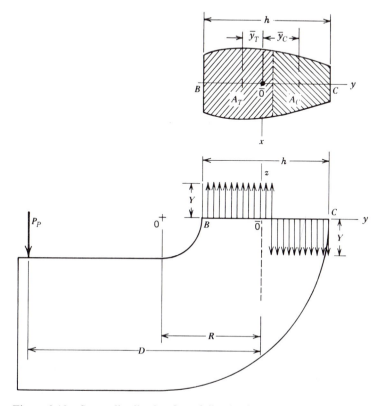

Figure 9.10 Stress distribution for a fully plastic load on a curved beam.

three unknowns A_T, A_C, and P_P are obtained from the equations of equilibrium and the fact that the sum of A_T and A_C must equal the cross-sectional area A, that is,

$$A = A_T + A_C \tag{9.38}$$

The equilibrium equations are (Fig. 9.10)

$$\sum F_z = 0 = A_T Y - A_C Y - P_P \tag{9.39}$$
$$\sum M_x = 0 = P_P D - A_T Y \bar{y}_T - A_C Y \bar{y}_C \tag{9.40}$$

In Eq. (9.40), $\bar{y}_T$ and $\bar{y}_C$ locate the centroids of A_T and A_C, respectively, as measured from the centroid $\bar{0}$ of the cross-sectional area of the curved beam (Fig. 9.10). Let M be the moment, about the centroid axis x, resulting from the stress distribution on section BC (Fig. 9.10). Then,

$$M = P_P D = A_T Y \bar{y}_T + A_C Y \bar{y}_C \tag{9.41}$$

The most convenient method of solving Eqs. (9.38), (9.39), and (9.40) for the magnitudes of A_T, A_C, and P_P is often a trial-and-error procedure, since $\bar{y}_T$ and $\bar{y}_C$ are not known until A_T and A_C are known (McWhorter et al., 1971).

The moment M [Eq. (9.41)] is generally less than the fully plastic moment M_P for pure bending. It is desirable to know the conditions under which M due to load P_P can be assumed equal to M_P, since for pure bending A_T is equal to A_C, and the calculations are greatly simplified. For some common sections, $M \approx M_P$, when $D > h$. For example, for $D = h$, we note that $M = 0.94M_P$ for curved beams with rectangular sections and $M = 0.96M_P$ for curved beams with circular sections. However, for curved beams with T-sections, M may be greater than M_P. Other exceptions are curved beams with I-sections and box sections, for which D should be greater than $2h$ in order for M to be approximately equal to M_P.

Fully Plastic vs Maximum Elastic Loads for Curved Beams

A linearly elastic analysis of a load-carrying member is required in order to predict the load-deflection relation for linearly elastic behavior of the member up to the load P_Y that initiates yielding in the member. The fully plastic load is also of interest since it is often considered to be the limiting load that can be applied to the member before the deformations become excessively large.

The fully plastic load P_P for a curved beam is often more than twice the maximum elastic load P_Y. Fracture loads for curved beams that are made of ductile metals and subjected to static loading may be four to six times P_Y. Dimensionless load-deflection experimental data for a uniform rectangular section hook made of a structural steel are shown in Fig. 9.11. The deflection is defined as the change in distance ST between points S and T on the hook. The hook does not fracture even for loads such that $P/P_Y > 5$. A computer program written by J. C. McWhorter, H. R. Wetenkamp, and O. M. Sidebottom (1971) gave the predicted curve in Fig. 9.11. The experimental data agree well with predicted results.

As noted in Fig. 9.11, the ratio of P_P to P_Y is 2.44. Furthermore, the load-deflection curve does not level off at the fully plastic load, but continues to rise.

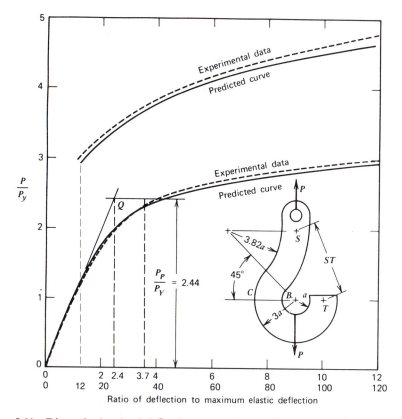

Figure 9.11   Dimensionless load-deflection curves for a uniform rectangular section hook made of structural steel.

This behavior may be attributed to strain hardening. Because of the steep stress gradient in the hook, the strains in the most strained fibers become so large that the material begins to strain harden before yielding can penetrate to sufficient depth at section BC in the hook to develop the fully plastic load.

The usual practice in predicting the deflection of a structure at the fully plastic load is to assume that the structure behaves in a linearly elastic manner up to the fully plastic load (point Q in Fig. 9.11) and multiply the deflection at this point by the ratio P_P/P_Y (in this case, 2.44). In this case, with this procedure (Fig. 9.11) the resulting calculated deflection [approximately calculated as 2.44(2.4) = 5.9] is greater than the measured deflection.

Usually, curved members such as crane hooks and chains are not subjected to a sufficient number of repetitions of peak loads during their life for fatigue failure to occur. Therefore, the working loads for these members are often obtained by application of a factor of safety to the fully plastic loads. It is not uncommon to have the working load as great as or greater than the maximum elastic load P_Y.

PROBLEMS
Section 9.2

9.1.   The frame shown in Fig. E9.1 has a rectangular cross section with a thickness of 10 mm and depth of 40 mm. The load P is located 120 mm from the centroid of section *BC*. The frame is made of steel having a yield stress of $Y = 430$ MPa. The frame has been designed using a factor of safety of $SF = 1.75$ against initiation of yielding. Determine the maximum allowable magnitude of *P*, if the radius of curvature at section *BC* is $R = 40$ mm.

9.2.   Solve Problem 9.1 for the condition that $R = 35$ mm.

Ans. $P = 3.174$ kN

9.3.   The curved beam in Fig. P9.3 has a circular cross section 50 mm in diameter. The inside diameter of the curved beam is 40 mm. Determine the stress at *B* for $P = 20$ kN.

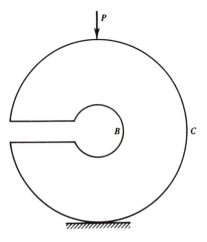

Figure P9.3

9.4.   Let the crane hook in Fig. E9.2 have a trapezoidal cross section as shown in row (*c*) of Table 9.2 with (see Fig. P9.4) $a = 45$ mm, $c = 80$ mm, $b_1 = 25$ mm, and $b_2 = 10$ mm. Determine the maximum load to be carried by the hook if the working stress limit is 150 MPa.

Ans. $P = 7.34$ kN

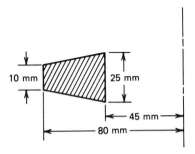

Figure P9.4

9.5.   A curved beam is built up by welding together rectangular and elliptical cross-section curved beams; the cross section is shown in Fig. P9.5. The center of curvature is located 20 mm from *B*. The curved beam is subjected to a positive bending moment M_x. Determine the stresses at points *B* and *C* in terms of M_x.

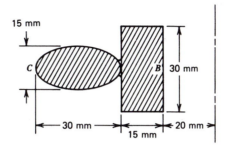

Figure P9.5

9.6. A commercial crane hook has the cross-sectional dimensions shown in Fig. P9.6 at the critical section that is subjected to an axial load $P = 100$ kN. Determine the circumferential stresses at the inner and outer radii for this load. Assume that area A_1 is half of an ellipse [see row (j) in Table 9.2] and area A_3 is enclosed by a circular arc.

Ans. $\sigma_{\theta\theta B} = 113.5$ MPa, $\sigma_{\theta\theta C} = -43.6$ MPa

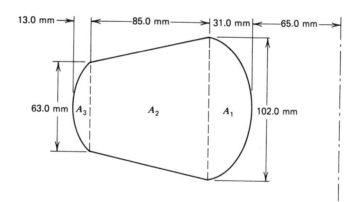

Figure P9.6

9.7. A crane hook has the cross-sectional dimensions shown in Fig. P9.7 at the critical section that is subjected to an axial load $P = 90.0$ kN. Determine

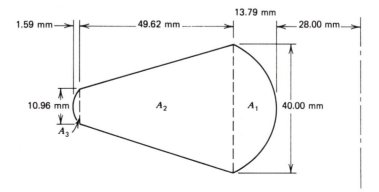

Figure P9.7

the circumferential stresses at the inner and outer radii for this load. Note that A_1 and A_3 are enclosed by circular arcs.

9.8.   The curved beam in Fig. P9.8 has a triangular cross section with the dimensions shown. If $P = 40$ kN, determine the circumferential stresses at B and C.

Ans. $\sigma_{\theta\theta B} = 297.8$ MPa, $\sigma_{\theta\theta C} = -238.1$ MPa

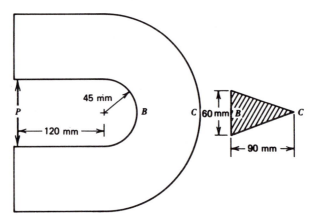

Figure P9.8

9.9.   A curved beam with a rectangular cross section strikes a 90° arc and is loaded and supported as shown in Fig. P9.9. The thickness of the beam is 50 mm. Determine the hoop stress $\sigma_{\theta\theta}$ along line A-A at the inside and outside radii and at the centroid of the beam.

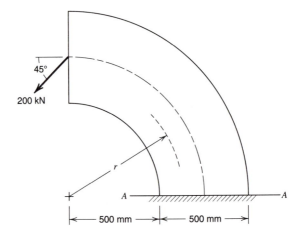

Figure P9.9

Section 9.3

9.10.   For the curved beam in Problem 9.5, determine the radial stress in terms of the moment M_x if the thickness of the web at the weld is 10 mm.

9.11. In Fig. P9.11 is shown a cast iron frame with a U-shaped cross section. The ultimate tensile strength of the case iron is $\sigma_u = 320$ MPa.

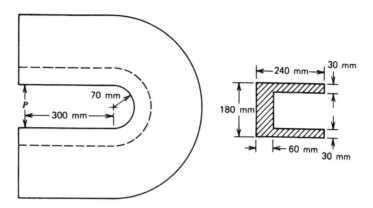

Figure P9.11

(a) Determine the maximum value of P based on a factor of safety $SF = 4.00$ which is based on the ultimate strength.

(b) Neglecting the effect of stress concentrations at the fillet at the junction of the web and flange, determine the maximum radial stress when this load is applied.

(c) Is the maximum radial stress less than the maximum circumferential stress?

Ans. **(a)** $P = 110.8$ kN, **(b)** $\sigma_{rr} = 42.4$ MPa **(c)** Yes

Section 9.4

9.12. A T-section curved beam has the cross section shown in Fig. P9.12. The center of curvature lies 40 mm from the flange. If the curved beam is subjected to a positive bending moment $M_x = 2.50$ kN·m, determine the stresses at the inner and outer radii. Use Bleich's correction factors. What is the maximum shear stress in the curved beam?

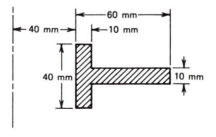

Figure P9.12

9.13. Determine the radial stress at the junction of the web and the flange for the curved beam in Problem 9.12. Neglect stress concentrations. Use the Bleich correction.

Ans. $\sigma_{rr} = 118.1$ MPa

9.14. A load $P = 12.0$ kN is applied to the clamp shown in Fig. P9.14. Determine the circumferential stresses at points B and C, assuming that the curved beam formula is valid at that section.

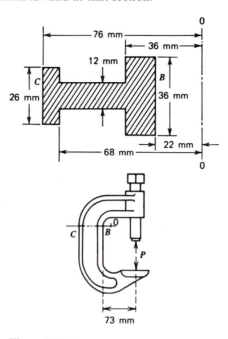

Figure P9.14

9.15. Determine the radial stress at the junction of the web and inner flange of the curved beam portion of the clamp in Problem 9.14. Neglect stress concentrations.

Ans. $\sigma_{rr} = 69.7$ MPa

Section 9.5

9.16. The curved beam in Fig. P9.16 is made of a steel ($E = 200$ GPa) that has a yield point stress $Y = 420$ MPa. Determine the magnitude of the bending moment $M_x = M_y$ required to initiate yielding in the curved beam, the angle change of the free end, and the horizontal and vertical components of the deflection of the free end.

9.17. Determine the deflection of the curved beam in Problem 9.3 at the point of load application. The curved beam is made of an aluminum alloy for which $E = 72.0$ GPa and $G = 27.1$ GPa. Let $k = 1.3$.

Ans. $\delta_P = 0.1629$ mm

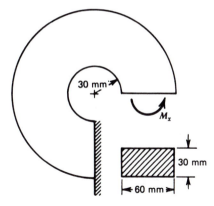

Figure P9.16

9.18. The triangular cross-section curved beam in Problem 9.8 is made of steel ($E = 200$ GPa, $G = 77.5$ GPa). Determine the separation of the points of application of the load. Let $k = 1.5$.

9.19. Determine the deflection across the center of curvature of the cast iron curved beam in Problem 9.11 for $P = 126$ kN. $E = 102.0$ GPa and $G = 42.5$ GPa. Let $k = 1.0$ with the area in shear equal to the product of the web thickness and the depth.

Ans. $\delta_Q = 0.3447$ mm

Section 9.6

9.20. The ring in Fig. P9.20 has an inside diameter of 100 mm, an outside diameter of 180 mm, and a circular cross section. The ring is made of a steel having a yield stress $Y = 520$ MPa. Determine the maximum allowable magnitude of P if the ring has been designed with a factor of safety $SF = 1.75$ against initiation of yielding.

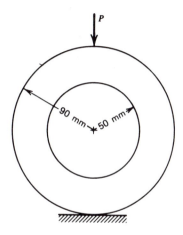

Figure P9.20

9.21. If $E = 200$ GPa and $G = 77.5$ GPa for the steel in Problem 9.20, determine the deflection of the ring for a load $P = 60$ kN. Let $k = 1.3$.

Ans. $\delta_p = 2.088$ mm

9.22. An aluminum alloy ring has a mean diameter of 600 mm and a rectangular cross section with 200 mm thickness and a depth of 300 mm (radial direction). The ring is loaded by diametrically opposite radial loads $P = 4.00$ MN. Determine the maximum tensile and compressive circumferential stresses in the ring.

9.23. If $E = 72.0$ GPa and $G = 27.1$ GPa for the aluminum alloy ring in Problem 9.22, determine the separation of the points of application of the loads. Let $k = 1.5$.

Ans. $\delta_p = 8.742$ mm

9.24. The link in Fig. P9.24 has a circular cross section and is made of a steel having a yield point stress of $Y = 250$ MPa. Determine the magnitude of P that will initiate yield in the link.

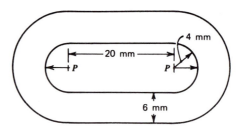

Figure P9.24

Section 9.7

9.25. Let the curved beam in Fig. 9.10 have a rectangular cross section with depth h and width b. Show that the ratio of the bending moment for fully plastic load P_P to the fully plastic moment for pure bending $M_P = Ybh^2/4$ is given by the relation

$$\frac{M}{M_P} = \frac{4D}{h} \sqrt{1 + \frac{4D^2}{h^2} - \frac{8D^2}{h^2}}$$

9.26. Let the curved beam in Problem 9.1 be made of a steel that has a flat top stress-strain diagram at the yield point stress $Y = 430$ MPa. From the answer to Problem 9.1, the load that initiates yielding is equal to $P_Y = SF(P) = 6.05$ kN. Since $D = 3h$, assume $M = M_P$ and calculate P_P. Determine the ratio P_P/P_Y.

Ans. $P_P = 14.33$ kN, $P_P/P_Y = 2.37$

9.27. Let the steel in the curved beam in Example 9.5 have a flat top at the yield point stress $Y = 280$ MPa. Determine the fully plastic moment for the curved beam. Note that the original cross section must be used. The distortion of the cross section increases the fully plastic moment for a positive moment.

REFERENCES

American Institute of Timber Construction (AITC) (1985). *Timber Construction Manual*, 3rd ed. New York: Wiley.

Anderson, C. G. (1950). Flexural Stresses in Curved Beams of I- and Box Sections. Presented to Inst. of Mech. Engineers, London, Nov. 3.

Bleich, H. (1933). Die Spannungsverteilung in den Gurtungen gekrümmter Stabe mit T- und I-formigem Querschnitt. *Der Stahlblau, Beilage zur Zeitschrift, Die Bautechnik*, **6**(1): 3–6.

Broughton, D. C., Clark, M. E., and Corten, H. T. (1950). Tests and Theory of Elastic Stresses in Curved Beams Having I- and T-Sections. *Exper. Mech.*, **8**(1): 143–155.

McWhorter, J. C., Wetenkamp, H. R., and Sidebottom, O. M. (1971). Finite Deflections of Curved Beams. *J. Eng. Mech. Div., Proc. ASCE*, **97**: 345–358.

Timoshenko, S. (1923). Bending Stresses in Curved Tubes of Rectangular Cross-Section. *Trans. ASME*, **45**: 135–140.

Timoshenko, S. and Goodier, J. (1970). *Theory of Elasticity*, 3rd ed. New York: McGraw-Hill.

von Kármán, T. (1911). *Zeitschrift des Vereines deutscher Ingenieure*, **55**: 1889.

Wang, C. C. (1985). A Unified Algorithm for Accurately Sizing Straight and Curved Beam Sections to Allowable Stress Limits. Presented at ASME Des. Eng. Div. Conf. and Exhibit on Mech. Vibration and Noise, Cincinnati, Ohio, Sept. 10–13, Paper 85-DET-102.

10

BEAMS ON ELASTIC FOUNDATIONS

In certain applications, a beam of relatively small bending stiffness is placed on an elastic foundation and loads are applied to the beam. The loads are transferred through the beam to the foundation. The beam and foundation must be designed to resist the loads without failing. Often, failure occurs in the beam before it occurs in the foundation. Accordingly, in this chapter we assume that the foundation has sufficient strength to prevent its own failure. Furthermore, we assume that the foundation resists the loads transmitted by the beam, in a linearly elastic manner; that is, the pressure developed at any point between the beam and the foundation is proportional to the deflection of the beam at that point. This type of foundation response is referred to as the Winkler foundation or Winkler model (Hetényi, 1946; Westergaard, 1948). Other types of foundation models are used, for example, the Vlasov model (Vlasov and Leont'ev, 1966). However, the Vlasov model is more complex than the Winkler model. Therefore, we employ the assumption that the foundation responds linearly with beam deflection. This assumption is fairly accurate for small deflections. However, if the deflections are large, the resistance of the foundation generally does not remain linearly proportional to the beam deflection. For large deflections, the stiffness of the foundation is larger than that for small deflections and is related in a nonlinear way to the beam deflection. The increased stiffness due to the nonlinear response of the foundation tends to reduce the deflections and stresses in the beam compared to those due to a linear foundation response. Since we consider small displacements, the solution presented in this chapter for the beam on an elastic foundation is generally conservative for the range of deflections treated. (However, for certain cases in which the load on the beam is distributed fairly uniformly, the Winkler model can lead to nonconservative design values; Vallabhan and Das, 1987). Furthermore, since we consider only a linear response of the foundation, we drop the term *linear* in our discussion.

The solution presented in this chapter for beams on elastic foundations can be used to obtain a simple approximate solution for beams supported by identical elastic springs that are spaced uniformly along the beam. Extensive surveys of studies on beams on elastic foundations have been given by Kerr (1964), Selvadurai (1979), and Scott (1981). Vallabhan and Das (1988) and Jones and Xenophontos (1977) have studied the Vlasov model. Design tables based on the Winkler model have been developed by Iyengar and Ramu (1979).

10.1

GENERAL THEORY

The response to loads of a beam resting on an elastic foundation is described by a single differential equation subject to different boundary conditions for the beam, depending on how the beam is supported at its ends. For instance, consider a beam of infinite length attached along its length to an elastic foundation (Fig. 10.1). Let the origin of coordinate axes (x, y, z) be located at the centroid of the beam cross section and let a concentrated lateral load P be applied to the beam at the origin of the (x, y, z) axes. The z axis coincides with the axis of the beam, the x axis is normal to the figure (directed toward the reader), and the y axis is normal to the elastic foundation. The load P causes the beam to deflect, which in turn displaces the elastic foundation. As a result, a distributed force is developed between the beam and the foundation. Thus, relative to the beam, the stiffness of the foundation produces a laterally distributed force q (force per unit length) on the beam

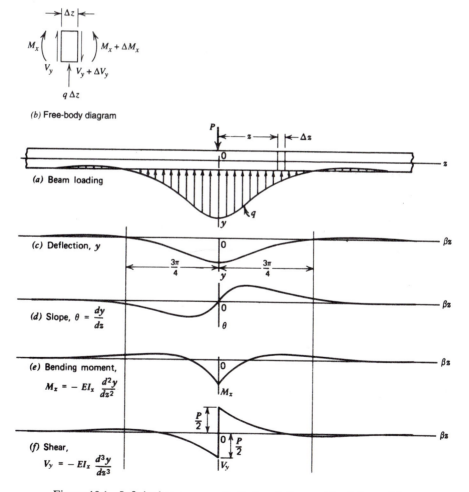

Figure 10.1   Infinite beam on an elastic foundation and loaded at origin.

(Fig. 10.1a). In the solution of the deflection problem, we shall see that in certain regions the deflection of the beam may be negative (upwards). Since the beam is assumed to be attached to the foundation, the foundation may in certain regions exert a tensile force on the beam.

A free-body diagram of an element Δz of the beam is shown in Fig. 10.1b, with positive sign conventions for the total shear V_y and moment M_x indicated. For the indicated sign convention and the condition of small displacements, we obtain the differential relations

$$\frac{dy}{dz} = \theta$$

$$EI_x \frac{d^2 y}{dz^2} = -M_x$$

$$EI_x \frac{d^3 y}{dz^3} = -V_y$$

$$EI_x \frac{d^4 y}{dz^4} = -q \tag{10.1}$$

where q is taken to be positive if it pushes up on the beam, that is, q is positive if it acts in the negative y direction.

For the linearly elastic foundation, the distributed load q is linearly proportional to the deflection y of the beam; thus,

$$q = ky \tag{10.2}$$

where the spring coefficient k may be written in the form

$$k = bk_0 \tag{10.3}$$

in which b is the beam width and k_0 the elastic spring constant for the foundation. The dimensions of k_0 are force/length^3. Substitution of Eq. (10.2) into the fourth of Eqs. (10.1) yields the differential equation of the bending axis of the beam on an elastic foundation.

$$EI_x \frac{d^4 y}{dz^4} = -ky \tag{10.4}$$

With the notation,

$$\beta = \sqrt[4]{\frac{k}{4EI_x}} \tag{10.5}$$

the general solution y of Eq. (10.4) may be expressed as

$$y = e^{\beta z}(C_1 \sin \beta z + C_2 \cos \beta z) + e^{-\beta z}(C_3 \sin \beta z + C_4 \cos \beta z) \tag{10.6}$$

Equation (10.6) represents the general solution for the response of an infinite beam on an elastic foundation subjected to a concentrated lateral load. The magnitudes

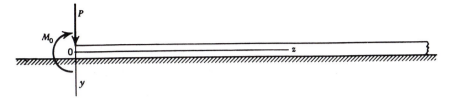

Figure 10.2   Semiinfinite beam on elastic foundation and loaded at the end.

of the constants of integration C_1, C_2, C_3, and C_4 are determined by the boundary conditions.

Solutions for the response of a beam supported by an elastic foundation and subjected to specific lateral loads can be obtained by the method of superposition, by employing the solution for an infinite beam loaded by a concentrated load (Fig. 10.1) and for a semiinfinite beam loaded at the end by a concentrated load P and moment M_0 as indicated in Fig. 10.2. In either of the cases shown in Figs. 10.1 and 10.2, the deflection of the beam goes to zero for large positive values of z. Consequently, the constants C_1 and C_2 in Eq. (10.6) are zero, and the equation for the displacement y of the bending axis of the beam reduces to

$$y = e^{-\beta z}(C_3 \sin \beta z + C_4 \cos \beta z), \qquad z \geq 0 \tag{10.7}$$

Because of symmetry, the displacement of the beam in Fig. 10.1 for negative values of z can be obtained from the solution for positive values of z, that is, $y(-z) = y(z)$. For the case of the semiinfinite beam (Fig. 10.2), $z \geq 0$, so that Eq. (10.7) applies directly.

10.2

INFINITE BEAM SUBJECTED TO A CONCENTRATED LOAD: BOUNDARY CONDITIONS

Consider a beam of infinite length, resting on an elastic foundation and loaded at the origin 0 of coordinate axes (y, z) with concentrated load P (Fig. 10.1). To determine the two constants of integration, C_3 and C_4 in Eq. (10.7), we employ the conditions (a) that the slope of the beam remains zero under the load because of symmetry and (b) that half of the load P must be supported by the elastic foundation under the half of the beam specified by positive values of z. The other half of P is supported by the elastic foundation where $z < 0$. Thus, we obtain the relations

$$\frac{dy}{dz} = 0 \quad \text{for } z = 0 \qquad \text{and} \qquad 2\int_0^\infty ky \, dz = P \tag{10.8}$$

The condition of vanishing slope at $z = 0$ yields, with Eq. (10.7),

$$C_3 = C_4 = C$$

Hence, Eq. (10.7) becomes

$$y = Ce^{-\beta z}(\sin \beta z + \cos \beta z) \tag{10.9}$$

Substituting Eq. (10.9) into the second of Eqs. (10.8), we obtain

$$C = \frac{P\beta}{2k} \tag{10.10}$$

Consequently, the equation of the deflected axis of the beam is

$$y = \frac{P\beta}{2k} e^{-\beta z}(\sin \beta z + \cos \beta z), \qquad z \geq 0 \tag{10.11}$$

Equation (10.11) holds for positive values of z. The deflections for negative values of z are obtained by the condition that $y(-z) = y(z)$, that is, by symmetry. Values for the slope, moment, and shear are obtained by substitution of Eq. (10.11) into Eqs. (10.1). Thus, we find

$$y = \frac{P\beta}{2k} A_{\beta z}, \qquad z \geq 0 \tag{10.12}$$

$$\theta = -\frac{P\beta^2}{k} B_{\beta z}, \qquad z \geq 0 \tag{10.13}$$

$$M_x = \frac{P}{4\beta} C_{\beta z}, \qquad z \geq 0 \tag{10.14}$$

$$V_y = -\frac{P}{2} D_{\beta z}, \qquad z \geq 0 \tag{10.15}$$

where

$$A_{\beta z} = e^{-\beta z}(\sin \beta z + \cos \beta z), \qquad B_{\beta z} = e^{-\beta z}\sin \beta z$$
$$C_{\beta z} = e^{-\beta z}(\cos \beta z - \sin \beta z), \qquad D_{\beta z} = e^{-\beta z}\cos \beta z \tag{10.16}$$

For convenience, values of $A_{\beta z}$, $B_{\beta z}$, $C_{\beta z}$, and $D_{\beta z}$ are listed in Table 10.1 for $0 \leq \beta z \leq 5\pi/2$.

Values of deflection, slope, bending moment, and shear at any point along the beam are given by Eqs. (10.12), (10.13), (10.14), and (10.15), respectively. By using the symmetry conditions, $y(-z) = y(z)$, $\theta(-z) = -\theta(z)$, $M_x(-z) = M_x(z)$, and $V_y(-z) = -V_y(z)$, these quantities are plotted vs βz in Figs. 10.1c, d, e, and f. Since all of these quantities approach zero as βz becomes large, the above solutions may be used as approximations for beams of finite length. In particular, in Table 10.1, we note that $A_{\beta z} = 0$ for $\beta z = 3\pi/4$; therefore, the beam has zero deflection at a distance $3\pi/(4\beta)$ from the load. A beam with a length $L = 3\pi/(2\beta)$ loaded at the center has a maximum deflection 5.5% greater (Hetényi, 1946) and a maximum bending moment 1.9% greater than for a beam with infinite length. Although the error in using the solution for a beam of length $L = 3\pi/(2\beta)$ is nonconservative, the error is not large; therefore, the infinite beam solution yields reasonable results for beams as short as $L = 3\pi/(2\beta)$ when loaded at the center. The infinite beam solution also

TABLE 10.1

βz	$A_{\beta z}$	$B_{\beta z}$	$C_{\beta z}$	$D_{\beta z}$
0	1	0	1	1
0.001	1.0000	0.0010	0.9980	0.9990
0.002	1.0000	0.0020	0.9960	0.9980
0.003	1.0000	0.0030	0.9940	0.9970
0.004	1.0000	0.0040	0.9920	0.9960
0.005	1.0000	0.0050	0.9900	0.9950
0.006	1.0000	0.0060	0.9880	0.9940
0.007	0.9999	0.0070	0.9861	0.9930
0.008	0.9999	0.0080	0.9841	0.9920
0.009	0.9999	0.0087	0.9821	0.9910
0.010	0.9999	0.0099	0.9801	0.9900
0.011	0.9999	0.0109	0.9781	0.9890
0.012	0.9999	0.0119	0.9761	0.9880
0.013	0.9998	0.0129	0.9742	0.9870
0.014	0.9998	0.0138	0.9722	0.9860
0.015	0.9998	0.0148	0.9702	0.9850
0.016	0.9997	0.0158	0.9683	0.9840
0.017	0.9997	0.0167	0.9663	0.9830
0.018	0.9997	0.0177	0.9643	0.9820
0.019	0.9996	0.0187	0.9624	0.9810
0.02	0.9996	0.0196	0.9604	0.9800
0.03	0.9991	0.0291	0.9409	0.9700
0.04	0.9984	0.0384	0.9216	0.9600
0.05	0.9976	0.0476	0.9025	0.9501
0.10	0.9906	0.0903	0.8100	0.9003
0.15	0.9796	0.1283	0.7224	0.8510
0.20	0.9651	0.1627	0.6398	0.8024
0.25	0.9472	0.1927	0.5619	0.7546
0.30	0.9267	0.2189	0.4888	0.7078
0.35	0.9036	0.2416	0.4204	0.6620
0.40	0.8784	0.2610	0.3564	0.6174
0.45	0.8515	0.2774	0.2968	0.5742
0.50	0.8231	0.2908	0.2414	0.5323
0.55	0.7934	0.3016	0.1902	0.4918
0.60	0.7628	0.3099	0.1430	0.4529
0.65	0.7315	0.3160	0.0996	0.4156
0.70	0.6997	0.3199	0.0599	0.3798
0.75	0.6676	0.3220	0.0237	0.3456
$\frac{1}{4}\pi$	0.6448	0.3224	0	0.3224
0.80	0.6353	0.3223	-0.0093	0.3131
0.85	0.6032	0.3212	-0.0391	0.2821
0.90	0.5712	0.3185	-0.0658	0.2527
0.95	0.5396	0.3146	-0.0896	0.2250
1.00	0.5083	0.3096	-0.1109	0.1987

(*Continues*)

TABLE 10.1 (*Continued*)

βz	$A_{\beta z}$	$B_{\beta z}$	$C_{\beta z}$	$D_{\beta z}$
1.05	0.4778	0.3036	−0.1294	0.1742
1.10	0.4476	0.2967	−0.1458	0.1509
1.15	0.4183	0.2890	−0.1597	0.1293
1.20	0.3898	0.2807	−0.1716	0.1091
1.25	0.3623	0.2719	−0.1815	0.0904
1.30	0.3355	0.2626	−0.1897	0.0729
1.35	0.3098	0.2530	−0.1962	0.0568
1.40	0.2849	0.2430	−0.2011	0.0419
1.45	0.2611	0.2329	−0.2045	0.0283
1.50	0.2384	0.2226	−0.2068	0.0158
1.55	0.2166	0.2122	−0.2078	0.0044
$\frac{1}{2}\pi$	0.2079	0.2079	−0.2079	0
1.60	0.1960	0.2018	−0.2077	−0.0059
1.65	0.1763	0.1915	−0.2067	−0.0152
1.70	0.1576	0.1812	−0.2046	−0.0236
1.75	0.1400	0.1720	−0.2020	−0.0310
1.80	0.1234	0.1610	−0.1985	−0.0376
1.85	0.1078	0.1512	−0.1945	−0.0434
1.90	0.0932	0.1415	−0.1899	−0.0484
1.95	0.0795	0.1322	−0.1849	−0.0527
2.00	0.0667	0.1230	−0.1793	−0.0563
2.05	0.0549	0.1143	−0.1737	−0.0594
2.10	0.0438	0.1057	−0.1676	−0.0619
2.15	0.0337	0.0975	−0.1613	−0.0638
2.20	0.0244	0.0895	−0.1547	−0.0652
2.25	0.0157	0.0820	−0.1482	−0.0663
2.30	0.0080	0.0748	−0.1416	−0.0668
2.35	0.0008	0.0679	−0.1349	−0.0671
$\frac{3}{4}\pi$	0	0.0671	−0.1342	−0.0671
2.40	−0.0056	0.0613	−0.1282	−0.0669
2.45	−0.0114	0.0550	−0.1215	−0.0665
2.50	−0.0166	0.0492	−0.1149	−0.0658
2.55	−0.0213	0.0435	−0.1083	−0.0648
2.60	−0.0254	0.0383	−0.1020	−0.0637
2.65	−0.0289	0.0334	−0.0956	−0.0623
2.70	−0.0320	0.0287	−0.0895	−0.0608
2.75	−0.0347	0.0244	−0.0835	−0.0591
2.80	−0.0369	0.0204	−0.0777	−0.0573
2.85	−0.0388	0.0167	−0.0721	−0.0554
2.90	−0.0403	0.0132	−0.0666	−0.0534
2.95	−0.0415	0.0100	−0.0614	−0.0514
3.00	−0.0422	0.0071	−0.0563	−0.0493
3.05	−0.0427	0.0043	−0.0515	−0.0472
3.10	−0.0431	0.0019	−0.0469	−0.0450
π	−0.0432	.0	−0.0432	−0.0432

TABLE 10.1 (*Continued*)

βz	$A_{\beta z}$	$B_{\beta z}$	$C_{\beta z}$	$D_{\beta z}$
3.15	−0.0432	−0.0004	−0.0424	−0.0428
3.20	−0.0431	−0.0024	−0.0383	−0.0407
3.25	−0.0427	−0.0042	−0.0343	−0.0385
3.30	−0.0422	−0.0058	−0.0306	−0.0365
3.35	−0.0417	−0.0073	−0.0271	−0.0344
3.40	−0.0408	−0.0085	−0.0238	−0.0323
3.45	−0.0399	−0.0097	−0.0206	−0.0303
3.50	−0.0388	−0.0106	−0.0177	−0.0283
3.55	−0.0378	−0.0114	−0.0149	−0.0264
3.60	−0.0366	−0.0121	−0.0124	−0.0245
3.65	−0.0354	−0.0126	−0.0101	−0.0227
3.70	−0.0341	−0.0131	−0.0079	−0.0210
3.75	−0.0327	−0.0134	−0.0059	−0.0193
3.80	−0.0314	−0.0137	−0.0040	−0.0177
3.85	−0.0300	−0.0139	−0.0023	−0.0162
3.90	−0.0286	−0.0140	−0.0008	−0.0147
$\frac{5}{4}\pi$	−0.0278	−0.0140	0	−0.0139
3.95	−0.0272	−0.0139	0.0005	−0.0133
4.00	−0.0258	−0.0139	0.0019	−0.0120
4.50	−0.0132	−0.0108	0.0085	−0.0023
$\frac{3}{2}\pi$	−0.0090	−0.0090	0.0090	0
5.00	−0.0046	−0.0065	0.0084	0.0019
$\frac{7}{4}\pi$	0	−0.0029	0.0058	0.0029
5.50	0.0000	−0.0029	0.0058	0.0029
6.00	0.0017	−0.0007	0.0031	0.0024
2π	0.0019	0	0.0019	0.0019
6.50	0.0018	0.0003	0.0012	0.0018
7.00	0.0013	0.0006	0.0001	0.0007
$\frac{9}{4}\pi$	0.0012	0.0006	0	0.0006
7.50	0.0007	0.0005	−0.0003	0.0002
$\frac{5}{2}\pi$	0.0004	0.0004	−0.0004	0

yields reasonable results for much longer beams for any location of the concentrated load as long as the distance from the load to either end of the beam is equal to or greater than $3\pi/(4\beta)$.

EXAMPLE 10.1
Diesel Locomotive Wheels on Rail

A railroad uses steel rails ($E = 200$ GPa) with a depth of 184 mm. The distance from the top of the rail to its centroid is 99.1 mm, and the moment of inertia of the rail is 36.9×10^6 mm⁴. The rail is supported by ties, ballast, and a road bed that together are assumed to act as an elastic foundation with spring constant $k = 14.0$ N/mm².

(a) Determine the maximum deflection, maximum bending moment, and maximum flexure stress in the rail for a single wheel load of 170 kN.

(b) A particular Diesel locomotive has three wheels per truck equally spaced at 1.70 m. Determine the maximum deflection, maximum bending moment, and maximum flexure stress in the rail if the load on each wheel is 170 kN.

SOLUTION

The equations for bending moment and deflection require the value of β. From Eq. (10.5), we find that

$$\beta = \sqrt[4]{\frac{k}{4EI_x}} = \sqrt[4]{\frac{14}{4(200 \times 10^3)(36.9 \times 10^6)}} = 0.000830 \text{ mm}^{-1}$$

(a) The maximum deflection and maximum bending moment occur under the load where $A_{\beta z} = C_{\beta z} = 1.00$. Equations (10.12) and (10.14) give

$$y_{max} = \frac{P\beta}{2k} = \frac{170 \times 10^3(0.000830)}{2(14)} = 5.039 \text{ mm}$$

$$M_{max} = \frac{P}{4\beta} = \frac{170 \times 10^3}{4(0.000830)} = 51.21 \text{ kN·m}$$

$$\sigma_{max} = \frac{M_{max}c}{I_x} = \frac{51.21 \times 10^6(99.1)}{36.9 \times 10^6} = 137.5 \text{ MPa}$$

(b) The deflection and bending moment at any section of the beam are obtained by superposition of the effects of each of the three wheel loads. With superposition, an examination of Figs. 10.1c and e indicates that the maximum deflection and maximum bending moment occur either under the center wheel or under one of the end wheels. Let the origin be located under one of the end wheels. The distance from the origin to the next wheel is $z_1 = 1.7 \times 10^3$ mm. Hence, $\beta z_1 = 0.000830(1.7 \times 10^3) = 1.411$. The distance from the origin to the second wheel is $z_2 = 2(1.7 \times 10^3)$ mm. Hence, $\beta z_2 = 0.000830(2)(1.7 \times 10^3) = 2.822$. From Table 10.1, we find

$$A_{\beta z1} = 0.2797, \qquad C_{\beta z1} = -0.2018$$
$$A_{\beta z2} = -0.0377, \qquad C_{\beta z2} = -0.0752$$

The deflection and bending moment at the origin (under one of the end wheels) are

$$y_{end} = \frac{P\beta}{2k}(A_{\beta z0} + A_{\beta z1} + A_{\beta z2}) = 5.039(1 + 0.2797 - 0.0377)$$

$$= 6.258 \text{ mm}$$

$$M_{end} = \frac{P}{4\beta}(C_{\beta z0} + C_{\beta z1} + C_{\beta z2}) = 51.20 \times 10^6(1 - 0.2018 - 0.0752)$$

$$= 37.02 \text{ kN·m}$$

Now, let the origin be located under the center wheel. The distance between the center wheel and either of the end wheels is $z_1 = 1.7 \times 10^3$ mm. Therefore,

$$y_{center} = \frac{P\beta}{2k}(A_{\beta z0} + 2A_{\beta z1}) = 5.039[1 + 2(0.2797)] = 7.858 \text{ mm}$$

$$M_{center} = \frac{P}{4\beta}(C_{\beta z0} + 2C_{\beta z1}) = 51.20 \times 10^6[1 - 2(0.2018)]$$

$$= 30.54 \text{ kN·m}$$

Thus, we find

$$y_{center} = y_{max} = 7.858 \text{ mm}$$
$$M_{end} = M_{max} = 37.02 \text{ kN·m}$$

and

$$\sigma_{max} = \frac{M_{max}c}{I_x} = \frac{37.02 \times 10^6(99.1)}{36.9 \times 10^6} = 99.4 \text{ MPa}$$

Beam Supported on Equally Spaced Discrete Elastic Supports

Long beams are sometimes supported by elastic springs equally spaced along the beam (Fig. 10.3a). Although coil springs are shown in Fig. 10.3, each spring support may be due to the resistance of a linearly elastic member or a structure such as a tension member, straight beam, or curved beam. It is possible to obtain an exact solution for the spring-supported beam of Fig. 10.3a by energy methods (see Sec. 5.5, Example 5.15); however, the computational work becomes prohibitive as the number of springs becomes large.

Alternatively, we may proceed as follows: Let each spring in Fig. 10.3a have the same constant K. The force R that each spring exerts on the beam is directly proportional to the deflection y of the beam at the section where the spring is attached. Thus, we write

$$R = Ky \tag{10.17}$$

We assume that the load R is distributed uniformly over a spacing l, a distance $l/2$ to the right and to the left of each spring. Thus, we obtain the stepped distributed loading shown in Fig. 10.3b. If the stepped distributed loading is approximated by a smooth average curve (dashed curve in Fig. 10.3b), the approximate distributed load is similar to the distributed load q of Fig. 10.1a. Since the dashed curve in Fig. 10.3b intersects each of the steps near its center, we assume that the dashed curve does indeed intersect each of the steps directly beneath the spring. Thus, we assume that an equivalent spring constant k exists, such that

$$k = \frac{K}{l} \tag{10.18}$$

Hence, substitution of Eq. (10.18) into Eq. (10.5) yields an equivalent β for the springs. Next, we assume that Eqs. (10.12) through (10.15) are valid for an infinite

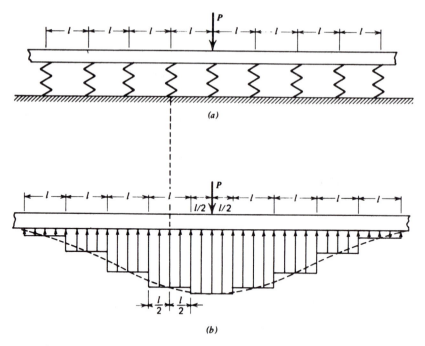

Figure 10.3   Infinite beam supported by equally spaced elastic springs.

beam supported by equally spaced elastic supports and loaded in the center. Obviously, the resulting approximate solution becomes more accurate as the spacing l between springs becomes small. However, we note that this approximate solution becomes greatly in error when the spacing l between springs becomes large. It has been found that the error in the solution is not excessive if we require that the spacing l between springs satisfies the condition

$$l \le \frac{\pi}{4\beta} \tag{10.19}$$

The magnitude of the error for spacing that satisfies Eq. (10.19) is discussed in the example problem that follows this section.

The approximate solution for a beam of infinite length, with equally spaced elastic supports, may be used to obtain a reasonable approximate solution for a sufficiently long finite-length beam. We note that the load exerted by each elastic spring has been assumed to be distributed over a distance l, the distribution being uniform over a distance $l/2$ to the left and right of the spring (Figs. 10.3a and b). Hence, consider a beam of length L supported by discrete elastic springs (Fig. 10.4a). In general, the end springs do not coincide with the ends of the beam but lie at some distance less than $l/2$ from the beam ends. Since the distributed effect of the end springs is assumed to act over length l, $l/2$ to the left and right of the end springs, we extend the beam of length L to a beam of length L'', where (Fig. 10.4b)

$$L'' = ml \tag{10.20}$$

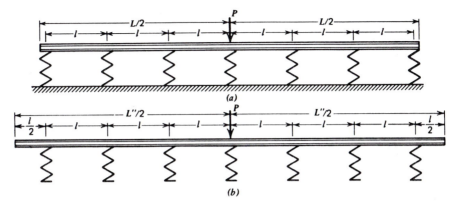

Figure 10.4

and the integer m denotes the number of spring supports. If $L'' \geq 3\pi/(2\beta)$, the approximate solution for a spring-supported infinite beam yields a reasonably good approximation for a spring-supported finite beam of length L.

EXAMPLE 10.2
Finite-Length Beam Supported by Seven Springs

An aluminum alloy I-beam (depth = 100 mm, $I_x = 2.45 \times 10^6$ mm^4, $E = 72.0$ GPa) has a length $L = 6.8$ m and is supported by seven springs ($K = 110$ N/mm) spaced at distance $l = 1.10$ m center to center along the beam. A load $P = 12.0$ kN is applied at the center of the beam over one of the springs. Using the approximate solution method described in Sec. 10.2, determine the load carried by each spring, the deflection of the beam under the load, the maximum bending moment, and the maximum bending stress in the beam. The exact solution of this problem has been presented in Example 5.15.

SOLUTION

The magnitude of the factor β is estimated by means of Eqs. (10.18) and (10.5). Thus, we find

$$k = \frac{K}{l} = \frac{110}{1.1 \times 10^3} = 0.100 \text{ N/mm}^2$$

$$\beta = \sqrt[4]{\frac{0.100}{4(72 \times 10^3)(2.45 \times 10^6)}} = 0.000614 \text{ mm}^{-1}$$

By Eqs. (10.19) and (10.20), we see that

$$l = 1.10 \times 10^3 < \frac{\pi}{4\beta} = \frac{\pi}{4(0.000614)} = 1279 \text{ mm}$$

$$L'' = 7(1.10 \times 10^3) = 7700 \text{ mm} > \frac{3\pi}{2\beta} = \frac{3\pi}{2(0.000614)} = 7675 \text{ mm}$$

Hence, the limiting conditions on l and L'' are satisfied. The maximum deflection and maximum bending moment occur under the load where $A_{\beta z} = C_{\beta z} = 1.00$. Equations (10.12) and (10.14) give

$$y_{max} = \frac{P\beta}{2k} = \frac{12 \times 10^3 (0.000614)}{2(0.10)} = 36.84 \text{ mm}$$

$$M_{max} = \frac{P}{4\beta} = \frac{12 \times 10^3}{4(0.000614)} = 4.886 \times 10^6 \text{ N·mm}$$

$$\sigma_{max} = \frac{M_{max} c}{I_x} = 99.7 \text{ MPa}$$

The deflection y_{max} ($y_D = y_{max}$ in Fig. E5.15b) occurs at the origin (at the center of the beam under the load). The magnitude of βz for the first, second, and third springs to the right and left of the load are $\beta l = 0.6754$, $2\beta l = 1.3508$, and $3\beta l = 2.0262$, respectively. From Table 10.1, $A_{\beta l} = 0.7153$, $A_{2\beta l} = 0.3094$, and $A_{3\beta l} = 0.0605$. The deflections of the springs C, B, and A (see Fig. E5.15b) are given by Eq. (10.12).

$$y_C = \frac{P\beta}{2k} A_{\beta l} = 36.84(0.7153) = 26.35 \text{ mm}$$

$$y_B = \frac{P\beta}{2k} A_{2\beta l} = 36.84(0.3094) = 11.40 \text{ mm}$$

$$y_A = \frac{P\beta}{2k} A_{3\beta l} = 36.84(0.0605) = 2.23 \text{ mm}$$

The reaction for each spring may be computed by means of Eq. (10.17). A comparison of the approximate solution presented here with the exact solution of Example 5.15 is given in Table E10.2. Although the reaction at A is considerably in error, the results in Table E10.2 indicate that the approximate maximum deflection is 5.50% less than the exact deflection, whereas the approximate maximum bending moment is 6.68% greater than the exact bending moment. These errors in the maximum deflection and maximum moment are not large when one considers the simplicity of the present solution compared to that of Example 5.15.

TABLE E10.2

Quantity	Exact Solution Example 5.15	Approximate Solution
Reaction A	−454 N	245 N
Reaction B	1216 N	1254 N
Reaction C	3094 N	2899 N
Reaction D	4288 N	4052 N
M_{max}	4.580 kN·m	4.886 kN·m
y_{max}	38.98 mm	36.84 mm

10.3

INFINITE BEAM SUBJECTED TO A DISTRIBUTED LOAD SEGMENT

The solution for the problem of a concentrated load at the center of an infinite beam on an infinite elastic foundation can be used to obtain solutions for distributed loads. Only segments of uniformly distributed loads are considered in this section. Consider an infinite beam resting on an infinite elastic foundation and subjected to a uniformly distributed load w over a segment of length L' (Fig. 10.5). The deflection, slope, bending moment, and shear of the beam can be determined with the solution presented in Sec. 10.2. Since the maximum values of these quantities generally occur within the segment of length L', we obtain the solution only in this segment.

Consider an infinitesimal length Δz of the beam within the segment of length L'. In this segment, the beam is subjected to a uniformly distributed load w (Fig. 10.5). Hence, a load $\Delta P = w \Delta z$ acts on the element Δz. We treat the load $\Delta P = w \Delta z$ as a concentrated load and choose the origin of the coordinate axes under load ΔP. Next, consider any point H at distance z from the load $\Delta P = w \Delta z$; note that H is located at distances a and b from the left and right ends of segment L', respectively. The deflection Δy_H at H due to the concentrated load $\Delta P = w \Delta z$ is given by Eq. (10.11) with $P = \Delta P = w \Delta z$. Thus, we have

$$\Delta y_H = \frac{w \Delta z \beta}{2k} e^{-\beta z}(\cos \beta z + \sin \beta z) \tag{10.21}$$

The total deflection y_H due to the distributed load over the entire length L' is obtained by superposition. It is the algebraic sum of increments given by Eq. (10.21). Hence, by the integration process, we obtain

$$y_H = \sum_{\lim \Delta z \to 0} \Delta y_H = \int_0^a \frac{w\beta}{2k} e^{-\beta z}(\cos \beta z + \sin \beta z)\, dz$$

$$+ \int_0^b \frac{w\beta}{2k} e^{-\beta z}(\cos \beta z + \sin \beta z)\, dz$$

$$= \frac{w}{2k}(2 - e^{-\beta a}\cos \beta a - e^{-\beta b}\cos \beta b) \tag{10.22}$$

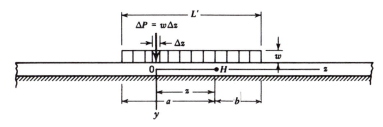

Figure 10.5   Uniformly distributed load segment on an infinite beam resting on an elastic foundation.

Values of slope, bending moment, and shear at point H may also be obtained by superposition. These expressions may be simplified by means of Eqs. (10.16). Thus, we obtain the results

$$y_H = \frac{w}{2k}(2 - D_{\beta a} - D_{\beta b}) \tag{10.23}$$

$$\theta_H = \frac{w\beta}{2k}(A_{\beta a} - A_{\beta b}) \tag{10.24}$$

$$M_H = \frac{w}{4\beta^2}(B_{\beta a} + B_{\beta b}) \tag{10.25}$$

$$V_H = \frac{w}{4\beta}(C_{\beta a} - C_{\beta b}) \tag{10.26}$$

Generally, the maximum values of deflection and bending moment are of greatest interest. The maximum deflection occurs at the center of segment L'. The maximum bending moment may or may not occur at the center of segment L'. In general, the location of the maximum bending moment depends on the magnitude of $\beta L'$.

$\beta L' \leq \pi$

For $\beta L'$ less than or equal to π, the data for $B_{\beta z}$ in Table 10.1 indicate that the maximum bending moment occurs at the center of segment L'.

$\beta L' \to \infty$

As $\beta L'$ becomes large,

$$\theta \to 0, \qquad M_x \to 0, \qquad V_y \to 0, \qquad \text{and} \qquad y \to \frac{w}{k} \tag{10.27}$$

everywhere, except near the ends of segment L'. The data in Table 10.1 indicate that the maximum bending moment occurs when either βa or βb is equal to $\pi/4$.

Intermediate Values of $\beta L'$

For $\beta L'$ greater than π, the location of the maximum bending moment may lie outside of segment L'. (See Problem 10.18 and Example 10.3.) However, the maximum moment value outside of segment L' for the example problem is only 3.0% greater than the maximum bending moment within segment L'. The location of the maximum bending moment can be obtained by trial and error; however, because of the small difference, sufficient accuracy can be obtained by taking the location of the maximum moment to be $\pi/(4\beta)$ from either end of the uniformly distributed load within length L'.

EXAMPLE 10.3
Uniformly Distributed Load on a Segment of Wood Beam

A long wood beam ($E = 10.0$ GPa) has a rectangular cross section with a depth of 200 mm and width of 100 mm. It rests on a soil foundation. The spring constant for the foundation is $k_0 = 0.040$ N/mm^3. A uniformly distributed load $w = 35.0$ N/mm extends over a length $L' = 3.61$ m of the beam (Fig. E10.3a). Determine the maximum deflection, maximum flexure stress, and maximum pressure between the beam and foundation. Take the origin of coordinates at the center of segment L'.

SOLUTION

The magnitude of β is obtained by means of Eqs. (10.3) and (10.5). Thus, we find

$$k = bk_0 = 100(0.040) = 4.00 \text{ N/mm}^2$$

$$I_x = \frac{bh^3}{12} = \frac{100(200)^3}{12} = 66.67 \times 10^6 \text{ mm}^4$$

$$\beta = \sqrt[4]{\frac{k}{4EI_x}} = \sqrt[4]{\frac{4}{4(10 \times 10^3)(66.67 \times 10^6)}} = 0.001107 \text{ mm}^{-1}$$

The magnitude of $\beta L'$, needed to determine where the maximum bending moment occurs, is $\beta L' = 0.001107(3.61 \times 10^3) = 4.00$. Since $\beta L'$ is greater than π, the maximum bending moment does not occur at the center of segment L'. With values of

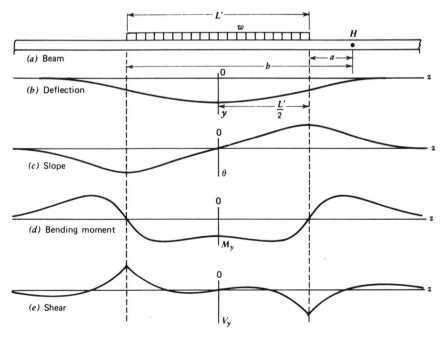

(a) Beam

(b) Deflection

(c) Slope

(d) Bending moment

(e) Shear

Figure E10.3

βa and $\beta b = 4.00 - \beta a$, values of the quantities y, θ, M_x, and V_y may be calculated by means of Eqs. (10.23), (10.24), (10.25), and (10.26), and the data from Table 10.1. These quantities are plotted in Figs. E10.3b, c, d, and e, respectively. Values are also shown for points in the beam outside the distributed load segment where $\beta b = \beta(a + L')$ and a is the distance from H to the nearest edge of the distributed load (Fig. E10.3a). Equations (10.24) and (10.26), for θ_H and V_H, respectively, are valid for points away from the distributed load; however, different equations are needed for y_H and M_H as indicated in Problem 10.18.

The maximum deflection occurs at the center of segment L', where $\beta a = \beta b = 2.00$. Equation (10.23), with $D_{\beta a} = -0.0563$ from Table 10.1, gives

$$y_{max} = \frac{w}{k}(1 - D_{\beta a}) = \frac{35}{4}(1 + 0.0563) = 9.243 \text{ mm}$$

The maximum pressure between the beam and foundation occurs at the point of maximum deflection; thus, we find that the maximum pressure is $p_{max} = y_{max}k_0 = 9.243(0.040) = 0.370$ MPa. There are four possible locations at which the largest bending moment may occur. They are located symmetrically with respect to the center of segment L'. Relative maximum bending moments occur at locations where $V_H = 0$. From Table 10.1, it is found that $V_H = 0$ ($C_{\beta a} = C_{\beta b}$) when $\beta a = 0.858$ and $\beta b = 3.142$ and also when $\beta a = 0.777$ and $\beta b = 4.777$. These conditions locate the position of relative maximum bending moments inside segment L' and outside of segment L', respectively. However, the value of the largest bending moment is located outside of segment L' and is given by the equation indicated in Problem 10.18. Thus, we find (outside of segment L')

$$M_{max} = \left| \frac{-w}{4\beta^2}(B_{\beta a} - B_{\beta b}) \right|$$

$$= \frac{35}{4(0.001107)^2}[0.3223 - (-0.0086)] = 2.363 \text{ kN·m}$$

This value is 3% greater than the bending moment calculated by means of Eq. (10.25) with $\beta a = 0.858$ and $\beta b = 3.142$. In practice, this difference is not especially significant.

The corresponding flexure stress is

$$\sigma_{max} = \frac{M_{max}c}{I_x} = \frac{2.363 \times 10^6(100)}{66.67 \times 10^6} = 3.544 \text{ MPa}$$

If the maximum bending moment is assumed to occur at $\pi/(4\beta)$ (see the end of Sec. 10.3), $\beta a = \pi/4$ and $\beta b = 4 - \pi/4$ (inside of segment L'). Substituting these values for βa and βb in Eq. (10.25), we find

$$M_H = \frac{w}{4\beta^2}(B_{\beta a} + B_{\beta b}) = \frac{35}{4(0.001107)^2}[0.3224 + (-0.0029)]$$

$$= 2.281 \text{ kN·m}$$

which is 3.5% less than the largest moment M_{max} computed above. Generally, the value M_H for $\beta a = \pi/4$ and $\beta b = \beta L' - \pi/4$ gives a good approximation of M_{max}.

10.4

SEMIINFINITE BEAM SUBJECTED TO LOADS AT ITS END

A semiinfinite beam resting on an infinite linearly elastic foundation is loaded at its end by a concentrated load P and positive bending moment M_0 (Fig. 10.2). The boundary conditions that determine the two constants of integration C_3 and C_4 in Eq. (10.7) are

$$EI_x \frac{d^2 y}{dz^2}\bigg|_{z=0} = -M_0$$

$$EI_x \frac{d^3 y}{dz^3}\bigg|_{z=0} = -V_y = P \tag{10.28}$$

Substitution of Eq. (10.7) into these boundary conditions yields two linear equations in C_3 and C_4. Solving these equations for C_3 and C_4, we obtain

$$C_3 = \frac{2\beta^2 M_0}{k}, \qquad C_4 = \frac{2\beta P}{k} - C_3 \tag{10.29}$$

Substituting these into Eq. (10.7), we find

$$y = \frac{2\beta e^{-\beta z}}{k}[P\cos\beta z - \beta M_0(\cos\beta z - \sin\beta z)] \tag{10.30}$$

Values of slope, bending moment, and shear are obtained by substitution of Eq. (10.30) into Eqs. (10.1). These equations are simplified with the definitions given by Eqs. (10.16). Thus, we have

$$y = \frac{2P\beta}{k} D_{\beta z} - \frac{2\beta^2 M_0}{k} C_{\beta z} \tag{10.31}$$

$$\theta = -\frac{2P\beta^2}{k} A_{\beta z} + \frac{4\beta^3 M_0}{k} D_{\beta z} \tag{10.32}$$

$$M_x = -\frac{P}{\beta} B_{\beta z} + M_0 A_{\beta z} \tag{10.33}$$

$$V_y = -PC_{\beta z} - 2M_0 \beta B_{\beta z} \tag{10.34}$$

These results are valid, provided that the beam is attached to the foundation everywhere along its length.

EXAMPLE 10.4
I-Beam Loaded at Its End

A steel I-beam ($E = 200$ GPa) has a depth of 102 mm, width of 68 mm, moment of inertia of $I_x = 2.53 \times 10^6$ mm^4, and length of 4 m. It is attached to a rubber foundation for which $k_0 = 0.350$ N/mm^3. A concentrated load $P = 30.0$ kN is applied at one end of the beam. Determine the maximum deflection, maximum flexure stress in the beam, and the location of each.

SOLUTION

The spring coefficient k is equal to the product of the beam width and the elastic spring constant k_0 for the foundation; that is, $k = 68(0.350) = 23.8$ N/mm^2. From Eq. (10.5), we find that

$$\beta = \sqrt[4]{\frac{k}{4EI_x}} = \sqrt[4]{\frac{23.8}{4(200,000)(2,530,000)}} = 0.001852 \text{ mm}^{-1}$$

Since

$$L = 4000 \text{ mm} > \frac{3\pi}{2\beta} = \frac{3\pi}{2(0.001852)} = 2540 \text{ mm}$$

the beam can be considered to be a long beam. Values for deflection y and moment M_x are given by Eqs. (10.31) and (10.33). The maximum deflection occurs at the end where load P is applied, since $D_{\beta z}$ is maximum where $\beta z = 0$. The maximum moment occurs at $z = \pi/4\beta$, where $B_{\beta z}$ is a maximum. Thus, the maximum deflection is

$$y_{\text{max}} = \frac{2P\beta}{k} = \frac{2(30,000)(0.001852)}{23.8} = 4.67 \text{ mm}$$

The location of y_{max} is at $z = 0$. The maximum moment is

$$M_{\text{max}} = -\frac{0.3224P}{\beta} = -\frac{0.3224(30,000)}{0.001852} = -5.22 \text{ kN·m}$$

and, therefore, the maximum stress is

$$\sigma_{\text{max}} = \frac{M_{\text{max}}c}{I_x} = \frac{5,220,000(51)}{2,530,000} = 105.3 \text{ MPa}$$

The location of σ_{max} is at $z = \pi/4\beta = 424$ mm.

10.5

SEMIINFINITE BEAM WITH CONCENTRATED LOAD NEAR ITS END

The solution for a semiinfinite beam resting on an infinite linearly elastic foundation with a concentrated load P near its end may be obtained from the solutions presented in Sec. 10.2 and 10.4. Consider a beam subjected to load P at distance a from its end (Fig. 10.6a). Let the beam be extended to infinity to the left as indicated by the dashed line. For the beam so extended, Eqs. (10.14) and (10.15) give

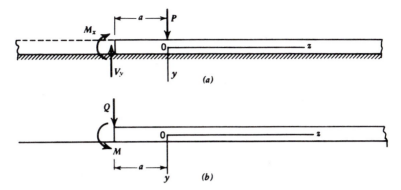

Figure 10.6   Semiinfinite beam on an elastic foundation loaded near its end.

magnitudes for $M_{x(z=-a)} = PC_{\beta a}/(4\beta)$ and $V_{y(z=-a)} = PD_{\beta a}/2$ at distance a to the left of the origin (Fig. 10.6a). Now let the beam (Fig. 10.6a) be loaded at the left end (Fig. 10.6b), by loads Q and M with magnitudes

$$Q = \frac{PD_{\beta a}}{2}, \qquad M = -\frac{PC_{\beta a}}{4\beta} \tag{10.35}$$

Since the origin of the coordinate axes is distance a to the right of the loaded end, the deflection and bending moment for this loading are given by Eqs. (10.31) and (10.33), respectively, if the coordinate z is replaced by $(a + z)$. Superposing the two loadings for the two beams in Fig. 10.6 cancels the moment and shear at the left end. Thus, superposition of the two results yields the solution for a semiinfinite beam loaded by a concentrated load P at distance a from the left end. Using Eqs. (10.12), (10.31), and (10.35), we obtain the deflection y for $z \geq -a$. Thus, we find for y the formula

$$y = \frac{P\beta}{2k}\left[A_{\beta z} + 2D_{\beta a}D_{\beta(a+z)} + C_{\beta a}C_{\beta(a+z)}\right] \tag{10.36}$$

Similarly, Eqs. (10.14), (10.33), and (10.35) give the bending moment M_x for $z \geq -a$ as follows:

$$M_x = \frac{P}{4\beta}\left[C_{\beta z} - 2D_{\beta a}B_{\beta(a+z)} - C_{\beta a}A_{\beta(a+z)}\right] \tag{10.37}$$

Since the quantities $A_{\beta z}$ and $C_{\beta z}$ in Eqs. (10.36) and (10.37) are symmetrical in z, for negative values of z $(-a \leq z \leq 0)$, we use the conditions $A_{\beta z}(-z) = A_{\beta z}(z)$ and $C_{\beta z}(-z) = C_{\beta z}(z)$.

EXAMPLE 10.5
I-Beam Loaded near One End

Let the load in Example 10.4 be moved to a location 500 mm from one end of the beam. Determine the maximum deflection, maximum flexure stress in the beam, and the location of each.

SOLUTION

From Example 10.4, $k = 23.8$ N/mm^2 and $\beta = 0.001852$ mm^{-1}. The deflection y and bending moment M_x are given by Eqs. (10.36) and (10.37). Since $\beta a = 0.001852(500) = 0.9260$, Table 10.1 gives $C_{\beta a} = -0.0782$ and $D_{\beta a} = 0.2383$. Hence,

$$y = \frac{P\beta}{2k}[A_{\beta z} + 2D_{\beta a}D_{\beta(a+z)} + C_{\beta a}C_{\beta(a+z)}]$$

$$= 1.1672[A_{\beta z} + 0.4766D_{\beta(a+z)} - 0.0782C_{\beta(a+z)}]$$

$$M_x = \frac{P}{4\beta}[C_{\beta z} - 2D_{\beta a}B_{\beta(a+z)} - C_{\beta a}A_{\beta(a+z)}]$$

$$= 4,050,000[C_{\beta z} - 0.4766B_{\beta(a+z)} + 0.0782A_{\beta(a+z)}]$$

By trial and error, it is found that the maximum deflection y_{max} occurs at 424 mm from the end of the beam, where $z = -76$ mm [$\beta z = 0.1408$ and $\beta(a + z) = \pi/4 = 0.7854$]. From Table 10.1, $A_{\beta z} = 0.9816$, $D_{\beta(a+z)} = 0.3224$, and $C_{\beta(a+z)} = 0$. Thus,

$$y_{max} = 1.1672[0.9816 + 0.4766(0.3224) - 0.0782(0)]$$

$$= 1.3251 \text{ mm}$$

By trial and error, it is found that the maximum bending moment M_{max} occurs at 500 mm from the end of the beam [$\beta z = 0$ and $\beta(a + z) = 0.9260$]. From Table. 10.1, $C_{\beta z} = 1.0000$, $A_{\beta(a+z)} = 0.5548$, and $B_{\beta(a+z)} = 0.3165$. Hence,

$$M_{max} = 4,050,000[1.0000 - 0.4766(0.3165) + 0.0782(0.5548)]$$

$$= 3,615,000 \text{ N·mm}$$

and, therefore,

$$\sigma_{max} = \frac{M_{max}c}{I_x} = \frac{3,615,000(51)}{2,530,000} = 72.9 \text{ MPa}$$

10.6

SHORT BEAMS

The solutions that have been presented in the foregoing sections are good approximations for a beam supported by an elastic foundation and with a length greater than $3\pi/(2\beta)$. However, for beams whose lengths are less than $3\pi/(2\beta)$, so called *short beams*, special solutions are required. The reader is referred to the book by M. Hetényi (1946) for a solution applicable for short beams. For the special case of a concentrated load located at the center of a short beam, the maximum deflection y_{max} and maximum bending moment M_{max} occur under the load; their

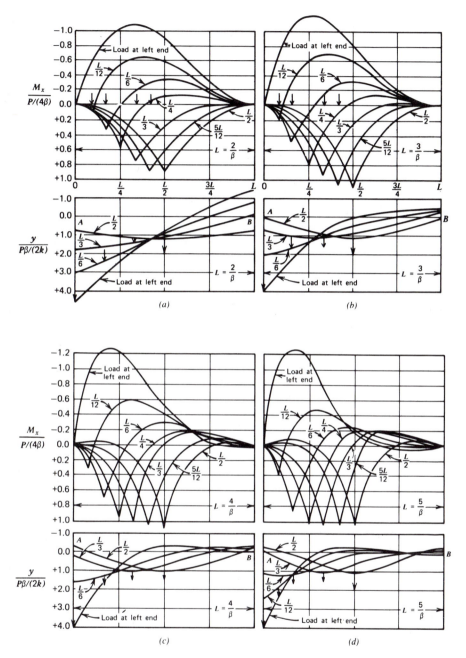

Figure 10.7 Bending moment diagrams and deflection curves for a short beam on elastic supports subjected to concentrated load located as shown on each curve. The ends of the beams are unrestrained (free). (a) Span = 2/β. (b) Span = 3/β. (c) Span = 4/β. (d) Span = 5/β.

magnitudes are given by the following equations:

$$y_{max} = \frac{P\beta}{2k} \frac{\cosh \beta L + \cos \beta L + 2}{\sinh \beta L + \sin \beta L} \qquad (10.38)$$

$$M_{max} = \frac{P}{4\beta} \frac{\cosh \beta L - \cos \beta L}{\sinh \beta L + \sin \beta L} \qquad (10.39)$$

in which L is the length of the beam. Magnitudes of the deflection y and bending moment M_x for other locations of the concentrated load are beyond the scope of the book. However, solutions have been calculated for several load locations for three short beams and one long beam. The results are presented in Fig. 10.7.

Design tables for finite beams with free ends on a Winkler foundation have been given by Iyengar and Ramu (1979). The cases of simply supported ends and clamped ends may be treated by appropriate superposition techniques. A solution for finite beams with elastic end restraints on a Winkler foundation has been given by Ting (1982). This solution can be used to simulate a beam on elastic foundations with various boundary conditions, including initial settlement of an end of the beam. The effect of other structural members connected to a beam on a Winkler foundation can also be assessed by using proper values of the elastic end restraints. The solution is in a form that can be coded easily into computer language.

10.7

THIN-WALL CIRCULAR CYLINDERS

The concept of a beam on an elastic foundation may be used to approximate the response of thin-wall circular cylinders subjected to loads that are rotationally symmetrical (Fig. 10.8). We use cylindrical coordinates r, θ, z for radial, circumferential, and axial directions. The dimensions of a long thin-wall cylinder may be represented by the mean radius a and wall thickness h. Let a long thin-wall cylinder be subjected to a ring load w having units N/mm, where the length dimension is measured in the circumferential direction. We show that the response

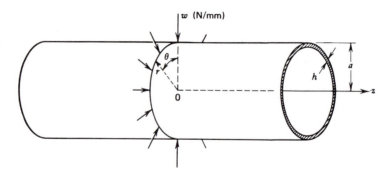

Figure 10.8   Ring load on a thin-wall cylinder.

of the cylinder is similar to that of a corresponding beam on an elastic foundation subjected to a concentrated load at its center (Sec. 10.2).

In developing an analogy between a thin-wall circular cylinder and a beam on an elastic foundation, we specify the analogous beam and elastic foundation as follows: Cut a longitudinal strip from the cylinder of width $a\,\Delta\theta$ (Fig. 10.9b). For convenience let the width $a\,\Delta\theta$ be unity. We consider this strip of length L and width $a\,\Delta\theta = 1$ as a beam. We consider the remainder of the cylinder to act as the elastic foundation. The spring constant k for the elastic foundation is obtained by imagining the open-ended cylinder to be subjected to an external pressure p_2. This pressure p_2 produces a uniaxial state of stress for which the only nonzero stress component is $\sigma_{\theta\theta} = ap_2/h$. Hence, by Hooke's law, the circumferential strain is $\epsilon_{\theta\theta} = \sigma_{\theta\theta}/E$. In turn, by strain-displacement relations, we can express $\epsilon_{\theta\theta}$ in terms of the radial displacement u as follows [see Eqs. (2.85)]:

$$u = a\epsilon_{\theta\theta} = \frac{a\sigma_{\theta\theta}}{E} = \frac{a^2 p_2}{Eh} \tag{10.40}$$

Since u is constant along the length of the cylinder, the magnitude of k is given by Eq. (10.27) where u replaces y and $w = p_2(a\,\Delta\theta) = p_2$, since $a\,\Delta\theta = 1$. Hence, we have

$$k = \frac{w}{u} = \frac{Eh}{a^2} \tag{10.41}$$

Note that the narrow strip (Fig. 10.9b), which represents the beam on an elastic foundation, has a different state of stress (and strain) than other beams considered in this chapter. The beam in Fig. 10.1 was assumed to be free to deform in the x direction, thus developing anticlastic curvature (Boresi and Chong, 1987). Each of the two sides of the beam in Fig. 10.9b lies in a radial plane of the cylinder; these sides are constrained to lie in the same planes after deformation much like a flat plate (see Chapter 13). Therefore, since the beam cannot deform anticlastically, EI_x in Eq. (10.5) must be replaced by $D = Eh^3/[12(1 - v^2)]$ [see Eq. (f) of Example 3.1].

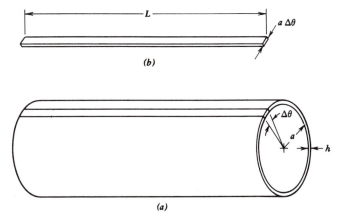

Figure 10.9   Thin-wall cylinder.

Replacing EI_x by D and using Eq. (10.41), we express β in the form

$$\beta = \sqrt[4]{\frac{3(1 - v^2)}{a^2 h^2}} \tag{10.42}$$

With the value of β given by Eq. (10.42), the solution for any of the loadings considered in Sec. 10.2 through 10.5 is applicable for thin-wall circular cylinders subjected to circumferential line loads. They may also be used to obtain estimates of the response of a thin-wall cylinder subjected to rotationally symmetric loads that vary along the axis of the cylinder.

Note: The analogous elastic foundation for the strip taken from a thin-wall circular cylinder is very stiff compared to the usual elastic foundation. Hence, the analogy is applicable even for a cylinder with length less than the radius a. If we assume that $v = 0.30$, the minimum length L for which the analogy is applicable is

$$L = \frac{3\pi}{2\beta} = 3.67a \sqrt{\frac{h}{a}} \quad \left(L = 0.82a \text{ for } \frac{h}{a} = \frac{1}{20}\right) \tag{10.43}$$

Generally for thin-wall cylinders, h/a is less than $1/20$, and the length L of the cylinder influenced by the concentrated ring load is less than $0.82a$. Often, the beam analogy can be employed to obtain estimates of the response of noncylindrical circular shell segments (for instance, conical shells) if the change in radius for a given length L is small compared to the average radius a in the length L.

EXAMPLE 10.6
Stresses in Storage Tank

A closed end thin-wall cylinder is used as an oil storage tank that rests on one of its ends (see Fig. E10.6). The tank has a diameter of 30 m, depth of 10 m, and wall thickness of 20 mm. The tank is made of steel for which $E = 200$ GPa and $v = 0.29$. Determine the maximum shear stress in the tank if it is filled with oil having a mass density of 900 kg/m^3 under the following different conditions:

(a) Assume that the bottom of the tank does not influence the circumferential stress in the cylindrical walls.

(b) Assume that the radial displacement of the junction between the cylinder and bottom remains zero during loading and that the bottom has infinite rotational stiffness.

(c) Assume that the radial displacement of the junction between the cylinder and bottom remains zero and that the bottom plate is sufficiently flexible that the moment at the junction can be considered to be zero.

SOLUTION

(a)   Choose cylindrical coordinates r, θ, and z. The pressure in the cylinder increases linearly with depth. If the bottom does not exert moments or radial

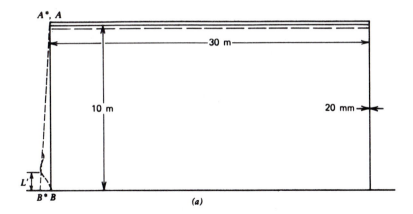

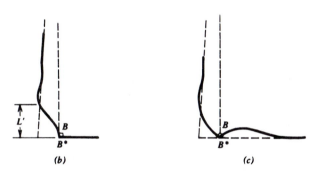

Figure E10.6   Thin-wall cylinder oil storage tank.

forces on the cylinder walls, wall AB in Fig. E10.6a deforms into the straight line $A*B*$. The stresses in the cylinder walls at $B*$ are σ_{rr}, $\sigma_{\theta\theta}$, and σ_{zz}. The radial stress σ_{rr} and longitudinal stress σ_{zz} at the bottom are small and are neglected compared to the circumferential stress $\sigma_{\theta\theta}$. By the solution for thin-wall cylinders, we find

$$\sigma_{\theta\theta} = \frac{pa}{h} = \frac{(10 \times 10^3)(9.807)(900 \times 10^{-9})(15 \times 10^3)}{20} = 66.20 \text{ MPa}$$

The maximum shear stress is given by Eq. (2.39)

$$\tau_{max} = \frac{\sigma_{max} - \sigma_{min}}{2} = \frac{\sigma_{\theta\theta}}{2} = 33.10 \text{ MPa}$$

(b)   In part (b) the bottom of the tank is assumed to have infinite stiffness. As indicated in Fig. E10.6b, the bottom prevents both a radial displacement and a change in slope of the cylinder wall at B. Although the cylinder is not uniformly loaded, we consider it to be a uniformly loaded long cylinder with a ring load w applied at its center. The cylinder center is taken as the junction between the cylinder and bottom, and it is cut at this line load. Hence, the bottom of the tank produces a ring load $w/2$ on the upper half of the long cylinder and bending moment to prevent rotation of the cut section. The associated magnitudes of k and β are given by Eqs. (10.41) and (10.42). Thus, we find

$$k = \frac{Eh}{a^2} = \frac{(200 \times 10^3)(20)}{(15 \times 10^3)^2} = 0.0178 \text{ N/mm}^2$$

$$\beta = \sqrt[4]{\frac{3(1-v^2)}{h^2 a^2}} = \sqrt[4]{\frac{3[1-(0.29)^2]}{(20)^2(15 \times 10^3)^2}} = 0.00235 \text{ mm}^{-1}$$

Since the cylinder is subject to internal pressure due to the oil, it is not uniformly loaded as assumed in the proposed solution. For the analogy to be valid, the minimum uniformly loaded length [Eq. (10.43)] needs to be

$$L' = \frac{L}{2} = \frac{3\pi}{4\beta} = \frac{3\pi}{4(0.00235)} = 1003 \text{ mm}$$

which corresponds to the distance L' in Figs. E10.5a and b. Thus, only 10% of the cylinder height needs to be uniformly loaded; the variation of pressure over this height is considered small enough to be neglected. The radial displacement u of the walls of the cylinder away from end effects is given by Eq. (10.40)

$$u = \frac{\sigma_{\theta\theta}a}{E} = \frac{66.20(15 \times 10^3)}{200 \times 10^3} = 4.965 \text{ mm}$$

Since the radial displacement of the bottom plate of the tank is assumed to be zero, the ring load w causes a radial displacement inward of 4.965 mm. The magnitude of w is obtained by substituting the known value of u (equal to y) into Eq. (10.12) for $\beta z = 0$. Hence, by

$$u = \frac{w\beta}{2k} = \frac{w(0.00235)}{2(0.0178)} = 4.965 \text{ mm}$$

we find

$$w = 75.21 \text{ N/mm}$$

The maximum bending moment is given by Eq. (10.14) for $\beta z = 0$. Thus, we obtain

$$M_{max} = \frac{w}{4\beta} = \frac{75.21}{4(0.00235)} = 8001 \text{ N·mm}$$

and

$$\sigma_{zz(max)} = \frac{M_{max}c}{I} = \frac{M_{max}(h/2)}{h^3/12} = \frac{8001(6)}{(20)^2} = 120.0 \text{ MPa}$$

The radial stress σ_{rr} is small and is neglected. Since the radial displacement of the cylindrical wall at the bottom is the same as for the unloaded cylinder, the average value of $\sigma_{\theta\theta}$ through the wall thickness is zero. However, due to bending, the ratio of $\sigma_{\theta\theta}$ to σ_{zz} is proportional to Poisson's ratio [see Eq. (a) of Example 3.1]. Therefore,

$$\sigma_{\theta\theta(max)} = v\sigma_{zz(max)} = 0.29(120.0) = 34.8 \text{ MPa}$$

The maximum shear stress at the junction between the bottom of the tank and cylindrical walls of the tank is

$$\tau_{max} = \frac{\sigma_{max} - \sigma_{min}}{2} = \frac{120.0}{2} = 60.0 \text{ MPa}$$

which is 81% greater than for part (a).

The radial displacement u for the junction between the bottom of the tank and cylindrical walls of the tank has been neglected. However, its magnitude may be computed by the following relation:

$$u_{bottom} = \frac{w(1 - v)a}{2Eh} = \frac{75.21(0.71)(15 \times 10^3)}{2(200 \times 10^3)(20)} = 0.100 \text{ mm}$$

This value is only 2% of the displacement of the unrestrained cylinder wall.

(c) If the bending moment at the junction of the cylindrical walls and tank bottom is zero, the thin-wall cylinder can be treated as a beam on an elastic foundation loaded at one end. The bottom of the tank is assumed to prevent a radial displacement as indicated in Fig. E10.6c. Let w be the ring load produced by the bottom of the tank. The radial displacement u is given by Eq. (10.31) for $\beta z = 0$.

$$u = \frac{2w\beta}{k} = \frac{2w(0.00235)}{0.0178} = 4.965 \text{ mm}$$

$$w = 18.80 \text{ N/mm}$$

The maximum moment occurs at a distance $\pi/(4\beta) = 334$ mm from the bottom and has a magnitude given by Eq. (10.33)

$$M_{max} = -\frac{w}{\beta}B_{\beta z} = -\frac{18.80(0.3224)}{0.00235} = -2579 \text{ N} \cdot \text{mm}$$

$$\sigma_{zz(max)} = \left|\frac{M_{max}c}{I}\right| = \left|-\frac{2579(6)}{20^2}\right| = |-38.69| \text{ MPa}$$

This bending stress causes a circumferential stress $\sigma_{\theta\theta 1}$, which is part of the resultant circumferential stress.

$$\sigma_{\theta\theta 1} = v\sigma_{zz(max)} = 0.29(-38.69) = -11.22 \text{ MPa}$$

Another part of the circumferential stress $\sigma_{\theta\theta 2}$ comes from the fact that the maximum bending stress occurs at a location ($\beta z = \pi/4$) where the displacement is not maximum. The radial displacement given by Eq. (10.31) is

$$u = \frac{2w\beta}{k}D_{\beta z} = 0.3224u_{max}$$

Since $\sigma_{\theta\theta(max)} = 66.20$ MPa is the uniform circumferential stress in the thin-wall cylinder, when $u = 0$, the average circumferential stress for $u = 0.3224u_{max}$ is

$$\sigma_{\theta\theta2} = (1 - 0.3224)\sigma_{\theta\theta(max)} = 0.6776(66.20) = 44.86 \text{ MPa}$$

The circumferential stress at the point where $\sigma_{zz(min)}$ occurs is

$$\sigma_{\theta\theta} = \sigma_{\theta\theta1} + \sigma_{\theta\theta2} = -11.22 + 44.86 = 33.64 \text{ MPa}$$

and

$$\tau_{max} = \frac{\sigma_{max} - \sigma_{min}}{2} = \frac{33.64 - (-38.69)}{2} = 36.17 \text{ MPa}$$

which is 9% greater than for part (a).

If the maximum shear-stress criterion of failure is used, the maximum shear stress indicates the severity of the loading conditions. If the bottom of the tank is rigid (one limiting condition), the maximum shear stress is 81% greater than that for unrestrained cylindrical walls. If the bottom does not offer any resistance to bending (a second limiting condition), the shear stress is 9% greater than that for unrestrained cylindrical walls. The actual condition of loading for most flat bottom tanks would be between the two limiting conditions but nearer to the condition of a rigid bottom. Some experimental measurements of the stresses, in what is reportedly the world's largest welded steel water-storage tank, have been given by James and Raba (1991).

PROBLEMS
Section 10.2

10.1. The ballast and roadbed under railroad rails may vary appreciably from location to location. If the magnitude of k is 50% less than the value in Example 10.1, determine the percentage increase in the maximum deflection and maximum bending moment for the rail for the same wheel load.

10.2. A steel I-beam ($E = 200$ GPa) has a depth of 127 mm, width of 76 mm, moment of inertia of $I_x = 5.12 \times 10^6$ mm^4, and length of 4 m. It rests on a hard rubber foundation. The value of the spring constant for the hard rubber is $k_0 = 0.270$ N/mm^3. If the beam is subjected to a concentrated load, $P = 60.0$ kN, at the center of the beam, determine the maximum deflection and maximum flexure stress at the center of the beam.

Ans. $y_{max} = 2.187$ mm, $\sigma_{max} = 124.4$ MPa

10.3. Solve Problem 10.2 if the steel beam is replaced by an aluminum alloy beam for which $E = 72.0$ GPa.

10.4. An infinitely long beam rests on an elastic foundation and is loaded by two equal forces P spaced at a distant L. The beam has bending stiffness EI and the foundation has a spring constant k.

(a) Find the distance L such that the deflection y under one of the forces is the same as the deflection midway between the two forces.

(b) For unit values of P, EI, and k, and with the origin midway between the two forces, write an expression for the deflection y as a function of position z. Evaluate your expression for deflection at $z = \pm L/2$.

(c) Plot the expression derived in (b) over the domain $-4.0 \leq \beta z \leq 4.0$.

10.5. A steel train rail ($E = 200\,\mathrm{GPa}$) has moment of inertia $I = 36.9 \times 10^6\,\mathrm{mm}^4$ and rests on a subgrade with $k = 14.0\,\mathrm{N/mm}^2$. Find the maximum wheel spacing for the train such that the rail never lifts from the subgrade between any two sets of wheels.

10.6. A heavy machine has a mass of 60,000 kg. Its mass center is equidistant from each of four ground supports located at the four corners of a square 1.5 m on a side. Before it is moved to its permanent location, temporary support must be designed to hold the machine on a level horizontal surface on the ground. The surface layer of the ground is silt above a thick layer of inorganic clay. By the theory of soil mechanics, it is estimated that the spring constant of the soil is $k_0 = 0.029\,\mathrm{N/mm}^3$. The machine is placed centrally on two long timber beams ($E = 12.4$ GPa), 200 mm wide and 300 mm deep. The beams are parallel to one another, with centers 1.50 m apart. Determine the maximum deflection of the beams, maximum flexure stress in the beams, and minimum required length L for the beams.

Ans. $y_{\mathrm{max}} = 13.27\,\mathrm{mm}$, $\sigma_{\mathrm{max}} = 14.84\,\mathrm{MPa}$, $L > 8.10\,\mathrm{m}$

10.7. A 60-kN capacity hoist may be moved along a steel I-beam ($E = 200$ GPa). The I-beam has a depth of 152 mm and moment of inertia, $I_x = 11.0 \times 10^6\,\mathrm{mm}^4$. The beam is hung from a series of vertical steel rods ($E = 200\,\mathrm{GPa}$) of length 2.50 m, of diameter 18.0 mm, and spaced 500 mm center to center.

(a) For capacity load at the center of the beam, located under one of the rods, determine the maximum stress in the beam and the rods.

(b) Does l satisfy Eq. (10.19)?

10.8. After installation of the I-beam of Problem 10.7, it becomes necessary to lower the I-beam 800 mm. This was done by adding 18.0-mm diameter aluminum alloy bars ($E = 72$ GPa) of length 800 mm to the steel bars. For a 60-kN load at the center of the beam located under one of the composite bars, determine the maximum stress in the beam and the rods.

Ans. $\sigma_{\mathrm{max(beam)}} = 82.8\,\mathrm{MPa}$ $\sigma_{\mathrm{max(rod)}} = 73.7\,\mathrm{MPa}$

10.9. A long wood beam ($E = 12.4$ GPa) of depth 200 mm and width 60 mm is supported by 100 mm rubber cubes placed equidistant along the beam at $l = 600$ mm. The cube edges are parallel and perpendicular to the axis of the beam. The rubber has a spring constant of $k_0 = 0.330\,\mathrm{N/mm}^3$. A load P is applied to the center of the beam located over one of the rubber cubes.

(a) If the wood has a yield stress of $Y = 40.0$ MPa, determine the magnitude of P based on a factor of safety $SF = 2.50$. What is the maximum pressure developed between the rubber and beam?

(b) Does l satisfy Eq. (10.19)?

10.10. A long 50-mm diameter steel bar ($E = 200$ GPa and $Y = 300$ MPa) is supported by a number of pairs of 2-mm diameter high-strength steel wires ($E = 200$ GPa and $Y = 1200$ MPa). An end view of the beam and wires is shown in Fig. P10.10. The pairs of wires are equally spaced at $l = 900$ mm. A load P is applied to the center of the long beam at the same location as one pair of wires.

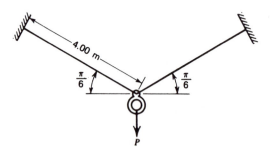

Figure P10.10

(a) Determine the magnitude of P if both the beam and wires are designed with factor of safety $SF = 2.00$.

(b) Does l satisfy Eq. (10.19)?

Ans.   **(a)** $P = 5.428$ kN,   **(b)** Yes

10.11. A long 40-mm diameter steel beam ($E = 200$ GPa) is supported by a number of semicircular curved beams. (See end view in Fig. P10.11.) The

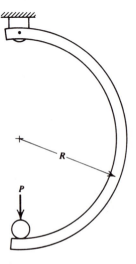

Figure P10.11

curved beams are spaced along the beam with spacing $l = 550$ mm. Each curved beam is made of steel, has a circular cross section of diameter 30 mm, and a radius of curvature $R = 300$ mm. A load $P = 3.00$ kN is applied to the center of the long beam located at one of the curved beams. Determine the maximum stress in the long beam and curved beams.

10.12. The beams in Fig. P10.12 are steel I-beams (203 mm deep, $I_x = 24.0 \times 10^6$ mm^4, $E = 200$ GPa). If a load $P = 90.0$ kN is applied to the center of the long beam located over one of the cross beams, determine the maximum flexure stess in the long beam and cross beams.

Ans. $\sigma_{max(long)} = 102.6$ MPa, $\sigma_{max(cross)} = 79.4$ MPa

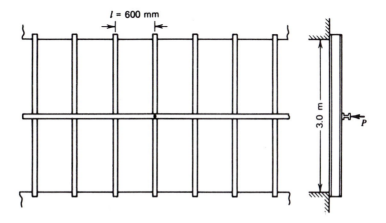

Figure P10.12

10.13. Let the curved beams in Problem 10.11 be made of an aluminum alloy ($E = 72.0$ GPa). Determine the maximum stress in the long beam and curved beams.

10.14. Let the long beam in Problem 10.12 be made of an aluminum alloy ($E = 72.0$ GPa). Determine the maximum flexure stress in the long beam and cross beams.

Ans. $\sigma_{max(cross)} = 79.4$ MPa, $\sigma_{max(long)} = 102.6$ MPa

10.15. For the beam on a linearly elastic foundation shown in Fig. 10.1, replace the concentrated load P by a concentrated (counterclockwise) moment M_0 at point 0. The beam has bending stiffness EI and the foundation has a spring constant k (force/area). Derive analytical expressions for the deflected shape $y(z)$, rotation $\theta(z)$, internal moment $M(z)$, and internal shear $V(z)$. Sketch each of the four expressions as is done in Fig. 10.1.

Section 10.3

10.16. Let the load of 60.0 kN in Problem 10.2 be uniformly distributed over a length of 1.00 m. Determine the maximum deflection and maximum flexure stress in the beam.

10.17. The long wood beam in Problem 10.9 is subjected to a distributed load w over length $L' = 3.00$ m. Determine the magnitude of w based on a factor of safety $SF = 2.00$.

Ans. $w = 123.2$ kN/m

10.18. Show that, for point H located outside segment L' (Fig. E10.3), the following equations are valid: $y_H = w(D_{\beta a} - D_{\beta b})/2k$ and $M_H = -w(B_{\beta a} - B_{\beta b})/(4\beta^2)$.

Section 10.4

10.19. Let load P be moved to one end of the beam in Problem 10.2. Determine the maximum deflection and maximum flexure stress in the beam and give the location of the maximum flexure stress.

10.20. Let the hoist in Problem 10.7 be moved to one end of the beam. Each rod supporting the I-beam is a spring exerting an influence over length l. If the end of the beam is $l/2 = 250$ mm from the nearest tension rod, determine the maximum stress in the rods and beam.

Ans. $\sigma_{max(rod)} = 149.3$ MPa, $\sigma_{max(beam)} = 60.7$ MPa

10.21. A long rectangular section brass beam ($E = 82.7$ GPa) has a depth of 20 mm, and a width of 15 mm and rests on a hard rubber foundation (Fig. P10.21). The value of the spring constant for the hard rubber foundation is 0.200 N/mm^3. If the beam is subjected to a concentrated load $P = 700$ N at the location shown, determine the maximum deflection of the beam and maximum flexure stress in the beam.

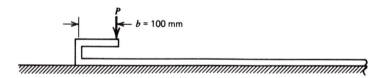

Figure P10.21

10.22. Solve Problem 10.21 for $b = 200$ mm.

Ans. $\sigma_{max} = 140$ MPa, $y_{max} = 0.833$ mm at $z = 159$ mm

10.23. A steel I-beam (depth $= 102$ mm, $I_x = 2.53 \times 10^6$ mm^4, $E = 200$ GPa) is long and supported by many springs ($K = 100$ N/mm) spaced at distance $l = 500$ mm center to center along the beam. A load $P = 3.50$ kN is applied to the left end of the beam at a distance of 2.00 m from the first spring. Determine the maximum flexure stress in the beam and maximum tension load and maximum compression load in the springs. *Hint:* $M_0 = -P(2000 - l/2)$.

10.24. Solve Problem 10.23 for the case where the steel beam is replaced by an aluminum alloy beam for which $E = 72.0$ GPa.

Ans. $\sigma_{max} = 141.1$ MPa, compression $= 4.23$ kN (1st spring), tension $= 720$ N (6th spring)

Section 10.5

10.25. Let the load $P = 60.0$ kN in Problem 10.2 be moved to one of the quarter points in the beam. Determine the maximum deflection and maximum flexure stress in the beam and locations for each.

10.26. Let the load $P = 60.0$ kN in Problem 10.2 be moved to a location 500 mm from one end of the beam. Determine the maximum deflection and maximum flexure stress in the beam and locations for each.

Ans. $y_{max} = 3.036$ mm at free end, $\sigma_{max} = 94.5$ MPa under load

10.27. Let the hoist in Problem 10.7 with a capacity load of 60 kN be located under the second rod from one end. Since each spring is assumed to exert an influence over a length $l = 500$ mm, the load acts at distance $a = 750$ mm from the end of the beam. Determine the maximum deflection of the beam, maximum flexure stress in the beam, maximum stress in the rods, and locations for each.

10.28. Let the hoist in Problem 10.7 with a capacity load of 60 kN be located under the first rod from one end. Since each spring is assumed to exert an influence over a length $l = 500$ mm, the load acts at distance $a = 250$ mm from the end of the beam. Determine the maximum deflection of the beam, maximum flexure stress in the beam, maximum stress in the rods, and locations for each.

Ans. $y_{max} = 2.80$ mm at free end; $\sigma_{max(beam)} = 39.2$ MPa at 880 mm from free end; $\sigma_{max(rod)} = 172.7$ MPa under load

10.29. A four-wheel car runs on steel rails ($E = 200$ GPa). The rails have a depth of 120 mm. The distance from the top of a rail to its centroid is 69 mm, and its moment of inertia is 17.07×10^6 mm^4. The rail rests on an elastic foundation with spring constant $k = 12.0$ N/mm^2. The two wheels on each side of the car are spaced 2.50 m center to center. If each wheel load is 80.0 kN, determine the maximum deflection and maximum flexure stress when a car wheel is located at one end of the rail and the other car wheel on the same rail is 2.50 m from the end.

Section 10.6

10.30. A steel I-beam ($E = 200$ GPa) has a length of $L = 3.00$ m, depth of 305 mm, flange width of 129 mm, and moment of inertia $I_x = 95.3 \times 10^6$ mm^4. The beam rests on a hard rubber elastic foundation whose spring constant is

$k_0 = 0.300$ N/mm^3. If the beam is subjected to a concentrated load $P = 270$ kN at its center, determine the maximum deflection and maximum flexure stress in the beam.

10.31. The magnitude of βL for the beam in Problem 10.30 is 2.532. Determine the maximum deflection and maximum flexure stress in the beam if the load is moved to one end of the beam. Use linear interpolation with the curves in Fig. 10.7.

Ans. $y_{max} = 12.37$ mm, $\sigma_{max} = 147.2$ MPa

Section 10.7

10.32. A steel ($E = 200$ GPa and $v = 0.29$) thin-wall cylinder has an inside diameter of 40 mm and a wall thickness of 1 mm. The cylinder may be considered fixed where it enters the stiffened end of a pressure vessel. The residual stress of installation may be considered negligible. Determine the bending stresses resulting from an internal pressure of 3 MPa.

10.33. A thin-wall cylinder is made of an aluminum alloy ($E = 72.0$ MPa and $v = 0.33$), has an outside diameter of 1 m, and a wall thickness of 5 mm. A split ring with square cross section 20 mm on a side is tightened on the cylinder until the stress in the split ring is 100 MPa. Assume that the split ring applies two line loads separated by the 20-mm dimension of the ring. Determine the principal stresses at the inner radius of the cylinder below the centerline of the split ring.

Ans. $\sigma_{zz} = 103.0$ MPa, $\sigma_{\theta\theta} = -62.1$ MPa

10.34. Let the split ring in Problem 10.33 be rounded on the inside surface so as to apply a line load at the center of the ring. Determine the maximum principal stresses at the inner radius of the cylinder.

10.35. A closed-end steel cylinder ($E = 200$ GPa and $v = 0.29$) has an inside radius $a = 2.00$ m, wall thickness $h = 10$ mm, and hemispherical ends. Since the state of stress is different for cylinder and hemisphere, their radial displacements will be different. Show that the length $L/2$ [see Eq. (10.43)] is small compared to a so that the short length of the hemisphere can be considered another cylinder. Determine the shear force w in terms of internal pressure p_1 at the junction of the cylinder and hemisphere (assumed to be another cylinder). Note that the bending moment at the junction is zero because of symmetry. Determine the maximum bending stress $\sigma_{zz(bending)}$ in the cylinder, axial stress σ_{zz}, and circumferential stress $\sigma_{\theta\theta}$ at the outside of the cylinder at the location where the maximum bending stress occurs, and the ratio of the maximum shear stress at that location to the maximum shear stress in the cylinder at a distance far from the junction.

Ans. $w = 13.73p_1$; $\sigma_{zz(bending)} = 29.17p_1$; $\sigma_{zz} = 129.2p_1$;
$\sigma_{\theta\theta} = 174.6p_1$;   ratio $= 0.874$

REFERENCES

Boresi, A. P. and Chong, K. P. (1987). *Elasticity in Engineering Mechanics*. New York: Elsevier.

Hetényi, M. (1946). *Beams on Elastic Foundations*. Ann Arbor: Univ. of Michigan Press.

Iyengar, K. T. S. R. and Ramu, S. A. (1979). *Design Tables for Beams on Elastic Foundations and Related Problems*. London: Elsevier Appl. Sci. Pub., Ltd.

James, R. W. and Raba, G. W. (1991). 'Behavior of Welded Steel Water-Storage Tank.' *J. Structural Eng.*, **117**, (1): 61–79.

Jones, R. and Xenophontos, J. (1977). 'The Vlasov Foundation Model.' *Int. J. Mech. Sci.*, **19**, (6): 317–324.

Kerr, A. D. (1964). 'Elastic and Viscoelastic Foundation Models.' *J. Appl. Mech., Trans. ASME*, **31**, Series E (3): 491–498.

Scott, R. F. (1981). *Foundation Analysis*. Englewood Cliffs, N. J. Prentice Hall.

Selvadurai, A. P. S. (1979). *Elastic Analysis of Soil Foundation Interaction*. London: Elsevier Appl. Sci. Publ., Ltd.

Ting, B.-Y. (1982). 'Finite Beams on Elastic Foundation with Restraints.' *J. Structural Div.*, **108**, (ST 3): 611–621.

Vallabhan, C. V. G. and Das, Y. C. (1987). 'A Note on Elastic Foundations.' *IDR Report*, Dept. of Civil Engr., Texas Tech. Univ., Lubbock, Texas.

Vallabhan, C. V. G. and Das, Y. C. (1988). 'Parametric Study of Beams on Elastic Foundations.' *ASCE J. Engr. Mech.*, **114**, (12): 2072–2082.

Vlasov, V. Z. and Leont'ev, N. N. (1966). *Beams, Plates and Shells on Elastic Foundations*. Washington, D. C.: NASA TT F-357 (translated from Russian).

Westergaard, H. M. (1948). 'New Formulas for Stresses in Concrete Pavements of Airfields.' *Trans. Amer. Soc. Civil Engineers*, **113**: 425–444.

11

THE THICK-WALL CYLINDER

11.1

BASIC RELATIONS

In this section, we derive basic relations for the axisymmetric deformation of a thick-wall cylinder. Thick-wall cylinders are used widely in industry as pressure vessels, pipes, gun tubes, etc. In many applications the cylinder wall thickness is constant, and the cylinder is subjected to a uniform internal pressure p_1, a uniform external pressure p_2, an axial load P, and a temperature change ΔT (measured from an initial uniform reference temperature; see Sec. 3.4) (Fig. 11.1). Often, the temperature change ΔT is a function of the radial coordinate r only (Fig. 11.1). Under such conditions, the deformations of the cylinder are symmetrical with respect to the axis of the cylinder (axisymmetric). Furthermore, the deformations at a cross section sufficiently far removed from the junction of the cylinder and its end caps (Fig. 11.1) are practically independent of the axial coordinate z. In particular, if the cylinder is open (no end caps) and unconstrained, it undergoes axisymmetric deformations due to pressures p_1 and p_2 and temperature change $\Delta T = \Delta T(r)$, which are independent of z. If the cylinder's deformation is constrained by supports or end caps, then in the vicinity of the supports or junction between the cylinder and end caps, the deformation and stresses will depend on the axial coordinate z. For example, consider a pressure tank formed by welding together hemispherical caps and a cylinder (Fig. 11.2). Under the action of an internal pressure p_1, the tank deforms as indicated by the dotted inside boundary and the long dashed outside boundary (the deformations are exaggerated in Fig. 11.2). If the cylinder were not constrained by the end caps, it would be able to undergo a larger radial displacement. However, at the junctions between the hemispherical caps and cylinder, the cylinder displacement is constrained by the stiff hemispherical caps. Consequently, the radial displacement (hence, the strains and stresses) at cylinder cross sections near the end cap junctions differs from those at sections far removed from the end cap junctions. In this section, we consider the displacement, strains, and stresses at locations far removed from the end caps. The determination of deformations, strains, and stresses near the junction of the thick-wall end caps and the thick-wall cylinder lies outside the scope of our treatment. This problem often is treated by experimental methods, since its analytical solution depends on a general three-dimensional study in the theory of elasticity (or plasticity). For thin-wall cylinders, the stress near the end cap junctions may be estimated by the procedure outlined in Sec. 10.7 (see Problem 10.35).

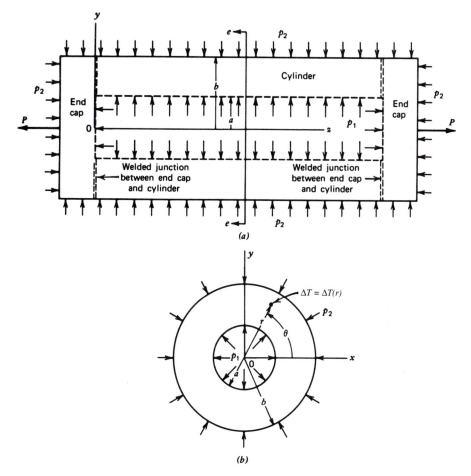

(a)

(b)

Figure 11.1 Closed cylinder with internal pressure, external pressure and axial loads. (a) Closed cylinder. (b) Section $e - e$.

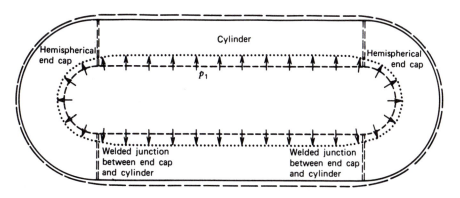

Figure 11.2 Closed cylinder with hemispherical ends.

Consequently, the solution presented in this chapter for thick-wall cylinders is applicable to locations sufficiently far from the end cap junctions where the effects of the constraints imposed by the end caps are negligible. The solution is also applicable to thick-wall cylinders that do not have end caps, so-called open cylinders. Since only axially symmetrical loads and constraints are admitted, the solution is axisymmetrical, that is, a function only of radial coordinate r.

We use cylindrical coordinates r, θ, z for radial, circumferential, and axial directions (Fig. 11.1). Let the cylinder be loaded as shown in Fig. 11.1. For analysis purposes, we remove a thin annulus of thickness dz from the cylinder (far removed from the end junctions) by passing two planes perpendicular to the z axis, a distance dz apart (Fig. 11.3a). The cylindrical volume element $dr(r\,d\theta)\,dz$ shown in Fig. 11.3b is removed from the annulus. Because of radial symmetry, no shear stresses act on the volume element and normal stresses are functions of r only. The nonzero stress components are principal stresses σ_{rr}, $\sigma_{\theta\theta}$, and σ_{zz}. The distributions of these stresses through the wall thickness are determined by the equations of equilibrium, compatibility relations, stress-strain-temperature relations, and material response data.

Equation of Equilibrium

We neglect body force components. Hence, the equations of equilibrium for cylindrical coordinates [Eqs. (2.50)] reduce to the single equation

$$r\frac{d\sigma_{rr}}{dr} = \sigma_{\theta\theta} - \sigma_{rr} \quad \text{or} \quad \frac{d}{dr}(r\sigma_{rr}) = \sigma_{\theta\theta} \tag{11.1}$$

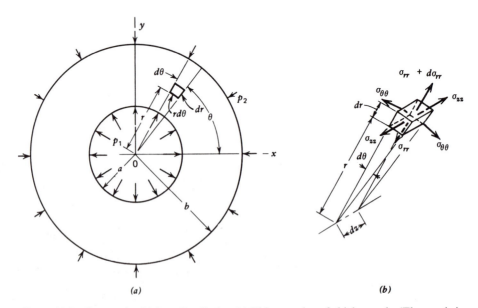

(a) (b)

Figure 11.3   Stresses in thick-wall cylinder. (a) Thin annulus of thickness dz. (The z axis is perpendicular to the plane of the figure.) (b) Cylindrical volume element of thickness dz.

Strain-Displacement Relations and Compatibility Condition

The strain-displacement relations for the thick-walled cylinder [Eqs. (2.85)] yield the three relations for extensional strains (since $v = 0$)

$$\epsilon_{rr} = \frac{\partial u}{\partial r}, \qquad \epsilon_{\theta\theta} = \frac{u}{r}, \qquad \epsilon_{zz} = \frac{\partial w}{\partial z} \tag{11.2}$$

where $u = u(r, z)$, $w = w(r, z)$ denote displacement components in the r and z directions, respectively. At sections far removed from the ends, the dependency on z in u and w is considered to be small. Hence, at sections far from the ends, the shear strain components are zero because of radial symmetry; furthermore, we assume that ϵ_{zz} is constant. Eliminating the displacement $u = u(r)$ from the first two of Eqs. (11.2), we obtain

$$r \frac{d\epsilon_{\theta\theta}}{dr} = \epsilon_{rr} - \epsilon_{\theta\theta}, \qquad \text{or} \qquad \frac{d}{dr}(r\epsilon_{\theta\theta}) = \epsilon_{rr} \tag{11.3}$$

Equation (11.3) is the strain compatibility condition for the thick-wall cylinder.

Stress-Strain-Temperature Relations

The material of the cylinder is taken to be isotropic and linearly elastic. Since, the stress-strain-temperature relations are [see Eqs. (3.38)]

$$\epsilon_{rr} = \frac{1}{E}[\sigma_{rr} - v(\sigma_{\theta\theta} + \sigma_{zz})] + \alpha \Delta T$$

$$\epsilon_{\theta\theta} = \frac{1}{E}[\sigma_{\theta\theta} - v(\sigma_{rr} + \sigma_{zz})] + \alpha \Delta T$$

$$\epsilon_{zz} = \frac{1}{E}[\sigma_{zz} - v(\sigma_{rr} + \sigma_{\theta\theta})] + \alpha \Delta T = \text{constant} \tag{11.4}$$

where E, v, and α denote the modulus of elasticity, Poisson's ratio, and the coefficient of linear thermal expansion, respectively. The term ΔT in Eq. (11.4) represents the change in temperature measured from a uniform reference temperature (constant throughout the cylinder initially); see Boresi and Chong (1987).

Material Response Data

For a cylinder made of isotropic linearly elastic material, the material response data are represented by the results of tests required to determine the elastic constants (modulus of elasticity E and Poisson's ratio v) and the coefficient of linear thermal expansion α. In order to determine the maximum elastic loads for the cylinder, the material data must include either the yield stress Y obtained from a tension test, or the shear yield stress τ_Y obtained from a torsion test of a hollow thin-wall tube. If the material response indicates that the material has a yield point (Fig. 1.5b), the value of either the yield point stress Y or shear yield point stress τ_Y is needed to calculate the fully plastic pressure for the cylinder.

11.2

STRESS COMPONENTS FOR A CYLINDER WITH CLOSED ENDS

In this section, we obtain expressions for the stress components σ_{rr}, $\sigma_{\theta\theta}$, σ_{zz}, for a cylinder with closed ends; the cylinder is subjected to internal pressure p_1, external pressure p_2, axial load P, and temperature change ΔT (Fig. 11.1).

We may express Eq. (11.3) in terms of σ_{rr}, $\sigma_{\theta\theta}$, σ_{zz} and their derivatives with respect to r, by substitution of the first two of Eqs. (11.4) into Eq. (11.3). Since ϵ_{zz} = constant, the last of Eqs. (11.4) may be used to express the derivative $d\sigma_{zz}/dr$ in terms of the derivatives of σ_{rr}, $\sigma_{\theta\theta}$, and ΔT with respect to r. By means of this expression, we may eliminate $d\sigma_{zz}/dr$ from Eq. (11.3) to rewrite Eq. (11.3) in terms of σ_{rr}, $\sigma_{\theta\theta}$, and derivatives of σ_{rr}, $\sigma_{\theta\theta}$, and ΔT. Since the undifferentiated terms in σ_{rr} and $\sigma_{\theta\theta}$ occur in the form $\sigma_{rr} - \sigma_{\theta\theta}$, Eq. (11.1) may be used to eliminate $\sigma_{rr} - \sigma_{\theta\theta}$. Hence, we obtain the differential expression

$$\frac{d}{dr}\left(\sigma_{rr} + \sigma_{\theta\theta} + \frac{\alpha E \, \Delta T}{1 - v}\right) = 0 \tag{11.5}$$

Incorporated in Eq. (11.5) is the equation of equilibrium, Eq. (11.1), the strain compatibility equation, Eq. (11.3), and the stress-strain-temperature relations, Eqs. (11.4).

Integration of Eq. (11.5) yields the result

$$\sigma_{rr} + \sigma_{\theta\theta} + \frac{\alpha E \, \Delta T}{1 - v} = 2C_1 \tag{11.6}$$

where $2C_1$ is a constant of integration (the factor 2 is included for simplicity of form in subsequent expressions). Elimination of the stress component $\sigma_{\theta\theta}$ between Eqs. (11.1) and (11.6) yields the following different expression for σ_{rr}:

$$\frac{d}{dr}(r^2\sigma_{rr}) = -\frac{\alpha E \, \Delta Tr}{1 - v} + 2C_1 r \tag{11.7}$$

Integration of Eq. (11.7) yields the result

$$\sigma_{rr} = -\frac{\alpha E}{r^2(1 - v)}\int_a^r \Delta Tr \, dr + \left(1 - \frac{a^2}{r^2}\right)C_1 + \frac{C_2}{r^2} \tag{11.8}$$

where the integration is carried out from the inner radius a of the cylinder (Fig. 11.1) to the radius r, and C_2 is a second constant of integration. Substitution of Eq. (11.8) into Eq. (11.6) yields the result

$$\sigma_{\theta\theta} = \frac{\alpha E}{r^2(1 - v)}\int_a^r \Delta Tr \, dr - \frac{\alpha E \, \Delta T}{1 - v} + \left(1 + \frac{a^2}{r^2}\right)C_1 - \frac{C_2}{r^2} \tag{11.9}$$

By Eqs. (11.8) and (11.9), we obtain

$$\sigma_{rr} + \sigma_{\theta\theta} = 2C_1 - \frac{\alpha E \, \Delta T}{1 - v} \tag{11.10}$$

Equation (11.10) serves as a check on the computations [see Eq. (11.6)]. The constants of integration C_1 and C_2 are obtained from the boundary conditions $\sigma_{rr} = -p_1$ at $r = a$ and $\sigma_{rr} = -p_2$ at $r = b$ (Fig. 11.1). Substituting these boundary conditions into Eq. (11.8), we find

$$C_1 = \frac{1}{b^2 - a^2}\left(p_1 a^2 - p_2 b^2 + \frac{\alpha E}{1 - v}\int_a^b \Delta T r\, dr\right), \qquad C_2 = -p_1 a^2 \quad (11.11)$$

Hence, Eq. (11.10) may be written as

$$\frac{\sigma_{rr} + \sigma_{\theta\theta}}{2} = \frac{p_1 a^2 - p_2 b^2}{b^2 - a^2} - \frac{\alpha E\, \Delta T}{2(1 - v)} + \frac{E}{(1 - v)(b^2 - a^2)}\int_a^b \Delta T r\, dr \quad (11.12)$$

To obtain σ_{zz}, we integrate each term of the last of Eqs. (11.4) over the cross-sectional area of the cylinder. Thus, we have

$$\int_a^b \epsilon_{zz} 2\pi r\, dr = \frac{1}{E}\int_a^b \sigma_{zz} 2\pi r\, dr - \frac{2v}{E}\int_a^b \frac{\sigma_{\theta\theta} + \sigma_{rr}}{2} 2\pi r\, dr + \alpha\int_a^b \Delta T\, 2\pi r\, dr \quad (11.13)$$

For sections far removed from the end section, ϵ_{zz} is a constant, and the integral of σ_{zz} over the cross-sectional area is equal to the applied loads. Hence, because of pressures p_1, p_2, and axial load P applied to an end plate (Fig. 11.4), overall equilibrium in the axial direction requires

$$\int_a^b \sigma_{zz} 2\pi r\, dr = P + \pi(p_1 a^2 - p_2 b^2) \quad (11.14)$$

If there is no axial load P applied to the closed ends, $P = 0$.

Since the temperature change ΔT does not appear in Eq. (11.14), the effects of temperature are self-equilibrating. With Eqs. (11.12), (11.13), and (11.14), the

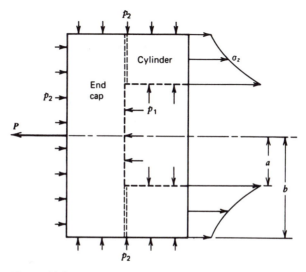

Figure 11.4

expression for ϵ_{zz} at a section far removed from the ends can be written in the form

$$\epsilon_{zz(\text{closed end})} = \frac{1 - 2v}{E(b^2 - a^2)}(p_1a^2 - p_2b^2) + \frac{P}{\pi(b^2 - a^2)E} + \frac{2\alpha}{b^2 - a^2}\int_a^b \Delta Tr\,dr$$

(11.15)

Substitution of Eq. (11.15) into the last of Eqs. (11.4), with Eq. (11.12), yields the following expression for σ_{zz} for a section far removed from the closed ends of the cylinder:

$$\sigma_{zz(\text{closed end})} = \frac{p_1a^2 - p_2b^2}{b^2 - a^2} + \frac{P}{\pi(b^2 - a^2)} - \frac{\alpha E \Delta T}{1 - v} + \frac{2\alpha E}{(1 - v)(b^2 - a^2)}\int_a^b \Delta Tr\,dr$$

(11.16)

Open Cylinder

If a cylinder has open ends and there is no axial load applied on its ends, overall equilibrium of an axial portion of the cylinder (Fig. 11.5) requires that

$$\int_a^b 2\pi r\sigma_{zz}\,dz = 0$$

(11.17)

Then by Eqs. (11.12), (11.13), and (11.17), the expression for ϵ_{zz} may be written in the form [also by Eqs. (11.14, 11.15, and 11.17)]

$$\epsilon_{zz(\text{open end})} = \frac{2v(p_2b^2 - p_1a^2)}{(b^2 - a^2)E} + \frac{2\alpha}{b^2 - a^2}\int_a^b \Delta Tr\,dr$$

(11.18)

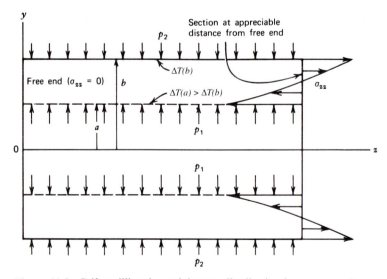

Figure 11.5   Self-equilibrating axial stress distribution in an open cylinder.

and for σ_{zz}, we obtain by Eqs. (11.4), (11.12), and (11.18),

$$\sigma_{zz(\text{open end})} = \frac{\alpha E}{1 - v}\left(\frac{2}{b^2 - a^2}\int_a^b \Delta Tr\,dr - \Delta T\right) \qquad (11.19)$$

We note, by Eq. (11.19), that if the temperature change $\Delta T = 0$, $\sigma_{zz} = 0$. However, $\epsilon_{zz} \neq 0$ [see Eq. (11.18)] when the Poisson ratio $v \neq 0$. Note that if $p_1 = p_2 = P = 0$ (temperature change still occurs), Eqs. (11.15) and (11.16) are identical to Eqs. (11.18) and (11.19), respectively.

11.3

STRESS COMPONENTS AND RADIAL DISPLACEMENT FOR CONSTANT TEMPERATURE

Stress Components
In the absence of temperature change, we set $\Delta T = 0$. Then Eqs. (11.8), (11.9), (11.10), (11.11), and (11.16) may be used to obtain the following expressions for the stress components in a closed cylinder (cylinder with end caps)

$$\sigma_{rr} = \frac{p_1 a^2 - p_2 b^2}{b^2 - a^2} - \frac{a^2 b^2}{r^2(b^2 - a^2)}(p_1 - p_2) \qquad (11.20)$$

$$\sigma_{\theta\theta} = \frac{p_1 a^2 - p_2 b^2}{b^2 - a^2} + \frac{a^2 b^2}{r^2(b^2 - a^2)}(p_1 - p_2) \qquad (11.21)$$

$$\sigma_{zz} = \frac{p_1 a^2 - p_2 b^2}{b^2 - a^2} + \frac{P}{\pi(b^2 - a^2)} = \text{constant} \qquad (11.22)$$

$$\sigma_{rr} + \sigma_{\theta\theta} = \frac{2(p_1 a^2 - p_2 b^2)}{b^2 - a^2} = \text{constant} \qquad (11.23)$$

For an open cylinder in the absence of axial force P, $\sigma_{zz} = 0$ by Eq. (11.19) with $\Delta T = 0$. Since the sum $\sigma_{rr} + \sigma_{\theta\theta}$ and stress σ_{zz} are constants through the thickness of the wall of the closed cylinder, by Eq. (11.13) or Eq. (11.15), we see that ϵ_{zz} is constant (extension or compression).

Radial Displacement for Closed Cylinder
For no temperature change, $\Delta T = 0$. Then the radial displacement u for a point in a thick-wall closed cylinder (cylinder with end caps) may be obtained by the second of Eqs. (11.2), the second of Eqs. (11.4), and Eqs. (11.20), (11.21), and (11.22). The resulting expression for u is

$$u_{(\text{closed end})} = \frac{r}{E(b^2 - a^2)}\left[(1 - 2v)(p_1 a^2 - p_2 b^2)\right.$$

$$\left. + \frac{(1 + v)a^2 b^2}{r^2}(p_1 - p_2) - v\frac{P}{\pi}\right] \qquad (11.24)$$

Radial Displacement for Open Cylinder

Of special interest are open cylinders (cylinder without end caps), since an open inner cylinder is often shrunk to fit inside an open outer cylinder to increase the strength of the resulting composite cylinder. For an open cylinder, in the absence of temperature changes ($\Delta T = 0$), Eq. (11.19) yields $\sigma_{zz} = 0$. Hence, proceeding as for the closed cylinder, we obtain

$$u_{(\text{open end})} = \frac{r}{E(b^2 - a^2)}\left[(1-v)(p_1a^2 - p_2b^2) \right.$$
$$\left. + \frac{(1+v)a^2b^2}{r^2}(p_1 - p_2)\right] \qquad (11.25)$$

EXAMPLE 11.1
Stresses in Hollow Cylinder

A thick-wall cylinder is made of steel ($E = 200$ GPa and $v = 0.29$), has an inside diameter of 20 mm, and outside diameter of 100 mm. The cylinder is subjected to an internal pressure of 300 MPa. Determine the stress components σ_{rr} and $\sigma_{\theta\theta}$ at $r = a = 10$ mm, $r = 25$ mm, and $r = b = 50$ mm.

SOLUTION

The external pressure $p_2 = 0$. Equations (11.20) and (11.21) simplify to

$$\sigma_{rr} = p_1\frac{a^2(r^2 - b^2)}{r^2(b^2 - a^2)}$$
$$\sigma_{\theta\theta} = p_1\frac{a^2(r^2 + b^2)}{r^2(b^2 - a^2)}$$

Substitution of values for r equal to 10 mm, 25 mm, and 50 mm, respectively, into these equations yields the following results:

Stress	$r = 10$ mm	$r = 25$ mm	$r = 50$ mm
σ_{rr}	− 300.0 MPa	− 37.5 MPa	0.0
$\sigma_{\theta\theta}$	325.0 MPa	62.5 MPa	25.0 MPa

EXAMPLE 11.2
Stresses and Deformations in Hollow Cylinder

A thick-wall closed-end cylinder is made of an aluminum alloy ($E = 72$ GPa and $v = 0.33$), has an inside diameter of 200 mm, and an outside diameter of 800 mm. The cylinder is subjected to an internal pressure of 150 MPa. Determine the principal stresses, maximum shear stress at the inner radius ($r = a = 100$ mm), and the increase in the inside diameter due to the internal pressure.

SOLUTION

The principal stresses are given by Eqs. (11.20), (11.21), and (11.22). For the conditions that $p_2 = 0$ and $r = a$, these equations give

$$\sigma_{rr} = p_1 \frac{a^2 - b^2}{b^2 - a^2} = -p_1 = -150 \text{ MPa}$$

$$\sigma_{\theta\theta} = p_1 \frac{a^2 + b^2}{b^2 - a^2} = 150 \frac{100^2 + 400^2}{400^2 - 100^2} = 170 \text{ MPa}$$

$$\sigma_{zz} = p_1 \frac{a^2}{b^2 - a^2} = 150 \frac{100^2}{400^2 - 100^2} = 10 \text{ MPa}$$

The maximum shear stress is given by Eq. (2.39).

$$\tau_{max} = \frac{\sigma_{max} - \sigma_{min}}{2} = \frac{170 - (-150)}{2} = 160 \text{ MPa}$$

The increase in the inside diameter due to the internal pressure is equal to twice the radial displacement given by Eq. (11.24) for the conditions $p_2 = P = 0$ and $r = a$.

$$u_{(r=a)} = \frac{p_1 a}{E(b^2 - a^2)} [(1 - 2v)a^2 + (1 + v)b^2]$$

$$= \frac{150(100)}{72,000(400^2 - 100^2)} [(1 - 0.66)100^2 + (1 + 0.33)400^2]$$

$$= 0.3003 \text{ mm}$$

The increase in the inside diameter due to the internal pressure is 0.6006 mm.

EXAMPLE 11.3
Stresses in a Composite Cylinder

Let the cylinder in Example 11.1 be a composite cylinder made by shrinking an outer cylinder on an inner cylinder. Before assembly, the inner cylinder has inner and outer radii of $a = 10$ mm and $c_i = 25.072$ mm, respectively. Likewise, the outer cylinder has inner and outer radii of $c_0 = 25.000$ mm and $b = 50$ mm, respectively. Determine the stress components σ_{rr} and $\sigma_{\theta\theta}$ at $r = a = 10$ mm, $r = 25$ mm, and $r = b = 50$ mm for the composite cylinder. For assembly purposes, the inner cylinder is cooled to a uniform temperature T_1 and the outer cylinder is heated to a uniform temperature T_2 in order for the outer cylinder to slide freely over the inner cylinder. It is assumed that the two cylinders will slide freely if we allow an additional 0.025 mm to the required minimum difference in radii of 0.072 mm. Determine how much the temperature (in degrees Celsius) must be raised in the outer cylinder above the temperature in the inner cylinder in order to freely assemble the two cylinders. $\alpha = 0.0000117$ per °C.

SOLUTION

After the composite cylinder has been assembled, the change in stresses due to the internal pressure $p_1 = 300$ MPa is the same as for the cylinder in Example 11.1. These stresses are added to the residual stresses in the composite cylinder caused by shrinking the outer cylinder on the inner cylinder.

The initial difference between the outer radius of the inner cylinder and the inner radius of the outer cylinder is 0.072 mm. After the two cylinders have been assembled and allowed to cool to their initial uniform temperature, a shrink pressure p_s is developed between the two cylinders. The pressure p_s is an external pressure for the inner cylinder and an internal pressure for the outer cylinder. The magnitude of p_s is obtained from the fact that the sum of the radial displacement of the inner surface of the outer cylinder and the radial displacement of the outer surface of the inner cylinder must equal 0.072 mm. Hence, by Eq. (11.25),

$$\frac{c_0}{E(b^2 - c_0^2)}[(1 - v)p_s c_0^2 + (1 + v)p_s b^2]$$

$$- \frac{c_i}{E(c_i^2 - a^2)}[-(1 - v)p_s c_i^2 - (1 + v)p_s a^2] = 0.072$$

Solving for p_s, we obtain

$$p_s = 189.1 \text{ MPa}$$

The pressure p_s produces stresses (so-called residual stresses) in the nonpressurized composite cylinder. For the inner and outer cylinders, the residual stresses σ_{rr}^R and $\sigma_{\theta\theta}^R$ at the inner and outer radii are given by Eqs. (11.20) and (11.21). For the inner cylinder, $p_1 = 0$, $p_2 = p_s$, $a = 10$ mm, and $b = 25$ mm. For the outer cylinder $p_1 = p_s$, $p_2 = 0$, $a = 25$ mm, and $b = 50$ mm. The residual stresses are found to be

	Inner Cylinder		Outer Cylinder	
Residual Stress	$r = 10$ mm	$r = 25$ mm	$r = 25$ mm	$r = 50$ mm
σ_{rr}^R	0	-189.1 MPa	-189.1 MPa	0
$\sigma_{\theta\theta}^R$	-450.2 MPa	-261.1 MPa	315.1 MPa	126.0 MPa

The stresses in the composite cylinder after an internal pressure of 300 MPa has been applied are obtained by adding these residual stresses to the stresses calculated in Example 11.1. Thus, we find

	Inner Cylinder		Outer Cylinder	
Stress	$r = 10$ mm	$r = 25$ mm	$r = 25$ mm	$r = 50$ mm
σ_{rr}	-300.0 MPa	-226.6 MPa	-226.6 MPa	0
$\sigma_{\theta\theta}$	-125.2 MPa	-198.6 MPa	377.7 MPa	151.0 MPa

A comparison of these stresses or the composite cylinder with those for the solid cylinder in Example 11.1 indicates that the stresses have been changed greatly. The determination of possible improvements in the design of the open end cylinder necessitates consideration of particular criteria of failure (see Sec. 11.4).

In order to have the inner cylinder slide easily into the outer cylinder during assembly, the difference in temperature between the two cylinders is given by the relation

$$\Delta T = T_2 - T_1 = \frac{u}{r\alpha} = \frac{0.072 + 0.025}{r\alpha} = \frac{0.097}{25(0.0000117)} = 331.6°C$$

since for uniform temperatures T_1, T_2, we have $\sigma_{rr} = \sigma_{\theta\theta} = \sigma_{zz} = 0$ in each cylinder, and since then Eqs. (11.2) and (11.4) yield $\epsilon_{\theta\theta} = u/r = \alpha\Delta T$, where $r = c_0 = c_i$.

11.4

CRITERIA OF FAILURE

The criterion of failure used in the design of a thick-wall cylinder depends on the type of material in the cylinder. As discussed in Sec. 4.3, the maximum principal stress criterion should be used in the design of members made of brittle isotropic materials if the principal stress of largest magnitude is a tensile stress. Either the maximum shear-stress or maximum octahedral shear-stress criterion of failure should be used in the design of members made of ductile isotropic materials (see Sec. 4.4).

Failure of Brittle Materials

If a thick-wall cylinder is made of a brittle material, the material property associated with fracture is the ultimate tensile strength σ_u. At the failure loads, the maximum principal stress in the cylinder is equal to σ_u. If the maximum principal stress occurs at the constrained ends of the cylinder, it cannot be computed using the relations derived in Sec. 11.2 and 11.3. At sections far removed from the ends, the maximum principal stress is either the circumferential stress $\sigma_{\theta\theta(r=a)}$ or axial stress σ_{zz}. If the cylinder is loaded so that the magnitude of the maximum compressive principal stress is appreciably larger than the magnitude of the maximum tensile principal stress, the appropriate criterion of failure to be used in design is uncertain. Such conditions are not considered in this book.

Failure of Ductile Materials

If excessive elastic deformation is not a design factor, failure of members made of ductile materials may be initiated as the result of inelastic deformation or fatigue (only high cycle fatigue is considered in this book; see Chapter 16). Failure of these members is predicted by either the maximum shear-stress criterion of failure or the maximum octahedral shear-stress criterion of failure. The failure of the member may be either a general yielding failure or a fatigue failure at a large number of stress cycles.

General Yielding Failure Thick-wall cylinders, which are subjected to static loads or peak loads only a few times during the life of the cylinder, are usually designed for the general yielding limit state. General yielding may be defined to occur when yielding is initiated in the member at some point other than at a stress concentration. This definition is used in examples at the end of this section (see also Sec. 4.6). However, yielding may be initiated in the region of stress concentrations at the ends of the cylinder or at an opening for pipe connections. Yielding in such regions in usually highly localized and subsequent general yielding is unlikely. However, the possibility of failure by fatigue still may exist (see Chapter 16). General yielding sometimes is considered to occur only after the member has yielded over an extensive region, such as occurs with fully plastic loads. Fully plastic loads for thick-wall cylinders are discussed in Sec. 11.5.

Fatigue Failure In practice, a thick-wall cylinder may be subjected to repeated pressurizations (loading and unloading) that may lead to fatigue failure. Since fatigue cracks often occur in the neighborhood of stress concentrations, every region of stress concentration must be considered in the design. In particular, the maximum shear stress must be determined in the region of stress concentrations, since fatigue cracking usually originates at a point where either the maximum shear stress or maximum octahedral shear stress occurs. The equations derived in Sec. 11.2 and 11.3 cannot be used to compute the design stresses, unless the maximum stresses occur at sections of the cylinder far removed from end constraints or other stress concentration regions.

Material Response Data for Design

If a member fails by general yielding, the material property associated with failure is the yield stress. This fact places a limit either on the value of the maximum shear stress, if the maximum shear-stress criterion of failure is used, or on the value of the maximum octahedral shear stress, if the maximum octahedral shear-stress criterion of failure is used. If the member fails by fatigue, the material property associated with the failure is the fatigue strength. For high cycle fatigue, both the maximum shear-stress criterion of failure and maximum octahedral shear-stress criterion of failure are used widely in conjunction with the fatigue strength (see Chapter 16, Example 16.1). The yield stress and fatigue strength may be obtained by tests of either a tension specimen or hollow thin-wall tube. It has been found that the values of these properties, as determined from tests of a hollow thin-wall tube in torsion, lead to a more accurate prediction of the material response for thick-wall cylinders than the values obtained from a tension specimen. This result is because the critical state of stress in the cylinder is usually at the inner wall of the cylinder, and for the usual pressure loading it is essentially one of pure shear (as occurs in the torsion test) plus a hydrostatic state of stress. Since in many materials a hydrostatic stress does not affect the yielding, the material responds (yields) as if it were subjected to a state of pure shear. Consequently, if the material properties are determined by means of a torsion test of a hollow thin-wall tube, the maximum shear-stress criterion and maximum octahedral shear-stress criterion predict failure loads that differ by less than 1% for either closed or open cylinders. The difference in these predictions may be as much as 15.5% if the material properties are obtained from tension specimen tests (Sec. 4.4). These conclusions pertain in general to most metals. However, the yield of most plastics is influenced by the hydrostatic state of stress. Hence, for most plastics, these conclusions may not generally hold.

The deviatoric state of stress (see Sec. 2.4) in a closed cylinder is identical to that for pure shear. Hence, the maximum shear-stress and octahedral shear-stress criteria of failure predict nearly identical factors of safety for the design of a closed cylinder if the yield stress for the material is obtained from torsion tests of hollow thin-wall tubes. Let the shear yield stress obtained from a torsion test of a thin-wall hollow tube specimen be designated as τ_Y. If the maximum shear stress for the inner radius of a closed cylinder is set equal to τ_Y, the pressure p_Y required to initiate yielding is obtained. (The reader is asked to derive the formula for p_Y in Problem 11.11.) For the special case of a closed cylinder with internal pressure only and with dimensions $b = 2a$, the yield pressure is found to be $p_Y = 0.75\,\tau_Y$; the corresponding dimensionless stress distribution is shown in Fig. 11.6.

Ideal Residual Stress Distributions for Composite Open Cylinders

It is possible to increase the strength of a thick-wall cylinder by introducing beneficial residual stress distributions. The introduction of beneficial residual stresses can be accomplished in several ways. In particular, there are two common ways of producing residual stresses in cylinders. One method consists of forming a composite cylinder from two or more open cylinders. For example, in the case of two cylinders, the inner cylinder has an outer radius that is slightly larger than the inner radius of the outer cylinder. The inner cylinder is slipped inside the outer cylinder after first heating the outer cylinder and/or cooling the inner cylinder. When the

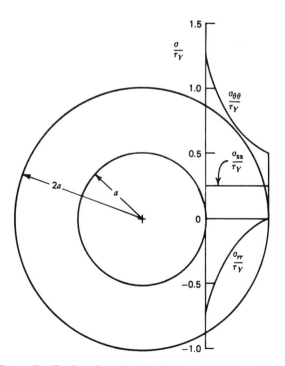

Figure 11.6   Stress distributions in a closed cylinder at initiation of yielding ($b = 2a$).

cylinders are allowed to return to their initially equal uniform temperatures (say, room temperature), a pressure (the so-called shrink pressure) is created between the cylinder surfaces in contact. This pressure introduces residual stresses in the cylinders. As a result, the strength of the composite cylinder under additional internal and external pressure loading is increased (Example 11.5). For more than two cylinders this process is repeated for each cylinder that is added to form the composite cylinder.

A second method consists of pressurizing a single cylinder until it deforms inelastically to some distance into the wall from the inner surface (the process is called autofrettage). When the pressure is removed, a beneficial residual stress distribution remains in the cylinder (see Sec. 11.5).

For a composite cylinder formed by two cylinders under a shrink fit and subject to internal pressure p_1, the most beneficial residual stress distribution is that which results in the composite cylinder failing (yielding or fracturing) simultaneously at the inner radii of the inner and outer cylinders. Consider, for example, a composite cylinder formed by inner and outer cylinders made of a *brittle* material whose stress-strain diagram remains linear up to its ultimate strength σ_u. The inner cylinder has inner radius r_1 and outer radius $1.5r_1 +$ (i.e., the outer radius is slightly larger than $1.5r_1$). The outer cylinder has an inner radius of $1.5r_1$ and outer radius of $3r_1$. See Fig. 11.7. Fracture of the brittle material occurs when the maximum principal stress reaches the ultimate strength σ_u. Since the maximum principal stress in the composite cylinder is the circumferential stress component $\sigma_{\theta\theta}$, for the most beneficial residual (dimensionless) stress distribution (Fig. 11.7a), failure of the composite cylinder occurs when $\sigma_{\theta\theta} = \sigma_u$, simultaneously at the inner radii of the inner and outer cylinders (Fig. 11.7b). The ideal residual stress distribution re-

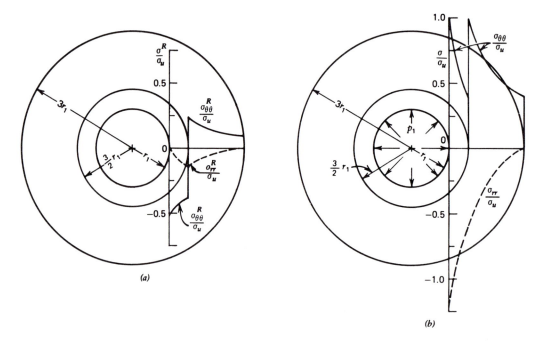

Figure 11.7   Stress distributions in composite cylinder made of brittle material that fails at inner radius of both cylinders simultaneously. (*a*) Residual stress distributions. (*b*) Total stress distributions.

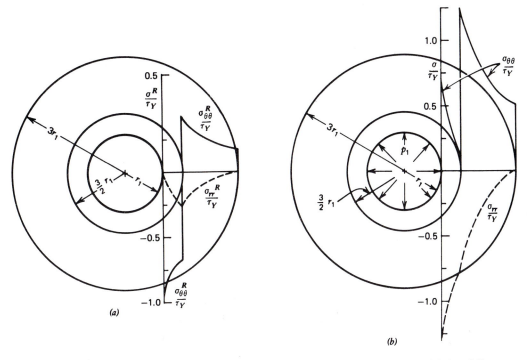

Figure 11.8   Stress distributions in composite cylinder made of ductile material that fails at inner radius of both cylinders simultaneously. (*a*) Residual stress distributions. (*b*) Total stress distributions.

quires a specific difference between the inner radius of the outer cylinder and the outer radius of the inner cylinder, which produces a shrink pressure p_s (see Problem 11.18). This shrink pressure produces a residual stress distribution (Fig. 11.7*a*) such that the application of an internal pressure p_1 produces the (dimensionless) stress distribution of Fig. 11.7*b* at failure.

If the composite cylinder is made of a *ductile* metal, either the maximum shear-stress criterion of failure or maximum octahedral shear-stress criterion of failure can be used. For example, let the composite cylinder of Fig. 11.8 be made of a ductile metal. Based on the maximum shear-stress criterion of failure, the ideal residual stress distribution due to the shrink pressure p_s is shown in Fig. 11.8*a*. (In this case, the interference fit is different from the cylinder of Fig. 11.7; see Problem 11.17.) For an internal pressure p_1 at failure of the cylinder, yield occurs simultaneously at the inner radii of the inner and outer cylinders, and the associated dimensionless stress distribution is shown in Fig. 11.8*b*.

EXAMPLE 11.4
Yield Failure of Thick-Wall Cylinder

The thick-wall cylinder in Example 11.1 is made of a ductile steel whose general yielding failure is accurately predicted by the octahedral shear-stress yield criterion. Determine the minimum yield stress for the steel for a factor of safety of $SF = 1.75$.

SOLUTION

The stress components calculated in Example 11.1 are for a cylinder that has been designed with a factor of safety of $SF = 1.75$. Yielding impends in the cylinder when the internal pressure is increased to $(SF)p_1 = 525$ MPa. The yield stress Y for the steel is obtained by setting the maximum octahedral shear stress in the cylinder [when the pressure in the cylinder is $(SF)p_1$] equal to the octahedral shear stress that occurs in a tension specimen made of the steel when the tension specimen axial stress is Y. The octahedral shear stress in the tension specimen is given by the relation [see Eqs. (2.22) and (4.24)],

$$\tau_{oct} = \frac{1}{3}\sqrt{(Y-0)^2 + (0-0)^2 + (0-Y)^2} = \frac{\sqrt{2}\,Y}{3} \tag{a}$$

The octahedral shear stress at any point in the thick-wall cylinder is given by the relation [see Eq. 2.22)]

$$\tau_{oct} = \frac{1}{3}\sqrt{(\sigma_{\theta\theta} - \sigma_{rr})^2 + (\sigma_{rr} - \sigma_{zz})^2 + (\sigma_{zz} - \sigma_{\theta\theta})^2} \tag{b}$$

For the open cylinder, the axial stress σ_{zz} is zero and the radial and circumferential stresses are

$$\sigma_{rr} = -1.75(300) = -525 \text{ MPa}$$
$$\sigma_{\theta\theta} = 1.75(325) = 568.8 \text{ MPa}$$

Substituting these stress components into Eq. (b) and setting Eq. (a) equal to Eq. (b), we obtain

$$Y = \frac{1}{\sqrt{2}}\sqrt{(568.8 + 525)^2 + (525)^2 + (568.8)^2} = 947.5 \text{ MPa}$$

EXAMPLE 11.5
Yield of Composite Thick-Wall Cylinder

The inner and outer cylinders of the composite thick-wall cylinder in Example 11.3 are made of the same ductile steel as the cylinder in Example 11.4. Determine the minimum yield stress for the steel in the composite cylinder for a factor of safety of $SF = 1.75$.

SOLUTION

Note: Equations (a) and (b) in Example 11.4 are valid for this problem also.

For the composite open cylinder, it is necessary to consider initiation of yielding for the inside of the inner cylinder, as well as for the inside of the outer cylinder. The axial stress σ_{zz} is zero for both cylinders. At the inside of the inner cylinder, the radial and circumferential stresses for a pressure $(SF)p_1$ are

$$\sigma_{rr} = (1.75)(300) = -525 \text{ MPa}$$
$$\sigma_{\theta\theta} = (1.75)(325) - 450.2 = 118.6 \text{ MPa}$$

Substituting these stress components into Eq. (b) and setting Eq. (a) equal to Eq. (b), we obtain

$$Y = \frac{1}{\sqrt{2}}\sqrt{(118.6 + 525)^2 + (525)^2 + (118.6)^2} = 593.3 \text{ MPa}$$

At the inside of the outer cylinder, the radial and circumferential stresses for a pressure $(SF)p_1$ are

$$\sigma_{rr} = -(1.75)(37.5) - 189.1 = -254.7 \text{ MPa}$$
$$\sigma_{\theta\theta} = (1.75)(62.5) + 315.1 = 424.5 \text{ MPa}$$

Substituting these stress components into Eq. (b) and setting Eq. (a) equal to Eq. (b), we find

$$Y = \frac{1}{\sqrt{2}}\sqrt{(424.5 + 254.7)^2 + (254.7)^2 + (424.5)^2}$$

$$= 594.3 \text{ MPa} > 593.3 \text{ MPa}$$

For the composite cylinder, the yield stress should be at least $Y = 594.3$ MPa. An ideal design for a composite cylinder should cause the required yield stress to be the same for the inner and outer cylinders. (Note that the above design is nearly ideal.)

A comparison of the required yield stress for the single cylinder in Example 11.4 and the required yield stress for the composite cylinders indicates the advantage of the composite cylinder. The yield stress of the single cylinder material must be 59.4% greater than that of the composite cylinder, if both cylinders are subjected to the same initial pressure and are designed for the same factor of safety against initiation of yielding.

11.5

FULLY PLASTIC PRESSURE. AUTOFRETTAGE

Thick-wall cylinders made of ductile material can be strengthened by introducing beneficial residual stress distributions. In Sec. 11.3 and 11.4, it was found that beneficial residual stress distributions may be produced in a composite cylinder formed by shrinking one cylinder on another. Beneficial residual stress distributions may also be introduced into a single cylinder by initially subjecting the cylinder to high internal pressure so that inelastic deformations occur in the cylinder. As a result, an increase in the load-carrying capacity of the cylinder occurs because of the beneficial residual stress distributions that remain in the cylinder after the high pressure is removed. The residual stress distribution in the unloaded cylinder depends on the depth of yielding produced by the high pressure, the shape of the inelastic portion of the stress-strain diagram for loading of a tensile specimen of the material, and the shape of the stress-strain diagram for unloading of the tensile specimen followed by compression loading of the specimen. If the material in the cylinder is a strain hardening material, a part (usually, a small part) of the increase in load-carrying capacity is due to the strengthening of the material, resulting from strain hardening of the material. If the material exhibits a flat top stress-strain diagram at the yield point (i.e., elastic-perfectly plastic), all the increase in load-carrying capacity is due to the beneficial residual stress distribution.

The process of increasing the strength of open and closed cylinders by increasing the internal pressure until the cylinder is deformed inelastically is called *autofrettage*. The beneficial effect of the autofrettage process increases rapidly with the spread of inelastic deformation through the wall thickness of the cylinder. Once yielding has spread through the entire wall thickness, any further improvement in load-carrying capacity resulting from additional inelastic deformation is due to strain hardening of the material. The minimum internal pressure p_1 required to produce yielding through the wall of the cylinder is an important pressure to be determined, since most of the increase in load-carrying capacity is produced below this pressure, and the deformation of the cylinder remains small up to this pressure. For the special case where the stress-strain diagram of the material is flat-topped at the yield point Y, the internal pressure p_1 is called the fully plastic pressure p_P.

We derive the fully plastic pressure by assuming that the maximum shear-stress criterion of failure is valid. If we assume that σ_{zz} is the intermediate principal stress $(\sigma_{rr} < \sigma_{zz} < \sigma_{\theta\theta})$ for the cylinder, $\sigma_{\theta\theta} - \sigma_{rr} = 2\tau_Y$, where τ_Y is the shear yield stress. This result may be substituted into the equation of equilibrium, Eq. (11.1), to obtain

$$d\sigma_{rr} = \frac{2\tau_Y}{r}\,dr \tag{11.26}$$

Integration yields

$$\sigma_{rr} = 2\tau_Y \ln r + C \tag{11.27}$$

The constant of integration C is obtained from the boundary condition that $\sigma_{rr} = -p_2$ when $r = b$. Thus, we obtain

$$\sigma_{rr} = -2\tau_Y \ln \frac{b}{r} - p_2 \tag{11.28}$$

the radial stress distribution at the fully plastic pressure p_P. The magnitude of p_P is given by Eq. (11.28) since the internal pressure is then $p_1 = p_P = -\sigma_{rr}$ at $r = a$. Thus, we obtain

$$p_P = 2\tau_Y \ln \frac{b}{a} + p_2 \tag{11.29}$$

In practice, p_2 is ordinarily taken equal to zero, since for $p_2 = 0$ the required internal pressure p_1 is smaller than for nonzero p_2. The circumferential stress distribution for the cylinder at the fully plastic pressure is obtained by substituting Eq. (11.28) into the relation $\sigma_{\theta\theta} - \sigma_{rr} = 2\tau_Y$ to obtain

$$\sigma_{\theta\theta} = 2\tau_Y\left(1 - \ln\frac{b}{r}\right) - p_2 \tag{11.30}$$

If the material in the cylinder is a *Tresca material*, that is, a material satisfying the maximum shear-stress criterion of failure, $\tau_Y = Y/2$, and the fully plastic pressure given by Eq. (11.29) is valid for cylinders subjected to axial loads in addition to internal and external pressures as long as σ_{zz} is the intermediate principal stress, that is, $\sigma_{rr} < \sigma_{zz} < \sigma_{\theta\theta}$. If the material in the cylinder is a *von Mises material*, that is, a material satisfying the maximum octahedral shear-stress criterion of failure $\tau_Y = Y/\sqrt{3}$ (see column 4, Table 4.2), the fully plastic pressure given by Eq. (11.29) is valid for closed cylinders subjected to internal and external pressures only. For this loading, the maximum octahedral shear-stress criterion of failure requires that the axial stress be given by the relation

$$\sigma_{zz} = \frac{\sigma_{\theta\theta} + \sigma_{rr}}{2} \tag{11.31}$$

The proof of Eq. (11.31) is left to the reader.

In many applications, the external pressure p_2 is zero. In this case, the ratio of the fully plastic pressure p_P [Eq. (11.29)] to the pressure p_Y that initiates yielding in the cylinder at the inner wall (see Problem 11.11) is given by the relation

$$\frac{p_P}{p_Y} = \frac{2b^2}{b^2 - a^2}\ln\frac{b}{a} \tag{11.32}$$

In particular, this ratio becomes large as the ratio b/a becomes large. For $b = 2a$, Eq. (11.32) gives $p_P = 1.85p_Y$; dimensionless radial, circumferential, and axial stress distributions for this cylinder are shown in Fig. 11.9. A comparison of these stress distributions with those at initiation of yielding (see Fig. 11.6) indicates that yielding throughout the wall thickness of the cylinder greatly alters the stress distributions. If the cylinder in Fig. 11.9 unloads elastically, the residual stress distributions can be obtained by multiplying the stresses in Fig. 11.6 by the factor 1.85 and subtracting them from the stresses in Fig. 11.9. For instance, the residual circumferential stress $\sigma_{\theta\theta}^R$ at the inner radius is calculated to be $\sigma_{\theta\theta}^R = -1.72\tau_Y$. This maximum circumferential residual stress can be expressed in terms of the tensile yield stress Y as follows: for a Tresca material $\sigma_{\theta\theta}^R = -0.86Y$ and for a von Mises material $\sigma_{\theta\theta}^R = -0.99Y$. However, one cannot always rely on the presence of this large compressive residual stress in the unloaded cylinder. In particular, all metals behave inelastically (due to the Bauschinger effect) when the cylinder is unloaded, resulting in a decrease in the beneficial effects of the residual stresses. For example, an investigation (Sidebottom et al., 1976) indicated that the beneficial effect of the residual stresses at the inside of the cylinder (when $b = 2a$) is decreased to about 50% of that calculated based on the assumption that the cylinder unloads elastically. Consequently, the cylinder will respond inelastically rather than elastically the next time it is loaded to the fully plastic pressure.

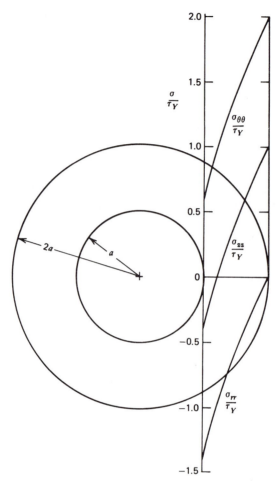

Figure 11.9 Stress distributions in a closed cylinder at fully plastic pressure ($b = 2a$). Cylinder made of von Mises material.

EXAMPLE 11.6
Fully Plastic Pressure for Cylinder

A closed cylinder has an inner radius of 20 mm and outer radius of 40 mm. It is made of steel that has a yield point stress of $Y = 450$ MPa and obeys the von Mises yield criterion.

(a) Determine the fully plastic internal pressure p_P for the cylinder.

(b) Determine the maximum circumferential and axial residual stresses when the cylinder is unloaded from p_P, assuming that the values based on linear elastic unloading are decreased by 50% because of inelastic deformation during unloading.

(c) Assuming that the elastic range of the octahedral shear stress has not been altered by the inelastic deformation, determine the internal pressure p_1 that can be applied to the cylinder based on a factor of safety $SF = 1.80$. For $SF = 1.80$, compare this result with the pressure p_1 for a cylinder without residual stresses.

SOLUTION

(a) The shear yield stress τ_Y for the von Mises steel is obtained using the octahedral shear-stress yield condition

$$\tau_Y = \frac{Y}{\sqrt{3}} = 259.8 \text{ MPa}$$

The magnitude of p_P is given by Eq. (11.29). Thus, we find

$$p_P = 2\tau_Y \ln\frac{b}{a} = 2(259.8)\ln\frac{40}{20} = 360.21 \text{ MPa}$$

The circumferential and axial stresses at the inner radius for fully plastic conditions are given by Eqs. (11.30) and (11.31). They are

$$\sigma_{\theta\theta} = 2\tau_Y\left(1 - \ln\frac{b}{a}\right) = 2(259.8)\left(1 - \ln\frac{40}{20}\right) = 159.4 \text{ MPa}$$

$$\sigma_{zz} = \frac{\sigma_{\theta\theta} + \sigma_{rr}}{2} = \frac{159.4 - 360.2}{2} = -100.4 \text{ MPa}$$

(b) Assuming linearly elastic unloading, we compute the circumferential and axial residual stresses at $r = a$ as

$$\sigma_{\theta\theta}^R = 159.4 - \frac{p_P(b^2 + a^2)}{b^2 - a^2} = 159.4 - \frac{360.2(40^2 + 20^2)}{40^2 - 20^2}$$

$$= 440.9 \text{ MPa}$$

$$\sigma_{zz}^R = -100.4 - \frac{p_P a^2}{b^2 - a^2} = -100.4 - \frac{360.2(20^2)}{40^2 - 20^2} = 220.5 \text{ MPa}$$

The actual residual stresses may be as much as 50% less than these computed values. Thus,

$$\sigma_{\theta\theta}^R = 0.50(-440.9) = -220.4 \text{ MPa}$$
$$\sigma_{zz}^R = 0.50(-220.5) = -110.2 \text{ MPa} \tag{a}$$

(c) Yielding is initiated in the cylinder at a pressure $(SF)p_1 = 1.80p_1$. If the residual stresses are neglected, the stresses at the inner radius due to pressure

$(SF)p_1$ are

$$\sigma_{rr} = -(SF)(p_1) = -1.80p_1$$

$$\sigma_{\theta\theta} = (SF)(p_1)\frac{b^2 + a^2}{b^2 - a^2} = (1.80)(p_1)\frac{40^2 + 20^2}{40^2 - 20^2}$$

$$= 3.000p_1 \qquad\qquad (b)$$

$$\sigma_{zz} = (SF)(p_1)\frac{a^2}{b^2 - a^2} = (1.80)(p_1)\frac{20^2}{40^2 - 20^2}$$

$$= 0.6000p_1$$

The actual stresses at the inner radius are obtained by adding the residual stresses given by Eq. (a) to those given by Eqs. (b). Thus,

$$\sigma_{rr} = -1.80p_1$$
$$\sigma_{\theta\theta} = 3.0000p_1 - 220.4 \qquad\qquad (c)$$
$$\sigma_{zz} = 0.6000p_1 - 110.2$$

The octahedral shear-stress yield condition requires that

$$\frac{\sqrt{2}\,Y}{3} = \frac{1}{3}\sqrt{(\sigma_{\theta\theta} - \sigma_{rr})^2 + (\sigma_{rr} - \sigma_{zz})^2 + (\sigma_{zz} - \sigma_{\theta\theta})^2} \qquad (d)$$

Substituting the values for the stress components given by Eq. (c) into Eq. (d), we find that

$$p_1 = 154.2 \text{ MPa}$$

is the working internal pressure for the cylinder that was preloaded to the fully plastic pressure. Substituting the values for the stress components given by Eq. (b) into Eq. (d), we obtain the working internal pressure for the cylinder without residual stresses

$$p_1 = 108.3 \text{ MPa}$$

Hence, the working pressure for the cylinder that is preloaded to the fully plastic pressure is 42.4% greater than the working pressure for the elastic cylinder without residual stresses.

11.6

CYLINDER SOLUTION FOR TEMPERATURE CHANGE ONLY

Consider the stress distribution in a thick-wall cylinder subjected to uniform internal and external pressures p_1 and p_2, axial load P, and temperature change ΔT that depends on the radial coordinate r only. The stress distribution may be obtained from Eqs. (11.8), (11.9), (11.10), (11.11), and (11.16). The special case of constant uniform temperature was considered in Sec. 11.3. In this section, the case of a cylinder subjected to a temperature change $\Delta T = T(r)$, in the absence of pressures and axial load, is treated. If internal and external pressures and temperature changes occur simultaneously, the resulting stresses may be obtained by superposition of the results of this section and Sec. 11.3. As in Sec. 11.3, the results of this section are restricted to the static, steady-state problem. Accordingly, the steady-state temperature change $\Delta T = T(r)$ is required input to the problem.

Steady-Steady Temperature Change (Distribution)

The temperature distribution in a homogeneous body in the absence of heat sources is given by Fourier's heat equation

$$\beta \nabla^2 T = \frac{\partial T}{\partial t} \tag{11.33}$$

in which β is the thermal diffusivity for the material in the body, where we consider $T = \Delta T$ to be the temperature change measured from the uniform reference temperature of the unstressed state, and t is the time. For steady-state conditions, $\partial T/\partial t = 0$, and Eq. (11.33) reduces to

$$\nabla^2 T = 0 \tag{11.34}$$

In cylindrical coordinates (r, θ, z), Eq. (11.34) takes the form

$$\frac{\partial^2 T}{\partial r^2} + \frac{1}{r}\frac{\partial T}{\partial r} + \frac{1}{r^2}\frac{\partial^2 T}{\partial \theta^2} + \frac{\partial^2 T}{\partial z^2} = 0 \tag{11.35}$$

Since T is assumed to be a function of r only, Eq. (11.35) simplifies to

$$\frac{d^2 T}{dr^2} + \frac{1}{r}\frac{dT}{dr} = 0 \tag{11.36}$$

The solution of Eq. (11.36) is

$$T = C_1 \ln r + C_2 \tag{11.37}$$

where C_1 and C_2 are constants of integration. With Eq. (11.37), the boundary conditions $T = T_b$ for $r = b$ and $T = T_a$ for $r = a$ are used to determine C_1 and C_2. The solution of Eq. (11.37) then takes the form

$$T = \frac{T_0}{\ln(b/a)} \ln \frac{b}{r} \tag{11.38}$$

where

$$T_0 = T_a - T_b$$

Stress Components

If $p_1 = p_2 = P = 0$, Eq. (11.38) can be used with Eqs. (11.8), (11.9), (11.10), (11.11), and (11.16) to obtain stress components for steady-state temperature distributions in a thick-wall cylinder. The results are

$$\sigma_{rr} = \frac{\alpha E T_0}{2(1 - \nu) \ln(b/a)} \left[-\ln\frac{b}{r} + \frac{a^2(b^2 - r^2)}{r^2(b^2 - a^2)} \ln\frac{b}{a} \right] \tag{11.39}$$

$$\sigma_{\theta\theta} = \frac{\alpha E T_0}{2(1 - \nu) \ln(b/a)} \left[1 - \ln\frac{b}{r} - \frac{a^2(b^2 + r^2)}{r^2(b^2 - a^2)} \ln\frac{b}{a} \right] \tag{11.40}$$

$$\sigma_{zz} = \sigma_{rr} + \sigma_{\theta\theta} = \frac{\alpha E T_0}{2(1 - \nu) \ln(b/a)} \left[1 - 2\ln\frac{b}{r} - \frac{2a^2}{b^2 - a^2} \ln\frac{b}{a} \right] \tag{11.41}$$

Thus, the stress distributions for linearly elastic behavior of a thick-wall cylinder subjected to a steady-state temperature distribution are given by Eqs. (11.39), (11.40), and (11.41). When $T_0 = T_a - T_b$ is positive, the temperature at the inner radius T_a is greater than the temperature at the outer radius T_b. For the case of positive T_0, dimensionless stress distributions for a cylinder with $b = 2a$ are shown in Fig. 11.10. Since for this case, the stress components $\sigma_{\theta\theta}$ and σ_{zz} are compressive, a positive temperature difference T_0 is beneficial for a cylinder that is subjected to a combination of internal pressure p_1 and temperature since the compressive stresses due to T_0 counteract tensile stresses due to p_1. The stresses in cylinders subjected to internal pressure p_1, external pressure p_2, axial load P, and steady-state temperature may be obtained as follows: The radial stress is given by adding Eq. (11.20) to Eq. (11.39), the circumferential stress is given by adding Eq. (11.21) to Eq. (11.40), and the axial stress is given by adding Eq. (11.22) to Eq. (11.41).

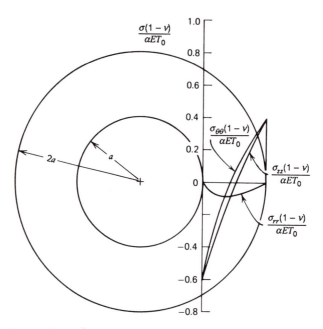

Figure 11.10   Stress distributions in a cylinder subjected to a temperature gradient ($b = 2a$).

PROBLEMS
Section 11.3

11.1. A long closed cylinder has an internal radius $a = 100$ mm and external radius $b = 250$ mm. It is subjected to an internal pressure $p_1 = 80.0$ MPa ($p_2 = 0$). Determine the maximum radial, circumferential, and axial stresses in the cylinder.

11.2. Determine the radial and circumferential stress distributions for the cylinder in Problem 11.1

11.3. Consider a 1-m length of the unloaded cylinder in Problem 11.1 at a location in the cylinder some distance from the ends. What are the dimensions of this portion of the cylinder after $p_1 = 80.0$ MPa is applied? The cylinder is made of a steel for which $E = 200$ GPa and $v = 0.29$.

11.4. A closed cylinder has an inside diameter of 20 mm and outside diameter of 40 mm. It is subjected to an external pressure $p_2 = 40$ MPa and internal pressure of $p_1 = 100$ MPa. Determine the axial stress and circumferential stress at the inner radius.

Ans. $\sigma_{zz} = -20.0$ MPa, $\sigma_{\theta\theta} = 60.0$ MPa

11.5. A composite aluminum alloy ($E = 72.0$ GPa and $v = 0.33$) cylinder is made up of an inner cylinder with inner and outer diameters of 80 and 120+ mm, respectively, and an outer cylinder with inner and outer diameters of 120 and 240 mm, respectively. The composite cylinder is subjected to an internal pressure of 160 MPa. What must the outside diameter of the inner cylinder be if the circumferential stress at the inside of the composite cylinder is equal to 130 MPa?

11.6. What must the outside diameter of the inner cylinder be for the composite cylinder in Problem 11.5 if the maximum shear stress at the inner radius of the inner cylinder is equal to the maximum shear stress at the inner radius of the outer cylinder? What are the values for the circumferential stress at the inside of the composite cylinder and the maximum shear stress?

Ans. Diameter $= 120.2271$ mm, $\sigma_{\theta\theta(a)} = 85.1$ MPa, $\tau_{max} = 122.6$ MPa

11.7. A gray cast iron ($E = 103$ GPa and $v = 0.20$) cylinder has an outside diameter of 160 mm and inside diameter of 40 mm. Determine the circumferential stress at the inner radius of the cylinder when the internal pressure is 60.0 MPa.

11.8. Let the cast iron cylinder in Problem 11.7 be a composite cylinder made up of an inner cylinder with inner and outer diameters of 40 and 80+ mm, respectively, and an outer cylinder with inner and outer diameters of 80 and 160 mm, respectively. What must the outside diameter of the inner cylinder be if the circumferential stress at the inside of the inner cylinder is equal to the circumferential stress at the inside of the outer cylinder? What is the magnitude of the circumferential stress at the inside of the composite cylinder?

Ans. Diameter $= 80.0287$ mm, $\sigma_{\theta\theta(a)} = 38.5$ MPa

11.9. A hollow steel cylinder ($E = 200$ GPa, $v = 0.3$), with an inner diameter of 100 mm and outer diameter of 300 mm, is press-fitted over a solid steel shaft of diameter 100.125 mm. Determine the maximum principal stress in the shaft and in the cylinder.

Section 11.4

11.10. **(a)** Derive the expression for the maximum shear stress in a thick-wall cylinder subjected to internal pressure p_1, external pressure p_2, and axial load P, assuming that σ_{zz} is the intermediate principal stress, that is, $\sigma_{rr} < \sigma_{zz} < \sigma_{\theta\theta}$.

(b) Derive an expression for the limiting value of the axial load P for which the expression in part (a) is valid.

11.11. Let σ_{zz} be the intermediate principal stress in a thick-wall cylinder ($\sigma_{rr} < \sigma_{zz} < \sigma_{\theta\theta}$). Using the maximum shear-stress criterion of failure, derive an expression for the internal pressure p_Y necessary to initiate yielding in the cylinder. The shear yield stress for the material is τ_Y.

Ans. $p_Y = \tau_Y(b^2 - a^2)/b^2 + p_2$

11.12. For a closed cylinder subjected to internal pressure p_1 only, show that the octahedral shear stress τ_{oct} at the inner radius is given by the relation $\tau_{oct} = \sqrt{2}p_1 b^2/[\sqrt{3}(b^2 - a^2)]$.

11.13. A closed cylinder is made of a ductile steel that has a yield stress $Y = 600$ MPa. The inside diameter of the cylinder is 80 mm. Determine the outside diameter of the cylinder if the cylinder is subjected to an internal pressure only, of $p_1 = 140$ MPa, and the cylinder is designed using a factor of safety of $SF = 1.75$ based on the maximum shear-stress criterion of failure.

11.14. Solve Problem 11.13 using the octahedral shear-stress criterion of failure.

11.15. A closed cylinder with inner and outer radii of 60 and 80 mm, respectively, is subjected to an internal pressure $p_1 = 30.0$ MPa and axial load $P = 650$ kN. The cylinder is made of a steel that has a yield stress of $Y = 280$ MPa. Determine the factor of safety SF used in the design of the cylinder based on

(a) the maximum shear-stress criterion of failure and

(b) the maximum octahedral shear-stress criterion of failure.

Ans.   **(a)** $SF = 1.96$,   **(b)** $SF = 2.00$

11.16. A closed cylinder with inner and outer diameters of 30 and 60 mm, respectively, is subjected to an internal pressure only. The cylinder is made of a brittle material having an ultimate strength of $\sigma_u = 160$ MPa. The outer diameter has been gradually reduced as we move away from each end so that stress concentrations at the ends can be neglected. Determine the magnitude of p_1 based on a factor of safety of $SF = 3.00$.

11.17. Two cylinders are slip-fitted together to form a composite open cylinder. Both cylinders are made of a steel having a yield stress $Y = 700$ MPa.

The inner cylinder has inner and outer diameters of 100 and 150+ mm, respectively. The outer cylinder has inner and outer diameters of 150 and 300 mm, respectively.

(a) Determine the shrink pressure p_s and maximum internal pressure p_1 that can be applied to the cylinder if it has been designed with a factor of safety of $SF = 1.85$ for simultaneous initiation of yielding at the inner radii of the inner and outer cylinders. Use the maximum shear-stress criterion of failure.

(b) Determine the outer diameter of the inner cylinder required for the design. For the steel $E = 200$ GPa and $v = 0.29$.

Ans. **(a)** $p_s = 91.2$ MPa, $p_1 = 247.0$ MPa; **(b)** diameter $= 150.292$ mm

11.18. Two cylinders are slip-fitted together to form a composite open cylinder. Both cylinders are made of a brittle material whose stress-strain diagram is linear up to the ultimate strength $\sigma_u = 480$ MPa. The inner cylinder has inner and outer radii of 50 and 75+ mm, respectively. The outer cylinder has inner and outer radii of 75 and 150 mm, respectively. Determine the shrink pressure p_s and maximum internal pressure p_1 that results in initiation of fracture simultaneously at the inner radii of both cylinders. Use the maximum principal stress criterion of failure.

Section 11.5

11.19. A thick-wall cylinder has an inside diameter of 180 mm and outside diameter of 420 mm. It is made of steel having a yield point stress of $Y = 460$ MPa and obeying the Tresca criterion. Determine the fully plastic pressure for the cylinder if $p_2 = 0$.

11.20. **(a)** Determine the working pressure p_1 for the thick-wall cylinder in Problem 11.19 if it is designed with a factor of safety of $SF = 3.00$ based on the fully plastic pressure.

(b) What is the factor of safety based on the maximum elastic pressure p_Y.

Ans. **(a)** $p_1 = 129.9$ MPa, **(b)** $SF = 1.45$

11.21. A composite open cylinder has an inner cylinder with inner and outer radii of 20 and 30 mm and is made of a steel with yield point stress $Y_1 = 400$ MPa. The outer cylinder has inner and outer radii of 30 and 60 mm and is made of a steel with yield point stress $Y_2 = 600$ MPa. Determine the fully plastic pressure for the composite cylinder if both steels obey the von Mises criterion.

11.22. The closed cylinder in Example 11.6 is made of a Tresca material instead of a von Mises material. Obtain the solution for the Tresca material.

Ans. $p_P = 311.9$ MPa, $p_1 = 133.5$ MPa (including residual stresses)
$p_1 = 93.8$ MPa (without residual stresses)

Section 11.6

11.23. An unloaded closed cylinder has an inner radius of 100 mm and outer radius of 250 mm. The cylinder is made of a steel for which $\alpha = 0.0000117$ per $°C$, $E = 200$ GPa, and $v = 0.29$. Determine the stress components at the inner radius for a steady-state temperature change with the temperature at the inner radius $100°C$ greater than the temperature at the outer radius.

11.24. Let the steel in the cylinder in Problem 11.23 have a yield stress of $Y = 500$ MPa. Determine the magnitude of T_0 necessary to initiate yielding in the cylinder based on the

(a) maximum shear-stress criterion of failure and

(b) maximum octahedral shear-stress criterion of failure.

Ans.   (a) $T_0 = 235.3°C$.   (b) $T_0 = 235.3°C$

11.25. The cylinder in Problem 11.23 is subjected to a temperature difference of $T_0 = 50°C$ and an internal pressure $p_1 = 100$ MPa. Determine the stress components at the inner radius.

11.26. A closed brass ($Y = 240$ MPa, $E = 96.5$ GPa, $v = 0.35$, $\alpha = 0.000020$ per $°C$) cylinder has an inside diameter of 70 mm and outside diameter of 150 mm. It is subjected to a temperature difference $T_0 = T_a - T_b = 70°C$. For this value of T_0,

(a) determine the magnitude p_1 of internal pressure required to initiate yield in the cylinder and

(b) the magnitude p_2 of external pressure required to initiate yield.

(c) Repeat parts (a) and (b) for the case $T_0 = 0$. Use the maximum shear-stress criterion of failure.

Ans.   (a) $p_1 = 135.9$ MPa,   (b) $p_2 = 51.9$ MPa,   (c) $p_1 = p_2 = 93.9$ MPa

REFERENCES

Boresi, A. P. and Chong, K. P. (1987). *Elasticity in Engineering Mechanics.* New York: Elsevier.

Sidebottom, O. M., Chu, S. C., and Lamba, H. S. (1976). 'Unloading of Thick-Walled Cylinders That Have Been Plastically Deformed.' *Exper. Mech.*, **16**(12): 454–460.

12

ELASTIC AND INELASTIC STABILITY OF COLUMNS

In this chapter, we consider columns subjected to axial compressive loads. The columns are considered sufficiently slender so that, at a critical compressive load, they may fail by sudden lateral deflection (buckling), rather than by yielding or crushing. If this lateral deflection continues to increase, a real column will undergo plastic deformation and possibly a catastrophic fracture or collapse. For very slender columns made of elastic-perfectly plastic materials, the critical load depends primarily on the modulus of elasticity and the properties of the column cross section. It is independent of the yield stress and ultimate strength. For moderately slender columns made of an elastic, strain-hardening material such as high-strength, alloy steels and aluminum alloys, the critical load will also depend on the inelastic stress-strain relationship of the material (see Sec. 12.6, "Inelastic Buckling of Columns"). Additionally, residual stresses due to nonuniform cooling of hot-rolled steel may cause the steel to behave as an elastic, strain-hardening material. Consequently, the critical load for a column made of such a steel may also be dependent on the inelastic stress-strain behavior of the material. A large part of our treatment is devoted to ideal columns. We define an ideal column to be one that remains elastic, is perfectly straight, is subjected to a compressive load that lies exactly along its central longitudinal axis, is not subjected to a bending moment or lateral force, and is weightless and free of residual stresses.

In general, the theoretical study of buckling is referred to as the theory of stability or theory of buckling. A comprehensive review of the general theory of buckling has been given by Langhaar (1958), including references to many works through mid–1958. More recently, the book by Bazant and Cedolin (1991) summarizes structural stability studies to 1990. Broadly speaking, the theory of buckling is the theory of stability of mechanical systems; that is, the theory of buckling deals principally with conditions for which equilibrium becomes unstable. The critical load at which a system may become unstable (buckle) may be determined by several methods. For conservative systems, these methods include (1) the equilibrium method, which leads to an eigenvalue problem, (2) the energy method, (3) snap-through theory, (4) imperfection theory, and (5) the dynamic method. Nonconservative (dissipative) systems require special attention (see Sec. 12.6, "Inelastic Buckling of Columns").

In Sec. 12.1, we introduce some basic concepts of column buckling. In Sec. 12.2, a physical description of the elastic buckling of columns for a range of lateral deflections, for both ideal and imperfect slender columns, is presented. In Sec. 12.3, the Euler formula for the critical load for elastic columns with pinned ends is derived. In Sec. 12.4, the effect of general end constraints on the elastic buckling load of columns is examined. In Sec. 12.5, the local buckling of thin-wall flanges of elastic columns with open cross sections (e.g., a channel) is discussed. Finally in Sec. 12.6, the topic of inelastic buckling of columns is introduced. For studies of the stability of other structural elements (e.g., plates and shells), the reader may refer to other works (Bleich, 1952; Timoshenko and Gere, 1961; Chajes, 1974; Brush and Almroth, 1975; Szilard, 1974; Calladine, 1988; Bazant and Cedolin, 1991).

12.1

INTRODUCTION TO THE CONCEPT OF COLUMN BUCKLING

When an initially straight, slender column with *pinned ends* is subjected to a large compressive load, theoretically, failure may occur by elastic buckling when the load exceeds the critical (buckling) load (Fig. 12.1; see also Sec. 12.3)

$$P_{cr} = \frac{\pi^2 EI}{L^2} \tag{12.1}$$

where E is the modulus of elasticity, I the moment of inertia of the cross section about the axis of bending, and L the length of the pinned end column. For an ideal pinned-end column, the load P may be increased beyond P_{cr}, along line OC, (Fig. 12.2). However, for loads $P \geq P_{cr}$, the column is in an unstable equilibrium state (Langhaar, 1989). More realistically, however, in a real structure, the line of action of force P does not lie *exactly* along the central axis of the column, and

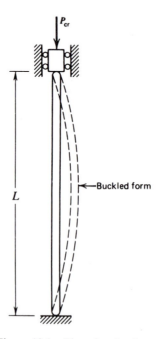

Figure 12.1   Pinned-end column.

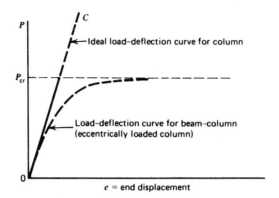

Figure 12.2   Load-deflection curves.

the column may not be exactly straight. Hence, in general, the force P produces a bending moment in the column. In other words, most real columns perform their load-carrying function as beam-columns (Bazant and Cedolin, 1991). Then, the load-deflection response of the column follows a curved path (Fig. 12.2) and the column eventually fails at a load $P \le P_{cr}$.

When an ideal column is subjected to a sufficiently small compressive force P, it remains in equilibrium in the straight position. If the column is subjected to an additional small lateral force, it will deflect laterally. However, when the lateral force is removed, the elastic restoring forces in the column are large enough to return the column to its original (straight) position. Thus, for sufficiently small loads P, the column is said to be in a stable equilibrium state. As P is increased, it reaches a magnitude for which the elastic restoring forces are only large enough to maintain equilibrium in a displaced lateral position. However, they are not capable of returning the column to its original straight position. At this critical load, the column is in a state of neutral equilibrium. For values of load greater than this critical load, the slightest lateral movement of the column from its straight position may result in a displacement that produces yield or fracture, or that exceeds the clearance tolerance of the column.*

As seen by Eq. (12.1), the critical load P_{cr} of a column with pinned ends depends on the moment of inertia I of the cross section, the elastic modulus E, and on the length L. Among structural materials, the value of E may vary by a factor as great as 15 or more (e.g., the modular ratio between steel and concrete is usually around 8 or 9). However, for structural steel columns used in modern construction, E is essentially constant (see Appendix A). For such columns, P_{cr} is governed by the column length L and the smallest moment of inertia I of the column cross section. Also, the magnitude of the buckling load of a column is influenced greatly by the nature of its end supports (Sec. 12.4). For given end conditions, buckling of slender columns is controlled most frequently by controlling L or selecting the column

---

* If a pinned-end column (Fig. 12.1) remains elastic for load $P \ge P_{cr}$, in the buckled form its ends may meet or if a column has one end free and one end fixed, it may become inverted for loads greater than its critical load. The problem of determining the deflections of such columns is called the *elastica* problem (Bazant and Cedolin, 1991). However, if an elastic column undergoes such a large displacement, it will probably exceed clearance tolerances of the system in which it is a part; that is, it will *jam*. In general, the problem of *jamming* may also occur in systems with small tolerances, such as required in electrical motors and generators, and internal combustion engines.

cross section to maximize I. To maximize I, closed-tube cross sections are commonly used in column design.

In summary, if an *ideal* slender elastic column is subjected to a compressive load P and is further subjected to a lateral disturbance, it can undergo destructive lateral displacements, when the magnitude of P equals or exceeds P_{cr}. It may fail by excessive, yet small, *stable* elastic deflections (jamming) if P is equal to P_{cr}. It may fail by elastic deflections, which may increase indefinitely, if P is increased to values larger than P_{cr}. In the latter case, since most real materials cannot withstand the strains associated with such large deflections without yielding or breaking, a real column will fail by plastic collapse or fracture when P is significantly larger than P_{cr}. In the next section, we consider the deflection response of columns to compressive loads that vary in magnitude from zero to greater than P_{cr}.

12.2

DEFLECTION RESPONSE OF COLUMNS TO COMPRESSIVE LOADS

In Sec. 12.1, we noted that a slender pinned-end column can buckle at the load $P = P_{cr}$ [Eq. (12.1)], if it is subjected to a lateral disturbance. For this load, the column may undergo a lateral elastic displacement. If this displacement exceeds the tolerance limits of the column, the column is said to have failed by excessive lateral deflection due to elastic buckling. If the compressive load on the column continues to increase beyond P_{cr}, the column will continue to deflect laterally, and unless it is extremely slender (an elastica), it will become completely unstable and undergo plastic collapse or fracture. Buckling, therefore, is frequently referred to as *structural instability*. Furthermore, in this type of failure of slender columns, the buckling load usually represents the maximum theoretical load that the column can be expected to resist. This is true even though, at this load, the stress in the column material is ordinarily less than the compressive proportional limit or yield stress. Hence, when elastic buckling is the mode of failure of the column, the problem is one of obtaining an expression for the critical (buckling) load. A brief physical description of the deflection response of a slender elastic column in the neighborhood of the critical load is given in the next subsection.

Caution: The effects of temperature (especially elevated temperature) and of time (such as in long loading periods resulting in creep) on buckling are not considered in the subsequent discussion. In practice, these effects may be extremely important (Manson, 1981; Moulin et al., 1989) and usually act to lower the critical load.

Elastic Buckling of an Ideal Slender Column

Consider an ideal slender pinned-end column that is straight, axially loaded, and made of a homogeneous material. The column will remain straight under any value of the axial load; that is, it will not bend. At a certain critical load, let a small lateral force be applied to the column to produce a small lateral deflection. As noted in Sec. 12.1, if when the lateral load is removed, the column remains in the slightly bent position (Fig. 12.1), the axial load is the critical load P_{cr}. If the load is increased slightly, the lateral deflection of the column increases rapidly; the deflection (and hence, the strain and stress in the column) is not proportional to the load, However, the material still acts elastically. This behavior of the column is represented by the region $0AB$ of the curve in Fig. 12.3a, where A represents the critical (buckling) load for the ideal slender column. Although the deflection increases in region AB, segment AB still represents a relatively small deflection [the larger deflections indicated by BCD (Fig. 12.3a) are discussed in the next subsection]; note that the magnitude of the deflection δ in the sketch is greatly exaggerated. As noted in Sec. 12.1, the value of the critical load is given by Eq. (12.1). Associated with the critical load P_{cr} is a critical stress $\sigma_{cr} = P_{cr}/A$, where σ_{cr} is the average normal stress in the column due to the load P_{cr} and A is the cross-sectional area of the column. Thus, we may write

$$P_{cr} = \frac{\pi^2 EI}{L^2}, \qquad \sigma_{cr} = \frac{P_{cr}}{A} = \frac{\pi^2 E}{(L/r)^2} \qquad (12.2)$$

where E is the modulus of elasticity of the material, L the length of the column, I the smallest moment of inertia of the cross section, r the least radius of gyration of the cross section ($r^2 = I/A$), and L/r the slenderness ratio. In order for the column to remain elastic, σ_{cr} must be less than the yield stress of the column material. In other words, the elastic buckling load has little physical significance, unless it is reached before the average stress equals or exceeds the yield stress. Equation (12.2) is called Euler's formula for a column with pinned ends. Further discussion and the derivation of Eq. (12.2) are given in Sec. 12.3.

Large Deflections. Southwell (1941) has shown that a very slender column can sustain a load greater than P_{cr} in a bent position, provided that the average stress is much less than the yield stress. The load-deflection relation for such a column is similar to the curves BCD in Fig. 12.3. For a real column, the yield stress of the material is exceeded at some deflection C, and the column yields at an outer fiber due to combined axial force and bending (Fig. 12.3a). At point C, the load-deflection

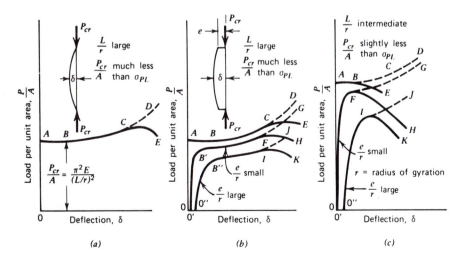

Figure 12.3   Relation between load and deflection for columns.

curve continues to rise slightly with increasing load, until the critical section of the column is fully yielded. Then, the load-deflection curve drops, as indicated by the region near point E, and the column fails, either by plastic collapse or fracture.

Imperfect Slender Columns

Columns that possess deviations from the ideal conditions assumed in the previous subsection are said to be *imperfect columns*. Real columns nearly always possess deviations from perfect (ideal) conditions. It is important, therefore, to consider the effects of these imperfections on the *elastic* buckling load. A common deviation is eccentricity of loading caused by misalignment of load or by initial crookedness of the column. If the load is slightly eccentric, a slender column will undergo lateral deflection as soon as a load is applied. This deflection will increase with increasing load (Fig. 12.2). However, before the load reaches the Euler load [Eq. (12.2)], the deflection of the column may be very large. Therefore, load-deflection relations for large deflections are needed to explain the effects of deviations from ideal conditions. Practically, however, unless a column is extremely slender, it will fail by yielding or fracture, before such large deflections occur.

An imperfect column may be considered to be equivalent to a straight slender column with an eccentricity e of load, as shown in the sketch in Fig. 12.3b. For small eccentricities, $0'B'FG$ represents the load-deflection curve of the column. The curve $0'B'FG$ approaches the curve $0ABCD$ of the ideal column if the material remains elastic. However, near point F, the yield stress is reached or exceeded. As for the ideal column, the load-deflection curve continues to rise until the critical section of the column is fully yielded, then drops (near point H). For a larger eccentricity, similar behavior occurs (curve $0''B''IJ$). For a small eccentricity, the maximum load within region FH does not differ much from P_{cr} (Fig. 12.3b). However, for large eccentricities, the maximum load within region IK may be much less than P_{cr}.

For a very slender [large L/r; see Eq. (12.2)] straight column subjected to a slightly eccentric load, Euler's formula [Eq. (12.2)] may give a fairly accurate estimate of the critical load. However, for a less slender column in which the critical stress $\sigma_{cr} = P_{cr}/A$ is only slightly less than the yield stress, the load-deflection curve is represented more like $0ABE$ in Fig. 12.3c instead of $0ABCD$ in Figs. 12.3a and b. This behavior is explained by the fact that at point B only a small amount of additional lateral deflection is required to cause yielding. The curves $0'FH$ and $0''IK$ in Fig. 12.3c show that a column with an intermediate L/r ratio is much more sensitive to eccentricity of load than is a very slender column (Fig. 12.3b). Hence, Euler's formula does not predict accurately the buckling load for eccentrically loaded columns of intermediate slenderness ratios.

Failure of Slender Columns. In view of the above discussion, the question arises as to whether it is appropriate to use the term buckling in describing the failure of a slender column that is loaded eccentrically. For example, if columns whose load-deflection relations are represented by the curves $0ABCE$, $0'B'FH$, and $0''B''IK$ in Fig. 12.3b are used as machine or structural parts, their failure is usually considered to occur by excessive deflection represented approximately by points B, B', and B''. However, at B, B', and B'', the loads are smaller than those required to cause instability or total collapse (regions CE, FH, and IK). Hence, failure of eccentrically loaded slender members (the condition that limits the maximum utilizable load) is not a condition of buckling in the usual sense of the term.

Failure of Columns of Intermediate Slenderness Ratio.   The load-deflection relations for columns of intermediate slenderness ratios are represented by the curves
in Fig. 12.3c. For such columns, a condition of instability is associated with points
B, F, and I. At these points, inelastic strain occurs and is followed, after only a small
increase in load, by instability and plastic collapse at relatively small lateral deflections. These deflections do not differ greatly from the deflections associated with
points B, B', and B'' of Fig. 12.3b.

Which Type of Failure Occurs?   As noted above, two types of failure of slender
columns are possible, depending on the slenderness ratio L/r, namely, (1) failure
by excessive deflection before plastic collapse or fracture and (2) failure by plastic
collapse or fracture. For a given value of L/r, it is difficult to determine which
type of failure will occur, except perhaps when L/r is very large. In other words,
for L/r values ordinarily employed in structural columns, the type of column failure is not easily determined. Furthermore, for responses indicated by curves $0'FH$
and $0''IK$ (Fig. 12.3c), it is uncertain how much of an increase in load is possible,
after inelastic strains occur and before collapse. To complicate matters further,
this increase in load depends on not only the value of L/r, but also the shape of
the cross section and the stress-strain diagram of the material (see Sec. 12.6). Thus,
a wholly rational method, or formula, for the failure of columns is difficult to
achieve. Hence, empirical methods are usually used in conjunction with analysis
to develop workable design criteria (Salmon and Johnson, 1990).

12.3

THE EULER FORMULA FOR COLUMNS WITH PINNED ENDS

In this section, we derive Euler's formula for the critical load of an axially compressed column with pinned ends (Fig. 12.4). As noted in the introductory remarks
for this chapter, several methods may be used to determine critical loads of a conservative system. By way of illustration, we derive the Euler formula using several
of these methods; namely, the equilibrium method, the imperfection method, and
the energy method. We do not demonstrate the snap-through method (since it is
more significant in the study of buckling of shells) or the vibration method (since
it lies outside the scope of this book; see Bazant and Cedolin, 1991). The definitive
study of buckling of columns has been attributed to Euler (1933), who studied the
buckling of a column clamped at the bottom and unrestrained at the top (the so-
called *flag-pole problem*). However, since the problem of a column with pinned
ends occurs frequently in practice, it is called the *fundamental problem of buckling
of a column*. Hence, we consider it first.

The Equilibrium Method
Consider the pin-ended column of Fig. 12.4. The free-body diagram of the lower
part AB of the column is shown in Fig. 12.5, where positive $M(x)$ is taken in the
clockwise sense. By equilibrium of moments about point A, we have $\sum M_A =
M(x) + Py = 0$, or

$$M(x) = -Py \tag{12.3}$$

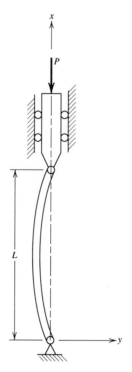

Figure 12.4   Column with pinned ends.

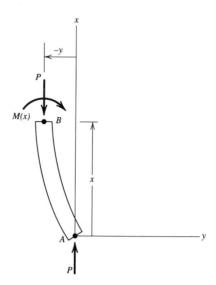

Figure 12.5   Free-body diagram of lower part of column.

Equation (12.3) represents a state of neutral equilibrium. By elementary beam theory, the moment $M(x)$ is related to the radius of curvature $R(x)$ of the centerline of the column in the displaced position by the relation

$$M(x) = \frac{EI}{R(x)} \tag{12.4}$$

where EI is the flexural stiffness for bending in the plane of Fig. 12.5. If the slope dy/dx of the displaced position is small, by the calculus formula for $R(x)$, we have

$$\frac{1}{R} = \pm \frac{(d^2y/dx^2)}{[1 + (dy/dx)^2]^{3/2}} \approx \pm \frac{d^2y}{dx^2} \tag{12.5}$$

Then, Eqs. (12.4) and (12.5) yield

$$M(x) = \pm EI \frac{d^2y}{dx^2} \tag{12.6}$$

where for given axes the plus sign corresponds to the case where $M(x)$ is chosen to produce a curvature with center 0 on the positive side of the y axis (Fig. 12.6a). Similarly, the minus sign is taken when a positive moment is chosen to produce a curvature with center 0 on the negative side of the y axis (Fig. 12.6b). For example, in Fig. 12.5, we have chosen positive $M(x)$ in the clockwise sense, and it produces

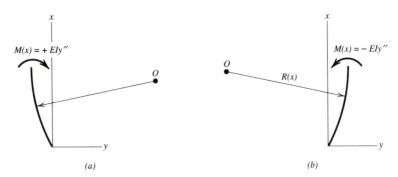

Figure 12.6 Sign convention for internal moment. (*a*) Positive moment taken clockwise. (*b*) Positive moment taken counterclockwise.

a curvature with center 0 on the positive side of the y axis. Therefore, we take the plus sign in Eq. (12.6). Thus, by Eqs. (12.3) and (12.6), we obtain, after dividing by EI,

$$\frac{d^2y}{dx^2} + k^2 y = 0 \tag{12.7}$$

where

$$k^2 = \frac{P}{EI} \tag{12.8}$$

The boundary conditions associated with Eq. (12.7) are

$$y = 0, \quad \text{for } x = 0 \quad \text{and} \quad x = L \tag{12.9}$$

For arbitrary values of P, Eqs. (12.7) and (12.9) admit only the trivial solution $y=0$. However, nontrivial solutions of Eq. (12.7) exist for specific eigenvalues k, as will now be shown.

The general solution to Eq. (12.7) is

$$y = A \sin kx + B \cos kx \tag{12.10}$$

where A and B are constants determined from the boundary conditions [Eq. (12.9)]. Therefore, Eqs. (12.9) and (12.10) yield,

$$A \sin kL = 0, \quad B = 0 \tag{12.11}$$

For a nontrivial solution ($A \neq 0$), Eq. (12.11) requires that $\sin kL = 0$, or

$$k_n = \sqrt{\frac{P_n}{EI}} = \frac{n\pi}{L}, \quad n = 1, 2, 3, \dots \tag{12.12}$$

For each value of n, by Eq. (12.10), for $k = k_n$, there exists an associated nontrivial solution (eigenfunction)

$$y_n = A_n \sin \frac{n\pi x}{L} \tag{12.13}$$

From Eq. (12.12), the corresponding Euler loads are

$$P_n = \frac{n^2 \pi^2 EI}{L^2}, \qquad n = 1, 2, 3, \ldots \tag{12.14}$$

The minimum P_n occurs for $n = 1$. This is the smallest load for which a nontrivial solution is possible; that is, it is the critical load for the column. By Eq. (12.12), with $n = 1$, we find

$$P_1 = \frac{\pi^2 EI}{L^2} = P_{cr} \tag{12.15}$$

As noted previously, Eq. (12.15) is the Euler formula for the buckling of a column with pinned ends [see Eqs. (12.1) and (12.2)]. The buckled shape of the column is, as given by Eq. (12.13) and shown in Fig. 12.4,

$$y_1 = A_1 \sin \frac{\pi x}{L} \tag{12.16}$$

However, the constant A_1 is indeterminant; that is, the maximum amplitude of the buckled column cannot be determined by this approach. It must be determined by the theory of the elastica. The interpretation of Eq. (12.14) for higher values of n follows.

Higher Buckling Loads; $n > 1$

A slender pinned-end elastic column (one with a large slenderness ratio L/r) has more than one buckling or critical load, Eq. (12.14). If the column is restrained from buckling into a single lobe by a lateral stop, it may buckle at a load higher than P_{cr}. For example, a column with pinned ends can buckle in the form of a single sine lobe at the critical load P_{cr} (Fig. 12.7a). If, however, it is prevented from bending in the form of one lobe by restraints at its midpoint, the load can increase until it buckles into two lobes ($n = 2$; Fig. 12.7b). By Eq. (12.14), for $n = 2$, the critical load and critical stress are

$$P_2 = 4 \frac{\pi^2 EI}{L^2} = 4P_{cr}, \qquad \sigma_{cr(2)} = \frac{P_2}{A} = 4 \frac{\pi^2 E}{(L/r)^2} \tag{12.17}$$

Similarly, for $n = 3$,

$$P_3 = 9 \frac{\pi^2 EI}{L^2} = 9P_{cr}, \qquad \sigma_{cr(3)} = \frac{P_3}{A} = 9 \frac{\pi^2 E}{(L/r)^2} \tag{12.18}$$

or for any number of lobes n,

$$P_n = n^2 \frac{\pi^2 EI}{L^2} = n^2 P_{cr}, \qquad \sigma_{cr(n)} = \frac{P_n}{A} = n^2 \frac{\pi^2 E}{(L/r)^2} \tag{12.19}$$

The buckling modes for $n = 1$, 2, and 3 are shown in Fig. 12.7.

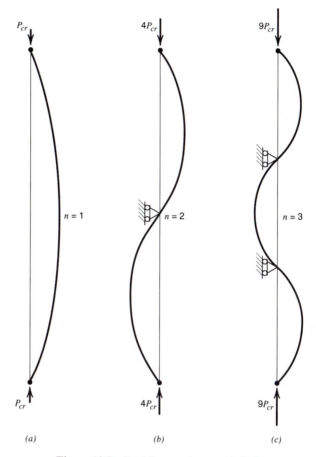

$$n = 1 \qquad n = 2 \qquad n = 3$$

(a) (b) (c)

Figure 12.7 Buckling modes: $n = 1, 2, 3$.

In summary, the equilibrium method is characterized by the question: What are the values of the load for which the perfect system admits nontrivial equilibrium configurations? As demonstrated above for the pin-ended column, an infinite number of such values (eigenvalues), $n = 1, 2, 3, \ldots, \infty$, exist. However, practically speaking, usually only the smallest value, P_{cr}, is significant.

The Imperfection Method

In the imperfection method, we acknowledge the fact that a real column usually is loaded eccentrically (the line of action of the compressive force does not lie exactly along the centroidal axis of the column or the column is not exactly straight). A common eccentricity occurs when the line of action of the load is displaced a distance e from the centerline of the column (Fig. 12.8a, c, and d). As an example, consider the column shown in Fig. 12.8a. A free-body diagram of a segment of the column in the displaced position is shown in Fig. 12.8b. By the equilibrium of moments about point A, we have $\sum M_A = M(x) + Pex/L + Py = 0$. Hence,

$$M(x) = -Py - Pe\frac{x}{L} \qquad (12.20)$$

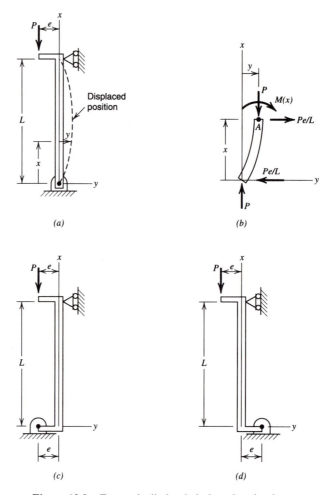

Figure 12.8 Eccentrically loaded pinned-end columns.

Thus, by Eqs. (12.6) and (12.20), we obtain $M(x) = +EI d^2y/dx^2 = -Py - Pex/L$, or dividing by EI, we obtain

$$\frac{d^2y}{dx^2} + k^2y = -\frac{k^2ex}{L}, \qquad k^2 = \frac{P}{EI} \tag{12.21}$$

The boundary conditions associated with Eq. (12.21) are

$$y = 0 \qquad \text{for } x = 0, L \tag{12.22}$$

The general solution of Eq. (12.21) is

$$y = A\sin kx + B\cos kx - \frac{ex}{L} \tag{12.23}$$

where A and B are constants determined by the boundary conditions. Hence,

Eqs. (12.22) and (12.23) yield the solution

$$y = e\left(\frac{\sin kx}{\sin kL} - \frac{x}{L}\right) \qquad (12.24)$$

As the load P increases (Fig. 12.8a), the deflection of the column increases. When P reaches the value for which $kL = n\pi$, $n = 1, 2, 3, \ldots$, the deflection given by Eq. (12.24) becomes infinitely large, since $\sin kL$ goes to zero as kL goes to $n\pi$. The term $\sin kL$ becomes zero for the first time when $n = 1$. Then, $P_1 = P_{cr} = \pi^2 EI/L^2$, the Euler load [Eq. (12.2)]. Thus, we see that the imperfection method yields the Euler buckling load for the pin-ended column. Other types of eccentricities, such as those shown in Figs. 12.8c and d, may also be analyzed by the method of imperfections (see Problems 12.8 and 12.9).

In summary, the imperfection method is characterized by the question: What is the value of the load for which the deflection of an imperfect system increases beyond any limit?

The Energy Method

The energy method is based on the first law of thermodynamics; namely, as a system is moved from one configuration to a second configuration, *the work that external forces perform on a system plus the heat energy that flows into the system equals the increase of internal energy of the system plus the increase of the kinetic energy of the system.* For a conservative (elastic) system, the increase in internal energy is equal to the change in strain energy (see Sec. 5.1). Thus, for a conservative system, the first law of thermodynamics is, in equation form,

$$\delta W + \delta H = \delta U + \delta K \qquad (12.25)$$

where δW is the work of the external forces as the system is moved from one configuration to a second configuration, δH the corresponding heat energy added to the system δU the corresponding change in the strain energy, and δK the corresponding change in kinetic energy. See also Sec. 3.1.

Application to Column Buckling. Consider an elastic column that is subjected to its critical axial load. A slight disturbance will cause the column to "kick out" or buckle into a new equilibrium position. Dynamically, the elastic column will vibrate about the buckled position indefinitely. This vibration represents an increase in kinetic energy. However, for an infinitesimal buckling displacement, the kinetic energy δK is of a higher order than the energies δW and δU; that is, $\delta K \ll \delta W$ or δU. In addition we assume an adiabatic system, so that $\delta H = 0$. Then, for an adiabatic elastic column, Eq. (12.25) may be approximated by the condition

$$\delta W = \delta U \qquad (12.26)$$

In the energy approach (Timoshenko and Gere, 1961), it is assumed that the column may buckle when the load first reaches a value for which $\delta W = \delta U$. If there exists a family of such loads, the critical load is the minimum load of the family.

To solve the problem for the lowest possible load by the Rayleigh method (which is an approximate solution; Langhaar, 1989), we may reduce the problem to a single degree of freedom by assuming a simple deflection form, say, $y = A \sin \pi x / L$. However, more generally, we may represent the buckled configuration by a Fourier series that satisfies the end conditions. For example, consider the pin-ended column subjected to an axial compressive load P, (Fig. 12.9). In the small displacement theory, it is assumed that the effect of the axial shortening of the column on the strain energy is of secondary importance. [This assumption is also fairly accurate for large displacements of the column (Kounadis and Mallis, 1988).] It is further assumed that the length L of the column remains essentially unchanged when it buckles. However, in the buckled position the load P is lowered a distance $(L - b)$ from its buckled position (Fig. 12.9).

Let us assume that the displacement y may be represented by the Fourier series

$$y(x) = \sum_{n=1}^{\infty} a_n \sin \frac{n\pi x}{L} \tag{12.27}$$

This equation satisfies the end conditions ($y = 0$ for $x = 0, L$) for the pin-ended column. Next consider the change in strain energy from the unbuckled position to the buckled position. By Eqs. (5.19) and (12.6), the strain energy due to bending of the column is

$$U_M = \frac{1}{2} \int_0^L EI(y'')^2 \, dx \tag{12.28}$$

where $y'' = d^2 y / dx^2$. Utilizing the orthogonality condition for the sine series,

$$\int_0^L \sin \frac{n\pi x}{L} \sin \frac{m\pi x}{L} \, dx = 0 \quad \text{for } m \neq n; \qquad = L \quad \text{for } m = n \tag{12.29}$$

we obtain by integration, with Eqs. (12.27) and (12.28)

$$U_M = \frac{\pi^4 EI}{4L^3} \sum_{n=1}^{\infty} n^4 a_n^2 \tag{12.30}$$

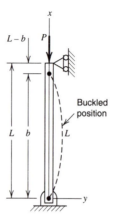

Figure 12.9

As noted previously, in the small displacement theory, we ignore the shortening of the column. Then, initially in the unbuckled position, $U = 0$. Therefore,

$$\delta U = U_M - 0 = \frac{\pi^4 EI}{4L^3} \sum_{n=1}^{\infty} n^4 a_n^2 \qquad (12.31)$$

To compute the work done by the load P, we recall that the bar will buckle without changing its length L significantly. Note also that an element ds of the column is given by the relation $ds^2 = dx^2 + dy^2$ or $ds = \sqrt{1 + (y')^2}\, dx$ and, therefore,

$$L = \int_0^L ds = \int_0^b \sqrt{1 + (y')^2}\, dx$$

By the binomial expansion,

$$\sqrt{1 + (y')^2} = [1 + (y')^2]^{1/2} = 1 + \tfrac{1}{2}(y')^2$$

to second-order terms in y'. Therefore,

$$L - b \approx \tfrac{1}{2}\int_0^b (y')^2\, dx \approx \tfrac{1}{2}\int_0^L (y')^2\, dx$$

Consequently, the work done by force P is

$$\delta W = P(L - b) = \frac{P}{2}\int_0^L (y')^2\, dx$$

or with Eq. (12.27)

$$\delta W = \frac{\pi^2 P}{4L} \sum_{n=1}^{\infty} n^2 a_n^2 \qquad (12.32)$$

With Eqs. (12.31) and (12.32), the criterion $\delta W = \delta U$ yields

$$P = \frac{\pi^2 EI}{L^2} \sum_{n=1}^{\infty} \frac{n^4 a_n^2}{n^2 a_n^2},$$

or

$$P = \frac{\pi^2 EI}{L^2} \frac{a_1^2 + 2^4 a_2^2 + 3^4 a_3^2 + \cdots}{a_1^2 + 2^2 a_2^2 + 3^2 a_3^2 + \cdots} \qquad (12.33)$$

If $a_1 \neq 0$ and $a_2 = a_3 = a_4 = \cdots = 0$, then $P = \pi^2 EI/L^2$. By inspection if any of the other a's are not zero, $P > \pi^2 EI/L^2$. Hence, the minimum buckling load is $P_1 = P_{cr} = \pi^2 EI/L^2$, with the corresponding displacement mode $y = a_1 \sin \pi x/L$. This value of P_1 and its displacement mode agree with those obtained by the imperfection method and by the equilibrium method. For $a_2 \neq 0$ and all other a's zero, we obtain the second mode and the buckling load $P_2 = 4\pi^2 EI/L^2 = 4P_{cr}$, as before. Similarly, for $a_3 \neq 0$ and all other a's zero, we get $P_3 = 9P_{cr}$, and so on for the other modes.

In summary, for an elastic column (or more generally, for any elastic structure), the energy method seeks an answer to the question: When does the load reach a value for which the work done by the external forces during the buckling deflection first equals the increase in the strain energy due to buckling? The equality yields an expression for the buckling load.

12.4

EULER BUCKLING OF COLUMNS WITH GENERAL END CONSTRAINTS

Consider a straight elastic column (Fig. 12.10) with linearly elastic end restraints. Let K_1, K_2 denote the elastic constants for the rotational springs, and k_1, k_2 denote the elastic constants for the extensional springs. Let an axial compressive load P be applied to the column. Initially, for small values of P, the column remains straight. When the load increases, a critical value is reached at which any small lateral disturbance will cause the column to displace laterally (to buckle). In the buckled position, the potential energy of the column-spring system is

$$V = \tfrac{1}{2}K_2(y_2')^2 + \tfrac{1}{2}k_2(y_2)^2 + \tfrac{1}{2}K_1(y_1')^2 + \tfrac{1}{2}k_1(y_1)^2$$
$$+ \tfrac{1}{2}EI \int_0^L (y'')^2 \, dx - \tfrac{1}{2}P \int_0^L (y')^2 \, dx \qquad (12.34)$$

where y_1, y_2 denote the displacements at $x=0$ and $x=L$, respectively, and primes denote derivatives with respect to x. The displaced equilibrium position of the column is given by the principle of stationary potential energy (see Sec. 5.1); that is, by the condition that the first variation δV of V, under a virtual displacement δy, vanishes identically (Langhaar, 1989). By Eq. (12.34), we set δV equal to zero.

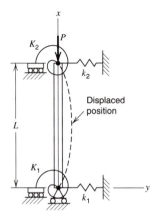

Figure 12.10

Thus, we find

$$\delta V = K_2 y_2' \, \delta y_2' + k_2 y_2 \, \delta y_2 + K_1 y_1' \, \delta y_1' + k_1 y_1 \, \delta y_1$$

$$EI \int_0^L y'' \, \delta y'' \, dx - P \int_0^L y' \, \delta y' \, dx = 0 \qquad (12.35)$$

Integration of Eq. (12.35) by parts yields

$$(K_2 y_2' + EI y_2'') \delta y_2' + (K_1 y_1' - EI y_1'') \delta y_1' + (k_2 y_2 - EI y_2''' - P y_2') \delta y_2$$

$$+ (k_1 y_1 + EI y_1''' + P y_1') \delta y_1 + \int_0^L (EI y'''' + P y'') \delta y \, dx = 0 \qquad (12.36)$$

The necessary and sufficient conditions that $\delta V = 0$ are that (see Langhaar, 1989)

$$EI y'''' + P y'' = 0 \qquad (12.37)$$

and

$$(K_2 y_2' + EI y_2'') \delta y_2' = 0$$
$$(K_1 y_1' - EI y_1'') \delta y_1' = 0$$
$$(k_2 y_2 - EI y_2''' - P y_2') \delta y_2 = 0$$
$$(k_1 y_1 + EI y_1''' + P y_1') \delta y_1 = 0 \qquad (12.38)$$

Equation (12.37) is the *Euler equation* for the column, and Eqs. (12.38) are the *boundary conditions.* Equations (12.38) include both *natural boundary conditions* and *forced boundary conditions*, as demonstrated in the following discussion.

The general solution of Eq. (12.37) is

$$y = A \sin kx + B \cos kx + Cx + D \qquad (12.39)$$

where $k^2 = P/EI$ and A, B, C, D are constants.

If y_1, y_2 and their first derivatives y_1', y_2' are arbitrary (not forced), their variations $\delta y_1, \delta y_2, \delta y_1'$, and $\delta y_2'$ are also arbitrary (nonzero). Then, Eqs. (12.38) and (12.39) yield the natural boundary conditions

$$\left(\cos kL - \frac{EIk}{K_2} \sin kL \right) A - \left(\sin kL + \frac{EIk}{K_2} \cos kL \right) B + \frac{C}{k} = 0$$

$$A + \frac{EIk}{K_1} B + \frac{C}{k} = 0$$

$$\left(\sin kL + \frac{EIk^3}{k_2} \cos kL - \frac{Pk}{k_2} \cos kL \right) A$$

$$+ \left(\cos kL - \frac{EIk^3}{k_2} \sin kL + \frac{Pk}{k_2} \sin kL \right) B + \left(L - \frac{P}{k_2} \right) C + D = 0$$

$$\left(\frac{Pk}{k_1} - \frac{EIk^3}{k_1} \right) A + B + \frac{P}{k_1} C + D = 0 \qquad (12.40)$$

Equations (12.40) are a set of four homogeneous algebraic equations in A, B, C, D. For a nontrivial solution of Eq. (12.37), the determinant Δ of the coefficients A, B, C, D must vanish; that is, $\Delta = 0$. This equation leads to the critical load (buckling) condition of the column.

If certain of the end displacements (y_1, y_2) and end slopes (y_1', y_2') of the column are forced (given), that is, they are not arbitrary, the associated variations must vanish. These specified conditions are called the forced boundary conditions (also called *geometric*, *kinematic*, or *essential* boundary conditions). For example, for pinned ends, we require that y_1 and y_2 be set (forced) equal to zero; that is, we require

$$y_1 = 0 \quad \text{for } x = 0 \quad \text{and} \quad y_2 = 0 \quad \text{for } x = L \tag{12.41}$$

Hence, the variations $\delta y_1 = \delta y_2 = 0$. Then, the last two of Eqs. (12.38) are identically satisfied. The first two of Eqs. (12.38) yield the natural (unforced) boundary conditions for pinned ends, since y_1' and y_2' and hence $\delta y_2'$ are arbitrary (nonzero). Also, for pinned ends, $K_1 = K_2 = 0$. Therefore, Eqs. (12.8) yield the natural boundary conditions (since $EI \neq 0$)

$$y_1'' = y_2'' = 0 \tag{12.42}$$

Equations (12.39), (12.41), and Eq. (12.42) yield $B = C = D = 0$, and $A \sin kL = 0$. Hence, $k_n L = n\pi$, $n = 1, 2, 3, \ldots$, and the critical load (for $n = 1$) is $P_{cr} = k_1^2 EI = \pi^2 EI / L^2$. This result agrees with that obtained in Sec. 12.3. In the following examples, columns with other end conditions are treated. See also Table 12.1, where a summary of column buckling loads and the corresponding modes for several end conditions are listed.

TABLE 12.1
Comparison of Boundary-Condition Effects

Boundary Conditions	Critical Load	Deflected Shape	Effective Length KL
Simple support-simple support	$\dfrac{\pi^2 EI}{L^2}$		L
Clamped-clamped	$\dfrac{4\pi^2 EI}{L^2}$		$\frac{1}{2}L$
Clamped-simple support	$\dfrac{2.04\pi^2 EI}{L^2}$		$0.70L$
Clamped-free	$\dfrac{\pi^2 EI}{4L^2}$		$2L$
Clamped-guided	$\dfrac{\pi^2 EI}{L^2}$		L
Simple support-guided	$\dfrac{\pi^2 EI}{4L^2}$		$2L$

EXAMPLE 12.1
Column with One End Clamped and the Other End Free

Consider a column clamped at one end ($x = 0$) and free at the other end ($x = L$); see Table 12.1. At $x = 0$, the forced boundary conditions are

$$y(0) = y_1 = 0, \qquad y'(0) = y'_1 = 0 \tag{a}$$

Hence, $\delta y_1 = \delta y'_1 = 0$. Also, at the free end, $K_2 = k_2 = 0$. Then, Eqs. (12.38) yield the natural boundary conditions

$$y''_2 = 0, \qquad EIy'''_2 + Py'_2 = 0 \tag{b}$$

Equations (a) and (b), with Eq. (12.39), yield the results

$$B + D = 0, \qquad kA + C = 0$$
$$A \sin kL + B \cos kL = 0, \qquad A \cos kL = 0 \tag{c}$$

The solution of Eqs. (c) is $A = C = 0$, $D = -B$, and $\cos kL = 0$. The condition $\cos kL = 0$ yields $k_n L = (2n - 1)\pi/2$. For $n = 1$, the eigenvalue is $k_1 = \pi/2$. Hence, the minimum buckling load is

$$P_1 = k_1^2 EI = \frac{\pi^2 EI}{(2L)^2} = \frac{\pi^2 EI}{(L_{\text{eff}})^2} = \tfrac{1}{4} P_{cr}$$

where

$$P_{cr} = \frac{\pi^2 EI}{L^2}$$

denotes the critical load of a pinned-end column of length L and where $L_{\text{eff}} = 2L$ denotes the *effective length* of the clamped-free ended column. The effective length of a given column is the length of an equivalent column with pinned ends that will buckle at the same critical load as the given column with its actual end conditions. The effective length L_{eff} is often denoted by KL, where K is the *effective length factor* (Table 12.1).

EXAMPLE 12.2
Column with Clamped (Fixed) Ends

Consider a column with clamped ends (Table 12.1). The specified (forced) boundary conditions are

$$\text{for } x = 0, \quad y_1 = y'_1 = 0; \qquad \text{for } x = L, \quad y_2 = y'_2 = 0 \tag{a}$$

Consequently, the variations $\delta y_1 = \delta y_2 = \delta y'_1 = \delta y'_2 = 0$, and Eqs. (12.40) are satisfied identically. Therefore, there are no natural boundary conditions for this problem; that is, all the boundary conditions are forced. Equations (a), with Eq. (12.39), yield

$$B + D = 0 \text{ (or } D = -B), \qquad kA + C = 0 \text{ (or } C = -kA)$$
$$A(\sin kL - kL) + B(\cos kL - 1) = 0, \qquad A(\cos kL - 1) - B \sin kL = 0 \tag{b}$$

The first two of Eqs. (b) express C and D in terms of A and B. For a nontrivial solution, the determinant of the coefficients of the last two of Eqs. (b) in A and B must vanish identically. This requirement yields the result

$$kL \sin kL = 2(1 - \cos kL) \tag{c}$$

Equation (c) is satisfied by the condition $k_n L = 2n\pi$. The minimum buckling load for $n = 1$ is

$$P_1 = k_1^2 EI = \frac{\pi^2 EI}{(L/2)^2} = \frac{\pi^2 EI}{(L_{\text{eff}})^2} = 4P_{cr} \tag{d}$$

where

$$P_{cr} = \frac{\pi^2 EI}{L^2}$$

and $L_{\text{eff}} = L/2$ denotes the *effective length* of the column with fixed ends.

For specific values of K_1, K_2, k_1, k_2 that are neither zero nor infinity, the buckling load is obtained by setting the determinant Δ of the coefficients A, B, C, D in Eq. (12.40) equal to zero. Then, in general, the determinantal equation $\Delta = 0$ must be solved numerically for the minimum buckling load P_1. In general, the minimum buckling loads for columns, with end conditions other than pinned, are given by the expression $P_1 = \pi^2 EI/(L_{\text{eff}})^2 = Q P_{cr}$, where $P_{cr} = \pi^2 EI/L^2$ (the Euler load for the pin-ended column), L_{eff} is the effective length of the column, and Q is a positive number determined by the values of K_1, K_2, k_1, k_2.

12.5

LOCAL BUCKLING OF COLUMNS

Consider a column with a cross section that is formed with several thin-wall parts (e.g., a channel, an angle, or a wide-flange I-beam). Depending on the relative cross-sectional dimensions of a flange or web, such a column may fail by local buckling of the flange or web, before it fails as an Euler column. For example, consider the test results of an aluminum column that has an equal-leg angle cross section of length b and wall thickness t, with dimensions as indicated in Fig. 12.11 (Bridget et al., 1934). The experimentally determined failure of the column exhibits distinctively different characteristics, depending on the ratio t/b. If the ratio t/b is relatively large, the column buckles as an Euler column. However, if t/b is relatively small, the column fails by buckling or wrinkling of one side or leg (Fig. 12.11) before it buckles as an Euler column. This type of failure is referred to by several names, namely, sheet buckling, plate buckling, crimping, wrinkling, or more generally, as *local buckling*, in contrast to Euler (*global*) buckling. Figure 12.12 shows a sketch of similar local buckling of the thin flanges of a channel section or one-half of an H-section (Stowell et al., 1951).

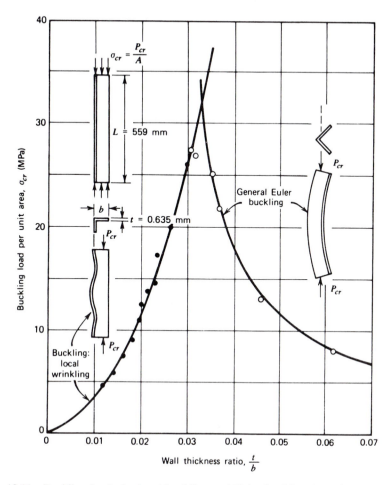

Figure 12.11   Buckling loads for local buckling and Euler buckling for columns made of 245 TR aluminum ($E = 74.5$ GPa).

Local buckling of a compressed thin-wall column may not cause immediate collapse of the column. However, it alters the stress distribution in the system, reduces the compressive stiffness of the column, and generally leads to collapse at loads less than the Euler load. The prediction of the occurrence of local buckling has been studied by a number of authors (see, e.g., Timoshenko and Gere, 1961; Rhodes, 1978). The study of compression of such columns beyond the local buckling load is less extensive (Rhodes and Harvey, 1976). In a sense, the effect of local buckling in a column is comparable to the effect of imperfections (Rhodes, 1978). In most instances, local buckling reduces the collapse load of the column. However, under certain conditions, load redistribution may not be particularly harmful, in that additional load may be applied before the strength of the column, as a whole, is exceeded (Rhodes and Harvey, 1976).

In the design of columns in building structures using hot-rolled steel, local buckling is controlled by selecting cross sections with t/b ratios such that the critical stress for local buckling will exceed the yield stress of the material (AISC, 1986). Hence, local buckling will not occur since the material will yield first. Local buckling is controlled in cold-formed steel members by the use of effective widths of

Figure 12.12   Local buckling (wrinkling) of thin flanges of channel section (or half of H-section; see Stowel et al., 1951).

the various compression elements (leg of an angle, flange of a channel, etc.), which account for the relatively small t/b ratio. These effective widths are then used to compute effective (reduced) cross-section properties, A, I, and so forth (AISI, 1986).

12.6

INELASTIC BUCKLING OF COLUMNS

In the previous discussion, we considered mainly elastic buckling of columns. In this section, we account for the fact that the average stress in the column may exceed the proportional limit stress of the material before buckling as an Euler column occurs.

Inelastic Buckling
As noted previously, buckling of columns that have intermediate slenderness ratios (L/r) is not solely affected by elastic action. For example, let the value of L/r be re-

latively small, so that the compressive stress in the column reaches the compressive proportional limit stress or the yield point stress (see Sec. 1.4) of the material before the load reaches the elastic buckling load. Then, the behavior of the column may be very similar to that of elastic buckling; that is, the column may undergo a rather abrupt lateral deflection at a fairly well-defined load. This behavior is called *inelastic* or *plastic buckling*.

The problem of determining the buckling load for a column that does not buckle before it is strained inelastically may be considered for two kinds of material behavior. In one case, the material in the column has a flat-top stress-strain curve. The initial nonzero slope of the compressive stress-strain curve changes abruptly to zero when the proportional limit stress is reached. Then, the stress-strain curve becomes horizontal and remains so until relatively large inelastic strains are developed (Fig. 1.5). For such materials, the proportional limit stress and yield point stress are essentially equal. Mild steel usually exhibits this type of stress-strain curve.

A column with a relatively small L/r (less than approximately 100) and made of structural steel having a yield point usually buckles when the average stress in the column is slightly less than the proportional limit of the material. When the average stress in the column is equal to the yield point, it is impossible for the column to develop a resisting moment until the column has deflected sufficiently to cause the most highly strained material in the column to strain harden. The column usually fails by excessive deflection before this happens.

In the second kind of material behavior, the compressive stress-strain curve changes in slope gradually as the stress is increased above the proportional limit (Figs. 1.3 and 1.4); aluminum alloys and some heat-treated steels, for example, exhibit this kind of behavior. Also, residual stresses cause this kind of behavior in rolled structural steel shapes (see Salmon and Johnson, 1990, p. 308). Inelastic buckling loads for columns made of this second kind of material behavior are considered in the following discussion.

Columns made of materials that exhibit this second type of stress-strain diagram (e.g. aluminum, alloy steels, and rolled structural steel shapes in which significant residual stresses exist) may have relative dimensions such that before buckling, small amounts of inelastic strains occur. These inelastic strains, however, may not be great enough to cause significant damage to the column. In fact, the maximum inelastic strain in the column, at impending buckling, is often of the same order of magnitude as that of the elastic strain at the proportional limit of the material. Usually, it is much less than the strain corresponding to the yield stress based on a 0.2% offset. By permitting this small amount of inelastic strain, a larger design load may be justified than if only strains within the elastic range were permitted. However, when the buckling load is finally reached, the deflection of the column may increase suddenly and result in a catastrophic collapse.

Two Formulas for Inelastic Buckling of an Ideal Column

The buckling load for a column is the axial load that holds the column in a slightly deflected position. Since an ideal column will not bend under axial load, a small lateral force must be applied to produce the initial deflection. This loading procedure may be carried out in any number of ways. For instance, the loading history might be (1) the lateral force may be applied first, the axial load required to hold the column in the slightly bent position may be applied next, and the lateral force is removed; or (2) the unknown critical load may be applied first, the lateral force may be applied next to cause a lateral deflection, and then the lateral force is removed. For *elastic* behavior of the column, the solution for the Euler buckling load is the same for the two procedures, since the physical process is conservative (*reversible*) and, hence, does not depend on the strain history.

For a system in which *inelastic* behavior occurs, the physical process is *irreversible*. Hence, the loading history (the order of force application) influences the resulting value of the buckling load. The main condition involved is whether a single-valued relationship exists between stress and strain for the loading/unloading process (see Sec. 4.2). The stress-strain relationship is single-valued if all strains are elastic or if inelastic strains increase monotonically. The stress-strain relationship is not single-valued when fibers in the member are allowed to unload elastically from an inelastic state.

Accordingly, two approaches have been developed to predict inelastic buckling loads for an ideal column. They differ in their assumptions regarding the loading history. In case (a), the lateral force and the last increment of the axial load are applied simultaneously so that the strains in all the fibers at any cross section increase, although they are not uniformly distributed on the section after the lateral force is applied. In case (b), an axial load equal to the buckling load is applied first, followed by application of a small lateral force that deflects the column. The bending in case (b) causes the strains in the fibers on the convex side to decrease (unload) and on the concave side to increase.

The essential difference between the two cases lies in the fact that, in case (b), the strains in some of the fibers on the convex side decrease elastically. Hence, the change in stress $\Delta\sigma$ accompanying the decrease in strain $\Delta\epsilon$ is given by $\Delta\sigma = E\,\Delta\epsilon$, in which E is the elastic modulus. However, in case (a), $\Delta\sigma = E_T\,\Delta\epsilon$, in which E_T is the tangent modulus corresponding to the inelastic stress σ ($\sigma = P/A$).

The buckling load in case (a) is called the *tangent-modulus load* and is given by the expression $P_T = \pi^2 E_T I/L^2$. In case (b), the buckling load is called the *double-modulus load* and is given by the expression $P_D = \pi^2 E_D I/L^2$, in which E_D is the effective modulus or double modulus, since it is expressed in terms of E and E_T.

The double-modulus theory was considered to be the more accurate theory of inelastic column buckling until Shanley (1946) showed that it represented a paradox requiring physically unattainable conditions. A development of the double-modulus theory is given by Bleich (1952); see also Bazant and Cedolin (1991). A comparison of predicted values of buckling loads by the tangent-modulus theory and double-modulus theory is given in the third edition of this book (Boresi et al., 1978). Note that neither of these theories applies to a column made of an elastic, perfectly plastic material (see Figs. 4.3a and 4.4a), since for such materials, both E_T and E_D are zero, which leads to $P_T = P_D = 0$.

A development of the tangent-modulus formula follows. We recommend use of the tangent-modulus equation, since it generally leads to a good, conservative estimate of the maximum (buckling) load that a real column having slight imperfections can be expected to safely resist.

Tangent-Modulus Formula for an Inelastic Buckling Load

Let Fig. 12.13a represent a column subjected to a gradually increasing axial load P. For convenience, let the column have a rectangular cross section. It is also assumed that the slenderness ratio L/r is sufficiently small to preclude elastic buckling. The load P may therefore attain a value that causes a uniform stress σ on the cross section of the column that is greater than the proportional limit σ_{PL} (Sec. 1.4) of the material. The stress-strain diagram for the material is shown in Fig. 12.13d. In this diagram, σ represents the uniformly distributed compressive stress on each cross section of the column and ϵ the corresponding strain; it is assumed that the value of the buckling load P_T (or buckling stress $\sigma_T = P_T/A$) and the corresponding inelastic strain are represented by a point in the neighborhood of C on the stress-strain curve. Hence, as previously noted, inelastic buckling involves relatively small inelastic strains.

The problem is to find the smallest load $P_T = A\sigma_T$ that will cause the ideal column to remain in a slightly bent position when a small lateral force is applied (simultaneously with the last increment of axial load) and then is removed. As increments of the axial load P are applied to the ideal column, the longitudinal strain at n-n increases but remains uniformly distributed as shown by the lines marked 1, 2, 3, and 4 in Fig. 12.13b. As P approaches the value P_T, which we wish to determine, let a small lateral force be applied *simultaneously* with the last increment of load as P attains the value P_T. The resulting distribution of strain is as shown by the line marked 5 in Fig. 12.13b in which the lateral bending is greatly exaggerated.

The resulting stress distribution on section n-n is shown in Fig. 12.13c by the sloping line AB and is obtained from Fig. 12.13d by taking the stresses corresponding to the strains. The assumption that line AB is straight is equivalent to the assumption that the slope of tangent line to the stress-strain curve, such as that shown at C in Fig. 12.13d, is constant during the change in strain from ϵ to $\epsilon + \Delta\epsilon$. This assumption is justified because the increment $\Delta\epsilon$ is small for the small lateral bending imposed on the column. The slope of the tangent line at any point such as C is called the *tangent modulus* at point C and is denoted by E_T; the increment of stress

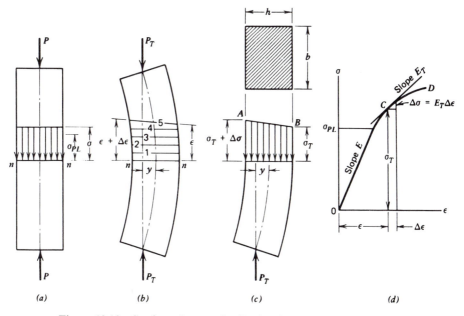

Figure 12.13   Strain and stress distribution for tangent-modulus load.

corresponding to $\Delta\epsilon$ is therefore $\Delta\sigma = E_T \Delta\epsilon$. The desired value P_T of the axial load P may now be found in the same manner that the Euler load for the beginning of elastic buckling is usually obtained. Thus, let Fig. 12.13c be a free-body diagram showing the forces acting on the lower half of the column. Equilibrium of the column requires that the external bending moment $P_T y$ for any cross section shall be equal and opposite to the resisting moment about the centroidal axis of the cross section of the internal forces on the section. This fact is expressed by the equation

$$P_T y = \frac{(\Delta\sigma/2)I}{h/2} \tag{12.43}$$

In Eq. (12.43), let $\Delta\sigma$ be replaced by $E_T \Delta\epsilon$ and, in turn, let $\Delta\epsilon$ be replaced by the expression h/R, which is obtained by relating the strain in the extreme fiber to the radius of curvature R of the column. Furthermore, for small deflections the curvature $1/R$ is given by the expression $1/R = -d^2y/dz^2$ [see Eq. (12.5)]. With these substitutions, Eq. (12.43) becomes

$$E_T I \frac{d^2 y}{dz^2} = -P_T y \tag{12.44}$$

The solution of this differential equation for a column with pinned ends leads to the buckling load (see the first subsection in Sec. 12.3)

$$P_T = \frac{\pi^2 E_T I}{L^2} \quad \text{or} \quad \sigma_T = \frac{P_T}{A} = \frac{\pi^2 E_T}{(L/r)^2} \tag{12.45}$$

in which P_T is the tangent modulus buckling load. It may be considered either the smallest load that will hold the ideal column in a slightly bent form or the largest load under which the ideal column will not bend. This formula is called the *tangent-modulus formula* or *Engesser's formula*.*

The solution of the tangent-modulus equation for a column of given material and dimensions involves a trial-and-error process for the reason that a value of E_T cannot be selected unless P_T is known. Furthermore, a stress-strain diagram for the given material must be available. The method of solution is illustrated in the following examples. An alternate procedure has been proposed by Rao (1991).

EXAMPLE 12.3
Tangent-Modulus Load P_T

A pinned-pinned column having a square cross section $b = 25.0$ mm on a side and a length of $L = 250$ mm is loaded axially through special bearing blocks that allow free rotation when bending of the column starts. The member is made of material for which the compressive stress-strain curve is shown by $0BC$ in Fig. E12.3. The tangent modulus for this stress-strain curve is shown by abscissas to the curve

---

* This theory is due in part to F. R. Shanley (1947, 1950).

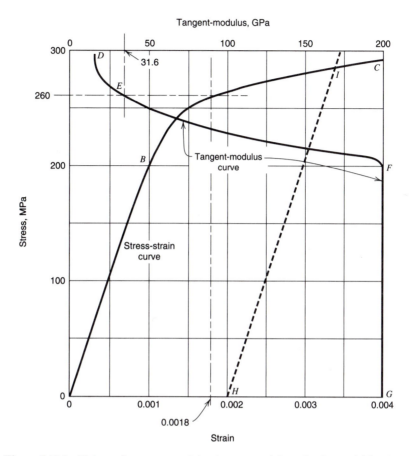

Figure E12.3   Values of tangent modulus for a material not having a yield point.

DEFG on the upper scale. (Point *F* corresponds to the proportional limit, point *B* on curve 0*BC*.) Compute the load P_T.

SOLUTION

By trial and error, we must select from the curves in Fig. E12.3 a set of corresponding values of stress σ_T and tangent modulus E_T that will satisfy Eq. (12.45). As a first trial value, select $E_T = 31.0$ GPa, which from curve *DEF* corresponds to $\sigma_T = P_T/A = 262$ MPa. For rectangular sections $r = b/\sqrt{12} = 7.217$ mm and $L/r = 34.64$. The right side of Eq. (12.45) is

$$\frac{\pi^2 E_T}{(L/r)^2} = \frac{\pi^2 (31 \times 10^3)}{(34.64)^2} = 255 \text{ MPa}$$

Since the left side of Eq. (12.45) is 262 MPa, a new trial is necessary. For the second trial assume that $E_T = 31.6$ GPa, which from curve *DEF* corresponds to $\sigma_T = 261$ MPa. The right side of Eq. (12.45) is now 260 MPa, which is sufficiently close to the assumed value. Hence, the buckling load is $P_T \cong 260(25)^2 = 163,000 \text{ N} = 163 \text{ kN}$.

EXAMPLE 12.4

In Example 12.3 the stress at the tangent-modulus load was found to be 260 MPa which from curve $0BC$ corresponds to a strain of 0.00180. If it is specified that the strain (or stress) in the column must not exceed the value corresponding to the yield stress of the material, based on 0.2% offset, will this stress and strain at the tangent modulus load be within the required limit?

SOLUTION

The yield stress based on 0.2% offset, as shown by the line HI in Fig. E12.3, is 288 MPa, which corresponds to a strain of 0.0034. Therefore, the stress and strain at the tangent-modulus load are less than values at the yield stress. It should be pointed out that the stress and strain at which the tangent-modulus load occurs in nearly all inelastic columns are smaller than the stress and strain values corresponding to the yield stress. This fact shows that, although the tangent-modulus formula is obtained on the assumption that some inelastic strain occurs, the inelastic strains that correspond to this load are smaller than the inelastic strains (the offset) that are usually assumed to be permissible without causing damage to the load-resisting behavior of the material or structure.

Direct Tangent-Modulus Method

The iterative method of applying the tangent modulus formula is suitable for calculating the buckling load for a given column. However, in design, one may wish to try several different column configurations to meet specifications most efficiently. Then, the procedure described below is more appropriate.

As noted in Example 12.3, the average tangent-modulus stress σ_T is determined by iterating between Eq. (12.45) and interpreted values from Fig. E12.3. This method is not satisfactory when more than one column configuration must be studied. Therefore, it is more expedient to determine a tangent-modulus stress curve as a function of the effective slenderness ratio of the column for stress greater than the proportional limit stress σ_{PL}. For this purpose, consider the compressive stress-strain diagram for 2024-T4 aluminum alloy, Fig. 12.14b. At several stress values $\sigma_T \geq \sigma_{PL}$, the corresponding slopes E_T of the stress-strain curve are determined from Fig. 12.14b. Corresponding values of σ_T and E_T are then substituted into Eq. (12.45) to determine the effective slenderness ratio L_{eff}/r. The values of L_{eff}/r are then plotted vs σ_T to obtain a tangent-modulus curve (Fig. 12.14a). The Euler curve CBG and tangent modulus curve DB intersect at point B at the proportional limit stress σ_{PL}. The curve DBG can be used to design both elastic and inelastic columns of 2024-T4 aluminum alloy. However, in practice, the procedure may be simplified further by approximating the tangent modulus curve DB and part of the Euler curve by a simpler empirical curve or equation (say, a straight line, a parabola, etc.); see Salmon and Johnson (1990).

Many empirical column formulas are represented in part by one of the following three types of column equations when the allowable stress design (ASD) method

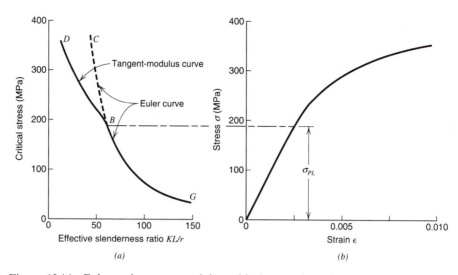

Figure 12.14   Euler and tangent-modulus critical stress for columns made of 2024-T4 aluminum alloy.

is used:

$$\sigma_{cr} = \frac{P_{cr}}{A} = \sigma_S - C_S\left(\frac{KL}{r}\right) \quad \text{(straight-line)} \tag{12.46}$$

$$\sigma_{cr} = \frac{P_{cr}}{A} = \sigma_P - C_P\left(\frac{KL}{r}\right)^2 \quad \text{(parabola)} \tag{12.47}$$

$$\sigma_{cr} = \frac{P_{cr}}{A} = \frac{\sigma_{GR}}{1 + C_{GR}(KL/r)^2} \quad \text{(Gordon–Rankine)} \tag{12.48}$$

where σ_{cr} and P_{cr} denote the critical stress and load for the column, respectively, and $KL = L_{\text{eff}}$ (see Table 12.1), σ_S, σ_P, σ_{GR}, C_S, C_P, and C_{GR} are empirical positive constants. These equations are usually referred to by the names in parentheses, and subscripts S, P, and GR refer to straight-line, parabola, and Gordon-Rankine formulas, respectively. The constants in Eqs. (12.46), (12.47), and (12.48) may be determined by fitting the equations to the tangent-modulus stress curve for materials like aluminum alloys, or by fitting the equations to experimental column buckling data for materials like structural steel.

For 2024-T4 aluminum alloy, we may approximate the curves of Fig. 12.14a as follows. The tangent-modulus curve and the portion of the Euler curve of interest are redrawn in Fig. 12.15. By inspection, the general shape of the tangent-modulus curve and a portion of the Euler curve may be approximated by a straight line. For example, the straight-line equation

$$\sigma_{cr} = 390 - 3.44\left(\frac{KL}{r}\right) \tag{12.49}$$

accurately approximates the curves for 2024-T4 aluminum alloy, provided that the effective slenderness ratio $KL/r \leq 80$, (Fig. 12.15). The Euler formula then may be

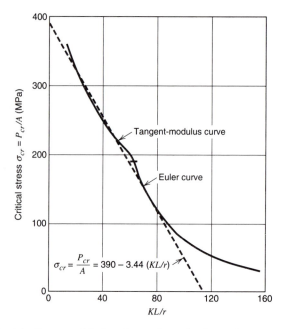

Figure 12.15   Straight-line empirical column formula for columns made of 2024-T4 aluminum alloy.

used for values of $KL/r \geq 80$. A straight-line equation is generally used for all aluminum alloys. However, the magnitudes of σ_S, C_S and the limiting value of KL/r are generally different for each alloy.

Design loads for columns depend on the safety factor used. The magnitude of the safety factor is specified by design codes, as is the range of KL/r for each column formula. For example, if a safety factor of 1.80 is used, the design load P for 2024-T4 aluminum alloy columns is given by the following design formulas:

$$\sigma = \frac{P}{A} = 216.6 - 1.91\left(\frac{KL}{r}\right), \qquad \frac{KL}{r} \leq 80 \tag{12.50}$$

$$\sigma = \frac{408,500}{(KL/r)^2}, \qquad \frac{KL}{r} \geq 80 \tag{12.51}$$

Design equations for steel columns are more complex. Generally, with appropriate constants, a parabolic equation is used for low values of KL/r (say, $KL/r \leq 120$), and the Euler equation is used for $KL/r \geq 120$ (Salmon and Johnson, 1990).

EXAMPLE 12.5
Buckling of a 2024-T4 Aluminum Alloy Column

The 2024-T4 aluminum alloy column shown in Fig. E12.5 has a length of 1.00 m and a solid rectangular cross section 40.0 mm by 75.0 mm. Its compressive stress-

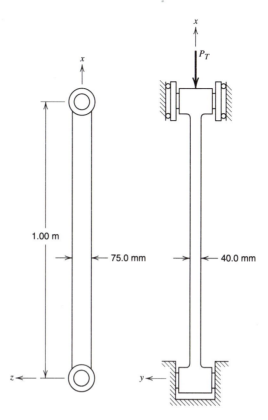

Figure E12.5

strain diagram is given in Fig. 12.14b. The column is supported at its ends by frictionless pins that allow rotation about the y axis and by rigid walls that prevent rotation about the z axis and lateral displacement of the ends in the y and z directions. Determine the design load for buckling of the column, based on a safety factor $SF = 2.50$.

SOLUTION

It is first necessary to determine how the column will buckle. The column will buckle in a direction perpendicular to the axis for which KL/r is a maximum, since the maximum slenderness ratio results in the smallest buckling load. If the column buckles in the z direction, about the y axis (Fig. E12.5), it rotates about the pinned ends ($K_y = 1$). Hence, the value of $(KL/r)_y$ may be computed as follows:

$$r_y = \sqrt{\frac{I_y}{A}} = \sqrt{\frac{bh^3}{12bh}} = \frac{h}{\sqrt{12}} = \frac{75.0}{\sqrt{12}} = 21.7 \text{ mm} \tag{a}$$

and

$$\left(\frac{KL}{r}\right)_y = \frac{(1)(1000)}{21.7} = 46.2 \tag{b}$$

If the column buckles in the y direction, about the z axis, it has fixed ends ($K_z = 0.50$). In this case,

$$r_z = \frac{h}{\sqrt{12}} = 11.5 \text{ mm}, \qquad \left(\frac{KL}{r}\right)_z = \frac{(0.50)(1000)}{11.5} = 43.5 \qquad \text{(c)}$$

Thus, the column buckles as a pin-ended column with $(KL/r)_y = 46.2$. With this value of KL/r, we obtain from Fig. 12.14a, $\sigma_{cr} = P_{cr}/A = 229$ MPa and, therefore, $P_{cr} = 687$ kN. The design load is $P = P_{cr}/SF = 687/2.50 = 275$ kN.

PROBLEMS
Section 12.3

12.1. An airplane compression strut 2.00 m long has an elliptical cross section with major and minor diameters of $b = 150$ mm and $a = 50.0$ mm, respectively. Calculate the slenderness ratio for pinned ends. If the member is made of spruce wood with $E = 11.0$ GPa, determine the load P that can be carried by the column based on a factor of safety of $SF = 1.50$, $I = \pi b a^3/64$, and $A = \pi ab/4$.

12.2. Three pinned-end columns each have a cross-sectional area of 2000 mm² and length of 750 mm. They are made of 7075-T5 aluminum alloy ($E = 72.0$ GPa and $\sigma_{PL} = 448$ MPa). One of the columns has a solid square cross section. A second column has a solid circular cross section. The third has a hollow circular cross section with an inside diameter of 30.0 mm. Determine the critical buckling load for each of the columns.

Ans. $P_{cr(square)} = 421$ kN, $P_{cr(circle)} = 402$ kN, $P_{cr(hollow\ circular)} = 686$ kN

12.3. An aluminum alloy ($E = 72.0$ GPa) extrusion has the cross section shown in Fig. P12.3. A 2.00-m length of the extrusion is used as a pin-ended column.

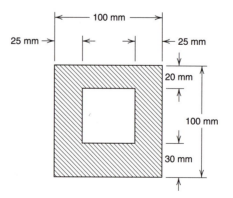

Figure P12.3

(a) Determine the minimum radius of gyration for the column cross and find the slenderness ratio of the column.

(b) Determine the buckling load for the column.

12.4. In Fig. P12.4, columns *AB* and *CD* have pinned ends, are made of an aluminum alloy ($E = 72.0$ GPa), and have equal rectangular cross sections 20 mm by 30 mm. Determine the magnitude of *P* that will first cause one of the columns to buckle. Assume elastic conditions.

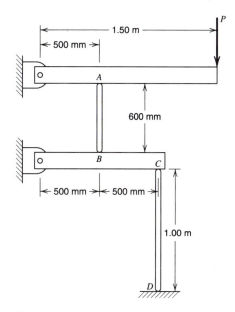

Figure P12.4

12.5. In Fig. P12.5, member *0A* can be considered rigid and weightless. Columns *BC* and *DF* are pin-ended with solid circular cross sections of diameter 10.0 mm. Column *BC* is made of structural steel ($E = 200$ GPa) and

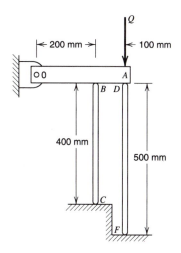

Figure P12.5

column DF is made of an aluminum alloy ($E = 72.0$ GPa). Determine the magnitude of Q that will first cause one of the columns to buckle.

12.6. In Fig. P12.6, member CD is a structural steel ($E = 200$ GPa) pipe with an outside diameter of 101.6 mm and inside diameter of 90.1 mm. The pipe has reinforced spherical end caps, so that the pipe acts as a pin-ended column. The column was used to support the weight (80 kN) of member AB while it was fixed in the position shown. In this position, the pipe supports the weight of AB. The exact location of member AB was obtained when the column had a uniform temperature of 20°C. After fixing member AB in position, the column temperature was increased uniformly by the sun and column buckled. Determine the temperature at which the column buckled. The coefficient of linear thermal expansion is $\alpha = 11.7 \times 10^{-6}$ per °C.

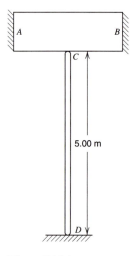

Figure P12.6

12.7. A long thin-wall steel ($E = 200$ GPa, $v = 0.29$) pressure vessel has a length of 9.00 m, an inside diameter of 100 mm, and a wall thickness of 2.00 mm. The ends of the vessel are reinforced hemispheres. The vessel is positioned between two rigid walls that touch each end of the vessel when the internal pressure is zero.

(a) Determine the internal pressure that will cause the vessel to buckle as a column. Assume that local buckling of the thin walls does not occur.

(b) What is the average axial stress in the vessel when it buckles?

12.8. Determine the Euler load for the column shown in Fig. 12.8c; see the discussion on the imperfection method in Sec. 12.3.

12.9. Determine the Euler load for the column shown in Fig. 12.8d; see the discussion on the imperfection method in Sec. 12.3.

Section 12.4

12.10. A steel ($E = 200$ GPa) column with a solid circular cross section has clamped ends, is 2.50 m long, and must support a load of 40.0 kN.

    (a) Determine the minimum required diameter using a factor of safety $SF = 2.00$.

    (b) What is the minimum value for the proportional limit in order for the column to buckle elastically?

12.11. An aluminum alloy ($E = 72.0$ GPa) column has the cross section shown in Fig. P12.11. It has a length of 9.00 m and is clamped at each end. A support at midlength of the column prevents deflection in the x direction, but does not prevent rotation of the section nor deflection in the y direction.

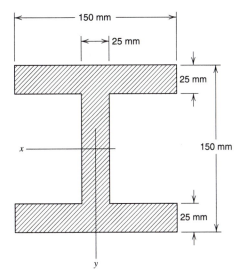

Figure P12.11

    (a) Determine the buckling load.

    (b) What is the minimum proportional limit in order for the column to buckle elastically?

12.12. In Fig. P12.12, bar AB is pinned at A. It can be considered rigid and weightless. The aluminum ($E = 72.0$ GPa) columns CD and FH have solid circular cross sections of diameter 100 mm. Column CD has pinned ends, and column FH is clamped at the bottom and pinned at the top. When the load $Q = 0$, bar AB is horizontal, and the two columns just are in contact with AB. Determine the least value of Q that will cause one of the columns to buckle.

12.13. Solve Problem 12.12 for the case where both columns are made of steel ($E = 200$ GPa). What is the minimum required proportional limit to ensure that the column will buckle elastically?

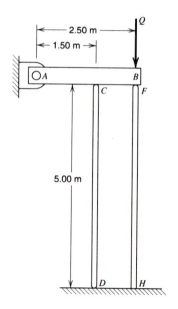

Figure P12.12

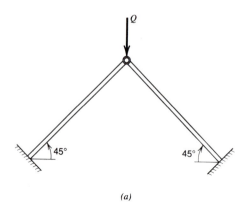

(a)

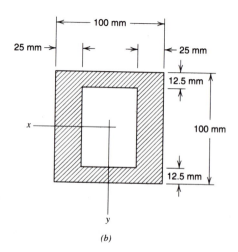

(b)

Figure P12.14

12.14. The two structural steel ($E = 200$ GPa) truss members in Fig. P12.14a are 5.00 m long and have the cross section shown in Fig. P12.14b. They are clamped at the bottom and are pinned together at the top. The members are free to deflect perpendicular to their plane; in which case, they both bend about the cross-section axis having the maximum moment of inertia. If they buckle in their plane, they both bend about the axis of minimum moment of inertia. Determine the magnitude of the load Q that will buckle the members.

Section 12.5

12.15. Assume that the dimensions of the angle section of the pinned-end column shown in Fig. 12.11 are $t = 0.635$ mm, $b = 12.0$ mm, and $L = 559$ mm. Show by Fig. 12.11 that the column fails as a unit by elastic buckling (not local buckling). Calculate the average stress in the column at the buckling load and compare it to the stress obtained experimentally (Fig. 12-11).

Section 12.6

12.16. Solve Example 12.3 for a column length of $L = 300$ mm.

12.17. A column made of the steel whose compressive stress-strain diagram is shown in Fig. E12.3 has a length of 1.00 m and solid circular cross section. The column must carry a design load of 810 kN with a factor of safety of $SF = 2.00$. Determine the required diameter of the column.

Ans. $D = 90.4$ mm

12.18. An aluminum alloy has a modulus $E = 72.0$ GPa and proportional limit $\sigma_{PL} = 310$ MPa. The equation $\sigma = A\epsilon^n$, where A and n are material constants, accurately approximates the compressive stress-strain diagram for small inelastic strains when it is made to coincide with the stress-strain diagram at the proportional limit and to pass through another test point whose coordinates are $\sigma = 370$ MPa and $\epsilon = 0.00600$.

(a) Determine values for A and n.

(b) A rectangular section column with dimensions 40 mm by 60 mm has fixed ends and is found to buckle at an average stress of $P_T/A = 345$ MPa; find the length L of the column.

12.19. For the aluminum alloy in Problem 12.18, determine two values of the slenderness ratio such that $P_T/A = \sigma_{PL}$.

Ans. $(L/r)_{max} = 47.9$ and $(L/r)_{min} = 35.0$

12.20. A rectangular cross-section (60.0 mm $\times$ 90.00 mm) column of 2024-T4 aluminum alloy is 1.00 m long. The top end of the column is pinned. The bottom of the column has a knife-edge support that runs perpendicular to the 90.0-mm dimension. The effect of the knife-edge support is to provide a clamped condition for deflection in the 90.0-mm direction and a pinned condition in the direction of the 60.0-mm dimension. Use Fig. 12.14 to determine the buckling load of the column.

12.21. A stainless steel column has the tangent-modulus curve shown in Fig. P12.21a and cross section illustrated in Fig. P12.21b. Its length is 2.20 m, and it has a clamped end at the top and pinned end at the bottom. Determine the design load for the column, based on a factor of safety $SF = 2.50$.

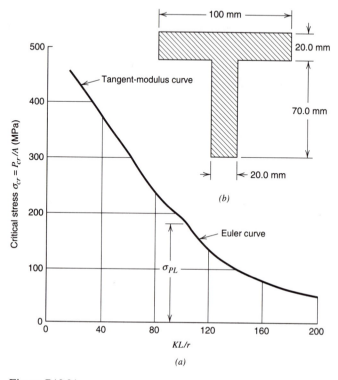

(b)

(a)

Figure P12.21

12.22. A 2024-T4 aluminum alloy column has the cross section shown in Fig. P12.22. It is 2.20 m long, with one end clamped and the other pinned.

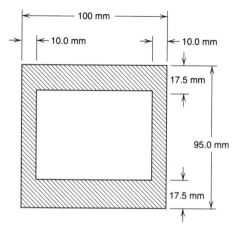

Figure P12.22

Determine the design load for the column, based on a factor of safety $SF = 1.80$ (see Fig. 12.14).

12.23. Columns *AB* and *CD* in Fig. P12.23*a* have identical rectangular cross sections 50.0 mm by 75.0 mm. The columns are made of a metal that has the stress-strain diagram shown in Fig. P12.23*b*. Column *AB* has clamped ends and column *CD* pinned ends. Assume that member *F* is rigid and weightless and that it is prevented from moving laterally. Also, neglect the rotation of member *F* due to a difference in shortening of the two columns. Determine the load *Q* that will first cause one of the columns to buckle.

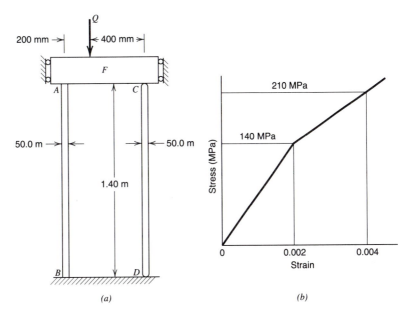

Figure P12.23

REFERENCES

American Institute of Steel Construction (AISC) (1986). *Manual of Steel Construction, Load and Resistance Factor Design*, 1st ed., Chicago, Ill.

American Iron and Steel Institute (AISI) (1986). *Specification for the Design of Cold-Formed Steel Structural Members*, Aug. 19, 1986 ed. Washington, D. C.

Bazant, Z. P. and Cedolin, L. (1991), *Stability of Structures*. New York: Oxford Univ. Press.

Bleich, F. (1952). *Buckling Strength of Metal Structures*. New York: McGraw-Hill, pp. 9–14.

Boresi, A. P., Sidebottom, O. M., Seely, F. B., and Smith, J. O. (1978). *Advanced Mechanics of Materials*, 3rd ed. New York: Wiley, pp. 649–653.

Bridget, F. J., Jerome, C. C., and Vosseller, A. B. (1934). Some New Experiments in Buckling of Thin-Walled Construction. *Trans. Amer. Soc. Mech. Engineers*, **56**: 569–578.

Brush, D. O. and Almroth, B. O. (1975). *Buckling of Bars, Plates, and Shells.* New York: McGraw-Hill.

Calladine, C. R. (1988). *Theory of Shell Structures.* Cambridge: Univ. Press.

Chajes, A. (1974). *Principles of Structural Stability Theory.* Englewood Cliffs, N.J.: Prentice Hall.

Euler, L. (1933). *Elastic Curves (Des Curvie Elasticis,* Lausanne and Geneva) translated and annotated by W. A. Oldfather, C. A. Ellis, and D. M. Brown. Reprinted from ISIS, No. 58, XX(1), 1774. Bruges, Belgium: Saint Catherine Press, Ltd.

Kounadis, A. N. and Mallis, J. (1988). On the Accuracy of Various Formulas for Establishing Large Axial Displacement of Columns. *Mech. Structures and Machines,* **16**(2): 123–146.

Langhaar, H. L. (1958). General Theory of Buckling. *Appl. Mech. Rev.,* **11**(11): 585–588.

Langhaar, H. L. (1989). *Energy Methods in Applied Mechanics.* Melbourne, Florida: Kreiger.

Manson, S. S. (1981). *Thermal Stress and Low-Cycle Fatigue.* Melbourne, Florida: Kreiger.

Moulin, D., Combescure, A., and Acker, D. (1989). A Review of Thermal Buckling Analysis Methods. *Nucl. Eng. Design,* **116**: 255–263.

Rao, B. N. (1991). A Simplified Procedure for Determining the Buckling Strength of Inelastic Columns. *Exper. Tech.* (Soci. of Exper. Mech.), **15**(2): 42–44.

Rhodes, J. (1978). Secondary Local Buckling in Thin-Walled Sections. *Acta Technica Academiae Scientiarum Hungaricae,* **87**(1-2): 143–153.

Rhodes, J. and Harvey, J. M. (1976). Plain Channel Section Struts in Compression and Bending Beyond the Local Buckling Load. *Int. J. Mech. Sci.,* **8**: 511–519.

Salmon, C. G. and Johnson, J. E. (1990). *Steel Structures Design and Behavior,* 3rd ed. New York: Harper and Row.

Shanley, F. R. (1946). The Column Paradox. *J. Aero. Sci.,* **13**(12): 678.

Shanley, F. R. (1947). Inelastic Column Theory. *J. Aero. Sci.,* **14**(5): 261–267.

Shanley, F. R. (1950). Applied Column Theory. *Trans. Amer. Soci. Civil Engineers,* **115**: 698–727.

Southwell, R. V. (1941). *Introduction to the Theory of Elasticity,* 2nd ed. London: Oxford Press, p. 434.

Stowell, E. Z., et al. (1951). Buckling Stresses for Flat Plates and Sections. *Proc. Amer. Soc. Civil Engineers,* **77**(7): 31.

Szilard, R. (1974). *Theory and Analysis of Plates.* Englewood Cliffs, N. J.: Prentice Hall.

Timoshenko, S. and Gere, J. M. (1961). *Theory of Elastic Stability,* 2nd ed. New York: McGraw-Hill.

PART III

SELECTED ADVANCED TOPICS

In Part III, Chapters 13 to 19, we introduce selected advanced topics of interest. In Chapter 13, the topic of linear plate theory is introduced. The effects of stress concentrations are discussed in Chapter 14. Fracture mechanics of metals and high cycle fatigue are treated in Chapters 15 and 16, respectively. Time-dependent deformation (creep) is addressed in Chapter 17, and contact stress problems are presented in Chapter 18. Finally, an introduction to and discussion of finite element applications are given in Chapter 19.

13

FLAT PLATES

In this chapter, we present the theory of elastic plates and a number of examples. We develop the basic equations (Sec. 13.1–13.7), and we outline methods of solution for rectangular plates (Sec. 13.8) and circular plates (Sec. 13.9) subjected to simple loading.

13.1

INTRODUCTION

A flat plate is a structural element or member whose middle surface lies in a flat plane. The dimension of a flat plate in a direction normal to the plane of its middle surface is called the *thickness* of the plate. A plate is characterized by the fact that its thickness is relatively small compared to the dimensions in the plane of the middle surface. As a consequence, the bending behavior of a plate depends strongly on the plate thickness, as compared to the in-plane dimensions of the plate.

Plates may be classified according to the magnitude of the thickness compared to the magnitude of the other dimensions and according to the magnitude of the lateral deflection compared to the thickness. Thus, we may speak of (1) relatively thick plates with small deflections, (2) relatively thin plates with small deflections, (3) very thin plates with large deflections, (4) extremely thin plates (membranes) that may undergo either large or small deflections, and so on. There are no sharp lines of distinction between these classifications; rather there are gradual transition regions between two categories, in which the response of the plate exhibits some of the characteristics of both categories.

Additional descriptions are applied to plates. For example, if the distance between the two surfaces (faces) of the plate is constant, the plate is said to be of constant thickness; if not, it is said to be of variable thickness. Further descriptions of a plate pertain, as we shall see, to the manner in which the plate edges are constrained and to the manner in which the plate material responds to load.

Some attention is given to anisotropic material behavior. However, the treatment presented here is largely a study of the small-deflection theory of thin constant thickness isotropic elastic plates with temperature effects included. In the general development of the theory, (x, y, z) coordinates are taken to be *orthogonal curvilinear plate coordinates*, where (x, y) are orthogonal *planar* curvilinear coordinates that lie in the middle surface of the plate and the z coordinate is perpendicular to the middle surface of the plate (Fig. 13.1).

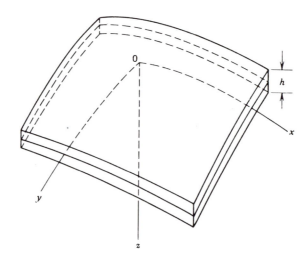

Figure 13.1   Flat plate coordinates.

In the development of the theory, kinematic relations and associated strain-displacement relations are presented. Stress resultants for a plate are defined, and the equations of equilibrium are derived by employing the principle of virtual work. Temperature effects are included in the elastic stress-strain relations. In turn, the stress resultant-displacement relations are derived including temperature effects. The boundary conditions for the plate follow directly from the principle of virtual work (Sec. 13.7).

The *small-deflection* theory of plates developed here is limited to cases in which the lateral displacement w of the plate in the z direction is less than about half of the plate thickness h. When $w < h/2$, small-deflection theory yields reasonably accurate estimates for plate behavior since the second-order effects that are omitted from the theory are negligible. However, when $w > h/2$, second-order effects become significant. Consequently, for a given lateral load, plate theory based upon small deflections yields displacements and stress resultants that are too large, relative to large-deflection theory. The primary effect of large deflections is to develop direct (membrane) tensile stresses that stiffen the plate. In other words, when displacements are large, in-plane tensile stresses are developed that influence both the stress resultants and the stiffness of the plate. Section 13.9 contains a more detailed discussion of the effects of these direct tensile stresses.

13.2

STRESS RESULTANTS IN A FLAT PLATE

The concept of stress and stress notation were introduced in Chapter 2. Although the major results were developed for rectangular coordinates, results were presented for orthogonal coordinates (x, y, z) [see Eqs. (2.46) and (2.47)]. In particular, we recall that σ_{xx} denotes the tensile stress on a plane element that is normal to an x coordinate line and $(\sigma_{xy}, \sigma_{xz})$ denote (y, z) components, respectively, of the shear stress that acts on a plane element normal to the x coordinate line. Similar inter-pretations apply for σ_{yy}, σ_{zz}, and σ_{yz}. As for rectangular coordinates, $\sigma_{yz} = \sigma_{zy}$, $\sigma_{xz} = \sigma_{zx}$, and $\sigma_{xy} = \sigma_{yx}$ for orthogonal curvilinear coordinates (x, y, z). For non-orthogonal curvilinear coordinates, the symmetry of shears does not hold (Boresi and Chong, 1987).

It is convenient to introduce special notation for in-plane forces (tractions), bending moments, twisting moments, and shears in a plate. Thus, with respect to orthogonal curvilinear coordinates (x, y, z), consider a differential element of the plate cut out by the surfaces $x = $ constant and $y = $ constant (Fig. 13.2), where (x, y) are orthogonal curvilinear coordinates in the middle plane of the plate and coor-dinate z is the straight-line coordinate perpendicular to the middle plane. The elements of area of these cross sections of a flat plate are

$$dA_y = \alpha \, dx \, dz, \qquad dA_x = \beta \, dy \, dz \qquad (13.1)$$

where

$$\alpha = \alpha(x, y), \qquad \beta = \beta(x, y), \qquad \gamma = 1$$

Let N_{xx} denote the tensile force on a cross-sectional face of the element $(x = $ constant), *per unit length of the y coordinate line* on the middle surface (Fig. 13.3). Then, the total tensile force on the *differential* element in the x direction is $N_{xx}\beta \, dy$. Hence, since $dA_x = \beta \, dy \, dz$, we have

$$N_{xx}\beta \, dy = dy \int_{-h/2}^{h/2} \beta\sigma_{xx} \, dz$$

where the middle surface has been taken as the reference surface. More generally, the reference surface $(z = 0)$ may be taken as any plane (e.g., the upper face of the plate). Then the integral in the above equation is determined by the thickness h and location of the reference surface. Thus, we have since $\beta = \beta(x, y)$,

$$N_{xx} = \int_{-h/2}^{h/2} \sigma_{xx} \, dz \qquad (a)$$

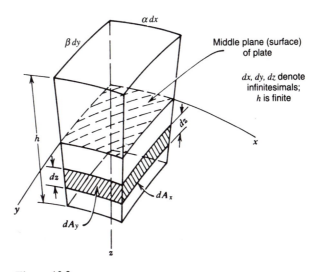

Figure 13.2

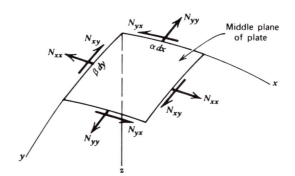

Figure 13.3   Resultant tractions on a reference surface.

In a similar manner, the traction N_{yy} *per unit length of the x coordinate line* on the middle surface (Fig. 13.3) is

$$N_{yy}\alpha\,dx = dx\int_{-h/2}^{h/2}\alpha\sigma_{yy}\,dz$$

or

$$N_{yy} = \int_{-h/2}^{h/2}\sigma_{yy}\,dz \tag{b}$$

Likewise, the shear force N_{xy} *per unit length of the y coordinate line* is given by

$$N_{xy}\beta\,dy = dy\int_{-h/2}^{h/2}\beta\sigma_{xy}\,dz$$

or

$$N_{xy} = \int_{-h/2}^{h/2}\sigma_{xy}\,dz \tag{c}$$

and for N_{yx}, the shear force *per unit length of the x coordinate line*,

$$N_{yx} = \int_{-h/2}^{h/2}\sigma_{xy}\,dz = N_{xy} \tag{d}$$

We let (Q_x, Q_y) be the transverse shears per unit length of a y coordinate line and x coordinate line, respectively. Hence, for the transverse shear forces (Q_x, Q_y) *per unit length of the coordinate line*, we find (Fig. 13.4) that

$$Q_x = \int_{-h/2}^{h/2}\sigma_{xz}\,dz$$

$$Q_y = \int_{-h/2}^{h/2}\sigma_{yz}\,dz \tag{e}$$

We let M_{xx} be the bending moment *per unit length of the y coordinate line*. Then by Fig. 13.4, we obtain, with positive directions indicated by the right-hand rule for moments (double-headed arrows),

$$M_{xx} = \int_{-h/2}^{h/2}z\sigma_{xx}\,dz \tag{f}$$

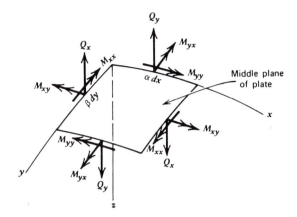

Figure 13.4   Resultant moments and shears on a reference surface.

For the twisting moment M_{xy} *per unit length of the y coordinate line*, we find

$$M_{xy} = \int_{-h/2}^{h/2} z\sigma_{xy}\, dz \tag{g}$$

Similarly for bending moment and twisting moment *per unit length of the x coordinate line*,

$$M_{yy} = \int_{-h/2}^{h/2} z\sigma_{yy}\, dz$$

$$M_{yx} = \int_{-h/2}^{h/2} z\sigma_{xy}\, dz = M_{xy} \tag{h}$$

In summary, we have the tractions $(N_{xx}, N_{yy}, N_{xy} = N_{yx})$, transverse shears (Q_x, Q_y), bending moments (M_{xx}, M_{yy}), and twisting moments $(M_{xy} = M_{yx})$ in the form (for the reference surface, $z = 0$, coincident with the plate middle surface)

$$N_{xx} = \int_{-h/2}^{h/2} \sigma_{xx}\, dz$$

$$N_{yy} = \int_{-h/2}^{h/2} \sigma_{yy}\, dz$$

$$N_{xy} = N_{yx} = \int_{-h/2}^{h/2} \sigma_{xy}\, dz$$

$$Q_x = \int_{-h/2}^{h/2} \sigma_{xz}\, dz$$

$$Q_y = \int_{-h/2}^{h/2} \sigma_{yz}\, dz$$

$$M_{xx} = \int_{-h/2}^{h/2} z\sigma_{xx}\, dz$$

$$M_{yy} = \int_{-h/2}^{h/2} z\sigma_{yy}\, dz$$

$$M_{xy} = M_{yx} = \int_{-h/2}^{h/2} z\sigma_{xy}\, dz \tag{13.2}$$

The positive senses of forces and moments are shown in Figs. (13.3) and (13.4). However, there is no universal agreement between authors on the sign conventions for the shears (Q_x, Q_y) and twisting moments $(M_{xy} = M_{yx})$.* Special attention must be paid to the notation M_{xx} and M_{yy}. For example, M_{xx} is a resultant moment per unit length of y coordinate due to the stress component σ_{xx}; it is not a moment about the x axis. Similar remarks pertain to M_{yy}. Also, as noted previously, M_{xy} and M_{yx} are twisting moments due to shear stresses acting on planes perpendicular to the x and y axes, respectively (Fig. 13.4).

13.3

KINEMATICS: STRAIN-DISPLACEMENT RELATIONS FOR PLATES

In this section, we let (U, V, W) be the components of the displacement vector, of any point P in the plate, on tangents to the local coordinate lines at P (Fig. 13.5). The notation (u, v, w) is reserved for the displacement components of the corresponding point P' on the middle surface of the plate (Fig. 13.5). Then, by Eqs. (2.84), for orthogonal coordinates [for a flat plate $\alpha = \alpha(x, y)$, $\beta = \beta(x, y)$, $\gamma = 1$; see Eq. (2.47)], we have the small-displacement strain-displacement relations for point P

$$\epsilon_{xx} = \frac{1}{\alpha}\left(U_x + \frac{\alpha_y V}{\beta}\right)$$

$$\epsilon_{yy} = \frac{1}{\beta}\left(V_y + \frac{\beta_x U}{\alpha}\right)$$

$$\epsilon_{zz} = W_z$$

$$2\epsilon_{xy} = \frac{U_y}{\beta} + \frac{V_x}{\alpha} - \frac{\beta_x V}{\alpha\beta} - \frac{\alpha_y U}{\alpha\beta}$$

$$2\epsilon_{xz} = U_z + \frac{W_x}{\alpha}$$

$$2\epsilon_{yz} = V_z + \frac{W_y}{\beta} \tag{13.3}$$

where the (x, y, z) subscripts on U, V, W, α, and β denote partial differentiation. Equations (13.3) are linear strain-displacement relations for the three-dimensional kinematical problem of the plate. However, the purpose of plate theory is to reduce the three-dimensional problem to a more tractable two-dimensional problem. The approximation that is usually used to achieve this reduction is due to Kirchhoff; namely, it is assumed that the straight-line normals to the undeformed middle plane (reference plane) of the plate remain straight, inextensional and normal to the

---

* Here, we follow the convention employed by H. L. Langhaar (1989). See also Marguerre and Woernle (1969).

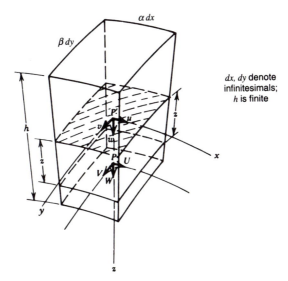

Figure 13.5

middle surface under the deformation of the plate. Since under this assumption line elements normal to the middle surface do not extend and no angular distortion occurs between normals and the reference surface, it follows that the Kirchhoff approximation is equivalent to assuming that $\epsilon_{zz} = \epsilon_{xz} = \epsilon_{yz} = 0$ (note these conditions match those of plane strain). The Kirchhoff assumption is not limited to problems of small displacements, since it is purely kinematical in form. It does not depend on material properties. Hence, it may be employed in plasticity studies of plates, etc. Since ϵ_{xz} and ϵ_{yz} are discarded, for an isotropic material, it implies that σ_{xz}, σ_{yz} are zero and, hence, the shears Q_x and Q_y are zero [(Eq. 13.2)]. If values of (Q_x, Q_y) are needed, they are reintroduced into the theory through the equations of equilibrium (Sec. 13.4). However, some inconsistencies are inevitable when the Kirchhoff approximation is employed in plates (such as the implied plane strain condition). It is nevertheless more accurate than the membrane theory (approximation) of plates, which requires not only Q_x and Q_y to be zero, but also that M_{xx}, M_{yy}, $M_{xy} = M_{yx}$ vanish.

The Kirchhoff approximation implies that U, V, W are linear functions of z, irrespective of the magnitude of the displacement. General expressions for (U, V, W) in terms of the displacement components (u, v, w) of the middle surface are very complicated for large displacements. The resulting nonlinear relations for the strains $(\epsilon_{xx}, \epsilon_{yy}, \dots \epsilon_{yz})$ in terms of (u, v, w) are even more complicated. However, it is feasible to derive general strain equations in terms of (u, v, w) for any plate, if we employ the strain-displacement relations for small displacements [Eqs. (13.3)].

Since, by the Kirchhoff approximation, the thickness is not changed by the deformation (Fig. 13.5), the difference $W - w$ is of second order. Therefore, we make the approximation $W = w$. Then by the last two of Eqs. (13.3), we obtain

$$2\epsilon_{xz} = \frac{w_x}{\alpha} + U_z$$

$$2\epsilon_{yz} = \frac{w_y}{\beta} + V_z \tag{13.4}$$

where (x, y, z) subscripts on U, V, and w denote partial differentiation. Since Kirchhoff's approximation implies that $\epsilon_{xz} = \epsilon_{yz} = 0$, Eqs. (13.4) yield

$$U_z + \frac{w_x}{\alpha} = 0, \qquad V_z + \frac{w_y}{\beta} = 0 \tag{13.5}$$

Integrations of Eqs. (13.5) yield

$$U = -\frac{w_x}{\alpha} z + f(x, y)$$

$$V = -\frac{w_y}{\beta} z + g(x, y) \tag{13.6}$$

The additive functions $f(x, y)$ and $g(x, y)$ are determined by the conditions $U = u$ and $V = v$ for $z = 0$. Then, by Eqs. (13.6)

$$U = u - z\frac{w_x}{\alpha}$$

$$V = v - z\frac{w_y}{\beta}$$

$$W = w \tag{13.7}$$

where (u, v, w) are functions of (x, y) only. Equation (13.7) determines how, U, V, W vary through the thickness of the plate in accord with Kirchhoff's approximation and small-displacement theory. Substitution of Eqs. (13.7) into Eqs. (13.3) yields

$$\epsilon_{xx} = \frac{1}{\alpha}\frac{\partial}{\partial x}\left(\frac{\alpha u - z w_x}{\alpha}\right) + \frac{\alpha_y}{\alpha\beta}\left(\frac{\beta v - z w_y}{\beta}\right)$$

$$\epsilon_{yy} = \frac{1}{\beta}\frac{\partial}{\partial y}\left(\frac{\beta v - z w_y}{\beta}\right) + \frac{\beta_x}{\alpha\beta}\left(\frac{\alpha u - z w_x}{\alpha}\right)$$

$$\epsilon_{zz} = 0$$

$$\gamma_{xy} = 2\epsilon_{xy} = \frac{\beta}{\alpha}\frac{\partial}{\partial x}\left(\frac{\beta v - z w_y}{\beta^2}\right) + \frac{\alpha}{\beta}\frac{\partial}{\partial y}\left(\frac{\alpha u - z w_x}{\alpha^2}\right)$$

$$\gamma_{xz} = \gamma_{yz} = 2\epsilon_{xz} = 2\epsilon_{yz} = 0 \tag{13.8}$$

Alternatively, Eqs. (13.8) may be written by separating the z terms as follows:

$$\epsilon_{xx} = \frac{u_x}{\alpha} + \frac{\alpha_y v}{\alpha\beta} - \frac{z}{\alpha}\left[\frac{\partial}{\partial x}\left(\frac{w_x}{\alpha}\right) + \frac{\alpha_y w_y}{\beta^2}\right] \tag{13.9a}$$

$$\epsilon_{yy} = \frac{v_y}{\beta} + \frac{\beta_x u}{\alpha\beta} - \frac{z}{\beta}\left[\frac{\partial}{\partial y}\left(\frac{w_y}{\beta}\right) + \frac{\beta_x w_x}{\alpha^2}\right] \tag{13.9b}$$

$$\gamma_{xy} = 2\epsilon_{xy} = \frac{\beta}{\alpha} \frac{\partial}{\partial x}\left(\frac{v}{\beta}\right) + \frac{\alpha}{\beta} \frac{\partial}{\partial y}\left(\frac{u}{\alpha}\right)$$

$$- \frac{z}{\alpha}\left[\frac{\partial}{\partial x}\left(\frac{w_y}{\beta}\right) - \frac{\beta_x w_y}{\beta^2}\right]$$

$$- \frac{z}{\beta}\left[\frac{\partial}{\partial y}\left(\frac{w_x}{\alpha}\right) - \frac{\alpha_y w_x}{\alpha^2}\right] \tag{13.9c}$$

$$\epsilon_{zz} = \gamma_{xz} = \gamma_{yz} = 0 \tag{13.9d}$$

Equations (13.8) and (13.9) are approximations of Eqs. (13.3) that result from application of the Kirchhoff approximation. These approximations form the basis of classical small-displacement plate theory.

Rotation of a Plate Surface Element

To obtain continuity conditions at the junction of two plates, it is sometimes necessary to compute the rotations of the plate (middle) surface at the junction. As noted in the theory of deformation (Boresi and Chong, 1987), the small-displacement rotation ω of a volume element is a vector quantity given by the relation

$$\omega = \tfrac{1}{2} \operatorname{curl} \mathbf{q} \tag{13.10}$$

where $\mathbf{q} = (U, V, W)$ is the displacement vector, and the operation $\operatorname{curl} \mathbf{q} = \nabla \times \mathbf{q}$ must be expressed in terms of the appropriate coordinate system (recall that here we employ orthogonal curvilinear coordinates). The expression for the curl in curvilinear orthogonal coordinates is (Newell, 1955)

$$\operatorname{curl} \mathbf{q} = \nabla \times \mathbf{q} = \frac{1}{\alpha\beta\gamma} \begin{vmatrix} \alpha\mathbf{i} & \beta\mathbf{j} & \gamma\mathbf{k} \\ \dfrac{\partial}{\partial x} & \dfrac{\partial}{\partial y} & \dfrac{\partial}{\partial z} \\ \alpha U & \beta V & \gamma W \end{vmatrix} \tag{13.11}$$

where the displacement vector $\mathbf{q}$ is

$$\mathbf{q} = U\mathbf{i} + V\mathbf{j} + W\mathbf{k} \tag{13.12}$$

and $(\mathbf{i}, \mathbf{j}, \mathbf{k})$ denote unit vectors tangent to (x, y, z) coordinate lines, respectively. With Eqs. (13.7) and (13.11), Eq. (13.10) yields with $\gamma = 1$ and with $z = 0$ (after differentiation)

$$\omega_x = \frac{w_y}{\beta}, \qquad \omega_y = -\frac{w_x}{\alpha}, \qquad \omega_z = \frac{1}{2\alpha\beta}\left[\frac{\partial}{\partial x}(\beta v) - \frac{\partial}{\partial y}(\alpha u)\right] \tag{13.13}$$

where $(\omega_x, \omega_y, \omega_z)$ are the projections of ω along tangents to the (x, y, z) coordinate lines, respectively.

In terms of (ω_x, ω_y), we may rewrite Eqs. (13.8) and (13.9) in the forms

$$\epsilon_{xx} = \frac{1}{\alpha} \frac{\partial}{\partial x} (u + z\omega_y) + \frac{\alpha_y}{\alpha\beta} (v - z\omega_x)$$

$$\epsilon_{yy} = \frac{1}{\beta} \frac{\partial}{\partial y} (v - z\omega_x) + \frac{\beta_x}{\alpha\beta} (u + z\omega_y)$$

$$\gamma_{xy} = 2\epsilon_{xy} = \frac{\beta}{\alpha} \frac{\partial}{\partial x} \left(\frac{v - z\omega_x}{\beta} \right) + \frac{\alpha}{\beta} \frac{\partial}{\partial y} \left(\frac{u + z\omega_y}{\alpha} \right) \tag{13.14}$$

and

$$\epsilon_{xx} = \frac{u_x}{\alpha} + \frac{\alpha_y v}{\alpha\beta} - \frac{z}{\alpha} \left(-\frac{\partial \omega_y}{\partial x} + \frac{\alpha_y}{\beta} \omega_x \right)$$

$$\epsilon_{yy} = \frac{v_y}{\beta} + \frac{\beta_x u}{\alpha\beta} - \frac{z}{\beta} \left(\frac{\partial \omega_x}{\partial y} - \frac{\beta_x \omega_y}{\alpha} \right)$$

$$\gamma_{xy} = 2\epsilon_{xy} = \frac{\beta}{\alpha} \frac{\partial}{\partial x} \left(\frac{v}{\beta} \right) + \frac{\alpha}{\beta} \frac{\partial}{\partial y} \left(\frac{u}{\alpha} \right) - \frac{z}{\alpha} \left(\frac{\partial \omega_x}{\partial x} - \frac{\beta_x \omega_x}{\beta} \right)$$

$$- \frac{z}{\beta} \left(-\frac{\partial \omega_y}{\partial y} + \frac{\alpha_y \omega_y}{\alpha} \right) \tag{13.15}$$

For rectangular coordinates, $\alpha = \beta = 1$ and $\alpha_x = \alpha_y = \beta_x = \beta_y = 0$. Then Eqs. (13.14) and (13.15) reduce to

$$\epsilon_{xx} = \frac{\partial}{\partial x} (u + z\omega_y), \qquad \epsilon_{yy} = \frac{\partial}{\partial y} (v - z\omega_x)$$

$$\gamma_{xy} = 2\epsilon_{xy} = \frac{\partial}{\partial x} (v - z\omega_x) + \frac{\partial}{\partial y} (u + z\omega_y) \tag{13.16}$$

and

$$\epsilon_{xx} = u_x + z \frac{\partial \omega_y}{\partial x}, \qquad \epsilon_{yy} = v_y - z \frac{\partial \omega_y}{\partial y}$$

$$\gamma_{xy} = 2\epsilon_{xy} = \frac{\partial v}{\partial x} + \frac{\partial u}{\partial y} - z \left(\frac{\partial \omega_x}{\partial x} - \frac{\partial \omega_y}{\partial y} \right) \tag{13.17}$$

where

$$\omega_x = w_y, \qquad \omega_y = -w_x \tag{13.18}$$

Alternatively in terms of (u, v, w), we may write

$$\epsilon_{xx} = u_x - zw_{xx}, \qquad \epsilon_{yy} = v_y - zw_{yy}$$
$$\gamma_{xy} = 2\epsilon_{xy} = v_x + u_y - 2zw_{xy} \tag{13.19}$$

where we recall that (x, y) subscripts on (u, v, w) denote partial differentiation.

The strain-displacement relations derived above are employed in the classical small-displacement theory of plates. For an alternative derivation of these relations, see Marguerre and Woernle (1969).

13.4

EQUILIBRIUM EQUATIONS FOR SMALL-DISPLACEMENT THEORY OF FLAT PLATES

The equations of equilibrium for a plate may be derived by several methods. For example, they may be derived (1) by considering the equilibrium requirements for an infinitesimal plate element (dx, dy, dz) (Fig. 13.2), (2) by integrating the point-wise equilibrium equations [Eqs. (2.45) or (2.46)] through the plate thickness and employing the definitions of Eqs. (13.2), or (3) by a direct application of the principal of virtual work. In the following derivation, we employ Method 2. Similar results have been obtained by Marguerre and Woernle (1969) by Methods 1 and 3 for rectangular coordinates.

We consider an element of the plate generated by all normals erected on an element $dx\, dy$ of the middle surface. This element may be subjected to external forces caused by gravity and by external shears and pressures applied to the faces of the plate. Since the area of the element $dx\, dy$ of the middle surface is $\alpha\beta\, dx\, dy$, the resultant external force on the element of the plate is denoted by $\mathbf{P}\,\alpha\beta\, dx\, dy$. The vector $\mathbf{P}$ is the resultant force per unit area of the middle surface. It is a function of the coordinates (x, y) of the middle surface. The vector $\mathbf{P}$ is considered to act at the middle surface of the plate, and it is resolved into components P_x, P_y, P_z along (x, y, z) coordinate lines, respectively. Often, the component P_z is denoted by p or q, since usually it results from normal pressures on the faces of the plate. In addition to the external force $\mathbf{P}\,\alpha\beta\, dx\, dy$, an external couple $\mathbf{R}\,\alpha\beta\, dx\, dy$ may act on the element of the plate. We consider a couple that results only from shear stresses on the external faces of the plate. Hence, relative to the midsurface $R_z = 0$, and

$$\alpha\beta R_x = -\alpha\beta z \sigma_{yz}\big|_{-h/2}^{h/2}$$

$$\alpha\beta R_y = \alpha\beta z \sigma_{xz}\big|_{-h/2}^{h/2} \tag{13.20}$$

where (R_x, R_y, R_z) are the (x, y, z) projections of couple $\mathbf{R}$.

To employ Method 2, we use the pointwise equilibrium equations. Thus, for $\alpha = \alpha(x, y)$, $\beta = \beta(x, y)$, $\gamma = 1$, we obtain by Eqs. (2.46)

$$\frac{\partial}{\partial x}(\beta\sigma_{xx}) + \frac{\partial}{\partial y}(\alpha\sigma_{xy}) + \frac{\partial}{\partial z}(\alpha\beta\sigma_{xz}) + \alpha_y\sigma_{xy} - \beta_x\sigma_{yy} + \alpha\beta B_x = 0$$

$$\frac{\partial}{\partial x}(\beta\sigma_{xy}) + \frac{\partial}{\partial y}(\alpha\sigma_{yy}) + \frac{\partial}{\partial z}(\alpha\beta\sigma_{yz}) + \beta_x\sigma_{xy} - \alpha_y\sigma_{xx} + \alpha\beta B_y = 0$$

$$\frac{\partial}{\partial x}(\beta\sigma_{xz}) + \frac{\partial}{\partial y}(\alpha\sigma_{yz}) + \frac{\partial}{\partial z}(\alpha\beta\sigma_{zz}) + \alpha\beta B_z = 0 \tag{13.21}$$

The force equilibrium equations for N_{xx}, N_{yy}, N_{xy}, Q_x, and Q_y are obtained by integrating the differential equations of equilibrium [Eqs. (13.21)] through the thickness h of the plate. For example, the first term in the first of Eqs. (13.21) is $\partial(\beta\sigma_{xx})/\partial x$. Integrating this term with respect to z between the limits $-h/2$ and $h/2$ and utilizing Eqs. (13.2), we obtain

$$\int_{-h/2}^{h/2} \frac{\partial}{\partial x}(\beta\sigma_{xx})\, dz = \frac{\partial}{\partial x}\int_{-h/2}^{h/2} \beta\sigma_{xx}\, dz = \frac{\partial}{\partial x}(\beta N_{xx})$$

The second term in Eq. (13.21) is integrated similarly. For the integral of the third term, we obtain

$$\int_{-h/2}^{h/2} \frac{\partial}{\partial z}(\alpha\beta\sigma_{xz})\,dz = \alpha\beta\sigma_{xz}\big|_{-h/2}^{h/2} = \alpha\beta P_x$$

The fourth integral obtained from Eq. (13.21) is

$$\int_{-h/2}^{h/2} \alpha_y\sigma_{xy}\,dz = \alpha_y N_{xy}$$

Similarly, the other terms can be integrated.

To obtain the moment equilibrium equations, we multiply Eqs. (13.21) by z and then integrate through the thickness and employ the definitions of Eq. (13.2).

The complete set of equilibrium equations obtained is thus

$$\frac{\partial}{\partial x}(\beta N_{xx}) + \frac{\partial}{\partial y}(\alpha N_{xy}) + \alpha_y N_{xy} - \beta_x N_{yy} + \alpha\beta P_x + \alpha\beta h B_x = 0$$

$$\frac{\partial}{\partial x}(\beta N_{xy}) + \frac{\partial}{\partial y}(\alpha N_{yy}) + \beta_x N_{xy} - \alpha_y N_{xx} + \alpha\beta P_y + \alpha\beta h B_y = 0$$

$$\frac{\partial}{\partial x}(\beta Q_x) + \frac{\partial}{\partial y}(\alpha Q_y) + \alpha\beta P_z + \alpha\beta h B_z = 0$$

$$\frac{\partial}{\partial x}(\beta M_{xx}) + \frac{\partial}{\partial y}(\alpha M_{xy}) + \alpha_y M_{xy} - \beta_x M_{yy} - \alpha\beta Q_x + \alpha\beta R_y = 0$$

$$\frac{\partial}{\partial x}(\beta M_{xy}) + \frac{\partial}{\partial y}(\alpha M_{yy}) + \beta_x M_{xy} - \alpha_y M_{xx} - \alpha\beta Q_y - \alpha\beta R_x = 0$$

$$N_{xy} = N_{yx} \tag{13.22}$$

For rectangular coordinates, $\alpha = \beta = 1$. Then Eqs. (13.22) yield

$$\frac{\partial N_{xx}}{\partial x} + \frac{\partial N_{xy}}{\partial y} + P_x + hB_x = 0$$

$$\frac{\partial N_{xy}}{\partial x} + \frac{\partial N_{yy}}{\partial y} + P_y + hB_y = 0$$

$$\frac{\partial Q_x}{\partial x} + \frac{\partial Q_y}{\partial y} + P_z + hB_z = 0$$

$$\frac{\partial M_{xx}}{\partial x} + \frac{\partial M_{xy}}{\partial y} - Q_x + R_y = 0$$

$$\frac{\partial M_{xy}}{\partial x} + \frac{\partial M_{yy}}{\partial y} - Q_y - R_x = 0$$

$$N_{xy} = N_{yx} \tag{13.23}$$

Equations (13.22) are exact relations, provided that (x, y, z) are orthogonal curvilinear plate coordinates for the *deformed* plate. They are approximations for the small-displacement theory of plates if (x, y, z) are orthogonal curvilinear plate coordinates in the *undeformed* plate, since Eqs. (13.21) are approximations for such axes.* Therefore, we shall use them as the equilibrium relations for the small-displacement theory of plates relative to orthogonal curvilinear plate axes in the undeformed plate.

The last of Eqs. (13.22) is an identity that follows from Eqs. (13.2). Often, R_x and R_y are zero; in any case, they may usually be discarded from Eqs. (13.22). However, if they are retained, we obtain from the third, fourth, and fifth of Eqs. (13.22), by the elimination of Q_x and Q_y,

$$\frac{\partial}{\partial x}\left\{\frac{1}{\alpha}\left[\frac{\partial}{\partial x}(\beta M_{xx}) + \frac{\partial}{\partial y}(\alpha M_{xy}) + \alpha_y M_{xy} - \beta_x M_{yy} + \alpha\beta R_y\right]\right\}$$
$$+ \frac{\partial}{\partial y}\left\{\frac{1}{\beta}\left[\frac{\partial}{\partial x}(\beta M_{xy}) + \frac{\partial}{\partial y}(\alpha M_{yy}) - \alpha_y M_{xx} + \beta_x M_{xy} - \alpha\beta R_x\right]\right\}$$
$$+ h\alpha\beta B_z + \alpha\beta P_z = 0 \tag{13.24}$$

Equation (13.24) is called the moment equilibrium equation of plates. For rectangular axes, $\alpha = \beta = 1$, and Eq. (13.24) reduces to (if we discard R_x and R_y)

$$\frac{\partial^2 M_{xx}}{\partial x^2} + 2\frac{\partial^2 M_{xy}}{\partial x\,\partial y} + \frac{\partial^2 M_{yy}}{\partial y^2} + hB_z + P_z = 0 \tag{13.25}$$

13.5

STRESS-STRAIN-TEMPERATURE RELATIONS FOR ISOTROPIC ELASTIC PLATES

The preceding equations, derived in Sec. 13.2, 13.3, and 13.4, are independent of material properties. Hence, they are equally applicable to problems of elasticity, plasticity, and creep, irrespective of the effects of temperature.

In conventional plate theory, it is assumed that the plate is in a state of plane stress; that is, $\sigma_{xz} = \sigma_{yz} = \sigma_{zz} = 0$. For isotropic elastic plates, the relations $\sigma_{xz} = \sigma_{yz} = 0$ are consistent with the Kirchhoff approximation, which signifies that $\epsilon_{xz} = \epsilon_{yz} = 0$. However, the Kirchhoff approximation has been criticized since it includes the approximation $\epsilon_{zz} = 0$. The condition $\epsilon_{zz} = 0$ conflicts with the assumption that $\sigma_{zz} = 0$. The condition $\epsilon_{zz} = 0$ is incorrect; however, the strain ϵ_{zz} has little effect on the strains ϵ_{xx}, ϵ_{yy}, ϵ_{xy}. Thus, the approximation $\epsilon_{zz} = 0$ is merely expedient. In the stress-strain relations, the condition of plane stress $\sigma_{zz} = 0$ is commonly used instead of $\epsilon_{zz} = 0$, and this circumstance is often regarded as an inconsistency. However, in approximations, the significant question is not the consistency of the assumptions, but rather the magnitude of the error that results, since nearly all approximations lead to inconsistencies. In plate theory, the values of ϵ_{zz} and σ_{zz} are not of particular importance. Viewed in this light, the Kirchhoff approximation merely implies that ϵ_{zz} has small effects on σ_{xx} and σ_{yy}, and that σ_{xz} and σ_{yz} are not significant. We observe further that the Kirchhoff approximation need not be restricted to linearly elastic plates; it is also applicable to studies of plasticity and creep of plates, and it is not restricted to small displacements.

---

* See Appendix 3C of Boresi and Chong (1987).

For linearly elastic isotropic materials and plane stress relative to the (x, y) plane $(\sigma_{zz} = \sigma_{xz} = \sigma_{yz} = 0)$, stress-strain-temperature relations are

$$\sigma_{xx} = \frac{E}{1 - v^2}(\epsilon_{xx} + v\epsilon_{yy}) - \frac{Ek\,\Delta T}{1 - v}$$

$$\sigma_{yy} = \frac{E}{1 - v^2}(v\epsilon_{xx} + \epsilon_{yy}) - \frac{Ek\,\Delta T}{1 - v}$$

$$\sigma_{xy} = 2G\epsilon_{xy} = G\gamma_{xy} \tag{13.26}$$

where E is Young's modulus, v Poisson's ratio, k the coefficient of linear thermal expansion [we use k instead of α (see Chapter 3) since α is used here as a metric coefficient; see Eq. (13.1)], G the shear modulus, and ΔT the temperature change measured relative to an arbitrary datum. It may be assumed without complication that k is a function of temperature change ΔT.

By Eqs. (13.9) and (13.26), $\sigma_{xx}, \sigma_{yy}, \sigma_{xy}$ may be expressed in terms of u, v, w, and ΔT. Then by Eqs. (13.2), the quantities $N_{xx}, N_{yy}, N_{xy}, M_{xx}, M_{yy}, M_{xy}$ may be expressed in terms of u, v, w, and ΔT. Then the first two of Eqs. (13.22) and Eq. (13.24) become differential equations in u, v, w. Thus, the equilibrium equations are expressed in terms of the displacement vector of the reference surface of the plate. For homogeneous plates, it is convenient to take the reference surface midway between the plate faces. However, for layered or reinforced plates, some other reference surface may be more appropriate. Then the integral limits $(-h/2, h/2)$ in Eqs. (13.2) would be modified accordingly. In the following, we take the reference surface as the middle surface of the plate. Hence, the faces of the plate are located at $z = \pm h/2$.

Although the Kirchhoff approximation implies that Q_x, Q_y vanish for isotropic linearly elastic plates, estimates of Q_x, Q_y may be obtained from the fourth and fifth of Eqs. (13.22).

Substitution of Eqs. (13.9) into Eqs. (13.26) and then substitution of the results into Eq. (13.2) yields

$$N_{xx} = \frac{Eh}{\alpha(1 - v^2)}\left(\frac{v\beta_x}{\beta}u + u_x + \frac{\alpha_y}{\beta}v + \frac{v\alpha}{\beta}v_y\right) - \frac{E}{1 - v}\int_{-h/2}^{h/2} k\,\Delta T\,dz$$

$$N_{yy} = \frac{Eh}{\alpha(1 - v^2)}\left(\frac{\beta_x}{\alpha}u + \frac{v\beta}{\alpha}u_x + \frac{v\alpha_y}{\alpha}v + v_y\right) - \frac{E}{1 - v}\int_{-h/2}^{h/2} k\,\Delta T\,dz$$

$$N_{xy} = Gh\left[\frac{\alpha}{\beta}\frac{\partial}{\partial y}\left(\frac{u}{\alpha}\right) + \frac{\beta}{\alpha}\frac{\partial}{\partial x}\left(\frac{v}{\beta}\right)\right]$$

$$M_{xx} = -\frac{D}{\alpha}\left[\frac{\partial}{\partial x}\left(\frac{w_x}{\alpha}\right) + \frac{\alpha_y w_y}{\beta^2} + \frac{v\alpha}{\beta}\frac{\partial}{\partial y}\left(\frac{w_y}{\beta}\right) + \frac{v\beta_x}{\alpha\beta}w_x\right]$$
$$- \frac{E}{1 - v}\int_{-h/2}^{h/2} zk\,\Delta T\,dz$$

$$M_{yy} = -\frac{D}{\beta}\left[\frac{\partial}{\partial y}\left(\frac{w_y}{\beta}\right) + \frac{\beta_x w_x}{\alpha^2} + \frac{v\beta}{\alpha}\frac{\partial}{\partial x}\left(\frac{w_x}{\alpha}\right) + \frac{v\alpha_y}{\alpha\beta}w_y\right] - \frac{E}{1 - v}\int_{-h/2}^{h/2} zk\,\Delta T\,dz$$

$$M_{xy} = -\frac{Gh^3}{6\alpha\beta}\left(w_{xy} - \frac{\alpha_y}{\alpha}w_x - \frac{\beta_x}{\beta}w_y\right) \tag{13.27}$$

where

$$G = \frac{E}{2(1 + v)}, \qquad D = \frac{Eh^3}{12(1 - v^2)} \tag{13.27a}$$

The quantity D is called the *flexural rigidity* of the plate.

Alternatively, with Eqs. (13.9) and (13.27), we may write

$$N_{xx} = \frac{Eh}{1 - v^2} (\epsilon_{xx}^0 + v\epsilon_{yy}^0 - T^0)$$

$$N_{yy} = \frac{Eh}{1 - v^2} (v\epsilon_{xx}^0 + \epsilon_{yy}^0 - T^0)$$

$$N_{xy} = 2Gh\epsilon_{xy}^0 = Gh\gamma_{xy}^0, \qquad \gamma_{xy}^0 = 2\epsilon_{xy}^0$$

$$M_{xx} = -D\left(\frac{\kappa_{xx}}{\alpha^2} + \frac{v\kappa_{yy}}{\beta^2} + T^1\right)$$

$$M_{yy} = -D\left(v\frac{\kappa_{xx}}{\alpha^2} + \frac{\kappa_{yy}}{\beta^2} + T^1\right)$$

$$M_{xy} = -\frac{(1 - v)D}{\alpha\beta}\kappa_{xy} \tag{13.28}$$

where ϵ_{xx}^0, ϵ_{yy}^0, ϵ_{xy}^0 are the strain components in the plate middle surface ($z = 0$),

$$T^0 = \frac{1 + v}{h} \int_{-h/2}^{h/2} k\,\Delta T\,dz$$

$$T^1 = \frac{12(1 + v)}{h^3} \int_{-h/2}^{h/2} zk\,\Delta T\,dz \tag{13.29}$$

are the zero-th and first moments of ΔT with respect to z, and

$$\kappa_{xx} = -\frac{\alpha_x w_x}{\alpha} + \frac{\alpha_y\alpha}{\beta^2}w_y + w_{xx}$$

$$\kappa_{yy} = \frac{\beta_x\beta}{\alpha^2}w_x - \frac{\beta_y}{\beta}w_y + w_{yy}$$

$$\kappa_{xy} = -\frac{\alpha_x w_x}{\alpha} - \frac{\beta_y}{\beta}w_y + w_{xy} \tag{13.30}$$

are the curvatures of the middle surface relative to the (x, y) axes. Hence,

$$\epsilon_{xx}^0 = \frac{1}{Eh}(N_{xx} - vN_{yy}) + \frac{T^0}{1 + v}$$

$$\epsilon_{yy}^0 = \frac{1}{Eh}(N_{yy} - vN_{xx}) + \frac{T^0}{1 + v}$$

$$\gamma_{xy}^0 = 2\epsilon_{xy}^0 = \frac{1}{Gh}N_{xy} \tag{13.31}$$

and

$$\kappa_{xx} = -\frac{12\alpha^2}{Eh^3}(M_{xx} - vM_{yy}) - \frac{\alpha^2 T^1}{1 + v}$$

$$\kappa_{yy} = -\frac{12\beta^2}{Eh^3}(M_{yy} - vM_{xx}) - \frac{\beta^2 T^1}{1 + v}$$

$$\kappa_{xy} = -\frac{12(1 + v)\alpha\beta}{Eh^3} M_{xy} \qquad (13.32)$$

For rectangular coordinates, $\alpha = \beta = 1$. Then, the moment curvature relations [the last three of Eqs. (13.28)] reduce to

$$M_{xx} = -D(\kappa_{xx} + v\kappa_{yy} + T^1)$$
$$M_{yy} = -D(v\kappa_{xx} + \kappa_{yy} + T^1)$$
$$M_{xy} = -(1 - v)D\kappa_{xy} \qquad (13.33)$$

where [by Eqs. (13.30)],

$$\kappa_{xx} = w_{xx}, \qquad \kappa_{yy} = w_{yy}, \qquad \kappa_{xy} = w_{xy} \qquad (13.34)$$

Stress Components in Terms of Tractions and Moments

Equations (13.9) and (13.26) lead to the conclusion that σ_{xx}, σ_{yy}, σ_{xy} vary linearly through the thickness of the plate; that is, $\sigma_{xx} = a + bz, \ldots, \ldots$. Hence, by Eqs. (13.2), $a = N_{xx}/h$, $b = 12M_{xx}/h^3$. Similarly, the coefficients in the linear expressions for σ_{yy} and σ_{xy} are determined. Thus, we find

$$\sigma_{xx} = \frac{N_{xx}}{h} + \frac{12zM_{xx}}{h^3}$$

$$\sigma_{yy} = \frac{N_{yy}}{h} + \frac{12zM_{yy}}{h^3}$$

$$\sigma_{xy} = \frac{N_{xy}}{h} + \frac{12zM_{xy}}{h^3} \qquad (13.35)$$

Pure Bending of Plates

If a plate is subjected to bending moments (M_{xx}, M_{yy}) only, we refer to the plate problem as one of *pure bending of plates*. In particular, for pure bending of plates, $N_{xx} = N_{yy} = N_{xy} = Q_x = Q_y = M_{xy} = 0$ and the preceding equations are simplified accordingly.

13.6

STRAIN ENERGY OF A PLATE

For plane stress theory, the strain energy density of a homogeneous isotropic elastic plate, referred to orthogonal plate coordinates, is (see Sec. 3.4).

$$U_0 = \frac{G}{1 - v}[\epsilon_{xx}^2 + \epsilon_{yy}^2 + 2v\epsilon_{xx}\epsilon_{yy} + 2(1 - v)\epsilon_{xy}^2 - 2(1 + v)(\epsilon_{xx} + \epsilon_{yy})k\,\Delta T] \qquad (13.36)$$

where U_0 has the dimensions of energy per unit volume. Since the volume element of a plate is $\alpha\beta\,dx\,dy\,dz$, the total strain energy U of the plate is

$$U = \iiint U_0 \alpha\beta\,dx\,dy\,dz \qquad (13.37)$$

The integrations with respect to x and y extend over the middle surface of the plate, whereas the integration with respect to z extends between the limits $-h/2$ and $h/2$. By Eqs. (13.9), (13.36), and (13.37), we find after integration with respect to z, the total strain energy

$$U = U_m + U_b + U_t \qquad (13.38)$$

where U_m, the *membrane energy* of the plate, is linear in the thickness h, and U_b, the *bending energy* of the plate, is cubic in h. The term U_t represents the strain energy that results from the temperature change ΔT. Hence, if G and v are taken independent of z, integration with respect to z yields

$$U_m = \iint \frac{Gh}{1-v}[(\epsilon_{xx}^0)^2 + (\epsilon_{yy}^0)^2 + 2v\epsilon_{xx}^0\epsilon_{yy}^0 + 2(1-v)(\epsilon_{yy}^0)^2]\alpha\beta\,dx\,dy$$

$$U_b = \iint \frac{Gh^3}{12(1-v)}\left[\left(\frac{\kappa_{xx}}{\alpha^2}\right)^2 + \left(\frac{\kappa_{yy}}{\beta^2}\right)^2 + 2v\left(\frac{\kappa_{xx}}{\alpha^2}\right)\left(\frac{\kappa_{yy}}{\beta^2}\right)\right.$$

$$\left. + 2(1-v)\left(\frac{\kappa_{xy}}{\alpha\beta}\right)^2\right]\alpha\beta\,dx\,dy$$

$$U_t = -\iint \frac{Eh}{1-v^2}\left[(\epsilon_{xx}^0 + \epsilon_{yy}^0)T^0 - \left(\frac{\kappa_{xx}}{\alpha^2} + \frac{\kappa_{yy}}{\beta^2}\right)\frac{h^2 T^1}{12}\right]\alpha\beta\,dx\,dy \qquad (13.39)$$

By means of Eqs. (13.9), with $z = 0$, and Eqs. (13.30), the strain energy [Eqs. (13.38) and (13.39)] may be expressed as a function of the middle surface displacement components (u, v, w). The strain energy is employed in conjunction with the Rayleigh–Ritz procedure to obtain approximate solutions of plate problems (Timoshenko and Woinowsky–Krieger, 1959; Szilard, 1974; Ugural, 1981). The strain energy also serves, by means of variational principals, to determine plate boundary conditions (Langhaar, 1989). In addition, the differential equations of equilibrium, in terms of (u, v, w), are obtained from the total *potential energy* expression by means of Euler's equation of the calculus of variations. In the next section, we employ the principle of stationary potential energy to determine boundary conditions for a plate.

13.7

BOUNDARY CONDITIONS FOR PLATES

In this section, we employ the principle of stationary potential energy (Sec. 5.1) to obtain boundary conditions for the classical theory of plates. For simplicity, we consider rectangular coordinates ($\alpha = \beta = 1$) and a rectangular plate that lies in the region $0 \leq x \leq a$, $0 \leq y \leq b$ (Fig. 13.6). Also, for purposes of demonstration, we

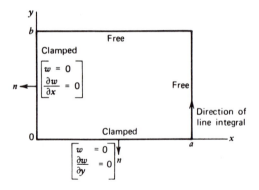

Figure 13.6   Boundary conditions at a reference surface edge.

discard temperature effects and consider the effects of tractions N_{xx}, N_{yy}, N_{xy} to be negligible compared to the moments M_{xx}, M_{yy}, M_{xy}. Furthermore, we recall that in Kirchhoff plate theory, the effects of Q_x, Q_y are also discarded.

The principle of stationary potential energy states

$$\delta W_e = \delta U \tag{13.40}$$

where the first variation δU of the strain energy is

$$\delta U = \int_0^b \int_0^a \delta \bar{U}\, dx\, dy \tag{13.41}$$

with

$$\delta \bar{U} = \int_{-h/2}^{h/2} (\sigma_{xx}\,\delta\epsilon_{xx} + \sigma_{yy}\,\delta\epsilon_{yy} + \sigma_{zz}\,\delta\epsilon_{zz} + 2\sigma_{xy}\,\delta\epsilon_{xy}$$
$$+ 2\sigma_{xz}\,\delta\epsilon_{xz} + 2\sigma_{yz}\,\delta\epsilon_{yz})\, dz \tag{13.42}$$

and for $(P_x = P_y = 0, P_z = p)$, $(R_x = R_y = 0)$, and $(B_x = B_y = B_z = 0)$ (see Sec. 13.4)

$$\delta W_e = \int_0^b \int_0^a p\,\delta w\, dx\, dy \tag{13.43}$$

Thus, Eqs. (13.40), (13.41) and (13.43) yield, with the Kirchhoff approximations $\epsilon_{zz} = \epsilon_{xz} = \epsilon_{yz} = 0$ (and hence, $\delta\epsilon_{zz} = \delta\epsilon_{xz} = \delta\epsilon_{yz} = 0$),

$$\int_0^b \int_0^a (\delta\bar{U} - p\,\delta w)\, dx\, dy = 0 \tag{13.44}$$

Since (N_{xx}, N_{yy}, N_{xy}) and temperature effects have been discarded, Eqs. (13.31) and (13.19) yield $\epsilon_{xx}^0 = u_x = 0$, $\epsilon_{yy}^0 = v_y = 0$ and $2\epsilon_{xy}^0 = v_x + u_y = 0$. Hence, by Eqs. (13.19),

$$\epsilon_{xx} = -zw_{xx}, \qquad \epsilon_{yy} = -zw_{yy}, \qquad \epsilon_{xy} = -zw_{xy} \tag{13.45}$$

Substitution of Eqs. (13.45) into Eq. (13.42) yields, with Eqs. (13.2) and (13.44),

$$\int_0^b \int_0^a (M_{xx}\,\delta w_{xx} + M_{yy}\,\delta w_{yy} + 2M_{xy}\,\delta w_{xy} + p\,\delta w)\,dx\,dy = 0 \qquad (13.46)$$

Now successive integration by parts of Eq. (13.46) yields (Boresi and Chong, 1987)

$$\int_0^b \int_0^a \left(\frac{\partial M_{xx}}{\partial x}\,\delta w_x + \frac{\partial M_{xy}}{\partial y}\,\delta w_y\right)dx\,dy - \oint (M_{xx}\,\delta w_x + M_{xy}\,\delta w_y)\,dy$$

$$+ \int_0^b \int_0^a \left(\frac{\partial M_{xy}}{\partial y}\,\delta w_x + \frac{\partial M_{yy}}{\partial y}\,\delta w_y\right)dx\,dy + \oint (M_{xy}\,\delta w_x + M_{yy}\,\delta w_y)\,dx$$

$$- \int_0^b \int_0^a p\,\delta w\,dx\,dy = 0$$

and

$$\int_0^b \int_0^a \left(\frac{\partial^2 M_{xx}}{\partial x^2} + 2\frac{\partial^2 M_{xy}}{\partial x\,\partial y} + \frac{\partial^2 M_{yy}}{\partial y^2} + p\right)\delta w\,dx\,dy$$

$$+ \oint \left[M_{xx}\,\delta w_x + M_{xy}\,\delta w_y - \left(\frac{\partial M_{xx}}{\partial x} + \frac{\partial M_{xy}}{\partial y}\right)\delta w\right]dy$$

$$+ \oint \left[-M_{xy}\,\delta w_x - M_{yy}\,\delta w_y + \left(\frac{\partial M_{xy}}{\partial x} + \frac{\partial M_{yy}}{\partial y}\right)\delta w\right]dx = 0 \qquad (13.47)$$

where the line integrals are taken along the boundary in a counterclockwise direction (Fig. 13.6).

We note that the integral over the area of the plate leads to the moment equilibrium equation [Eq. (13.25), with $B_z = 0$ and $P_z = p$].

To be specific, consider the rectangular plate to be *clamped* along the edges $y = 0$ and $x = 0$. Let the edges $x = a$ and $y = b$ be *free* of forces and moments. Then, we have the *forced boundary conditions*

$$w = 0 \quad \text{and} \quad \frac{\partial w}{\partial n} = 0 \quad \text{for } x = 0 \quad \text{and} \quad y = 0 \qquad (13.48)$$

where n denotes the normal direction to the edge. Since the variations must satisfy the forced boundary conditions, we also have

$$\delta w = \delta \frac{\partial w}{\partial n} = 0 \quad \text{for } x = 0 \quad \text{and} \quad y = 0 \qquad (13.49)$$

Consequently, Eq. (13.47) reduces to

$$\int_0^b \left[M_{xx}\,\delta w_x + M_{xy}\,\delta w_y - \left(\frac{\partial M_{xx}}{\partial x} + \frac{\partial M_{xy}}{\partial y}\right)\delta w\right]dy\,\bigg|^{x=a}$$

$$+ \int_0^a \left[M_{xy}\,\delta w_x + M_{yy}\,\delta w_y - \left(\frac{\partial M_{xy}}{\partial x} + \frac{\partial M_{yy}}{\partial y}\right)\delta w\right]dx\,\bigg|^{y=b} = 0 \qquad (13.50)$$

The line integrals of Eq. (13.50) lead to additional boundary conditions (natural conditions) for the free edge after further integration by parts. In this regard, we note that for $x = a$, the functions $\delta w_x(y)$ and $\delta w(y)$ are independent, where we recall that $w_x = \partial w / \partial x$. However, the functions $\delta w_y = \delta(\partial w / \partial y)$ and δw are not independent for $x = a$. Hence, the second term of the integral in dy must again be integrated by parts. Thus, integrating by parts and noting that $\partial M_{xx}/\partial x + \partial M_{xy}/\partial y = Q_x$ [see the fourth of Eqs. [(13.23), with $R_y = 0$], we obtain

$$\int_0^b \left[M_{xx} \delta w_x - \left(Q_x + \frac{\partial M_{xy}}{\partial y} \right) \delta w \right] dy \bigg|^{x=a} + M_{xy} \delta w \big|^{x=a, y=b}$$

and similarly for $y = b$,

$$\int_0^a \left[M_{yy} \delta w_x - \left(Q_y + \frac{\partial M_{xy}}{\partial x} \right) \delta w \right] dx \bigg|^{y=b} + M_{xy} \delta w \big|^{x=a, y=b}$$

Hence, for the free edges $x = a$ and $y = b$, we must have the natural boundary conditions

$$M_{xx} = 0, \qquad V_x = Q_x + \frac{\partial M_{xy}}{\partial y} = 0 \qquad \text{for } x = a$$

$$M_{yy} = 0, \qquad V_y = Q_y + \frac{\partial M_{xy}}{\partial x} = 0 \qquad \text{for } y = b \qquad (13.51)$$

In addition, at the corner of two free edges, we have the additional natural boundary condition

$$M_{xy} = 0 \qquad \text{for } x = a, \, y = b \qquad (13.52)$$

Consequently, at a free edge of a classical plate, say, $x = a$, the shear Q_x and twisting moment M_{xy} do not vanish separately, but rather the combination $Q_x + \partial M_{xy}/\partial y = V_x$, the so-called Kirchhoff shear, vanishes. Alternatively, we may express V_x and V_y in the form [with the fourth and fifth of Eqs. (13.23), with $R_x = R_y = 0$]

$$V_x = \frac{\partial M_{xx}}{\partial x} + 2 \frac{\partial M_{xy}}{\partial y}$$

$$V_y = \frac{\partial M_{yy}}{\partial y} + 2 \frac{\partial M_{xy}}{\partial x} \qquad (13.53)$$

In summary, in terms of the displacement w and its derivatives, we may write [see Eqs. (13.33)]

$$M_{xx} = -D(w_{xx} + v w_{yy})$$
$$M_{yy} = -D(w_{yy} + v w_{xx})$$
$$M_{xy} = -(1 - v)D w_{xy}$$
$$V_x = -D[w_{xxx} + (2 - v)w_{xyy}]$$
$$V_y = -D[w_{yyy} + (2 - v)w_{xxy}] \qquad (13.54)$$

Consequently, substitution for M_{xx}, M_{xy}, M_{yy}, in terms of w, in Eq. (13.25) yields, with $B_z = 0$ and $P_z = p$,

$$\nabla^2\nabla^2 w = \nabla^4 w = \frac{p}{D} \tag{13.55}$$

where $\nabla^2\nabla^2 w = \nabla^4 w = w_{xxxx} + 2w_{xxyy} + w_{yyyy}$. Equation (13.55) is the plate equation; it is one of the main results of classical plate theory. It is a fourth-order partial differential equation. Hence, the plate problem is to find solutions of Eq. (13.55) that satisfy the boundary conditions (clamped, free, simply supported, etc.) at the edges of the plate. Fortunately, the most important plate shapes are rectangular and circular, which may be treated most readily.

13.8

SOLUTION OF RECTANGULAR PLATE PROBLEMS

A large collection of solved rectangular plate problems has been presented by Timoshenko and Woinowsky-Krieger (1959). [See also Marguerre and Woernle (1969), in which isotropic and orthotropic plate solutions are presented for rectangular and circular plates for a wide variety of boundary conditions.] Marguerre and Woernle have presented a systematic treatment that clarifies the effects of shear deformation and hence clarifies the boundary conditions for the classical plate, in which shear deformation is discarded. In addition, the treatment by Marguerre and Woernle emphasizes the orthotropic plate, which is more interesting and more important practically than the isotropic plate. Naruoka (1981) has presented an extensive bibliography on the theory of plates indexed by author and subject matter.

In this section, initially, we treat the small displacement theory of simply supported rectangular plates for certain simple loadings. Thus, initially, we consider bending effects only, since in the case of small displacements, these effects dominate. Fourier series methods of solutions are employed. We also present results of an approximate solution due to Westergaard and Slater (1921).

Solution of $\nabla^2\nabla^2 w = \dfrac{p}{D}$. Rectangular Plate

In Sec. 13.7, when bending effects are dominant, we obtained the plate equation

$$\nabla^2\nabla^2 w = \frac{p}{D} \tag{13.56}$$

where p denotes lateral pressure and D is the flexural rigidity. The plate theory based on Eq. (13.56) is often referred to as the flexural (or bending) theory of plates. For this case, the solution of the plate problem requires that the lateral displacement w satisfies Eq. (13.56) and appropriate boundary conditions. We note that since $\nabla^2\nabla^2$ is an invariant vector operator, Eq. (13.56) holds for all coordinate

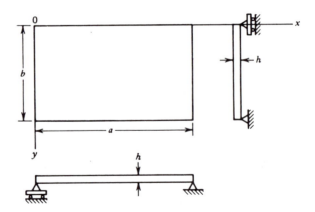

Figure 13.7

systems, provided that proper expressions for $\nabla^2\nabla^2$ are employed (Boresi and Chong, 1987).

For simplicity, we consider here a simply supported rectangular plate of thickness h and in-plane dimensions a and b (Fig. 13.7). Then, we observe that any function (Levy, 1899)

$$w(x, y) = X_n(x) \sin \frac{n\pi y}{b} \tag{13.57a}$$

where n is an integer, satisfies the simple support boundary conditions at $y = 0$ and $y = b$*

$$\left. \begin{array}{l} w = 0 \\ M_{yy} = -D(w_{yy} + v w_{xx}) = 0 \end{array} \right\} \text{ at } y = 0, b$$

Similarly, we may also write $w(x, y)$ in the form

$$w(x, y) = Y_n(y) \sin \frac{n\pi x}{a} \tag{13.58a}$$

which, in turn, satisfies the simple support boundary conditions at $x = 0$ and $x = a$; that is,

$$\left. \begin{array}{l} w = 0 \\ M_{xx} = -D(w_{xx} + v w_{yy}) = 0 \end{array} \right\} \text{ at } x = 0, a \tag{13.58b}$$

For our purposes here, we employ Eq. (13.57a). Thus, substitution into Eq. (13.56) yields an ordinary fourth-order differential equation for $X_n(x)$. Its solu-

---

* One advantage of this single-series method (the Levy method) is that the subsequent series solution [see Eq. (13.63)] converges quite rapidly compared to a double-series representation for w (the Navier method), that is, a solution form of the type

$$w = \sum_{m=1}^{\infty} \sum_{n=1}^{\infty} A_{mn} \sin \frac{m\pi x}{a} \sin \frac{n\pi y}{b}$$

tion contains four constants of integration, which may be selected to satisfy the remaining four boundary conditions at the edges $x = 0$ and $x = a$ (two at $x = 0$ and two at $x = a$). However, before this procedure may be carried out, the lateral pressure p must be expressed in appropriate form. Corresponding to the solution form [Eq. (13.57a)], we express p in the form

$$p(x, y) = p_0 \sum_{n=1}^{\infty} f_n(x) \sin \frac{n\pi y}{b} \tag{13.59}$$

In many practical cases, p may be written in the product form

$$p(x, y) = p_0 f(x) g(y) \tag{13.60}$$

Then, Eqs. (13.59) and (13.60) yield

$$p(x, y) = f(x) \sum_{n=1}^{\infty} p_n \sin \frac{n\pi y}{b} \tag{13.61}$$

where

$$p_n = \frac{2p_0}{b} \int_0^b g(y) \sin \frac{n\pi y}{b} dy \tag{13.62}$$

Consequently, to satisfy Eq. (13.56), we must generalize $w(x, y)$ to

$$w(x, y) = \sum_{n=1}^{\infty} X_n(x) \sin \frac{n\pi y}{b} \tag{13.63}$$

Then substitution of Eqs. (13.61) and (13.63) into Eq. (13.56) yields the set of ordinary differential equations

$$D\left[X_n'''' - 2\left(\frac{n\pi}{b}\right)^2 X_n'' + \left(\frac{n\pi}{b}\right)^4 X_n \right] = p_n f(x), \qquad n = 1, 2, \ldots \tag{13.64}$$

for the functions $X_n(x)$. The solution of Eq. (13.64) for the X_n and substitution into Eq. (13.63) yield the solution of the simply supported rectangular plate subjected to pressure p [Eq. (13.61)]. The resulting series solution gives good results (converges well) for $a > b$, and often for $a = b$. If $a < b$, it is better to use the series form of Eq. (13.58a) or simply interchange the labels a, b, so that again $a > b$.

In the treatment of Eq. (13.64), for simplicity, we take $f(x) = 1$. Then, Eq. (13.64) yields

$$X_n''''(x) - 2\left(\frac{n\pi}{b}\right)^2 X_n''(x) + \left(\frac{n\pi}{b}\right)^4 X_n(x) = \frac{p_n}{D} \tag{13.65}$$

By the theory of ordinary differential equations, the general solution of Eq. (13.65) is

$$X_n(x) = \frac{p_n}{D}\left(\frac{b}{n\pi}\right)^4 \left[1 + (A_{1n} + x A_{2n}) \cosh \frac{n\pi x}{b} \right.$$

$$\left. + (B_{1n} + x B_{2n}) \sinh \frac{n\pi x}{b} \right], \qquad n = 1, 2, \ldots \tag{13.66}$$

The constants A_{1n}, A_{2n}, B_{1n}, B_{2n} are selected to satisfy the four boundary conditions

$$\left.\begin{array}{l} w = 0 \\ M_{xx} = -D(w_{xx} + vw_{yy}) = 0 \end{array}\right\} \text{ at } x = 0, a \qquad (13.67)$$

Substitution of Eqs. (13.66) into Eq. (13.63) and then substitution of the results into Eq. (13.67) yield, after considerable algebra (Marguerre and Woernle, 1969),

$$X_n(x) = \frac{p_n}{D}\left(\frac{b}{n\pi}\right)^4\left\{1 - \cosh\frac{n\pi x}{b} + \frac{n\pi x}{b}\sinh\frac{n\pi x}{b}\right.$$

$$+ \frac{1}{1 + \cosh\dfrac{n\pi a}{b}}\left[\left(\sinh\frac{n\pi a}{b} - \frac{n\pi a}{b}\right)\sinh\frac{n\pi x}{b}\right.$$

$$\left.\left. - \frac{n\pi a}{b}\sinh\frac{n\pi a}{b}\cosh\frac{n\pi x}{b}\right]\right\} \qquad (13.68)$$

With $X_n(x)$ and hence $w(x, y)$ known, Eqs. (13.54) may be used to compute M_{xx}, M_{yy}, M_{xy}, V_x, V_y.

EXAMPLE 13.1
Square Plate Subject to Sinusoidally Distributed Pressure

A square plate is simply supported on all edges (Fig. 13.7) and is loaded by gravel such that

$$p(x, y) = p_0 \sin\frac{\pi x}{a}\sin\frac{\pi y}{b}, \qquad a = b \qquad (a)$$

(a) Determine the maximum deflection and its location.
(b) Determine the maximum values of the moments M_{xx}, M_{yy}.
(c) Determine the maximum values of the Kirchhoff shear forces V_x, V_y.

SOLUTION

The boundary conditions for simply supported edges are

$$\begin{array}{lll} w = 0, & M_{xx} = 0 & \text{for } x = 0, a \\ w = 0, & M_{yy} = 0 & \text{for } y = 0, b \end{array} \qquad (b)$$

Since $w = 0$ around the plate boundary, $\partial^2 w/\partial x^2 = 0$ for edges parallel to the x axis and likewise $\partial^2 w/\partial y^2 = 0$ for edges parallel to the y axis. Hence, noting the expressions for M_{xx}, M_{yy} in Eq. (13.54), we may rewrite the boundary conditions,

Eqs. (b) in the form (note that $b = a$)

$$w = 0, \qquad \frac{\partial^2 w}{\partial x^2} = 0 \qquad \text{for } x = 0, a$$

$$w = 0, \qquad \frac{\partial^2 w}{\partial y^2} = 0 \qquad \text{for } y = 0, a \tag{c}$$

(a) Equations (c) may be satisfied by taking w in the form

$$w = w_0 \sin\frac{\pi x}{a} \sin\frac{\pi y}{a} \tag{d}$$

where w_0 is a constant that must be chosen to satisfy the plate equation [Eq. (13.56)], namely, with Eq. (a),

$$\frac{\partial^4 w}{\partial x^4} + 2\frac{\partial^4 w}{\partial x^2 \partial y^2} + \frac{\partial^4 w}{\partial y^4} = \frac{p_0}{D} \sin\frac{\pi x}{a} \sin\frac{\pi y}{a} \tag{e}$$

Substitution of Eq. (d) into Eq. (e) yields

$$w_0 = \frac{p_0 a^4}{4\pi^4 D} \tag{f}$$

By Eq. (d), we see that the maximum deflection of the plate occurs at $x = y = a/2$. Thus, the maximum deflection of the plate is

$$w_{\max} = w_0 = \frac{p_0 a^4}{4\pi^4 D} \qquad \text{at } x = y = \frac{a}{2} \tag{g}$$

(b) To determine the maximum values of moments M_{xx}, M_{yy}, we find from Eqs. (13.54) with Eqs. (d) and (f)

$$M_{xx} = M_{yy} = \frac{p_0 a^2 (1 + v)}{4\pi^2} \sin\frac{\pi x}{a} \sin\frac{\pi y}{a} \tag{h}$$

It is seen that the maximum values of M_{xx} and M_{yy} occur at $x = y = a/2$. Thus,

$$M_{xx(\max)} = M_{yy(\max)} = \frac{p_0 a^2 (1 + v)}{4\pi^2} \qquad \text{at } x = y = \frac{a}{2} \tag{i}$$

(c) To calculate the Kirchhoff shear forces, we have by Eqs. (13.54) with Eqs. (d) and (f)

$$V_x = \frac{p_0 a}{4\pi}(3 - v)\cos\frac{\pi x}{a} \sin\frac{\pi y}{a} \tag{j}$$

$$V_y = \frac{p_0 a}{4\pi}(3 - v)\sin\frac{\pi x}{a} \cos\frac{\pi y}{a} \tag{j}$$

We see that the maximum values of V_x, V_y occur along the edges of the plate. Thus, by Eqs. (j),

$$V_{x(\max)} = \frac{p_0 a}{4\pi}(3 - v) \qquad \text{at } y = \frac{a}{2}, \qquad x = 0, a$$

$$V_{y(\max)} = \frac{p_0 a}{4\pi}(3 - v) \qquad \text{at } x = \frac{a}{2}, \qquad y = 0, a \qquad \text{(k)}$$

Westergaard Approximate Solution for Rectangular Plates. Uniform Load

The solution of the simply supported rectangular plate subjected to pressure was indicated above. By the results of the bending (flexural) theory of plates for uniform pressure, it may be shown that, at the center of the plate, the stress is always greater in the direction of the shorter span than in the direction of the larger span (Fig. 13.8). This fact may be made plausible by physical considerations. For example, consider the two strips EF and GH (Fig. 13.8). The deflections of the two strips at the center of the plate are, of course, equal. However, the shorter strip (GH), being the stiffer, carries the greater load, and hence, a greater stress is developed in it.

Rectangular Plate with Simply Supported Edges.   In Fig. 13.9, the bending moment *per unit width across the diagonal* at the corner (denoted by M_{diag}), the bending moment per unit width at the center of the strip GH (Fig. 13.8) in the short span b (denoted by M_{bc}) and the bending moment per unit width at the center of the strip EF (Fig. 13.8) in the long span a (denoted by M_{ac}) are plotted.

The curves and equations in Fig. 13.9 were obtained by Westergaard and Slater (1921) with slight modifications in the results obtained from the theory of flexure of plates. The modifications were made in order to obtain relatively simple expressions and, in doing so, allowance was made for some redistribution of stress accompanying slight yielding in the regions of high (and more or less localized) stresses. Note that the moment coefficient for a square slab ($b/a = 1$) is $1/24 = 0.0417$, and that for a long narrow slab ($b/a = 0$) the moment coefficient for the short span is $1/8 = 0.125$. The factor $1/8$ is the same as for a simply supported beam. For intermediate values of b/a, the moment coefficient is always greater in the short span than elsewhere, and its value is intermediate between the limiting values of $1/24$ and $1/8$.

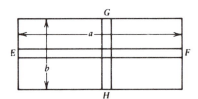

Figure 13.8

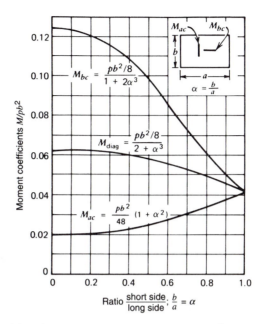

Figure 13.9   Ratio of bending moment M per unit width to pb^2 in rectangular plates with simply supported edges. Poisson's ratio v is assumed to be zero.

Rectangular Plate with Fixed Edges.   If the plate is rigidly held (fixed) at the edges and is subjected to a uniformly distributed load, the maximum moment per unit width occurs at the centers of the long edges, that is, at the fixed ends of the central strip of the short span.

Two limiting cases of fixed-edged rectangular slab will be considered first. If the plate is very long and narrow ($b/a = 0$), the forces at the short ends of the plate will have negligible effect on the moment in the central part of the plate and, hence, the plate may be considered a fixed-end beam with a span equal to the short dimension of the plate; therefore, the negative moment per unit width M_{be} at the fixed edges of the short span is $pb^2/12$, and the positive moment M_{bc} at the center of the short span is $pb^2/24$. The other limiting case is that of the square slab ($b/a = 1$) for which the moment coefficient at the center of the edges is approximately 0.05 and the moment coefficient at the center is 0.018.

For plates having other values of b/a, the maximum negative moment M_{be} and the maximum positive moment M_{bc} are given in Fig. 13.10. These values were obtained by Westergaard and Slater (1921) by simplifying the results obtained from the theory of flexure of flat plates. Owing to the advantageous redistribution of stresses accompanying slight yielding of the plate at points of maximum stress, the plate is somewhat stiffer than is indicated by the results obtained from the theory.

For plates made of ductile metal, the maximum moment used in design should probably be about the average of the values of M_{be} and M_{bc} given in Fig. 13.10. Bach (1920), from the results of experiments, recommends the moment coefficients given by the dotted line in Fig. 13.10. Experimental results for steel plates 0.61 m by 1.22 m ($b/a = 0.5$) with the thicknesses varying from 3 to 19 mm indicate the maximum moment per unit width to be approximately $0.042\,pb^2$. Results indicate that there is not much difference in the value of the stress at the center and at the end of the short span.

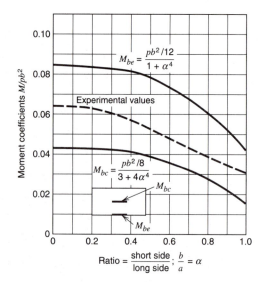

Figure 13.10 Ratio of bending moment M per unit width to pb^2 in rectangular plates with fixed edges. Poisson's ratio v is assumed to be zero.

Other Types of Edge Conditions. Formulas obtained by Westergaard and Slater (1921), giving approximate values of the moments per unit width in rectangular plates, including some of the formulas discussed in the preceding articles, are shown in Table 13.1. These formulas give results fairly close to those found from the theory of flexure of slabs, in which for convenience the value of Poisson's ratio $v = 0$ has been assumed. The effect of Poisson's ratio is to increase the bending moment per unit width in the plate. Let M_{acv} and M_{bcv} represent the values of the bending moments at the center of a rectangular plate when the material has a Poisson's ratio v not assumed to be zero. Approximate values of these bending moments are given by the expressions

$$M_{acv} = M_{ac} + v M_{bc}$$
$$M_{bcv} = M_{bc} + v M_{ac} \tag{13.69}$$

in which M_{ac} and M_{bc} are values of the bending moments as given in Table 13.1, or subsequent tables, in which v has been assumed to be zero. In using these formulas for plates made of ductile material, it should be borne in mind that they give results that probably err somewhat on the side of safety.

Deflection of Rectangular Plate; Uniformly Distributed Load

The differential equation for plates has been solved only for relatively simple shapes of plates and certain simple types of loading. From the solution of this equation for rectangular plates subjected to uniformly distributed loads, the maximum deflection w_{max} at the center of the plate is given by the equation

$$w_{max} = C(1 - v^2)\left(\frac{pb^4}{Eh^3}\right) \tag{13.70}$$

TABLE 13.1
Formulas Obtained by the Theory of Flexure of Slabs, Giving Approximate Values of Bending Moments per Unit Width and Maximum Deflections in Rectangular and Elliptical Slabs Under Uniform Load (Given by Westergaard)[a]

	Moments in Span b		Moments in Span a		Maximum Deflection $w_{max} = C(1-v^2) \times (pb^4/Eh^3)$
	At Center of Edge $-M_{be}$	At Center of Slab M_{bc}	At Center of Edge $-M_{ae}$	Along Center Line of Slab M_{ac}	Values of C
Four edges simply supported	0	$\dfrac{\frac{1}{8}pb^2}{1+2\alpha^3}$	0	$\dfrac{pb^2}{48}(1+\alpha^2)$	$\dfrac{0.16}{1+2.4\alpha^3}$
Span b fixed; span a simply supported	$\dfrac{\frac{1}{12}wb^2}{1+0.2\alpha^4}$	$\dfrac{\frac{1}{24}pb^2}{1+0.4\alpha^4}$	0	$\dfrac{pb^2}{80}(1+0.3\alpha^2)$	$\dfrac{0.032}{1+0.4\alpha^3}$
Span a fixed; span b simply supported	0	$\dfrac{\frac{1}{8}pb^2}{1+0.8\alpha^2+6\alpha^4}$	$\dfrac{\frac{1}{8}pb^2}{1.08\alpha^4}$	$0.015pb^2\left(\dfrac{1+3\alpha^2}{1+\alpha^4}\right)$	$\dfrac{0.16}{1+\alpha^2+5\alpha^4}$
All edges fixed	$\dfrac{\frac{1}{12}wb^2}{1+\alpha^4}$	$\dfrac{\frac{1}{8}pb^2}{3+4\alpha^4}$	$\dfrac{1}{24}wb^2$	$0.009pb^2(1+2\alpha^2-\alpha^4)$	$\dfrac{0.032}{1+\alpha^4}$
Elliptical slab with fixed edges; axes a and b b/a = α	$\dfrac{\frac{1}{12}wb^2}{1+\frac{2}{3}\alpha^2+\alpha^4}$	$\dfrac{\frac{1}{24}pb^2}{1+\frac{2}{3}\alpha^2+\alpha^4}$	$\dfrac{\frac{1}{12}pb^2\alpha^2}{1+\frac{2}{3}\alpha^2+\alpha^4}$	$\dfrac{\frac{1}{24}pb^2\alpha^2}{1+\frac{2}{3}\alpha^2+\alpha^4}$	

Rectangular Slabs

[a] Poisson's ratio $v = 0$ [see Eq. (13.69)]. b = shorter side; a = longer side; $b/a = \alpha$.

where p is the uniformly distributed load per unit of area, b the short span length, E the modulus of elasticity of the material in the plate, h the plate thickness, v Poisson's ratio, and C a dimensionless constant whose value depends on the ratio b/a of the sides of the plate and on the type of support at the edge of the plate.

Several investigators have computed values of the constant C in Eq. (13.70); some of the values are as follows: For a uniformly loaded square ($b/a = 1$) plate simply supported at its edges, $C = 0.047$; if the plate is very long and narrow ($b/a = 0$, approximately), $C = 0.16$. Thus, the deflection of a long narrow plate is more than three times that of a square plate having the same thickness as the narrow plate; in fact, the supports at the short ends of a narrow plate ($b/a < \frac{1}{3}$) have very little effect in preventing deflection at the center of the plate. If all the edges of a uniformly loaded square plate are fixed, the constant in Eq. (13.70) is $C = 0.016$. A comparison of this value of C with the value 0.047 for simply supported edges shows that if the edges of a square plate are fixed, the deflection at the center of the plate is about one-third the deflection for simply supported edges. However, the edges of a plate are seldom if ever rigidly clamped and, therefore, the deflection at the center of a plate having partial restraint at its edges would be given by a value of C between 0.016 and 0.047.

Values of the constant C in Eq. (13.70) for various ratios of b/a and various conditions at the supports are given in Table 13.1. From experiments on plates 0.61 m by 1.22 m with the edges carefully clamped, the measured deflections on relatively thin plates ($h/b \leq 0.02$) agree very closely up to values of deflections not greater than about one-half the plate thickness with those given by the formulas for deflections in Table 13.1. The formulas for deflections in this table give values that are too large when the direct tensile stresses in the plate are appreciable; this condition begins when the maximum deflection of the plate reaches a value of about one-half the thickness of the plate. The stiffening effect of the direct (membrane) tensile stresses also serves to reduce the bending stresses in the plate.

EXAMPLE 13.2
Water Tank

A water tank 3.60 m deep and 2.70 m square is to be made of structural steel plate. The sides of the tank are divided into nine panels by two vertical supports (or stiffeners) and two horizontal supports; that is, each panel is 0.90 m wide and 1.20 m high, and the average head of water on a lower panel is 3.00 m (Fig. E13.2).

(a) Determine the thickness of the plate for the lower panels, using a working stress limit of $\sigma_w = 124.0$ MPa.

(b) Calculate the maximum deflection of the panel.

SOLUTION

The mean pressure on a bottom panel is $p = (3.00 \text{ m})(9.80 \text{ kPa/m}) = 29.4 \text{ kPa}$. We assume this pressure to be uniformly distributed over the panel. We also assume that the edges of the panel are fixed.

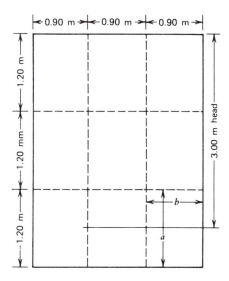

Figure E13.2

(a) For fixed edges, by Fig. 13.10 with $b/a = 0.75$, we have approximately, using the experimental curve,

$$M = 0.042\ pb^2 = (0.042)\left(29.4 \times 10^3\ \frac{\text{N}}{\text{m}^2}\right)(0.90^2\text{m}^2)$$

$$= 1000\frac{\text{N}\cdot\text{m}}{\text{m}}$$

and hence,

$$\sigma = M\frac{c}{I} = \frac{6M}{h^2}$$

thus

$$h = \sqrt{\frac{6M}{\sigma_w}} = \sqrt{\frac{6(1000)}{124}} = 6.96\ \text{mm}$$

(b) To find displacement, we have from Table 13.1, for fixed edges, $C = 0.032/[1 + (0.75)^4] = 0.0243$. With $v = 0.29$ and $E = 200$ GPa, we find

$$w_{\text{max}} = 0.0243(1 - 0.29^2)\frac{(29.4 \times 10^3\ \text{Pa})(900\ \text{mm})^4}{(200 \times 10^9\ \text{Pa})(6.96\ \text{mm})^3}$$

or

$$w_{\text{max}} = 6.37\ \text{mm}$$

This deflection is more than one-half the thickness of the plate. Hence, direct tensile stress would probably reduce the value of w_{max}. See Sec. 13.9.

13.9

SOLUTION OF CIRCULAR PLATE PROBLEMS

In this section, we consider solutions for circular plates undergoing small elastic displacements. We also present some results for large elastic deflections of circular plates; that is, for maximum deflections that are large compared to the plate thickness h. In the case of large deflections, direct tensile forces (tractions) that, though small for deflections less than one-half the plate thickness, become relatively large for deflections greater than the thickness.

Solution of $\nabla^2\nabla^2 w = \dfrac{p}{D}$. Circular Plate

For the circular plate with radius a and thickness h, we employ polar coordinates with origin at the center of the plate (Fig.13.11). Then, Eq. (13.56) may be written in the form (Boresi and Chong, 1987)

$$\nabla^2\nabla^2 w = \left(\frac{\partial^2}{\partial r^2} + \frac{1}{r}\frac{\partial}{\partial r} + \frac{1}{r^2}\frac{\partial^2}{\partial\theta^2}\right)\left(\frac{\partial^2 w}{\partial r^2} + \frac{1}{r}\frac{\partial w}{\partial r} + \frac{1}{r^2}\frac{\partial^2 w}{\partial\theta^2}\right) = \frac{p}{D} \quad (13.71)$$

The general solution of Eq. (13.71) is presented by Marguerre and Woernle (1969). Here, we consider only the axisymmetric case, in which the plate is loaded and supported symmetrically with respect to the z axis. Then, Eq. (13.71) reduces to (since dependency on θ vanishes),

$$\nabla^2\nabla^2 w = \left(\frac{d^2}{dr^2} + \frac{1}{r}\frac{d}{dr}\right)\left(\frac{d^2 w}{dr^2} + \frac{1}{r}\frac{dw}{dr}\right) = \frac{p}{D} \quad (13.72)$$

The solution of Eq. (13.72), with $p = p_0 = $ constant, is

$$w = \frac{p_0 r^4}{64D} + A_1 + A_2\ln r + B_1 r^2 + B_2 r^2\ln r \quad (13.73)$$

where A_1, A_2, B_1, B_2 are constants of integration. The constants A_1, A_2, B_1, B_2 are determined by the boundary conditions at $r = a$ and the regularity conditions

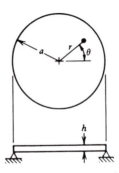

Figure 13.11

that w, ω_r [see Eq. (13.13)], M_{rr}, and V_r must be finite at the center of the plate [origin $r = 0$ of the (r, θ) coordinate system].

Analogous to the expressions for the rectangular plate, we have [Eqs. (13.13), (13.28), (13.30) with $\alpha = 1$, $\beta = r$; see also Eq. (13.51)], in general

$$M_{rr} = -D\left[w_{rr} + v\left(\frac{w_r}{r} + \frac{w_{\theta\theta}}{r^2}\right)\right]$$

$$M_{\theta\theta} = -D\left[\frac{w_r}{r} + \frac{w_{\theta\theta}}{r^2} + vw_{rr}\right]$$

$$M_{rr} + M_{\theta\theta} = -D(1 + v)\nabla^2 w$$

$$M_{r\theta} = -D(1 - v)\frac{\partial}{\partial r}\left(\frac{w_\theta}{r}\right)$$

$$V_r = -D\left[\frac{\partial}{\partial r}(\nabla^2 w) + (1 - v)\frac{1}{r}\frac{\partial}{\partial r}\left(\frac{w_{\theta\theta}}{r}\right)\right]$$

$$V_\theta = -D\left[\frac{1}{r}\frac{\partial}{\partial\theta}(\nabla^2 w) + (1 - v)\frac{\partial^2}{\partial r^2}\left(\frac{w_\theta}{r}\right)\right]$$

$$\omega_r = \frac{1}{r}w_\theta, \qquad \omega_\theta = -w_r \tag{13.74}$$

where subscripts (r, θ) on w denote partial differentiation. Accordingly, for the solid plate, by Eqs. (13.73) and (13.74), we conclude that $A_2 = B_2 = 0$ for axisymmetric conditions.

Circular Plates with Simply Supported Edges

For a solid circular plate simply supported at the edge $r = a$, the boundary conditions are, with Eqs. (13.73) and (13.74) with $A_2 = B_2 = 0$,

$$w(a) = A_1 + B_1 a^2 + \frac{p_0 a^4}{64D} = 0$$

$$-\frac{1}{D}M_{rr}(a) = 2(1 + v)B_1 + (3 + v)\frac{p_0 a^2}{16D} = 0$$

Hence, solving these equations for A_1 and B_1, we obtain with Eqs. (13.73) and (13.74) the following results for the simply supported solid circular plate with uniform lateral pressure $p = p_0$:

$$w = \frac{p_0 a^4}{64D}\left[1 - \left(\frac{r}{a}\right)^2\right]\left[\frac{5 + v}{1 + v} - \left(\frac{r}{a}\right)^2\right]$$

$$M_{rr} = \frac{p_0 a^2}{16}(3 + v)\left[1 - \left(\frac{r}{a}\right)^2\right]$$

$$M_{\theta\theta} = \frac{p_0 a^2}{16}\left[3 + v - (1 + 3v)\left(\frac{r}{a}\right)^2\right] \tag{13.75}$$

Circular Plates with Fixed Edges

For a solid circular plate with fixed edge at $r = a$, the boundary conditions with $A_2 = B_2 = 0$ and Eqs. (13.73) and (13.74) are

$$w(a) = A_1 + B_1 a^2 + \frac{p_0 a^4}{64D} = 0$$

$$w_\theta(a) = -w_r(a) = -2B_1 a - \frac{p_0 a^3}{15D} = 0$$

Solving these equations for A_1 and B_1, we obtain by Eqs. (13.73) and (13.74) the following results for the solid circular plate with fixed edge at $r = a$, subject to uniform lateral pressure $p = p_0$:

$$w = \frac{p_0 a^4}{64D}\left[1 - \left(\frac{r}{a}\right)^2\right]^2$$

$$M_{rr} = \frac{p_0 a^2}{16}\left[1 + v - (3 + v)\left(\frac{r}{a}\right)^2\right]$$

$$M_{\theta\theta} = \frac{p_0 a^2}{16}\left[1 + v - (1 + 3v)\left(\frac{r}{a}\right)^2\right] \tag{13.76}$$

Equations (13.73), (13.74), (13.75), and (13.76) summarize the bending theory of simply supported and clamped circular plates subject to uniform lateral pressure. Numerous solutions for other types of plates, loadings, and boundary conditions have been presented by Marguerre and Woernle (1969). In particular, Marguerre and Woernle have presented extensive results for orthotropic plates.

Circular Plate with Circular Hole at the Center

For a simply supported circular plate of radius a with circular hole of radius b at the center and subjected to uniform lateral pressure $p = p_0$, the boundary conditions are [see Eqs. (13.73) and (13.74)]

$$V_r(b) = -D\left(\frac{4B_2}{b} + \frac{p_0 b}{2D}\right) = 0$$

$$M_{rr}(b) = -D\left\{-(1 - v)\frac{A_2}{b^2} + 2B_1(1 + v)\right.$$

$$\left. + B_2[3 + v + 2(1 + v)\ln b] + \frac{(3 + v)p_0 b^2}{16D}\right\} = 0 \tag{13.77}$$

and

$$w(a) = A_1 + A_2 \ln a + B_1 a^2 + B_2 a^2 \ln a + \frac{p_0 a^4}{64D} = 0$$

$$M_{rr}(a) = -D\left\{-(1 - v)\frac{A_2}{a^2} + 2B_1(1 + v)\right.$$

$$\left. + B_2[3 + v + 2(1 + v)\ln a] + \frac{(3 + v)p_0 a^2}{16D}\right\} = 0 \tag{13.78}$$

Solving these equations for A_1, A_2, B_1, and B_2, we obtain

$$
A_1 = -\frac{p_0 a^4}{4D}\left\{\frac{(1+v)\ln\frac{a}{b}\ln a}{(1-v)\left(\frac{a}{b}\right)^2\left[\left(\frac{a}{b}\right)^2-1\right]} - \frac{(5-v)\ln a}{4(1-v)\left(\frac{a}{b}\right)^2}\right.
$$

$$
\left. + \frac{\left(\frac{a}{b}\right)^2\ln a - \ln b}{2\left(\frac{a}{b}\right)^2\left[\left(\frac{a}{b}\right)^2-1\right]} - \frac{(3+v)\left[\left(\frac{a}{b}\right)^2-1\right]}{8(1+v)\left(\frac{a}{b}\right)^2} + \frac{1}{16}\right\}
$$

$$
A_2 = \frac{p_0 a^4}{4D}\left\{\frac{(1+v)\ln\frac{a}{b}}{(1-v)\left(\frac{a}{b}\right)^2\left[\left(\frac{a}{b}\right)^2-1\right]} - \frac{(3+v)}{4(1-v)\left(\frac{a}{b}\right)^2}\right\}
$$

$$
B_1 = \frac{p_0 a^2}{8D}\left\{\frac{\left(\frac{a}{b}\right)^2\ln a - \ln b}{\left(\frac{a}{b}\right)^2\left[\left(\frac{a}{b}\right)^2-1\right]} - \frac{(3+v)\left[\left(\frac{a}{b}\right)^2-1\right]}{4(1+v)\left(\frac{a}{b}\right)^2}\right\}
$$

$$
B_2 = -\frac{p_0 b^2}{8D} \tag{13.79}
$$

With these coefficients and Eqs. (13.73) and (13.74), the displacement and stress resultants may be computed.

For example, for $a/b = 2$ and $v = 0.30$, the maximum displacement is

$$
w(b) = w_{max}
$$

$$
= 0.682\frac{p_0 a^4}{Eh^3} \tag{13.80}
$$

Except for simple types of loading and shapes of plates, such as a circular shape, the method of finding the bending moment by solving the plate equation [Eq. (13.56)] is somewhat complicated. However, the results obtained can be reduced to tables or curves of coefficients for the maximum bending moments per unit width of a plate and for the maximum deflections of the plate. Some of these results are presented below.

The bending theory of elastic plates, however, does not make allowance for adjustments that take place when slight local yielding at portions of high stress causes a redistribution of stress. This redistribution of stress, in turn, may result in additional strength of the plate, which may often be incorporated into the design of plates, particularly plates of ductile material. We also observe that the bending theory of plates based on Eq. (13.56) does not take into account the added resistance of the plate resulting from direct tensile stresses that accompany relatively large deflections.

Summary for Circular Plates with Simply Supported Edges

Consider a circular plate with simply supported edges, so that no displacement occurs at the edge. The lateral displacement w and bending moments M_{rr}, $M_{\theta\theta}$ for uniform lateral pressure p are given by Eqs. (13.75). The maximum displacement occurs at the center of the plate ($r = 0$). The maximum stress σ_{max} also occurs at the center of the plate. The value of σ_{max} is tabulated in Table 13.2. Results are given in Table 13.2 also for the case of a spot load ($P = \pi r_0^2 p$) at the center of the plate, where the solution is reasonably accurate, provided r_0 is a sufficiently small (nonzero) value.

Summary for Circular Plates with Fixed Edges

Consider a circular plate rigidly held (fixed) so that no rotation or displacement occurs at the edge. We observe that under service conditions the edges of plates are seldom completely "fixed," although usually they are subject to some restraint; furthermore, a slight amount of yielding at the fixed edge may destroy much of the effect of the restraint and thereby transfer the moment to the central part of the plate. For these reasons, the restraint at the edges of a plate is considered of less

TABLE 13.2

Formulas for Values of the Maximum Principal Stresses and Maximum Deflections in Circular Plates as Obtained by Theory of Flexure of Plates[a]

Support and Loading	Principal Stress, σ_{max}	Point of Maximum Stress	Maximum Deflection, w_{max}
Edge simply supported; load uniform ($r_0 = a$)	$\dfrac{3}{8}(3 + v)p\dfrac{a^2}{h^2}$	Center	$\dfrac{3}{16}(1 - v)(5 + v)\dfrac{pa^4}{Eh^3}$
Edge fixed; load uniform ($r_0 = a$)	$\dfrac{3}{4}p\dfrac{a^2}{h^2}$	Edge	$\dfrac{3}{16}(1 - v^2)\dfrac{pa^4}{Eh^3}$ [b]
Edge simply supported; load at center. $P = \pi r_0^2 p$ $r_0 \to 0$, but $r_0 > 0$	$\dfrac{3(1 + v)}{2\pi h^2}P\left(\dfrac{1}{v + 1}\right.$ $\left. + \ln\dfrac{a}{r_0} - \dfrac{1 - v}{1 + v}\dfrac{r_0^2}{4a^2}\right)$	Center	$\dfrac{3(1 - v)(3 + v)Pa^2}{4\pi Eh^3}$
Fixed edge; load at center. $P = \pi r_0^2 p$ $r_0 \to 0$, but $r_0 > 0$	$\dfrac{3(1 + v)P}{2\pi h^2}\left(\ln\dfrac{a}{r_0} + \dfrac{r_0^2}{4a^2}\right)$ a must be $> 1.7 r_0$	Center	$\dfrac{3(1 - v^2)Pa^2}{4\pi Eh^3}$

[a] a = radius of plate; r_0 = radius of central loaded area; h = thickness of plate; p = uniform load per unit area; v = Poisson's ratio.
[b] For thicker plates ($h/r > 0.1$), the deflection is $w_{max} = C(\frac{3}{16})(1 - v^2)(pa^4/Eh^3)$, where the constant C depends on the ratio h/a as follows: $C = 1 + 5.72(h/a)^2$.

importance, particularly if the plate is made of relatively ductile material, than would be indicated by the results of the theory of flexure of plates with fixed edges. In general, an actual medium-thick plate with a fixed edge will be intermediate in stiffness between the plate with a simply supported edge and the plate with an ideally fixed edge.

Formulas are given in Table 13.2 for the maximum deflection of clamped circular plates of ideal, elastic material (Morley, 1935). Experiments have verified the formulas for uniformly distributed loads and a simply supported edge. These experiments with fixed-edged plates under uniformly distributed loads show that the formula for the deflection is correct for thin and medium-thick plates [$(h/a) < 0.1$] for deflections not larger than about one-half the plate thickness. For thicker plates the measured values of deflection are much larger than those computed by the formula. Two reasons for this discrepancy exist: (1) lack of ideal fixity at the edge and (2) additional deflection in the thicker plates due to the shear stresses. These experiments suggested that for thicker [$(h/a) > 0.1$] circular plates with fixed edges subjected to uniform loads, the values of w_{max} given in Table 13.2 be multiplied by a factor that depends on the ratio of the thickness h to the radius r. This factor is $C = 1 + 5.72(h/a)^2$. Experiments on plates with edges securely clamped gave deflections that agreed closely with values computed by the use of the bending theory formula and the constant C.

Formulas for deflections by the bending theory give values that are too large for thin to medium-thick plates when the deflections are larger than about one-half the plate thickness.

Summary for Stresses and Deflections in Flat Circular Plates with Central Holes

Circular plates of radius a with circular holes of radius r_0 at their center are commonly used in engineering systems. For example, they occur in thrust-bearing plates, telephone and loudspeaker diaphragms, steam turbines, diffusers, piston heads, etc. Several cases of practical importance have been studied by Wahl and Lobo (1930). In all these cases, the maximum stress is given by simple formulas of the type

$$\sigma_{max} = k_1 \frac{pa^2}{h^2} \quad \text{or} \quad \sigma_{max} = \frac{k_1 P}{h^2} \quad (13.81)$$

depending on whether the applied load is uniformly distributed over the plate or concentrated along the edge of the central hole. Likewise, the maximum deflections are given by simple formulas of the type

$$w_{max} = k_2 \frac{pa^4}{Eh^3} \quad \text{or} \quad w_{max} = k_2 \frac{Pa^2}{Eh^3} \quad (13.82)$$

Wahl and Lobo have calculated numerical values for k_1 and k_2 for several values of the ratio a/r_0 and for a Poisson's ratio of $v = 0.30$. The cases that they studied are shown in Fig. 13.12 and the corresponding values of k_1 and k_2 are tabulated in Table 13.3. For other solutions for symmetrical bending of circular plates, the interested reader is referred to Timoshenko and Woinowsky-Krieger (1959).

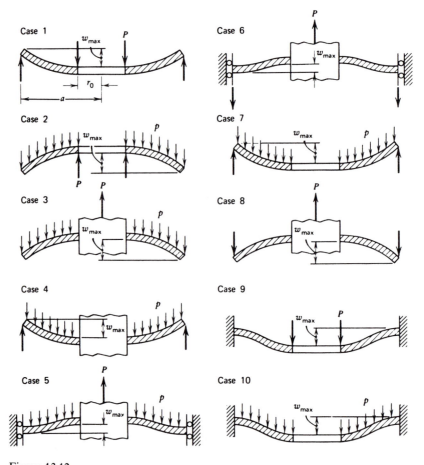

Figure 13.12

EXAMPLE 13.3
Circular Plate Fixed at Edges

A mild steel plate ($E = 200$ GPa, $v = 0.29$, $Y = 315$ MPa) has a thickness $h = 10$ mm and covers a circular opening having a diameter of 200 mm. The plate is fixed at the edges and is subjected to a uniform pressure p.

(a) Determine the magnitude of the yield pressure p_Y and deflection w_{max} at the center of the plate when this pressure is applied.

(b) Determine a working pressure based on a factor of safety of $SF = 2.00$ relative to p_Y.

SOLUTION

(a) The maximum stress in the plate is a radial flexure stress at the outer edge of the plate given either by Eq. (13.76) and the flexure formula or the appropriate

TABLE 13.3

Coefficients k_1 and k_2 Eqs. (13.81) and (13.82) for the Ten Cases Shown in Fig. 13.12: Poisson's Ratio $\nu = 0.30$.

$\dfrac{a}{r_0} =$	1.25		1.5		2		3		4		5	
Case	k_1	k_2	k_1	k_2	k_1	k_2	k_1	k_2	k_1	k_2	k_1	k_2
1	1.10	0.341	1.26	0.519	1.48	0.672	1.88	0.734	2.17	0.724	2.34	0.704
2	0.66	0.202	1.19	0.491	2.04	0.902	3.34	1.220	4.30	1.300	5.10	1.310
3	0.135	0.00231	0.410	0.0183	1.04	0.0938	2.15	0.293	2.99	0.448	3.69	0.564
4	0.122	0.00343	0.336	0.0313	0.74	0.1250	1.21	0.291	1.45	0.417	1.59	0.492
5	0.090	0.00077	0.273	0.0062	0.71	0.0329	1.54	0.110	2.23	0.179	2.80	0.234
6	0.115	0.00129	0.220	0.0064	0.405	0.0237	0.703	0.062	0.933	0.092	1.13	0.114
7	0.592	0.184	0.976	0.414	1.440	0.664	1.880	0.824	2.08	0.830	2.19	0.813
8	0.227	0.00510	0.428	0.0249	0.753	0.0877	1.205	0.209	1.514	0.293	1.745	0.350
9	0.194	0.00504	0.320	0.0242	0.454	0.0810	0.673	0.172	1.021	0.217	1.305	0.238
10	0.105	0.00199	0.259	0.0139	0.480	0.0575	0.657	0.130	0.710	0.162	0.730	0.175

equation in Table 13.2

$$\sigma_{max} = \frac{3}{4} p_Y \frac{a^2}{h^2} = \frac{3 p_Y (100)^2}{4(10)^2} = 75 p_Y$$

The magnitude of p_Y by the maximum shear-stress theory of failure is obtained by setting σ_{max} equal to Y

$$p_Y = \frac{Y}{75} = \frac{315}{75} = 4.20 \text{ MPa}$$

The maximum deflection of the plate when this pressure is applied is given by the appropriate equation in Table 13.2. Thus,

$$w_{max} = \frac{3}{16}(1 - v^2)\frac{p_Y a^4}{Eh^3} = \frac{3(1 - 0.29^2)(4.20)(100)^4}{16(200 \times 10^3)(10)^3}$$

$$= 0.361 \text{ mm}$$

(b) Let p_w be the working pressure; its value is based on p_Y

$$p_w = \frac{p_Y}{SF} = \frac{4.20}{2.00} = 2.10 \text{ MPa}$$

Summary for Large Elastic Deflections of Circular Plates. Clamped Edge and Uniformly Distributed Load

Consider a circular plate of radius a and thickness h (Fig. 13.13a). Let the plate be loaded by lateral pressure p that causes a maximum deflection w_{max} that is large compared to the thickness h (Fig. 13.13c). Let the edge of the plate be clamped so that rotation and radial displacement are prevented (Fig. 13.13b). In Fig. 13.13d a diametral strip of one unit width is cut from the plate to show the bending moments per unit of width and the direct tensile forces that act in this strip at the edge and center of the plate. The direct tensile forces arise from two sources: First, the fixed support at the edge prevents the edge at opposite ends of a diametral strip from moving radially, thereby causing the strip to stretch as it deflects. Second, if the plate is not clamped at its edge but is simply supported as shown in Figs. 13.13e, f, and g, radial stresses arise out of the tendency for outer concentric rings of the plate, such as shown in Fig. 13.13h, to retain their original diameter as the plate defects. In Fig. 13.13h the concentric ring at the outer edge is shown cut from the plate. This ring tends to retain the original outside diameter of the unloaded plate; the radial tensile stresses acting on the inside of the ring, as shown in Fig. 13.13h, cause the ring diameter to decrease, and in doing so they introduce compressive stresses on every diametral section such as x-x. These compressive stresses in the circumferential direction sometimes cause the plate to wrinkle or buckle near the edge, particularly if the plate is simply supported. The radial stresses are usually larger in the central portion of the plate than near the edge.

Thus, when the plate is deflected more than about one-half the thickness, there are direct tensile stresses in addition to bending stresses; as will be indicated later,

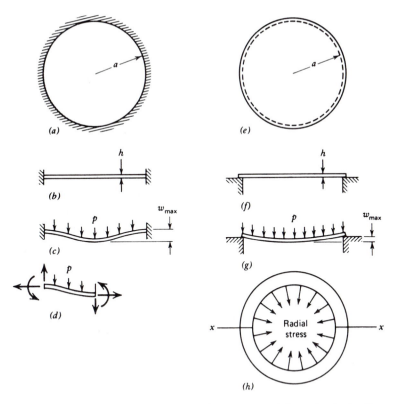

Figure 13.13 Thin plates having large deflections in which tension is significant.

the significant values of these stresses occur either at the edge or center of the plate. Let the bending stresses in a radial plane at the edge and center of the plate be designated by σ_{be} and σ_{bc}, respectively, and let the corresponding direct tensile stresses be σ_{te} and σ_{tc}, respectively. Values of these stresses for a plate with clamped edges having a radius a and thickness h and made of a material having a modulus of elasticity E are given in Fig. 13.14. In Fig. 13.14 the ordinates are values of the stress multiplied by the quantity a^2/Eh^2 (to make dimensionless ordinates), and the abscissas are values of the maximum deflection w_{max} divided by the thickness h (Prescott, 1946). Note that the dimensionless ordinates and abscissas make it possible to use the curves for plates of any dimensions, provided that other conditions are the same. Also note that the bending stress σ_{be} at the fixed edge is the largest of these four stresses. The direct tensile stresses, though small for small deflections (deflections less than about one-half the plate thickness), become relatively large as the deflection increases. For example, if the deflection is equal to twice the plate thickness, the direct tensile stress σ_{tc} at the center of the plate is equal to the bending stress σ_{bc} at the center; if the deflection is four times the thickness, the stress σ_{tc} is twice σ_{bc}.

Significant Stress; Edges Clamped

The maximum stress in the plate is at the edge and is the sum of the values of the bending stress σ_{be} and the direct tensile stress σ_{te} associated with the curves in

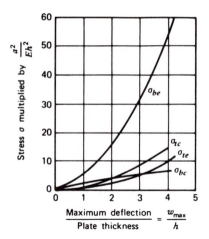

Figure 13.14   Stresses in thin plates having large deflections; circular plate with clamped edges.

Fig. 13.14. Values of this maximum stress σ_{max} multiplied by the quantity a^2/Eh^2 are shown as ordinates to the upper curve in Fig. 13.15a. The values of σ_{max} at points in the plate a short distance radially from the edge are very much smaller than at the edge; a minimum value occurs near the edge, and the stresses gradually approach another maximum value that occurs at the center of the plate. The maximum stress at the center of the plate is indicated by the lower curve in Fig. 13.15a, which represents the sum of the stresses σ_{bc} and σ_{tc} as given by the curves in Fig. 13.14. If failure of the plate is by general yielding, the maximum stress at the center is the significant stress, since the effect of the maximum stress at the edge is localized. However, if the failure of the plate is by fatigue crack growth resulting from repeated applications of loads, or if the plate is made of brittle material and

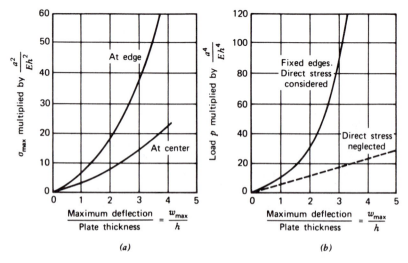

Figure 13.15   Maximum stresses and deflections in thin plates having large deflections; circular plate with clamped edges.

hence fails by sudden fracture under static loads, the stress at the edge would be the significant stress.

Load on Plate; Edges Clamped

In Fig. 13.15b the values of the load p on the plate with fixed edges multiplied by the quantity a^4/Eh^4 are represented as ordinates, and maximum deflections divided by the plate thickness are abscissas, thus giving a dimensionless curve. The dashed line represents values of load and maximum deflection as computed by neglecting the effect of direct tensile stresses. A significant increase in the load p is indicated by the upward trend of the curve above the straight line for deflection larger than about one-half the plate thickness, which shows that the plate is much stiffer than is indicated by the analysis in which the stiffening effect contributed by the direct tensile stress is neglected.

The relation between the load p and stresses in the plate is obtained by using Figs. 13.15a and b jointly. For example, if the dimensions and modulus of elasticity of the plate and load p are given, the quantity pa^4/Eh^4 can be computed. In Fig. 13.15b the abscissa w_{max}/h corresponding to this value of pa^4/Eh^4 is found from the curve. The value of w_{max}/h thus found is now used as the abscissa in Fig. 13.15a, and the stress at the center or edge of the plate is found by reading the ordinate corresponding to this abscissa to the appropriate curve in Fig. 13.15a and dividing it by a^2/Eh^2. This procedure is used in the following example.

EXAMPLE 13.4
Large Deflection of a Uniformly Loaded Circular
Plate with Clamped Edge

A circular plate of aluminum alloy is 500 mm in diameter and 5 mm thick. The plate is subjected to a uniformly distributed pressure p and fixed at its edge. The maximum pressure that the plate can support is assumed to be that pressure that causes a significant tensile stress equal to the tensile yield stress of the material (say, 288 MPa).

(a) Determine the allowable magnitude of the pressure p that develops not more than one-half the maximum pressure that the plate can support.

(b) Compute the maximum deflection corresponding to this allowable pressure. The modulus of elasticity of the aluminum alloy is $E = 72.0$ GPa.

SOLUTION

(a) We note by Fig. 13.15, that neither pressure p nor stress σ are linearly proportional to the deflections. In addition, the stress σ, either at the edge or center of the plate, is not linearly proportional to the pressure p. We must therefore apply the reduction factor (factor of safety $SF = 2$) to the load rather than the stress.

The factor of safety is applied to the failure pressure for the plate. The plate is assumed to fail by general yielding. As indicated in Fig. 13.15a, yielding initiates at the edge of the plate when σ_{max} at the edge is equal to Y. We assume that general yielding failure occurs shortly after the maximum stress at the center of the plate reaches the yield stress of the material; the pressure-deflection curve in Fig. 13.15b is assumed not to be influenced by the localized yielding at the edge. Hence, we seek the value of the pressure p that will cause a stress of 288 MPa at the center of the plate; this value of p is then to be reduced by the factor $SF = 2$.

Accordingly, we compute the factor

$$\frac{\sigma a^2}{Eh^2} = \frac{288(250^2)}{72 \times 10^3(5^2)} = 10$$

With the value 10 as the ordinate in Fig. 13.15a for the curve $\sigma_{\max}$ at the center of the plate, we read the corresponding abscissa $w_{\max}/h = 2.4$. By Fig. 13.15b, with the abscissa equal to 2.4, we find

$$\frac{pa^4}{Eh^4} = 50$$

The value of p determined from this ratio ($p = 576$ kPa) represents the maximum pressure that the plate can support without yielding over a large portion of its volume. Therefore, $p/2$ or 288 kPa is considered the allowable magnitude of pressure that the plate may support.

(b)   To determine the maximum deflection, we compute first the quantity

$$\frac{pa^4}{Eh^4} = \frac{0.288(250^4)}{72 \times 10^3(5^4)} = 25$$

By Fig. 13.15b, we find the corresponding abscissa $w_{\max}/h = 1.8$. Hence, the deflection of the center of the plate is $w_{\max} = 1.8h = 9.00$ mm.

Summary for Large Elastic Deflections of Circular Plates. Simply Supported Edge and Uniformly Distributed Load

It was found that when the edge of a circular plate as shown in Fig. 13.13 is fixed and the plate is subjected to a uniformly distributed load, there exist direct radial tensile stresses in addition to the bending stresses. If a circular plate has its edge simply supported instead of fixed, the direct tensile stresses have somewhat smaller magnitudes, but they are still effective in increasing the load resistance of the plate, particularly when the deflections are large relative to the thickness of the plate.

In Fig. 13.16a the ordinates to the curve marked σ_{tc} represent the direct tensile stresses at the center of the simply supported plate where these stresses are a maximum, and the ordinates to the curve marked σ_{bc} represent the bending stresses at the center of the plate that also have a maximum value at the center. The axes of the curves in Figs. 13.16a and b have the same meaning as those for Figs. 13.15a and b for a plate whose edge is fixed. In Fig. 13.16a the ordinates to the curve marked $\sigma_{\max}$ represent the sum of the stresses σ_{tc} and σ_{bc} that occur on the tensile side of the plate at the center. In Fig. 13.16b the curve represents the relation between the

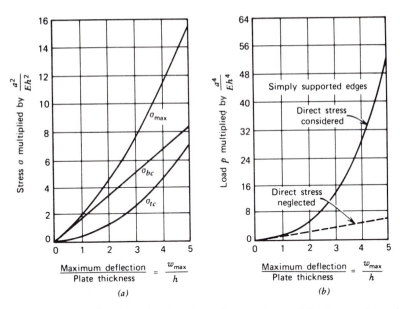

Figure 13.16 Stresses in thin circular plates having large deflections; edges simply supported.

load and maximum deflection, and the dashed line represents this relationship if the direct tensile stresses are neglected in the analysis. The solid curve in Fig. 13.16b, which rises above the dashed line when the maximum deflection becomes greater than one-half to one times the thickness of the plate, shows the influence of the direct tensile stress in increasing the stiffness, especially of relatively thin plates for which the deflections are likely to be large in comparison with the thickness. Figures 13.16a and b are used in solving problems in a manner similar to the use of Figs. 13.15a and b as described in Example 13.4.

Rectangular or Other Shaped Plates with Large Deflections

The general behavior described for circular plates when the deflections are large also applies to rectangular, elliptical, or other shapes of plates. Curves giving data for rectangular plates similar to those given in Figs. 13.15 and 13.16 for circular plates are given by Ramberg et al. (1942).

PROBLEMS
Section 13.8

13.1. Repeat Example 13.1 for the case of a rectangular plate $a \neq b$.

13.2. Repeat Example 13.1 for the case $p = p_0 \sin \dfrac{m\pi x}{a} \sin \dfrac{n\pi y}{a}$, where m and n are integers.

13.3. Repeat Problem 13.2 for the case of a rectangular plate $a \neq b$.

13.4.   Determine the twisting moment M_{xy} and stress σ_{xy} for the plate of Example 13.1.

13.5.   Compute the stresses σ_{xx}, σ_{yy}, σ_{xy} for the plate of Example 13.1.

13.6.   Let a simply supported rectangular plate be subjected to load $p(x, y)$ given in the form of a double trigonometric series

$$p(x, y) = \sum_{m=1}^{\infty} \sum_{n=1}^{\infty} A_{mn} \sin\frac{m\pi x}{a} \sin\frac{n\pi y}{b} \tag{a}$$

Let the displacement $w(x, y)$ be represented in terms of a double trigonometric series

$$w(x, y) = \sum_{m=1}^{\infty} \sum_{n=1}^{\infty} W_{mn} \sin\frac{m\pi x}{a} \sin\frac{n\pi y}{b} \tag{b}$$

This double-series method was used by Navier (the Navier method; see also, Timoshenko and Woinowsky-Krieger, 1959) in a lecture presented to the French Academy in 1820.

(a)   Show that

$$W_{mn} = \frac{1}{\pi^4 D} \frac{A_{mn}}{(m^2/a^2 + n^2/b^2)^2} \tag{c}$$

(b)   For the case $p(x, y) = p_0$, show by the method of Fourier series that

$$A_{mn} = \frac{4p_0}{ab} \int_0^a \int_0^b \sin\frac{m\pi x}{a} \sin\frac{n\pi y}{b} \, dx \, dy$$

$$= \frac{16p_0}{\pi^2 mn} \tag{d}$$

(c)   and hence, that

$$w = \frac{16p_0}{\pi^6 D} \sum_{m=1}^{\infty} \sum_{n=1}^{\infty} \frac{\sin\dfrac{m\pi x}{a} \sin\dfrac{n\pi y}{b}}{mn(m^2/a^2 + n^2/b^2)^2} \tag{e}$$

(d)   and that the maximum deflection is given by

$$w_{max} = \frac{16p_0}{\pi^6 D} \sum_{m=1}^{\infty} \sum_{n=1}^{\infty} \frac{(-1)^{(m+n-2)/2}}{mn(m^2/a^2 + n^2/b^2)^2}$$

Ans.   This series converges extremely rapidly. Using only the first term for a square plate ($a = b$), we obtain

$$w_{max} \cong \frac{4p_0 a^4}{\pi^6 D} \cong 0.0042\frac{p_0 a^4}{D}$$

13.7. A rectangular steel plate ($E = 200$ GPa, $v = 0.29$, $Y = 280$ MPa) has a length of 2 m, width of 1 m, and fixed edges. The plate is subjected to a uniform pressure $p = 270$ kPa. Assume that the design pressure for the plate is limited by the maximum stress in the plate; this would be the case for fatigue loading, for instance. For a working stress limit $\sigma_w = Y/2$, determine the required plate thickness and maximum deflection.

13.8. If the pressure for the plate in Problem 13.7 is increased, yielding will be initiated by moment M_{be} at the fixed edge of the plate; however, the pressure-deflection curve for the plate will remain nearly linear until after the pressure has been increased to initiate yielding due to bending at the center of the plate. Determine the required plate thickness and maximum deflection for the plate in Problem 13.7 if the plate has a factor of safety $SF = 2.00$ against initiation of yielding at the center of the plate.

> *Ans.* $h = 22.1$ mm, $w_{max} = 3.45$ mm

13.9. A square structural steel trap door ($E = 200$ GPa, $v = 0.29$, $Y = 240$ MPa) has a side length of 1.50 m and thickness of 15 mm. The plate is simply supported and subjected to a uniform pressure. Determine the yield pressure p_Y and maximum deflection when this pressure is applied.

> *Ans.* $p_Y = 74.4$ kPa, $w_{max} = 24.1$ mm

Section 13.9

13.10. Verify Eq. (13.73).

13.11. With Eqs. (13.73) and (13.74) and the boundary conditions for a solid circular plate simply supported at the outer edge, $r = a$, derive the results of Eqs. (13.75).

13.12. Repeat Problem 13.11 for the case of the solid circular plate with fixed edge at $r = a$; that is, derive Eqs. (13.76).

13.13. Derive Eqs. (13.79) and, hence, verify Eq. (13.80).

13.14. The cylinder of a steam engine is 400 mm in diameter, and the maximum steam pressure is 690 kPa. Find the thickness of the cylinder head that is a flat steel plate, assuming that the working stress is $\sigma_w = 82.0$ MPa. Determine the maximum deflection of the cylinder head. The plate has fixed edges. For the steel, $E = 200$ GPa and $v = 0.29$.

13.15. A cast-iron disk valve is a flat circular plate 300 mm in diameter and is simply supported. The plate is subjected to uniform pressure supplied by a head of 60 m of water (9.80 kPa/m). Find the thickness of the disk using a working stress of $\sigma_w = 14$ MPa. Determine the maximum deflection of the plate. For cast iron, $E = 100$ GPa and $v = 0.20$.

> *Ans.* $h = 33.7$ mm, $w_{max} = 0.061$ mm

13.16. A circular plate is made of steel ($E = 200$ GPa, $v = 0.29$, and $Y = 276$ MPa), has a radius $a = 250$ mm, and thickness $h = 25$ mm. The plate is simply supported and subjected to a uniform pressure $p = 1.38$ MPa.

(a) Determine the maximum bending stress in the plate and maximum deflection.

(b) Determine the pressure p_Y that is required to initiate yielding in the plate and the factor of safety against initiation of yielding in the plate.

13.17. A circular steel plate with a central hole is fixed at the central hole and uniformly loaded as indicated in Case 3 of Fig. 13.12. For the plate, $a = 300$ mm, $r_0 = 100$ mm, $h = 10$ mm, $p = 100$ kPa, $E = 200$ GPa, and $Y = 290$ MPa.

(a) Determine the maximum bending stress and maximum deflection.

(b) What is the factor of safety against initiation of yielding?

Ans. **(a)** $\sigma_{max} = 194$ MPa, $w_{max} = 1.19$ mm   **(b)** $SF = 1.50$

13.18. A circular opening in the flat end of a nuclear reactor pressure vessel is 254 mm in diameter. A circular steel plate 2.54 mm thick, with tensile yield stress $Y = 241$ MPa, is used as a cover for the opening. When the cover plate is inserted in the opening, its edges are clamped securely. Determine the maximum internal pressure to which the vessel may be subjected if it is limited by the condition that it must not exceed one-third the pressure that will cause general yielding of the cover plate. $E = 200$ GPa for steel.

13.19. A circular plate made of aluminum alloy ($E = 72.0$ GPa and $Y = 276$ MPa) is to have a 254-mm diameter. The edge of the plate is clamped and a pressure of $p = 73.8$ kPa is applied. Determine the required thickness h of the plate, so that this pressure (73.8 kPa) is two-thirds of the pressure that will cause the plate to just reach yield.

Hint: Here, the stress at the edge of the plate is the significant stress, since no yield of the plate is permitted. Use Figs. 13.15a and b to solve for h by trial and error, with a value $p = \frac{3}{2} \times 73.8$ kPa = 110.7 kPa and $\sigma = 276$ MPa.

Ans. $h = 2.0$ mm

13.20. A circular steel plate whose diameter is 2.54 m and thickness is 12.7 mm is simply supported at its edge and subjected to a uniformly distributed pressure p. The tensile yield point stress of the steel is 207 MPa. Determine the pressure p_Y that produces a maximum stress in the plate equal to the tensile yield point stress. Determine the maximum deflection for this pressure.

13.21. In Problem 13.20, determine the pressure p that produces a maximum stress at the center of the plate equal to one-half the yield point stress. Compare this pressure to that determined in Problem 13.20? Explain the result.

Ans. $p = 14.0$ kPa

13.22. Rework Problem 13.18 for the case of a simply supported edge.

13.23. Let the aluminum plate in Problem 13.19 have a thickness of 2.0 mm and simply supported edges. Determine the magnitude of the internal pressure p that can be applied to the plate if the pressure is two-thirds the pressure that will cause the plate to just reach yield.

Ans. $p = 154 \, \text{kPa}$

REFERENCES

Bach, C. (1920). *Elastizität und Festigkeit*, 8th ed. Berlin: Springer-Verlag, p. 598.

Boresi, A. P. and Chong, K. P. (1987). *Elasticity in Engineering Mechanics*. New York: Elsevier.

Langhaar, H. L. (1989). *Energy Methods in Applied Mechanics*. Malabar, Florida: Krieger.

Levy, M. (1899). *Comptes Rendes*, **129**: 535–539.

Marguerre K. and Woernle, H.-T. (1969). *Elastic Plates*. Waltham, Mass.: Blaisdell.

Morley, A. (1935). Strength of Materials, 8th ed. London: Longmans, Green.

Naruoka, M. (1981). *Bibliography on Theory of Plates*, 1st ed. Tokyo: Gihodo.

Newell, H. E., Jr. (1955). *Vector Analysis*. New York: McGraw-Hill.

Prescott, J. (1946). *Applied Elasticity*. New York: Dover, pp. 455–469.

Ramberg, W. A., McPherson, A. E., and Levy, S. (1942). Normal Pressure Tests of Rectangular Plates. Rept. 748, Nat. Advisory Comm. for Aero., Washington, D.C.

Szilard, R. (1974). *Theory and Analysis of Plates*. Englewood Cliffs, N.J.: Prentice-Hall.

Timoshenko, S. and Woinowsky-Krieger, S. (1959). *Theory of Plates and Shells*, 2nd ed. New York: McGraw-Hill.

Ugural, A. C. (1981). *Stresses in Plates and Shells*. New York: McGraw-Hill.

Wahl, A. M. and Lobo, G. (1930). Stresses and Deflections in Flat Circular Plates with Central Holes. *Trans. Amer. Soc. Mech. Eng.*, **52**: 29–43.

Westergaard, H. M. and Slater, W. A. (1921). Moments and Stresses in Slabs. *Proc. Amer. Concrete Inst.*, **17**: 415–538.

14

STRESS CONCENTRATIONS

As noted in previous chapters, the formulas for determining stresses in simple structural members and machine elements are based on the assumption that the distribution of stress on any section of a member can be expressed by a mathematical law or equation of relatively simple form. For example, in a tension member subjected to an axial load the stress is assumed to be distributed uniformly over each cross section; in an elastic beam the stress on each cross section is assumed to increase directly with the distance from the neutral axis; etc.

The assumption that the distribution of stress on a section of a simple member may be expressed by relatively simple laws may be in error in many cases. The conditions that may cause the stress at a point in a member, such as a bar or beam, to be radically different from the value calculated from simple formulas include effects such as

1. abrupt changes in section such as occur at the roots of the threads of a bolt, at the bottom of a tooth on a gear, at a section of a plate or beam containing a hole, at the corner of a keyway in a shaft

2. contact stress at the points of application of the external forces, as, for example, at bearing blocks near the ends of a beam, at the points of contact of the wheels of a locomotive and the rail, at points of contact between gear teeth or between ball bearings and the races

3. discontinuities in the material itself, such as nonmetallic inclusions in steel, voids in concrete, pitch pockets and knots in timber, or variations in the strength and stiffness of the component elements of which the member is made, such as crystalline grains in steel, fibers in wood, aggregate in concrete

4. initial stresses in a member that result, for example, from overstraining and cold working of metals during fabrication or erection, from heat treatment of metals, from shrinkage in castings and in concrete, or from residual stress resulting from welding operations

5. cracks that exist in the member, which may be the result of fabrication, such as welding, cold working, grinding, or other causes.

The conditions that cause the stresses to be greater than those given by the ordinary stress equations of mechanics of materials are called *discontinuities* or *stress raisers*. These discontinuities cause sudden increases in the stress (*stress peaks*) at points near the stress raisers. The term *stress gradient* is used to indicate the rate of increase of stress as a stress raiser is approached. The stress gradient may have an influence on the damaging effect of the peak value of the stress.

Often, large stresses due to discontinuities are developed in only a small portion of a member. Hence, these stresses are called *localized stresses* or simply *stress concentrations*. In many cases, particularly in which the stress is highly localized, a mathematical analysis is difficult or impracticable. Then, experimental, numerical, or mechanical methods of stress analysis are used.

Whether the *significant stress* (stress associated with structural damage) in a metal member under a given type of loading is the localized stress *at a point*, or a somewhat smaller value representing the *average stress* over a small area including the point, depends on the internal state of the metal such as grain type and size, state of stress, stress gradient, temperature, and rate of straining; all these factors may influence the ability of the material to make local adjustments in reducing somewhat the damaging effect of the stress concentration at the point.

The solution for the values of stress concentrations by the theory of elasticity applied to members with known discontinuities or stress raisers requires in general the solution of differential equations that are difficult to solve. However, the elasticity method has been used with success to evaluate stress concentrations in members containing changes of section, such as that caused by a circular hole in a wide plate (see Sec. 14.2). In addition, the use of numerical methods, such as finite elements (see Chapter 19) has lead to *approximate* solutions to a wide range of stress concentration problems. Experimental methods of determining stress concentrations may also prove of value in cases for which the elasticity method becomes excessively difficult to apply.

Some experimental methods are primarily mechanical methods of solving for the significant stress; see, for example, the first three of the list of methods given in the next paragraph. These three methods tend to give values comparable with the elasticity method. Likewise the elastic strain (strain-gage) method, when a very short gage length is used over which the strain is measured with high precision, gives values of stress concentration closely approximating the elasticity value. In the other methods mentioned, the properties of the materials used in the models usually influence the stress concentration obtained, causing values somewhat less than the elasticity values.

Each experimental method, however, has limitations, but at least one method usually yields useful results in a given situation. Some experimental methods that have been used to evaluate stress concentrations are (1) photoelastic (polarized light), (2) elastic membrane (soap film), (3) electrical analogy, (4) elastic strain (strain gage), (5) brittle coating, (6) Moiré methods, and (7) repeated stress, and so on; see Hetényi (1950); Peterson (1974); Kobayashi (1988); Doyle and Phillips (1989).

In this chapter, we consider large stress gradients that arise in the vicinity of holes, notches, and cracks in a structural member or solid. In many practical engineering situations, the failure of a structural member or system is due to the propagation of a crack or cracks that occur in the presence of large stress gradients. The state of stress in the neighborhood of such geometrical irregularities is usually three-dimensional in form, thus increasing the difficulty of obtaining complete analytical solutions. Generally, powerful mathematical methods are required to describe the stress concentrations. We present some general concepts and basic techniques of stress concentrations calculations. For more explicit and more advanced solutions, the reader should refer to specialized works.

The results for computation of stress gradients play a fundamental role in the analysis of fracture and the establishment of fracture criteria. In particular, stress concentrations coupled with repeated loading (fatigue loading; Chapter 16) cause a large number of the failures in structures. The reason for this fact is fairly clear, since stress concentrations lead to local stresses that exceed the nominal or average stress by large amounts.

The concept of a stress concentration factor is often employed by designers to account for the localized increase in stress at a point, the nominal stress being multiplied by a stress concentration factor to obtain an estimate of the local stress at the point. Examples of the use of stress concentration factors are given in the following sections.

14.1

NATURE OF A STRESS CONCENTRATION PROBLEM. STRESS CONCENTRATION FACTOR

In the tension test of an isotropic homogeneous bar of constant cross-sectional area A, the stress σ is assumed to be uniformly distributed over the cross section, provided the section is sufficiently far removed from the ends of the bar, where the load may be applied in a nonuniform manner (Fig. 14.1a). At the end sections, ordinarily the stress distribution is not uniform. Nonuniformity of stress may also occur because of geometric changes (holes or notches) in the cross section of a specimen (Figs. 14.1b and c). This nonuniformity in stress distribution may result in a maximum stress σ_{max} at a section that is considerably larger than the average stress ($\sigma_n = P/A$, where P is the total tension load).* The ratio S_c defined as

$$S_c = \frac{\sigma_{max}}{\sigma_n} \tag{14.1}$$

is called the *stress concentration factor* for the section (point); the more abrupt the cross-sectional area transition in the tension specimen, the larger the stress concentration factor (Fig. 14.1d).

If σ_{max} is the *calculated* value σ_c of the localized stress as found from the theory of elasticity, or experimental methods, S_c is given an additional subscript c and is written S_{cc}. Then, S_{cc} is called the *calculated stress concentration factor*; it is also sometimes referred to as a *form factor*. If, on the other hand, σ_{max} is the *effective* value σ_e found *from tests of the actual material under the conditions of use*, as, for example, under repeated stress by determining first the effective stress σ_e (fatigue strength) from specimens that contain the abrupt change in section or notch and then obtaining the fatigue strength from specimens free from the notch, S_c is given the additional subscript e. Then, S_{ce} is called the *effective* or *significant stress concentration factor*; the term *strength reduction factor* is also used, especially in connection with repeated loads (fatigue). Thus, we may write

$$\sigma_c = S_{cc}\sigma_n \quad \text{and} \quad \sigma_e = S_{ce}\sigma_n \tag{14.2}$$

---

* When the dimension of the hole or notch is small compared to the width of the bar, the area A is considered to be the cross-sectional area of the bar away from the load application region or from the hole or notch in the member. For bars with relatively small widths, we take A to be the area of the bar at the hole or notch section. When the width of the bar is large compared to the diameter of the hole, the difference in the two definitions of A is small.

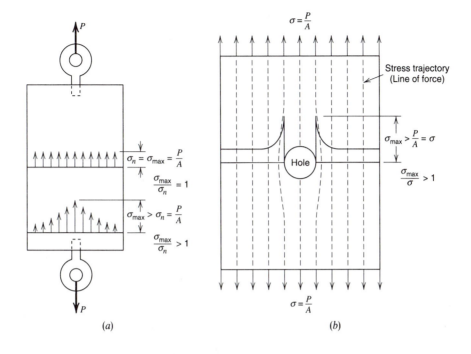

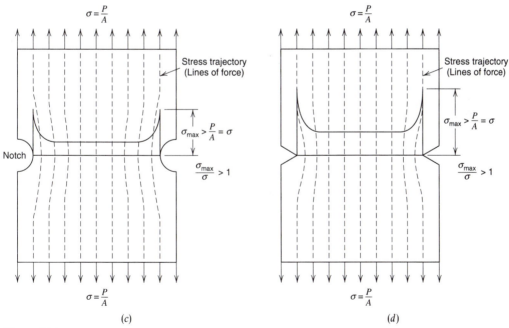

Figure 14.1

The significance of values of S_{ce} is discussed in Sec. 14.5. Analytical and experimental values for S_{cc} are presented in Sec. 14.2, 14.3, and 14.4.

The values of calculated stress concentrations given in this chapter are not meant to be exhaustive, but rather illustrative of the effects of different discontinuities as computed by the various methods of determining calculated stress concentrations or localized stresses.

A pictorial representation of stress trajectories (Figs. 14.1b, c, d) is often employed as an approximate model in the physics of solids to explain the nature of the strain (stress) in the neighborhood of a geometrical discontinuity (crack, dislocation, etc.) in a solid. This representation is based on the analogy between magnetic lines of forces and stress trajectories.

For example, analogous to magnetic lines of forces, the *stress trajectories*, whose paths must lie in the material, cluster together in passing around a geometric hole or discontinuity. In doing so, the average spacing between the lines of force is reduced and, therefore, there results a stress concentration (stress gradient) or an increase in local stress (more lines of force are squeezed into the same area). To expand this idea further, consider a geometrical discontinuity (crack) and sketch the hypothetical local arrangement of atoms around the tip of the crack (Fig. 14.2). The lines of force may be considered to be transmitted from one row of atoms to another. Therefore, the transmission of force around the tip of the crack (say, a small crack in an infinite plate) entails heavy loading and straining of the bonds (AB, CD, AC, etc.). Smaller loads and strains are carried by bonds away from the crack (the strain of bond MN is much less than that of AB). For bonds sufficiently far removed from AB, for example, bond MN, the associated stress is essentially $\sigma = P/A$. The conceptual model of Fig. 14.2 leads to the conclusion that for bond AB to be extended, bonds AC and BD also must be extended. Hence, the uniaxial loading of the plate causes the region around the crack tip to have not only a high tensile strain in the y direction but also a high tensile strain in the x direction. The concept of lines of force also suggests a redistribution of strain energy from regions above or below the crack (regions R and Q in Fig. 14.2) to the highly strained region at the crack tip (see also Figs, 14.1b, c, d). Also because of the distortion of rectangular elements (Fig. 14.2), high shear stresses exist in the neighborhood of a stress concentration.

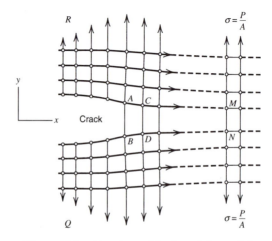

Figure 14.2   Atomic model of crack in a solid.

In practical problems of stress concentrations, the state of stress in the neighborhood of the crack is three-dimensional in nature. For such complex situations, few complete analytical solutions exist. Indeed, the majority of mathematical solutions to stress concentration problems are at best approximate two-dimensional solutions of plane stress cases, the case of plane strain being derived from the plane stress case (Savin, 1961). Consequently, experimental methods of determining stress concentration factors are often employed to supplement or verify analytical predictions. Unfortunately, experimental methods are also limited in accuracy and particularly in generality. For this reason, stress concentration factors are usually determined by several methods.

Stress concentrations may also arise because of concentrated loads such as point loads, line loads, spot loads, etc. (see Sec. 14.3, 14.4. and Chapter 18).

14.2

STRESS CONCENTRATION FACTORS. THEORY OF ELASTICITY

Circular Hole in an Infinite Plane Under Uniaxial Tension

Consider first the case of an infinite plate or sheet with a small circular hole of radius a under uniaxial tension σ (Fig. 14.3).

With respect to polar coordinates (r, θ), the plane stress components at any point P are given by the formulas (Boresi and Chong, 1987)

$$\sigma_{rr} = \frac{\sigma}{2}\left(1 - \frac{a^2}{r^2}\right) + \frac{\sigma}{2}\left(1 - \frac{a^2}{r^2}\right)\left(1 - \frac{3a^2}{r^2}\right)\cos 2\theta$$

$$\sigma_{\theta\theta} = \frac{\sigma}{2}\left(1 + \frac{a^2}{r^2}\right) - \frac{\sigma}{2}\left(1 + \frac{3a^4}{r^4}\right)\cos 2\theta$$

$$\sigma_{r\theta} = -\frac{\sigma}{2}\left(1 - \frac{a^2}{r^2}\right)\left(1 + \frac{3a^2}{r^2}\right)\sin 2\theta \tag{14.3}$$

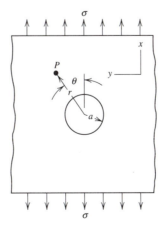

Figure 14.3 Infinite plate with a small circular hole.

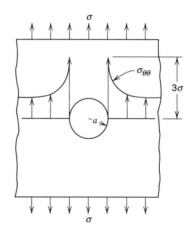

Figure 14.4 $\sigma_{\theta\theta}$ distribution for $\theta = \pi/2, 3\pi/2$.

We note that the stress state given by Eqs. (14.3) satisfies the boundary conditions at $r = a$ ($\sigma_{rr} = \sigma_{r\theta} = 0$ for all θ) and at $r = \infty$ ($\sigma_{xx} = \sigma_{rr} = \sigma$, $\sigma_{xy} = \sigma_{r\theta} = 0$ for $\theta = 0$, π and $\sigma_{yy} = \sigma_{rr} = 0$, $\sigma_{xy} = \sigma_{r\theta} = 0$ for $\theta = \pi/2, 3\pi/2$). For $r = a$,

$$\sigma_{\theta\theta} = \sigma(1 - 2\cos 2\theta) \tag{14.4}$$

Hence, for $\theta = \pi/2$, $3\pi/2$, $\sigma_{\theta\theta}$ attains its maximum value of $\sigma_{\theta\theta(\max)} = 3\sigma$. For $\theta = 0$, π, $\sigma_{\theta\theta}$ attains a compressive value $-\sigma$. Thus, $\sigma_{\theta\theta}$ attains a maximum tensile value of three times the uniformly distributed stress σ, at the hole $r = a$ for $\theta = \pi/2$, $3\pi/2$ (Fig. 14.4) This value (3σ) is the largest normal stress that occurs in the plate. Hence, the stress concentration factor at the hole [Eq. (14.2)] is $S_{cc} = 3$. Figure 14.4 shows the fact that as r increases ($> a$), the maximum value of $\sigma_{\theta\theta}$ decreases rapidly [see Eqs. (14.3)]. Thus, the high stress gradient or stress concentration is quite localized in effect. For this reason, Eqs. (14.3) are often used to estimate the stress concentration effect of a hole in a plate of finite width in the direction normal to the direction of tension σ. However, when the diameter of the hole is comparable to the width of the plate, Eqs. (14.3) are considerably in error. Several authors have studied the problem of a plate strip with a circular hole by theoretical and experimental (photo-elastic and strain-gage) methods. The results are summarized by the formula

$$S_{cc} = \frac{\sigma_{\max}}{\sigma_n} = \frac{3\kappa - 1}{\kappa + 0.3} \tag{14.5}$$

where κ is the ratio (width of strip/diameter of hole) and σ_n the average stress over the weakened cross-sectional area (the cross-sectional area of the plate remaining at the section containing the hole).

Elliptic Hole in an Infinite Plate Stressed in Direction Perpendicular to Major Axis of the Hole
Consider an infinite plate or sheet with an elliptic hole of major axis $2a$ and minor axis $2b$ (Fig. 14.5). A uniform tensile stress σ is applied at a large distance from the hole and is directed perpendicular to the major axis of the elliptical hole; that is, $\sigma_{yy} = \sigma$ at infinity. For this problem, it is desirable to express the stress components

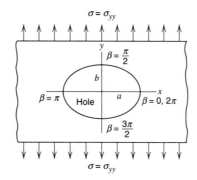

Figure 14.5   Elliptical hole in an infinite plate.

relative to orthogonal curvilinear coordinates (Boresi and Chong, 1987) (elliptic coordinates, Fig. 14.6). In terms of elliptic coordinates (α, β), the equation of an ellipse is

$$\frac{x^2}{\cosh^2 \alpha} + \frac{y^2}{\sinh^2 \alpha} = c^2 \tag{14.6}$$

where for the ellipse with semiaxes (a, b), we have (Fig. 14.5)

$$a = c \cosh \alpha_0, \qquad b = c \sinh \alpha_0 \tag{14.7}$$

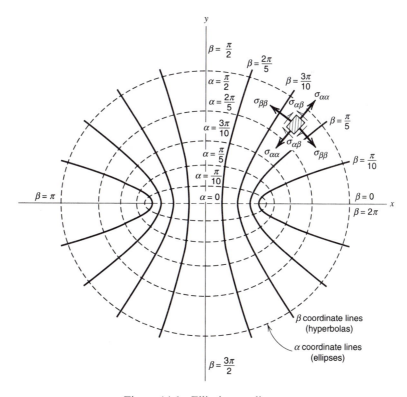

Figure 14.6   Elliptic coordinates.

Thus, in the limit as $\alpha_0 \to 0$, the elliptical hole becomes a sharp crack (an ellipse of zero height and length $2a = 2c$). Because of this condition, the solution for the stresses in a plate with an elliptical hole is employed to study the stresses in a plate with a narrow crack of length $2a$.

The elastic stress distribution in a plate with an elliptical hole has been determined by Inglis (1913) by the method of complex potentials (see also Savin, 1961; Timoshenko and Goodier, 1970). For uniaxial tension stress, perpendicular to the major axis of the elliptical hole, the sum of the stress components $\sigma_{\alpha\alpha}$, $\sigma_{\beta\beta}$ is given by the formula

$$\sigma_{\alpha\alpha} + \sigma_{\beta\beta} = \sigma e^{2\alpha_0}\left[\frac{(1 + e^{-2\alpha_0})\sinh 2\alpha}{\cosh 2\alpha - \cos 2\beta} - 1\right] \tag{14.8}$$

Since the stress $\sigma_{\alpha\alpha} = 0$ at the hole ($\alpha = \alpha_0$), Eq. (14.8) yields the stress $\sigma_{\beta\beta}$ at the hole as

$$\sigma_{\beta\beta}|_{\alpha=\alpha_0} = \sigma e^{2\alpha_0}\left[\frac{(1 + e^{-2\alpha_0})\sinh 2\alpha_0}{\cosh 2\alpha_0 - \cos 2\beta} - 1\right] \tag{14.9}$$

where (α, β) are elliptic coordinates ($\alpha = \alpha_0$ at the hole) and by Eqs. (14.7)

$$\tanh \alpha_0 = \frac{b}{a} \tag{14.10}$$

where a is the semimajor axis of the ellipse and b the semiminor axis. Therefore, by Eq. (14.9), the maximum value of $\sigma_{\beta\beta}$ is (for $\beta = 0, \pi$; $\cos 2\beta = 1$; this occurs at the ends of the major axis)

$$\sigma_{\beta\beta(\text{max})} = \sigma(1 + 2\coth \alpha_0) = \sigma\left(1 + \frac{2a}{b}\right) \tag{14.11}$$

Thus, the maximum value of $\sigma_{\beta\beta}$ increases without bound as $b/a \to 0$, that is, as the semiminor axis b becomes smaller and smaller relative to a. It is noteworthy that for $a = b$ (a circular hole), the maximum value of $\sigma_{\beta\beta}$ is 3σ, which agrees with the results given by Eq. (14.4). The distribution of $\sigma_{\beta\beta}$ around a circular hole ($a/b = 1$) is shown in Fig. 14.7. The distribution of $\sigma_{\beta\beta}$ at the hole for $a/b = 5$ is shown in Fig. 14.8. By geometry, the radius of curvature of an ellipse at the end of the major axis is [Eq. (14.6)]

$$\rho = \frac{b^2}{a} \tag{14.12}$$

where (a, b) are major and minor semiaxes lengths, respectively. Hence, Eqs. (14.11) and (14.12) yield

$$\sigma_{\beta\beta(\text{max})} = \sigma\left(1 + 2\sqrt{\frac{a}{\rho}}\right) \tag{14.13}$$

Also by Eq. (14.9), the minimum value of $\sigma_{\beta\beta}$ is $\sigma_{\beta\beta(\text{min})} = -\sigma$ (at the ends of the minor axis, where $\beta = \pi/2, -\pi/2$).

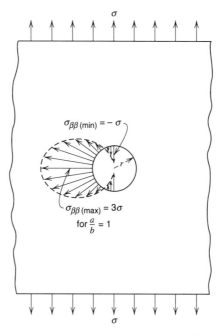

Figure 14.7 Circumferential stress distribution around an edge of a circular hole in an infinite plate.

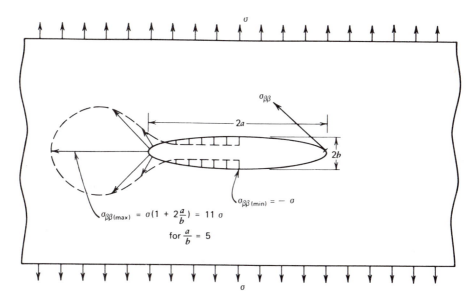

Figure 14.8 Distribution of $\sigma_{\beta\beta}$ around an elliptical hole in an infinite plate loaded perpendicular to the major axis.

Elliptical Hole in Infinite Plate Stressed in Direction Perpendicular to the Minor Axis of the Hole

Let the plate be subjected to stress σ as indicated in Fig. 14.9, where, as above, the dimensions (a, b) are very small compared to the length and width dimensions of the plate. By a transformation of Eq. (14.9), the value of $\sigma_{\beta\beta}$ at any point on the perimeter of the hole is

$$\sigma_{\beta\beta}|_{\alpha = \alpha_0} = \sigma \left(\frac{1 + \sinh 2\alpha_0 - e^{2\alpha_0} \cos 2\beta}{\cosh 2\alpha_0 - \cos 2\beta} \right) \tag{14.14}$$

For $\beta = \pi/2, -\pi/2$, $\sigma_{\beta\beta}$ attains the maximum value

$$\sigma_{\beta\beta(\text{max})} = \sigma(1 + 2 \tanh \alpha_0) = \sigma \left(1 + \frac{2b}{a} \right) \tag{14.15}$$

at the ends of the minor axis. Again as above, for $\beta = 0, \pi$, $\sigma_{\beta\beta}$ attains the minimum value $\sigma_{\beta\beta(\text{min})} = -\sigma$ (which now occurs at the ends of the major axis). The distribution of $\sigma_{\beta\beta}$ is given in Fig. 14.9 for $a/b = 5$.

Crack in a Plate

As $b \to 0$, the elliptical hole in an infinite plate becomes very flat and approaches the shape of a line crack (see Chapter 15). The maximum value of $\sigma_{\beta\beta}$ may become quite large compared to the applied stress for nonzero values of b as $b \to 0$, depend-

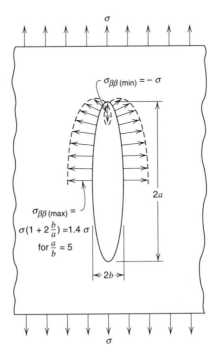

Figure 14.9 Distribution of $\sigma_{\beta\beta}$ around an elliptical hole in an infinite plate loaded perpendicular to the minor axis.

ing on the nature of the load. For example, for the case of Fig. 14.5, Eq. (14.11) yields with $a/b = 100$, $(\sigma_{\beta\beta})_{max} = 201\sigma$, which corresponds to a stress concentration factor of $S_{cc} = 201$. For the loading case of Fig. 14.9, with $a/b = 100$, Eq. (14.15) yields $(\sigma_{\beta\beta})_{max} = 1.02\sigma$ or $S_{cc} = 1.02$. The case $b = 0$ leads to a special study of stress singularities. The practical significance of very large stress concentrations is discussed in Chapter 15.

Ellipsoidal Cavity

In a member subjected to axial tension, the theoretical stress at the edge of an internal cavity having the shape of an ellipsoid has been obtained by Sadowsky and Sternberg (1949). The stress concentration factors for two special cases of such an internal discontinuity will be considered; namely for ellipsoids of revolution of the prolate spheroid type (football shape) and the oblate spheroid type (door-knob shape). The data for a prolate spheroid are given in Table 14.1. For this case, the semimajor axis a of the ellipsoid, which is the axis of revolution, is oriented so that it is perpendicular to the direction of the axial pull in the member, and the semi-minor axis b always lies in a plane parallel to the axial pull. Dimensions a and b are considered to be very small compared to the cross-sectional dimensions of the axial member. If the nominal (average) stress in the member is σ_n, the maximum stress occurs at the end of the semimajor axis a and has values for various ratios of b/a as given in Table 14.1.

The ellipsoid of revolution having the shape of the oblate spheroid has its semiminor axis b, which is the axis of revolution, oriented in the direction of the uniaxial pull in the member, and the semimajor axis a always lies in a plane per-pendicular to the load. If the nominal (average) stress in the member is σ_n, the max-imum stress occurs at the end of a semimajor axis a and has values for various ratios of b/a as given in Table 14.2. These values of the calculated maximum elastic stress show that an internal flaw or cavity of spherical shape such as a gas bubble (an ellipsoid for which $b/a = 1$) raises the stress from σ_n to $2.05\sigma_n$; a long, narrow, stringlike internal flaw or cavity ($b/a = 0$) oriented in a direction perpendicular to the load raises the stress from σ_n to $2.83\sigma_n$; and a very flat, round cavity oriented so that the flat plane is perpendicular to the load raises the stress from σ_n to values as high or higher than $13.5\sigma_n$ if the material remains elastic; this value is compa-rable to the value for a narrow elliptical hole as given by Eq. (14.11).

TABLE 14.1
Stress at End of Semimajor Axis a of Internal Ellipsoidal Cavity of Prolate Spheroid Shape

Ratio b/a	1.0	0.8	0.6	0.4	0.2	0.1
Calculated stress	$2.05\sigma_n$	$2.17\sigma_n$	$2.33\sigma_n$	$2.52\sigma_n$	$2.70\sigma_n$	$2.83\sigma_n$

TABLE 14.2
Stress at End of Semimajor Axis a of Internal Ellipsoidal Cavity of Oblate Spheroid Shape

Ratio b/a	1.0	0.8	0.6	0.4	0.2	0.1
Calculated stress	$2.05\sigma_n$	$2.50\sigma_n$	$3.3\sigma_n$	$4.0\sigma_n$	$7.2\sigma_n$	$13.5\sigma_n$

Grooves and Holes

The values of the calculated stress concentration factors for grooves as shown in Figs. *A* through *D* of Table 14.3 may be obtained from the diagram given by Neuber (1958) (Fig. 14.10).

Consider first the construction of Fig. 14.10. For example, let it be assumed that a member contains the groove shown in Fig. *A* of Table 14.3 and is subjected to an axial load *P*.

Let the calculated stress concentration factor be S_{cs} when the groove is very shallow. Then from Neuber (1958)

$$S_{cs} = 1 + 2\sqrt{\frac{t}{\rho}} \tag{14.16}$$

TABLE 14.3
Directions for Use of Fig. 14.10 (Neuber) in Finding Calculated Stress Concentration Factor S_{cc} in Bars

Type of Notch	Type of Load	Formula for Nominal Stress	Scale for $\sqrt{\dfrac{t}{\rho}}$	Curve for Finding S_{cc}
	Tension	$\dfrac{P}{2bh}$	f	1
	Bending	$\dfrac{3M}{2b^2h}$	f	2
	Tension	$\dfrac{P}{bh}$	f	3
	Bending	$\dfrac{6M}{b^2h}$	f	4
	Tension	$\dfrac{P}{2bh}$	f	5
	Bending	$\dfrac{3Mt}{2h(c^3 - t^3)}$	e	5
	Tension	$\dfrac{P}{\pi b^2}$	f	6
	Bending	$\dfrac{4M}{\pi b^3}$	f	7
	Direct shear	$\dfrac{1.23V}{\pi b^2}$	e	8
	Torsional shear	$\dfrac{2T}{\pi b^3}$	e	9

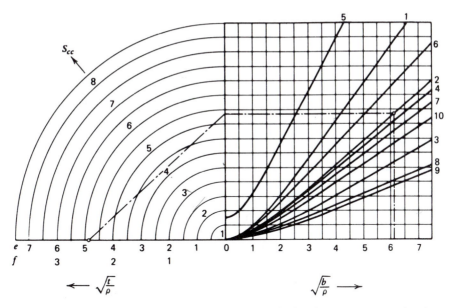

Figure 14.10 Neuber's diagram (nomograph) for a calculated stress concentration factor at the root of a notch.

Let the calculated stress concentration factor be S_{cd} when the groove is very deep. Then from Neuber (1958),

$$S_{cd} = \frac{2[(b/\rho) + 1]\sqrt{b/\rho}}{[(b/\rho) + 1]\arctan\sqrt{b/\rho} + \sqrt{b/\rho}} \tag{14.17}$$

Let S_{cc} represent the calculated stress concentration for any depth of groove. Then, according to Neuber, an approximate, and usually quite accurate, value of S_{cc} is given by the following equation:

$$S_{cc} = 1 + \frac{(S_{cs} - 1)(S_{cd} - 1)}{\sqrt{(S_{cs} - 1)^2 + (S_{cd} - 1)^2}} \tag{14.18}$$

When the groove is very shallow, Eq. (14.18) reduces to $S_{cc} = S_{cs}$, and when the groove is very deep, Eq. (14.18) reduces to $S_{cc} = S_{cd}$. Curve number 1 in Fig. 14.10 has been plotted by making use of Eqs. (14.16), (14.17), and (14.18). The other curves were obtained in a similar manner.

To show how Fig. 14.10 is used, assume that $\rho = 6.35$ mm, $t = 38.0$ mm, and $b = 241.0$ mm in Fig. A of Table 14.3 and that the bar is subjected to a bending moment M. From these values, $\sqrt{t/\rho} = 2.45$ and $\sqrt{b/\rho} = 6.16$. As indicated in Table 14.3, scale f applies for $\sqrt{t/\rho}$ and curve 2 for $\sqrt{b/\rho}$. Thus, to find the value of the calculated stress concentration factor, we enter Fig. 14.10 with $\sqrt{b/\rho} = 6.16$, proceed vertically upward to curve 2, then horizontally to the left to the axis of ordinates. We join this point to the point $\sqrt{t/\rho} = 2.45$ on the left-hand axis of abscissas (on which scale f is applicable) by a straight line. This line is tangent to the circle corresponding to the appropriate calculated stress concentration factor; thus, $S_{cc} = 4.25$.

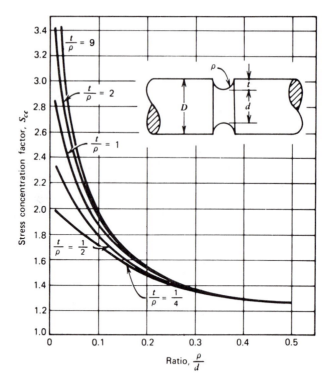

Figure 14.11   Calculated stress concentration factors for semicircular grooves in a cylindrical member subjected to bending only as obtained for Neuber's diagram. (From the unpublished results of Moore and Jordan.)

Some values of a calculated stress concentration factor for bending obtained from Neuber's diagram (Fig. 14.10) as found by Moore and Jordan are given in Fig. 14.11.

14.3

STRESS CONCENTRATION FACTORS. COMBINED LOADS

In Sec. 14.2, we discussed stress concentrations for several types of notches for simple loading of members made of an isotropic material that is assumed to behave in a linearly elastic manner. Because of the linearity of the response, if these same conditions prevail when such a member is subjected to more complex loading, the loads in some cases may be resolved into simple component parts, for which the results of Sec. 14.2 hold. Then by means of the principle of superposition, the results may be combined to yield the effect of complex loading.

Infinite Plate with Circular Hole

Consider an infinite plate, with a circular hole, subjected to stress $\sigma = \sigma_1$ on two parallel edges far removed from the hole (Fig. 14.3) and to stress $\sigma = \sigma_2$ on the other distant parallel edges. The stress distribution may be derived from Eqs. (14.3) by superposition. One need merely set $\sigma = \sigma_1$ and $\theta = 0$ in Eqs. (14.3) to obtain stresses due to σ_1. Then set $\sigma = \sigma_2$ and $\theta = \theta + \pi/2$ in Eqs. (14.3) to obtain stresses due to σ_2 and add the stresses so obtained to those due to σ_1. Special results are obtained for $\sigma = \sigma_1 = \sigma_2$, the case of uniform tension in all directions [then $\sigma_{\theta\theta(\max)} = 2\sigma$] and for $\sigma = \sigma_1 = -\sigma_2$, the case of pure shear [then $\sigma_{\theta\theta(\max)} = 4\sigma$ for $\theta = \pm\pi/2$]. Thus, for uniform tension $S_{cc} = 2$ and for uniform shear $S_{cc} = 4$.

Elliptical Hole in Infinite Plate Uniformly Stressed in Directions of Major and Minor Axes of the Hole

Analogous to the circular hole case, the stresses for the state of uniform tension σ on the boundary ($\sigma_{xx} \to \sigma$ for $x \to \infty$ and $\sigma_{yy} \to \sigma$ for $y \to \infty$) may be computed for the elliptical hole. The results are (Neuber, 1958)

$$\sigma_{\alpha\alpha} + \sigma_{\beta\beta} = \frac{2\sigma \sinh 2\alpha}{\cosh 2\alpha - \cos 2\beta} \tag{14.19}$$

Again since $\sigma_{\alpha\alpha} = 0$ for $\alpha = \alpha_0$ (at the hole),

$$\sigma_{\beta\beta}\big|_{\alpha=\alpha_0} = \frac{2\sigma \sinh 2\alpha_0}{\cosh 2\alpha_0 - \cos 2\beta} \tag{14.20}$$

and

$$\sigma_{\beta\beta(\max)} = 2\sigma\left(\frac{a}{b}\right) \tag{14.21}$$

which, for $a/b = 1$, becomes equal to 2σ as derived previously for the circular hole.

Pure Shear Parallel to Major and Minor Axes of the Elliptical Hole

Let an infinite plate be subjected to uniform shear stress τ as shown in Fig. 14.12. The stress state due to this case of pure shear parallel to the (x, y) axes may be found by superposition of the two cases for uniform tension $\sigma(=\tau)$ at $\beta = \pi/4$ and $-\sigma(=-\tau)$ at $\beta = 3\pi/4$; see Figs. 14.5 and 14.6 and also Eqs. (14.9) and (14.14). The value of $\sigma_{\beta\beta}$ on the perimeter of the hole ($\alpha = \alpha_0$) may be found in this manner to be

$$\sigma_{\beta\beta}\big|_{\alpha=\alpha_0} = -\frac{2\tau e^{2\alpha_0} \sin 2\beta}{\cosh 2\alpha_0 - \cos 2\beta} \tag{14.22}$$

By differentiation of $\sigma_{\beta\beta}$ with respect to β, we may show that the maximum value of $\sigma_{\beta\beta}$ occurs when

$$\tan \beta = -\tanh \alpha_0 = -\frac{b}{a} \tag{14.23}$$

and the maximum value of $\sigma_{\beta\beta}$ is

$$\sigma_{\beta\beta(\max)} = \tau \frac{(\cosh \alpha_0 + \sinh \alpha_0)^2}{\sinh \alpha_0 \cosh \alpha_0} = \tau \frac{(a + b)^2}{ab} \tag{14.24}$$

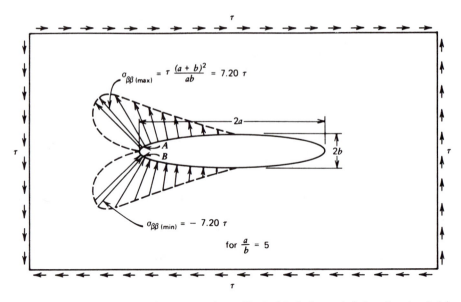

Figure 14.12   Distribution of $\sigma_{\beta\beta}$ around an elliptical hole in an infinite plate loaded in pure shear.

For the case $a/b = 5$, the distribution of $\sigma_{\beta\beta}$ around the hole is given in Fig. 14.12, where point A locates the maximum value. Analogously, the minimum (compressive) value of $\sigma_{\beta\beta}$ is

$$\sigma_{\beta\beta(min)} = -\tau \frac{(a + b)^2}{ab} \tag{14.25}$$

where $\tan \beta = \tanh \alpha_0 = b/a$ (point B in Fig. 14.12).

Solutions for the stress distribution around an elliptical hole in a plane isotropic sheet have been obtained for other loadings, for example, pure bending in the plane, as well as for other shapes of holes. (Neuber, 1958).

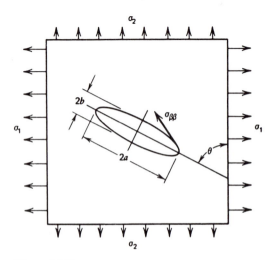

Figure 14.13

Elliptical Hole in Infinite Plate with Different Loads in Two Perpendicular Directions

Consider an infinite plate with an elliptical hole (Fig. 14.13). Let the plate be subjected to uniformly distributed stresses $\sigma_1 > \sigma_2$ along straight-line edges far removed from the hole. Let the major axis of the hole form an angle θ with the edge on which stress σ_1 acts. We wish to compute the maximum value of $\sigma_{\beta\beta}$ at the perimeter of the hole.

The solution to the above problem may be obtained by superposing the loadings of Figs. 14.8, 14.9, and 14.12. By the rules of transformation of stress [see Chapter 2, Eqs. (2.31)], we compute normal and shear stresses on planes parallel to the major and minor axes of the ellipse (Fig. 14.13), as shown in Fig. 14.14. Thus, we obtain

$$\sigma_{\mathrm{I}} = \frac{\sigma_1 + \sigma_2}{2} + \frac{\sigma_1 - \sigma_2}{2}\cos 2\theta$$

$$\sigma_{\mathrm{II}} = \frac{\sigma_1 + \sigma_2}{2} - \frac{\sigma_1 - \sigma_2}{2}\cos 2\theta$$

$$\tau_{\mathrm{I,II}} = \frac{\sigma_1 - \sigma_2}{2}\sin 2\theta \tag{14.26}$$

Then the substitutions $\sigma = \sigma_{\mathrm{I}}$ into Eq. (14.9), $\sigma = \sigma_{\mathrm{II}}$ into Eq. (14.14), and $\tau = \tau_{\mathrm{I,II}}$ into Eq. (14.22) and addition of the results yield

$$\sigma_{\beta\beta} = [(\sigma_1 + \sigma_2)\sinh 2\alpha_0 + (\sigma_1 - \sigma_2)(e^{2\alpha_0}\cos 2\beta - 1)\cos 2\theta$$
$$- (\sigma_1 - \sigma_2)e^{2\alpha_0}\sin 2\beta \sin 2\theta]/(\cosh 2\alpha_0 - \cos 2\beta) \tag{14.27}$$

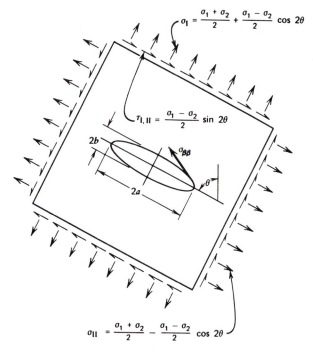

Figure 14.14

For a given value of θ, Eq. (14.27) gives $\sigma_{\beta\beta}$ as a function of β. Hence, by setting the derivative of $\sigma_{\beta\beta}$ with respect to β equal to zero, we may compute the values of β that give extreme values of $\sigma_{\beta\beta}$. The values of β are solutions of the equation

$$1 - \cos 2\beta \cosh 2\alpha_0 - \sin 2\beta \cot 2\theta \sinh 2\alpha_0$$

$$= \left(\frac{\sigma_1 + \sigma_2}{\sigma_1 - \sigma_2}\right)\left(\frac{\sinh 2\alpha_0}{e^{2\alpha_0}}\right)\left(\frac{\sin 2\beta}{\sin 2\theta}\right) \tag{14.28}$$

In general, Eq. (14.28) is satisfied by two values of β, depending on the quantities σ_1, σ_2, and α_0 ($\tanh \alpha_0 = b/a$), for each value of θ. One value of β is associated with $\sigma_{\beta\beta(\max)}$ and the other with $\sigma_{\beta\beta(\min)}$. Because of symmetry, for given β, the significant values of $\sigma_{\beta\beta}$ may be determined by considering values of θ between 0 and $\pi/2$.

Case 1. $\theta = 0$ or $\theta = \pi/2$. By Eq. (14.28), for $\theta = 0$ or $\theta = \pi/2$, we obtain the values $\beta = 0$ and $\beta = \pi/2$. Hence, we find for $\theta = 0$ and $\beta = 0$

$$\sigma_{\beta\beta} = \sigma_{\beta\beta(\max)} = \sigma_1\left(1 + 2\frac{a}{b}\right) - \sigma_2 \tag{14.29}$$

at the ends of the major axis. For $\theta = \pi/2$ and $\beta = 0$

$$\sigma_{\beta\beta} = \sigma_{\beta\beta(\max)} = -\sigma_1 + \sigma_2\left(1 + 2\frac{a}{b}\right) \tag{14.30}$$

at the ends of the major axis. Likewise, we find for $\theta = 0$ and $\beta = \pi/2$

$$\sigma_{\beta\beta} = \sigma_{\beta\beta(\min)} = -\sigma_1 + \sigma_2\left(1 + 2\frac{b}{a}\right) \tag{14.31}$$

at the ends of the minor axis. For $\theta = \pi/2$ and $\beta = \pi/2$

$$\sigma_{\beta\beta} = \sigma_{\beta\beta(\min)} = \sigma_1\left(1 + 2\frac{b}{a}\right) - \sigma_2 \tag{14.32}$$

at the ends of the minor axis.

Case 2. $0 < \theta < \pi/2$. For a fixed value of β, the extreme values of $\sigma_{\beta\beta}$ occur at values of θ determined by setting the derivative of $\sigma_{\beta\beta}$ [Eq. (14.27)] with respect to θ equal to zero. Thus, we obtain

$$\tan 2\theta = \frac{e^{2\alpha_0}\sin 2\beta}{1 - e^{2\alpha_0}\cos 2\beta} \tag{14.33}$$

Consequently, by Eqs. (14.28) and (14.33), we find that extreme values of $\sigma_{\beta\beta}$ are obtained when

$$\left(\frac{\sigma_1 + \sigma_2}{\sigma_1 - \sigma_2}\right)\cos 2\theta = \pm\left[\left(\frac{\sigma_1 + \sigma_2}{\sigma_1 - \sigma_2}\right)^2 \sinh 2\alpha_0 - \frac{1}{2}(\sinh 2\alpha_0 + \cosh 2\alpha_0)\right] \tag{14.34}$$

and

$$e^{2\alpha_0}\cos 2\beta = e^{2\alpha_0}\cosh 2\alpha_0 - 2\left(\frac{\sigma_1 + \sigma_2}{\sigma_1 - \sigma_2}\right)^2 \sinh^2 2\alpha_0 \tag{14.35}$$

provided that

$$\frac{1 + \coth 2\alpha_0}{2\coth \alpha_0} \le \left(\frac{\sigma_1 + \sigma_2}{\sigma_1 - \sigma_2}\right)^2 \le \frac{1 + \coth 2\alpha_0}{2\tanh \alpha_0} \tag{14.36}$$

where Eq. (14.36) follows from Eqs. (14.34) and (14.35) and the conditions

$$-1 \le \cos 2\theta \le 1, \qquad -1 \le \cos 2\beta \le 1 \tag{14.37}$$

By Eqs. (14.34), (14.35), and (14.27), we find the two values of $\sigma_{\beta\beta}$

$$\sigma_{\beta\beta 1} = -\frac{(\sigma_1 - \sigma_2)^2}{2(\sigma_1 + \sigma_2)}(1 + \coth 2\alpha_0) \tag{14.38}$$

$$\sigma_{\beta\beta 2} = \frac{3(\sigma_1 - \sigma_2)^2}{2(\sigma_1 + \sigma_2)}(1 + \coth 2\alpha_0) \tag{14.39}$$

Depending on the sign of the applied stresses, the maximum value of $\sigma_{\beta\beta}$ is given by either the value of $\sigma_{\beta\beta 1}$ or $\sigma_{\beta\beta 2}$, depending on which is larger. For example, assume that the values of σ_1, σ_2, and α_0 are such that Eq. (14.36) is satisfied. Let the elliptical hole be oriented at angle θ (Fig. 14.13) given by Eq. (14.34). Under these conditions, the value of $\sigma_{\beta\beta 2}$ from Eq. (14.39) is never greater than the value of $\sigma_{\beta\beta(\text{max})}$ given by Eqs. (14.29) and (14.30). However, the stress $\sigma_{\beta\beta 1}$ is a tensile stress when $\sigma_1 + \sigma_2 < 0$. The values of $\sigma_{\beta\beta 1}$ may exceed the maximum tensile stress that can exist for $\theta = 0$ or $\pi/2$. Hence, when σ_1 and σ_2 are both negative (compressive stresses), a tensile stress $\sigma_{\beta\beta 1}$ exists on the perimeter of the elliptical hole. When θ is equal to the value given by Eq. (14.34), with the positive sign, $\sigma_{\beta\beta 1}$ is the largest tensile stress that exists for any other value of θ that may be chosen for this state of stress [values of σ_1, σ_2, α_0 that satisfy Eq. (14.36)]. Consequently the presence of an elliptical hole in a flat plate (even for the case $b/a \approx 0$) may result in a tensile stress on the perimeter of the hole, even when the plate is subjected to negative stresses σ_1 and σ_2 (compression) on its edges (Fig. 14.13).

EXAMPLE 14.1
Narrow Elliptical Hole in Plate

Consider an elliptical hole in a plate with ratio $a/b = 100$ (Fig. 14.13). For this large value of a/b, the hole appears as a very narrow slit (crack) in the plate. Let compressive stresses $\sigma_1 = -20$ MPa and $\sigma_2 = -75$ MPa be applied to the plate edges.

(a) Determine the orientation of the hole (value of θ) for which the tensile stress at the perimeter of the hole is a maximum.

(b) Calculate the value of this tensile stress.

(c) Calculate the associated value of β (location of the point) for which this tensile stress occurs.

SOLUTION

Since $a/b = 100$, Eq. (14.10) indicates that $\coth \alpha_0 = 1/\tanh \alpha_0 = a/b = 100$. Hence, $\alpha_0 = 0.0100$ rad, $\sinh 2\alpha_0 = 0.0200$, $\cosh 2\alpha_0 = 1.000$, $\coth 2\alpha_0 = 50.0$. For these values of σ_1, σ_2, and α_0, Eq. (14.36) is satisfied.

(a)   The value of θ is given by Eq. (14.34). Hence,

$$\cos 2\theta = \left[\frac{-1 - 0.0200 + \left(\dfrac{-20 - 75}{-20 + 75}\right)^2 (0.020)}{2\left(\dfrac{-20 - 75}{-20 + 75}\right)} \right] = 0.2607$$

or

$$\theta = 0.6535 \text{ rad}$$

(b)   The maximum value of the tensile stress is given by Eq. (14.38). Thus,

$$\sigma_{\beta\beta(\text{max})} = \sigma_{\beta\beta 1} = -\frac{(-20 + 75)^2}{2(-20 - 75)}(1 + 50) = 812 \text{ MPa tension}$$

(c)   This tensile stress is located on the perimeter of the hole at a value of β given by Eq. (14.35)

$$\cos 2\beta = 1 - 2\left(\frac{-20 - 75}{-20 + 75}\right)^2 \frac{(0.020)^2}{1.020} = 0.9977$$

or

$$\beta = 0.0342 \text{ rad}$$

This small value of β means that the maximum tensile stress occurs very near the end of the major axis of the elliptical hole (see Figs. 14.13 and 14.5).

The above computation shows that a slender elliptical hole (long narrow crack) in a plate may result in a high tensile stress concentration even when the applied edge stresses are compressive.

Stress Concentration at a Groove in a Circular Shaft

Consider a machine element consisting of a circular shaft in which a circumferential circular groove (notch) is cut (Fig. 14.15 and Fig. D, Table 14.3). In practice, the shaft is subjected to an axial force P, bending moment M, and twisting moment (torque) T. We wish to compute the maximum principal stress in the cross section of the shaft at the root of the notch. In addition, a shear V may act on the shaft (Fig. D, Table 14.3). However, this shear has only a small effect on the maximum stress at the root of the notch (Neuber, 1958). Hence, we do not consider its effect.

The maximum principal stress at the root of the notch occurs at point A in Fig. 14.15. The stress components at A are σ_{zz} and σ_{zx}. Hence, by Eq. (2.37), the

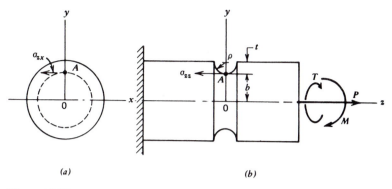

Figure 14.15

maximum principal stress is

$$\sigma_{\max} = \tfrac{1}{2}\sigma_{zz} + \tfrac{1}{2}\sqrt{\sigma_{zz}^2 + 4\sigma_{zx}^2} \tag{14.40}$$

The stress component σ_{zz} is produced by the axial load P and bending moment M. Hence,

$$\begin{aligned}
\sigma_{zz} &= S_{cc}^{(P)}\frac{P}{A} + S_{cc}^{(M)}\frac{Mc}{I} \\
&= S_{cc}^{(P)}\frac{P}{\pi b^2} + S_{cc}^{(M)}\frac{4M}{\pi b^3}
\end{aligned} \tag{14.41}$$

where $S_{cc}^{(P)}$ and $S_{cc}^{(M)}$ are the calculated stress concentration factors for axial load and bending moment, respectively. These stress concentration factors are determined from curves 6 and 7 in Fig. 14.10. The stress σ_{zx} is given by the relation

$$\sigma_{zx} = S_{cc}^{(T)}\frac{Tc}{J} = S_{cc}^{(T)}\frac{2T}{\pi b^3} \tag{14.42}$$

where $S_{cc}^{(T)}$ is the calculated stress concentration factor for torque and is determined from curve 9 of Fig. 14.10. For a given set of dimensions of the shaft (Fig. 14.15), Eqs. (14.40), (14.41) and (14.42) yield the value of $\sigma_{\max}$.

14.4

STRESS CONCENTRATION FACTORS. EXPERIMENTAL TECHNIQUES

Photoelastic Method
The values of calculated stress concentration factors found by the photoelastic method agree well with the results obtained from the theory of elasticity. Thus, the photoelastic method may be used as a check, and it may be applied also to some members in which the stress cannot be obtained mathematically; however, the technique of obtaining reliable results with the photoelastic method is acquired

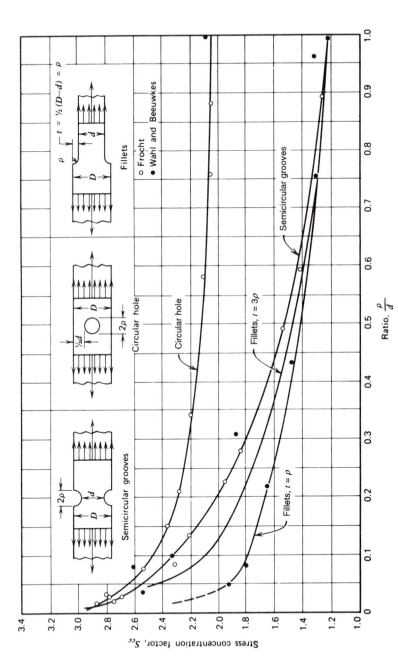

Figure 14.16   Stress concentration factors obtained by use of the photoelastic method.

only after considerable experience. In particular, special care must be exercised to obtain trustworthy results when the radius of the notch is very small (Kobayashi, 1988).

Values of the calculated stress concentration factors obtained by the photoelastic method for three forms of abrupt changes in section in flat specimens are shown as reported by Frocht (1936) in Fig. 14.16. In each specimen, the stress distribution is uniform at distant sections on either side of the abrupt change; when the stress distribution is variable on either side of the abrupt change in section, as in bending, the calculated stress concentration factor is found to be somewhat smaller. These curves show that the value of S_{cc} varies with the ratio ρ/d. However, S_{cc} also depends on the ratio D/d. For the particular groove, hole, and fillet shown in Fig. 14.16, the values of ρ/d and D/d are related by the equation $D/d = 1 + 2\rho/d$.

The values of S_{cc} for the hole and groove in Fig. 14.16 can be found also by Neuber's solution, as obtained from Fig. 14.10 for various values of ρ/d. These values obtained from Neuber's nomograph agree satisfactorily with those found by the photoelastic method. The elasticity solution for the calculated stress concentration factor for the fillet is achieved by a numerical method that is an approximation. Hence, the photoelastic method is of special value for this type of discontinuity. For the fillet in Fig. 14.16 for which $t = \rho$, the curve marked $t = \rho$ gives values of S_{cc}. For members in which t is not equal to ρ, the values of S_{cc} will be different as shown, for example, by the curve marked $t = 3\rho$. The influence of t/ρ on the values of S_{cc} for a fillet subjected to axial tension and to bending has been studied by Frocht (1936).

The distribution of stress shown in Fig. 14.17 was obtained by Coker and Filon (1957) by the photoelastic method. The maximum stress at the edge of the groove

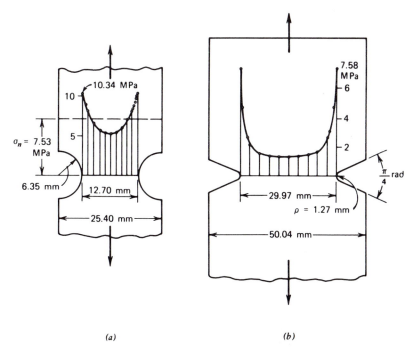

(a)   (b)

Figure 14.17   Stress distribution at notches found by the photoelastic method. (From Coker and Filon, 1957.)

in Fig. 14.17*a* is 1.37 times the average stress on the reduced section, that is, $S_{cc} = 1.37$, by the photoelastic method. The value as found by using Neuber's nomograph is $S_{cc} = 1.45$. In Fig. 14.17*b* the groove has a much smaller radius and the plate is much wider. The photoelastic method gives a maximum stress of 7.58 MPa, whereas the nominal or average stress was 1.59 MPa, that is, $S_{cc} = 4.77$. The value as found by Neuber's nomograph is $S_{cc} = 5.50$. The rather sharp notch gives a high concentration of stress. However, the stress concentration depends on the relative depth of the notch. For example, if in Fig. 14.17*b* the notch geometry and dimensions are kept as shown and the outer width of the plate is reduced to 29.97 mm (the width of the root section is then 9.90 mm), the value of $S_{cc} = 2.6$ is obtained from Neuber's solution.

Strain-Gage Method

Two examples are presented to indicate the use of the strain-gage method to determine calculated stress concentration factors for a hole in a shaft and to determine the effect of a concentrated load or the strain (stress) distribution in a beam at the section where the load is applied.

Transverse Hole in a Shaft.   By using a specially designed, mechanical strain gage that measured elastic strains in a 2.54-mm gage length, Peterson and Wahl (1936) obtained elastic stress concentration factors for a shaft containing a transverse hole and subjected to bending loads. Their results are shown in Fig. 14.18. With the same instrument, they obtained the stress at a fillet in a large steel shaft tested as a beam. These values checked closely with the values found by Frocht by the photoelastic method for fillets of the same proportions (Fig. 14.16).

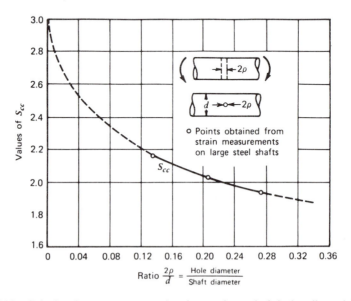

Figure 14.18   Calculated stress concentration factors for a shaft in bending with a transverse hole as found by elastic strain method. (From Peterson and Wahl, 1936.)

Effect of Local Pressure on Strain (Stress) Distributions in a Beam.   The effect on the longitudinal bending strains (stresses) in a beam caused by the bearing pressure of a concentrated load applied at the midspan section of a steel rail beam is shown in the upper part of Fig. 14.19. The load was applied approximately along a line across the top of the rail section. The effect of the bearing pressure on the longitudinal stress extends well below mid-depth of the rail. The point of zero longitudinal stress is about 25 mm above the calculated position of the neutral axis for the section beneath the load, and the strain (stress) on the cross section does not vary directly with the distance from the neutral axis, as is usually assumed for such a beam. The results for the section underneath the load, however, are approximate because relatively long gage lengths were used and the two-dimensional aspect of the state of stress was neglected.

If, however, the same beam is loaded as shown in the lower part of Fig. 14.19, the strains (stresses) in the central portion, which is subjected to constant bending moment free from the influence of the bearing pressure of the loads, are in agreement with the usual assumptions for simple bending.

Elastic Torsional Stress Concentration at Fillet in Shaft

If all cross sections of a shaft are circular but the shaft contains a rather abrupt change in diameter, a localized stress occurs at the abrupt change of section.

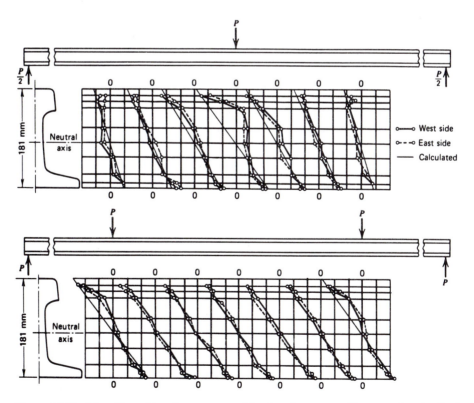

Figure 14.19   The effect of bearing pressure of load at the center of beam on longitudinal strains in beam. [From A. N. Talbott (1930). *Bull. A. R. E. A.*, **31**.]

Jacobsen (1925) investigated the concentration of torsional shear stress at a fillet, where the diameter of a shaft changes more or less abruptly, depending on the radius of the fillet.

The results of the investigation are given in Fig. 14.20. For example, if the radius of a circular shaft changes from 52 to 39 mm by means of a fillet with radius of 3.25 mm, $R/r = 1.33$, and $\rho/r = 1/12 = 0.083$; the maximum elastic shear stress at the fillet as given by Fig. 14.20 is approximately 1.7 times the maximum shear stress in the small shaft as found by the equation $\tau = Tr/J$, where T is the twisting moment and J the polar moment of inertia of the cross section of the smaller shaft ($J = \pi r^4/2$).

Elastic Membrane Method. Torsional Stress Concentration

Griffith and Taylor (1917), by using a soap film as the elastic membrane (see Sec. 6.4), found the torsional shear stress in a hollow shaft at the filleted corner of

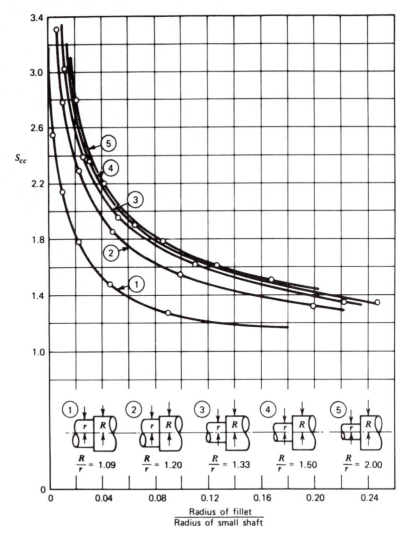

Figure 14.20 Torsional shear-stress concentration at fillet in shaft of two diameters.

a keyway and also at the center of the flat bottom of the keyway. The external and internal diameters were $2a = 254$ mm and 147 mm, respectively, and the keyway was 25.4 mm deep and 63.5 mm wide.

Figure 14.21 shows the value of the ratios of the maximum torsional shear stress at the fillet for various radii r of fillet to the maximum shear stress that would be developed in the shaft if the shaft had no keyway. In other words, the ordinates to the curve give the elastic calculated stress concentration factors S_{cc} due to the keyway.

Ordinates to the dotted line in Fig. 14.21 are the elastic stress concentration factors for the shear stress at the center of the bottom of the keyway; the stress at this point is approximately twice as great as would be the maximum shear stress in the shaft if it had no keyway.

Torsional Stress at Fillet in Angle Section. The torsional shear stress at a sharp internal corner of a bar subject to torque is infinitely large if the material does not yield when the stress becomes sufficiently high. If the corner is rounded off by means of a fillet, the stress is reduced; the amount of reduction corresponding to fillets of different radii in an angle section was found by Griffith and Taylor (1917) by use of the soap-film method. They used a section 25.4 mm wide (Fig. 14.22) and the straight portions or arms of the section were long.

The ratios of the maximum shear stress at the fillet to the shear stress in the straight portion or arm of the angle section for various radii of fillets are given in Table 14.4. These values show that a small fillet has a large influence in reducing the stress at the corner, and that practically no advantage is gained by making the radius of the fillet larger than about 6 mm.

Note: The stress concentration factors given in the foregoing discussions are for particular forms of discontinuities. Values of stress concentration factors

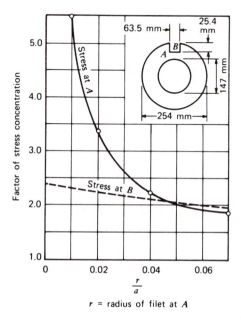

Figure 14.21   Factors of torsional shear-stress concentration at keyway in hollow shaft.

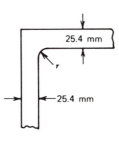

Figure 14.22

TABLE 14.4

Radius r of Fillet, mm (see Fig. 14.22)	Ratio: Maximum Stress / Stress in Arm
2.54	1.89
5.08	1.54
7.62	1.48
10.16	1.44
12.70	1.43
15.24	1.42
17.78	1.41

for many other forms of discontinuities are available in the technical literature (Peterson, 1974).

Beams with Rectangular Cross Sections

In Chapter 7, the normal stress distribution for elastically loaded beams was assumed to be given by the flexure formula [Eq. (7.1)]. There are many conditions that arise in practice that produce stress distributions that differ from the one given by the flexure formula. For example, three such conditions are the following. First, residual stresses that alter the stress distribution may be present in the beam before loading. Second, concentrated loads (large loads applied over a small area; see Chapter 18) cause contact stresses that also alter the distribution of flexure stress (see Fig. 14.19 and the associated discussion). Third, stress concentrations caused by abrupt changes in the cross section of the beam produce normal stress distributions that differ from those predicted by the flexure formula.

Consider a simply supported beam with rectangular cross section loaded as shown in Fig. 14.23. The portion of the beam between the loads P is subjected to pure bending with moment $M = Pd$. For elastic bending, the normal stress distribution at sections far removed from the stress concentration is given by the flexure formula. Hence, the nominal value for the flexure stress is $\sigma_n = Mc/I$, where $c = h/2$, (Fig. 14.23). The maximum stress at the base of the fillet is given by Eq. (14.2) as $\sigma_{max} = S_{cc}\sigma_n = S_{cc}Mc/I$, where S_{cc} is the stress concentration factor for bending. The magnitude of S_{cc} depends on the ratio of the radius ρ of the fillet to the beam depth h and the ratio of H to h (Fig. 14.23). The magnitude of S_{cc} is larger for sharp notches, that is, for cases where H is large compared to h and ρ is small compared to h. Values of S_{cc} for fillets in rectangular section beams are given in Fig. 14.24 as functions of H/h and ρ/h; also, stress concentration factors for grooves in rectangular section beams are given in Fig. 14.25 as functions of H/h and ρ/h.

Design of Beams Having Stress Concentrations. If a beam is made of ductile material and it is not subjected to a large number of repeated loads (fatigue loading; see Chapter 16), the effects of stress concentrations are usually disregarded. Then, the flexure formula is used in the design of the beam. However, if the beam is made

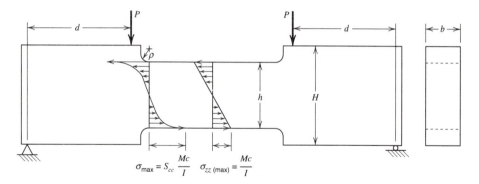

$$\sigma_{max} = S_{cc}\,\frac{Mc}{I} \qquad \sigma_{zz\,(max)} = \frac{Mc}{I}$$

Figure 14.23   Stress concentration at fillet in rectangular section beam. (The beam depth is exaggerated for clarity.)

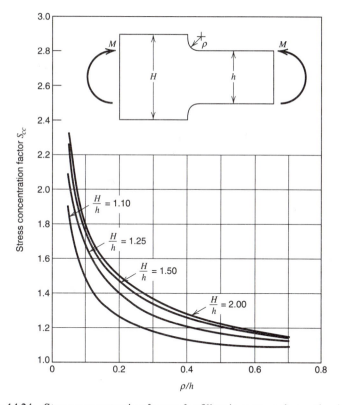

Figure 14.24   Stress concentration factors for fillets in rectangular section beams.

of a brittle material or the beam is subjected to fatigue loading, the effect of stress concentrations must be included in the design. In the case of a beam made of a brittle material, failure loads for the beam are estimated, based on Eq. (14.2), with σ_c being equal to the ultimate strength of the material and σ_n to the nominal flexure stress. The design of beams subjected to fatigue loading is considered in Chapter 16.

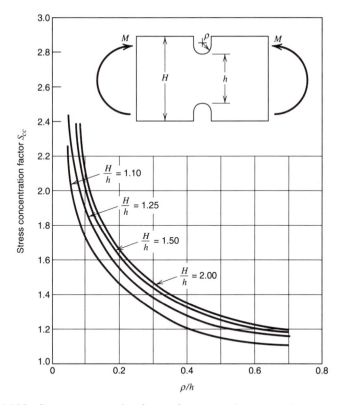

Figure 14.25   Stress concentration factors for grooves in rectangular section beams.

EXAMPLE 14.2
Fracture Load of a Filleted Tension Member

The filleted tension member in Fig. 14.16 is made of brittle material whose stress-strain diagram remains essentially linear up to its ultimate strength, $\sigma_u = 420$ MPa. Assuming that fracture of the member will occur at the base of the fillet, determine the magnitude of the tension working load P that can be applied to the member based on a safety factor $SF = 4.00$. The member has a width $w = 20.0$ mm perpendicular to the plane of the figure; also, $D = 110.0$ mm, $d = 50.0$ mm, and $\rho = 10.0$ mm.

SOLUTION

By Fig. 14.16, with the given dimensions, $t = (110.0 - 50.0)/2 = 30$ mm, $t/\rho = 3$, and $\rho/d = 0.20$. Thus, by Fig. 14.16, $S_{cc} = 1.83$. At fracture, $\sigma_c = \sigma_u = S_{cc}P_F/A = S_{cc}\sigma_n$, where σ_c is the calculated stress at the base of the fillet, S_{cc} is the stress concentration factor, P_F is the load at fracture, $A = wd = 20 \times 50 = 1000$ mm^2, and σ_n is the nominal stress P_F/A. Hence, by Eq. (14.2), 420 MPa $= 1.83P_F/0.001$ or $P_F = 420(0.001)10^6/1.83 = 229.5$ kN. Therefore, the working load is $P = P_F/SF = 229.5/4.00 = 57.4$ kN.

EXAMPLE 14.3
Beam with Stress Concentration

The beam in Fig. E14.3 is made of cast iron (σ_u = 250 MPa), a material considered to be brittle.

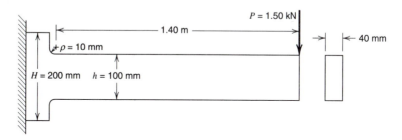

Figure E14.3

(a) If the design working load is $P = 1.50$ kN, determine the maximum stress in the beam at the fillet.

(b) What factor of safety was used in the design of the beam?

SOLUTION

Since $H = 200$ mm, $h = 100$ mm, and $\rho = 10$ mm, we have $H/h = 2.00$ and $\rho/h = 0.10$. With these values, the stress concentration factor is, by Fig. 14.24, $S_{cc} = 1.77$.

(a) The maximum flexure design stress is $\sigma_{max(D)}$ at the base of the fillet, and it is given by $\sigma_{max(D)} = S_{cc}PLc/I = (1.77)(1500)(1400)(50)(12)/[(40)(100)^3] = 55.8$ MPa.

(b) To obtain the fracture load P_F, the working load P is multiplied by the factor of safety. Hence, the maximum flexural fracture stress is $\sigma_{max(F)} = \sigma_u = S_{cc}(SF)PLc/I$. Therefore, the factor of safety is $SF = \sigma_u I/(S_{cc}PLc) = (250)(40)(100)^3/[(1.77)(1500)(1400)(50)(12)] = 4.48$. Since the stress-strain relationship for cast iron is essentially linear to fracture, the stress is proportional to the load. Therefore, the factor of safety could also have been obtained by dividing σ_u by $\sigma_{max(D)}$.

14.5

EFFECTIVE STRESS CONCENTRATION FACTORS

Definition of Effective Stress Concentration Factor
As noted in Sec. 14.1 and 14.2, calculated stress concentration factors apply mainly to ideal, elastic materials, and they depend mainly on the geometry or form of the abrupt change in section. For these reasons, they are often called *form factors*.

However, in applications involving real materials, the significance of a stress concentration factor is not indicated satisfactorily by the calculated value. Rather, it is found through experience that the *significant* or *effective stress value* that indicates impending structural damage (failure) of a member depends on the characteristics of the material and the nature of the load, as well as the geometry or form of the stress raiser. Consequently, in practice, the significant (or effective) value of the stress concentration is obtained by multiplying the nominal stress by a *significant* or *effective stress concentration factor*,* S_{ce}. Often, the nominal stress is computed from an elementary stress formula, such as $\sigma_n = P/A$, $\sigma_n = Mc/I$, etc. Usually, the magnitude of S_{ce} is less than the magnitude S_{cc} of the calculated stress concentration factor for a given stress raiser.

The magnitude of S_{ce} is always obtained experimentally, in contrast to the calculated value S_{cc}. Ordinarily, S_{ce} is obtained by testing two or more samples or sets of specimens of the actual material. One specimen (or set of specimens) is prepared without the presence of the discontinuity or stress raiser, so that the nominal stress is the significant or effective stress. A second specimen (or set of specimens) is prepared with the discontinuity or stress raiser built in. The second set of specimens is tested in the same manner as the first set. For simple members, such as axial rods, beams, or torsion bars, the stress in each set of specimens is usually calculated by means of elementary formulas.

One may assume that damage (failure) in the two sets of specimens is initiated when the significant stress quantities in the specimens attain the same critical value. The loads causing these equal stress quantities are unequal. The damaging stress in the specimens with the stress raiser is, of course, caused by a smaller load. Hence, the effective stress concentration factor may be defined as the ratio of these two loads $S_{ce} = P_n/P_e$, where P_n is the load that causes failure for the specimens without the stress concentration (the nominal load) and P_e is the load that causes failure for the specimens with the stress concentration (the effective load).

Alternatively, one may recognize that the effective stress concentration factor S_{ce} is usually less than the calculated stress concentration factor S_{cc}. Hence, S_{ce} may be defined in terms of an effective stress quantity σ_e that is greater than the nominal stress quantity σ_n but is less than the calculated localized stress quantity σ_c. Consider a condition in which a stress concentration exists in a member. Using the theory of elasticity or other methods, we may determine the calculated localized stress σ_c. By experimental means, the effective stress quantity σ_e is found. Finally, the nominal stress quantity σ_n is computed by using an appropriate elementary stress formula. Then σ_c may be expressed as the nominal stress *plus* some proportion q of the *increase* in the calculated stress caused by the stress concentration. That is, $\sigma_e = \sigma_n + q(\sigma_c - \sigma_n)$. This expression can be written in terms of the nominal stress and the two stress concentration factors.

$$S_{ce}\sigma_n = \sigma_n + q(S_{cc}\sigma_n - \sigma_n)$$

Thus,

$$S_{ce} = 1 + q(S_{cc} - 1) \tag{14.43}$$

and

$$q = \frac{(S_{ce} - 1)}{(S_{cc} - 1)} \tag{14.44}$$

---

* The term strength reduction factor is sometimes used. However, one should note that the strength of the material is not reduced by the stress raiser, but rather the load-carrying capacity of the member is reduced.

For example, if the stress concentration is caused by a small hole in the center of a plate subjected to an axial tensile load, then the nominal stress is $\sigma_n = P/A$, and the calculated localized stress quantity $\sigma_c = 3P/A$. Suppose that, under experimental conditions, yielding around the hole in the plate begins at an effective stress of $\sigma_e = 2P/A$. For these conditions, the proportion of the increase in calculated stress is $q = (\sigma_e - \sigma_n)/(\sigma_c - \sigma_n) = 0.5$ and the effective stress concentration factor is $S_{ce} = 2.0$.

The ratio q is called the *notch sensitivity index* of the material for the given form of discontinuity and the given type of loading. For example, in Eq. (14.43), if $q = 0$, $S_{ce} = 1$, and the material and member are said to be insensitive to the effects of the stress concentration, whereas if $q = 1$, $S_{ce} = S_{cc}$, and the member is said to be fully sensitive to the effects of the stress concentration. The value of S_{ce} (and hence q) is determined from tests as described above. It has been found from such tests (Fig. 14.26) that the values of S_{ce} and q depend mainly on the ability of the material and member to make adjustments or accommodations, such as local yielding, that reduce the damaging effects of the localized stress. The ability of the material to make these adjustments or accommodations depends, in turn, on the type of loading applied to the member (whether static, repeated, impact, etc.); the existence in the member of initial or residual stresses; the character of the internal structure of the material; the temperature of the member; the surface finish at the abrupt change of section; the stress gradient in the region of the stress concentration, etc. These factors are discussed briefly below.

Static Loads; Ductile Material

At abrupt changes of section in members made of *ductile** materials (especially metals) and subjected to static loads at ordinary temperatures, the localized stresses at the abrupt change of section are relieved to a large degree by localized yielding of the material that occurs largely, in metals, as slip across intercrystalline planes

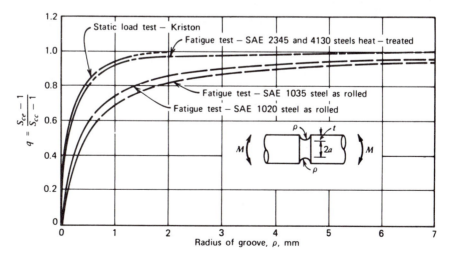

Figure 14.26   The influence of radius of groove on notch sensitivity index.

---

* See Chapter 1; ductile materials exhibit yield stresses and undergo large plastic strains before fracture.

(see Sec. 1.5). Because of this action, the value of q for the conditions specified is very low and lies usually in the range from 0 to 0.1. However, if the use or function of the member is such that the amount of inelastic strain required for this relieving action must be restricted, the value of q may approach 1.0 (Fig. 14.26). If the temperature of a metal member is very low when subjected to static loads, slip in the crystals seems to be reduced and is likely to be less effective in relieving the concentrated stress; hence, the value of q may be as much as 0.5 or even greater.

If the metal member is subjected to static load while at an elevated temperature, the mechanism (creep, Chapter 17) by which localized yielding occurs may cause the value of q to vary from nearly zero to nearly unity. This situation arises from the fact that the creep of metals may be the result of either one or both of two different inelastic mechanisms, depending on the temperature and stress imposed: (1) Creep may be caused mainly by intercrystalline slip, especially at the lower range of creep temperatures and relatively high stresses, this type of creep relieves the stress concentration to a large degree ($q = 0$, nearly); or (2) creep may be due to viscous flow of the unordered (so-called amorphous) grain boundary material, especially at higher temperatures and lower stresses, and stress concentration is relieved very little by such inelastic deformation ($q = 1$, nearly).

Static Loads; Brittle Material

If a member that contains an abrupt change in cross section is made of a relatively brittle material and subjected to static loads, q will usually have a value in the range from 0.5 to unity, except for certain materials that contain many internal stress raisers inherent in the internal structure of the material such as graphite flakes in gray cast iron. An external stress raiser in the form of an abrupt change in section in such a material as gray cast iron has only a small additional influence on the strength of the member and, hence, the value of q is relatively small.

Repeated Loads

If a member has an abrupt change in section and is subjected to a load that is repeated many times, the mode of failure is one of progressive fracture, even though the material is classified as ductile. Under these conditions, the ability of the material to make adjustments or accommodations by localized yielding is greatly reduced. This type of fracture, known as fatigue (Chapter 16), is illustrated in Fig. 14.27. The herringbone gear in Fig. 14.27 has several regions of stress concentration located at the root of each tooth and at the inside corners of the keyway. Due to the relative proximity of the keyway to the root of the teeth, a pair of fatigue cracks developed and joined. Normally, there is little or no evidence of yielding during fatigue crack growth. In this case, after the fatigue cracks became sufficiently long, average stress levels in the vicinity of the crack increased above yield. The subsequent yielding is commonly noticed in rough and noisy operation of the equipment. The gear was removed from service just prior to complete fracture of the part. Because of yielding, the cracks remained open.

The minimal influence of yielding under repeated loads leads to a relatively large value of q, usually between 0.5 and unity. The value of unity is approached in general for the harder, heat-treated metals (including tools and machine parts) and the lower value (0.5) is approached for metals in their softer condition (such as structural steel). Furthermore, the internal structure of metals, especially

(a)

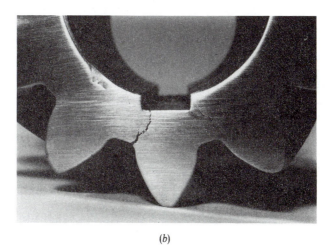

(b)

Figure 14.27   Fatigue cracks due to repeated (fatigue) loading of herringbone gear. (a) Perspective view of failed gear. (b) End view of fatigue cracks.

of steel, has some influence on the value of q. If the pearlitic grain size in steel is very fine, q is near unity. If the grain size is very coarse, the value of q is less.

Residual Stresses

The presence of initial or residual stresses in a member at an abrupt change in section also may influence the value of q. If the member is made of ductile metal and subjected to static loads at room temperature, localized yielding relieves the effects of residual stresses. Generally, in this case, it is assumed that q is not altered by the residual stresses. On the other hand, if the member is made of brittle material and the residual stresses act along the same directions as the load stresses, the effects of the residual stresses may either add or subtract from the effects of the load stresses, depending on the relative signs of the load stresses and the residual stresses. Correspondingly, the magnitude of q is increased or decreased. If, however, the member is made of a ductile metal and subjected to repeated loads, the influence of initial or residual stresses is uncertain. The relatively large inelastic deformation that occurs (in a small volume of the member surrounding the stress concentration) in low cycle fatigue is assumed generally to negate any effect of residual stress on the magnitude of q. However, in the case of high cycle fatigue ($N > 10^6$, Chapter 16), inelastic deformation in the region of a stress concentration is ordinarily minimal and residual stresses are assumed generally to alter the magnitude of q; the magnitude of q may either increase or decrease, depending on the sign of the residual stresses.

Very Abrupt Changes in Section. Stress Gradient

Let the change in section of a member be very abrupt; that is, let the hole, fillet, or groove, etc., forming the abrupt change in section have a very small characteristic dimension compared to the dimensions of the section, so that the calculated stress gradient is steep in the region of stress concentration. The value of S_{cc} for such a stress raiser is large, but the value of S_{ce} found from tests of such members, under either static or repeated loads, is usually much smaller than S_{cc}; that is, the value of q is smaller than would be found from tests of members of the same material with less abrupt changes of section. Figure 14.26 gives the results of tests of specimens having an abrupt change of section caused by a circumferential groove that show the foregoing facts. In this figure, the value of q is plotted as ordinates, and the radius of the groove at the abrupt change of section is plotted as abscissas.

The results of these tests are represented by smooth curves drawn through points (not shown) representing the test data. The data used for each curve were obtained by testing specimens of the same material, the specimens being identical except for the size of the groove radius. The upper curve is for static load tests of specimens of Kriston (a plastic), which is a very brittle material. The other curves are for repeated bending-load tests of steels. In these tests, the unpublished results of Moore, Jordan, and Morkovin, the values of a/ρ and t/ρ were kept constant, which means that the value of S_{cc} was kept constant (see Neuber's nomograph, Fig. 14.10). However, the groove radius ρ was varied and all these curves show that when the groove radius approaches very small values, q is quite small, but when the groove radius is relatively large, the value of q approaches unity.

The results of these tests indicate that the damaging effects on a member from notches having small radii at the roots of the notches such as scratches, small holes, grooves, fillets, or small inclusions, are considerably less than would be indicated by the large values of the theoretical stress at such stress raisers; in other words, q (and hence, S_{ce}) is relatively small. Much of the available data for the value of S_{ce} and q have been obtained by conducting repeated load tests of specimens with cross sections of relatively small dimensions containing fillets, grooves, holes, etc., having small radii. These data furnish valuable information for computing significant stresses in a member having such discontinuities within the range of conditions used in the tests, but the values of q are probably unrealistically small for use in computing S_{ce} by Eq. (14.43) for holes, fillets, grooves, etc., whose radii are relatively large.

Significance of Stress Gradient

The question naturally arises as to why the value q for a given material under a given type of loading should depend on the value of the root radius of the notch when it is small, as indicated by the curves of Fig. 14.26. Much discussion of this question is found in the technical literature, but no completely satisfactory reason can be given. A possible explanation is as follows: At one or more points on the surface of the member at the root of the notch, the stress concentration will have its highest value, but at nearby points in the member in any direction from the root of the notch, the values of the stress diminish. For most notches, the highest rate (stress gradient) at which the stress diminishes occurs at points in a cross section of the member at the notch root. Let S be the stress gradient at the root of the notch, that is, S is the slope of a line that is tangent at the root of the notch to the curve of stress distribution on the cross section at the root of the notch. This slope gives the rate at which the stress is diminishing at points just underneath the root of the notch. If S is large, the stress magnitude will diminish rapidly so that the stress at a point just underneath the root of the notch will be only slightly larger than the value given at this point by the ordinary (nominal) stress equation.

It may be shown that S for notches such as holes, fillets, and grooves is given approximately by the following equation:

$$S = \frac{2.5\sigma_{max}}{\rho} = \frac{1.5S_{cc}\sigma_n}{\rho} \tag{14.45}$$

From Eq. (14.45), it is seen that, for a given value of nominal stress σ_n and S_{cc}, if ρ becomes small, the value of S becomes very large. When ρ is small and S large, the magnitude of the concentrated stress diminishes so rapidly that only a very thin layer of material at the root of the notch is subjected to the stress concentration. This means that the so-called adjustments or accommodations that take place in the material and that tend to relieve high stresses can take place more easily since such a small amount of material is involved. Furthermore, the machining and polishing of a specimen at the root of the notch will frequency result in an increase of the ability of the material in this thin layer (by work hardening) to resist stress. The greater apparent ability of this thin layer of material to resist higher stress plus the fact that the unchanged material (parent material) under this layer is not required to resist the highly concentrated stress also help explain why q becomes so much smaller as ρ becomes very small.

The foregoing discussion of stress gradient applies mainly to so-called mechanical notches such as holes and fillets, rather than chemical notches such as corrosion pits (see Chapter 16).

Impact or Energy Loading

If machine parts and structural members are subjected to impact or energy loading, for example, if a member is required to absorb energy delivered to it by a body having a relatively large velocity when it comes in contact with the member, localized stresses have, in general, a large influence in decreasing the load-carrying capacity of the member. As discussed in Chapter 3 [Eq. (3.33) with $\sigma_{xx} = \sigma$ and all other stress components equal to zero], the energy absorbed per unit volume by a material when stressed within the elastic strength is $\sigma^2/(2E)$; that is, the energy absorbed by a material is proportional to the square of the stress in the material. This means that the small portions of a member where the high localized stresses occur absorb an excessive amount of energy before the main portion of the member can be stressed appreciably and, hence, before the main portion can be made to absorb an appreciable share of the energy delivered to the member. As a result, the small portion where the localized stress occurs is likely to be stressed above the yield stress of the material. Then the energy required to be absorbed may be great enough to cause rupture even if the material is relatively ductile. (A familiar method of breaking a bar of ductile metal is to file a V-notch in one side of the bar to create a local stress raiser and then clamp one end of the bar in a vise, with the notch close to the face of the vise, and strike the bar near the other end a sharp blow with a hammer so that the bar is bent with the notch on the tension side of the bar.)

Tests widely used to measure the effects of a notch under impact loads are the Charpy and Izod impact tests. However, neither of these notched-bar single-blow impact tests gives a quantitative value of S_{ce}. These tests are important primarily in determining whether or not a material of known history of manufacture and treatment is substantially the same as a similar material that has proved to be satisfactory in service. There is no satisfactory test or method for determining a value of q for stress raisers in members subjected to impact loading. The effects of repeated loads and certain other influences on stress concentration factors are discussed in Chapter 16.

14.6

EFFECTIVE STRESS CONCENTRATION FACTORS. INELASTIC STRAINS

Consider a flat plate of width l and thickness d, with symmetrically placed edge notches of radius ρ (Fig. 14.28a). The elastic stress concentration factor S_{cc} for this case may be found from Fig. 14.10 for given values of ρ, t, and a. The tensile stress-strain curve for the material is shown in Fig. 14.28b. We consider here the problem of determining the maximum stress $\sigma_{\max}$ and maximum strain $\epsilon_{\max}$ at the roots of the edge notches for the case where the axial load P produces inelastic deformation in the material surrounding the notches. Before we present a solution to this problem, we define certain quantities and state a theorem that is employed in obtaining this solution.

In Fig. 14.28a, we assume that the stress distribution may be represented by the curve CD. The nominal stress on the cross section at the notch is σ_n. One of the quantities we wish to determine is the maximum stress $\sigma_{\max}$, where in terms of the

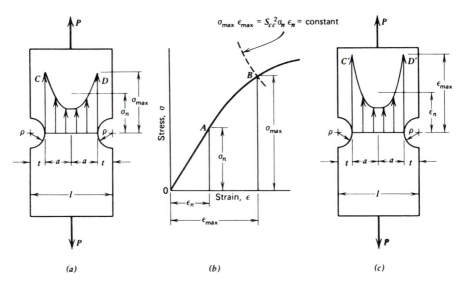

Figure 14.28   (a) Stress distribution. (b) Stress-strain curve. (c) Strain distribution.

significant stress concentration factor S_{ce} and nominal stress σ_n, we have

$$\sigma_{\max} = S_{ce}\sigma_n \tag{14.46}$$

Corresponding to the nominal stress σ_n, we have the nominal strain ϵ_n, where (σ_n, ϵ_n) are the coordinates of point A in Fig. 14.28b. In general, the strain distribution across the specimen and maximum strain $\epsilon_{\max}$ in the specimen are not known; see curve $C'D'$ in Fig. 14.28c. Corresponding to the effective stress concentration factor S_{ce}, we define a strain concentration factor E_{ce} by the relation

$$E_{ce} = \frac{\epsilon_{\max}}{\epsilon_n} \tag{14.47}$$

We wish to determine the values of both $\sigma_{\max}$ and $\epsilon_{\max}$. From curve $0AB$ of Fig. 14.28b, we note that $(\sigma_{\max}, \epsilon_{\max})$ are the coordinates of point B. For this purpose, we employ a theorem due to Neuber (1961a).

Neuber's Theorem
For relatively sharp notches, the following relation between S_{ce}, S_{cc}, and E_{ce} exists:

$$S_{ce} \cdot E_{ce} = S_{cc}^2 \tag{14.48}$$

where S_{ce} and E_{ce} are defined by Eqs. (14.46) and (14.47), respectively, and S_{cc} is the calculated (theoretical) stress concentration factor.

Equation (14.48) holds for $\sigma_{\max}$ above and below the elastic limit of the material. For example, when $\sigma_{\max}$ is below the elastic limit, $S_{ce} = S_{cc}$ and also $E_{ce} = S_{cc}$. Hence, Eq. (14.48) is satisfied identically. When $\sigma_{\max}$ is above the elastic limit, $S_{ce} < S_{cc}$. Hence, by Eq. (14.48), $E_{ce} > S_{ce}$. Substituting Eqs. (14.46) and (14.47) into

Eq. (14.48), we find

$$\sigma_{max}\epsilon_{max} = S_{cc}^2 \sigma_n \epsilon_n \qquad (14.49)$$

Equation (14.49) may be used to determine the values of σ_{max} and ϵ_{max} (coordinates of points B, Fig. 14.28b), since in a typical problem the values of S_{cc}, σ_n, and ϵ_n are usually available; that is, the load, dimensions of the member, stress-strain curve of the material, and value of S_{cc} are known or obtainable. Thus, with these values known, Eq. (14.49) may be written

$$\sigma_{max}\epsilon_{max} = \text{constant} \qquad (14.50)$$

Equation (14.50) represents a hyperbola in (σ, ϵ) space (Fig. 14.28b). The intersection of this hyperbola with the stress-strain curve occurs at point B (Fig. 14.28b). Thus, by plotting Eq. (14.50) in Fig. 14.28b, we locate point B and, hence, we may read the values of σ_{max} and ϵ_{max} as the coordinates of point B. Then, the substitution of σ_{max} and ϵ_{max} values into Eqs. (14.46) and (14.47) yields values of S_{ce} and E_{ce}, respectively.

EXAMPLE 14.4
Application of Neuber's Theorem

Consider a low carbon steel with the stress-strain diagram shown in Fig. E14.4a. Let the nominal stress in a notched specimen (Fig. 14.28a) be $\sigma_n = 105$ MPa. From Fig. E14.4a, we find $\epsilon_n = 0.0005$. Also, let $S_{cc} = 2.43$. By Eq. (14.49) we obtain

$$\sigma_{max}\epsilon_{max} = 0.31$$

This curve intersects the stress-strain curve at point B (Fig. E14.4a). For point B, we find $\sigma_{max} = 236.3$ MPa and $\epsilon_{max} = 0.0013$. Hence, by Eq. (14.46).

$$S_{ce} = \frac{\sigma_{max}}{\sigma_n} = \frac{236.3}{105} = 2.25$$

This value of S_{ce} corresponds to the value of the ordinate of point C, (Fig. E14.4b) with an abscissa value of $\sigma_n = 105$ MPa. Proceeding in a similar manner, we may plot a continuous curve FCG of values of S_{ce} as shown in Fig. E14.4b. For values of $\sigma_n < 58$ MPa (abscissa of point F), $S_{ce} = \sigma_{max}/\sigma_n = 2.43 = S_{cc}$. For values of $\sigma_n > 58$ MPa, S_{ce} decreases from the value of 2.43 to 1.5 at point G. In this region of decreasing value of S_{ce} (from point F to point G), $S_{ce} < S_{cc}$. In this region, S_{ce} is the significant (effective) stress concentration factor rather than S_{cc}.

Three other curves for values of S_{cc} equal to 2.06, 1.88, and 1.60, respectively, are also plotted in Fig. E14.4b, employing the method for $S_{cc} = 2.43$. In addition, experimental values obtained by Neuber (1961b) are also shown.

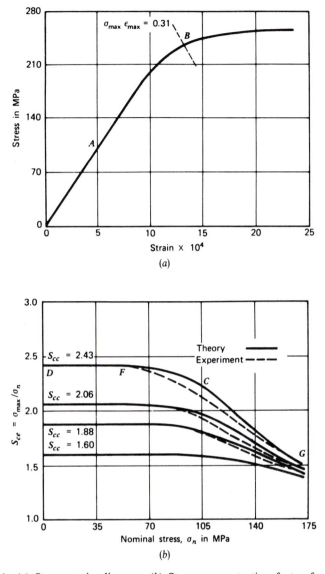

Figure E14.4 (a) Stress-strain diagram. (b) Stress concentration factor for low carbon steel of (a). Experimental data from Neuber (1961b).

PROBLEMS
Section 14.2

14.1. For the flat bar in Fig. C of Table 14.3, let $b = 16t$, $c = 17t$, $\rho = t$ (circular hole). By means of Neuber's nomograph (Fig. 14.10), show that for the bar loaded in tension, S_{cc} is approximately 3. Note that for this case, the half-width c of the bar is large compared to the radius of the hole.

14.2. With $\rho = t$, solve Problem 14.1 for

(a) $b = 4t$ and $c = 5t$ and

(b) $b = t$ and $c = 2t$.

Ans. (a) $S_{cc} = 2.7$, (b) $S_{cc} = 2.3$

14.3. For the flat bar in Fig. A of Table 14.3, let $t = 4\rho$ and $b = 16\rho$. By Neuber's nomograph (Fig. 14.10), determine the value of S_{cc} for the cases where the bar is subjected to

(a) axial tensile load and

(b) bending.

14.4. A cylindrical shaft has a circular groove, the depth of the groove is $t = 6.00$ mm, and the radius at the root of the groove is $\rho = 2.20$ mm. (See Fig. D, Table 14.3.) The radius of the cross section at the root of the groove is $b = 60$ mm. By Neuber's nomograph (Fig. 14.10), determine the value of S_{cc} for the cases where the shaft is subjected to

(a) axial tensile load,

(b) bending, and

(c) torsion.

Ans. (a) $S_{cc(P)} = 3.7$, (b) $S_{cc(M)} = 3.3$, (c) $S_{cc(T)} = 2.1$

14.5. The rectangular section tension member in Fig. P14.5 is made of a brittle material for which $\sigma_u = 300$ MPa. The member dimension into the plane of the figure is 20 mm. Determine the design load P of the member using a safety factor $SF = 3.50$.

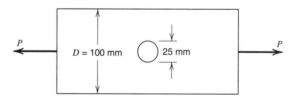

Figure P14.5

14.6. A 100-mm diameter steel ($E = 200$ GPa) tension member has a semicircular groove with depth equal to the radius $\rho = 5$ mm (Fig. P14.6). Short-gage-length electrical strain gages are cemented to the bottom of the groove and to the member at a location more than 100 mm from the groove. An axial load produces axial strain readings of 0.00100 at the bottom of the groove and 0.00032 at the location away from the groove. Assuming elastic material behavior, determine the stress concentration factor for the groove and magnitude of the axial load P. Assume that the state of stress at the bottom of the groove is uniaxial.

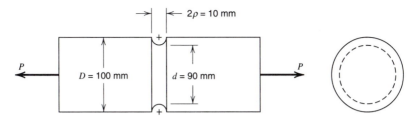

Figure P14.6

14.7. The stress concentration factor for a hole in a beam is approximately 3.00, if the diameter of the hole is small compared to the depth of the beam. Two 10.0-mm diameter holes are drilled through the beam, Fig. P14.7, at equal distances from the neutral surface. Using the moment of inertia for the net section through the holes and $S_{cc} = 3.00$, determine the magnitude of bending moment M for the case in which the limiting flexure stress is 120 MPa

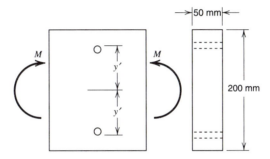

Figure P14.7

(a) for $y' = 50.0$ mm to the outer edges of the holes and
(b) for $y' = 75.0$ mm.

Section 14.3

14.8. In Example 14.1, let the hole be circular rather than elliptical. For stresses, σ_1, σ_2 as given,

(a) determine the maximum tensile stress in the plate.
(b) Determine the maximum compressive stress in the plate.

14.9. In Example 14.1, let the ratio $b/a = 5$. Determine the orientation of the hole (value of θ) for which the tensile stress at the perimeter of the hole is a maximum, the magnitude of the maximum tensile stress in the plate, and its location.

Ans. $\sigma_{max} = 57.3$ MPa, $\theta = 0.6406$ rad, $\beta = 0.5837$ rad

14.10. A thin-wall cylindrical tank, of diameter D and wall thickness h, is subjected to internal pressure p. A small circular hole exists in the wall of the cylinder. By means of Eqs. (14.29) through (14.32), derive expressions for the maximum stresses σ_A and σ_B at the hole, on longitudinal and transverse sections of the tank, respectively. Assume that the material remains elastic.

14.11. A thin-wall cylindrical pressure vessel, of diameter D and wall thickness h, is filled with a fluid whose weight density is γ (force per unit volume). The fluid is pressurized until the average pressure in the vessel is p (force per unit area). Assume that the difference in pressure between the top and bottom of the tank is small enough to be neglected. The vessel is supported near its ends on horizontal supports a distance l apart. Design considerations require that a small circular hole be drilled into the vessel at midspan at either the top (point A) or bottom (point B) (Fig. P14.11).

Show that if the hole is drilled at point A, the maximum stress at the hole due to the fluid and pressure is $\sigma_A = (5pD/4h) + (\gamma l^2/8h)$, and if at point B, $\sigma_B = (5pD/4h) - (\gamma l^2/8h)$. The weight of the vessel is neglected in estimating the bending stresses at section AB, and the bending stress is assumed to be smaller than the circumferential stress.

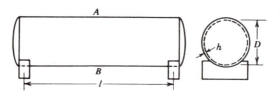

Figure P14.11

14.12. Let the shaft of Problem 14.4 be subjected simultaneously to a bending moment $M = 15.0 \text{ kN·m}$ and torsional moment $T = 30.0 \text{ kN·m}$. With the stress concentration factors determined in Problem 14.4, compute

(a) the maximum principal stress,

(b) the maximum shear stress, and

(c) the maximum octahedral shear stress that occur in the shaft at the root of the groove.

Ans.   **(a)** $\sigma_{max} = 382 \text{ MPa}$,   **(b)** $\tau_{max} = 236 \text{ MPa}$,   **(c)** $\tau_{oct} = 205 \text{ MPa}$

Section 14.4

14.13. A rectangular section tension member has semicircular grooves as shown in Fig. P14.13. The thickness of the member is 40 mm. The member is made

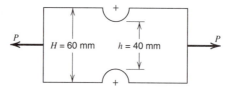

Figure P14.13

of a ductile metal that has a yield strength $Y = 350$ MPa. Determine the failure load for static loading.

14.14. The tension member in Fig. P14.14 has a rectangular cross section with a thickness of 20 mm. If $P = 80$ kN, determine the maximum normal stress at a section through the hole and at the section through the base of the fillets.

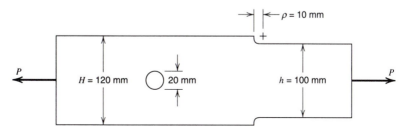

Figure P14.14

14.15. A rectangular section beam has grooves as shown in Fig. 14.25. If $H = h + 2\rho$ and $\rho = 0.20h$, determine the stress concentration factor.

14.16. The beam in Fig. P14.16 is made of steel ($E = 200$ GPa), has a diameter of 60.0 mm over the length of 600 mm, and has a fillet to the larger diameter. The magnitude of the stress concentration factor S_{cc} for the fillet is determined by strain readings from a strain gage cemented to the top of the beam at the base of the fillet. A strain reading of 0.00080 was recorded when $P = 3.00$ kN. What is the magnitude of S_{cc} for the fillet?

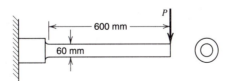

Figure P14.16

14.17. The rectangular section beam in Fig. P14.17 is made of an aluminum alloy ($E = 72.0$ GPa). Strain gages are used to determine the stess concentration factor for the grooves. One strain gage is located at the bottom of the groove and another strain gage is located some distance from the stress concentration as shown. Strain-gage readings at the groove and away from the stress concentration were recorded as 0.00250 and 0.00100, respectively. Determine the magnitudes of P and the stress concentration factor S_{cc}.

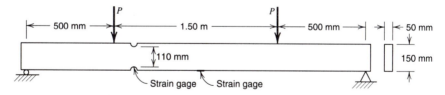

Figure P14.17

14.18. The beam in Fig. P14.18 is made of a brittle material that has an ultimate strength $\sigma_u = 450$ MPa. If $h = 125$ mm and $\rho = 15.0$ mm, determine the magnitude of P based on a safety factor $SF = 3.50$. Assume that the material is linearly elastic up to the ultimate strength.

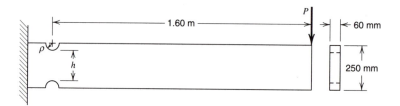

Figure 14.18

Section 14.5

14.19. A flat bar of a relatively brittle material has a fillet as shown in Fig. 14.24. The thickness of the bar is 24 mm, $\rho = 3$ mm, $h = 50$ mm, and $H = 68$ mm. The bar is subjected to a static moment M. If the allowable (working) stress for the material under this condition of loading is $\sigma_w = 14.0$ MPa, compute the maximum allowable moment M_{max} for the member. Assume that $q = 0.80$ [see Eq. (14.44)].

14.20. In Problem 14.12, let the twisting moment T and bending moment M be repeatedly applied through completely reversed cycles. Assume that $q = 0.80$ for the material in the shaft under repeated load. Compute the significant value of the maximum principal stress and maximum shear stress in the shaft at a cross section through the root of the groove.

Ans. $\sigma_{max} = 333.8$ MPa, $\tau_{max} = 208.3$ MPa

Section 14.6

14.21. In Fig. 14.28a, let $a = 30$ mm, $t = \rho = 5$ mm, and the thickness of the plate be $d = 12.5$ mm. Let the load $P = 110$ kN. If the stress-strain curve of the material is given by Fig. E14.4a, determine the stress concentration factor.

14.22. Let the stress-strain diagram for Problem 14.21 be flat-topped at a stress of 258 MPa. What is the magnitude of the strain at the root of the notch when $S_{ce} = 1.10$?

Ans. $\epsilon_{max} = 0.0065$

REFERENCES

Boresi, A. P. and Chong, K. P. (1987). *Elasticity in Engineering Mechanics*. New York: Elsevier.

Coker, E. G. and Filon, L. N. G. (1957). *A Treatise on Photoelasticity* (revised by H. T. Jessop). London: Cambridge Univ. Press, pp. 563–596.

Doyle, J. F. and Phillips, J. W. (1989). *Manual on Experimental Stress Analysis.* Bethel, Conn.: Soc. for Experimental Mech.

Frocht, M. M. (1936). Photoelastic Studies in Stress Concentrations. *Mech. Eng.*: 485.

Frocht, M. M. (1948). *Photoelasticity,* Vols. 1 and 2. New York: Wiley.

Griffith, A. A. and Taylor, G. I. (1917). The Use of Soap Films in Solving Torsional Problems. In *Proceedings of the Institute of Mechanical Engineers.* London: Oct.–Dec., p. 755.

Hetényi, M. (1950). *Handbook of Experimental Stress Analysis,* New York: Wiley.

Inglis, C. E. (1913). Stresses in the Plate due to the Presence of Cracks and Sharp Corners. *Trans. Inst. Naval Arch.,* **60**: 219.

Jacobsen, L. S. (1925). Torsional-Stress Concentrations in Shafts of Circular Cross-Section and Variable Diameter. *Trans. Amer. Soc. Mech. Eng.* 47: 619.

Kobayashi, A. S. (1988). *Handbook on Experimental Mechanics.* Bethel, Conn.: Soc. for Experimental Mech.

Neuber, H. (1958), *Kerbspannungslehre,* 2nd ed. Berlin: Springer-Verlag.

Neuber, H. (1961a). Theory of Stress Concentration for Shear Strained Prismatic Bodies with Nonlinear Stress-Strain Law. *J. Appl. Mech.,* **28**, Series E (4): 544–550.

Neuber, H. (1961b). Research on the Distribution of Tension in Notched Construction Parts. WADD Rept. 60-906, Jan.

Peterson, R. E. (1974). *Stress Concentrations Factors for Design.* New York: Wiley.

Peterson, R. E. and Wahl, A. M. (1936). Two- and Three-Dimensional Cases of Stress Concentration, and Comparison with Fatigue Tests. *Trans. Amer. Soc. Mech. Eng.,* **47**: 619; **59**: A15–A22.

Sadowsky, M. A. and Sternberg, E. (1949). Stress Concentrations Around a Triaxial Ellipsoidal Cavity., *J. Appl. Mech.,* **71**: 149.

Savin, G. N. (1961). *Stress Concentrations Around Holes.* New York: Pergamon Press.

Timoshenko, S. P. and Goodier, J. N. (1970). *Theory of Elasticity,* 3rd ed. New York: McGraw-Hill.

15

FRACTURE MECHANICS

Unexpected failure of weapons, buildings, bridges, ships, trains, airplanes, and various machines has occurred throughout the industrial world. A number of these failures have been due to poor design. However, it has been discovered that many failures have been caused by preexisting flaws in materials that initiate cracks that grow and lead to fracture. This discovery has, in a sense, lead to the field of study known as fracture mechanics. The field of fracture mechanics is extremely broad. It includes applications in engineering, studies in applied mechanics (including elasticity and plasticity), and materials science (including fracture processes, fracture criteria, and crack propagation). The successful application of fracture mechanics requires some understanding of the total field. Unfortunately, this vast subject cannot be treated fully in the few pages allotted to it here. Therefore, we have focused mainly on certain fundamental concepts. In particular, we briefly examine criteria for crack initiation and the three modes of crack extension or crack propagation (Sec. 15.1). Using the concept of an elliptical hole in a thin flat plate, we examine the stress distribution at the leading edge of a blunt crack (Sec. 15.2). By letting the minor axis of the elliptical hole go to zero, we estimate the stress distribution at the leading edge of a sharp crack and introduce the concept of stress intensity factor for the opening or tension mode of crack propagation (Sec. 15.3). Although fracture is influenced by many factors, such as plasticity (Chapter 4), temperature, corrosion, fatigue (Chapter 16), creep (Chapter 17), etc., we restrict our study mainly to linear elastic fracture mechanics (LEFM) of materials that fracture in a brittle manner. A brief discussion of some of these effects is given in Sec. 15.4. The reader who is interested in pursuing these topics is referred to the literature (see, e.g., the following excellent references: Broek, 1985 and 1988; Barsom and Rolfe, 1987; Ewalds and Wanhill, 1986; Knott 1973; Kanninen and Popelar, 1984; Liebowitz, 1968–1971). For testing methods in fracture mechanics, the reader is referred to Ewalds and Wanhill (1986) and Curbishley (1988).

15.1

FAILURE CRITERIA. FRACTURE

As noted in Chapter 1, failure of a structural system may occur by excessive deflection, yield, or fracture. Unfortunately, these modes of failure do not occur in a singular fashion, since prior to failure, say, by fracture, yielding of a member may occur. Furthermore, a member may undergo considerable deflection before it fails, say, by extensive yielding. Consequently, failure criteria are usually based on the

dominant failure mode. Thus, for yield-dominant failure of ductile metals the criteria of octahedral shear stress (also distortional strain energy density) and maximum shear stress seem most appropriate. For fracture-dominant failure, several types of fracture must be considered. For example, fracture may occur in a "sudden" manner (brittle materials at ordinary temperatures or structural steels at low temperatures), it may occur as brittle fracture of cracked or flawed members, it may occur in progressive stages (so-called fatiguing; Chapter 16) at general levels of stress below yield, and finally, it may occur with time at elevated temperatures (creep rupture). In contrast to yield dominant failures, different types of failure criteria are applicable to different types of fracture-dominant failures.

Material defects are of significance in all kinds of failures. However, different types of defects influence various modes of failure differently. For example, for initiation of yielding, the significant defects tend to distort and interrupt crystal lattice planes and interfere with easy glide of dislocations. These defects are of the nature of dislocation entanglements, interstitial atoms, out-of-size substitutional atoms, grain-boundary spacings, bounded precipitate particles, etc. In general, these defects provide resistance to yielding that is essential to the proper performance of high-strength metals. On the other hand, little resistance to yielding is provided by larger defects such as inclusions, porosity, surface scratches, and small cracks, although such defects may alter the net load-bearing section.

For failure by fracture before extensive yielding of the section (fracture-dominant failure), the significant defects (size scale) depend principally on the *notch toughness* of the material. Notch toughness is a measure of the ability of a material to absorb energy in the presence of a flaw. Unfortunately, there is no clear boundary between yielding (ductile-type material) failures and fracture-dominant (brittle-type material) failures. Indeed, classification of many materials as ductile or brittle is meaningless unless physical factors such as temperature, state of stress, rate of loading, and chemical environment are specified. For example, many materials can be made to behave in a ductile manner for a given set of conditions and in a brittle manner for another set of conditions. To be more precise, one should speak of a material being in a *brittle* or *ductile state*. However, here too difficulties arise, since there is not always a clear demarcation between brittle and ductile states. Nevertheless, it is fortuitous that for an important range of materials and conditions in either the ductile state or the brittle state, time effects, temperature, stress gradients, microstructural features, and size effects, for example, are of secondary importance. For the ductile state, it is possible to postulate failure criteria based on concepts of macroscopic states of stress that define critical values of quantities for which yielding begins (Chapter 4).

Under similar circumstances, it is also possible to postulate reasonable failure criteria based on macroscopic stress concepts for the onset of brittle fracture. In general, in contrast to materials in the ductile state, failure (fracture) states for materials in the brittle state are sensitive to both the magnitude and sign of the mean stress. The fracture states for isotropic materials in the brittle state are frequently (conveniently) represented by pyramid–like surfaces in principal stress space, which are cut (limited) by suitable tension (critical value) cutoffs (Liebowitz, 1968; Paul, 1968).

Brittle fracture problems that we consider are subdivided into three types as follows: (1) brittle fracture of members free of cracks and flaws under static loading conditions, (2) brittle fracture originating at cracks and flaws in members under static loading conditions, and (3) brittle fracture resulting from high cycle fatigue

loading (Chapter 16). Another type of brittle fracture, which we do not consider, occurs at elevated temperatures after long-time creep in which small deformations occur as creep in grain boundaries (see Smith and Nicolson, 1971). This type of fracture is sometimes referred to as *creep fracture*.

Brittle Fracture of Members Free of Cracks and Flaws

So-called brittle materials (such as glass, gray cast iron, and chalk) exhibit nearly linear tensile stress-strain diagrams up to their ultimate strengths. If, at the fracture location in a member made of a brittle material, the principal stress of maximum absolute magnitude is tensile in nature, fracture will occur on the plane on which this principal stress acts. Then, the maximum principal stress criterion of failure is considered to be valid for design purposes. When mean stresses at the fracture locations in members are tensile, the brittle materials in these members are considered to be loaded in a brittle state. It may be possible for the same materials in other members to be loaded in a ductile state if the mean stresses in these members are large compressive. See, for instance, the Mohr–Coulomb and Drucker–Prager criteria presented in Sec. 4.5.

Brittle Fracture of Cracked or Flawed Members

Cracks may be present in members before loading, created (initiated) by flaws (high stress concentrations) at low nominal stress levels, or initiated and made to propagate with a large number of cycles because of fatigue loading (Chapter 16). Failure by fracture (complete fracture) results when a crack propagates sufficiently far through the member so that the member is unable to support the load and hence fractures into two or more pieces.* In general, brittle fracture consists of at least two stages: crack initiation and crack extension or crack propagation. Once a crack has been initiated, subsequent crack propagation may occur in several ways, depending on the relative displacements of the particles in the two faces (surfaces) of the crack. Three basic modes of crack surface displacements are Mode I, the opening mode (Fig. 15.1a), Mode II, the (edge) sliding mode (Fig. 15.1b), and Mode III, the tearing mode (Fig. 15.1c). In Mode I, the opening mode, the crack surfaces move directly apart. In Mode II, the sliding mode, the crack surfaces move (slide) normal to the crack tip and remain in the plane of the crack. In the tearing mode, Mode III, the crack surfaces move parallel to the crack tip and again remain in the plane of the crack. The most general case of crack surface displacements, so called mixed-mode is obtained by superposition of these basic three modes. We follow the convention of adding Roman numeral subscripts I, II, III to symbols associated with quantities that describe Modes I, II, III, respectively.

In isotropic materials, brittle fracture usually occurs in Mode I. Consequently, we confine our attention mainly to Mode I in establishing fracture criteria for sudden fracture of flawed members when the materials in these members are loaded in the brittle state. Although fractures induced by sliding (Mode II) and tearing

---

* In pipe systems, which carry fluids and other materials, failure may occur when a crack propagates through the pipe thickness, allowing fluid to escape from the pipe. Such failures may be extremely harmful to life and property if the liquid is a dangerous chemical or is contaminated, say, by nuclear fission products such as those that occur in nuclear reactor pipe loop systems (see Clarke and Gordon, 1973).

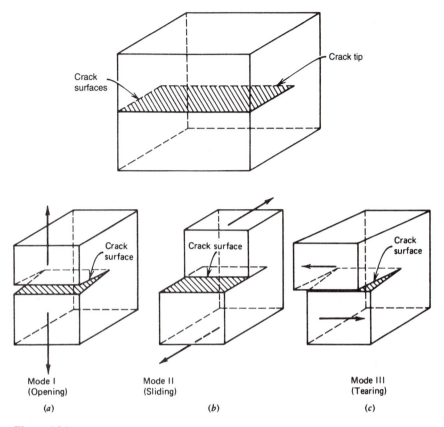

Figure 15.1

(Mode III) do occur, their frequency is much less than the opening mode fracture (Mode I). Although the combined influence of two or three modes of crack extension has been studied (Paris and Sih, 1965), we do not consider such problems here. However, we do note that improvement of fracture resistance for Mode I usually results in improved resistance to mixed mode crack extension.

The crack surfaces, which are stress-free boundaries in the neighborhood of the crack tip, strongly influence the distribution of stress around the crack. More remote boundaries and remote loading affect mainly the intensity of the stress field at the crack tip. Elastic stress analysis of cracks leads to the concept of *stress intensity factor K*, which is employed to describe the elastic stress field surrounding the crack tip. As noted above, the motion of crack surfaces can be divided into three types, with corresponding stress fields. Hence, three stress intensity factors K_I, K_{II}, and K_{III} are employed to characterize the stress fields for these three modes. The dimensions of stress intensity factor K are [stress] $\times$ [length]$^{1/2}$. The factor K depends on specimen dimensions and loading conditions. In general, K is proportional to [average stress] $\times$ [crack length]$^{1/2}$. When K is known for a given mode (say, K_I), stresses and displacements in the neighborhood of the crack tip can be calculated (Sec. 15.2). The stresses are inversely proportional to the square of the distance from the crack tip, becoming infinite at the tip. In general, fracture criteria for brittle fracture are based on critical values of the stress intensity factor, for which the crack rapidly propagates (leading to fracture).

In order to determine the load or loads required to cause brittle fracture of a cracked member or structure, it is necessary that relations be developed so that K can be determined for the member or structure and that the critical value K_C be determined for the material. Test specimens have been developed to measure the critical value of K for the opening mode (Mode I); when certain test conditions are satisfied, the critical value is designated as K_{IC}, and it is called *fracture toughness*. Fracture toughness K_{IC} is considered to be the material property measure of resistance to brittle fracture (Srawley and Brown, 1965).

The designs of test specimens recommended in ASTM and British standards to determine values of K_{IC} are indicated in Fig. 15.2. The relative dimensions of the two test specimens in Fig. 15.2 are specified by the magnitude of W. The minimum magnitude of W depends on the values of material properties K_{IC} and the yield

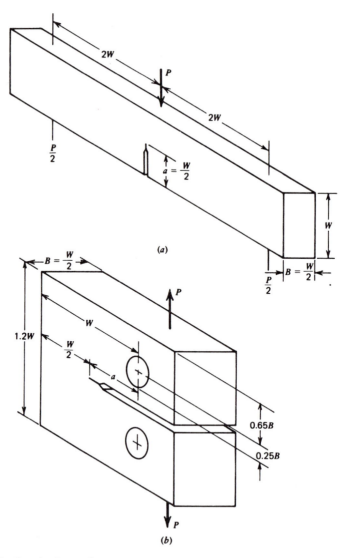

Figure 15.2 Standard toughness specimens. (*a*) Single edge-cracked bend specimen. (*b*) Compact tension specimen.

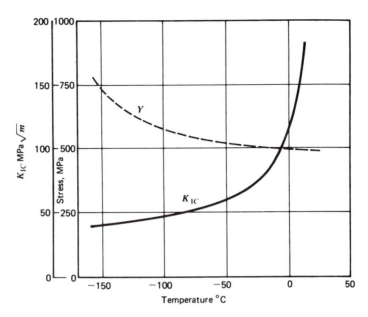

Figure 15.3   Temperature dependence of K_{IC} for A533B steel.

stress Y of the material. The derivation of K is based on a linear elastic solution. Therefore, the stress state at the crack tip of the test specimen used to determine K_{IC} should approximate linear elastic conditions as closely as possible. If the magnitude of the thickness $B = W/2$ (Fig. 15.2) is small, appreciable inelastic deformation occurs at the tip of the crack and the measured value of K_C is larger than K_{IC}. Values of toughness that exceed K_{IC} are defined as K_C. As B is increased in magnitude, the measured value of K approaches a minimum value obtained when the inelastic deformation at the tip of the crack is held to a minimum, that is, for nearly linear elastic conditions. This will occur when the state of stress at the tip of the crack over most of the width of the specimen is that associated with plane strain. In order to insure plane strain conditions over the major length of the crack tip, it has been recommended (Knott, 1973) that the magnitude of B satisfy the relation

$$B \geq 2.5\left(\frac{K_{IC}}{Y}\right)^2 \tag{15.1}$$

Because of the large size of specimens required to satisfy Eq. (15.1) for some materials, the expense of specimen preparation and of testing may be large. Wessel (1969) has obtained values of K_{IC} for an A533B steel at several temperatures as indicated in Fig. 15.3. At temperatures of -150 and $10°C$, Eq. (15.1) gives values for B of 7.4 mm and 242.0 mm, respectively. The large variation of K_{IC} with temperature indicated in Fig. 15.3 is typical of relatively low-strength structural steels. Nonferrous alloys and very high-strength steels show a rather small variation of K_{IC} with temperature. Room-temperature values of K_{IC} for several metals are listed in Table 15.1. Except for the data for A533B steel that were taken from Fig. 15.3, the other data in Table 15.1 were taken from published papers.*

---

* H. D. Greenberg, E. T. Wessel, and W. H. Pryle, 1970; E. A. Steigerwald, 1969; W. A. Logsdon, 1975; F. G. Nelson, P. E. Schilling, and J. G. Faufman, 1972; C. N. Freed, 1968.

TABLE 15.1
K_{IC} Critical Stress Intensity Factor (Fracture Toughness) (Room Temperature Data)

Material	σ_u (MPa)	Y (MPa)	K_{IC} (MPa$\sqrt{m}$)	Minimum Values for B, a, t (mm)
Alloy Steels				
A533B	—	500	175	306.0
2618 Ni Mo V	—	648	106	66.9
V1233 Ni Mo V	—	593	75	40.0
124 K 406 Cr Mo V	—	648	62	22.9
17-7PH	1289	1145	77	11.3
17-4PH	1331	1172	48	4.2
Ph 15-7Mo	1600	1413	50	3.1
AISI 4340	1827	1503	59	3.9
Stainless Steel				
AISI 403	821	690	77	31.1
Aluminum Alloys				
6061-T651	352	299	29	23.5
2219-T851	454	340	32	22.1
7075-T7351	470	392	31	15.6
7079-T651	569	502	26	6.7
2024-T851	488	444	23	6.7
Titanium Alloys				
Ti-6Al-4Zr-2Sn-0.5Mo-0.5V	890	836	139	69.1
Ti-6Al-4V-2Sn	852	798	111	48.4
Ti-6.5Al-5Zr-1V	904	858	106	38.2
Ti-6Al-4Sn-1V	889	878	93	28.0
Ti-6Al-6V-2.5Sn	1176	1149	66	8.2

In order to use values of K_{IC} from Table 15.1 in design, it is necessary that formulas for K_I be derived for typical load-carrying members. A few formulas for K_I for several geometric configurations and loads are given in Table 15.2. These formulas along with others may be found in Paris and Sih (1965). We assume that the dimensions of each member are such that the state of stress at the crack tip over most of the thickness of the member is linearly elastic so that $K_I = K_{IC}$ at initiation of crack propagation. In order to insure that the state of stress is linearly elastic for each of the cases in Table 15.2, it is assumed that the magnitudes of a and thickness t satisfy the relation

$$a, t \geq 2.5\left(\frac{K_{IC}}{Y}\right)^2 \tag{15.2}$$

Fracture mechanics analysis is also employed in establishing failure criteria for general yielding as well as fracture criteria for materials loaded in the ductile state (Knott, 1973). These topics are beyond the scope of this book.

TABLE 15.2
Stress Intensity Factors K_I

Case 1. Infinite Sheet with Through-Thickness Crack and Uniform Tension at Infinity. Griffith's Crack.

$$K_I = \sigma\sqrt{\pi a}$$

Case 2. Periodic Array of Through-Thickness Cracks in Infinite Sheet with Uniform Tension at Infinity

$$K_I = \sigma\sqrt{\pi a}\,f(\lambda); \;\lambda = \frac{a}{c}$$

λ	$f(\lambda)$
0.1	1.00
0.2	1.02
0.3	1.04
0.4	1.08
0.5	1.13
0.6	1.21

Case 3. Central Crack in Finite-Width Strip Subjected to Uniform Tension at Infinity

$$K_I = \sigma\sqrt{\pi a}\,f(\lambda); \;\lambda = \frac{a}{c}$$

λ	$f(\lambda)$
0.1	1.01
0.2	1.03
0.3	1.06
0.4	1.11
0.5	1.19
0.6	1.30

Case 4. Single-Edge Crack in Finite-Width Sheet

$$K_I = \sigma\sqrt{\pi a}\,f(\lambda); \;\lambda = \frac{a}{c}$$

λ	$f(\lambda)$
$0(c \to \infty)$	1.12
0.2	1.37
0.4	2.11
0.5	2.83

Case 5. Double-edge Crack in Finite-Width Sheet

$$K_I = \sigma\sqrt{\pi a}\,f(\lambda); \;\lambda = \frac{a}{c}$$

λ	$f(\lambda)$
$0(c \to \infty)$	1.12
0.2	1.12
0.4	1.14
0.5	1.15
0.6	1.22

Case 6. Edge Crack in Beam in Bending

$$K_I = \sigma\sqrt{\pi a}\,f(\lambda)$$
$$\lambda = \frac{a}{2c}$$
$$\sigma = \frac{3M}{2tc^2}$$

λ	$f(\lambda)$
0.1	1.02
0.2	1.06
0.3	1.16
0.4	1.32
0.5	1.62
0.6	2.10

EXAMPLE 15.1
Longitudinal Cracks in Pressurized Pipes

General experience in nondestructive testing of pressurized pipes made of various materials indicates that longitudinal cracks of maximum length of 10 mm may be present. There is concern that the pipe will undergo sudden fracture. Hence, an estimate of the maximum allowable pressure is required. Consider two cases, one for which 17-4PH precipitation hardening steel heat-treated to the properties in Table 15.1 is used and the other for which Ti-6Al-4Sn-1V titanium alloy heat-treated to the properties in Table 15.1 is used.

SOLUTION

By fracture mechanics concepts, unstable crack growth (crack propagation) occurs at a load level for which the potential energy available for crack growth exceeds the work done in extending the crack (creating additional crack surface). For the pressurized pipe, the stress state of the crack corresponds to that of Case 1 of Table 15.2; see also Fig. E15.1. Thus, $K_1 = \sigma \sqrt{\pi a}$, where a is the crack half-length, $\sigma = pr/t$, where p is the internal pressure, r the pipe inner radius, and t the pipe thickness. By fracture mechanics concepts, $K_1 = K_{IC}$ for unstable crack growth.

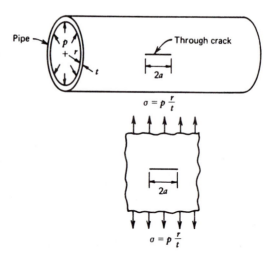

Figure E15.1

Case A (17-4PH Precipitation Hardening Steel)
By Table 15.1, $K_{IC} = 48 \text{ MPa}\sqrt{m}$. We find the maximum allowable pressure to be

$$p_{max} = \frac{t}{r}\frac{K_{IC}}{\sqrt{\pi a}} = \frac{48\sqrt{1000}}{\sqrt{5\pi}}\frac{t}{r} = 382\frac{t}{r} \qquad (a)$$

If, on the other hand, the pressure p is fixed, the critical value of the ratio t/r is

$$\left(\frac{t}{r}\right)_{critical} = 0.00261p \qquad (b)$$

These results assume that the thickness t is greater than 4.2 mm so that Eq. (15.2) is satisfied. If the crack half-length is decreased in magnitude, the fracture pressure given by Eq. (a) increases in magnitude; the computed value is no longer valid for $a < 4.2$ mm.

Case B (Ti-6Al-4Sn-1V Titanium Alloy)

By Table 15.1, $K_{IC} = 93$ MPa $\sqrt{m}$. As in Case A, by fracture mechanics

$$p_{max} = \frac{t}{r} \frac{K_{IC}}{\sqrt{\pi a}} = \frac{93\sqrt{1000}}{\sqrt{5\pi}} \frac{t}{r} = 742 \frac{t}{r}$$

This pressure probably would not cause brittle fracture. Even if the thickness t was equal to or greater than 28 mm, the crack half-length $a = 5$ mm is much less than that required (28 mm) to satisfy Eq. (15.2).

15.2

THE STATIONARY CRACK

The solution to the stress concentration problem for an elliptical hole may be used to obtain an estimate of the stress distribution in the neighborhood of the tip of a crack (either blunt or sharp). As noted in Sec. 14.2, the tangential stress component $\sigma_{\beta\beta}$ around an elliptical hole in an infinite plate (sheet) subjected to uniform tensile stress σ in a direction perpendicular to the major axis of the hole depends on the ratio a/b [Eq. (14.11)]. Hence, as $a/b \to \infty$ (the elliptical hole becomes a crack), the maximum value of $\sigma_{\beta\beta}$ becomes very large. For example, for $a/b = 100$, $\sigma_{\beta\beta(max)} = 201\sigma$; for $a/b = 1000$, $\sigma_{\beta\beta(max)} = 2001\sigma$; etc. For sufficiently large ratios of a/b, the radius ρ of curvature at the edge of the major axis of the elliptical hole decreases but remains finite. If we take the radius ρ very small (but nonzero), the elliptical hole solution (at the end of the major axis of the hole) is used as the solution for a blunt crack.

As the ratio $b/a \to 0$, we consider that $\rho \to 0$, and we are led to the case of a sharp crack of length $2a$ in an infinite plate with uniform stress σ applied at infinity in a direction perpendicular to the length $2a$. The stress distribution in the neighborhood of a *sharp crack tip* may be obtained directly from the elliptical hole problem by considering the case $b \to 0$.

In terms of ρ, the stress concentration factor S_{cc} is [Eq. (14.13)]

$$S_{cc} = \frac{\sigma_{\beta\beta(max)}}{\sigma} = 1 + 2\sqrt{\frac{a}{\rho}} \tag{15.3}$$

Thus, since many geometrical holes, notches, flaws, cracks, etc., may be approximated by an elliptical hole, it is to be expected that as $\rho \to 0$, $S_{cc} \to \infty$. All tabulated solutions of the crack problem exhibit this behavior. Most of the studies of fracture mechanics are directed toward the behavior of the stress solution in the neighborhood of a crack tip, as $\rho \to 0$.

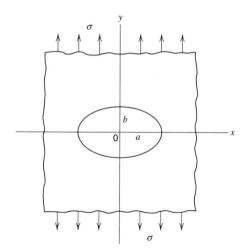

Figure 15.4

To examine the normal stresses in the neighborhood of a crack tip, it is convenient to represent the stress components in terms of (x, y) axes (Fig. 15.4). Thus, for stresses along the major axes of the elliptical hole, $y = 0$, $x > a$, we obtain by transforming the stress components relative to the (α, β) axes (Fig. 14.6) into stress components relative to the (x, y) axes (Inglis, 1913)

$$\sigma_{xx} = F_1(s) - F_2(s)$$
$$\sigma_{yy} = F_1(s) + F_2(s) \tag{15.4}$$

where s is a distance parameter

$$s = \frac{x}{2B} + \sqrt{\left(\frac{x}{2B}\right)^2 - m} \tag{15.5}$$

and

$$F_1(s) = \frac{\sigma}{2}\left[1 + \frac{2(1 + m)}{s^2 - m}\right]$$

$$F_2(s) = \frac{\sigma}{2}\left\{1 + \frac{m^2 - 1}{s^2 - m}\left[1 + \left(\frac{m - 1}{s^2 - m}\right)\left(\frac{3s^2 - m}{s^2 - m}\right)\right]\right\} \tag{15.6}$$

with

$$B = \frac{1}{2}(a + b), \qquad m = \frac{a - b}{a + b} \tag{15.7}$$

By means of Eqs. (15.4), the stresses along the major axes in the neighborhood of the end of the major axes of an elliptical hole may be examined. By assuming that the radius of curvature ρ of a crack may be approximated by the radius of curvature of an equivalent elliptical hole [Eq. (14.12)], approximations of the stress in the neighborhood of the crack may be obtained, provided $\rho \neq 0$, that is, for a blunt crack.

Blunt Crack

For the elliptical hole, let the radius of curvature be $\rho \ll a$ at the end of the major axis. Let $r = x - a$ be the distance from the end of the major axis (in the major axis direction, Fig. 15.5). In the neighborhood of $x = a, r \ll a$. Hence, in terms of r and ρ, we may write, with Eqs. (14.12), (15.4), (15.5), (15.6), and (15.7),

$$2B = a\left(1 + \sqrt{\frac{\rho}{a}}\right), \qquad m = \frac{1 - \sqrt{\rho/a}}{1 + \sqrt{\rho/a}}$$

$$s = 1 - \sqrt{\frac{\rho}{a}} + \sqrt{\frac{(2r + \rho)}{a}}$$

$$F_1(s) = F_1(r) \approx \frac{\sigma}{\sqrt{(2r + \rho)/a}}$$

$$F_2(s) = F_2(r) \approx \frac{\sigma(\rho/a)}{[(2r + \rho)/a]^{3/2}} \tag{15.8}$$

Hence, in the neighborhood of the tip of the crack,

$$\sigma_{yy} = \frac{\sigma\sqrt{a}}{\sqrt{2r}} \frac{1 + (\rho/r)}{[1 + (\rho/2r)]^{3/2}} \tag{15.9}$$

At the tip of the crack, $r = 0$, and then Eq. (15.9) reduces to

$$\sigma_{yy} = 2\sqrt{\frac{a}{\rho}}\,\sigma \tag{15.10}$$

which agrees with Eq. (15.3) for $a \gg \rho$.

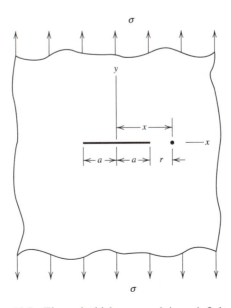

Figure 15.5   Through-thickness crack in an infinite plate.

Sharp Crack

For the sharp crack, we may estimate the stress distribution from that of the elliptical hole by letting $b \to 0$. Then, we have

$$2B = a, \qquad m = 1, \qquad s = \frac{x}{a} + \sqrt{\left(\frac{x}{a}\right)^2 - 1}$$

$$x = r + a, \qquad r \ll a, \qquad b = 0 \tag{15.11}$$

and by Eqs. (15.6)

$$F_1(s) = F_1(r) \approx \frac{\sigma}{\sqrt{2r/a}}$$

$$F_2(s) = F_2(r) \approx \frac{\sigma}{2} \tag{15.12}$$

Hence,

$$\sigma_{yy} = \frac{\sigma\sqrt{a}}{\sqrt{2r}} \tag{15.13}$$

Clearly at the crack tip ($r = 0$), the stress is singular ($\sigma_{yy} \to \infty$ as $r \to 0$).

Alternatively in terms of x, it may be shown that

$$\sigma_{yy} = \frac{\sigma x}{\sqrt{x^2 - a^2}}$$

$$\sigma_{xx} = \sigma\left(\frac{x}{\sqrt{x^2 - a^2}} - 1\right) \tag{15.14}$$

Again at the crack tip ($x = a$), σ_{yy} (and σ_{xx}) become infinite. For large values of x, $\sigma_{yy} \to \sigma$, and $\sigma_{xx} \to 0$ as expected (Fig. 15.5).

As we shall see in Sec. 15.3 in describing crack propagation, it is conventional to introduce the combination $\sigma_{yy}\sqrt{2r}$ since this factor remains finite as $r \to 0$. In addition, a factor π is introduced so that

$$K_{\mathrm{I}} = \sigma_{yy}\sqrt{2\pi r} = \sigma\sqrt{\pi a} \tag{15.15}$$

The factor K_{I} is called the *stress intensity factor*. In certain fracture theories, it is assumed that the material fractures (the crack propagates) if K_{I} exceeds a critical value

$$K_{\mathrm{I}} = \sigma_c\sqrt{\pi a} \geq K_{\mathrm{IC}} \tag{15.16}$$

where σ_c is the corresponding critical tensile stress. The term stress intensity factor should not be confused with the term stress concentration factor [Eq. (14.1)], which represents the ratio between the maximum stress in a region of stress concentration and the average stress.

The results of this section are of importance in fracture mechanics, failure theories, and crack propagation studies. See Sec. 15.3. One might question how, in the presence of such a high stress ($\sigma_{yy} \rightarrow \infty$, as $r \rightarrow 0$), the material can remain elastic, permitting the application of linear elastic fracture mechanics. In fact, at the tip of a sharp crack, it is possible for a state of high hydrostatic stress to exist (in a sufficiently thick member), whereas the maximum shear stress and distortional energy density remain low. Thus, yielding may not occur prior to brittle fracture.

15.3

CRACK PROPAGATION. STRESS INTENSITY FACTOR

Elastic Stress at the Tip of a Sharp Crack

In Sec. 14.2, we noted that the maximum stress at the ends of an elliptical hole in an infinite plate may be quite large. For example, when the plate is subjected to an edge tensile stress σ in the direction perpendicular to the major axis of the elliptic hole, the stress at the edge of the hole is given by Eq. (14.11). Hence, if the ratio $a/b = 100$, the value of $\sigma_{\beta\beta(\text{max})}$ is 201σ; if $a/b = 1000$, $\sigma_{\beta\beta(\text{max})} = 2001\sigma$; etc. The elliptical hole becomes very narrow and approaches the shape of an internal line crack as $a/b \rightarrow \infty$. In this case, $\sigma_{\beta\beta(\text{max})} \rightarrow \infty$, and we can no longer utilize the concept of stress concentration factor in describing the behavior around the crack tip.

Physically, one might expect that when loads are applied to a member that contains a line crack, the extremely large stress at the tips of the crack will cause the crack to extend or propagate. Experiments bear out this expectation in that it has been observed that the crack may* propagate when the load attains a critical nominal value. In general, under lower values of the applied stress σ, the crack may propagate slowly a short distance and stop, whereas under higher values of σ, the crack may propagate rapidly and continuously until a catastrophic separation of parts of the member occurs.

For a given member made of a given material, cracks may propagate under conditions such that the material is in the ductile state or under conditions such that the material is in the brittle state (Sec. 15.1). If the dimensions of the member are such that the state of stress over most of the length of the crack tip is plane strain, the crack will propagate with minimum plastic deformation occurring at the crack tip. The material in such members is considered to be loaded in the brittle state. The state of stress outside the small plastic zone is assumed to be characterized by the elasticity solution presented in Sec. 15.2. In the discussion that follows, we assume that the materials are loaded in the brittle state.

Investigators have attempted to explain the mechanism of crack propagation in terms of the distribution of stress in the neighborhood of the crack tip. However, in addition, to help explain the crack propagation, another concept is required. Earlier investigators, particularly Griffith, introduced the concept of strain energy release rate, G. The quantity G represents the amount of strain energy lost by the member per unit area of the newly formed crack area as the crack propagates. This strain energy is used up in forming the new surface area of the crack. In other words, the energy required to form the surface area of the extended crack is obtained from the strain energy of the body. Since the dimensions of G may be

---

* The capacity of the material to absorb relatively large amounts of energy per unit volume by plastic flow before fracture determines the level of nominal stress at which the crack propagates. In mild steel, a crack may not propagate until catastrophic fracture is imminent.

written $[F/L]$, G is referred to as the crack extension force. Hence, for a given geometry and material, a crack will propagate when the load reaches a level that produces a critical value G_{IC} of G.

Stress Intensity Factor. Definition and Derivation

To examine the stress distribution near the tip of a crack in a flat plate, consider a crack of length $2a$, which is very small compared to the width and length of the plate (Fig. 15.6). Let the plate be subjected to uniformly distributed stress in a direction perpendicular to the crack length $2a$. As noted in Sec. 15.2, the elastic stress at the tip of the crack becomes infinitely large as the radius of curvature ρ at the tip goes to zero [Eq. (15.10)]. As shown in Sec. 15.2, [see Eq. (15.15) and Figs. 15.5 and 15.6], the stress σ_{yy} along the extension of the major axis (the expected path of crack propagation) is given by

$$\sigma_{yy} = \frac{K_I}{\sqrt{2\pi r}} \tag{15.17}$$

where r is the distance from the crack tip measured along the x axis and K_I the stress intensity factor for a mode I crack. (See Sec. 15.2 and Fig. 15.1.)

Following Irwin (1957), we define the stress intensity factor by means of the following limit:

$$K_I = \lim_{\rho \to 0} \frac{\sqrt{\pi\rho}}{2} \sigma_{max} \tag{15.18}$$

where ρ is the radius of curvature at the crack tip (Fig. 15.6), and σ_{max}, the maximum stress at the crack tip, is a function of ρ; see Eqs. (15.9) and (14.13) for the case of an elliptical hole. Consequently, if we consider the line crack to be the limit-

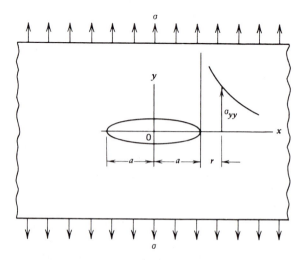

Figure 15.6

ing case of an elliptical hole, as $b \to 0$, we obtain by Eqs. (14.13) and (15.18)

$$K_1 = \lim_{\rho \to 0} \frac{\sqrt{\pi\rho}}{2}\left[\sigma\left(1 + 2\sqrt{\frac{a}{\rho}}\right)\right] = \sigma\sqrt{\pi a} \qquad (15.19)$$

for mode I propagation of an internal crack in a flat plate; see also Eq. (15.15). Values of K_1 for some other types of cracked members are listed in Table 15.2.

Crack Extension Force G. Derivation

Following concepts proposed by Griffith, one may derive a relationship between the crack extension force G and stress intensity factor K for various modes. For example, for a Griffith crack of length $2a$, centrally located in a plate subjected to a uniformly distributed stress σ at edges far removed from the crack (Fig. 15.5), the surfaces of the crack undergo a relative displacement of magnitude $2v$ under a Mode I separation (Fig. 15.7). For a condition of plane strain (Paul, 1968), it may be shown that (Knott, 1973)

$$2v = 4(1 - v^2)\frac{\sigma}{E}\sqrt{a^2 - x^2}, \qquad x \le a \qquad (15.20)$$

where v is Poisson's ratio and E the modulus of elasticity. The problem is to calculate the strain energy released when a crack of half-length a is extended to a half-length $(a + \delta a)$. For constant load σ, the release of potential energy is equal to the release of strain energy as $\delta a \to 0$. Alternatively, we may calculate the change in energy in the plate as a whole, by calculating the work done by the surface forces at the crack tip acting across the length δa when the crack is closed from length $(a + \delta a)$ to length a. In other words, we may employ the principle of virtual work.

In terms of the crack extension force G, the energy change may be expressed in the form

$$G\,\delta a = \int_a^{a + \delta a} \sigma_{yy} v\, dx \qquad (15.21)$$

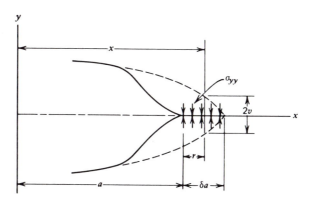

Figure 15.7

where the plate thickness is taken as unity and, by Eq. (15.14),

$$\sigma_{yy} = \frac{\sigma x}{\sqrt{x^2 - a^2}}, \qquad a \le x \le a + \delta a \qquad (15.22)$$

and v is given by Eq. (15.20), where we let $a \to a + \delta a$ and $a \le x \le a + \delta a$. Letting $r = x - a$, we may write for sufficiently small r (Fig. 15.7)

$$\sigma_{yy} = \frac{\sigma \sqrt{\pi a}}{\sqrt{2\pi r}} = \frac{K_{\mathrm{I}}}{\sqrt{2\pi r}}$$

$$2v = 4(1 - v^2)\frac{\sigma}{E}\sqrt{2a}\sqrt{\delta a - r} \qquad (15.23)$$

and, hence, with Eq. (15.20) we have

$$G\,\delta a = 2(1 - v^2)\frac{\sigma^2 a}{E} \int_0^{\delta a} \left(\frac{\delta a - r}{r}\right)^{1/2} dr \qquad (15.24)$$

Integration of Eq. (15.24) with the convenient substitution $r = \delta a \sin^2 \omega$ yields for *plane strain*

$$G = \frac{(1 - v^2)\pi a \sigma^2}{E} = (1 - v^2)\frac{K_{\mathrm{I}}^2}{E} \qquad (15.25)$$

where $K_{\mathrm{I}} = \sigma\sqrt{\pi a}$ is the stress intensity factor for a Mode I opening (Fig. 15.1, Table 15.2).

Critical Value of Crack Extension Force

As noted in Sec. 15.1, under certain conditions of loading, a crack in a structural member may gradually increase in length as the load is increased. This period of gradual increase in crack length may be followed by a rapid (catastrophic) propagation of the crack, resulting in complete separation of two parts of the member. In certain fracture mechanics hypotheses, this rapid propagation of the crack is associated with a *critical crack length* a_c. Alternatively, since G, the Griffith crack extension force, is related to the crack length a [Eq. (15.25)], the rapid propagation of the crack (say for Mode I) may also be associated with G_{IC}, a *critical crack extension force*, defined by

$$G_{\mathrm{IC}} = \frac{(1 - v^2)\pi a_c \sigma^2}{E} = (1 - v^2)\frac{K_{\mathrm{IC}}^2}{E} \qquad (15.26)$$

where analogous to G_{IC}, the factor

$$K_{\mathrm{IC}} = \sigma\sqrt{\pi a_c} \qquad (15.27)$$

is called the *critical stress intensity factor* for Mode I opening of the crack (Sec. 15.1) . The factor K_{IC} is also referred to as the *fracture toughness* (Knott,

1973). Typical values of K_{IC} for several metals are listed in Table 15.1. These values have been obtained for elastic (plane strain) conditions according to ASTM standards and can be used in the design of the members shown in Table 15.2. If the fracture loads for these members are to be calculated with reasonable accuracy, it is necessary that the state of stress at the tip of the crack is plane strain over most of the length of the crack tip. This is insured (Knott, 1973) by specifying that the crack half-length a (or crack length a when applicable) and thickness t of the cracked member satisfy the relation [see also Sec. 15.1, Eq. (15.2)]

$$a, t \geq 2.5 \left(\frac{K_{IC}}{Y} \right)^2 \tag{15.28}$$

The magnitude of the right side of Eq. (15.28) for each metal is listed (in millimeters) in Table 15.1. If the crack half-length a (or crack length a when applicable) is appreciably less than the value indicated by Eq. (15.28), the computed fracture load may be greater than the failure load for another mode of failure (yielding failure, for instance). If the thickness t of the member is small compared to the value given by the right side of Eq. (15.28), the state of stress approaches plane stress, appreciable yielding may occur at the crack tip, and the actual fracture load may be as much as several hundred percent greater than the value calculated using K_{IC}.

EXAMPLE 15.2
Brittle Fracture for Combined Tension and Bending

A hook similar to that shown in Fig. E15.2 is part of a scarifier used to dig up old road beds before replacing them. Let the tool be made of AISI 4340 steel and heat-treated to the properties indicated in Table 15.1. The dimensions of the tool are $d = 250$ mm, $2c = 60$ mm, and the width of the rectangular cross section is $t = 25$ mm. Determine the magnitude of the fracture load P for a crack length of **(a)** $a = 5$ mm and **(b)** $a = 10$ mm.

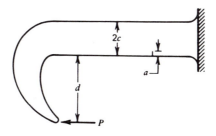

Figure E15.2

SOLUTION

Note that both the width t and crack length a satisfy Eq. (15.28); therefore, the hook is assumed to be loaded in the brittle state. At a section through the crack, the hook is subjected to combined axial load (Case 4 of Table 15.2) and bending

(Case 6 of Table 15.2). Since a linear elastic analysis is assumed, the state of stress for combined loading can be obtained by superposition of the two states of stress for the two types of loading. Thus,

$$\sigma f(\lambda) = \sigma_1 f_1(\lambda) + \sigma_2 f_2(\lambda)$$

(a)   When $a = 5$ mm, $\lambda = a/2c = 0.083$. From Table 15.2 for Cases 4 and 6, we obtain $f_1(\lambda) = 1.22$ and $f_2(\lambda) = 1.02$.

$$\sigma f(\lambda) = \frac{P}{25(60)}1.22 + \frac{3(280P)}{2(25)(30)^2}1.02 = 0.0195P$$

This is substituted in the relation

$$K_{\text{IC}} = \sigma f(\lambda)\sqrt{\pi a}$$

$$P = \frac{59\sqrt{1000}}{0.0195\sqrt{5\pi}} = 24{,}100 \text{ N} = 24.1 \text{ kN}$$

(b)   When $a = 10$ mm, $\lambda = 0.167$

$$\sigma f(\lambda) = \frac{P}{25(60)}1.33 + \frac{3(280P)(1.05)}{2(25)(30)^2} = 0.0205P$$

$$P = \frac{59\sqrt{1000}}{0.0205\sqrt{10\pi}} = 16{,}200 \text{ N} = 16.2 \text{ kN}$$

15.4

FRACTURE: OTHER FACTORS

In Sec. 15.1 through 15.3, concepts related to the stationary crack and initiation of crack propagation in metallic materials were introduced for brittle fracture [linear elastic fracture mechanics (LEFM)]. As noted, however, many other factors that influence fracture are being researched today. Many of the factors pertain to secondary effects. Indeed, as noted by Broek (1988), "Too many 'refinements' in engineering (fracture) solutions pertain to secondary errors; they increase the complexity, but do not improve the solution." However, some of these factors are of primary importance. For example, the topics of elastic-plastic fracture mechanics (EPFM; see Parton and Morozov, 1989), crack-growth analysis (fatigue, variable amplitude loading, etc.), load spectra and stress histories (statistical models; see Hermann and Roux, 1990), testing and experimental data interpretation (Broek, 1988; Curbishley, 1988; Ewalds and Wanhill, 1986; Herrmann and Roux, 1990), environmental effects (Liebowitz, 1968, Vol. III, 1971; Parton and Morozov, 1989), surface flaws, and residual stresses play important roles in fracture. The ultimate uses of the results of fracture mechanics studies are to control fracture and assess damage tolerances (Broek, 1988; Kachanov, 1986), and to develop fracture control plans that help to minimize the potential for brittle fracture (Barsom and Rolfe, 1987).

Elastic-Plastic Fracture Mechanics

In previous sections of this chapter, we examined conditions in which a crack may propagate. However, cracks or flaws differ in nature. At one extreme, the fracture of a crystal grain initiates from a submicroscopic crack produced when two atomic layers move apart. At the other extreme, the fracture of a pressure vessel may originate from a crack whose length and width are measured in millimeters (macroscopic crack and macroscopic fracture). Macroscopic fracture is due to macroscopic cracks whose dimensions are several orders of magnitude larger than the largest structural constituent of the material. Consequently, the problem can be treated as a problem in continuum mechanics. Thus, if we assume that the material is continuous, homogeneous, isotropic, and linearly elastic, we can use either the theory of linear elasticity or LEFM to compute the stresses in the neighborhood of a crack (Sec. 14.3 and 15.2). However, the theory of linear elasticity leads to the conclusion that the stresses for a sharp crack grow without bound at the crack tip. Thus, for a real, brittle material, fracture will occur almost instantaneously at a critical stress. For a real, ductile material, plastic deformation will usually occur before fracture, and then, LEFM is not applicable. In this case, elastic-plastic fracture mechanics (EPFM) must be used (see Parton and Morozov, 1989; Liebowitz, 1968, Vol. III, 1971, Chapter 2; Broek, 1988, Chapter 4; Clausmeyer et. al, 1991; Hwang et. al, 1990). The topic of EPFM is beyond the scope of this book.

Crack-Growth Analysis

The initiation of crack growth (crack propagation) is predicted by LEFM at a load for which the stress intensity factor K_I reaches a critical value (K_{IC} for the crack opening mode, see Sec. 15.3; see also Broek, 1988; Liebowitz, 1968, Vol. III, 1971, Chapter 1; Parton and Morozov, 1989, Chapters 1 and 2). However, crack initiation and growth are strongly affected by several factors such as repeated loads (fatigue, see Chapter 16; also Ewald and Wanhill, 1986), high temperatures (creep, Chapter 17), plasticity, environmental conditions (see Liebowitz, 1968, Vol. III, Broek 1988, Chapters 6 and 7).

Load Spectra and Stress History

As employed in Sec. 15.1 through 15.3, LEFM predicts a critical value of load that initiates crack propagation. More generally, members of structures are subjected to load spectra and subsequent stress histories. The word *spectra* means any statistical representation of loads or stresses (Herrmann and Roux, 1990). In the determination of damage due to crack propagation, one must establish reliable prediction methods to estimate the number of load cycles that will result in a maximum permissible crack size for a given stress history (Broek, 1988, Chapter 6). The type of stress histories (constant amplitude, repeated loads, variable amplitude loads, impulse loads, etc.) and methods of obtaining load spectra [say, records obtained by strain-gage readings over long periods of time, power spectrum density analysis, exceedance diagrams* from counts (peak, mean-cross peak, etc.) of

---

* Exceedance diagrams give the number of times that a quantity, say, stress, exceeds a certain level during a given time interval. Exceedance diagrams may be for one year of operation or any other appropriate time, depending on the application. The exceedance diagram may also be for a range of stress (or load, etc.).

the records] as well as other factors enter into the design problem of fracture control and the establishment of damage tolerance criteria. To simplify the process somewhat, many industries have established standard spectra for general use for a variety of structures (Broek, 1988).

Testing and Experimental Data Interpretation

As noted in Sec. 15.1, the experimentally determined critical stress intensity K_C depends on the specimen thickness. For a very thin specimen (<1 mm thick), the value of K_C is not exactly certain. For a relatively thin specimen, a state of plane stress exists approximately, and the value of K_C determined experimentally is fairly large. As the thickness of the specimen is increased, there is a transition from a plane stress state to a plane strain state (Fig. 15.8). For a given test temperature and loading rate, the value of K_C tends to a limiting (minimum) value. This minimum value of K_C is called the *plane strain fracture toughness* (e.g., K_{IC} for the opening crack mode). The plane strain fracture toughness is considered to be a material property; however, it is dependent on temperature (Fig. 15.3) and loading rate. After several years of testing (in the 1960–1970 period), a standardized plane strain K_{IC} test method was developed by the American Society for Testing Materials (ASTM) for metallic materials. Two standard test specimens, the single-edge cracked bend specimen and compact tension specimen, were proposed (Fig. 15.2). The method was first published in 1970, and it is described in ASTM standard E399 (ASTM, 1984). However, like most standards, this standard employs a number of compromises and approximations of data. The interpretation and use of plane strain fracture toughness and problems associated with the use of toughness data are discussed by Broek (1988, Chapter 7). The user of toughness data would be well advised to refer to Broek's discussion (see also Barsom and Rolfe, 1987; Curbishley, 1988, Chapter 6; Ewalds and Wanhill, 1986, Chapter 5; Herrmann and Roux, 1990).

In summary, fracture toughness testing and the use of fracture toughness data are still in the developmental stage. Much work still remains before structural integrity and damage tolerance assessments will be fully understood. Because of the limited space allotted here to these topics, our treatment is brief and incomplete. Indeed, the effects of environmental factors, residual stresses, and many current research topics are omitted. For example, fracture mechanics of ceramics, concrete, rock, and masonry materials are not discussed (Mihashi et al., 1989). Nevertheless,

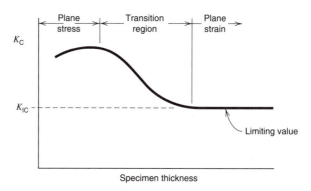

Figure 15.8   Variation of K_C with specimen thickness.

the interested reader may refer to the excellent references cited. In addition, much of the current research is documented in technical journals devoted exclusively to fracture mechanics. These include *Engineering Fracture Mechanics, International Journal of Fracture,* and *Theoretical and Applied Fracture Mechanics.* In addition, other journals contain articles on fracture mechanics, for example, *Computer Methods in Applied Mechanics and Engineering, Experimental Mechanics, International Journal of Fatigue,* and *International Journal of Plasticity.*

PROBLEMS
Section 15.1

15.1. A circular shaft is made of gray cast iron, which may be considered to be linearly elastic up to its ultimate strength $\sigma_u = 145$ MPa. The shaft is subjected to a moment $M = 5.50$ kN·m and torque $T = 5.00$ kN·m. Determine the diameter d of the shaft if the factor of safety against brittle fracture is $SF = 4.00$.

15.2. A piece of chalk of diameter d is subjected to an axial load P and a torque T. Assume that the chalk remains linearly elastic up to the ultimate strength σ_u. The axial load is $P = \sigma_u \pi d^2/12$. Determine the magnitude of the torque T that will cause brittle fracture.

Ans. $T = \pi d^3 \sigma_u/8\sqrt{6}$

15.3. A 50-mm diameter shaft is made of a brittle material. The shaft is subjected to a static torque $T = 1.20$ kN·m. A bending moment M is increased in magnitude until fracture. The fracture surface is found to make an angle of 1.000 rad with a longitudinal line drawn on the shaft. If the maximum principal stress criterion of failure is valid for this material and loading, determine the magnitude of M and ultimate strength σ_u for the material.

15.4. A circular shaft is made of gray cast iron, which may be considered to be linearly elastic up to its ultimate strength $\sigma_u = 150$ MPa. The shaft has a diameter of 125 mm and is subjected to a bending moment $M = 7.50$ kN·m. Determine the maximum value for torque T that can be applied to the shaft if it has been designed with a factor of safety of 3.00 for both M and T.

Ans. $T = 8.98$ kN·m

15.5. Let the bending moment for Problem 15.4 be applied by a dead load that is known to remain constant with time. The variation of torque T with time is unknown. Determine the limiting value for T if the factor of safety for M remains 3.00, whereas the factor of safety for T is increased to 5.00.

15.6. A long strip of aluminum alloy 2024-T851 has a width of 150 mm and thickness of 8.00 mm. An edge crack of length $a = 9.00$ mm (Case 4 of Table 15.2) is located at one edge of the strip near the center of its length. Determine the magnitude of the axial load P that will cause brittle fracture.

Ans. $P = 137.3$ kN

15.7. The long strip in Problem 15.6 has a double-edge crack with $a = 9.00$ mm (Case 5 in Table 15.2). Determine the magnitude of the axial load P that will cause brittle fracture.

15.8. A 2024-T851 aluminum alloy pipe is used as a tension member. The pipe has an outside diameter of 100 mm and wall thickness of 8.00 mm. An inspection of the pipe locates a circumferential through-thickness crack having a length of 15.0 mm (Case 3 of Table 15.2). If the axial load P is increased to failure, will the failure be brittle fracture? What is the failure load?

Ans.   Yes. $P = 342.7$ kN

15.9. Let the tension member in Problem 15.8 be made of AISI 403 stainless steel. If the axial load P is increased to failure, will the failure be brittle fracture? What is the lower limit for the failure load?

15.10. A 60.0-mm square beam is made of AISI 4340 steel that has been heat-treated to give the properties indicated in Table 15.1. On the tension side of the beam, a transverse crack has a depth of 8.00 mm (Case 6 of Table 15.2). Determine the magnitude of the moment M that will cause brittle fracture.

Ans. $M = 12.97$ kN·m

15.11. A simple beam has a span of 4.00 m, depth of 250 mm, and width of 100 mm. The beam is made of 6061-T651 aluminum alloy and is loaded by a concentrated load P at midspan. The design load for the beam has been calculated using a factor of safety of 3.00 and assuming the general yielding theory of failure. Determine the magnitude of P. An inspection of the beam located a transverse crack a distance of 1.50 m from one end. The crack has a depth of 24.0 mm. What is the factor of safety for the beam against brittle fracture?

Section 15.3

15.12. Solve Example 15.2 for the condition that $d = 200$ mm and $2c = 50$ mm.

15.13. A rectangular section beam has a depth $2c = 150$ mm, width $t = 25$ mm, and length $L = 2.00$ m. The beam is loaded as a simply supported beam with a concentrated load P at the center. A notch is machined into the beam on the tension side opposite the point of application of P. The depth of the notch was increased by fatigue loading until $a = 15$ mm. The beam is made of 17-7PH precipitation hardening steel heat-treated to yield the properties indicated in Table 15.1.

(a) Determine whether or not plane strain conditions are satisfied for the beam.

(b) Determine the fracture load P.

Ans.   **(a)** Conditions are satisfied   **(b)** $P = 65.2$ kN

15.14. Solve Problem 15.13 if the beam is made of 2024-T851 aluminum alloy.

15.15. A closed-end cylinder is made of 7079-T651 aluminum alloy. The cylinder has an inside diameter $D = 1000$ mm, a wall thickness $h = 20$ mm, and is subjected to an internal pressure $p = 6.00$ MPa. Determine the length of crack ($2a$) required to cause fracture at this pressure if plane strain conditions are assumed to be satisfied. The inside of the cylinder is covered with a thin layer of rubber to prevent leakage. Determine whether or not conditions are satisfied for a plane strain state of stress.

Ans. $2a = 19.1$ mm, yes

15.16. If the crack with length $2a = 19.1$ mm is circumferential instead of longitudinal for the cylinder in Problem 15.15, determine the internal pressure that will cause fracture.

15.17. A bar of titanium alloy (Ti-6Al-4Sn-1V) has a rectangular cross section ($t = 30$ mm and $2c = 300$ mm), a length $L = 1$ m, and is heat-treated to give the material properties in Table 15.1. The bar is subjected to an axial load P whose action line is at the centerline of one of the 30-mm edges. If a transverse edge crack at the center of length L has the minimum length required for a plane strain state of stress, determine the magnitude of P to cause brittle fracture.

Ans. $P = 656$ kN

15.18. Solve Problem 15.17 if the action line of load P is at the center of the tip of the crack.

15.19. In a plane strain compact specimen test (Fig. 15.2), a value of fracture toughness $K_C = 55$ MPa$\sqrt{m}$ is estimated. The yield strength of the material is $Y = 689$ MPa, and the specimen thickness is 12.7 mm.

(a) Is the value $K_C = 55$ MPa$\sqrt{m}$ a valid value for K_{IC}?

(b) What is the maximum value of fracture toughness that can be measured with this specimen?

(c) If the fracture toughness determined in the test is not valid, estimate the plane strain fracture toughness with an accuracy of 6% or less.

(d) What is the required thickness of the specimen to measure plane strain fracture toughness for all conditions?

REFERENCES

American Society of Testing Materials (ASTM) (1984). Standard Test Method for Plane-Strain Fracture Toughness of Metallic Materials. *In Annual Book of ASTM Standards*, Vol. 03.01, ASTM Std. E399-83. Philadelphia, Pa.: pp. 519–554.

Barsom, J. M. and Rolfe, S. T. (1987). *Fracture and Fatigue Control in Structures*, 2nd ed. Englewood Cliffs, N. J.: Prentice-Hall.

Broek, D. (1985). *Elementary Engineering Fracture Mechanics*, 4th ed. London: Martinus Nijhoff.

Broek, D. (1988). *The Practical Use of Fracture Mechanics*. London: Kluwer Academic Pub.

Clarke, W. L. and Gordon, G. M. (1973). Investigation of Stress Corrosion Cracking Susceptibility of Fe-Ni-Cr Alloys in Nuclear Reactor Water Environments. *Corrosion* **29**, Part 1: 1–12.

Clausmeyer, H., Kussmaul, K., and Roos, E. (1991). Influence of Stress State on the Failure Behavior of Cracked Components Made of Steel. *Appl. Mech. Rev.*, **44**, (2): 77–92.

Curbishley, I. (1988). *Mechanical Testing: Characterization of High-Temperature Materials*, **3**. London: Inst. of Metals.

Ewalds, H. L. and Wanhill, R. J. H. (1986). *Fracture Mechanics*. London: Edward Arnold.

Freed, C. N. (1968). A Comparison of Fracture Toughness Parameters for Titanium Alloys. *Eng. Fracture Mech*, **1** (1): 175–190.

Greenberg, H. D., Wessel, E. T., and Pryle, W. H. (1970). Fracture Toughness of Turbine Generator Rotor Forgings. *Eng. Fracture Mech*, **1** (4): 653–674.

Herrmann, H. J. and Roux, S. (1990). *Statistical Models for the Fracture of Disordered Media*. New York: Elsevier.

Hwang, K. C., Yu, S. W., and Yang, W. (1990). Theoretical Study of Crack-Tip Singularity Fields in China, *App. Mech. Rev.*, **43**, (3).

Inglis, C. E. (1913). Stresses in the Plate due to the Presence of Cracks and Sharp Corners. *Trans. Inst. Naval Arch.*, **60**. 219.

Irwin, G. R. (1957). Analyses of Stresses and Strains Near the End of a Crack Traversing a Plate. *J. Appl. Mech.*, **24**: 361–364.

Kachanov, L. M. (1986). *Introduction to Continuum Damage Mechanics*. Hingham, Mass.: Kluwer Academic Publ.

Kanninen, M. F. and Popelar, C. H. (1984). *Advanced Fracture Mechanics*, Oxford Engineering Science Series, Vol. 15. London: Oxford Univ. Press.

Knott, J. F. (1973). *Fundamentals of Fracture Mechanics*. New York: Wiley.

Liebowitz, H. (1968). *Fracture: An Advanced Treatise*, Vol. II. New York: Academic Press, Chapter 4. (See also Vol. I, 1968, and Vols. III through VII, 1971–1972.)

Logsdon, W. A. (1975). An Evaluation of the Crack Growth and Fracture Properties of AISI 403 Modified 12 Cr. Stainless Steel. *Eng. Fracture Mech.*, **7** (1): 23–40.

Mihashi, H., Takahashi, H., and Wittmann, F. (1989). *Fracture Toughness and Fracture Energy: Test Methods for Concrete and Rock*. Brookfield, Vermont: A. A. Balkema Publ.

Nelson, F. G., Schilling, P. E., and Faufman. J. G. (1972). The Effect of Specimen Size on the Results of Plane Strain Fracture Toughness Tests. *Eng. Fracture Mech.* **4** (1): 33–50.

Paris, P. G. and Sih, G. C. (1965). Stress Analysis of Cracks. In *Fracture Toughness Testing and Its Applications*, STP 381. Philadelphia, Pa.: ASTM, p. 30.

Parton, V. Z. and Morozov, E. M. (1989). *Mechanics of Elastic-Plastic Fracture*, 2nd ed. (revised). New York: Hemisphere Publ.

Paul, B. (1968). Generalized Pyramidal Fracture and Yield Criteria. *Internat. J. Solids and Structures*, **4**: 175–196.

Smith, A. I. and Nicolson, A. M. (1971). *Advances in Creep Design*. New York: Halsted Press Div., Wiley.

Srawley, J. E. and Brown, W. F., Jr. (1965). Fracture Toughness Methods. In *Fracture Toughness Testing and Its Applications*, STP 381. Philadelphia, Pa.: ASTM, p. 133.

Steigerwald, E. A. (1969). Plane Strain Fracture Toughness of High Strength Materials. *Eng. Fracture Mech.*, **1** (3): 473–494.

Wessel, E. T. (1969). Practical Fracture Mechanics for Structural Steel. Paper H, London: UKAEA/Chapman and Hall.

16

FATIGUE: PROGRESSIVE FRACTURE

Fatigue has been defined as "the progressive localized permanent structural change that occurs in a material subjected to repeated or fluctuating strains at stresses having a maximum value less than the tensile strength of the material" (ASM, 1975). As noted in Chapter 15, failures occur in many mechanical systems. It has been estimated that between 50 and 90% of these failures are due to fatigue (Fuchs and Stephens, 1980). Failures due to fatigue culminate in cracks or fracture after a sufficient number of fluctuations of load.

Fracture of a structural member due to repeated cycles of load or fluctuating loads is commonly referred to as a *fatigue failure* or *fatigue fracture*. The corresponding number of load cycles or the time during which the member is subjected to these loads before fracture occurs is referred to as the *fatigue life* of the member. The fatigue life of a member is affected by many factors (ASM, 1975). For example, it is affected by (1) the type of load (uniaxial, bending, torsion), (2) the nature of the load-displacement curve (linear, nonlinear), (3) the frequency of load repetitions or cycling, (4) the load history [cyclic load with constant or variable amplitude, random load, etc. (Gauthier and Petrequin, 1989; Buxbaum et. al, 1991)], (5) the size of the member, (6) the material flaws, (7) the manufacturing method (surface roughness, notches), (8) the operating temperatures (high temperature that results in creep, low temperature that results in brittleness), (9) the environmental operating conditions (corrosion, see Clarke and Gordon, 1973). In practice, accurate estimates of fatigue life are difficult to obtain, because for many materials, small changes in these conditions may strongly affect fatigue life. The designer may therefore be forced to rely on experience or testing of full-scale members under in-service conditions. Testing of full-scale members is time-consuming and costly. Therefore, data from laboratory tests of small material specimens are used to establish fatigue failure criteria, even though these data may not be sufficient to determine the fatigue life of the real member. Nevertheless, laboratory tests are useful in determining the effect of load variables on fatigue life, in comparing the relative fatigue resistance of various materials and establishing the importance of fabrication methods, surface finish, environmental effects, etc., on fatigue life predictions (Buch, 1988; Fuchs and Stephens, 1980).

The total period of fatigue life (total life) may be considered to consist of three phases: (1) initial fatigue damage that produces crack initiation, (2) propagation of a crack or cracks that results in partial separation of a cross section of a member, until the remaining uncracked cross section is unable to support the applied load, and (3) final fracture of the member. Traditionally, fatigue life data have been expressed as the number of stress cycles required to initiate a fatigue crack that will grow large enough to produce fracture (e.g., breaking of a test specimen into two pieces). Alternatively, fatigue data may also be expressed in terms of crack-growth rate (Fuchs and Stephens, 1980, p. 82). Early investigators of fatigue life assumed that total fatigue life consisted mainly of the time required to initiate a minute fatigue crack and that the time required for the crack to grow (propagate) was an insignificant portion of the total life. However, with the development of more accurate methods of crack detection and tracking (Skelton, 1988), it was discovered that microscopic cracks develop very early in the fatigue life and grow at various rates until fracture occurs. This fact has led to the use of crack initiation and crack-growth rates to more accurately predict fatigue life (Fuchs and Stephens, 1980; Knott, 1973). In this chapter, we estimate fatigue life on the basis of experimental fatigue data (stress-cycle data) of test specimens subject to appropriate conditions, including the effects of stress concentrations. The reader interested in analytical methods of fatigue life predictions is referred to the literature (Buch et. al, 1986; Buch, 1988; Kliman, 1985; Socie, 1977; Fuchs and Stephens, 1980; Weronski and Hejwowski, 1991; see also current journals, e.g., the *International Journal of Fatigue*).

Ductile Fracture Resulting from Low Cycle Fatigue Loading

Many current studies of fatigue are devoted to problems of low cycle fatigue of members made of ductile materials. For such problems, large plastic strains occur at the section of the member where fracture finally occurs. Consequently, we consider the material in members that undergo low cycle fatigue to be in a ductile state. Failure resulting from low cycle fatigue is beyond the scope of this book, and the reader is referred to the literature (Manson, 1981; Sandor, 1972). Fatigue failures may occur with only small plastic strains. Such failures are called brittle failures due to high cycle fatigue. For members made of ductile metals, high cycle fatigue failure occurs after about 10^6 cycles.

16.1

PROGRESSIVE FRACTURE (HIGH CYCLE FATIGUE FOR NUMBER OF CYCLES $N > 10^6$)

A basic concept in fracture predictions by fracture mechanics analysis is the existence of a critical crack size for a given geometry and load. In some practical applications, the size of the critical crack or defect is so large that the effect can usually be detected and corrected before the part is put into service or during maintenance of the part in service. However, most parts contain subcritical cracks or flaws. These subcritical cracks may, during operation, grow to critical size and cause catastrophic failures. Several mechanisms of subcritical flaw growth exist. Of particular importance in practical problems are the mechanisms of fatigue and stress corrosion cracking. Here we briefly consider fatigue criteria associated with subcritical flaw (crack) growth by the mechanism of fatigue. The mechanism of stress corrosion cracking is left to more specialized works (Clarke and Gordon, 1973; Fuchs and Stephens, 1980; Chapter 11). However, one should note that fatigue crack growth processes cannot be fully explained unless effects of environment (corrosion) are considered.

Encouraged by the success of linear elastic fracture mechanics (LEFM) in explaining sudden brittle fracture, several investigators have attempted to describe subcritical crack growth in terms of LEFM parameters. The objective of the fracture mechanics approach is essentially to replace uncertainties (i.e., the degree of ignorance) in conventional design factors by more reliable quantitative parameters that are more direct measures of the material fracture resistance. Early results seem to indicate that a material exhibiting high static fracture toughness also gives good resistance to subcritical crack propagation due to fatigue.

By fatigue failure (progressive fracture), we mean one that occurs after a number of cycles under alternating stresses, with peak stresses below the ultimate strength of the material in a simple tension test. We restrict our discussion to ordinary (room) temperatures. Fatigue fracture at high temperature (thermal fatigue) has been treated in the literature (Smith and Nicolson, 1971; Manson, 1981). To simplify our discussion, we further divide our treatment of the fatigue growth of subcritical cracks into the initiation of cracks as microcracks and the propagation of cracks as macrocracks to fracture.

For example, consider a smooth shaft rotating in bearings and subjected to loads that produce bending moments. As the shaft rotates, the maximum fiber stresses alternate between tension and compression. In turn, these cyclic components at a surface point set up alternating shear stresses, maximum on 45° planes with the tension-compression direction. If these stresses locally exceed the elastic limit, alternating plastic deformation (strain) is produced in the surface grains. Since the plastic deformation is not fully reversible, at least two effects result: (1) a general strain hardening of the surface grains that localizes the deformation along active slip-bands inclined roughly at 45° to the direction of the maximum principal stress and (2) a nonreversible flow at the surface producing *extrusions* that pile up material on the surface and associated *intrusions* that act as microcracks along the active slip-bands. An intrusion initially propagates along an active slip-band as a so-called *stage I* crack until it reaches a length sufficiently large with respect to the member for the crack tip stress field to become dominant. Under continued repeated loading, the intrusion then propagates as a *stage II* crack, normal to the maximum principal (tensile) stress until the member breaks by a fast tensile fracture. During stage II propagation, *striations* or *ripples* occur on portions of the fatigue crack surface perpendicular to the tensile direction. The growth of the crack from intrusion to the stage II propagation is a rapidly accelerating process. Hence, the process is strongly controlled by the initiation of the intrusion. Fairly large amounts of alternating plastic deformations are required to form intrusions and extrusions on an initially smooth surface. Consequently, rather large alternating stresses are needed to precipitate fatigue fracture. It follows that once a crack has been initiated in any initially smooth surface, it propagates rapidly due to the high stress concentration.

Conventional fatigue (endurance) testing has been concerned primarily with the testing of specimens with smooth surfaces under conditions of rotating-bending or uniaxial tension-compression cycling. The results of these tests are presented in the form of plots of stress (applied alternating stress magnitude $\pm \sigma$) *vs* the number N of stress cycles (usually represented as $\log N$) required to cause fracture. These plots are called σ-N diagrams (also called *S*-*N* diagrams in the literature) (Fig. 16.1). Wöhler (Anonymous, 1967) discovered that the steel in the railroad car axles he tested exhibited a behavior called an *endurance limit*: a stress level below which a material can undergo repeated cycling of stress indefinitely and show no evidence of fracture. However, later investigators found that many materials did not exhibit

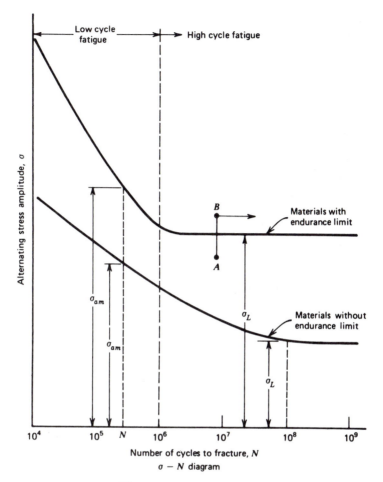

Figure 16.1 $\sigma - N$ diagram.

the endurance limit response, but rather continued to exhibit fracture, provided that the repetition of load was continued for a sufficiently large number of cycles (Fig. 16.1). Thus in general, under fatigue testing of smooth specimens, materials exhibit one of two types of responses. In mild steel or certain other steels, an endurance limit is observed, below which the specimen seems to last indefinitely. On the other hand, many materials do not exhibit a clear-cut endurance limit, but the σ-N curves continue downward as N increases. For these materials (e.g., most nonferrous materials), it is customary to *define* the stress to cause failure in a given number of cycles (say, $N = 10^8$) as the endurance limit stress σ_L (Fig. 16.1).

The endurance limit σ_L is an important material property for members subjected to fatigue loading as long as the number of cycles of loading approaches the number associated with σ_L. It should be noted that other fatigue properties for a given material can be obtained from the σ-N curve. Many members are subjected to fewer cycles than are associated with the endurance limit. For each value of N in Fig. 16.1, there is a stress σ_{am}, the fatigue strength, where the subscripts *am* denote alternating maximum; a specimen subjected to completely reversed cycles of stress at σ_{am} will fracture after N cycles. Note that $\sigma_{am} = \sigma_L$ at the endurance limit.

Typical σ-N curves for completely reversed loading of smooth specimens of a structural steel, a stainless steel, and aluminum alloy are shown in Fig. 16.2. If a large number of specimens of one of the metals in Fig. 16.2 were tested at one stress level, the data would indicate appreciable scatter. The σ-N curve usually reported for a given metal (Fig. 16.2) is often taken to represent a 50% probability of failure curve. That is, if a large number of fatigue specimens of one of the metals in Fig. 16.2 were tested at a given fatigue strength σ_{am}, approximately 50% of the specimens would be expected to fail prior to N cycles of load corresponding to the given σ_{am}. The statistical nature (Fuchs and Stephens, 1980) of fatigue data may be represented either as a series of σ-N curves representing different probabilities of failure or a σ-N band (Fig. 16.3). Because of the large expense involved, σ-N probability curves or σ-N bands (Fig. 16.3) are seldom obtained.

The experimental σ-N curves in Figs. 16.1 and 16.2 remain fairly valid for constant amplitude ($\pm\sigma_{am}$) tests. However, deviations from constant amplitude alternating stress may alter the σ-N curve. For example, if a steel is subjected to cyclic stress of constant amplitude for a sufficiently long time below the endurance limit (point A in Fig. 16.1), its endurance limit may be increased (point B). This process, known as *coaxing*, is sometimes employed to improve resistance to fatigue fracture.

In addition to coaxing, various other factors affect the fatigue strength. For example, the fatigue strength of a material may be altered by such factors as frequency of cycling, cold working of the material, temperature, corrosion, residual stresses, surface finish, and mean stress.

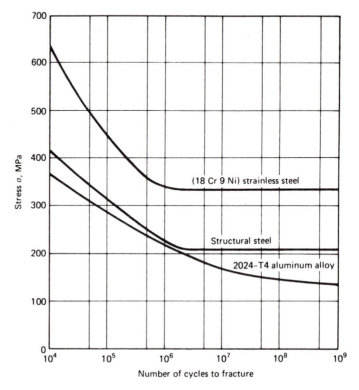

Figure 16.2 $\sigma - N$ diagrams for three metals.

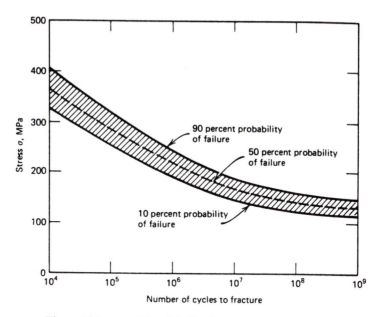

Figure 16.3 $\sigma - N$ band indicating scatter of fatigue data.

As noted above, the σ-N curve gives the fatigue strength σ_{am} for specified N for members subjected to completely reversed loading (loading under the condition of zero mean stress). Nonzero mean stresses have a marked effect on the fatigue strength. There have been several relations proposed to describe the effects of mean stress. Three such relations are (for one-dimensional testing)

(a) Soderberg Relation

$$\frac{\sigma_a}{\sigma_{am}} + \frac{\sigma_m}{Y} = 1 \tag{16.1}$$

(b) Gerber Relation

$$\frac{\sigma_a}{\sigma_{am}} + \left(\frac{\sigma_m}{\sigma_u}\right)^2 = 1 \tag{16.2}$$

(c) Goodman Relation

$$\frac{\sigma_a}{\sigma_{am}} + \frac{\sigma_m}{\sigma_u} = 1 \tag{16.3}$$

where σ_a is the stress amplitude, σ_{am} the fatigue strength for given N for zero mean stress, Y the yield stress, σ_m the mean stress, and σ_u the ultimate strength. The relation between σ_a and σ_m for cyclic loading with unequal stresses is indicated in Fig. 16.4. For most metals, the Soderberg relation yields conservative estimates of critical stress amplitude σ_a (or range of stress $2\sigma_a$). The Goodman relation gives reasonably good results for brittle materials, whereas it is conservative for ductile materials. The Gerber relation yields fairly good estimates for σ_a for ductile materials. The Soderberg relation, Gerber relation, and Goodman relation are interpreted in Fig. 16.5. For any mean stress σ_m, the ordinate to a particular curve

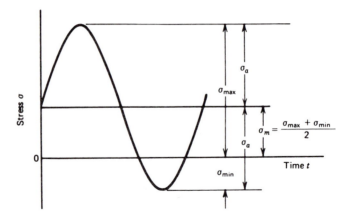

Figure 16.4

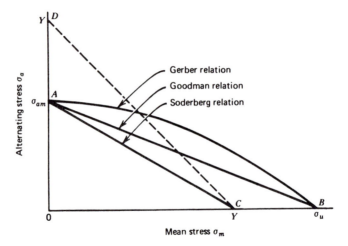

Figure 16.5

gives the magnitude of σ_a for that relation. The dashed line CD is generally used along with the Gerber and Goodman relations since failure by general yielding is assumed to occur along CD.

In the case of fatigue loading, design applications in this book are limited to high cycle fatigue (Fig. 16.1) of members made of ductile metals and subjected to cyclic loading with constant mean stress and constant amplitude of alternating stress. The material property (fatigue strength) is assumed to be obtained from smooth specimens (free of stress concentrations) subject to completely reversed loading under a uniaxial state of stress (tension-compression specimens or rotating bending specimens). The fatigue strength for a specified number of cycles N, where $N > 10^6$, is specified by the magnitude of σ_{am}. The effect of mean stress σ_m is assumed to be given by either the Soderberg relation [Eq. (16.1)], Gerber relation [Eq. (16.2)], or Goodman relation [Eq. (16.3)]. The criteria of failure for members subject to multiaxial states of stress are assumed to be the same as for general yielding failure. Both the maximum shear-stress criterion and maximum octahedral shear-stress criterion of failure are widely used in the design for high cycle fatigue.

In the case of low cycle fatigue, criteria of failure are often formulated in terms of the total strain range (Manson, 1981).

Stress Concentrations

Stress concentrations (Chapter 14) greatly increase the stresses in the neighborhood of the stress concentrations and generally limit design loads when the member is subjected to fatigue loading. The effect of stress concentrations in fatigue loading is discussed in Sec. 16.2 and 16.3.

EXAMPLE 16.1
Fatigue of Torsion-Bending Member

The member in Fig. E16.1 is made of steel ($Y = 345$ MPa and $\sigma_u = 586$ MPa), has a diameter $d = 20$ mm, lies in the plane of the paper, and has a radius of curvature $R = 800$ mm. The member is simply supported at A and B and is subjected to a cyclic load P at C normal to the plane of the member.

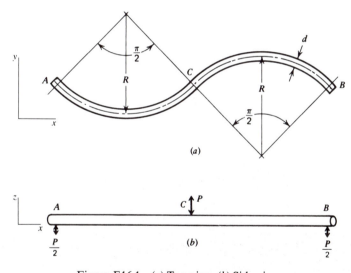

Figure E16.1 (*a*) Top view. (*b*) Side view.

(a) The load varies from P_{max} to $P_{min} = -5P_{max}/6$. The endurance limit for $N = 10^7$ for the steel is $\sigma_{am} = 290$ MPa. Determine the magnitude of P_{max} based on a factor of safety $SF = 1.80$ against failure at $N = 10^7$ cycles. Use the maximum octahedral shear-stress criterion of failure. Assume that the Gerber relation [Eq. (16.2)] is valid.

(b) Obtain the solution for $P_{min} = -P_{max}/2$.

SOLUTION

(a) The magnitude of the alternating component of stress σ_a is obtained by Eq. (16.2). For linearly elastic behavior, $\sigma_{min} = -5\sigma_{max}/6$. However,

$\sigma_{\max} - \sigma_{\min} = 2\sigma_a$ and $\sigma_{\max} = \sigma_m + \sigma_a$. Hence, $\sigma_m = \sigma_a/11$. Substituting this value of σ_m and values of σ_{am} and σ_u in Eq. (16.2), we obtain $\sigma_a = 289$ MPa. Thus, $\sigma_{\max} = \frac{12}{11}\sigma_a = 315$ MPa and $\sigma_{\min} = -263$ MPa. This result indicates that a smooth fatigue specimen cycled between these stress levels would not fracture before 10^7 cycles. Since $\sigma_{\max}$ is less than Y, failure would be by fatigue and not general yielding.

The load P on member ACB in Fig. E16.1 can be cycled from $P_{\max}(SF)$ to $P_{\min}(SF)$ through 10^7 cycles before fracture by fatigue. The reactions at A and B when $P_{\max}(SF)$ is applied are equal to $P_{\max}(SF)/2 = 0.90P_{\max}$. The reaction $0.90P_{\max}$ produces a moment and torque of equal magnitude at the critical section at C. Thus,

$$M = T = 0.90P_{\max}R = 720P_{\max}$$

The bending stress σ due to M and shear stress τ due to T at C are

$$\sigma = \frac{Mc}{I} = \frac{720P_{\max}(10)(4)}{\pi(10)^4} = 0.917P_{\max}$$

$$\tau = \frac{Tc}{J} = \frac{720P_{\max}(10)(2)}{\pi(10)^4} = 0.458P_{\max}$$

The magnitude of $P_{\max}$ is obtained by means of Eq. (4.44) when $\sigma_{\max} = 315$ MPa is substituted for Y. Thus,

$$3\left(\frac{0.458P_{\max}}{315}\right)^2 + \left(\frac{0.917P_{\max}}{315}\right)^2 = 1$$

$$P_{\max} = 260 \text{ N} \quad \text{and} \quad P_{\min} = -216 \text{ N}$$

(b) For $\sigma_{\min} = -\sigma_{\max}/2$, we obtain $\sigma_m = \sigma_a/3$. Substitution of this value of σ_m along with values for Y and σ_u into Eq. (16.2) gives $\sigma_a = 282$ MPa and $\sigma_{\max} = 376$ MPa. Since $\sigma_{\max}$ is greater than Y, failure of the member occurs by general yielding and not fatigue. Substitution of values of σ, τ, and Y into Eqs. (4.44) gives $P_{\max} = 285$ N and $P_{\min} = -142$ N.

16.2

EFFECTIVE STRESS CONCENTRATION FACTORS: REPEATED LOADS

The value of the stress concentration factor S_{ce} for a notched member subjected to completely reversed repeated loads (fatigue) is obtained by comparison of data taken from two sets of test specimens. One set of specimens (5 to 10 specimens) is notch-free, the other set notched. The significant stress in the notch-free specimens is the nominal stress as computed with an elementary stress formula. For the notched specimens, the nominal stress is again computed with the same elementary stress formula as for the notch-free specimens. Both sets of specimens are subjected to the same type of repeated load or fatigue test (say, bending).

It is assumed that the failure (fracture) in each set of specimens for a specified number of completely reversed loading cycles N occurs when the stress attains the same value in each set. Since the notch causes a stress concentration, the load required to cause the fracture stress is less for the notched specimens.

To illustrate the method of determining S_{ce} for bending fatigue loading, consider the σ-N diagrams of Fig. 16.6. The nominal stress is computed by the equation $\sigma_n = Mc/I$ and plotted as ordinates in Fig. 16.6; the abscissa is the number of cycles of bending stress to which the specimen is subjected. For a given value of N, say,

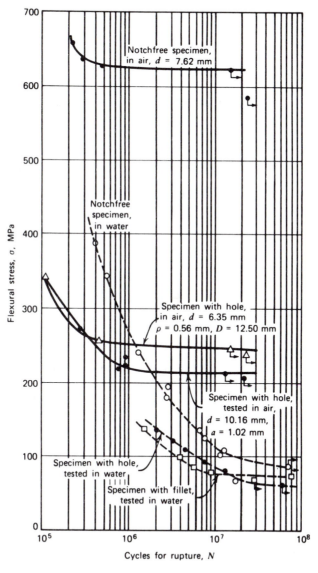

Figure 16.6 $\sigma - N$ diagrams showing the effect of abrupt changes in cross section and corrosion on the resistance of steel to repeated cycles of completely reversed bending stress. Quenched and tempered SAE 3140 steel was used for all tests. [From T. J. Dolan (1937). Urbana-Champaign, Ill.: Univ. Ill., Bull. 293, Eng. Exper. Station.]

300,000 cycles, the value of S_{ce} is computed by taking the ratio of the failure stress of the notch-free specimen to the failure stress of the notched specimen. Thus, for $N = 300,000$ cycles, by Fig. 16.6, we find

$$S_{ce} = \frac{634.4}{268.9} = 2.36$$

Likewise, for $N = 10^7$ cycles we find

$$S_{ce} = \frac{620.6}{248.2} = 2.50$$

Thus, the value of S_{ce} varies with N. By Fig. 16.6, the value of S_{ce} remains relatively constant for $N > 10^7$, since the curves at $N = 10^7$ are changing very slowly. In fatigue testing, the value of S_{ce} is often based on the endurance limit of the steel (stress at $N = 10^7$).

For the specimen with a transverse hole (Fig. 14.16), the calculated (elastic) stress concentration factor is $S_{cc} = 3.00$. However, as shown by the results of fatigue tests for these specimens, the significant stress concentration factor is $S_{ce} = 2.5$. Thus, there is a difference between the value of S_{ce} and S_{cc}, the value of S_{ce} being smaller. In the following section, additional conditions that influence the value of S_{ce} are discussed.

16.3

EFFECTIVE STRESS CONCENTRATION FACTORS: OTHER INFLUENCES

Corrosion Fatigue

Figure 16.6 shows the damaging effects that mechanical notches such as holes and fillets and so-called chemical notches such as corrosion pits are likely to have on the resistance of steel to repeated stress, particularly of alloy steels heat-treated to give high strength. The effect of corrosion that takes place while the material is being repeatedly stressed is much more damaging to the fatigue strength of steel than corrosion that takes place prior to stressing (called stressless corrosion). The main reason for this fact seems to be that in the absence of stress the products of the corrosion tend to form a protecting film that excludes the corroding agent from contacting the metal. If, however, the rather brittle film is repeatedly stressed in the presence of the corroding agent, it cracks and allows the corroding agent to continue to attack the metal underneath the film. The effect of corrosion on the fatigue strength of steel is shown by the σ-N diagrams in Fig. 16.6. For example, the quenched and tempered SAE 3140 steel tested indicated an endurance limit of approximately 620.5 MPa when tested in air (a relatively noncorrosive medium), and this was reduced to about 68.94 MPa when the specimens were tested in the presence of water; the presence of a small hole caused little further decrease in fatigue strength. Also the shape of the σ-N diagram for stresses above the endurance limit was influenced greatly by the corrosion.

Effect of Range of Stress

In Sec. 16.2, it was assumed that the member or specimen was subjected to repeated cycles of completely reversed stress, that is, in each stress cycle the stress varied from a given tensile stress to an equal compressive stress. If a specimen in which the stress is concentrated is subjected to repeated cycles of stress in which the stress is not completely reversed, it is convenient to consider the cycle or range of stress to be made up of a steady stress and a completely reversed (alternating) stress superimposed on the steady stress (see Sec. 16.1). There is considerable evidence (Eisenstadt, 1971) indicating that the damaging effect of the stress concentration in such a repeated cycle of stress is associated only with the completely reversed (alternating) component of the stress cycle and not with the mean stress in the cycle. Thus, the stress concentration factor for the particular discontinuity is applied only to the alternating stress component (see Example 16.4).

Methods of Reducing Harmful Effects of Stress Concentrations

A problem that frequently arises in engineering is that of reducing the value of a stress concentration below the minimum value that will cause a fatigue fracture to occur or of raising the fatigue strength of the material so that fracture is avoided, rather than that of calculating the effective stress concentration. Some of the methods that have been employed in an attempt to reduce the damaging effects of localized stresses are the following:

1. Reducing the value of the stress concentration by decreasing the abruptness of the change in cross section of the member by use of fillets, etc., either by adding or removing small amounts of material.

2. Reducing the value of the stress concentration by making the portion of the member in the neighborhood of the stress concentration less stiff; this may be done, for example, by removing material in various ways, as indicated in Fig. 16.7. Sometimes it may be done by substituting a member made of material with a lower modulus of elasticity, such as replacing a steel nut on a steel bolt by a bronze nut to reduce the stress concentration at the threads of the steel bolt.

3. Increasing the fatigue strength of the material by cold-working the portions of the members where the stress concentrations occur; for example, by the cold rolling of fillets and bearing surfaces on axles, or by the shot blasting or shot peening of surfaces of machine parts. The increased fatigue strength of a member caused by local cold-working of the metal at the region of stress concentration in some cases may be due primarily to residual compressive stresses set up in the cold-worked metal by the surrounding elastic material as this elastic material attempts to return to its original position when the cold-working tool is removed, especially if the repeated cycle of stress is not reversed. Likewise, overstraining of the outer fibers of a beam or the inner fibers of a thick-walled pressure vessel or pipe may create favorable initial (residual) stresses (see Chapter 11).

4. Increasing the fatigue strength of the material by alloying and heat-treating portions of steel members that resist the high stress, by case hardening, nitriding, flame hardening, etc. In such treatments, however, care must be taken to avoid inducing tensile residual stresses.

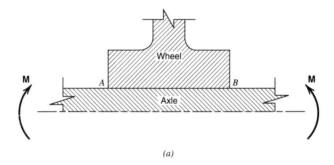

(a)

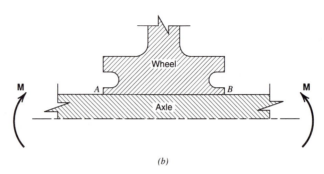

(b)

Figure 16.7   Wheel shrunk on an axle subjected to bending (top half of a symmetric cross section). (a) Stress concentration at A and B. (b) Stress concentration reduced by removing material in the vicinity of A and B.

5. Reducing the stress concentration by removing surface scratches, tool marks, small laps, and similar stress raisers to create a smooth surface by polishing.

6. Reducing the stress concentration by the prevention of minute surface corrosion pits by protecting the surface from acid fumes or moisture through the use of a corrosion-resisting covering, as, for example, by encasing the member in grease or paint.

EXAMPLE 16.2
Slot in Cantilever Beam

A cantilever beam is made of a flat bar of hot-rolled SAE 1020 steel. The beam contains a slot, Fig. C of Table 14.3, with dimensions $b = 10$ mm, $t = 50$ mm, $\rho = 5$ mm, $h = 25$ mm, and $c = 60$ mm.

Let the beam be subjected at its free end to a large number of completely reversed cycles of bending moment of maximum amplitude M.

(a) Compute the significant value of the stress at the top or bottom of the slot (i.e., at the root of the notch) in terms of M; assume q is given by Fig. 14.26.

(b) The maximum utilizable stress for this material under completely reversed cycles of stress is 172 MPa. Compute the allowable moment M based on a factor of safety of $SF = 4.0$.

SOLUTION

(a) From Fig. 14.10, $S_{cc} = 2.8$, and from Fig. 14.26 the value of q, the ordinate of the test data for SAE 1020 steel, is $q = 0.94$. Hence, by Eq. (14.43),

$$S_{ce} = 1 + 0.94(2.8 - 1) = 2.70$$

The nominal stress σ_n at the root of the notch (see Table 14.3, Fig. C) is

$$\sigma_n = \frac{3Mt}{2h(c^3 - t^3)} = \frac{3M(50)}{2(25)(60^3 - 50^3)} = 32.97 \times 10^{-6}\ M$$

Hence, the significant (or effective) stress is

$$\sigma_e = S_{ce}\sigma_n = 2.70 \times 32.97 \times 10^{-6}\ M = 89.0 \times 10^{-6}\ M$$

(b) The allowable (working) stress is $\sigma_w = \sigma_{max}/SF = 172/4.0 = 43.0$ MPa. Hence, $43.0 = 89.0 \times 10^{-6}\ M$ or $M = 483,100$ N·mm $= 483.1$ N·m.

EXAMPLE 16.3
Long Narrow Slot in Cantilever Beam

Let the cantilever beam of Example 16.2 be unchanged, except that $\rho = 0.75$ mm. Then the slot approaches a long crack in the bar. Compute the significant stress at the root of the notch.

SOLUTION

(a) From Fig. 14.10, we find that $S_{cc} = 6.1$. Figure 14.26 shows that for fatigue tests of SAE 1020 steel, $q = 0.69$ when $\rho = 0.75$ mm. Hence,

$$S_{ce} = 1 + 0.69(6.1 - 1) = 4.52$$

and the significant stress is

$$\sigma_e = S_{ce}\sigma_n = 4.52(32.97 \times 10^{-6}\ M) = 149.0 \times 10^{-6}\ M$$

Comparison with the results of Example 16.2 shows that the value of S_{cc} is increased 118%, whereas the value of S_{ce} is increased 67%, which corresponds to the increase in the significant stress σ_e. These facts indicate that as S_{cc} increases with a decrease in ρ, so does S_{ce} but to a lesser degree.

(b) The allowable stress is $\sigma_w = 43.0$ MPa. Hence, $43.0 = 149.0 \times 10^{-6}\ M$ or $M = 288,600$ N·mm $= 288.6$ N·m.

EXAMPLE 16.4
Fillet in Bar Subjected to Range of Load

The filleted tension member in Fig. 14.16 is made of 2024-T4 aluminum alloy ($E = 72.0$ GPa, $v = 0.33$, $\sigma_u = 470$ MPa, $Y = 330$ MPa, and $\sigma_{am} = 190$ MPa for $N = 10^6$ cycles). Perpendicular to the figure, the thickness of the member is 10 mm. The other dimensions are $D = 59$ mm, $d = 50$ mm, and $t = \rho = 3.00$ mm. The member is subjected to a tensile load ranging from $P_{min} = 20.0$ kN to P_{max}. Assuming that $q = 0.95$, determine the magnitude of P_{max} to produce fracture of the tension member in 10^6 cycles. Also determine σ_{max} and σ_{min}.

SOLUTION

The calculated stress concentration factor S_{cc} for the fillet can be read from the curve, fillet $t = \rho$, in Fig. 14.16, with $\rho/d = 0.06$. As read from the curve,

$$S_{cc} = 1.90$$

Since $q = 0.95$, Eq. (14.43) gives

$$S_{ce} = 1 + 0.95(1.9 - 1) = 1.86$$

Experimental evidence (Smith, 1942) indicates that S_{ce} should be applied only to the alternating part of the stress. Therefore, it is convenient to work with nominal values of the stresses as follows: nominal minimum stress $\sigma_{n(min)} = P_{min}/A$, nominal maximum stress $\sigma_{n(max)} = P_{max}/A$, nominal alternating stress σ_{na}, nominal mean stress σ_{nm}, and nominal fatigue strength $\sigma_{nam} = \sigma_{am}/S_{ce}$.

Since 2024-T4 aluminum alloy is a ductile metal, we assume that the Gerber relation, Eq. (16.2), is valid when written in terms of nominal stress values.

$$\frac{\sigma_{na}}{\sigma_{nam}} + \left(\frac{\sigma_{nm}}{\sigma_u}\right)^2 = 1 \tag{a}$$

Equation (a) can be interpreted graphically in Fig. 16.5; each ordinate to curve *AB* is reduced in magnitude by the factor $1/S_{ce}$. Nominal stress relations are defined as follows:

$$\sigma_{n(min)} = \frac{P_{min}}{A} = \frac{20.0 \times 10^3}{50(10)} = 40.0 \text{ MPa} \tag{b}$$

$$\sigma_{nm} = \sigma_{n(min)} + \sigma_{na} \tag{c}$$

$$\sigma_{n(max)} = \frac{P_{max}}{A} = \sigma_{n(min)} + 2\sigma_{na} \tag{d}$$

$$\sigma_{nam} = \frac{\sigma_{am}}{S_{ce}} = \frac{190}{1.86} = 102.2 \text{ MPa} \tag{e}$$

Substitution of Eqs. (c) and (e) into Eq. (a) gives

$$\sigma_{na} = 93.9 \text{ MPa}$$

which when substituted into Eq. (d) gives

$$\sigma_{n(max)} = 40.0 + 2(93.9) = 227.8 \text{ MPa} = \frac{P_{max}}{A}$$

$$P_{max} = 227.8(50)(10) = 113,900 \text{ N} = 113.9 \text{ kN}$$

The assumption that the effective stress concentration factor should be applied only to the alternating part of the stress defines the maximum and minimum stresses in the member at the stress concentration as

$$\sigma_{max} = \sigma_{nm} + S_{ce}\sigma_{na} = 133.9 + 1.86(93.9) = 308.6 \text{ MPa}$$

$$\sigma_{min} = \sigma_{nm} - S_{ce}\sigma_{na} = -40.8 \text{ MPa} \tag{f}$$

Since the load cycles between a tensile load of 20.0 kN and a tensile load of 114 kN, the negative sign for σ_{min} may be suspect. The maximum and minimum stresses given by Eqs. (f) give the correct range in stress at the stress concentration for linearly elastic conditions; however, the values given by Eqs. (f) can be only a rough approximation of their true magnitudes. If residual stresses are not present at the stress concentration, linearly elastic analysis gives the maximum stress at the stress concentration as $S_{cc}P_{max}/A = 432.8$ MPa, which exceeds the yield stress of the material. Plasticity theories and experimental evidence indicate that plastic deformation in the region of the stress concentration produces residual stresses that result in a reduction of the mean stress at the stress concentration. The residual stresses at the stress concentration have a sign opposite to the sign of the stresses that caused the inelastic deformation. Thus, the residual stresses decrease the magnitudes of both σ_{max} and σ_{min} so that the values given by Eqs. (f) are realistic.

PROBLEMS
Section 16.1

16.1. A tension member is cycled an indefinitely large number of times from $P_{min} = -10.0$ kN to $P_{max} = 16.0$ kN. The member is made of steel ($\sigma_u = 700$ MPa, $Y = 450$ MPa, and $\sigma_{am} = \sigma_L = 350$ MPa). Using the Gerber relation, determine the diameter of the rod for a factor of safety $SF = 2.20$.

16.2. Let the tension member in Problem 16.1 be cycled an indefinitely large number of times between $P_{min} = 0$ and $P_{max} = 16.0$ kN. Determine the diameter of the tension member for a factor of safety $SF = 2.20$. What is the mode of failure?

Ans. $d = 9.98$ mm, general yielding mode of failure

16.3. A cast iron I-beam has a depth of 150 mm, width of 100 mm, and equal flange and web thicknesses of 20 mm. The beam is subjected to 10^6 cycles of loading from $M_{min} = 5.00$ kN·m to M_{max}. Consider the cast iron to be a brittle material ($\sigma_u = 200$ MPa and $\sigma_{am} = 90.0$ MPa for

$N = 10^6$). Using the Goodman relation, determine M_{max} based on a factor of safety $SF = 2.50$.

16.4.   A thin-wall cylinder is made of 2024-T4 aluminum alloy ($\sigma_u = 430$ MPa, $Y = 330$ MPa, $\sigma_{am} = 190$ MPa for $N = 10^6$). The cylinder has an inside diameter of 300 mm and a wall thickness of 8.00 mm. The ends are strengthened so that fatigue failure is assumed to occur in the cylinder away from the ends. The pressure in the cylinder is cycled 10^6 times between $p_{min} = -2.00$ MPa and $p_{max} = 7.00$ MPa. What is the factor of safety against fatigue failure if design is based on the Gerber relation?

Ans. $SF = 2.13$

16.5.   A shaking mechanism of a machine has a crank shown in Fig. P16.5. The crank is made of a stress-relieved cold-worked SAE 1040 steel ($\sigma_u = 830$ MPa, $Y = 660$ MPa, and $\sigma_{am} = \sigma_L = 380$ MPa). A completely reversed load $P = 500$ N is to be applied for up to 10^8 cycles to the crank pin, normal to the plane of the crank. Determine the diameter d of the shaft based on a factor of safety $SF = 1.75$ using the octahedral shear-stress criterion of failure.

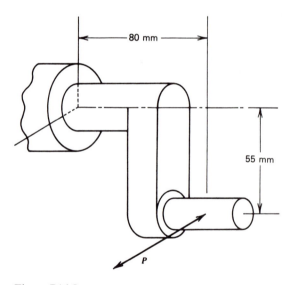

Figure P16.5

16.6.   The crank in Problem 16.5 has a diameter $d = 15.0$ mm and is made of 2024-T4 aluminum alloy ($\sigma_u = 430$ MPa, $Y = 330$ MPa, $\sigma_{am} = 160$ MPa for $N = 10^8$). Let the load vary between 0 and $P_{max} = 500$ N for 10^8 cycles. Assume that the Gerber relation [Eq. (16.2)] is valid.

(a)   Is the design governed by general yielding or fatigue failure?

(b)   Determine the magnitude of the safety factor SF used in the design based on the octahedral shear-stress criterion of failure.

Ans.   **(a)** Fatigue failure,   **(b)** $SF = 2.03$

16.7. Let the shaft in Problem 4.21 be made of SAE 1040 steel ($\sigma_u = 830$ MPa, $Y = 660$ MPa, $\sigma_{am} = \sigma_L = 380$ MPa). If the shaft is rotated under constant load an indefinitely large number of times, determine the diameter of the shaft for a factor of safety $SF = 2.00$.

16.8. A 30.0-mm diameter shaft is subjected to cyclic combined bending and torsion loading such that $M = 200\,P$ and $T = 150\,P$, where the magnitude of P varies from $P_{min} = -0.60\,P_{max}$ to P_{max} and length is measured in millimeters. The shaft is made of a stress-relieved cold-worked SAE 1060 steel ($\sigma_u = 810$ MPa, $Y = 620$ MPa, $\sigma_{am} = \sigma_L = 410$ MPa). Using a factor of safety $SF = 1.80$, determine P_{max} for 10^7 cycles of loading. Use the octahedral shear-stress criterion of failure and Gerber relation.

Ans. $P_{max} = 3.12$ kN

Sections 16.2–16.3

16.9. The crank shaft in Fig. P16.5 has a diameter $d = 13.0$ mm. Let a small-diameter hole be drilled in the crank shaft at a location 50 mm from the load P (measured along the axis of the shaft). Determine the magnitude of the completely reversed load P that can be cycled 10^8 times based on a factor of safety $SF = 1.75$. Assume that $q = 1.00$. Material properties are given in Problem 16.5.

16.10. The load P in Problem 16.9 is cycled from zero to P_{max}. Determine the magnitude of P_{max} for 10^8 cycles based on a factor of safety $SF = 1.75$. Material properties are given in Problem 16.5. Assume that $q = 0.90$ and the Gerber relation is valid when expressed in terms of nominal stress values [see Eq. (a) of Example 16.4]. Since the state of stress at the hole is uniaxial, we define S_{cc} for the stress concentration as the ratio of σ_{max} for the crank shaft with the hole to σ_{max} for the crank shaft without the hole.

Ans. $S_{ce} = 2.98$, $P_{max} = 494.1$ N

16.11. A crank shaft has a fillet and minimum diameter of 30 mm. The critical section of the crank shaft is subjected to a bending moment $M = 200P$ and a torque $T = 180P$, where P is a completely reversed load. The calculated stress concentrations for the fillet are $S_{cc}^{(M)} = 2.50$ for bending and $S_{cc}^{(T)} = 2.00$ for torsion. The shaft is made of a stress-relieved cold-worked SAE 1060 steel ($E = 200$ GPa, $v = 0.29$, $\sigma_u = 810$ MPa, $Y = 620$ MPa, and $\sigma_{am} = 410$ MPa for 10^7 cycles). Determine the completely reversed load P that can be applied 10^7 times based on a factor of safety of $SF = 2.20$. Assume $q = 0.85$.

16.12. Let the crank shaft in Problem 16.11 be subjected to a range of load from zero to P_{max}. Determine the completely reversed load P_{max} that can be applied 10^7 times based on a factor of safety $SF = 2.20$. Assume that the Gerber relation is valid when expressed in terms of nominal stress values [see Eq. (a) of Example 16.4]. Use the maximum shear-stress criterion of failure; therefore, assume that S_{ce} for the fillet is the ratio τ_{max} in the crank shaft with the fillet to τ_{max} in the crank shaft without the fillet.

Ans. $P_{max} = 1.66$ kN

16.13. A tension member has a hole drilled at its center, similar to the member in Fig. 14.16. The dimensions of the member are $w = 20$ mm perpendicular to the figure, $D = 60$ mm, and $\rho = 5.0$ mm. The member is subjected to a tension load that cycles between $P_{min} = 30.0$ kN and P_{max}. The member is made of 2024-T4 aluminum alloy, with properties $E = 72.0$ GPa, $v = 0.33$, $\sigma_u = 470$ MPa, $Y = 330$ MPa, and $\sigma_{am} = 190$ MPa (by testing a smooth specimen to $N = 10^6$ cycles).

   (a) Determine the magnitude of P_{max} to produce general fatigue fracture of the member in 10^6 cycles (see Example 16.4). Assume that fracture occurs at the hole section.

   (b) Determine the maximum and minimum stress at the edge of the hole, assuming that the stress concentration factor need be applied only to the alternating component of stress (see Example 16.4).

16.14. A rectangular cross-section tension member made of a ductile material ($Y = 350$ MPa and $\sigma_{am} = \sigma_L = 280$ MPa at 10^8 cycles of completely reversed axial load) has semicircular grooves (Fig. P16.14). The thickness of the member is 40 mm.

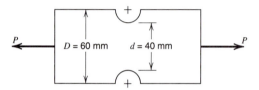

Figure P16.14

   (a) Determine the failure load for static loading.

   (b) Determine the failure load for 10^8 cycles of completely reversed axial load.

16.15. For the member in Problem 16.14, determine P_{max} if the member fails in 10^8 cycles of load from $P_{min} = 0$ to P_{max}.

16.16. For the member in Problem 16.14, determine P_{max} if the member fails in 10^8 cycles of load from $P_{min} = -100$ kN to P_{max}.

16.17. A rectangular cross-section cantilever beam is made of steel ($E = 200$ GPa, $\sigma_u = 590$ MPa, and $\sigma_{am} = 220$ MPa for 10^6 completely reversed cycles of load); see Fig. P16.17. Determine the magnitude of load P to cause

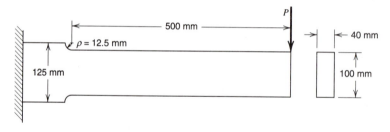

Figure P16.17

failure at 10^6 cycles of completely reversed load, based on a factor of safety of 2.20.

16.18. Solve Problem 16.17 for the case where the load is cycled from 0 to P_{max}. (See Example 16.4.)

16.19. The beam in Fig. P16.19 is made of 2024-T4 aluminum alloy ($E = 72.0\,\text{GPa}$, $Y = 330\,\text{MPa}$, $\sigma_u = 470\,\text{MPa}$, and $\sigma_{am} = 170\,\text{MPa}$ for 10^7 cycles of completely reversed load). The beam is subjected to 10^7 completely reversed cycles of load P. If $h = 200\,\text{mm}$ and $\rho = 25.0\,\text{mm}$, determine the magnitude of P, based on a factor of safety of 1.80.

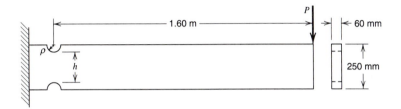

Figure P16.19

16.20. Solve Problem 16.19 for the case where P is cycled 10^7 times from 0 to P_{max}. (See Example 16.4.)

16.21. The tension member in Fig. P16.21 has a thickness of 60 mm. The member is made of 2024-T4 aluminum alloy. The member is subjected to 10^6 completely reversed cycles of loading. The fatigue strength for $N = 10^6$ is $\sigma_{am} = 220\,\text{MPa}$. What is the design load based on a factor of safety $SF = 2.20$?

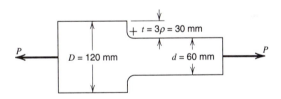

Figure P16.21

16.22. Solve Problem 16.21 for the case where the load is cycled between $P_{min} = 0$ and P_{max}.

REFERENCES

Anonymous (1967). Wöhler's Experiments on the Strength of Materials. *Eng.*, **207**: 10.

American Society for Metals (ASM) (1975). *Metals Handbook*, 8th ed., Vol. 10, Failure Analysis and Prevention. Metals Park, Ohio: pp. 95–125.

Buch, A. (1988). *Fatigue Strength Calculation*, Materials Science Surveys, No. 6. Brookfield, Vermont: Trans Tech Publications, Brookfield Publ.

Buch, A., Seeger, T., and Vormwald, M. (1986). Improvement of Fatigue Life Prediction Accuracy for Various Realistic Loading Spectra by Use of Correction Factors. *Internat. J. Fatigue*, **8**: 175.

Buxbaum, O., Klätschke, H., and Opperman, H. (1991). Effect of Loading Sequence on the Fatigue Life of Notched Specimens Made from Steel and Aluminum Alloys. *App. Mech. Rev.*, **44** (1).

Clarke, W. L. and Gordon, G. M. (1973). Investigation of Stress Corrosion Cracking Susceptibility of Fe-Ni-Cr Alloys in Nuclear Reactor Water Environments. *Corrosion*, **29**, Part 1: 1–12.

Eisenstadt, M. M. (1971). *Introduction to Mechanical Properties of Metals*, New York: Macmillan.

Fuchs, H. O. and Stephens, R. I. (1980). *Metal Fatigue in Engineering*. New York: Wiley.

Gauthier, J. P. and Petrequin, P. (1989). High Cycles Fatigue of Austenitic Stainless Steels Under Random Loading. *Nucl. Eng. Design*, **116**: 343–353.

Kliman, V. (1985). Fatigue Life Estimation Under Random Loading Using Energy Criterion. *Internat. J. Fatigue*, **7**: 39.

Knott, J. F. (1973). *Fundamentals of Fracture Mechanics*. New York: Wiley.

Manson, S. S. (1981). *Thermal Stress and Low-Cycle Fatigue*. Melbourne, Florida: Kreiger.

Sandor, B. I. (1972). *Fundamentals of Cyclic Stress and Strain*. Madison, Wis.: Univ. Wis. Press.

Skelton, R. P. (1988). Fatigue Crack Growth. In *Mechanical Testing* (Curbishley, I., ed.). Brookfield, Vermont: Inst. of Metals, Chapter 3.

Smith, A. I. and Nicolson, A. M. (1971). *Advances in Creep Design*, New York: Halsted Press Division of Wiley.

Smith, J. O. (1942). Effect of Range of Stress on the Fatigue Strength of Metals. Urbana-Champaign, Ill.: Univ. Ill., Bull. 334, Eng. Exper. Station.

Socie, D. F. (1977). Fatigue Life Prediction Using Local Stress-Strain Concepts. *Exper. Mech.*, **17** (2): 50.

Weronski, A. and Hejwowski, T. (1991). *Thermal Fatigue of Metals*. New York: Marcel Dekker.

17

CREEP. TIME-DEPENDENT DEFORMATION

17.1

DEFINITION OF CREEP. THE CREEP CURVE

At ordinary temperatures, say, 0°C to 50°C (32°F to 122°F), and in the absence of a corrosive environment, a properly designed member will support its static design load for an unlimited time. However, at so-called elevated temperatures, the life of the member may be severly limited, even for loads less than the design load. At elevated temperatures, a sustained load may produce inelastic strain in the material that increases with time. Hence, the material is said to *creep*, creep being defined as time-dependent inelastic strain under sustained load and elevated temperature. If creep is maintained for a sufficiently long time, excessive deflection (creep-failure) or fracture (creep-fracture) occurs. The combination of temperature, load, and time that produces creep and possibly creep-failure or creep-fracture of a member depends on the material and the environment. Consequently, creep, creep-failure, and creep-fracture of a member may occur over a wide range of temperature and load.

Creep will occur in any metal subjected to a sustained load at a temperature slightly above its recrystallization temperature. At this temperature, the atoms become quite mobile. As a result, time-dependent alterations of the metal's structure occur. It is often stated that "elevated temperature" for creep behavior of a metal begins at about one-half the melting temperature T_m of a metal measured in degrees Kelvin. However, this is a rule of thumb that greatly oversimplifies a very complex behavior. The temperature at which a member's function is limited by creep rather than, say, yield strength is not directly related to T_m. In reality, the meaning of elevated temperature must be determined individually for each material on the basis of its behavior. As noted in the American Society for Metals' Handbook (ASM, 1976), elevated temperature behavior for various metals occurs over a wide range of temperature, for example, at 205°C (400°F) for aluminum alloys, 315°C (600°F) for titanium alloys, 370°C (700°F) for low-alloy steels, 540°C (1000°F) for austenitic, iron-based high-temperature alloys, 650°C (1200°F) for nickel-based and cobalt-based high-temperature alloys, and 980°C to 1540°C (1800°F to 2800°F) for refractory metals and alloys. Whereas for certain plastics, asphalt, concrete, lead, and lead alloys, elevated temperatures for creep behavior may lie in the range of "ordinary temperatures," say, from 0°C to 50°C (32°F to 122°F).

The original observations of creep in materials are lost in antiquity. A simple form of creep in nature is the slow, almost imperceptible downslope movement of soil particles and rock debris under the influence of gravity. Early mankind may have observed the creep of rocks that formed the sides and roofs of caves or of ice in the walls and roofs of igloos. The fact that windows in old churches in Europe are thinner at the top than at the bottom has been attributed to the creep of the glass under the effect of gravity. Today, engineers and scientists confront creep in energy-producing systems such as power-generating plants (coal, gas, and nuclear plants). Creep also occurs in energy conversion systems, such as thermionic converters, and in modern-day applications of electronic packaging that involve the heat transfer and cooling of microcircuits [microelectronic chips, electronic circuit boards, surface-mounted electronic components, solder joints, etc.; see, e.g., the *Journal of Electronic Packaging*, published quarterly by the American Society of Mechanical Engineers (ASME)]. Unfortunately, we cannot provide in the space available here an exhaustive historical development of creep analysis, even for metals, let alone other materials. However, such developments may be found in previously published treatises. For example, the creep of wires of hardened iron at room temperature was observed and studied quantitatively as long ago as 1834 by the French engineer L. J. Vicat (1834). He observed, among other things, the first part (primary range) of the classical form of the strain-time plot (creep curve; Fig. 17.1). Vicat's interest was focused mainly on the use of wire for load-carrying members in suspension bridges. However, it was not until the beginning of this century that the entire creep curve of Fig. 17.1 was developed for iron wire and for several other materials (Phillips, 1905; Andrade, 1910).

Much of the history of creep of metals has its origins in the industrial revolution that led to the operation of machines at the highest possible temperatures to

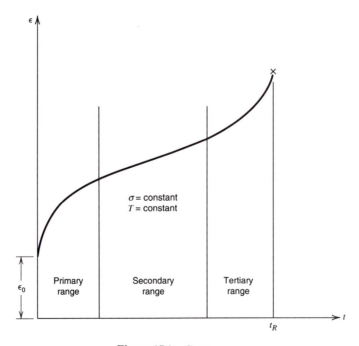

Figure 17.1   Creep curve.

achieve the greatest thermal efficiencies. One of the first comprehensive studies of the purposes and techniques of stress analysis and design in creep problems was made by Bailey (1935). In the late 1950s and 60s, a great increase in studies of creep occurred, due in part to the interest in nuclear reactor power generation and the high temperatures used in such systems. During this period, several books dealing with stress analysis of creep were published in various countries. One of these books (Finnie and Heller, 1959) summarized many of the earlier creep studies and the associated stress analysis techniques. Other publications during this period include books by Kachanov (1960), Lubahn and Felgar (1961), Odqvist (1966), Hult (1966), Rabotnov (1969), and Penny and Marriott (1971). Several surveys were also published during this period (e.g., ASTM, 1959; ASTM, 1965). A review of creep of metals under multiaxial states of stress, including an extensive list of references up to 1971, was given by Boresi and Sidebottom (1972). More recently, the book by Kraus (1980) presents an introduction to design problems of creep; the book by Boyle and Spence (1983) treats basic methods of stress analysis of creep; the book edited by Bernasconi and Piatti (1979) covers a wide range of creep problems from basic concepts to experimental techniques; Cadek (1988) explores high-temperature creep as affected by metallurgical mechanisms (e.g., motion of dislocations, dislocation structure, dislocation creep in pure metals, etc.); the volume edited by Curbishley (1988) discusses tensile testing (T. G. F. Gray, Chapter 1) and creep testing (M. S. Loveday, Chapter 2).

In this chapter, we are concerned mainly with mathematical equations used to represent creep strain (Fig. 17.1) as a function of stress, temperature, and time, and the use of these equations to study the effects of creep. Therefore, we assume that an elevated temperature for a material exists so that the material may creep. A phenomenological approach is taken to describe the physical processes that alter the metallurgical structure of a material, allowing creep to occur. Furthermore, we do not attempt to describe in any detail various creep models (viscoelastic, elastic-plastic, etc.) that have been proposed. Rather, we base our study on the typical creep curve (plot of strain *vs* time, Fig. 17.1) and mathematical modeling of the creep curve. Creep curves are ordinarily obtained by tests of bars subjected to sustained axial tension (Loveday, 1988). Standards for creep tests have been established by several technical organizations [International Standards Organization (ISO), 1987; British Standards Institute (BSI), 1987; ASTM, 1983].

In Sec. 17.2, we briefly describe the tension creep test for metals. In Sec. 17.3 and 17.4, we present one-dimensional creep formulas for metals subjected to stress and elevated temperature. In Sec. 17.5 and 17.6, the creep of metals subjected to multidimensional states of stress is considered. Some applications to simple problems in creep of metals are discussed in Sec. 17.7. In Sec. 17.8, a few observations relative to creep of nonmetals are given.

17.2

THE TENSION CREEP TEST FOR METALS

The creep behavior of various materials is often based on a one-dimensional (tension) test. Various standards for creep testing specify the geometric design of test specimens (ASTM, 1983; BSI, 1987; ISO, 1987). Careful control of machined dimensions is specified (Loveday, 1988). During the test, the tension specimen is subjected to sufficiently high stress σ and temperature T to produce time-dependent inelastic strain (creep). Highly sensitive creep-testing systems have been developed to measure load (stress) and temperature to ensure accurate creep data over long periods of time (Loveday, 1988). In the creep test, the strain in the specimen varies with time. For an appropriate constant stress and elevated temperature, a strain-time plot (creep curve) is shown in Fig. 17.1. This creep curve exhibits three distinct ranges. Beginning at time $t = 0$, the strain is ϵ_0 due to the initially loading. The strain ϵ_0 may be partly elastic and partly plastic, depending on the level of load and temperature. In the first interval of time, the primary range of the creep curve, the strain rate (the slope of the creep curve), decreases, until it reaches some minimum rate. During the next interval of time, the secondary range, this minimum rate is maintained, more or less, until a time at which the strain rate begins to increase, the beginning of the tertiary range. In the tertiary range, the strain rate continues to increase under the sustained stress and temperature until, at time $t = t_R$, the specimen is pulled apart (point x in Fig. 17.1). In the following section, various formulas that have been used to approximate one-dimensional curves are discussed.

17.3

ONE-DIMENSIONAL CREEP FORMULAS FOR METALS SUBJECTED TO CONSTANT STRESS AND ELEVATED TEMPERATURE

As noted in Sec. 17.1, by creep we mean the inelastic strain that occurs when the relationship between stress, strain, and elevated temperature is time-dependent. In general, creep behavior is a function of the material, stress, temperature, time, stress history, and temperature history. Creep behavior includes the phenomenon of *relaxation*, which is characterized by the reduction of stress in a member, with time, while total strain remains constant. It also includes *recovery*, which is characterized by the reduction of inelastic strain with time after the stress has been removed (see Kraus, 1980, Chapter 3). However, the topics of relaxation and recovery lie outside the scope of our discusssion.

Creep of metals at elevated temperatures is characterized by the fact that most of the deformation is irreversible; that is, only a small part of the strain is recovered after removal of load. In addition, the dependence of *creep rate* on stress is quite nonlinear. As a consequence, linear theories of viscoelasticity do not ordinarily apply to metals (Rabotnov, 1969). Thus, the theory of creep of metals, the objective of which is to describe time-dependent irreversible deformation, is patterned after the general theory of plasticity. At elevated temperatures, plastic deformation of metals is usually accompanied by creep. Therefore, in real situations of creep, the concepts of creep and plasticity intertwine. However, in the representation of creep data by empirical formulas, creep deformation is separated from plastic, and likewise elastic, deformation.

For one-dimensional states, the classical creep curve for a material is obtained from a tensile test at constant stress σ and temperature T (Fig. 17.1). As noted in Sec. 17.1, the creep curve for metals usually exhibits three regions in which the creep deformation takes on a different character (curve C_1, Fig. 17.2). With reference to curve C_1, interval OA represents the instantaneous deformation that occurs *immediately* as the load is applied. This strain is denoted by $\epsilon_0(\sigma, T)$. Depending on the stress level σ and temperature T, ϵ_0 may include both elastic and plastic parts. The interval AB represents the primary (initial) stage of creep deformation. In this interval, the creep deformation is changing (is transient) at a *decreasing strain rate*. It is for this reason that investigators often represent the creep behavior in the primary time stage by formulas that express the creep strain rate $\dot{\epsilon}_c$ as a function of stress σ, temperature T, and time t, where the dot denotes derivative with respect to time. The interval BC represents the second stage of creep in which the creep rate reaches a minimum value. If in this region the creep rate remains constant, the creep strain rate is a function of stress σ and temperature T only; that is, $\dot{\epsilon}_c = f(\sigma, T)$, and the creep strain ϵ_c is a linear function of time t. The interval CD represents the third (tertiary) creep stage, in which the creep strain rate increases rapidly. If the load is sustained in this region, creep rupture will occur (point D on curve $ABCD$).

Although the division of the creep curve into three intervals is conventional for many metals, depending on the metal, stress, and temperature, a variety of creep curves may be obtained as a consequence of the complexity of the metallurgical processes involved. For example, for materials different from that used to generate curve C_1 and for the same load and temperature for curve C_1 (Fig. 17.2), the strain-time response may be given by curves C_0, C_2, or C_3. Or if the material used to generate curve C_1 is subjected to a lower load and/or temperature,

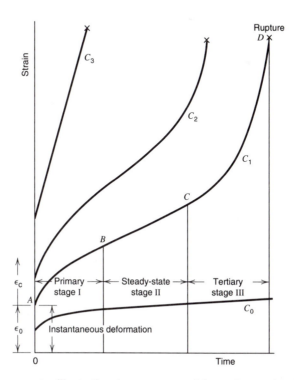

Figure 17.2 Creep curve illustrating instantaneous deformation and primary, steady-state (secondary), and tertiary stages.

its response may be given by curve C_0, for which the tertiary range of creep is never reached. If the material of curve C_1 is subjected to a higher stress and/or temperature, its response may be given by curve C_2 (for which the primary range of creep is suppressed) or curve C_3 (for which both the primary and secondary ranges of creep are bypassed and for which fracture occurs in a relatively short time). For example, for annealed SAE 1035 steel at a constant temperature of 524°C, Fig. 17.3 illustrates the change in the creep curve that is produced by increasing the stress level in steps of approximately 20 MPa from 83 to 164 MPa. Similar increases in the creep strain also occur with temperature for constant stress. Additionally, it is difficult to determine the instantaneous deformation ϵ_0 precisely, since it depends on the method of loading. Much of the published creep data ignore this quantity, and creep curves are simply plots of ϵ_C vs time (part *ABCD* of curve C_1, Fig. 17.2).

Because of the extreme complexity of creep behavior, the analysis of creep problems is often based on curve-fitting of experimental creep data. These representations generally attempt to represent the creep strain ϵ_C or creep strain rate $\dot{\epsilon}_C$ as functions of stress σ, temperature T, and time t. Usually, such equations have been developed by one of three methods: (1) by deriving empirical formulas that model experimental data (Rabotnov, 1969, Chapter IV); (2) by deriving equations based on metallurgical creep mechanisms (Dorn, 1962; Cadek, 1988); or (3) combinations of methods (1) and (2) (Kennedy, 1962). In these methods, attempts have been made to separate various influences for each stage of creep. For example, in the first method, one may represent any one of the three stages separately by an

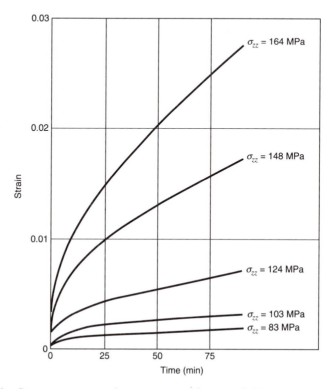

Figure 17.3   Constant-stress tension creep curves for annealed SAE 1035 steel at 524°C.

empirical formula; or one may represent two stages by a single formula, say, the primary and secondary stages; or all three stages by a single formula. Alternatively, in method (2), one may consider the effect of a particular creep-producing mechanism, say, dislocations (Cadek, 1988), on the various stages of creep and attempt to relate various parameters in the strain-stress-temperature-time relations to the properties of the dislocations. In Table 17.1, we list a number of formulas that have been used to represent one-dimensional creep curves. Many of these formulas are discussed in the references listed in the table. Following Kennedy (1962), we separate the equations into time-, temperature-, and stress-dependent parts. We also indicate combinations of time, stress, and temperature components. The time-dependence formulas are sometimes of the form $\epsilon = \epsilon_0 + \epsilon_C$, where $\epsilon_C = \epsilon_{PC} + \epsilon_{SC} + \epsilon_{TC}$, ϵ is the total strain, ϵ_0 is the instantaneous strain, ϵ_C is creep strain, and ϵ_{PC}, ϵ_{SC}, ϵ_{TC} denote primary, secondary, and tertiary creep, respectively. The various components of the total strain ϵ are selected to fit creep data at constant stress and temperature (method 1). If practical interest does not extend to the tertiary stage, the effects of ϵ_{TC} are not included in some of the formulas. Generally, ϵ_0 is a constant, $\dot{\epsilon}_{PC}$ a monotonically decreasing function of time, ϵ_{SC} a linear function of time, and $\dot{\epsilon}_{TC}$ a monotonically increasing function of time, where a dot denotes derivative with respect to time.

The temperature dependency of creep is often related to thermodynamics and rate processes of solid-state physics (Dorn, 1962; Cadek, 1988). Consequently, the temperature dependency is often of exponential form. Also, experimental evidence indicates that the creep rate in the secondary stage of creep increases more rapidly with temperature increases than does the creep rate in the primary stage of creep. Finally, since stress is a tensor (Chapter 2), whereas temperature and time are

TABLE 17.1
Empirical One-Dimensional Creep Formulas

Equation Form	References	Equation
Time Dependence		
Rational		
$\epsilon_C = at/(1 + bt)$	(Freundenthal, 1936)	(a)
Logarithmic		
$\epsilon = a + b\ln(t)$	(Phillips, 1905)	(b)
$\epsilon = a + b\ln(1 + ct)$	Modification of (b)	(c)
Exponential		
$\epsilon = a + bt - c\exp(-dt)$	(McVetty, 1934)	(d)
$\epsilon_C = at + b[1 - \exp(-ct)]$	(McVetty, 1934) (Söderberg, 1936)	(e)
Power		
$\epsilon_C = bt^n; \quad 1/3 < n < 1/2$	(Bailey, 1935)	(f)
Power series		
$\epsilon_C = at^m + bt^n; \quad m > 1, \quad 0 < n < 1$	(de Lacombe, 1939)	(g)
$\epsilon_C = at_m + bt_n + ct_p \dots$	(Graham, 1953)	(h)

(continues)

TABLE 17.1 (*Continued*)

Equation Form	References	Equation
Time Dependence (continued)		
Combined exponential-power $\epsilon_C = a(1 + bt^{1/3})\exp(kt) - a$	(Andrade, 1910)	(i)
Combined logarithmic-power $\epsilon_C = a\ln(t) + bt^n + ct$	(Wyatt, 1953)	(j)
Temperature Dependence		
Exponential		
$\dot{\epsilon}_C = a\exp(-Q/RT)$	(Mott, 1953)	(k)
$\epsilon_C = a[t\exp(-Q/RT)]$	(Dorn, 1962)	(l)
$\dot{\epsilon}_C = aT\exp(-Q/RT)$	(Stowell, 1957)	(m)
Rational		
$\epsilon_C = aT^{2/3}f(t)$	(Mott and Nabarro, 1948)	(n)
$\epsilon_C = aTf(t)$	(Smith, 1948)	(o)
$\epsilon_C = f\{T[a + \ln(t)]\}$	(Larson and Miller, 1952)	(p)
$\epsilon_C = f[(T - a)/\ln(t - b)]$	(Manson and Haferd, 1954)	(q)
Hyperbolic-exponential		
$\dot{\epsilon}_C = a\exp(-Q/RT)\sinh(b/RT)$	(Feltham, 1953)	(r)
Other		
$\epsilon_C = cf[t(T - T')^{-B}]$	(Warren, 1967)	(s)
Stress Dependence		
Exponential		
$\epsilon_C = af(t)\exp(b\sigma)$	(Dorn, 1962)	(t)
$\dot{\epsilon}_C = a\exp(b + c\sigma)$	(Nadai, 1931)	(u)
$\dot{\epsilon}_C = a[\exp(b\sigma) - 1]$	(Söderberg, 1936)	(v)
Power		
$\epsilon_C = af(t)\sigma^b$	(Dorn, 1962)	(w)
$\epsilon_C = at^n\sigma^b; \quad 0 < n < 1, \quad b > 1;$ Bailey–Norton law	(Bailey, 1935) (Norton, 1929)	(x)
Hyperbolic		
$\dot{\epsilon}_C = a\sinh(b\sigma)$	(Ludwik, 1908) (McVetty, 1943)	(y)
$\dot{\epsilon}_C = a\sinh(b\sigma/RT)$	(Feltham, 1953)	(z)
Other		
$\dot{\epsilon}_C = a\sigma\exp[\,f(\sigma)]$	(Kanter, 1938)	(aa)
Combined Time-Temperature-Stress Dependencies		
$\dot{\epsilon}_C = T\exp(-a/T - b + c\sigma)$	(Nadai, 1931)	(bb)
$\epsilon_C = a\exp(-A/T)\sigma^n t^k$	(Pickel et al, 1971)	(cc)
$\epsilon_C = a\exp(-A/T)\sinh(a\sigma)t^k$	(Pickel et al., 1971)	(dd)
$\epsilon_C = a\exp(-A/T) \times [\sinh(b\sigma)]^m t^k$	(Pickel et al., 1971)	(ee)
$\epsilon_C = a\exp(-A/T)(\sigma/b)^c + (\sigma/d)^e t$	(Odqvist, 1953)	(ff)
$\epsilon_C = \sum_{i=1}^{n} C_i\sigma^{a_i}\phi^{b_i}; \qquad \phi = t(T' - T)^{-A}$	(Graham and Walles, 1955)	(gg)

scalars, the introduction of stress dependency into creep formulas is more difficult. Therefore, more than one function of stress has to be employed, if creep behavior over a wide range of stress is to be fitted accurately. In general, to model creep curves (Figs. 17.1 through 17.3), one needs expressions of the form $\epsilon_C = f(t, T, \sigma)$, where f is a general function of time t, temperature T, and stress σ. It is customary to assume that the effects of t, T, and σ are separable. Then, in general, f may be taken as a sum of n products of t, T, and σ, and ϵ_C may be written in the form

$$\epsilon_C = \sum_{i=1}^{n} f_i(t) g_i(T) h_i(\sigma)$$

Thus, experiments for one-dimensional creep behavior are usually run allowing only one of the variables (t, T, σ) to change (see Fig. 17.3, where σ is varied for constant temperature T and a given time t). The formulas in Table 17.1 reflect the separation of stress, temperature, and time.

It should be noted that, in practice, the determination of the time dependence of creep for a complex metal (alloy) that exhibits large structural change with time at elevated temperature is very difficult. It requires extensive curve-fitting procedures (see Conway, 1968; 1969; Penny and Marriott, 1971; Kraus, 1980). Consequently, Eqs. (a) through (j) are not generally applicable. Any one of them may hold for a certain metal and for certain test conditions. However, many of them are not conveniently adapted to include effects of temperature and stress.

By combining the time, temperature, and stress representations, Eqs. (bb) through (gg) in Table 17.1, the entire functional behavior of creep may be approximated by a single equation. An extensive discussion of the application of Eq. (gg) is given by Kennedy (1962). The successful use of these equations generally requires numerical methods (Kraus, 1980) such as finite element methods (see Chapter 19). Many of the formulas in Table 17.1 may be applied to multiaxial stress states through the use of the concept of effective stress (Sec. 4.3) and the corresponding concept of effective strain rate (Sec. 17.6). Kennedy (1962) also discusses at length strain-stress-temperature-time relations based on quasi-empirical methods and metallurgical (microstructure) observations. The distinguishing feature of microstructure formulations is that an attempt is made to relate the parameters in the creep relation to creep-producing microstructure mechanisms such as grain boundary displacement, slip, and subgrain size (see Cadek, 1988).

In Table 17.1, ϵ denotes total strain, ϵ_C creep strain, σ stress, T temperature, t time, ln the natural logarithm, exp the exponential e, and $a, b, c, \ldots, A, B, C, \ldots$ parameters that may be functions of σ, t, T or they may be constants. Time derivative is denoted by a dot over a symbol (e.g., $\dot{\epsilon}_C$). The notation $f(x)$ denotes a function of x.

17.4

ONE-DIMENSIONAL CREEP OF METALS SUBJECTED TO VARIABLE STRESS AND TEMPERATURE

Preliminary Concepts

As indicated by Loveday (1988), there are many designs for creep-testing machines. They differ mainly in the method of measuring deformation and the methods of heating and controlling temperature. Several of these machines are of the lever type that ensures constant load. However, during a constant load creep test, a creep specimen enlongates and its cross-sectional area decreases. Consequently, the stress is not truly constant; that is, for a constant applied force, the stress increases during the test. Some attention has been devoted to the fact that the stress is not constant, and various machines have been designed to maintain constant stress (Loveday, 1988).

In the design of these machines, it is assumed that the volume of the creep specimen remains constant and that the change in cross-sectional area can be determined from the elongation of the specimen. Lubahn and Felgar (1961, Chapter 6) give an interesting discussion of the difference between constant load and constant stress creep tests. Also, Andrade (1910) found that the elongation of lead wire for constant load testing was considerably larger than for constant stress loading, particularly at stress levels near the ultimate strength. He also found that under constant load, the creep curve for lead wire exhibited three stages of creep (primary, secondary, and tertiary), whereas for constant stress, the creep curve exhibited only the primary and secondary stages (Fig. 17.4). Lubahn and Felgar (1961, p. 136) observe that for a metallurgically stable material subject to a tensile creep test at constant stress, the creep rate continues to decrease indefinitely, this effect being caused by strain hardening (see Sec. 4.2).

If the load is kept constant, the creep rate will tend to increase after sufficient creep has occurred, because of the reduction of cross-sectional area of the specimen. Since the strain-hardening effect tends to decrease with increasing deformation and the reduction of cross section increases with increasing deformation, a strain is eventually reached at which the effect of reduction of area dominates, the strain rate begins to increase rapidly, and the creep curve exhibits an inflection point (the start of the tertiary stage). This balance between reduction of cross section and strain hardening is similar to the balance of necking down and strain hardening in the static tension test (Chapter 1). Therefore, it may be regarded as a

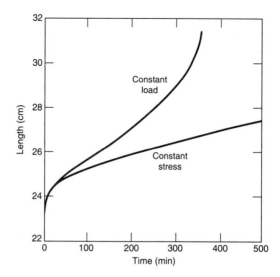

Figure 17.4 Creep tests of lead wire. Initial loads and initial lengths were the same. (From Andrade, 1910.)

structural instability that leads to fracture. For many metals this instability occurs at about the same strain in the creep test, as it does in the tension test. At temperatures where creep is important, metals are often metallurgical unstable; that is, their properties change either gradually or abruptly with time. Because of these metallurgical changes, or the lack of them, the creep rate of a metal may remain constant for long periods of time, it may increase continuously from the beginning of the creep test, or it may remain constant for a brief time and then increase rapidly (even for a constant stress creep test) to fracture. Thus, one should not assume that all metals always exhibit three distinct regions of primary, secondary, and tertiary creep.

In machines and structures, ordinarily only small creep strains on the order of 1% or 2% are permitted. In addition, high-strength, heat-resistant alloys fracture at relatively small deformations. Consequently, much creep testing is restricted to small strains. For constant stress creep tests of a metallurgically stable metal, creep strain in the secondary stage appears to vary almost linearly with time (Fig. 17.1); that is, the creep rate is constant. For this reason, the secondary stage of creep is referred to as *steady-state creep*, and straight-line approximations of this stage are used. However, early published results of tests (Robinson, 1943) that lasted *12 years* (100,000 hr) showed that the creep rate changed continuously throughout this time. Nevertheless, over a particular period of time, experimentally determined straight-line approximations can be used with sufficient accuracy for many purposes. More recently, Evans and Wilshire (1985) questioned the steady-state ideas that have dominated theoretical studies of creep for over the past half-century. Although Evans and Wilshire acknowledge that the concept of steady-state creep has led to progress, they believe that further development depends on new concepts based on the view that the steady-state creep condition is not attained even at high temperatures. These new concepts are based on experimental data obtained using high-precision constant-stress machines, rather than constant-load machines. Evans and Wilshire (1985) explored the idea that most materials exhibit a *minimum* rather than a *steady-state* creep rate and have developed new formulas and computer programs to study creep. Their work may be particularly important in the case of large creep deformations that lead to rupture. However, these concepts and procedures lie outside the scope of our study here. Therefore, we employ the concept of steady-state creep and consider mainly the primary and secondary stages of creep.

Similarity of Creep Curves

The similarity of creep curves has been used by many authors in the development of phenomenological theories of creep (Boresi and Sidebottom, 1972; Rabotnov, 1969). Similarity of creep curves means that the creep deformation is representable in the form

$$\epsilon_c = F(\sigma)f(t) \tag{17.1}$$

Thus, for a given temperature, the stress dependency $F(\sigma)$ is separate from the time dependency $f(t)$. For example, for isothermal conditions a common form of Eq. (17.1) is the power form [the Bailey–Norton equation; Eq. (x), Table 17.1]

$$\epsilon_c = at^n\sigma^b \tag{17.2}$$

Experimentally, the separation of time and stress dependencies is fairly well justified for the initial part of the creep curve. Indeed, for many metals, Eq. (17.2) is valid
for the initial part of the creep curve.

A series of creep curves may be considered as a graphical representation of the
equation

$$\epsilon = \epsilon(\sigma, t) \qquad (17.3)$$

with one (ϵ, t) curve for each value of σ (Fig. 17.5). Alternatively, the relation among
ϵ, σ, and t may be expressed by plotting σ vs ϵ for given times $t_1, t_2, \dots$ (Fig. 17.6).
Curves of (σ, ϵ), for given times $t_1, t_2, \dots$, are called *isochronous creep curves*.
For some materials, isochronous creep curves are similar. Thus, by analogy to
Eq. (17.1), isochronous creep curves may be represented by the relation

$$\sigma = G(\epsilon)g(t) \qquad (17.4)$$

However, the conditions of similarity of isochronous creep curves are very different from ordinary creep curves; see Eq. (17.1) (see Rabotnov, 1969).

In Eq. (17.4), if we set $g(0) = 1$, then $\sigma = G(\epsilon)$ is the instantaneous deformation
relation. In published experimental data on creep, the data for the initial stage of
creep are sometimes unreliable, since the instantaneous strain ϵ_0 that occurs on

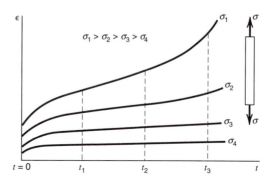

Figure 17.5   Constant-stress creep-time curves.

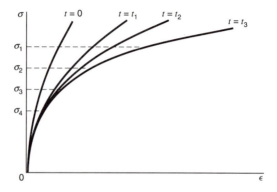

Figure 17.6   Isochronous stress-strain diagram.

applying the load is not recorded accurately. Consequently, creep curves for very small values of t frequently are not accurate (Evans and Wilshire, 1985). By selecting $g(t)$ appropriately, we can obtain the instantaneous curve $\sigma = G(\epsilon)$ by extrapolation. However, we have no guarantee that the actual instantaneous curve is obtained. Rabotnov (1969) has represented the function $g(t)$ by the formula

$$g(t) = \frac{1}{1 + at^b} \tag{17.5}$$

and from experimental data has estimated that $b \approx 0.3$.

By comparison of analytical and experimental creep results, it is found that very small changes in stress produce a large change in creep rate, and consequently, the time required to achieve a particular strain. Small differences in micro-structure or chemical composition of material specimens also greatly affect creep rate. Consequently, in experiments to determine the time required to attain a given creep strain at a fixed stress level, the scatter of results for a given set of specimens may be quite large. However, if we determine experimentally the stress at which a particular strain is reached in a given time, the scatter is small. On this basis, Rabotnov (1969) noted that there is some experimental evidence of accurately predicted creep deformation that confirms the similarity of isochronous creep curves [Eq. (17.4)]. However, much experimental evidence does not fully confirm Eq. (17.4). In some cases, similarity is clearly not valid, but if it exists, it can be used to simplify calculations.

Temperature Dependency

The effect of temperature on creep response may be summarized by noting that an increase in temperature results in an increase of creep rate. Hence, at a given stress level, a given strain is obtained more quickly at higher rather than at lower temperatures. In general, all creep parameters are affected by temperature. Consequently, a correlation of the temperature effects on the parameters that occur in the numerous creep equations that have been proposed (Table 17.1) is not feasible. Physical concepts are usually employed to predict the temperature dependence of creep curves (Cadek, 1988; Evans and Wilshire, 1985). The results have been reasonably successful. Formulas that have been used to predict the temperature dependency of creep curves of metals are listed in Table 17.1 [Eqs. (k) through (s)]. Ordinarily, in practice, the temperature dependency of creep curves is required mainly for interpolation over a fairly narrow range of temperatures; extrapolation far outside of the range is not particularly important to the engineer. In general, it is reasonable to use a simple formula for temperature dependency, even though the formula may have no physical relationship to a physically based formula that requires complicated calculations.

Variable Stress and Temperature

The objective of elementary creep theory based on a phenomenological model is to determine the strain as a function of time, given stress and temperature as functions of time. By elementary, we mean theories that include all the results of one-dimensional creep tests at constant stress and temperature. Alternatively, the objective is to develop an equation or a system of equations that accurately relate measured values of stress, strain, temperature, and time. The form of the stress-strain-temperature-time creep relation may be chosen to fit only certain parts of the creep curve. For example, for a creep test in the secondary range of creep, the creep rate may be approximately constant for constant stress and temperature. If the creep test is run at a higher stress level, the creep rate increases. Thus, for constant temperature tests in the secondary range, the creep rate $\dot{\epsilon}_C$ may be expressed as a function of stress level σ

$$\dot{\epsilon}_C = \dot{\epsilon}_{SC}(\sigma) \tag{17.6}$$

where $\dot{\epsilon}_{SC}$ is the secondary-stage creep rate. Equation (17.6) is sometimes applied to creep of metals that undergo long-time use in which most of the creep occurs at a constant rate (see curve C_0, Fig. 17.2), or to short-time creep at very high temperature and very high stress (curve C_3, Fig. 17.2). The use of Eq. (17.6) ignores the primary and tertiary stages of creep (Fig. 17.2). Hence, the creep deformation ϵ_C is approximated by straight lines (Fig. 17.7). Creep models that employ Eq. (17.6) are called *steady-state creep models*. Steady-state creep models are not capable of describing relaxation phenomena, since they do not include unloading effects properly. There have been several modifications of steady-state models. For example, to allow for the effect of initial elastic deformation on steady-state creep, Söderberg (1936) proposed the equation

$$\dot{\epsilon} = \frac{\dot{\sigma}}{E} + \dot{\epsilon}_{SC}(\sigma) \tag{17.7}$$

where $\dot{\epsilon}$ is the total strain rate, $\dot{\sigma}$ is the stress rate, $\dot{\epsilon}_{SC}$ is the steady-state creep rate, and E is the modulus of elasticity. Equation (17.7) ignores primary creep. Experiments and the predictions of Eq. (17.7) do not ordinarily agree well, since the primary creep strain is often as large or larger than the elastic strain. Odqvist (1953) proposed an equation for steady-state creep that approximates the effect of elastic strain, instantaneous plastic strain, and primary creep on the secondary creep rate, It is

$$\dot{\epsilon} = [\epsilon_0'(\sigma)]\dot{\sigma} + \dot{\epsilon}_{SC}(\sigma) \tag{17.8}$$

where ϵ_0 is elastic strain plus instantaneous plastic strain plus primary stage creep, and the prime denotes derivative with respect to σ. Equation (17.8) generally can

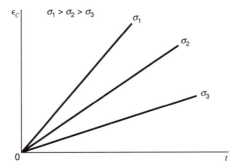

Figure 17.7   Straight-line approximation of creep curves.

be made to agree well with experimental creep curves. The creep model curve associated with Eq. (17.8) starts at time $t = 0$ and is a straight line, asymptotic to the steady-state creep curve (Fig. 17.8).

One may represent the creep curve for constant temperature as a relationship of the type

$$\epsilon = f(\sigma, t) \tag{17.9}$$

that is a general form of Eq. (17.1). It is tempting to assume that this relationship holds true when stress varies with time. However, on the basis of invariance relative to time, Eq. (17.9) leads to contradictions (Rabotnov, 1969). Models in which stress, strain, temperature, and time are related functionally as in Eqs. (17.1) and (17.9) are called *aging (time) models*. Since these models are applied mainly to primary and secondary ranges of creep and in these regions the creep rate decreases (i.e., the resistance to creep increases or the material hardens), they are also called *time-hardening models*. In a modified form of aging (time hardening), the creep strain rate $\dot{\epsilon}_C$ is taken as a function of stress and time. In particular, in a form analogous to Eq. (17.9), the creep rate is taken as

$$\dot{\epsilon}_C = \dot{\epsilon} - \frac{\dot{\sigma}}{E} = f(\sigma, t) \tag{17.10}$$

The time-hardening model of Eq. (17.10) is more logically acceptable than that of Eq. (17.9), since instantaneous change in stress does not produce an instantaneous change in creep strain; rather, it produces an instantaneous change in creep strain rate. Predictions based on the time-hardening model of Eq. (17.10) also agree well with experiments for small changes in stress levels. The time-hardening model of Eq. (17.10) is particularly easy to apply for similar creep curves [Eq. (17.1)]. Then,

$$\dot{\epsilon}_C = \dot{\epsilon} - \frac{\dot{\sigma}}{E} = F(\sigma)f(t) \tag{17.11}$$

If we change the time scale and take $f(t)$ as the independent variable rather than t, we may write Eq. (17.11) in the form

$$\frac{d\epsilon_C}{df} = \frac{d\epsilon}{df} - \frac{1}{E}\frac{d\sigma}{df} = F(\sigma) \tag{17.12}$$

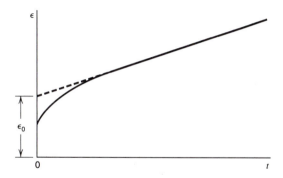

Figure 17.8   Approximation of steady-state creep stage.

This form is particularly suited to the relaxation problem, since then $\epsilon = $ constant (refer to the first paragraph in Sec. 17.3). For $\epsilon = $ constant, Eq. (17.12) yields

$$f(t) = -\frac{1}{E} \int \frac{d\sigma}{F(\sigma)} \qquad (17.13)$$

For thermally stable materials (i.e., for materials whose structure and properties do not change under prolonged exposure to test temperatures in the absence of load), it is natural to assume the existence of an equation of state that relates the creep rate to the applied stress and accumulated creep strain. Such an assumption uniquely relates the degree of strain hardening to the amount of plastic deformation in a manner analogous to the way that work hardening in the theory of plasticity is related to plastic deformation (Sec. 4.2). Thus, for a strain-hardening hypothesis, the equation of state is represented as

$$F(\dot{\epsilon}_C, \sigma, \epsilon_C, T) = 0 \qquad (17.14)$$

Equations of state were employed early in the study of plasticity (Ludwik, 1908; Nadai, 1931). However, Davenport (1938) was one of the first to introduce the concept of strain hardening. Strain-hardening theory can be checked most simply by a creep test in which stepwise changes in stress are made (Pickel et al., 1971). Generally speaking, stepwise changes in stress (load) are easy to make. Also, theoretical predictions of various hardening theories show wider differences for stepwise loading than they do for relaxation tests. Consequently, stepwise loading tests serve as a better check on the accuracy of hardening models than relaxation tests (Boyle and Spence, 1983).

For example, let us examine creep tests with a stepwise change in stress. To illustrate the differences between predictions of a time-hardening model and strain-hardening model, consider two creep curves for two stress levels: σ_1, σ_2; and $\sigma_1 < \sigma_2$ (Fig. 17.9). Assume that the creep curves can be represented by the relation $\epsilon_C = a\sigma^b t^n$ [Eq. (17.2)], a common representation of creep in the primary and secondary

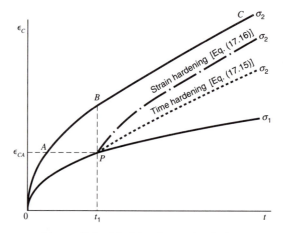

Figure 17.9 Models of creep hardening.

creep ranges for isothermal conditions. The creep rate form of this equation is

$$\dot{\epsilon}_C = na\,\sigma^b t^{n-1} \tag{17.15}$$

Equation (17.15) is a time-hardening model of the creep rate, since $0 < n < 1$ and therefore the creep rate decreases (creep resistance hardens) with time. Equation (17.15) can be written in a form independent of time t by eliminating t between Eqs. (17.2) and (17.15); then,

$$\dot{\epsilon}_C = \frac{na^{(1/n)}\sigma^{(b/n)}}{\epsilon_C^{(1-n)/n}} \tag{17.16}$$

Equation (17.16) indicates that the creep strain rate decreases (resistance to creep strain hardens) with increasing creep strain ϵ_C. Thus, Eq. (17.16) is called a strain-hardening model, since the hardening phase is modeled using the creep parameter ϵ_C. Time- and strain-hardening models can be developed similarly for more complex creep functions. Equations (17.15) and (17.16) give the same strain rate for constant stress.

Let us assume that Eq. (17.2) can be applied to varying stress conditions, although it was developed for constant stress states. For varying stress, we find that time and strain hardening give different creep rates. For example, let us subject a metal specimen first to stress σ_1. The creep is predicted by the lower curve in Fig. 17.9. At time $t = t_1$, let the stress level be increased instantaneously to σ_2. Deleting the instantaneous elastic strain (since we plot creep strain ϵ_C), we find that the new creep curve leaves the σ_1 creep curve at point P. Depending on the hardening model used, predictions for the new curve differ. For example, if the time-hardening model is used [Eq. (17.15)], the creep rate at point P depends solely on the time t_1 and stress level σ_2. Hence, by the time-hardening model, the new creep curve will leave point P with a creep rate of curve σ_2 at time t_1 (point B) and continue parallel to BC of curve σ_2. On the other hand, for a strain-hardening model, the new creep rate depends on the stress level σ_2 and accumulated creep ϵ_{CA}. Hence, by the strain-hardening model, the new creep curve will leave point P with the creep rate of curve σ_2 at point A and will continue parallel to ABC of curve σ_2. Thus, as observed in Fig. 17.9, appreciable differences in the two predictions are apparent.

Quite often, strain-hardening models give more accurate predictions of experimental results for stepwise changes of stress. Unfortunately, strain-hardening models do not always yield accurate predictions, particularly when several step changes in stress occur in the same test (Rabotnov, 1969). Furthermore, the strain-hardening model is unable to accurately predict results due to structural instabilities (Pickel et al., 1971). Nevertheless, for structurally stable metals, generally predictions by the strain-hardening model are fairly reliable and the model is relatively easy to use.

Finally, the equation-of-state approach [Eq. (17.14)] has been used by many authors to describe various processes in the behavior of metals. As in the theory of plasticity, it has been assumed that ϵ_C is the total plastic strain and $\dot{\epsilon}_C$ is therefore the rate of plastic strain. If one accepts these assumptions, for a fixed temperature T, Eq. (17.4) may be considered a surface in space with coordinates $\dot{\epsilon}_C$, ϵ_C, and σ; then the instantaneous plastic deformation curve should be given by the intersection of this surface with the plane $\dot{\epsilon}_C = \infty$. Rabotnov (1969) has shown that this procedure is not generally valid, and the problem of instantaneous plastic strain must be solved by other means. Accordingly, the equation-of-state approach to hardening [Eq. (17.14)] is best suited to those situations involving structurally stable metals in which a predominate creep mechanism exists and in which only instantaneous elastic deformation σ/E exists. Then the total strain is $\epsilon = \sigma/E + \epsilon_C$. See Evans and Wilshire (1985) for another approach to varying stress effects on creep that does not directly use either time- or strain-hardening models.

A more general hardening hypothesis is that the creep rate depends on stress, temperature, and a number of parameters $(p_1, p_2, \ldots, p_N)$ which characterize the creep processes (Rabotnov, 1963). Then, Eq. (17.14) becomes

$$F(\dot{\epsilon}_C, \sigma, T, p_i) = 0, \qquad i = 1, 2, \ldots, N \tag{17.17}$$

The time-hardening hypothesis is characterized by $N = 1$, $p_1 = t$, and the strain-hardening hypothesis is given by $N = 1$, $p_1 = \epsilon_C$. By selecting $N > 1$ and different p_i, other hardening models may be obtained (Taira, 1962).

In applications, many special forms of Eqs. (17.14) or (17.17) have been used. In particular, analytical forms have been chosen to represent Eq. (17.14), with the objective of including the effect of the principal mechanism of creep. Often, the condition of similarity of creep curves is assumed. Then, for constant temperature, Eq. (17.14) is represented by [see Eq. (17.1)]

$$\dot{\epsilon}_C = F(\sigma)f\left[\frac{\epsilon_C}{S(\sigma)}\right] \tag{17.18}$$

If f is a power function of ϵ_C/S, we may write

$$\dot{\epsilon}_C = \epsilon_C^{-\beta}h(\sigma) \tag{17.19}$$

where β is a constant. Integration of Eq. (17.19) yields (Rabotnov, 1969, p. 210)

$$\epsilon_C = g(\sigma)t^n \tag{17.20}$$

where $n = 1/(1 + \beta)$. Various forms of $g(\sigma)$ have been used; for example, a power function

$$g(\sigma) = A\sigma^b; \qquad b > \frac{1}{n} \tag{17.21}$$

or an exponential function

$$g(\sigma) = B\exp\left(\frac{\sigma}{c}\right) \tag{17.22}$$

Equation (17.22) is not suitable for small values of stress, since $g(\sigma)$ must tend to zero as σ goes to zero. An improvement

$$g(\sigma) = C\left[2\sinh\left(\frac{\sigma}{B}\right)\right]^b \tag{17.23}$$

was suggested by Garofalo (1965), but this expression is more complicated to use in practice.

More generally, to adequately describe both the primary and secondary creep of metals, it is necessary to broaden Eq. (17.18) into the form

$$\dot{\epsilon}_C = H(\epsilon_C)h(\sigma) \tag{17.24}$$

where $H(\epsilon_C)$ behaves like $\epsilon_C^{-\beta}$ for small values of ϵ_C and tends to a constant for the secondary range of creep. Rabotnov (1969) suggests the form

$$H(\epsilon_C) = \epsilon_C^{-\beta} + C_1 \tag{17.25}$$

where C_1 is a constant. For $\beta = 2$, Eqs. (17.24) and (17.25) are equivalent to Andrade's law in the primary range, since creep deformation is then proportional to the 1/3 power of time [see Eq. (i), Table 17.1].

Techniques for the experimental determination of the constants in creep equations, such as Eqs. (17.20) through (17.25), are discussed by Conway (1968) and by Evans and Wilshire (1985). The creep parameters A and b in Eq. (17.21) or B and c in Eq. (17.22) generally depend on temperature T; see, for example, Eq. (cc), Table 17.1. If the temperature range is sufficiently small, b (or c) is practically constant. However, A (or B) varies considerably with T. To express the effects of temperature T, the parameters A and B are often taken in the form

$$A = A_0 \exp\left(\frac{-U}{RT}\right), \qquad B = B_0 \exp\left(\frac{-U}{RT}\right) \tag{17.26}$$

where U is activation energy (Cadek, 1988) and R the universal gas constant. Thus, a general form of the strain hardening law that includes temperature effects is

$$\dot{\epsilon}_C = \epsilon_C^{-\beta} f(\sigma) \exp\left(\frac{-U}{RT}\right) \tag{17.27}$$

where $f(\sigma)$ and the parameter β are considered to be independent of temperature T.

17.5

CREEP UNDER MULTIAXIAL STATES OF STRESS

General Discussion

Most engineering systems (machines, structures, aircraft, etc.) operate under multiaxial stress conditions. Creep tests of members in such systems are very difficult and expensive to perform. In addition, a large amount of experimental and analytical data must be accumulated to make meaningful comparisons between experimental results and analytical predictions. Digital computer representations of analytical predictions in the form of field maps (Boresi and Sidebottom, 1972) or finite element programs (Chapter 19) overcome some of this difficulty by displaying the entire analytical calculation in a single diagram or map.

Metallurgical models of multiaxial stress creep are lacking. Therefore, models of multi-axial creep are mainly phenomenological in form (Gooch and How, 1986; Boyle and Spence, 1983; Kraus, 1980). These models are based mainly on concepts from the theory of plasticity of metals at normal (room) temperatures, where time effects are negligible or absent. Multiaxial plasticity theories predict reasonably accurate results for proportional loading. However, for arbitrary loading paths, considerable differences between experimental results and analytical predictions may occur. In addition, experimental data does not absolutely indicate the validity of any particular theory of plasticity. Since the extension of a single plasticity theory to creep may proceed in several ways, the number of possible creep models is quite large. Fortunately, in a number of engineering creep problems, the stress state ordinarily varies slowly with time. Consequently, different creep models may predict rather similar results.

The multiaxial creep problem is far more complex than the uniaxial case because of the fact that one-dimensional quantities (scalars) must now be replaced by tensor quantities. Hence, instead of a single creep strain ϵ_C and strain rate $\dot{\epsilon}_C$, now a creep strain tensor ϵ_{ij}^C and creep strain rate tensor $\dot{\epsilon}_{ij}^C$ $(i, j = 1, 2, 3)$ enter. Thus, the equation of state becomes a relationship among the creep strain rate tensor $\dot{\epsilon}_{ij}^C$, stress tensor σ_{ij}, and hardening parameters that may be scalars or more generally tensors of any order. When the hardening parameters are scalars, the hardening is said to be *isotropic*. Engineering phenomenological models of multiaxial creep have been based primarily on isotropic hardening assumptions, even though predictions so obtained may, in some cases, disagree with experiments.

The simplest case of multiaxial creep is that in which the stress state is homogeneous (constant from point to point) and constant with time. As expected, most available experimental results are for this case. From this basis, a number of methods are used to extend the analysis to nonhomogeneous stress states that vary with time. As noted above, however, the number of possibilities is much larger than in plasticity theory. Indeed, it is possible to describe a large number of creep models treating special effects such as steady-state creep (creep deformation rate constant with constant stress), creep with isotropic hardening, creep with anisotropic hardening, and so on.

As noted in the one-dimensional creep theory (Sec. 17.3), it is difficult if not impossible to determine the end of the instantaneous deformation (elastic and plastic) and the beginning of creep deformation in a creep test. However, the error introduced into the analysis by this unknown is generally small. Likewise, it is not possible to distinguish precisely between the primary and secondary stages of creep. The determination of the transition from the primary to secondary stage of creep has been attempted in several ways. For example, in the one-dimensional case, it is often assumed that as creep deformation ϵ_C increases, the function $H(\epsilon_C)$ in Eq. (17.24) tends to a definite limit that is attained either for a particular value of ϵ_C or as $\epsilon_C \to \infty$ [Eq. (17.25)]. Another method based on metallurgical concepts is to assume that primary and secondary creep are controlled by different micromechanisms that coexist simultaneously, but independently. In this case, the total deformation at any time consists of the instantaneous strain ϵ_0 (elastic and plastic), primary creep strain ϵ_{PC}, and secondary creep strain ϵ_{SC}. The primary creep (transient creep) is described by an equation of state like Eq. (17.14), namely,

$\dot{\epsilon}_{PC} = F_P(\sigma, \epsilon_C, T)$, which dampens out with time so that $\dot{\epsilon}_C \to 0$. Other elaborate schemes have been devised. Depending on the scheme employed for the transition from the primary to secondary stage of creep, the analysis may proceed by widely differing paths. In engineering problems, issues such as simplicity and convenience often dictate methods. On this basis, the first method noted above is used most frequently.

Two cases of steady-state creep are prevalent. Under one set of conditions, strain hardening may be negligible from the initiation of loading. Under other conditions, the creep rate becomes constant only after some time as the material becomes fully strain-hardened and cannot undergo further hardening as creep continues. For example, if the temperature and stress levels are sufficiently high (as in short-term creep tests), strain hardening is negligible. However, at relatively low temperatures, the creep rate may become constant only after long times, steady-state creep being preceded by a period of strain hardening. The difference in responses in the two cases is clear under variable loading. If the material does not strain harden, the instantaneous creep rate depends only on the instantaneous stress. If steady-state creep is preceded by a transient (primary) period, when a step change in load is applied in the steady-state region, the steady-state creep rate does not change instantaneously to a new value, but rather the transient period of the creep curve is repeated (more or less), until finally after some time the creep rate is again constant (Fig. 17.10).

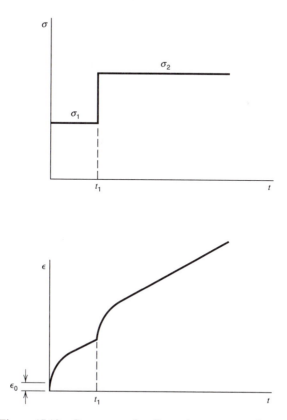

Figure 17.10   Creep curve for discontinuous stress change.

For creep under multiaxial states of stress, the effects of strain hardening are more important, since strain hardening may create anisotropy in the material relative to subsequent creep after a change in loading path. Consequently, in steady-state creep, as well as nonsteady creep, under multiaxial stress states, one must distinguish between conditions of isotropic and anisotropic behavior. These conditions are influenced by the initial state of the material, strain hardening or structural changes caused by instantaneous plastic creep (which may occur due to large instantaneous loads or due to creep if the stresses are nonuniform and become redistributed), and strain hardening that occurs in the transient creep phase.

As noted above, time-dependent inelastic theories for multiaxial states of stress are based on idealized models of material behavior. The models are similar to those used in the theory of plasticity (time-independent, inelastic behavior; Chapter 4). As noted in Sec. 4.3, the mathematical theory of plasticity is based on three postulates (assumptions):

1. There exists an initial yield state defined by a point on a yield surface specifying the states of stress for which plastic flow begins, the yield criterion.

2. There exists a rule relating the increment of the plastic state of strain to a specified increment in the state of stress, the flow rule.

3. There exists a rule specifying the modification of the yield surface during the course of the plastic flow, the hardening rule.

In addition, it is often assumed that the yield surfaces are independent of hydrostatic states of stress.

In the case of time-dependent deformation, the concept of a yield surface has no meaning. However, Drucker (1959) has pointed out that one may speak of surfaces $\phi(\sigma_{ij}) = $ constant in stress space for both time-dependent and time-independent inelastic behavior. For von Mises and Tresca materials, these surfaces correspond to the condition $\phi = \sigma_e = $ constant, where σ_e denotes the effective stress (Sec. 4.3). For time-independent inelastic deformation, the surface $\sigma_e = $ constant is a yield surface. For time-dependent inelastic deformation, the surface $\sigma_e = $ constant is interpreted by Drucker as a surface along which the rate of dissipation of energy is a constant; that is, $\sigma_e \dot{\epsilon}_{eC}$, where $\dot{\epsilon}_{eC} = d\epsilon_{eC}/dt$ is the effective creep strain rate defined by

$$\dot{\epsilon}_{eC} = \frac{\sqrt{2}}{3} \sqrt{(\dot{\epsilon}_{xx}^C - \dot{\epsilon}_{yy}^C)^2 + (\dot{\epsilon}_{yy}^C - \dot{\epsilon}_{zz}^C)^2 + (\dot{\epsilon}_{zz}^C - \dot{\epsilon}_{xx}^C)^2 + 6(\dot{\epsilon}_{xy}^C)^2 + 6(\dot{\epsilon}_{yz}^C)^2 + 6(\dot{\epsilon}_{zx}^C)^2}$$

(17.28)

where here the superscript C denotes creep.

The mathematical theory of inelasticity for time-dependent (creep) deformations, analogous to plasticity theory, is based on the following three conditions:

1. There exist surfaces for which $\phi = $ constant that are independent of hydrostatic states of stress.

2. There exists a rule relating $\dot{\epsilon}_{eC}$ to σ_e, temperature, and stress histories, the creep flow rule.

3. There exists a rule specifying the modification of the surface $\phi = $ constant during the process of inelastic deformation, the creep hardening rule.

Condition (1) is generally assumed valid for metals, and isotropic hardening is often assumed in condition (3). The flow rule (condition 2) for uniaxial states of stress is obtained from the tension test. However, for time-dependent inelastic deformation, even for the uniaxial case, the flow rule is not known precisely, since the strain rate is generally a function of not only stress and temperature, but also stress and temperature histories (Sec. 17.3). We consider the question of selection of a flow rule in Sec. 17.6.

17.6

FLOW RULE FOR CREEP OF METALS SUBJECTED TO MULTIAXIAL STATES OF STRESS

The flow rule for multiaxial states of stress generally is based on the existence of a flow rule for the uniaxial state of stress (the tension test or the torsion test of a thin-wall circular cross-section cylinder; see Chapter 4). Often, the multiaxial flow rule is obtained from the flow rule for tension specimens by replacing the tensile stress σ by the effective stress σ_e (Chapter 4) and the tensile creep strain rate $\dot{\epsilon}_C$ by the effective creep strain $\dot{\epsilon}_{eC}$ [Eq. (17.28)].

Since an equation of state does not truly exist for metals that creep, the flow rule is not known for the uniaxial state of stress. However, as discussed in Sec. 17.4, a number of approximate flow rules have been proposed. Most of them are based on a family of constant stress creep curves obtained from tension specimens tested at the temperature of interest. If a temperature gradient is to be included in the analysis, a family of constant stress creep curves is obtained at each of two or more temperatures in the range of interest. The family of constant stress creep curves is incorporated into the multiaxial flow rule in a number of different ways.

Steady-State Creep

As noted in Sec. 17.5, in the case of multiaxial states of creep, the presence or absence of strain-hardening effects is most important. This is because strain hardening that results from creep or plastic deformation produces anisotropic changes in the material properties that alter subsequent time-dependent deformation (creep). The effects of anisotropy are well known in rolled members of steel or aluminum alloy. For example, the elastic properties of such members may vary 10 to 20% between the direction of rolling and the direction transverse to rolling. However, the creep rate in these directions, for a given stress and temperature, may vary by a factor of 2 or more. It has also been observed that the effects of anisotropy due to strain hardening during creep may cause even greater variations in the creep rate.

If one wishes to describe multiaxial creep in the secondary stage of creep (steady-state creep stage), one must distinguish between isotropic and anisotropic creep. In particular, one must be aware of the fact that anisotropy may be produced by several effects; for example, anisotropy may be caused by the manufacturing process, strain hardening due to plastic deformation at the instant of loading, strain hardening due to plastic deformation caused by changes in load during creep, or by strain hardening that occurs in the primary stage of creep. Anisotropy may also arise in steady-state creep, if the creep is of sufficient duration and magnitude to result in structural (physical) changes in the metal. The study of the effects of creep under anisotropic conditions lies outside the scope of our study here. However, it is a topic of current research in metals and composites (Sullivan, 1991; Ohno, 1990; Pan, 1991).

We restrict our discussion to isotropic creep in which the principal axes of stress and strain coincide during the creep process. We have seen in Chapters 2 and 3 that the general state of stress at any point in a body may be characterized by three principal stresses σ_1, σ_2, and σ_3, in three mutually perpendicular principal stress directions (axes). Likewise, the state of strain at the point may be defined in terms of three principal strains ϵ_1, ϵ_2, ϵ_3, in three mutually perpendicular principal strain directions (axes). For linear elastic isotropic material properties, the principal axes of stress and strain coincide (see Chapter 3). In terms of principal axes, the stress-strain relations of a linear elastic isotropic material may be written in the form

$$\epsilon_1 = \frac{1}{E}[\sigma_1 - v(\sigma_2 + \sigma_3)]$$

$$\epsilon_2 = \frac{1}{E}[\sigma_2 - v(\sigma_3 + \sigma_1)]$$

$$\epsilon_3 = \frac{1}{E}[\sigma_3 - v(\sigma_1 + \sigma_2)] \tag{17.29}$$

where E is the modulus of elasticity and v Poisson's ratio. In linear elastic theory, the history of loading is neglected. However, as noted previously, in inelastic deformations, such as plasticity and creep, the history of loading affects the deformations. Also, as noted in Sec. 17.5, it is therefore useful to employ creep strain-rate-stress relations in the study of creep deformation [see Eq. (17.28)].

In general, to derive a creep strain-rate-stress relation (flow rule) for creep problems, we employ the fact that the deformation in a creep process is largely inelastic. It has been noted experimentally that inelastic deformation does not involve volumetric changes, the volumetric change being principally elastic in form. Accordingly, as in the theory of plasticity, we assume that the volumetric change due to inelastic (creep) deformation is zero. The total volumetric strain is (Boresi and Chong, 1987)

$$e = \bar{I}_1 - 2\bar{I}_2 + 4\bar{I}_3 \tag{17.30}$$

where $\bar{I}_1$, $\bar{I}_2$, and $\bar{I}_3$ are the strain invariants [see Eq. (2.80)]. For small strains, relative to principal axes, Eq. (17.30) may be approximated as

$$e \approx \bar{I}_1 = \epsilon_1 + \epsilon_2 + \epsilon_3 \tag{17.31}$$

since $\bar{I}_2$, $\bar{I}_3$ are higher-order terms in the principal strains ($\epsilon_1, \epsilon_2, \epsilon_3$). We may separate the strain into elastic and inelastic (creep) parts. Thus,

$$\epsilon_1 + \epsilon_2 + \epsilon_3 = (\epsilon_1 + \epsilon_2 + \epsilon_3)_{\text{elastic}} + (\epsilon_1 + \epsilon_2 + \epsilon_3)_{\text{creep}} \tag{17.32}$$

By Eq. (17.29),

$$(\epsilon_1 + \epsilon_2 + \epsilon_3)_{\text{elastic}} = \frac{1 - 2v}{E}(\sigma_1 + \sigma_2 + \sigma_3) \tag{17.33}$$

With the assumption that the volume change due to creep is zero, the volumetric creep strain is zero; that is,

$$(\epsilon_1 + \epsilon_2 + \epsilon_3)_{creep} = \epsilon_{1C} + \epsilon_{2C} + \epsilon_{3C} = 0 \qquad (17.34)$$

where the subscript C denotes creep strain.

For isotropic creep deformation in which the stress distribution does not change with time, it follows that the principal axes of stress and strain remain coincident and do not rotate during the creep deformation. Then, we can differentiate Eq. (17.34) with respect to time to obtain

$$\dot\epsilon_{1C} + \dot\epsilon_{2C} + \dot\epsilon_{3C} = 0 \qquad (17.35)$$

as the condition of constant volume under creep deformation. Although Eq. (17.35) has been obtained on the basis that the principal strain directions do not rotate, it appears that is also gives good results in certain cases in which the principal strain directions do rotate (see Pickel et al., 1971, where solid circular bars of SAE 1035 steel were tested in torsion).

A second assumption, analogous to that used in plasticity theory (see Lubahn and Felgar, 1961, Chapter 8), is that the maximum shear strain rates [i.e., $\dot\epsilon_{ij} = (\dot\epsilon_i - \dot\epsilon_j)/2$] are proportional to the maximum shear stresses [see Eq. (4.14)]. Thus, we write

$$\frac{\dot\epsilon_{1C} - \dot\epsilon_{2C}}{\sigma_1 - \sigma_2} = \frac{\dot\epsilon_{2C} - \dot\epsilon_{3C}}{\sigma_2 - \sigma_3} = \frac{\dot\epsilon_{3C} - \dot\epsilon_{1C}}{\sigma_3 - \sigma_1} = C(x, y, z, t) \qquad (17.36)$$

where $C(x, y, z, t)$ is a function of location (x, y, z) in the body and time t. For steady-state creep, $C(x, y, z, t) \to C(x, y, z)$; that is C remains constant in time. In transient creep, C will change with time, since the creep strain rates change with time.

Solving Eqs. (17.35) and (17.36) for $\dot\epsilon_{1C}$, $\dot\epsilon_{2C}$, and $\dot\epsilon_{3C}$, we obtain

$$\dot\epsilon_{1C} = \tfrac{2}{3}C[\sigma_1 - \tfrac{1}{2}(\sigma_2 + \sigma_3)]$$
$$\dot\epsilon_{2C} = \tfrac{2}{3}C[\sigma_2 - \tfrac{1}{2}(\sigma_3 + \sigma_1)]$$
$$\dot\epsilon_{3C} = \tfrac{2}{3}C[\sigma_3 - \tfrac{1}{2}(\sigma_1 + \sigma_2)] \qquad (17.37)$$

To determine the parameter C, the creep behavior of the material must be known for given strain rates and stresses. For this purpose, we employ concepts analogous to plasticity theory (Sec. 4.3) and define an effective stress, an effective strain, and an effective strain rate. First, we note that the yielding of many metals has been shown to be predicted by either the Tresca [Eq. (4.12)] or von Mises criterion [Eq. (4.22)]. Accordingly, say, for the von Mises criterion, we write for the effective stress [see Eq. (4.23)]

$$\sigma_e^M = \frac{1}{\sqrt{2}}[(\sigma_1 - \sigma_2)^2 + (\sigma_2 - \sigma_3)^2 + (\sigma_3 - \sigma_1)^2]^{1/2} \qquad (17.38)$$

where superscript M denotes von Mises. Similarly, for the effective strain and effective strain rate, we write

$$\epsilon_{eC}^M = \frac{\sqrt{2}}{3}[(\epsilon_{1C} - \epsilon_{2C})^2 + (\epsilon_{2C} - \epsilon_{3C})^2 + (\epsilon_{3C} - \epsilon_{1C})^2]^{1/2} \qquad (17.39)$$

and

$$\dot{\epsilon}_{eC}^{M} = \frac{\sqrt{2}}{3}[(\dot{\epsilon}_{1C} - \dot{\epsilon}_{2C})^2 + (\dot{\epsilon}_{2C} - \dot{\epsilon}_{3C})^2 + (\dot{\epsilon}_{3C} - \dot{\epsilon}_{1C})^2]^{1/2} \qquad (17.40)$$

Returning to the case of steady-state creep, we consider a structural member subjected to constant stress and constant temperature for a long period of time. If the deformations are sufficiently small to preclude tertiary creep, the creep deformation-time diagram for the member corresponds approximately to the solid curve in Fig. 17.11. The instantaneous deformation OA may be entirely elastic or partly elastic and partly plastic. The primary creep range AB is followed by the secondary (steady-state) creep range BC. For long times and for a relatively brief period of primary creep, it may be sufficiently accurate to approximate the steady-state creep deformation by the dashed straight line OS (taken parallel to line DBC). In derivations of a steady-state creep theory based on the straight line OS, the temperature and stress components of each volume element in the member are usually assumed to remain constant with time. Since the primary creep effect is neglected, the flow rule is given by a relation that approximates the creep strain rate for steady-state conditions. Then for a given temperature, the creep rate is a function of stress only, and the flow rule takes the form [see Eq. (17.6)]

$$\dot{\epsilon}_{eC}^{M} = F(\sigma_e) \qquad (17.41)$$

Several relations have been proposed for the function F [see Eqs. (t) through (aa), Table 17.1). A widely used formula proposed by Bailey (1935) is

$$\dot{\epsilon}_{eC}^{M} = B(\sigma_e^{M})^n \qquad (17.42)$$

Equation (17.42) has the merit that it is simple to use. The factor $1/\sqrt{2}$ in Eq. (17.38) is such that in a simple tension test with stress σ_1 and strain rate $\dot{\epsilon}_{1C}$, we have $\sigma_e^{M} = \sigma_1$ and $\dot{\epsilon}_{eC}^{M} = \dot{\epsilon}_{1C}$ by Eq. (17.40), since $\dot{\epsilon}_{2C} = \dot{\epsilon}_{3C} = -(1/2)\dot{\epsilon}_{1C}$ by Eq. (17.37). Then, by the first of Eqs. (17.37) and Eq. (17.41), we obtain $\dot{\epsilon}_{eC}^{M} = 2C\sigma_e^{M}/3$, or

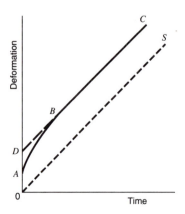

Figure 17.11   Deformation-time diagram: $0A$=instantaneous deformation; AB=primary stage; BC=secondary stage; $0S$=straight-line approximation ignoring $0D$; DBC = Odqvist approximation of ABC.

$C = 3\dot{\epsilon}_{eC}^{M}/(2\sigma_{e}^{M})$. With this value of C and Eq. (17.42), we can write Eq. (17.37) in the form (flow rule)

$$\dot{\epsilon}_{1C} = B(\sigma_{e}^{M})^{n-1}[\sigma_{1} - \tfrac{1}{2}(\sigma_{2} + \sigma_{3})]$$
$$\dot{\epsilon}_{2C} = B(\sigma_{e}^{M})^{n-1}[\sigma_{2} - \tfrac{1}{2}(\sigma_{3} + \sigma_{1})]$$
$$\dot{\epsilon}_{3C} = B(\sigma_{e}^{M})^{n-1}[\sigma_{3} - \tfrac{1}{2}(\sigma_{1} + \sigma_{2})] \qquad (17.43)$$

Alternatively, if we take the Tresca yield criterion, the effective stress and effective strain rate are defined as [see Eq. (4.14) with $\sigma_{1} > \sigma_{2} > \sigma_{3}$]

$$\sigma_{e}^{T} = \sigma_{1} - \sigma_{3}, \qquad \dot{\epsilon}_{eC}^{T} = \dot{\epsilon}_{1C} - \dot{\epsilon}_{3C} \qquad (17.44)$$

where superscript T denotes Tresca. Then in place of Eq. (17.43), we have

$$\dot{\epsilon}_{1C} = \frac{2}{3}\frac{\dot{\epsilon}_{eC}^{T}}{\sigma_{e}^{T}}[\sigma_{1} - \tfrac{1}{2}(\sigma_{2} + \sigma_{3})]$$

$$\dot{\epsilon}_{2C} = \frac{2}{3}\frac{\dot{\epsilon}_{eC}^{T}}{\sigma_{e}^{T}}[\sigma_{2} - \tfrac{1}{2}(\sigma_{3} + \sigma_{1})]$$

$$\dot{\epsilon}_{3C} = \frac{2}{3}\frac{\dot{\epsilon}_{eC}^{T}}{\sigma_{e}^{T}}[\sigma_{3} - \tfrac{1}{2}(\sigma_{1} + \sigma_{2})] \qquad (17.45)$$

In cases where the order $\sigma_{1} > \sigma_{2} > \sigma_{3}$ is retained during loading, Eq. (17.45) may be easier to use than Eq. (17.43); see Finnie and Heller (1959, Sec. 7.3) for a discussion of the differences between the results predicted by Eqs. (17.43) and (17.45).

Nonsteady Creep

Multiaxial creep models have been used to predict stresses or deformations of structural members subjected to prescribed stress or deformation histories and to prescribed temperature histories, including temperature gradients. Often, the flow rule for these problems is based on an equation that approximates a family of constant-stress creep curves, if only one temperature is considered, or that approximates families of constant stress creep curves over a range of temperatures. As a generalization of Eq. (17.27), the flow rule takes the form [see also Eqs. (bb) through (gg), Table 17.1]

$$\epsilon_{eC} = F(\sigma_{e})G(T_{a})\phi(t) \qquad (17.46)$$

where $F(\sigma_{e})$, $G(T_{a})$, and $\phi(t)$ are functions of effective stress σ_{e}, absolute temperature T_{a}, and time t, respectively. A number of investigators have used the form [see Eqs. (x) and (cc), Table 17.1]

$$\epsilon_{eC} = B\sigma_{e}^{n}[e^{(-A/T_{a})}]t^{k}, \qquad 0 < k < 1 \qquad (17.47)$$

where B, n, A, and k are material constants. Frequently, better accuracy is obtained by replacing σ_{e}^{n} by $(\sinh b\sigma_{e})^{n}$, where b is an additional material constant [see Eq. (ee), Table 17.1].

Forms such as Eq. (17.47) are ordinarily used for constant stress and temperature conditions. The effect of varying stress or temperature on the effective strain rate $\dot{\epsilon}_{eC}$ is included in the creep model by the introduction of a hardening rule. As noted in Sec. 17.4, two hardening rules, the time-hardening and strain-hardening [see Eqs. (17.15) and (17.16)] concepts, are commonly used. The time-hardening rule assumes that the creep strain rate $\dot{\epsilon}_C$ depends on stress, temperature, and time. With the time-hardening rule, the creep strain rate is obtained by taking the time derivative of Eq. (17.46). For the specific form of Eq. (17.47), we have

$$\dot{\epsilon}_{eC} = kB\sigma_e^n[e^{(-A/T_a)}]t^{(k-1)} \tag{17.48}$$

The strain-hardening rule states that $\dot{\epsilon}_C$ depends on stress, temperature, and strain. The strain-hardening form is obtained by eliminating time between Eqs. (17.47) and (17.48). Thus, we get

$$\dot{\epsilon}_{eC} = k\{B\sigma_e^n[e^{(-A/T_a)}]\}^{1/k}\epsilon_C^{(k-1)/k} \tag{17.49}$$

Numerous applications of time- and strain-hardening flow rules have been given in the literature (Rabotnov, 1969; Boresi and Sidebottom, 1972; Boyle and Spence, 1983; Kraus, 1980). Because of the complexity of the creep problem, approximate solutions are obtained by numerical techniques, such as iteration methods (e.g., successive elastic solutions) and finite element methods (Zienkiewicz, 1977; see also Chapter 19). Boresi and Sidebottom (1972) have given several comparisons of multiaxial flow rules by the method of successive elastic solutions.

Relatively few problems in creep analysis admit closed-form solutions. One such problem is the steady-state creep of a thick cylinder; other cases include the problem of the steady-state, small-strain creep of an infinite rectangular plate with a small circular hole (see Sec. 14.2) and certain steady-state creep problems in the membrane theory of shells (Boyle and Spence, 1983, Chapter 4; Kraus, 1980, Chapter 3).

17.7

A SIMPLE APPLICATION OF CREEP OF METALS

One of the simplest problems for which a closed-form solution of creep is possible is that of a thin-wall metal tube with closed ends, subjected to constant internal pressure and temperature (see Chapter 11). By equilibrium, the stresses in the radial, circumferential, and axial directions are

$$\sigma_{rr} = 0, \qquad \sigma_{\theta\theta} = \frac{pr}{h}, \qquad \sigma_{zz} = \frac{pr}{2h} \tag{17.50}$$

where r is the inner radius of the tube, h is the tube wall thickness, and (r, θ, z) refer to the radial, circumferential, and axial directions, respectively. For small strains, the stresses remain constant for constant pressure, since the changes in geometry are negligible.

Consider a Tresca material. Then, if we use the time-hardening formulation given by Eq. (17.48), we get

$$\dot{\epsilon}_{eC}^T = kB(\sigma_e^T)^n [e^{(-A/T_a)}] t^{(k-1)} \tag{17.51}$$

Also, by Eqs. (17.50) and the first of Eqs. (17.44), we have

$$\sigma_e^T = \sigma_1 - \sigma_3 = \sigma_{\theta\theta} - \sigma_{rr} = \frac{pr}{h} \tag{17.52}$$

By Eqs. (17.45), (17.51), and (17.52), we obtain

$$\dot{\epsilon}_{1C} = \dot{\epsilon}_{\theta\theta C} = \frac{1}{2} \frac{\dot{\epsilon}_{eC}^T}{\sigma_e^T} \left(\frac{pr}{h}\right) = \frac{1}{2} kB \left(\frac{pr}{h}\right)^n [e^{(-A/T_a)}] t^{(k-1)}$$

$$\dot{\epsilon}_{2C} = \dot{\epsilon}_{zzC} = 0$$

$$\dot{\epsilon}_{3C} = \dot{\epsilon}_{rrC} = -\frac{1}{2} \frac{\dot{\epsilon}_{eC}^T}{\sigma_e^T} \left(\frac{pr}{h}\right) = -\frac{1}{2} kB \left(\frac{pr}{h}\right)^n [e^{(-A/T_a)}] t^{(k-1)} \tag{17.53}$$

As a result, we see that the tube grows radially in diameter, but maintains its length. Considering the radial displacement due to creep, $u_C = r\epsilon_{\theta\theta C}$, we have

$$u_C = r\epsilon_{\theta\theta C} = r \int_0^t \dot{\epsilon}_{\theta\theta C} \, d\tau \tag{17.54}$$

If we assume that initial strain is entirely elastic, the total radial displacement is $u = u_{\text{elastic}} + u_C$, where u_{elastic} is given by Hooke's law [Eq. (17.29)]

$$u_{\text{elastic}} = r\epsilon_{\theta\theta\text{elastic}} = \frac{r}{E} [\sigma_{\theta\theta} - v(\sigma_{zz} + \sigma_{rr})] = \frac{pr^2}{Eh} \left(1 - \frac{v}{2}\right) \tag{17.55}$$

Hence, the total radial displacement is, after carrying out the integration of Eq. (17.54),

$$u = \frac{pr^2}{Eh} \left(1 - \frac{v}{2}\right) + \frac{1}{2} Br \left(\frac{pr}{h}\right)^n [e^{(-A/T_a)}] t^k \tag{17.56}$$

Summary

Since $0 < k < 1$, the creep rate $\dot{\epsilon}_{eC}^T \to 0$ as time t becomes large [see Eq. (17.51)]. The radial displacement u continues to increase with time, and the circumference of the tube increases. Since the volumetric change is zero, the thickness of the tube decreases.

17.8

CREEP OF NONMETALS

Under appropriate conditions, most materials will creep. For example, nonmetallic materials such as glass, polymers, portland cement paste, and so on, creep when subjected to sufficiently high temperatures and stresses. As pointed out by Finnie and Heller (1959), the mechanical behavior of many nonmetallic materials during creep is somewhat simpler than that of metals. This has been attributed to the fact that nonmetallic materials like glass, polymers, and cements are more nearly isotropic than metals and large creep strains are required to induce anisotropy in them. The creep of glass and polymers is often treated by the theory of linear viscoelasticity. The creep behavior of other nonmetals, such as concrete, asphalt, and wood, is very complex. Nevertheless, one of the first applications of the theory of linear viscoelasticity was in the study of creep in concrete (Rabotnov, 1969). Concrete is a material that undergoes an aging process, such that under sustained load the modulus of elasticity changes with time. Generally, the properties of concrete depend on its age; that is, property changes occur that are independent of deformation. Aging is a phenomenon that alters creep of concrete. It is caused mainly by cement hydration, a process that continues for a long time after the initial hardening period. Aging changes the rate of creep and, hence, must be accounted for. This fact increases the difficulty of predicting the creep behavior of concrete. Aging effects have been discussed by Bazant (1977), and Bazant and Prasannan (1989a, b). Asphalt, a widely used pavement material, acts much like a viscoelastic material. However, the creep behavior of asphalt resembles that of concrete. In the following, we give a brief description of the creep behavior of asphalt, concrete and wood.

Asphalt

The early work of Van der Poel (1954) discusses asphaltic mixtures and their applications to road design. A large number of references to 1953 are listed by Van der Poel. More recently, Bolk (1981) published a manual on the creep test, which summarizes much of the work on the creep of asphalt conducted by The Netherlands Government Highway Engineering Laboratory. In this study, the representation of creep behavior was examined from the viewpoint that asphalt possesses elastic, viscous, and plastic properties dependent on the temperature and duration of loads. At low temperatures and/or for short duration of load, asphalt behaves in an almost linear elastic manner. At high temperatures and/or long duration of loads, asphalt responds in a viscous manner. Asphalt responds plastically at high levels of loads or under localized high stress (even at low loads). Consequently, rheological modeling is employed (Hills, 1973) and, for theoretical analyses, the various rheological components are assumed to be independent of one another. In a test, it is difficult, if not impossible, to separate the measured deformation into its rheological components. Also in creep tests, deformation behavior over long periods of time is of primary interest. Nevertheless, the permanent deformation (the deformation that remains after reversible deformation is recovered; see Monismith and Tayeboli, 1988) of asphalt is composed of viscous, plastic, and visco-plastic components. These effects change in importance with the duration of load and temperature. For example, during a creep test of asphalt, the binder film that exists between mineral particles becomes thinner. As a result, mineral particle-to-particle contact gradually occurs. Because these particles are relatively dry, the shear force required to maintain a shear strain rate gradually increases; that is, the asphalt appears to strain-harden (Prendergast, 1992). Ordinarily, creep tests are performed in a temperature range for which asphalt does not behave as a purely viscous material, and in these tests, the viscous component of deformation is nonlinear in time. Thus, even in a well-controlled creep test, the separation of the permanent deformation of asphalt into various linear rheological components is not feasible. Consequently, in practice, a phenomenological approach is taken, much as in the case of metals (Bolk, 1981). However, the material constants in the creep rate equations exhibit a stronger dependency on temperature and time than they do for metals (see Table 17.1).

The relation of creep to the engineering properties of asphalt (e.g., rutting of pavements, total deformation, strength, etc.) has been examined extensively (see, e.g., Bolk, 1981; Eckmann, 1989; Monismith and Tayeboli, 1988). Bolk (1981) has concluded that the correlation between the deformation measured in a creep test and that measured in a rutting test of asphalt hardly changes when the permanent strain in the flow rule is replaced by the total strain measured in the creep test. Hence, for practical purposes, it is sufficient to measure only total strain. The measurement of the reversible strain is therefore optional. Bolk (1981) concluded that the creep test is a valuable tool for prediction of engineering properties, such as rutting of asphalt pavements. He also gives recommendations for conducting uniaxial static creep tests with asphalt test specimens and recommends the logarithmic flow rule, $\epsilon = A + B \log t$, where A and B are material constants determined by the tests. A relatively low value of B indicates low viscous behavior; a high value of B suggests mainly viscous behavior.

Concrete

The creep of concrete is affected by a large number of factors. For example, water-reducing admixtures tend to increase creep rates, as do retarding admixtures and accelerating admixtures (Mindess and Young, 1981). Many other experimental variables affect the creep of concrete, for example, paste parameters (porosity, age, etc.), concrete parameters (aggregate stiffness, aggregate/cement content, volume to surface ratio), and environmental parameters (applied stress, duration of load, humidity, etc.). Usually, the creep of concrete is influenced more by paste properties, since the aggregate tends to retard creep rate. Since creep data on paste are limited, creep data on concrete are relied on to assess the influence of various parameters. Concrete is often treated as an isotropic material, and creep flow formulas similar to those of metals are used to represent concrete creep. For example, creep dependence on stress and temperature is often represented by the formula

$$\epsilon_C = C \sinh\left(\frac{V\sigma}{RT}\right) \tag{17.57}$$

where C is a constant, V the activation volume, R the universal gas constant, and T the absolute temperature [see Eq. (z), Table 17.1]. Equation (17.57) has been used with success to represent experimental data. Since the parameter $V\sigma/RT$ is generally small, the creep strain-stress-temperature relation is approximately linear in the stress range ordinarily used in concrete. From a practical viewpoint, the creep strain-stress relation in concrete is commonly taken to be

$$\epsilon_C = \phi\sigma \tag{17.58}$$

where ϕ is called the *specific creep*. The concept of specific creep is useful for comparing the creep of different concrete specimens at different stress levels. A typical value of ϕ is approximately 150 μ/MPa, $\mu = 10^{-6}$. Although Eqs. (17.57) or

(17.58) are frequently used to estimate creep in concrete, many empirical equations (some simple, some complex) have been used to predict creep in concrete. These equations are used in the same manner as for metal creep. There is considerable disagreement regarding a specific equation to represent different aspects of concrete creep, because of the difficulty in separating effects such as shrinkage from the more conventional effects. Nevertheless, under the assumption that the initial instantaneous strain, creep strain, and shrinkage strain are independent and additive, the American Concrete Institute (ACI, 1991) has developed a simplified creep equation of the form

$$\frac{\epsilon_C}{\epsilon_e} = \frac{t^{0.6}}{B + t^{0.6}} C_{\text{ult}}$$

(17.59)

where t denotes time, B is a constant that depends on the age of the concrete before loading (B is taken to be 10 when the concrete is more than 7 days old before loading), and C_{ult} is the ultimate creep coefficient. The value of C_{ult} is difficult to determine, as it may vary considerably (for 40% relative humidity C_{ult} may range between 1.30 and 4.5). ACI recommends a value of $C_{\text{ult}} = 2.35$, if experimental data are not available. ACI also recommends certain correction factors to adjust C_{ult} for different conditions of humidity and age at loading. The interested reader is referred to the *ACI Manual of Concrete Practice* (ACI, 1991) for details.

Wood

Hearmon (1954) gave one of the early reviews of creep data of wood up to 1953. More recently, Bodig and Jayne (1982) wrote a comprehensive treatise on the mechanics of wood and wood composites. They approached the creep of wood from a rheological (flow) point of view, as a study of the time-dependent stress-strain behavior of materials. Since wood is highly anisotropic, the magnitude of creep strain depends on a large number of factors. The most critical conditions include the alignment of the orthotropic axes of wood relative to the load, the magnitude and type of stress, the rate of load, the duration of load, moisture content, and temperature. There are many practical situations in which the creep of wood is particularly important: for example, the deflection of wood beams and other types of load-carrying wood members under long durations of load; the reduction of pressure between layers of glulam members due to creep relaxation, resulting in loss of bonding; creep rupture of wood members at sustained loads less than the ultimate static load; and so on. In spite of the complexity of wood, many of the same concepts employed in the study of creep in metals are used, and the study of creep in wood rests heavily on curve-fitting of experimental data to obtain approximate flow rules. Because of the nature of the manufacturing process of wood composites, creep relaxation may strongly affect the serviceability of wood composites (Bodig and Jayne, 1982). In contrast to creep models of metals that are described in terms of three stages of creep, Bodig and Jayne consider the total creep deformation to consist of elastic and viscoelastic parts. For example, consider a tension specimen subjected to an instantaneously applied load P at time t_0 (Fig. 17.12). The deformation at some later time t_1 is taken to be

$$\delta_1 = \delta_e + \delta_{de} + \delta_v$$

(17.60)

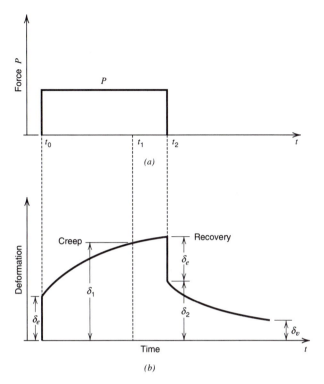

Figure 17.12   Creep curve. (a) Load-time function. (b) Components of creep.

where δ_e is the instantaneous elastic deformation due to the instantaneous application of load, δ_{de} a delayed elastic deformation, and δ_v a viscous component. To determine the delayed elastic part, one imagines that the load is instantaneously reduced to zero at time t_2. Then, the elastic deformation δ_e is instantaneously recovered. As time continues, additional deformation is recovered (Fig. 17.12). Since the process of recovery is irreversible, a residual deformation remains. The nonrecoverable (viscous) deformation developed to time t_2 is δ_v. At time t_2, the total deformation is δ_2. It consists of the viscous deformation δ_v and the delayed elastic deformation δ_{de}. For a particular time $t > t_2$, δ_v can be obtained from the data (Fig. 17.12). Therefore,

$$\delta_{de} = \delta_2 - \delta_v \qquad (17.61)$$

Tests show that for many wood composites, δ_v increases linearly with time. By subtracting the viscous contribution from the total deformation-time curve, the delayed elastic component can be determined. Guided by this simplistic one-dimensional model and employing linear rheological models (e.g., linear viscoelastic models, namely, the Maxwell body, Kelvin body, and Burger body), Bodig and Jayne (1982, Sec. 5.4) derived creep formulas that are the synthesis of linear viscoelastic models and experimental data. For example, for a beam made of a flakeboard-veneer composite, they derived the strain-time relation

$$\epsilon = \frac{\sigma}{10^6}\left(\frac{1}{0.84} + \frac{1 - e^{-t/1.2557}}{5.48} + \frac{t}{72.37}\right) \qquad (17.62)$$

where σ is in lb/in.^{2} and t in hours. Note that the first term is the instantaneous elastic strain, the second term is the delayed elastic strain, and the third is the viscous strain [see Eq. (17.59)]. A similar approach for a multiaxial stress state may be carried out following the concepts presented in Sec. 17.6.

REFERENCES

American Concrete Institute (ACI) (1991). *Manual of Concrete Practice*, Part I: Materials and General Properties of Concrete. Detroit, Mich.

American Society for Metals (ASM) (1976). *Metals Handbook; Failure Analysis and Prevention*, Vol. 10. Metals Park, Ohio: p. 249.

American Society for Testing and Materials (ASTM) (1959). Effects of Nonsteady Load and Temperature Conditions on the Creep of Metals, STP 260. Philadelphia, Pa.

ASTM (1965). Literature Survey on Creep Damage in Metals, STP 391. Philadelphia, Pa.

ASTM (1983). Recommended Practice for Conducting Creep, Creep-Rupture and Stress-Rupture Tests of Metallic Materials, Std. E139. Philadelphia, Pa.

Andrade, E. N. da C. (1910). The Viscous Flow in Metals and Allied Phenomena. *Proc. Roy. Soc.* Ser. A, **84**: 1.

Bailey, R. W. (1935). The Utilization of Creep Test Data in Engineering Design. *Proc. Inst. Mech. Eng.*, **131**: 131.

Bazant, Z. P. (1977). Viscoelasticity of Solidifying Porous Material — Concrete. *J. Eng. Mech., ASCE*, **103** (6): 1049–1067.

Bazant, Z. P. and Prasannan, S. (1989a). Solidification Theory for Concrete Creep I: Foundation. *J. Eng. Mech., ASCE*, **115** (8): 1691–1703.

Bazant, Z. P. and Prasannan, S. (1989b), Solidification Theory for Concrete Creep II: Verification and Application. *J. Eng. Mech. ASCE*, **115** (8): 1704–1725.

Bernasconi, G. and Piatti, G. (eds.) (1979). *Creep of Engineering Materials and Structures*. Essex, England: Elsevier.

Bodig, J. and Jayne, B. A. (1982). *Mechanics of Wood and Wood Composites*. New York: Van Nostrand Reinhold.

Bolk, Ir. H. J. N. (1981). *The Creep Test*. The Netherlands: Study Ctr. for Road Construction.

Boresi, A. P. and Chong, K. P. (1987). *Elasticity in Engineering Mechanics*. New York: Elsevier.

Boresi, A. P. and Sidebottom, O. M. (1972). Creep of Metals Under Multiaxial States of Stress. *Internat. J. Nucl. Eng. and Design*, **18**: 415–456.

Boyle, J. T. and Spence, J. (1983). *Stress Analysis for Creep*. London: Butterworth.

British Standard Institute (BSI) (1987). Methods for Creep and Rupture Testing of Metals, Tensile Creep Testing, BS3500, Part 6. London.

Cadek, J. (1988). *Creep in Metallic Materials*. New York: Elsevier.

Conway, J. B. (1968). *Numerical Methods for Creep and Rupture Analysis*. New York: Gordon and Breach.

Conway, J. B. (1969). *Stress-Rupture Parameters: Origin, Calculation and Use*. New York: Gordon and Breach.

Curbishley, I. (ed.) (1988). *Mechanical Testing*. London: Ins. of Metals.

Davenport, C. C. (1938). Correlation of Creep and Relaxation Properties of Copper. *J. Appl. Mech.*, **60**: A55 – A60.

de Lacombe, J. (1939). A Method of Representing Creep Curves. *Rev. Metallurgy*, **36**: 178.

Dorn, J. E. (1962). Progress in Understanding High-Temperature Creep, H. W. Gillet Mem. Lecture. Philadelphia, Pa.: ASTM.

Drucker, D. C. (1959). A Definition of Stable Inelastic Material. *J. Appl. Mech. Trans. ASME*, **81**: 101–106.

Eckmann, B. (1989). Exxon Research in Pavement Design—Moebius Software: A Case Study Reduction of Creep Through Polymer Modification. *Proc. Assoc. Asphalt Paving Technologist*, **58**: 337–361.

Evans, R. and Wilshire, B. (1985). *Creep of Metals and Alloys*. London: Inst. of Metals.

Feltham, P. (1953). *Proc. Phys. Soc.*, London, **66**: 865.

Finnie, I. and Heller, W. R. (1959). *Creep of Engineering Materials*. New York: McGraw-Hill.

Freundenthal, A. M. (1936). Theory of Wide-Span Arches in Concrete and Reinforced Concrete. *Internat. Assoc. Bridge and Structural Engineers*, **4**: 249.

Garofalo, F. (1965). *Fundamental of Creep and Creep Rupture in Metals*. New York: MacMillan.

Gooch, D. J. and How, I. M. (eds.) (1986). *Techniques for Multiaxial Creep Testing*. New York: Elsevier.

Graham, A. (1953). The Phenomenological Method in Rheology. *Res.*, **6**: 92.

Graham, A. and Walles, K. F. A. (1955). Relations Between Long and Short Time Properties of a Commercial Alloy. *J. Iron and Steel Inst.*, **179**: 105–120.

Gray, T. G. F. (1988). Tensile Testing. In *Mechanical Testing*, I., Curbishley, ed.). London: Inst. of Metals, Chapter 1.

Hearmon, R. F. S. (1954). Wood. In *Building Materials, Their Elasticity and Inelasticity*, (M. Reiner, ed.). Amsterdam: North Holland, Chapter 5.

Hills, J. F. (1973). The Creep of Asphalt Mixes. *J. Inst. Petrol.*, **59** (570): 247–262.

Hult, J. (1966). *Creep in Engineering Structures*. Waltham, Mass.: Blaisdell Publ.

International Standards Organization (ISO) (1987). *Metallic Materials— Verifications of Extensometers Used in Uniaxial Testing*. Geneva, Switzerland: DIS 9513.

Kachanov, L. M. (1960). *Theory of Creep*. London: UK Nat. Lending Library for Sci. and Technol. (translation 1967).

Kanter, J. (1938). Problem of Temperature Coefficient of Tensile Creep Rate. *Trans. AIME*, **131**: 385–418.

Kennedy, A. J. (1962). *Creep and Stress Relaxation in Metals*. London: Oliver and Boyd.

Kraus, H. (1980). *Creep Analysis*. New York: Wiley.

Larson, F. R. and Miller, J. (1952). A Time-Temperature Relationship for Rupture and Creep Stresses. *Trans. ASME*, **74**: 765.

Loveday, M. S. (1988). Creep Testing. In *Mechanical Testing*, (I. Curbishley, ed.). London: Inst. of Metals, Chapter 2.

Lubahn, J. D. and Felgar, R. P. (1961). *Plasticity and Creep of Metals*. New York: Wiley.

Ludwik, P. (1908). *Elemente der Technologischen Mechanik*: Berlin: Springer-Verlag.

Manson, S. S. and Haferd, A. M. (1954). Washington, D.C.: NACA, Tech. Note 3159.

McVetty, P. G. (1934). Working Stresses for High Temperature Service, *Mech. Eng.*, **56**: 149.

McVetty, P. G. (1943). Creep of Metals at Elevated Temperatures—The Hyperbolic Sine Relation Between Stress and Creep Rate. *Trans. ASME*, **65**: 761.

Mindess, S. and Young, J. F. (1981). *Concrete*. Englewood Cliffs, N.J.: Prentice-Hall.

Monismith, C. L. and Tayeboli, A. A. (1988). Permanent Deformation (Rutting) Considerations in Asphalt Concrete Pavement Sections. *Proc. Assoc. Asphalt Paving Technologists*, **57**: 414–463.

Mott, N. F. (1953). A Theory of Work-Hardening of Metals II. *Phil. Mag.*, **44**: 742.

Mott, N. F. and Nabarro, F. R. N. (1948). Dislocation Theory and Transient Creep: Report on a Conference on the Strength of Solids. *Proc. Phys. Soc.*, London, Series B, **64**: 1–19.

Nadai, A. (1931). *Plasticity*. New York: McGraw-Hill.

Norton, F. H. (1929). *The Creep of Steel at High Temperature*. New York: McGraw-Hill.

Odqvist, F. K. G. (1953). Influence of Primary Creep on Stresses in Structural Parts. *Teckniska Hogskolan—Handlingar*, Stockholm (66): 18 pp.

Odqvist, F. K. G. (1966). *Mathematical Theory of Creep and Creep Rupture*, 2nd ed. (1974). London: Clarendon Press.

Odqvist, F. K. G. and Hult, J. (1962). *Creep Strength of Metallic Structures*. Berlin: Springer-Verlag. (in German).

Ohno, N. (1990). Recent Topics In Constitutive Modeling of Cyclic Plasticity and Viscoplasticity. *Appl. Mech. Rev.*, **43** (11): 283–295.

Pan, T.-Y. (1991). Thermal Cycling Induced Plastic Deformation in Solder Joints—Part 1: Accumulated Deformation in Surface Mount Joints. *J. Elec. Packaging, Trans. ASME*, **113** (1): 8–15.

Penny, R. K. and Marriott, D. L. (1971). *Design for Creep*. New York: McGraw-Hill.

Phillips, F. (1905). The Slow Stretch in India Rubber, Glass and Metal Wire When Subjected to a Constant Pull. *Phil. Mag.*, **9**: 513.

Pickel, T. W., Jr., Sidebottom, O. M., and Boresi, A. P. (1971). Evaluation of Creep Laws and Flow Criteria for Two Metals Subjected to Stepped Load and Temperature Changes. *Exper. Mech.*, **11** (5): 202–209.

Prendergast, J. (1992). A European Road Comes to the U.S. *Civil Eng.*, **62** (5): 52–54.

Rabotnov, Yu. N. (1963). On the Equations of State of Creep. *Proc. Joint. Internat. Conf. Creep*, **2**: 117–122.

Rabotnov, Yu. N. (1969). *Creep Problems in Structural Members*. Amsterdam: North-Holland.

Robinson, E. L. (1943). 100,000 Hour Creep Test. *Mech. Eng.* **65** (3): 166–168.

Smith, C. L. (1948). A Theory of Transient Creep in Metals. *Proc. Phys. Soc.*, Series B, London, **61**: 201–205.

Söderberg, C. R. (1936). The Interpretation of Creep Tests for Machine Design. *Trans ASME*, **58** (8): 733–743.

Stowell, E. Z. (1957). Phenomenological Relation Between Stress, Strain, and Temperature for Metals at Elevated Temperatures. Washington, D.C.: NACA, Tech. Note 4000.

Sullivan, J. L. (1991). Measurements of Composite Creep. *Exper. Mech.*, **15** (5): 32–37.

Taira, S. (1962). Lifetime of Structures Subjected to Varying Load and Temperature. In *Creep in Structures*. Berlin: Springer-Verlag, pp. 96–124.

Van der Poel, C. (1954). Road Asphalt. In *Building Materials, Their Elasticity and Inelasticity*, (M. Reiner, ed.). Amsterdam: North Holland, Chapter 9.

Vicat, L. J. (1834). Note sur l'allongemont progressif du fil de fer soumis à diverses tensions. *Ann. ponts et chausées, Mem. et Doc.* 7 (1): 40.

Warren, J. W. L. (1967). A Survey of the Mechanics of Uniaxial Creep Deformation of Metals. London: Aero. Res. Council.

Wyatt, O. H. (1953). Transient Creep in Pure Metals. *Proc. Phys. Soc.*, London, Series B, **66**: 459.

Zienkiewicz, O. C. (1977). *The Finite Element Method*, 3rd ed. New York: McGraw-Hill.

18

CONTACT STRESSES

18.1

INTRODUCTION

Contact stresses are caused by the pressure of one solid on another over limited areas of contact. Most load-resisting members are designed on the basis of stress in the main body of the member, that is, in portions of the body not affected by the localized stresses at or near a surface of contact between bodies. In other words, most failures (by excessive elastic deflection, yielding, and fracture) of members are associated with stresses and strains in portions of the body far removed from the points of application of the loads.

In certain cases, however, the contact stresses created when surfaces of two bodies are pressed together by external loads are the significant stresses; that is, the stresses on or somewhat beneath the surface of contact are the major cause of failure of one or both of the bodies. For example, contact stresses may be significant at the area (1) between a locomotive wheel and the railroad rail; (2) between a roller or ball and its race in a bearing; (3) between the teeth of a pair of gears in mesh; (4) between the cam and valve tappets of a gasoline engine; etc.

We note that in each of these examples, the members do not necessarily remain in fixed contact. In fact, the contact stresses are often cyclic in nature and are repeated a very large number of times, often resulting in a fatigue failure that starts as a localized fracture (crack) associated with localized stresses. The fact that contact stresses frequently lead to fatigue failure largely explains why these stresses may limit the load-carrying capacity of the members in contact and hence may be the significant stresses in the bodies. For example, a railroad rail sometimes fails as a result of "contact stresses"; the failure starts as a localized fracture in the form of a minute transverse crack at a point in the head of the rail somewhat beneath the surface of contact between the rail and locomotive wheel, and progresses outwardly under the influence of the repeated wheel loads until the entire rail cracks or fractures. This fracture is called a *transverse fissure failure*.

On the other hand, bearings and gear teeth sometimes fail as a result of formation of pits (pitting) at the surface of contact. The bottom of such a pit is often located approximately at the point of maximum shear stress. Steel tappets have been observed to fail by initiation of microscopic cracks at the surface that then spread and cause flaking. Chilled cast-iron tappets have failed by cracks that start beneath the surface, where the shear stress is highest, and spread to the surface, causing pitting failure.

The principal stresses at or on the contact area between two curved surfaces that are pressed together are greater than at a point beneath the contact area, whereas the maximum shear stress is usually greater at a point a small distance beneath the contact surface.

The problem considered here initially is to determine the maximum principal (compressive) and shear "contact stresses" on and beneath the contact area between two ideal *elastic* bodies having curved surfaces that are pressed together by external loads. Several investigators have attempted to solve this problem. H. Hertz* (1895) was the first to obtain a satisfactory solution, although his solution gives only principal stresses in the contact area.

18.2

THE PROBLEM OF DETERMINING CONTACT STRESSES

Two semicircular disks made of elastic material are pressed together by forces P (Fig. 18.1). The two bodies are initially in contact at a single point. Sections of the boundaries of the two bodies at the point of contact are smooth curves before the loads are applied. The principal radii of curvature of the surface of the upper solid at the point of contact are R_1 and R_1'. Likewise, R_2 and R_2' are the principal radii of curvature of the surface of the lower solid at the point of contact. The intersection of the planes in which the radii R_1 and R_2 (or the radii R_1' and R_2') lie form an angle α. Elevation and plan views, respectively, of the two solids are shown in

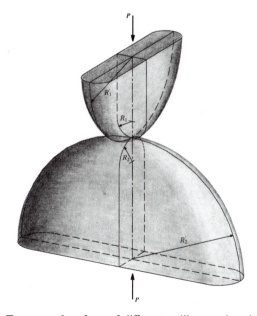

Figure 18.1 Two curved surfaces of different radii pressed against each other.

---

* In 1881 Hertz published a paper entitled "On Contact of Elastic Solids" and the following year, "On the Contact of Rigid Elastic Solids and on Hardness." See (Hertz, 1895).

Figs. 18.2a and b. The lines v_1 and v_2, which form the angle α, lie in the plane sections containing the radii R_1 and R_2, respectively. The line of action of load P lies along the axis that passes through the centers of curvature of the solids and through the point of contact. Hence, the line of action of force P is perpendicular to a plane that is tangent to both solids at the point of contact. In other words, it is assumed that there is no tendency for one body to slide with respect to the other and, hence, no friction force is present. The effect of a friction force is discussed in Sec. 18.9.

The effect of the load P is to cause the surface of the solids to deform elastically over a region surrounding the initial point of contact, thereby bringing the two bodies into contact over a small area in the neighborhood of the initial point of contact (Fig. 18.2b). The problem is to determine a relation between the load P and the maximum compressive stress on this small area of contact and to determine the principal stresses at any point in either body on the line of action of the load, designated as the z axis. The principal stresses σ_{xx}, σ_{yy}, and σ_{zz} acting on a small cube at a point on the z axis are shown in Fig. 18.2c. The maximum shear stress at the point is $\tau_{\max} = \frac{1}{2}(\sigma_{zz} - \sigma_{yy})$, where σ_{zz} and σ_{yy} are the maximum and minimum principal stresses at the point.

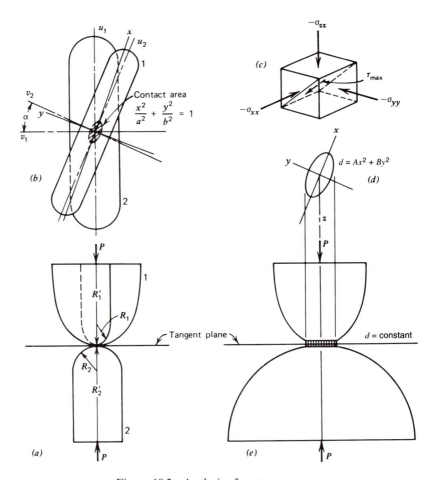

Figure 18.2 Analysis of contact stresses.

The detailed development of the solution of the problem will not be presented here. However, the main assumptions made in the solution are given so that the limitations on the use of the results may be understood. A brief discussion is given to explain and justify the assumptions.

18.3

GEOMETRY OF THE CONTACT SURFACE

Fundamental Assumptions
The solution of the problem of the contact stresses in the neighborhood of the point of contact of two bodies is based on the following two assumptions.

(a) Properties of Materials. The material of each body is homogeneous, isotropic, and elastic in accordance with Hooke's law, but the two bodies are not necessarily made of the the same material.

(b) Shape of Surfaces near Point of Contact, Before Loading. If two bodies are in contact at a point, there is a common tangent plane to the surfaces at the point of contact. In the solution for contact stresses, an expression for the distance between corresponding points on the surfaces near the point of contact is required; corresponding points are points that lie on the surfaces of the bodies and on a line perpendicular to the common tangent plane. Equations that express the two distances z_1, z_2 from corresponding points to the common tangent plane are needed to determine the deformations of the two bodies near the initial point of contact. In the analysis, an equation that *approximates* the total distance $d = z_1 + z_2$ between corresponding points on any two surfaces is used. This equation is

$$d = Ax^2 + By^2 \tag{18.1}$$

in which x and y are coordinates with respect to the y and x axes with origin at the point of contact; these coordinates lie in the tangent plane, and A and B are (positive) constants (Hertz, 1895) that depend on the principal radii of curvature of the surfaces at the point of contact. The derivation of Eq. (18.1) is discussed later in this section. Figures 18.2d and e illustrate the fact that the curve representing Eq. (18.1) for a given value of d is an ellipse. This fact will be important in considering the shape of the area of contact between the two bodies.

Contact Surface Shape After Loading
When the loads P are applied to the bodies, their surfaces deform elastically near the point of contact so that a small area of contact is formed. It is *assumed* that, as this small area of contact forms, points that come into contact are points on the two surfaces that originally were equal distances from the tangent plane. According to Eq. (18.1), such equidistant points on the two surfaces lie on an ellipse. Hence the boundary line of the area of contact is assumed to be an ellipse whose

equation is

$$\frac{x^2}{a^2} + \frac{y^2}{b^2} = 1 \tag{18.2}$$

where x and y are coordinates referred to the same axes as were specified for Eq. (18.1). The contact area described by Eq. (18.2) is shown in Fig. 18.2b. Equation (18.1) is of sufficient importance to warrant further discussion of its validity, particularly since a method of determining the constants A and B is required in the solution of the problem of finding contact stresses.

Justification of Eq. (18.1)

In order to obtain Eq. (18.1), an expression is derived first for the perpendicular distance z_1 from the tangent plane to any point on the surface of body 1 near the point of contact, by assuming that the bodies are free from loads and in contact at a point. A portion of body 1 showing the distance z_1 is illustrated in Fig. 18.3a. Let the points considered lie in the planes of principal radii of curvature. Let u_1 and v_1 be axes in the tangent plane that lie in the planes of principal radii of curvature of body 1. The distance z_1 to point C or D is found as follows. From triangle ODD'

$$z_1 = u_1 \tan \tfrac{1}{2}\beta = \tfrac{1}{2} u_1 \beta \tag{18.3}$$

since the angle β is small. From triangle HKD

$$\tan \beta = \beta = \frac{KD}{HK} = \frac{u_1}{R'_1} \tag{18.4}$$

since the radius R'_1 is approximately equal to HK. Substitution of the value of β from Eq. (18.4) into Eq. (18.3) gives

$$z_1 = \frac{u_1^2}{2R'_1} \tag{18.5}$$

In a similar manner, the distance z_1 to the points E and F lying in the plane of radius R_1 is found to be

$$z_1 = \frac{v_1^2}{2R_1} \tag{18.6}$$

On the basis of these results, it is assumed that the distance z_1 to any point G not lying in either plane of principal curvature may be approximated by

$$z_1 = \frac{u_1^2}{2R'_1} + \frac{v_1^2}{2R_1} \tag{18.7}$$

This assumption seems justified by the fact that Eq. (18.7) reduces to Eq. (18.6) for $u_1 = 0$, and Eq. (18.5) for $v_1 = 0$. In particular, we note that if z_1 is constant for all points G, Eq. (18.7) is an equation for an ellipse.

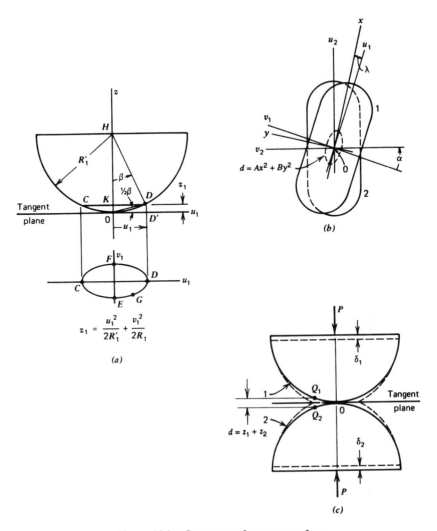

Figure 18.3   Geometry of contact surface.

Attention is directed now to the second body. The distance z_2 from the tangent plane to any point in the surface of body 2 near the point of contact is obtained in the same way as z_1 in Eq. (18.7). It is

$$z_2 = \frac{u_2^2}{2R_2'} + \frac{v_2^2}{2R_2} \tag{18.8}$$

where u_2 and v_2 are coordinates with respect to axes lying in the tangent plane and also in the planes of the principal radii of curvature R_2' and R_2, respectively. The locations of the axes u_1, v_1 and u_2, v_2 are shown in Fig. 18.3b, which is the same view of the bodies as in Fig. 18.2b. The axes v_1 and v_2 subtend the angle α, that is, the angle between the lines v_1 and v_2 of the bodies as shown in Fig. 18.2b.

The distance d between points on the two surfaces near the point of contact is the numerical sum of z_1 and z_2 given by Eqs. (18.7) and (18.8). Hence, we find

$$d = z_1 + z_2 = \frac{u_1^2}{2R_1'} + \frac{v_1^2}{2R_1} + \frac{u_2^2}{2R_2'} + \frac{v_2^2}{2R_2} \tag{18.9}$$

Equation (18.9) may now be transformed into the form of Eq. (18.1). The first transformation is the elimination of the coordinates u_2 and v_2 by the relationships

$$u_2 = u_1 \cos \alpha + v_1 \sin \alpha$$
$$v_2 = -u_1 \sin \alpha + v_1 \cos \alpha \tag{18.10}$$

When Eqs. (18.10) are substituted into Eq. (18.9), there results

$$d = A'u_1^2 + 2H'u_1v_1 + B'v_1^2 \tag{18.11}$$

where

$$2A' = \frac{1}{R_1'} + \frac{1}{R_2'}\cos^2 \alpha + \frac{1}{R_2}\sin^2 \alpha$$

$$2H' = \left(\frac{1}{R_2'} - \frac{1}{R_2}\right) \sin \alpha \cos \alpha$$

$$2B' = \frac{1}{R_1} + \frac{1}{R_2'}\sin^2 \alpha + \frac{1}{R_2}\cos^2 \alpha \tag{18.12}$$

Equation (18.11) is the equation of an ellipse, as shown in Fig. 18.3b, with center at point 0. To find the equation of the ellipse referred to axes x and y, which coincide with the major and minor axes of the ellipse, the value of the angle λ through which the axes u_1 and v_1 must be rotated in order to eliminate the product term u_1v_1 in Eq. (18.11) is required. The transformation is

$$u_1 = x \cos \lambda - y \sin \lambda$$
$$v_1 = x \sin \lambda + y \cos \lambda \tag{18.13}$$

If Eqs. (18.13) are substituted into Eq. (18.11) and the value of the angle λ taken to eliminate the product term u_1v_1, Eq. (18.11) becomes

$$d = Ax^2 + By^2 \tag{18.14}$$

which is identical in form to Eq. (18.1). In the process of making the transformation, it is found that A and B are the roots of a quadratic equation and have the following values:

$$B = \frac{1}{4}\left(\frac{1}{R_1} + \frac{1}{R_2} + \frac{1}{R_1'} + \frac{1}{R_2'}\right) \tag{18.15}$$
$$+ \frac{1}{4}\sqrt{\left[\left(\frac{1}{R_1} - \frac{1}{R_1'}\right) + \left(\frac{1}{R_2} - \frac{1}{R_2'}\right)\right]^2 - 4\left(\frac{1}{R_1} - \frac{1}{R_1'}\right)\left(\frac{1}{R_2} - \frac{1}{R_2'}\right)\sin^2 \alpha}$$

$$A = \frac{1}{4}\left(\frac{1}{R_1} + \frac{1}{R_2} + \frac{1}{R_1'} + \frac{1}{R_2'}\right) \tag{18.16}$$
$$- \frac{1}{4}\sqrt{\left[\left(\frac{1}{R_1} - \frac{1}{R_1'}\right) + \left(\frac{1}{R_2} - \frac{1}{R_2'}\right)\right]^2 - 4\left(\frac{1}{R_1} - \frac{1}{R_1'}\right)\left(\frac{1}{R_2} - \frac{1}{R_2'}\right)\sin^2 \alpha}$$

The constants A and B depend on the principal radii of curvature of the two bodies at the point of contact and on the angle α between the corresponding planes of the principal curvatures.

Note: With the definition for radii of curvature in Fig. 18.2, it is possible for the ratio B/A to be greater than or less than 1. If $\alpha = 0$ and $B/A > 1$, the *major* axis of the ellipse of contact is along the x axis and is said to be in the direction of rolling If $\alpha = 0$ and $B/A < 1$, the *minor* axis of the ellipse of contact is along the x axis (in the direction of rolling). In this latter case, it is convenient to let R_1 and R_2 be radii of curvature in the plane of rolling [the x-z plane, Fig. 18.2e] so that $B/A > 1$.

Brief Discussion of Solution

It was pointed out earlier in this section that Eq. (18.1) is used to estimate the displacement of points on the surfaces of the two bodies that eventually lie within the contact area. In Fig. 18.3c the solid outline shows the two bodies of Fig. 18.1 in contact at one point, before the loads are applied, and the dashed lines show the new positions of the two bodies after the loads P are applied and the two bodies are in contact over an area around the original point of contact 0. The centers of the bodies move toward each other by amounts of δ_1 and δ_2, respectively, which means that the distance between points on the bodies not affected by the local deformation near 0 is decreased by an amount $\delta_1 + \delta_2 = \delta$.

Let w_1 denote the displacement, due to local compression, of point Q_1, Fig. 18.3c. We take w_1 positive in the direction away from the tangent plane, assumed to remain immovable during local compression. Similarly, let w_2 denote the displacement, due to local compression, of point Q_2, where w_2 is taken positive in the direction away from the tangent plane. These positive directions of w_1 and w_2 conform to the positive directions of displacement in a small loaded region on a part of the boundary of a semiinfinite solid, that is, the positive displacement is directed into the solid. Hence, the distance between two points, such as Q_1 and Q_2 in Fig. 18.3, will diminish by $\delta - (w_1 + w_2)$. If, finally, due to the local compression, points Q_1 and Q_2 come in contact, we have

$$\delta - (w_1 + w_2) = z_1 + z_2 = d$$

With the expression for d, given by Eq. (18.14), we may write

$$w_1 + w_2 = \delta - Ax^2 - By^2 \tag{18.17}$$

Equation (18.17) has been obtained from geometrical considerations only. To compute the displacements (w_1, w_2), local deformation at the surface of contact must be considered. Under the assumption that the surface of contact is very small compared to the radii of curvatures of the bodies, the solution obtained for semiinfinite bodies subjected to spot loads may be employed to determine $w_1 + w_2$ (Timoshenko and Goodier, 1970; Thomas and Hoersch, 1930). Hertz noted that Eq. (18.17) has the same form as that of the Newtonian potential equation for the attraction of a homogeneous mass M in the shape of an ellipsoid on a unit of mass concentrated at a point P some distance from the ellipsoid. This Newtonian potential function satisfies the same differential equations that are required to be satisfied by the theory of elasticity. The problem is solved by placing into the potential equation the stresses at the contact surface instead of the mass, etc., and the constants are evaluated (Thomas and Hoersch, 1930). The solution is given in terms of elliptic integrals. The results are summarized in the following sections.

18.4

NOTATION AND MEANING OF TERMS

The following notation and interpretations of terms are needed for an understanding of subsequent equations:

P = total force exerted by body 1 on body 2, and vice versa.

E_1, E_2 = tensile (or compressive) moduli of elasticity for bodies 1 and 2.

v_1, v_2 = Poisson's ratio for bodies 1 and 2.

a = semimajor axis of ellipse of contact.

b = semiminor axis of ellipse of contact.

$k = b/a = \cos\theta; k \leq 1$.

$k' = \sqrt{1 - k^2} = \sin\theta$.

R_1, R_1' = principle values of the radii, respectively, of the surface of body 1 at the point of contact. The plane sections in which R_1, R_1' lie are perpendicular to each other. See Fig. 18.1. The signs of R_1 and R_1' are determined as follows. If the center of curvature lies inside the body (i.e., if the body surface is convex at the point of contact), the radius is positive. If the center of curvature lies outside the body (i.e., if the body surface is concave at the point of contact), the radius is negative. For example, in Fig. 18.1, the radii R_1, R_1' are positive.

R_2, R_2' = same as R_1, R_1', but for body 2; in Fig. 18.12c, R_2 is negative.

α = angle between planes of principal curvatures at point of contact (see Fig. 18.2b).

$k(z/b)$ = relative depth below the surface of contact to a point on the z axis at which stresses are to be calculated. The depth is expressed in terms of $k(z/b)$ rather than by z directly so that, in evaluating the integrals obtained in the mathematical solution of the problem, the term $k(z/b)$ can conveniently be replaced by a trigonometric function. Thus, $\cot\phi = k(z/b)$.

z_s = distance from surface to point on z axis at which maximum shear stress occurs in either body.

In the expressions for the principal stresses, two integrals (called elliptic integrals) are found that involve ϕ, θ, and k'. These integrals are denoted as $F(\phi, k')$ and $H(\phi, k')$. Likewise, two integrals involving k' alone, denoted as $K(k')$ and $E(k')$, are required. These elliptic integrals are

$$F(\phi, k') = \int_0^\phi \frac{d\theta}{\sqrt{1 - k'^2 \sin^2\theta}}$$

$$H(\phi, k') = \int_0^\phi \sqrt{1 - k'^2 \sin^2\theta}\, d\theta$$

$$K(k') = F\left(\frac{\pi}{2}, k'\right) = \int_0^{\pi/2} \frac{d\theta}{\sqrt{1 - k'^2 \sin^2 \theta}}$$

$$E(k') = H\left(\frac{\pi}{2}, k'\right) = \int_0^{\pi/2} \sqrt{1 - k'^2 \sin^2 \theta} \, d\theta$$

These integrals have been tabulated and are readily available in most mathematical handbooks.

18.5

EXPRESSIONS FOR PRINCIPAL STRESSES

The analysis involving the assumptions and limitations indicated in Sec. 18.3 yields the following expressions for the principal stresses σ_{xx}, σ_{yy}, and σ_{zz} at a point on the z axis. The point is at the distance z from the origin, which lies on the surface of contact of the two elastic bodies. The stresses act on orthogonal planes perpendicular to the x, y, and z axes, respectively. The solution of this problem is (Thomas and Hoersch, 1930)

$$\sigma_{xx} = [M(\Omega_x + v\Omega_x')]\frac{b}{\Delta} \tag{18.18}$$

$$\sigma_{yy} = [M(\Omega_y + v\Omega_y')]\frac{b}{\Delta} \tag{18.19}$$

$$\sigma_{zz} = -\left[\frac{M}{2}\left(\frac{1}{n} - n\right)\right]\frac{b}{\Delta} \tag{18.20}$$

in which

$$M = \frac{2k}{k'^2 E(k')}, \qquad n = \sqrt{\frac{k^2 + k^2(z/b)^2}{1 + k^2(z/b)^2}}$$

$$\Delta = \frac{1}{A+B}\left(\frac{1 - v_1^2}{E_1} + \frac{1 - v_2^2}{E_2}\right) \tag{18.20a}$$

where A and B are constants given by Eqs. (18.15) and (18.16), and where

$$\Omega_x = -\frac{1 - n}{2} + k\frac{z}{b}[F(\phi, k') - H(\phi, k')]$$

$$\Omega_x' = -\frac{n}{k^2} + 1 + k\frac{z}{b}\left[\left(\frac{1}{k^2}\right)H(\phi, k') - F(\phi, k')\right]$$

$$\Omega_y = \frac{1}{2n} + \frac{1}{2} - \frac{n}{k^2} + k\frac{z}{b}\left[\frac{1}{k^2}H(\phi, k') - F(\phi, k')\right]$$

$$\Omega_y' = -1 + n + k\frac{z}{b}[F(\phi, k') - H(\phi, k')]$$

Also, $v = v_1$ for a point in body 1 and $v = v_2$ for a point in body 2. We note that the stresses depend on the variables, A, B, k, k', v_1, v_2, E_1, E_2, b, and z. The first four variables depend only on the shape of the surfaces near the point of contact. Of these four, A and B are found from Eqs. (18.15) and (18.16), and from Sec. 18.4, $k' = \sqrt{1 - k^2}$. Therefore, one additional equation is needed for determining the value of k. This equation is

$$\frac{B}{A} = \frac{(1/k^2)E(k') - K(k')}{K(k') - E(k')} \tag{18.21}$$

The second group of four variables, v_1, v_2, E_1, and E_2, depend only on the properties of the two bodies in contact and are found by tests of the materials. The variable b, the semiminor axis of the area of contact, depends on the eight variables previously listed, but it is important to note that it also depends on the load P. The equation expressing this fact is

$$b = \sqrt[3]{\frac{3kE(k')}{2\pi}(P\Delta)} = ka \tag{18.22}$$

Values of the variable z, which represent the distance of a point *from the surface of contact*, may be chosen. Then the three principal stresses at any point on the z axis may be obtained.

18.6

METHOD OF COMPUTING CONTACT STRESSES

Principal Stresses

In Sec. 18.5 it is noted that the values of A and B must be computed first, and that in Eq. (18.21) the ratio B/A determines the value of k (and k'). It should be remembered that the values of A and B are related to the geometric shape and configuration of the two bodies. For example, if two cylinders are crossed so that they are in point contact with their longitudinal axes perpendicular, the value of $B/A = 1$, but if these cylinders are arranged so that their longitudinal axes are parallel (line contact), $B/A = \infty$. With the values of the four quantities A, B, k, and k' known, the terms in the brackets in Eqs. (18.18), (18.19), and (18.20) can be evaluated for a selected value of Poisson's ratio v. Fortunately, the value of v in these bracket terms has only a small influence on the final values of the stresses. Consequently, we take a value of $v = \frac{1}{4}$ to compute these terms. The actual values of v_1 and v_2 of the two bodies are used later in computing Δ. Thus, since the terms within the brackets do *not* depend strongly on the elastic constants of the two bodies or the load P, their magnitudes can be computed and tabulated for use as coefficients in the determination of the ratio b/Δ. For example, let a value of the ratio $B/A = 1.24$ be chosen. From Eq. (18.21), $k = 0.866$ and hence, $k' = 0.5$. For specified values of the ratio kz/b, required coefficients can be found for determination of the stresses at a distance z from the area of contact. The results of these computations are given in Fig. 18.4, in which the coefficients of b/Δ are plotted as abscissas, and the values of kz/b to the point at which the stresses occur are plotted as ordinates.

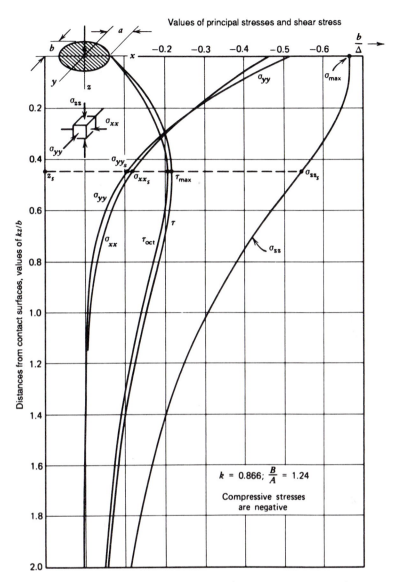

Figure 18.4   Curves showing variation in principal stresses, maximum shear stress, and octahedral shear stress with variation in distance from the contact surface; $v = 0.25$.

The curves representing σ_{xx}, σ_{yy}, and σ_{zz} show that their largest magnitudes occur when $z = 0$ (at the center of the surface of contact) and that all three stresses decrease as z increases. The principal stress having the greatest magnitude is σ_{zz} and, hence, at each point, $\sigma_{max} = \sigma_{zz}$. In this example, in which $B/A = 1.24$, the value of $\sigma_{max} = 0.67b/\Delta$. The coefficient 0.67 of b/Δ is found at $z = 0$ from the curve σ_{zz}.

Maximum Shear Stress

The maximum shear stress at any point is $\tau_{max} = \frac{1}{2}(\sigma_{max} - \sigma_{min})$. In Fig. 18.4 the curves show that the magnitudes of σ_{xx} and σ_{yy} decrease more rapidly than that of

σ_{zz} at points just beneath the surface of contact. Because of this fact, the maximum shear stress at points just beneath the surface of contact increases in magnitude and reaches its maximum value $\frac{1}{2}(\sigma_{zz_s} - \sigma_{yy_s})$ at z_s, as shown by the curve marked τ. In this example in which $B/A = 1.24$, the value of $\tau_{max} = 0.22b/\Delta$, and the depth $kz_s/b = 0.44$, that is, $z_s = 0.44b/0.866 = 0.51b$. The coefficient 0.22 of b/Δ is the ordinate to the τ curve at the depth z_s.

Maximum Octahedral Shear Stress

The octahedral shear stress τ_{oct} [see, Eq. (2.22)] is

$$\tau_{oct} = \frac{1}{3}\sqrt{(\sigma_{xx} - \sigma_{yy})^2 + (\sigma_{yy} - \sigma_{zz})^2 + (\sigma_{zz} - \sigma_{xx})^2}$$

The values of τ_{oct} have been computed by this equation for several points along the z axis and are plotted as ordinates to the curved marked τ_{oct}. In this example, for $B/A = 1.24$, the maximum value of the octahedral shear stress $\tau_{oct(max)} = 0.21b/\Delta$, and it occurs at the same $z_s = 0.51b$ as the maximum shear stress. The coefficient 0.21 of b/Δ is the ordinate to the τ_{oct} curve at z_s.

Maximum Orthogonal Shear Stress

As noted above, the maximum shear stress and maximum octahedral shear stress occur in the interior of the contacting bodies at points located equidistant from the tangent plane on a line perpendicular to the center of the contacting area. These maximum values are considered to be the significant stresses in certain failure criteria associated with initiation of yielding (Chapter 4). Other shear stress components that are considered to be significant in the fatigue failure of bearings and other rolling elements (e.g., cylinders) in contact are shear stresses that occur on planes perpendicular and parallel to the plane tangent to the contact area. For example, with reference to Figs. 18.2b and d, orthogonal shear stress components σ_{xz} and σ_{yz} act on a plane perpendicular to the z axis; they are called orthogonal shear stresses since they act also on planes x-y and y-z that are perpendicular (orthogonal) to the plane of contact. These shear stress components are zero on the z axis, where τ_{max} occurs. We choose the x axis in the direction of rolling and consider only those problems for which the x axis coincides with either the major or minor axes of the contacting ellipse. The maximum orthogonal shear stress τ_0 is defined as $\sigma_{xz(max)}$, which occurs at points in the interior of the contacting bodies located in the (x, z) plane equidistant from the z axis at some distance from the contacting surface.

Although τ_0 is always smaller than τ_{max}, τ_0 for a given point in a contacting body changes sign as the rolling element (the contact area) approaches and leaves the region above the point. Therefore, the range of the maximum orthogonal shear stress is $2\tau_0$, and for most applications this range is greater than the range of the maximum shear stress τ_{max}. Note that $\sigma_{yz(max)}$ may be greater than τ_0; however, $\sigma_{yz(max)}$ does not change sign during rolling so that the range of σ_{yz} is equal to $\sigma_{yz(max)}$, which is less than $2\tau_0$. The range in shear stress is considered to be important (Moyar and Morrow, 1964) in studies of fatigue failure due to rolling contact.

The location of the point (perpendicular distance to the point from the tangent plane at the contact area) at which τ_0 occurs, as well as the magnitude of τ_0, is a function of the ellipticity (values of a, b) and orientation of the contact ellipse with respect to the rolling direction (Moyar and Morrow, 1964) In particular, for toroids under radial loads (Fig. 18.5) Fessler and Ollerton (1957) have derived expressions for the orthogonal shear stress components σ_{xz}, σ_{yz} (Fig. 18.6). Their results are, with the notations of Sec. 18.4 and 18.5,

$$\sigma_{xz} = -\frac{3PQ}{2\pi a^2} = -\frac{b}{\Delta}\left[\frac{kQ}{E(k')}\right]$$

$$\sigma_{yz} = -\frac{3PR}{2\pi b^2} = -\frac{b}{\Delta}\left[\frac{R}{kE(k')}\right] \tag{18.23}$$

where

$$Q = \frac{\left(\dfrac{x}{a}\right)\left(\dfrac{z}{a}\right)^2\left[\left(1+\dfrac{c^2}{a^2}\right)\dfrac{c^2}{a^2}\right]^{-3/2}\left(k^2+\dfrac{c^2}{a^2}\right)^{-1/2}}{\left[\left(\dfrac{ax}{a^2+c^2}\right)^2+\left(\dfrac{ay}{b^2+c^2}\right)^2+\left(\dfrac{az}{c^2}\right)^2\right]} \tag{18.24}$$

and

$$R = \frac{\left(\dfrac{y}{b}\right)\left(\dfrac{z}{b}\right)^2\left[\left(1-\dfrac{c^2}{b^2}\right)\dfrac{c^2}{b^2}\right]^{-3/2}\left(\dfrac{1}{k^2}+\dfrac{c^2}{b^2}\right)^{-1/2}}{\left[\left(\dfrac{bx}{a^2+c^2}\right)^2+\left(\dfrac{by}{b^2+c^2}\right)^2+\left(\dfrac{bz}{c}\right)^2\right]} \tag{18.25}$$

where c^2 is the positive root of the equation

$$\frac{x^2}{a^2+c^2}+\frac{y^2}{b^2+c^2}+\frac{z^2}{c^2}=1 \tag{18.26}$$

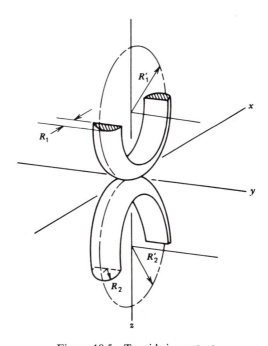

Figure 18.5   Toroids in contact.

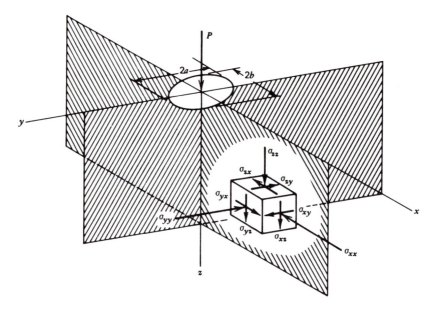

Figure 18.6   State of stress of an element. Compressive stresses are negative.

The maximum values of σ_{xz} and σ_{yz} occur in the planes of symmetry, $y = 0$ and $x = 0$, respectively. For the plane of symmetry $y = 0$, we get

$$\sigma_{xz} = -\frac{3PQ(y = 0)}{2\pi a^2} = -\frac{b}{\Delta}\left[\frac{kQ(y = 0)}{E(k')}\right]$$

$$\sigma_{yz} = 0 \tag{18.27}$$

and for the plane of symmetry $x = 0$, we have

$$\sigma_{xz} = 0$$

$$\sigma_{yz} = -\frac{3PR(x = 0)}{2\pi b^2} = -\frac{b}{\Delta}\left[\frac{R(x = 0)}{kE(k')}\right] \tag{18.28}$$

In particular, we note that along the z axis ($x = y = 0$), $\sigma_{xz} = \sigma_{yz} = 0$.

As noted earlier, in rolling contact problems, the perpendicular distance from the tangent plane at the contact area to the point at which the maximum range of τ_0 occurs (as the roller passes over a given region) and the magnitude of τ_0 depend on the ellipticity and orientation of the contact ellipse with respect to the rolling direction. Consequently, the geometric configuration of the contacting rollers is an important factor. For example, for contacting toroids (Fig. 18.5), the distance z_0, from the contact surface, at which the maximum orthogonal shear stress $\tau_0 = \sigma_{zx(\text{max})}$ occurs is plotted in Fig. 18.7; note that the contact ellipse semiaxis in the rolling direction is equal to b for $e < 1$ and a for $e > 1$. The magnitude of τ_0 is plotted in Fig. 18.8. (Fessler and Ollerton, 1957).

The range of the maximum orthogonal shear stress $2\tau_0$ is considered to be important in rolling fatigue problems as long as $2\tau_0$ is greater than the range of the maximum shear stress τ_{max}. Fessler and Ollerton (1957) found that $2\tau_0 > \tau_{\text{max}}$ for

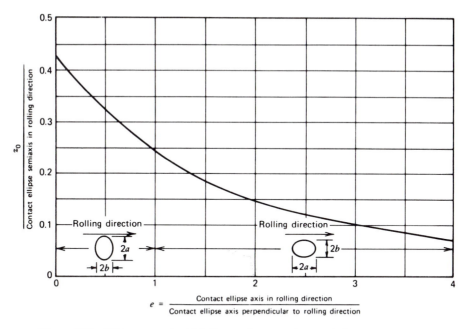

Figure 18.7   Distance z_0 at which the maximum orthogonal shear stress occurs.

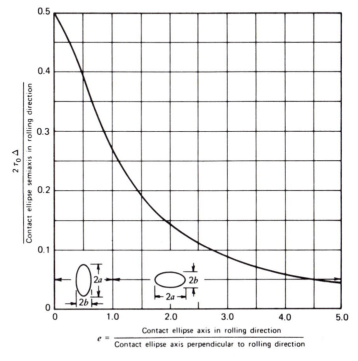

Figure 18.8   Magnitude of maximum orthogonal shear stress.

$e < 2.25$ when $v = 0.25$ for both bodies and that $2\tau_0 > \tau_{max}$ for $e < 2.85$ when $v = 0.50$ for both bodies. Poisson's ratio for most isotropic materials falls between these two values. Furthermore, the contour surfaces for most rolling contact problems result in values of e that are less than 2.25.

Fatigue failure of members under rolling contact depends on the magnitude of $2\tau_0$ for $e < 2.25$; however, the magnitude of the range of the orthogonal shear stress $2\tau_0$ for fatigue failure for a specified number of cycles is not equal to the range of maximum shear stress (for the same number of cycles of loading) in specimens (tension-compression or reversed torsion) that are generally used to obtain fatigue properties (Moyar and Morrow, 1964). The high hydrostatic state of compressive stress in contact problems strengthens the material against fatigue failure. The usual procedure is to obtain the magnitude of τ_0 at which fatigue failure initiates for a specified number of cycles for one orientation of the contact ellipse and one value of e and assume that fatigue failure for another rolling contact stress problem having a different value of e will occur for the same number of cycles when τ_0 has the same magnitude. Since σ_{max} is generally calculated for all contact stress problems, it is not necessary that τ_0 be calculated if the ratio $2\tau_0/\sigma_{max}$ is plotted vs e (Fig. 18.9). The curve in Fig. 18.9 has been plotted for $v = 0.25$. Since σ_{max} depends on v, whereas τ_0 does not, the curve in Fig. 18.9 is moved slightly for other values of v. The ratios $2\tau_0/\tau_{max}$ or $2\tau_0/\tau_{oct(max)}$ vs e could also be plotted; however, only one plot is needed and σ_{max} is more often calculated than τ_{max} or $\tau_{oct(max)}$.

To illustrate the use of Fig. 18.9 in the analysis of rolling contact stress fatigue problems, let fatigue properties for rolling contact stress be determined using two cylinders that roll together. Since the major axis of the contact ellipse is large compared to the minor axis b, e is assumed to be equal to zero. Let σ_{max1} be the maximum principal stress in the cylinders when fatigue failure of the contact surface occurs after N cycles of load. From Fig. 18.9 we read $2\tau_0 = 0.50\sigma_{max1}$. Let the toroids in Fig. 18.5 be made of the same material as the cylinders. Let the radii of curvature be such that $e = 1.3$; for this value of e we read $2\tau_0 = 0.40\sigma_{max2}$ from Fig. 18.9. The magnitude of σ_{max2} such that fatigue failure of the toroids occurs after the same number of cycles N as the two cylinders is obtained by setting the two values of $2\tau_0$ equal. Thus, we obtain $\sigma_{max2} = 1.25\sigma_{max1}$. This result is based on the assumption that the fatigue strength of the material is the same for both types

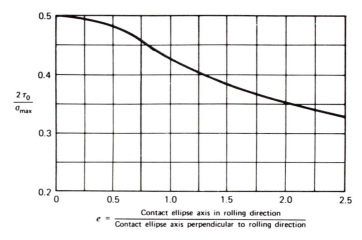

Figure 18.9   Ratio of range of maximum orthogonal shear stress to σ_{max} ($v = 0.25$).

of loading. Moyar and Morrow (1964) indicate that the fatigue strength is not the same because of the size effect. Because of the larger volume of material under stress for the cylinders than the toroids, the fatigue strength for the toroids should be larger than for the cylinders. Thus, the magnitude of $2\tau_0$ for the toroids should be larger than for the cylinders in order to produce fatigue failure in the same number of cycles. Our result that $\sigma_{max2} = 1.25\sigma_{max1}$ is therefore conservative. If fatigue properties for materials are obtained for $e = 0$, the use of Fig. 18.9 without correction for the size effect predicts conservative results.

Curves for Computing Stresses for Any Value of B/A

The above example in which $B/A = 1.24$ ($k = 0.866$) shows that for a value of B/A (or k), a set of curves may be drawn representing the values of the principal stresses σ_{xx}, σ_{yy}, and σ_{zz} along the z axis at small distances z from the surface of contact. These curves may be used to find the magnitude and location of the maximum shear stress and maximum octahedral shear stress. Curves may also be constructed for a wide range of values of the ratio B/A. For each value of B/A, the maximum values of stresses may be found from the equations

$$\sigma_{max} = -c_\sigma\left(\frac{b}{\Delta}\right)$$

$$\tau_{max} = c_\tau\left(\frac{b}{\Delta}\right)$$

$$\tau_{oct(max)} = c_G\left(\frac{b}{\Delta}\right) \tag{18.29}$$

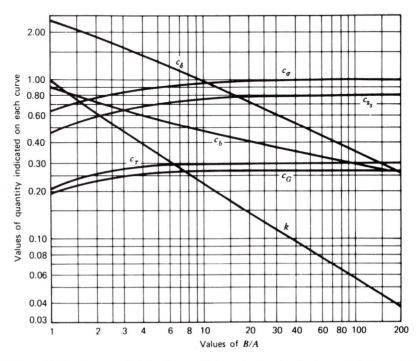

Figure 18.10   Stress and deflection coefficients for two bodies in contact at a point.

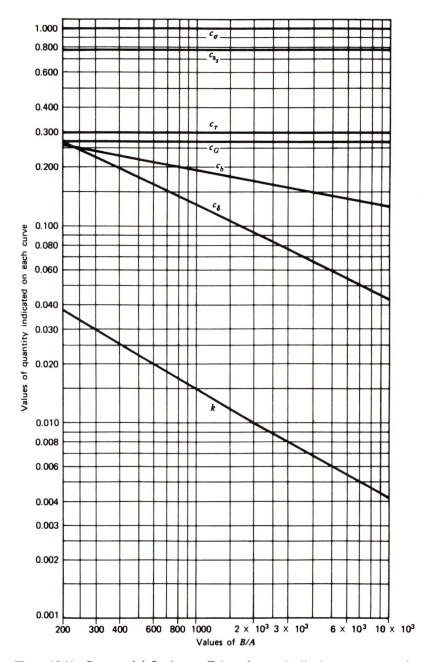

Figure 18.11   Stress and deflection coefficients for two bodies in contact at a point.

where the values of the coefficients c_σ, c_τ, and c_G may be read from the curves shown in Fig. 18.10. In the example in which $B/A = 1.24$, the values of the coefficients were given as $c_\sigma = 0.67$, $c_\tau = 0.22$, and $c_G = 0.21$. In Figs. 18.10 and 18.11, values of these coefficients for use in Eq. (18.29) are given as ordinates to the curves marked c_σ, c_τ, and c_G for a range of values B/A from 1 to 10,000. The values of k that are required in computing the semimajor axis a and semiminor axis b of the area of contact are given as ordinates to the curved marked k. The value of b that may be computed by using Eq. (18.22) is found as follows. Equation (18.22) is rewritten as

$$b = c_b \sqrt[3]{P\Delta} \tag{18.30}$$

in which $c_b = \sqrt[3]{3kE(k')/2\pi}$. The values of k (and k') as found from the curve marked k are used to compute the value of the coefficient c_b. These values of c_b are given as ordinates to the curve marked c_b. The length of the semimajor axis is $a = b/k$. The distance z_s from the surface of contact to the location on the z axis of the point at which the maximum stresses τ_{max} and $\tau_{oct(max)}$ occur is

$$z_s = c_{z_s} b \tag{18.31}$$

The coefficient c_{z_s} is plotted in Figs. 18.10 and 18.11. The examples following Sec. 18.7 illustrate the use of Figs. 18.10 and 18.11.

18.7

DEFLECTION OF BODIES IN POINT CONTACT

An expression for the distance $\delta = \delta_1 + \delta_2$ through which two bodies in contact at a point move toward each other when acted on by a load P (Fig. 18.3c) is given by Eq. (18.17). The deflection δ is sometimes called the *approach* because it expresses the sum of the "deflections" of the two bodies as they approach each other. The expression for the value of δ is given by the following equation due to Hertz (1895).

$$\delta = \frac{3kPK(k')}{2\pi}\left(\frac{A + B}{b/\Delta}\right) \tag{18.32}$$

where P is the load, $K(k')$ is the complete elliptic integral described in Sec. 18.4, and A, B, k, Δ, and b are defined in Sec. 18.4 and 18.5. For convenient use of Eq. (18.32), the substitution of

$$c_\delta = \frac{3kK(k')}{2} \tag{18.33}$$

is made to obtain

$$\delta = c_\delta \frac{P}{\pi}\left(\frac{A + B}{b/\Delta}\right) \tag{18.34}$$

In Eq. (18.33), the value c_δ depends only on k (and k'), and since from Eq. (18.21) there is a value of B/A corresponding to each value of k, there is a value of c_δ corresponding to each value of B/A. In Figs. 18.10 and 18.11 values of this coefficient have been computed by Eq. (18.33) and plotted as ordinates for the curve marked c_δ. Equation (18.34) gives approximate results since the elastic strains in the two bodies away from the contact region are neglected.

EXAMPLE 18.1
Contact Stresses Between Two Semicircular Disks

Let the two semicircular disks in Fig. 18.1 be made of steel ($E_1 = E_2 = 200$ GPa and $v_1 = v_2 = 0.29$). The radii of curvature of the two surfaces at the point of contact are $R_1 = 60$ mm, $R_1' = 130$ mm, $R_2 = 80$ mm, and $R_2' = 200$ mm. The angle α between the planes of minimum curvature is $\pi/3$ rad. If the load $P = 4.50$ kN, determine the maximum principal stress, maximum shear stress, and maximum octahedral shear stress in the disks and state the location of the point where each of these stresses occur. Determine the approach δ for the two disks because of load P.

SOLUTION

All stress and displacement calculations require first that values be obtained for B, A, and Δ; these are given by Eqs. (18.15), (18.16), and (18.20a).

$$B = \frac{1}{4}\left(\frac{1}{60} + \frac{1}{80} + \frac{1}{130} + \frac{1}{200}\right) + \frac{1}{4}\left[\left(\frac{1}{60} - \frac{1}{130} + \frac{1}{80} - \frac{1}{200}\right)^2\right.$$
$$\left. - 4\left(\frac{1}{60} - \frac{1}{130}\right)\left(\frac{1}{80} - \frac{1}{200}\right)\sin^2\frac{\pi}{3}\right]^{1/2}$$
$$= 0.01255 \text{ mm}^{-1}$$

$$A = \frac{1}{4}\left(\frac{1}{60} + \frac{1}{80} + \frac{1}{130} + \frac{1}{200}\right) - \frac{1}{4}\left[\left(\frac{1}{60} - \frac{1}{130} + \frac{1}{80} - \frac{1}{200}\right)^2\right.$$
$$\left. - 4\left(\frac{1}{60} - \frac{1}{130}\right)\left(\frac{1}{80} - \frac{1}{200}\right)\sin^2\frac{\pi}{3}\right]^{1/2}$$
$$= 0.00838 \text{ mm}^{-1}$$

$$\Delta = \frac{2(1 - v^2)}{(A + B)E} = \frac{2[1 - (0.29)^2]}{(0.01255 + 0.00838)(200 \times 10^3)}$$
$$= 438 \times 10^{-6} \text{ mm}^3/\text{N}$$

$$\frac{B}{A} = 1.50$$

The coefficients needed to calculate b, σ_{max}, τ_{max}, $\tau_{oct(max)}$, z_s, and the deflection are read from Fig. 18.10 for the ratio $B/A = 1.50$. Hence, by Fig. 18.10 we obtain the following coefficients: $c_b = 0.77$, $c_\sigma = 0.72$, $c_\tau = 0.24$, $c_G = 0.22$, $c_{z_s} = 0.53$, and $c_\delta = 2.10$. For known values of c_b, P, and Δ, Eq. (18.30) gives

$$b = c_b\sqrt[3]{P\Delta} = 0.77\sqrt[3]{(4.5 \times 10^3)(438 \times 10^{-6})} = 0.965 \text{ mm}$$

from which

$$\frac{b}{\Delta} = \frac{0.965}{438 \times 10^{-6}} = 2203 \text{ MPa}$$

Values of σ_{max}, τ_{max}, $\tau_{oct(max)}$, z_s, and δ are obtained by substituting known values of the coefficients into Eqs. (18.29), (18.31), and (18.34).

$$\sigma_{max} = -c_\sigma \frac{b}{\Delta} = -0.72(2203) = -1586 \text{ MPa}$$

$$\tau_{max} = c_\tau \frac{b}{\Delta} = 0.24(2203) = 529 \text{ MPa}$$

$$\tau_{oct(max)} = c_G \frac{b}{\Delta} = 0.22(2203) = 485 \text{ MPa}$$

$$z_s = c_{z_s} b = 0.53(0.965) = 0.51 \text{ mm}$$

$$\delta = c_\delta \frac{P}{\pi} \left(\frac{A+B}{b/\Delta} \right) = \frac{2.10(4.5 \times 10^3)}{\pi} \left(\frac{0.01255 + 0.00838}{2203} \right)$$

$$= 0.029 \text{ mm}$$

The maximum Hertz stress $\sigma_{max} = -1586$ MPa occurs at the contact surface under the load. The maximum shear stress and maximum octahedral shear stress occur at $z_s = 0.51$ mm from the surface of contact.

EXAMPLE 18.2
Contact Stresses in a Steel Ball Bearing

A steel ball bearing consisting of an inner race, an outer race, and 12 balls is shown in Fig. E18.2 ($E = 200$ GPa, $v = 0.29$, and $Y = 1600$ MPa). A rated load of $P_0 = 4.2$ kN is given in a manufacturer's handbook for this bearing when operated at 3000 rpm. An empirical relation (Allen, 1945) is used to determine the load P on the topmost ball that bears the largest portion of the load; $P = 5P_0/n = 1.75$ kN in which n is the number of balls.

(a) At the region of contact between the inner race and topmost ball, determine the maximum principal stress, maximum shear stress, maximum octahedral shear stress, dimensions of the area of contact, maximum orthogonal shear stress, and distance from the point of contact to the point where these stresses occur.

(b) What is the factor of safety against initiation of yielding based on the octahedral shear stress criterion of failure?

SOLUTION

(a) Let the ball be designated as body 1 and the inner race as body 2 so $R_1 = R'_1 = 4.76$ mm, $R_2 = -4.86$ mm, and $R'_2 = 18.24$ mm. We substitute these values in Eqs. (18.15) and (18.16) to obtain values for A and B. The following results are obtained:

$$B = 0.13245 \text{ mm}^{-1}, \qquad A = 0.00216 \text{ mm}^{-1}, \qquad \frac{B}{A} = 61.3$$

$$\Delta = \frac{2}{A+B} \frac{1-v^2}{E} = \frac{2(1 - 0.29^2)}{(0.13246 + 0.00216)(200 \times 10^3)}$$

$$= 68.0 \times 10^{-6} \text{ mm}^3/\text{N}$$

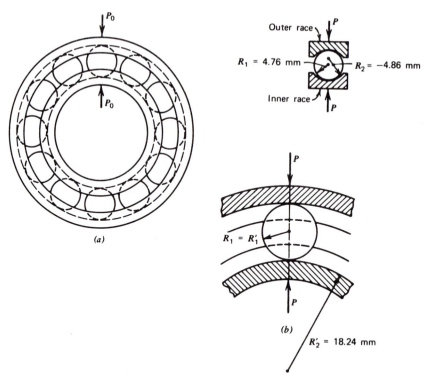

Figure E18.2   Contact load in ball bearing.

By Fig. 18.10, with $B/A = 61.3$, we obtain the following values for the coefficients: $c_b = 0.32$, $k = 0.075$, $c_\sigma = 1.00$, $c_\tau = 0.30$, $c_G = 0.27$, and $c_{z_s} = 0.78$. Hence,

$$b = c_b \sqrt[3]{P\Delta} = 0.32 \sqrt[3]{1.75 \times 10^3 (68.0 \times 10^{-6})} = 0.1574 \text{ mm}$$

$$a = \frac{b}{k} = \frac{0.1574}{0.075} = 2.099 \text{ mm}$$

$$\frac{b}{\Delta} = \frac{0.1574}{68.0 \times 10^{-6}} = 2315 \text{ MPa}$$

$$\sigma_{max} = -c_\sigma \frac{b}{\Delta} = -1.00(2315) = -2315 \text{ MPa}$$

$$\tau_{max} = c_\tau \frac{b}{\Delta} = 0.30(2315) = 695 \text{ MPa}$$

$$\tau_{oct(max)} = c_G \frac{b}{\Delta} = 0.27(2315) = 625 \text{ MPa}$$

The maximum principal stress occurs under the load at the contact tangent plane. The maximum shear stress and maximum octahedral shear stress are located at a distance

$$z_s = c_{z_s} b = 0.78(0.1574) = 0.123 \text{ mm}$$

from the contact tangent plane directly under the load. The magnitude of the maximum orthogonal shear stress is obtained from Fig. 18.8. Since b is in the direction of rolling, $e = b/a = k = 0.075$. From Fig. 18.8, we obtain

$$\tau_0 = \frac{0.486b}{2\Delta} = \frac{0.486(0.1574)}{2(68.0 \times 10^{-6})} = 556 \text{ MPa}$$

The location of the maximum orthogonal shear stress from the contact tangent plane is obtained from Fig. 18.7.

$$z_0 = 0.41b = 0.41(0.1574) = 0.065 \text{ mm}$$

(b) Since contact stresses are not linearly related to load P, the safety factor is not equal to the ratio of the maximum octahedral shear stress in the specimen used to obtain material properties and the maximum octahedral shear stress in the ball bearing. The magnitude of the yield load P_Y for a single ball may be obtained from the relation

$$(\tau_{\text{oct}})_Y = \frac{\sqrt{2}}{3} Y = \frac{\sqrt{2}}{3} 1600 = c_G \frac{b}{\Delta} = c_G c_b \frac{\sqrt[3]{P_Y \Delta}}{\Delta}$$

$$= 0.27(0.32) \frac{\sqrt[3]{P_Y(68.0 \times 10^{-6})}}{68.0 \times 10^{-6}}$$

from which

$$P_Y = 3076 \text{ N}$$

The safety factor SF is equal to the ratio of P_Y to P

$$SF = \frac{P_Y}{P} = \frac{3076}{1750} = 1.76$$

Significance of Stresses

In the preceding examples, the magnitude of the maximum principal stress is quite large in comparison with the value of this stress usually found in direct tension, bending, and torsion. In these problems, as in all contact stress problems, the three principal stresses at the point of maximum values are all compressive stresses. As a result, the maximum shear stress and maximum octahedral shear stress are always less than one-half the maximum principal stress; we recall that for a state of uniaxial stress (one principal stress), the maximum shear stress is one-half the principal stress. In fact by a comparison of the values of c_σ, c_τ, and c_G for various values of B/A in Figs. 18.10 and 18.11, we see that when $B/A = 1$, $c_\tau = 0.32c_\sigma$, and $c_G = 0.30c_\sigma$ and when $B/A = 100$ or larger, $c_\tau = 0.30c_\sigma$ and $c_G = 0.27c_\sigma$. Thus, τ_{max} and $\tau_{\text{oct(max)}}$ are always slightly smaller than one-third of the maximum principal stress σ_{max}. This fact is of special importance if the maximum shear stress or the octahedral shear stress is considered to be the cause of structural damage (failure) of the member; if the shear stresses are relatively small in comparison to the maximum principal stress, very high principal stresses can occur. However, the maximum utilizable values of the maximum shear stress or maximum octahedral shear stress are not easily determined, because in many problems involving two bodies under pressure at a small area of contact, such as occurs in rolling bearings, there are additional factors that affect the behavior of the material, for example, sliding friction, the effect of a lubricant, the effect of repeated loads, the effect of variation in the metal properties near the surface of contact such as that due to case hardening, and the effects of metallurgical changes that often occur in parts such as the races of ball bearings due to the repeated stressing.

18.8

STRESS FOR TWO BODIES IN LINE CONTACT. LOADS NORMAL TO CONTACT AREA

If two cylindrical surfaces are in contact, the contact region is approximately along a straight-line element before loads are applied. Figure 18.12a illustrates contact between two circular cylinders, the line of contact being perpendicular to the paper. Figure 18.12b also shows a line contact of a circular cylinder resting on a plane. Figure 18.12c shows a line contact of a small circular cylinder resting inside a larger hollow cylinder. In these cases, the radii R'_1 and R'_2, which lie in a plane perpendicular to the paper, are each infinitely large so that $1/R'_1$ and $1/R'_2$ each vanish identically and the angle $\alpha = 0$ (Fig. 18.3b). Therefore, from Eqs. (18.15) and (18.16) the expressions for B and A are

$$B = \frac{1}{2}\left(\frac{1}{R_1} + \frac{1}{R_2}\right), \qquad A = 0, \qquad \frac{B}{A} = \infty$$

where R_1 and R_2 are the radii of curvature of the cylindrical surfaces (Fig. 18.12). Note that R_2 is negative in Fig. 18.12c. Hence, the value of the ratio B/A is infinitely

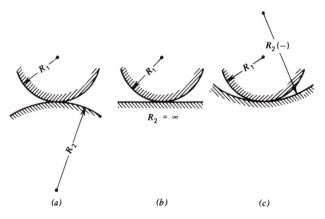

(a)   (b)   (c)

Figure 18.12   Line contact between cylindrical bodies.

large, and from Eq. (18.21) the corresponding value of k approaches zero. However, k is the ratio of the semiminor axis b of the area of contact to semimajor axis a [Eq. (18.22)] and, therefore, a must be infinitely large, which is the case of contact along a line between two bodies. The area of contact, when a distributed load of w (force per unit length) is applied, is a long narrow rectangle of width $2b$ in the x direction and length $2a$ in the y direction. When $k = 0$, Eqs. (18.18), (18.19), and (18.20) for the stresses at points on the z axis at various distances z/b from the contact surface do not involve elliptic functions. In this case,

$$\sigma_{yy} = -2v\left[\sqrt{1 + \left(\frac{z}{b}\right)^2} - \frac{z}{b}\right]\frac{b}{\Delta} \tag{18.35}$$

$$\sigma_{xx} = -\left[\frac{(\sqrt{1 + (z/b)^2} - z/b)^2}{\sqrt{1 + (z/b)^2}}\right]\frac{b}{\Delta} \tag{18.36}$$

$$\sigma_{zz} = -\left[\frac{1}{\sqrt{1 + (z/b)^2}}\right]\frac{b}{\Delta} \tag{18.37}$$

The value of b from Eq. (18.22) for the limiting case in which $k = 0$ is

$$b = \sqrt{\frac{2w\,\Delta}{\pi}} \tag{18.38}$$

in which w is the load per unit length of the contact area. The value of Δ is [Eq. (18.20a)]

$$\Delta = \frac{1}{(1/2R_1) + (1/2R_2)}\left(\frac{1 - v_1^2}{E_1} + \frac{1 - v_2^2}{E_2}\right) \tag{18.39}$$

where R_1 and R_2 are the radii of curvature of the cylindrical surfaces as shown in Fig. 18.12. The values of the stresses at a point on the line of contact are obtained from Eqs. (18.35), (18.36), and (18.37) by setting $z = 0$.

Maximum Principal Stresses: $k = 0$

It is seen from Eqs. (18.35), (18.36), and (18.37) that the principal stresses σ_{xx}, σ_{yy}, and σ_{zz} have their maximum numerical value when $z/b = 0$, that is, at the surface of contact. These stresses are

$$\sigma_{xx} = -\frac{b}{\Delta}, \qquad \sigma_{yy} = -2v\left(\frac{b}{\Delta}\right), \qquad \sigma_{zz} = -\frac{b}{\Delta} \tag{18.40}$$

Maximum Shear Stress: $k = 0$

The shear stress at any point on the z axis is $\tau = \frac{1}{2}(\sigma_{xx} - \sigma_{zz})$. If the expressions for σ_{xx} and σ_{zz} from Eq. (18.36) and (18.37) are substituted in this equation for τ and the first derivative of τ with respect to z is equated to zero, the value of z (or z/b) found from the resulting equation is the distance from the contact surface at which the greatest value τ_{max} of the shear stress occurs. The value thus found is $z_s/b = 0.7861$. At this point, the principal stresses are, from Eqs. (18.35–18.37),

$$\sigma_{xx} = -0.1856\left(\frac{b}{\Delta}\right), \qquad \sigma_{yy} = -0.9718\frac{vb}{\Delta}$$

$$\sigma_{zz} = -0.7861\left(\frac{b}{\Delta}\right) \tag{18.41}$$

Hence,

$$\tau_{max} = \tfrac{1}{2}(\sigma_{xx} - \sigma_{zz}) = 0.300\left(\frac{b}{\Delta}\right) \tag{18.42}$$

At $z_s/b = 0.7861$, the magnitude of σ_{xx} is smaller than that of σ_{yy} for values of v greater than about 0.19.

Maximum Octahedral Shear Stress: $k = 0$

The maximum octahedral shear stress occurs at the same point as the maximum shear stress and is found by substituting the values of σ_{xx}, σ_{yy}, and σ_{zz} from Eq. (18.41) into Eq. (2.22). The result is

$$\tau_{oct(max)} = 0.27\frac{b}{\Delta} \tag{18.43}$$

We note that the coefficients for determining the quantities σ_{max}, τ_{max}, $\tau_{oct(max)}$, and z_s as obtained from Figs. 18.10 and 18.11 for values of B/A greater than about 50 are 1.00, 0.30, 0.27, and 0.78, respectively, and these are the same coefficients found for the case of line contact between two bodies. This fact means that when the ratio B/A is about 50 or larger, the area of contact between the two bodies is very nearly a long narrow rectangle.

18.9

STRESSES FOR TWO BODIES IN LINE CONTACT. LOADS NORMAL AND TANGENT TO CONTACT AREA

In the preceding sections, the contact stresses in two elastic bodies held in contact by forces normal to the area of contact were found. Frequently, the normal force is accompanied by a tangential (frictional) force in the contact area such as occurs when the teeth of spur gears come into contact or when a shaft rotates in a bearing. The frictional force that results from the sliding contact lies in the plane of the area of contact in a direction perpendicular to the normal force. The presence of frictional force causes the maximum values of the contact stresses in the two elastic bodies to become substantially larger than those produced by a normal force acting alone. Furthermore, the presence of a frictional force combined with a normal force causes certain changes in the nature of the stresses. For example, when a normal force acts alone, the three principal stresses are compressive stresses at every point in the body near the contact area, and this fact makes it difficult to understand how a crack can form and progressively spread to cause a separation type of failure such as occurs in pitting failures of some bearing surfaces. However, when a frictional force is introduced, two of the three principal stresses are changed into tensile stresses in the region immediately behind the frictional force (see Figs. 18.14b and 18.14c). If the coefficient of friction for the two surfaces of contact is sufficiently large, these tensile stresses are relatively large. However, if these tensile stresses are nominally small, as they probably are on well-lubricated surfaces, their values may be raised by stress concentration that results from surface irregularities or small microscopic cracks that usually exist in the surfaces of real materials. These tensile stresses, when considered in conjunction with the many other factors involved, such as wear, nonhomogeneity of the material, and type of lubrication, help in explaining why a crack may develop and progressively spread in the surface of contact of such parts as gear teeth, roller bearings, etc.

The addition of a frictional force to a normal force on the contact surface also causes a change in the shear stresses in the region of the contact surface. One important change is that the location of the point at which the maximum shear stress occurs moves toward the contact area. In fact, when the coefficient of friction is 0.10 or greater, this point is located in the contact surface. The foregoing remarks also apply to the maximum octahedral shear stress.

The facts described above may be illustrated for an elastic cylindrical roller pressed against the plane surface of another elastic body.

Roller on Plane

Let Fig. 18.13a represent the cross section of a long roller of elastic material that rests on a flat surface of a thick, solid elastic body. The roller is subjected to a distributed load w (force per unit length), which presses it against the body over a long

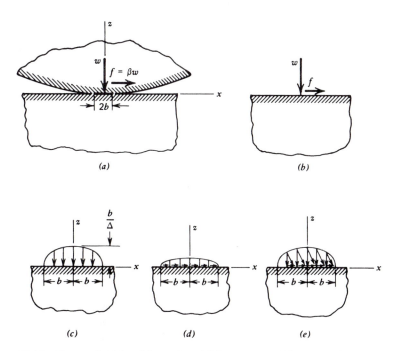

Figure 18.13   Tangential (shear) forces in addition to normal forces on the contact area.

narrow area of contact whose width is $2b$. A lateral distributed load f (force per unit length) causes the roller to slide on the body. If the coefficient of sliding friction is designated as β, then $f = \beta w$. In Fig. 18.13b, a part of the solid body is shown with the distributed loads w and f acting on the contact area. In Fig. 18.13c, which is an enlarged view of the part near the contact area, the ordinates to the ellipse show the distribution of normal stresses over this area and the maximum stress is $\sigma_{zz} = -b/\Delta$ [Eq. (18.40)]. R. D. Mindlin (1949) has found that when sliding occurs, the shear stress on the contact area due to the frictional force f are distributed as ordinates to an ellipse as shown in Fig. 18.13d, and the maximum shear stress σ_{zx} at the center is $\sigma_{zx} = \beta(b/\Delta)$. Figure 18.13e shows the distribution of the combined normal and friction stresses on the contact surface. C. K. Liu (Smith and Liu, 1953) has derived the equations for the stresses σ_{zz}, σ_{xx}, σ_{yy}, and σ_{zx} at any point in the body. These equations are

$$\sigma_{zz} = -\frac{b}{\pi \Delta} [z(b\phi_1 - x\phi_2) + \beta z^2 \phi_2]$$

$$\sigma_{xx} = -\frac{b}{\pi \Delta} \left\{ z\left(\frac{b^2 + 2z^2 + 2x^2}{b}\phi_1 - \frac{2\pi}{b} - 3x\phi_2\right) \right.$$
$$\left. + \beta\left[(2x^2 - 2b^2 - 3z^2)\phi_2 + \frac{2\pi x}{b} + 2(b^2 - x^2 - z^2)\frac{x}{b}\phi_1\right] \right\}$$

$$\sigma_{yy} = -\frac{2vb}{\pi \Delta} \left\{ z\left(\frac{b^2 + x^2 + z^2}{b}\phi_1 - \frac{\pi}{b} - 2x\phi_2\right) \right.$$
$$\left. + \beta\left[(x^2 - b^2 - z^2)\phi_2 + \frac{\pi x}{b} + (b^2 - x^2 - z^2)\frac{x}{b}\phi_1\right] \right\}$$

$$\sigma_{zx} = -\frac{b}{\pi \Delta} \left\{ z^2\phi_2 + \beta\left[(b^2 + 2x^2 + 2z^2)\frac{z}{b}\phi_1 - 2\pi\frac{z}{b} - 3xz\phi_2\right] \right\} \quad (18.44)$$

where ϕ_1, ϕ_2 are

$$\phi_1 = \frac{\pi(M + N)}{MN\sqrt{2MN + 2x^2 + 2z^2 - 2b^2}}$$

$$\phi_2 = \frac{\pi(M - N)}{MN\sqrt{2MN + 2x^2 + 2z^2 - 2b^2}}$$

where $M = \sqrt{(b + x)^2 + z^2}$ and $N = \sqrt{(b - x)^2 + z^2}$. The values of stress as given by Eq. (18.44) do not depend on y because it is assumed that either a state of plane strain or plane stress exists relative to the (x, z) plane.

Principal Stresses

In Eq. (18.44) σ_{yy} is a principal stress, say, σ_3, but σ_{zz} and σ_{xx} are not principal stresses because of the presence of the shear stress σ_{zx} that acts on these planes. Let the other two principal stresses at any point be designated by σ_1 and σ_2. These two stresses may be found from stress theory (Sec. 2.4) with the values of σ_{zz}, σ_{xx}, and

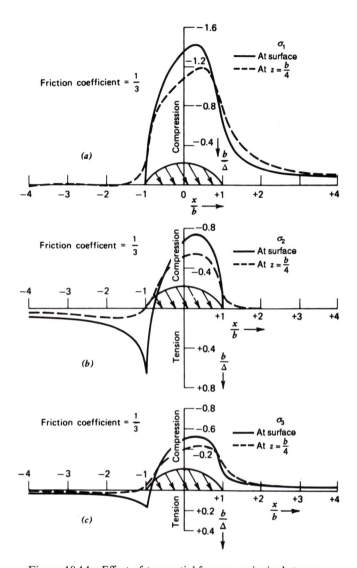

Figure 18.14 Effect of tangential force on principal stresses.

σ_{zx} for the point. The principal stresses σ_1, σ_2, and σ_3 for points on the surface*
and at a distance $z = b/4$ from the surface have been computed by this theory for a
value of friction coefficient of $\frac{1}{3}$, and their values have been plotted in Figs. 18.14a,
b, and c. Each principal stress has its maximum value in the surface of the body at
a distance of about $0.3b$ from the center of the area of contact in the direction of
the frictional force. These maximum values, all of which occur at the same point,
are $\sigma_1 = -1.4b/\Delta$, $\sigma_2 = -0.72b/\Delta$, and $\sigma_3 = -0.53b/\Delta$. These values may be compared with $\sigma_1 = -b/\Delta$, $\sigma_2 = -b/\Delta$, and $\sigma_3 = -0.5b/\Delta$, as found from Eq. (18.40)

---

* A special method of evaluating Eq. (18.44) may be used when solving for the stresses on the surface
where $z = 0$ (Smith and Liu, 1953).

for the normal distributed load w only. This comparison shows that the frictional force corresponding to a coefficient of friction of $\frac{1}{3}$ increases the maximum principal stress by 40%. Furthermore, the curves in Fig. 18.14 show that the principal stresses σ_2 and σ_3 are tension stresses near the edge of the contact area opposite the direction of the frictional force. The largest magnitudes of these stresses are $0.667(b/\Delta)$ and $0.167(b/\Delta)$, respectively, but these values are sometimes quite large. The presence of the tensile stresses on the surfaces aids in understanding the occurrence of fatigue failure by pitting, etc., of bearing surfaces subjected to repeated loads.

Maximum Shear Stress

From the values of maximum and minimum principal stresses at a point in the surface of contact, the maximum shear stress at the point on the surface is found to be

$$\tau_{max} = \frac{1}{2}\left(-\frac{1.4b}{\Delta} + \frac{0.53b}{\Delta}\right) = -0.43\left(\frac{b}{\Delta}\right) \tag{18.45}$$

To determine whether or not this value of the shear stress is the maximum value occurring in the body, it is necessary to compute the maximum shear stress at all other points, and especially at points inside the body under the contact area, since in all previous results presented in this chapter the maximum shear stress was found to be a subsurface shear stress. The values of shear stress at points on the surface and from the surface a distance of $z = b/4$ (where the maximum subsurface shear occurs) have been computed by making use of the principal stresses in Fig. 18.14 and are represented as ordinates to the curves in Figs. 18.15a, b, and c. There are three extreme values of shear stresses at each point; they are

$$\tau_1 = \frac{1}{2}(\sigma_1 - \sigma_3)$$
$$\tau_2 = \frac{1}{2}(\sigma_1 - \sigma_2)$$
$$\tau_3 = \frac{1}{2}(\sigma_2 - \sigma_3) \tag{18.46}$$

From Figs. 18.15a and c, we see that the ordinates to the curves representing τ_1 and τ_3 at distance $z = b/4$ from the surface are everywhere smaller than at the surface. This fact is true of the curves for these values at all distances from the surface. However, in Fig. 18.15b, the curve for τ_2 at $z = b/4$ rises above the curve representing values of τ_2 at the surface. Such curves for values of τ_2 have been plotted for several different distances from the surface, and it is found that the largest value of τ_2 is $0.36b/\Delta$. This value occurs at a distance of about $b/4$ from the surface. Therefore, the value of $\tau_1 = -0.43b/\Delta$ as given by Eq. (18.46) is the maximum shear stress, and it occurs at a point in the contact area about $0.3b$ from the center of the area. In Eq. (18.46) the maximum value of τ_2, which always occurs away from the surface, does not exceed τ_1 until the coefficient of friction has a value less than $\frac{1}{10}$.

Maximum Octahedral Shear Stress

In Fig. 18.16 the ordinates to the curves represent the values of the octahedral shear stresses τ_{oct} that have been computed at each point from Eq. (2.22) by substitution of values of the principal stresses obtained from Fig. 18.14. The maximum value is $\tau_{oct(max)} = 0.37b/\Delta$, and this value occurs on the contact area at the same

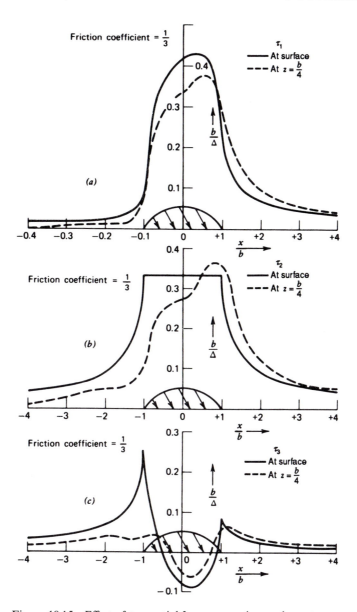

Figure 18.15 Effect of tangential force on maximum shear stresses.

point that the maximum principal stress and maximum shear stresses occur (Figs. 18.14 and 18.15).

Effect of Magnitude of Friction Coefficient

The magnitude of the coefficient of friction determines the size of the frictional distributed load f for a given value of w and, therefore, of the values of the maximum principal stresses, maximum shear stresses, and maximum octahedral shear stress. The changes in the maximum contact stresses with the coefficient of friction are

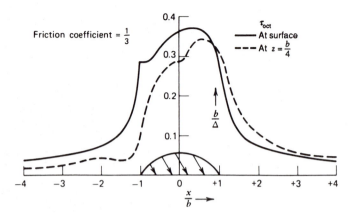

Figure 18.16   Effect of tangential force on octahedral shear stress.

given by Table 18.1. The increases in the maximum values of the tensile and compressive principal stresses caused by the frictional distributed load are very nearly proportional to the increases in the friction coefficient. For small values of the friction coefficient, the values of shear stress are increased only slightly by an increase in the friction coefficient, whereas there is a small decrease in octahedral shear up to a friction coefficient of $\frac{1}{6}$. For the case of a disk in rolling contact with a cylinder and for $\beta = \frac{1}{3}$, Mitsuda (1965) has noted that the range of maximum octahedral shear stress is 26% larger than the range of maximum orthogonal shear stress.

TABLE 18.1
Values of Contact Stresses Between Two Long Cylindrical Bodies Sliding Against Each Other While in Line Contact (Normal and Friction Forces)

Coefficient of Friction	0	$\dfrac{1}{12}$	$\dfrac{1}{9}$	$\dfrac{1}{6}$	$\dfrac{1}{3}$
Kind of Stress and Its Location	\multicolumn{5}{c}{**Values of Stress in Terms of b/Δ Corresponding to the Above Friction Coefficients**}
Maximum tensile principal stress that occurs in surface at $x = -b$	0	$\dfrac{2}{12}\dfrac{b}{\Delta}$	$\dfrac{2}{9}\dfrac{b}{\Delta}$	$\dfrac{2}{6}\dfrac{b}{\Delta}$	$\dfrac{2}{3}\dfrac{b}{\Delta}$
Maximum compressive principal stress that occurs in the surface between $x = 0$ and $x = 0.3b$	$-\dfrac{b}{\Delta}$	$-1.09\dfrac{b}{\Delta}$	$-1.13\dfrac{b}{\Delta}$	$-1.19\dfrac{b}{\Delta}$	$-1.40\dfrac{b}{\Delta}$
Maximum shear stress[a]	$0.300\dfrac{b}{\Delta}$	$0.308\dfrac{b}{\Delta}$	$0.310\dfrac{b}{\Delta}$	$0.339\dfrac{b}{\Delta}$	$0.435\dfrac{b}{\Delta}$
Maximum octahedral shear stress[a]	$0.272\dfrac{b}{\Delta}$	$0.265\dfrac{b}{\Delta}$	$0.255\dfrac{b}{\Delta}$	$0.277\dfrac{b}{\Delta}$	$0.368\dfrac{b}{\Delta}$

[a] Note that these stresses occur at the surface when the friction coefficient is $\frac{1}{10}$ or larger.

Range of Shear Stress for One Load Cycle

The *magnitude* of the maximum shear stress or maximum octahedral shear stress serves to indicate if yielding has taken place or to determine the factor of safety against impending yielding. However, in the case of fatigue loading, the *range* of shear stress on a given plane in a given direction in that plane is more commonly used to indicate the severity of a given loading. In the absence of friction, the *range* of the orthogonal shear stress for two cylinders in rolling contact is $2\tau_0 = 0.50b/\Delta[\tau_0 = \sigma_{zx(max)}]$; this range is greater than either the corresponding range of maximum shear stress $(0.30b/\Delta)$ or the corresponding range of octahedral shear stress $(0.27b/\Delta)$.

TABLE 18.2
Principal Stresses at Fixed Point 0 as Contact Surface Moves Relative to 0

Position of Contact Surface Relative to Fixed Point 0	Direction of Principle Stresses of Fixed Point 0

The presence of friction has little influence on the range of the orthogonal shear stress. However, the ranges of shear stress on certain planes increase with the coefficient of friction; the maximum range is as large as $0.67b/\Delta$ for $\beta = \frac{1}{3}$. Furthermore, this range occurs for points at the free surface of the contacting bodies where fatigue failures are more likely to be initiated.

For rolling cylinders in contact, Smith and Liu (1953) have determined the principal state of stress for a point in the surface of one of the rolling cylinders for the special case of the coefficient of friction $\beta = \frac{1}{3}$. The principal stresses are indicated in Table 18.2 for a volume element located at a point 0 for several locations of point 0 relative to the contact surface. For each location of point 0, the shear stress and octahedral shear stress attain maximum values on certain planes passing through point 0.

The largest values of these maxima [τ_{max} and $\tau_{oct(max)}$] occur for the volume element located at point 0 in Fig. (D) of Table 18.2. Smith and Liu noted the two sets of planes on which the largest values of τ_{max} and $\tau_{oct(max)}$ occur and determined the magnitude and sense of the shear stress acting on these planes for various locations of point 0. They also defined the maximum range of shear stress on either set of planes to be the magnitude of the maximum diameter of the shear stress envelope. (These maximum ranges were found to be $0.53b/\Delta$ and $0.63b/\Delta$.) (Smith and Liu, 1953).

Note that in Figs. (B) of Table 18.2 the maximum shear stress for these two states of stress occur on the same planes and they are opposite in sign; the range of shear stress for these planes is $0.67b/\Delta$. Only four sets of the infinite number of planes through point 0 [including planes perpendicular to the x and z axes on which $\tau_0 = \sigma_{zx(max)}$ occurs] were investigated; therefore, the range of shear stress ($0.67b/\Delta$) may not be the largest that occurs. However, the results do indicate that a tangential component of force at the contact surface increases the probability of fatigue failure, particularly if the coefficient of friction approaches a value of $\frac{1}{3}$.

EXAMPLE 18.2
Contact Stress in Cylinders with Friction

The fatigue testing machine described in Problem 18.13 has two identical steel disks ($E = 200\,\text{GPa}$ and $\nu = 0.29$) rolling together. The identical disks have a radius of curvature of 40 mm and width $h = 20$ mm. For rolling without friction, a load $P = 24.1$ kN produces the following stresses: $\sigma_{max} = 1445$ MPa, $\tau_{max} = 433$ MPa and $\tau_{oct(max)} = 361$ MPa. Let the cylinders be subjected to a load $P = 24.1$ kN and be rotated at slightly different speeds so that the roller surfaces slide across each other. If the coefficient of sliding friction is $\frac{1}{9}$, determine σ_{max}(tension), σ_{max}(compression), τ_{max}, and $\tau_{oct(max)}$.

SOLUTION

From Table 18.1 the value of the stresses are found as follows:

$$\sigma_{max}(\text{tension}) = \frac{2}{9}\frac{b}{\Delta}$$

$$\sigma_{max}(\text{compression}) = -1.13\frac{b}{\Delta}$$

$$\tau_{max} = 0.310\frac{b}{\Delta}$$

$$\tau_{oct(max)} = 0.255\frac{b}{\Delta}$$

The magnitudes of Δ and b are given by Eqs. (18.39) and (18.38)

$$\Delta = 2R\left(\frac{1-v^2}{E}\right) = \frac{2(40)(1-0.29^2)}{200 \times 10^3} = 0.0003664$$

$$b = \sqrt{\frac{2P\Delta}{h\pi}} = \sqrt{\frac{2(24.1 \times 10^3)(0.0003664)}{20\pi}} = 0.5301 \text{ mm}$$

$$\frac{b}{\Delta} = 1447 \text{ MPa}$$

Therefore, we have the following results:

$$\sigma_{max}(\text{tension}) = \frac{2}{9}(1447) = 322 \text{ MPa}$$

$$\sigma_{max}(\text{compression}) = -1.13(1447) = -1635 \text{ MPa}$$

$$\tau_{max} = 0.310(1447) = 449 \text{ MPa}$$

$$\tau_{oct(max)} = 0.255(1447) = 369 \text{ MPa}$$

The friction force (coefficient of sliding friction is $\frac{1}{9}$) increasses the maximum compression stress by 13.1%, maximum shear stress by 3.7%, and maximum octahedral shear stress by 2.2%.

PROBLEMS
Section 18.7

18.1. A steel railway car wheel may be considered a cylinder with a radius of 440 mm. The wheel rolls on a steel rail whose top surface may be considered another cylinder with a radius of 330 mm. For the steel wheel and steel rail, $E = 200$ GPa, $v = 0.29$, and $Y = 880$ MPa. If the wheel load is 110 kN, determine σ_{max}, τ_{max}, $\tau_{oct(max)}$, $2\tau_0$, and the factor of safety against initiation of yielding based on the maximum shear-stress criterion.

18.2. Determine the vertical displacement of the center of the wheel in Problem 18.1 due to the deflections in the region of contact.

Ans. $\delta = 0.116$ mm

18.3. In terms of P compute the maximum principal stress, maximum shear stress, and maximum octahedral shear stress in two steel balls ($E = 200$ GPa and $v = 0.29$) 200 mm in diameter pressed together by a force P.

18.4.   Solve Problem 18.3 for the condition that a single steel ball is pressed against a thick flat steel plate. (Length in mm)

Ans. $\sigma_{max} = -61\sqrt[3]{P}, \quad \tau_{max} = 20\sqrt[3]{P},$

$\tau_{oct(max)} = 18\sqrt[3]{P}$

18.5.   Solve Problem 18.3 for the condition that a single steel ball is pressed against the inside of a thick spherical steel race of inner radius 200 mm.

18.6.   A feed roll (a device used to surface-finish steel shafts) consists of two circular cylindrical steel rollers, each 200 mm in diameter and arranged so that their longitudinal axes are parallel. A cylindrical steel shaft (60 mm in diameter) is fed between the rollers in such a manner that its longitudinal axis is perpendicular to that of the rollers. The total load P between the shaft and rollers is 4.5 kN. Determine the values of the maximum principal stress and maximum shear stress in the shaft. Determine the distance from the plane of contact to the point of maximum shear stress, $E = 200$ GPa and $v = 0.29$.

Ans. $\sigma_{max} = -1589$ MPa, $\quad \tau_{max} = 517$ MPa, $\quad z_s = 0.515$ mm

18.7.   The longitudinal axes of the two feed rollers in Problem 18.6 are rotated in parallel planes until they form an angle of $\pi/6$ radians. The steel shaft is then fed between the two rollers at an angle of $\pi/12$ radians with respect to each of the rollers; again, $P = 4.5$ kN. Determine the maximum principal stress, maximum shear stress, and distance from the plane of contact to the maximum shear stress.

18.8.   A cast-iron push rod ($E = 117$ GPa, $v = 0.20$) in a valve assembly is operated by a steel cam ($E = 200$ GPa, $v = 0.29$) (Fig. P18.8). The cam is cylindrical in shape and has a radius of curvature of 5.00 mm at its tip. The surface of the push rod that contacts the cam is spherical in shape with

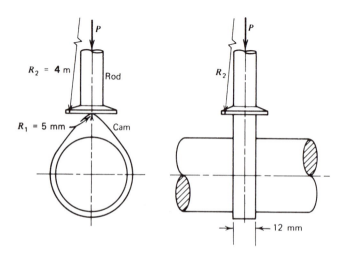

Figure P18.8   Contact load in valve tappet.

a radius of curvature 4.00 m so that the rod and cam are in point contact. If the allowable maximum principal stress for cast iron is -1400 MPa, determine the maximum load P that may act on the rod.

Ans. $P = 5.58$ kN

18.9. A fatigue testing machine used to determine fatigue life under rolling contact consists of a steel toroid (body 2) rolling on a steel cylinder (body 1), where $R_1 = 32$ mm, $R_1' = \infty$, $R_2 = 32$ mm, and $R_2' = 20$ mm. For steel, $E = 200$ GPa, $v = 0.29$.

  (a) Determine an expression for σ_{max} in terms of P.
  (b) Fatigue test results indicate that fatigue failure occurs at approximately $N = 10^9$ cycles with $\sigma_{max} = -2758$ MPa. Determine the applied load P. Since $\alpha = 0$ and R_1 and R_2 lie in the x-z plane (see Fig. 18.5), b (the minor semiaxis of the contact eclipse) is in the direction of rolling.

18.10. In the fatigue testing machine of Problem 18.9, the same cylinder is used, but the toroid is replaced by a second toroid, where $R_1 = \infty$, $R_1' = 32$ mm, $R_2 = 12.8$ mm, and $R_2' = 32$ mm. For the same steel properties as in Problem 18.9,

  (a) determine σ_{max} for fatigue failure at approximately $N = 10^9$ cycles. Neglect size effects and assume that fatigue failure is governed by the maximum range $2\tau_0$ of the orthogonal shear stress.
  (b) Determine the required load P. Since $\alpha = 0$ and R_1' and R_2' lie in the x-z plane (see Fig. 18.5), a (the major semiaxis of the ellipse) lies in the direction of rolling.

Ans. $P = 2.60$ kN

18.11. A hard steel ball ($E = 200$ GPa, $v = 0.29$) of diameter 50 mm is pressed against a thick aluminum plate ($E = 72.0$ GPa, $v = 0.33$, and $Y = 450$ MPa). Determine the magnitude of load P_Y required to initiate yield in the aluminum plate according to the maximum octahedral shear-stress criterion of failure.

18.12. For a safety factor $SF = 1.75$,

  (a) recalculate the required load in Problem 18.11.
  (b) For this load, determine the displacement (approach) δ of the ball relative to the plate.

Ans. $P = 196$ N, $\delta = 0.006$ mm

Section 18.8

18.13. A fatigue testing machine rolls together two identical steel disks ($E = 200$ GPa, $v = 0.29$), with radii 40 mm and thickness $h = 20$ mm. In terms of the applied load P, determine σ_{max}, τ_{max}, $\tau_{oct(max)}$, and τ_0.

18.14. Test data for the disks of Problem 18.13 indicate that fatigue failure occurs at approximately 10^8 cycles of load for $\sigma_{max} = -1380$ MPa.

(a) Determine the corresponding value of load P.

(b) For a fatigue failure at 10^8 cycles and a factor of safety $SF = 2.50$, determine the value of σ_{max}.

Ans. **(a)** $P = 21.9$ kN,   **(b)** $\sigma_{max} = -872$ MPa

18.15. The rail in Problem 18.1 wears in service until the top of the rail is flat with a width $h = 100$ mm.

(a) For other conditions given in Problem 18.1 remaining constant, determine the values of σ_{max} and τ_{max}.

(b) Using the maximum shear-stress criterion of failure, determine the safety factor SF against initiation of yield.

18.16. A cylindrical steel roller, with diameter 30 mm, is used as a follower on a steel cam. The surface of the cam at the contact region is cylindrical with radius of curvature 6 mm. Under no load, the follower and cam are in line contact over a length of 15 mm. For a value $\sigma_{max} = -1000$ MPa, determine the corresponding applied load P ($E = 200$ GPa, $v = 0.29$).

Ans. $P = 1.85$ kN

Section 18.9

18.17. Two cylindrical steel rollers ($E = 200$ MPa and $v = 0.29$), each 80 mm in diameter and 150 mm long, are mounted on parallel shafts and loaded by a force $P = 80$ kN. The two cylinders are rotated at slightly different speeds so that the roller surfaces slide across each other. If the coefficient of sliding friction is $\beta = \frac{1}{3}$, determine the maximum compressive principal stress, maximum shear stress, and maximum octahedral shear stress.

18.18. The two cylinders in Problem 18.17 are hardened. It is found that fatigue failures occur in the cylinders after 10^9 cycles for $\sigma_{max} = -1500$ MPa when $\beta = 0$.

(a) Determine the load P that can be applied to the cylinders to cause fatigue failure after 10^9 cycles ($\beta = 0$).

(b) Determine the load P that can be applied to the cylinders to cause fatigue failure at approximately 10^9 cycles for $\beta = \frac{1}{3}$. Assume that σ_{max} to cause fatigue failure is inversely proportional to the range of shear stress on a given plane and that the maximum range of shear stress is $0.67b/\Delta$ for $\beta = \frac{1}{3}$.

Ans.   **(a)** $P = 194$ kN,   **(b)** $P = 55.2$ kN

REFERENCES

Allen, R. K. (1945), *Rolling Bearings*. London.: Pitman.

Fessler, H. and Ollerton, E. (1957). Contact Stresses in Toroids Under Radial Loads. *Brit. J. Appl. Phys.*, **8** (10): 387.

Hertz, H. (1895). *Gesammetlte Werke*, Vol. 1. Lepzig. [For an English translation, see Hertz, H. (1896). *Miscellaneous Papers*. New York: Macmillan.

Mindlin, R. D. (1949). Compliance of Elastic Bodies in Contact. *J. App. Mech.*, **16** (3): 259.

Mitsuda, T. (1965). An Investigation of Pitting and Shelling Failure in Rolling Contact. Urbana/Champaign, Ill.: Univ. Ill., Depart. Theoretical and Appl. Mech., Ph. D. thesis, p. 16 and Table 1.

Moyar, G. J. and Morrow, J. D. (1964). Surface Failure of Bearings and Other Roller Elements. Urbana/Champaign, Ill.: Univ. Ill., Eng. Exper. Station, Bull. 468.

Smith, J. O. and Liu, C. K. (1953). Stresses due to Tangential and Normal Loads on an Elastic Solid with Applications to Some Contact Stress Problems. *J. Appl. Mech.*, **20** (2): 157.

Thomas, H. R. and Hoersch, V. A. (1930). Stresses due to the Pressure of One Elastic Solid on Another. Urbana/Champaign, Ill.: Univ. Ill., Eng. Exper. Station, June 15, Bull. 212.

Timoshenko, S. and Goodier, J. (1970). *Theory of Elasticity*, 3rd ed., New York: McGraw-Hill.

19

THE FINITE ELEMENT METHOD

19.1

INTRODUCTION

The finite element method* is the most powerful numerical technique available today for the analysis of complex structural and mechanical systems. It is used to obtain numerical solutions to a wide range of problems. The finite element method is used to analyze both linear and nonlinear systems. Nonlinear analysis includes material yielding, creep or cracking; aeroelastic response; buckling and postbuckling response; contact and friction; etc. The finite element method is used for both static and dynamic analyses. In its most general form, the method is not restricted to structural (or mechanical) systems. It has been applied to problems in fluid flow, heat transfer, and electric potential. This versatility is a major reason for the popularity of the method.

A complete study of finite element methods is beyond the scope of this book. So, the objective of this chapter is to outline the basic formulation for problems in linear elasticity. The formulation for plane elasticity is presented first. Then, the use of the method to analyze framed structures is examined. Finally, accuracy, convergence, and modeling techniques are discussed. Advanced topics, such as analysis of plate bending and shell problems, three-dimensional problems, and dynamic and nonlinear analysis, are left to more specialized texts.

Analytical Perspective

The classical method of analysis in elasticity involves the study of an infinitesimal element of an elastic body (continuum or domain). Relationships among stress, strain, and displacement for the infinitesimal element are developed (see Chapters 1–3) that are usually in the form of differential (or integral) equations that apply to each point in the body. These equations must be solved subject to appropriate boundary conditions. In other words, the approach is to define and solve a classical boundary value problem in mathematics (Boresi and Chong, 1987). Problems in engineering usually involve very complex shapes and boundary conditions. Consequently, for such cases, the equations cannot be solved exactly, but must finally be solved by approximate methods; for example, by truncated series, finite differences,

---

* The discovery of the method is often attributed to Courant (1943). The use of the method in structural (aircraft) analysis was first reported by Turner et al. (1956). The method received its name from Clough (1960).

numerical integration, etc. All these approximate methods require some form of discretization of the solution.

By contrast, the formulation of finite element solutions recognizes at the outset that discretization is likely to be required. The first step in application of the method is to discretize the domain into an assemblage of a finite number of finite size *elements* (or subregions) that are connected at specified node points. The quantities of interest (usually nodal displacements) are assumed to vary in a particular fashion over the element. This *assumed* element behavior leads to relatively simple integral equations for the individual elements. The integral equations for an element are evaluated to produce algebraic equations (in the case of static loading) in terms of the displacements of the node points. The algebraic equations for all elements are assembled to achieve a system of equations for the structure as a whole. Appropriate numerical methods are then used to solve this system of equations.

In summary, using the classical approach, we often are confronted with differential (or integral) equations that cannot be solved in closed form. This is due to the complexity of the geometry of the domain or boundary conditions. Consequently, we are forced to use numerical methods to obtain an approximate solution. These numerical methods always involve some type of discretization. In the finite element method, the discretization is performed at the outset. Then, further approximations, either in the formulation or in the solution may not be necessary.

Sources of Error

There are three sources of error in the finite element method: errors due to approximation of the domain (discretization error), errors due to approximation of the element behavior (formulation error), and errors due to use of finite precision arithmetic (numerical error).

Discretization error is due to the approximation of the domain with a finite number of elements of fixed geometry. For instance, consider the analysis of a rectangular plate with a centrally located hole (Fig. 19.1a). Due to symmetry, it is sufficient to model only one-quarter of the plate. If the region is subdivided into triangular elements (a triangular mesh or grid), the circular hole is approximated by a series of straight lines. If a few large triangles are used in a coarse mesh, (Fig. 19.1b), greater discretization error results than if a large number of small elements are used in a fine mesh, (Fig. 19.1c). Other geometric shapes may be chosen for the elements. For example, with quadrilateral elements that can represent curved sides, the circular hole is more accurately approximated (Fig. 19.1d). Hence, discretization error may be reduced by grid refinement. The grid can be refined by using more elements of the same type but of smaller size (*h-refinement*, Cook et al., 1989) or by using elements of a different type (*p-refinement*).

Formulation error results from the use of finite elements that do not precisely describe the behavior of the continuum. For instance, a particular element might be formulated on the assumption that displacements vary linearly over the domain. Such an element would contain no formulation error when used to model a prismatic bar under constant tensile load; in this case, the assumed displacement matches the actual displacement. If the same bar were subjected to uniformly distributed body force, then the actual displacements vary quadratically and formulation error would exist. Formulation error can be minimized by proper selection of element type and appropriate grid refinement.

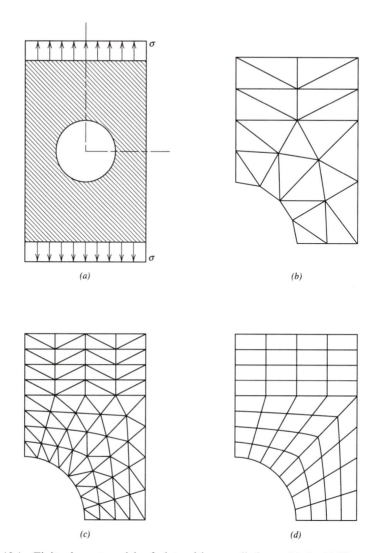

Figure 19.1   Finite element models of plate with centrally located hole. (*a*) Plate geometry and loading. (*b*) Coarse mesh of triangles. (*c*) Fine mesh of triangles. (*d*) Mesh of quadrilaterals with curved edges.

Numerical error is a consequence of round-off during floating-point computations and the error associated with numerical integration procedures. This source of error is dependent on the order in which computations are performed in the program and the use of double or extended precision variables and functions. The use of bandwidth minimization* can help control numerical error. Generally, in a well-designed finite element program, numerical error is small relative to formulation error.

---

* See Sec. 19.6 for a discussion of bandwidth minimization.

19.2

FORMULATION FOR PLANE ELASTICITY

Elasticity Concepts

One approach for developing the algebraic equations of the finite element method is to use energy principles. Fundamental energy expressions for an elastic solid are presented in Chapters 3 and 5. For plane elasticity, these expressions are simplified appropriately. The first law of thermodynamics states that for a two-dimensional body in equilibrium and subjected to arbitrary virtual displacements $(\delta u, \delta v)$, the variation in work of the external forces δW is equal to the variation of internal energy δU. Since *virtual displacements* are imposed, we define δW as the *virtual work of the external loads* and δU as the *virtual work of the internal forces*. The virtual work of the external loads δW can be divided into the work δW_S of the surface tractions, the work δW_B of the body forces, and the work δW_C* of the concentrated forces. For a two-dimensional body, these quantities are [see Eqs. (3.1a), (3.3), (3.4), (3.7) and (3.8) for definitions of the terms used]

$$\delta W - \delta U = \delta W_S + \delta W_B + \delta W_C - \delta U = 0 \tag{19.1}$$

$$\delta W_S = \int_S (\sigma_{Px}\, \delta u + \sigma_{Py}\, \delta v)\, dS \tag{19.2}$$

$$\delta W_B = \int_V (B_x\, \delta u + B_y\, \delta v)\, dV \tag{19.3}$$

$$\delta W_C = \sum F_{ix}\, \delta u_i + \sum F_{iy}\, \delta v_i \tag{19.4}$$

$$\delta U = \int_V (\sigma_{xx}\, \delta\epsilon_{xx} + \sigma_{yy}\, \delta\epsilon_{yy} + \sigma_{xy}\, \delta\gamma_{xy})\, dV \tag{19.5}$$

where (F_{ix}, F_{iy}) are (x, y) components of the concentrated force F_i at point i, $(\delta u_i, \delta v_i)$ are (x, y) components of the virtual displacement at point i, and $\gamma_{xy} = 2\epsilon_{xy}$. In matrix† notation,

$$\delta W_S = \int_S \{\delta u\}^T \{F_S\}\, dS \tag{19.6}$$

$$\delta W_B = \int_V \{\delta u\}^T \{F_B\}\, dV \tag{19.7}$$

$$\delta W_C = \sum \{\delta u_i\}^T \{F_i\} \tag{19.8}$$

$$\delta U = \int_V \{\delta\epsilon\}^T \{\sigma\}\, dV \tag{19.9}$$

---

* Concentrated forces were not discussed in Chapter 3 or 5, but are included here for completeness.
† In this chapter, vector quantities are denoted with braces { }. Two-dimensional arrays are contained in brackets [ ].

where

$$\{\delta u\} = [\delta u \quad \delta v]^T$$
$$\{\delta u_i\} = [\delta u_i \quad \delta v_i]^T$$
$$\{F_S\} = [\sigma_{Px} \quad \sigma_{Py}]^T$$
$$\{F_B\} = [B_x \quad B_y]^T$$
$$\{F_i\} = [F_{ix} \quad F_{iy}]^T$$
$$\{\delta\epsilon\} = [\delta\epsilon_{xx} \quad \delta\epsilon_{yy} \quad \delta\gamma_{xy}]^T$$
$$\{\sigma\} = [\sigma_{xx} \quad \sigma_{yy} \quad \sigma_{xy}]^T$$

In matrix form, the two-dimensional linear-elastic stress-strain relations are, by appropriate simplification of Eq. (3.32),

$$\{\sigma\} = [D]\{\epsilon\} \tag{19.10}$$

where $\{\epsilon\} = \{\epsilon_{xx} \quad \epsilon_{yy} \quad \gamma_{xy}\}^T$ and $[D]$ is the matrix of elastic coefficients. For plane stress,

$$[D] = \frac{E}{1 - v^2}\begin{bmatrix} 1 & v & 0 \\ v & 1 & 0 \\ 0 & 0 & \dfrac{1 - v}{2} \end{bmatrix} \tag{19.11}$$

and for plane strain,

$$[D] = \frac{E}{(1 + v)(1 - 2v)}\begin{bmatrix} 1 - v & v & 0 \\ v & 1 - v & 0 \\ 0 & 0 & \dfrac{1 - 2v}{2} \end{bmatrix} \tag{19.12}$$

Similarly, the two-dimensional, small displacement, strain-displacement relations are [see Eq. (2.81)],

$$\{\epsilon\} = [L]\{u\} \tag{19.13}$$

where $\{u\} = [u(x, y) \quad v(x, y)]^T$ and $[L]$ is a matrix of linear differential operators

$$[L] = \begin{bmatrix} \dfrac{\partial}{\partial x} & 0 \\ 0 & \dfrac{\partial}{\partial y} \\ \dfrac{\partial}{\partial y} & \dfrac{\partial}{\partial x} \end{bmatrix} \tag{19.14}$$

Displacement Interpolation: The Constant Strain Triangle

Consider a plane elasticity problem such as that shown in Fig. 19.1a. As discussed above, the first step in applying the finite element method is the discretization of the domain into a finite number of elements. Consider triangular elements as shown in Figs. 19.1b and c. If the entire domain is in equilibrium, then too is each element. Hence, the above virtual work concepts can be applied to an individual triangular element.

A typical triangular element is shown in Fig. 19.2 with corner nodes 1, 2, and 3 numbered in a counterclockwise order. The (x, y) displacement components at the nodes are (u_1, v_1), (u_2, v_2), and (u_3, v_3) as shown. The *nodal* displacements are the primary variables (unknowns) that are to be determined in the analysis. In general, for plane elasticity elements, node i has two degrees of freedom (DOF), u_i and v_i, where the subscript identifies the node at which the DOF exist. Quantities that are continuous over the element (those not associated with a particular node) are denoted without a subscript. A single triangular element with three nodes has six nodal DOF. These DOF are ordered according to node numbering as

$$\{u_i\} = [u_1 \quad v_1 \quad u_2 \quad v_2 \quad u_3 \quad v_3]^T \tag{19.15}$$

Displacements (u, v) at any point P within the element are continuous functions of the spatial coordinates (x, y). A fundamental approximation in the finite element method (that leads to formulation error) is that the displacement (u, v) at any point P in the element can be written in terms of the nodal displacements. Specifically, the displacement (u, v) at point P within the element is *interpolated* from the displacements of the nodes using interpolation polynomials. The order of the interpolation depends on the number of DOF in the element. For the three-node triangular element, the displacement is assumed to vary linearly over the element.

$$u(x, y) = a_1 + a_2 x + a_3 y$$
$$v(x, y) = a_4 + a_5 x + a_6 y \tag{19.16}$$

The coefficients a_i are constants (sometimes called generalized displacement coordinates) that are evaluated in terms of the nodal displacements. Before making this evaluation, we consider some properties of the linear displacement approximation.

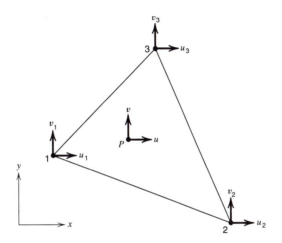

Figure 19.2   Constant strain triangle element.

1. Substitution of Eq. (19.16) into Eq. (19.13) yields $\epsilon_{xx} = a_2$, $\epsilon_{yy} = a_6$, and $\gamma_{xy} = a_3 + a_5$. Thus, the strain components in the element are constant; hence, the name *constant strain triangle* (CST) element. Since the stress-strain relations are linear [Eq. (19.10)], stress components are also constant in the element.

2. If $a_2 = a_3 = a_5 = a_6 = 0$, then $u(x, y) = a_1$ and $v(x, y) = a_4$. Constant values of u and v displacement indicate that the element can represent rigid-body translation.

3. If $a_1 = a_2 = a_4 = a_6 = 0$ and $a_3 = -a_5$, then $u(x, y) = a_3 y$ and $v(x, y) = -a_3 x$. Thus, for small strains and small rotations, the element can represent rigid-body rotation.

These three element characteristics ensure that the solution will converge monotonically as the mesh is refined (see Sec. 19.6 for a discussion of convergence).

To express the continuous displacement field in terms of the nodal displacements, Eq. (19.16) is evaluated at each node. The resulting equations are then solved for the coefficients a_i. Consider first the u displacement.

$$u(x_1, y_1) = a_1 + a_2 x_1 + a_3 y_1 = u_1$$
$$u(x_2, y_2) = a_1 + a_2 x_2 + a_3 y_2 = u_2$$
$$u(x_3, y_3) = a_1 + a_2 x_3 + a_3 y_3 = u_3$$

In matrix form, these equations are written as

$$[A]\{a\} = \{u_i\} \tag{19.17}$$

where

$$[A] = \begin{bmatrix} 1 & x_1 & y_1 \\ 1 & x_2 & y_2 \\ 1 & x_3 & y_3 \end{bmatrix}, \quad \{a\} = \begin{Bmatrix} a_1 \\ a_2 \\ a_3 \end{Bmatrix}, \quad \text{and} \quad \{u_i\} = \begin{Bmatrix} u_1 \\ u_2 \\ u_3 \end{Bmatrix}$$

Solution of Eq. (19.17) for $\{a\}$ and substitution into Eq. (19.16) yields

$$u(x, y) = \frac{1}{2A}(\alpha_1 + \beta_1 x + \gamma_1 y)u_1$$

$$+ \frac{1}{2A}(\alpha_2 + \beta_2 x + \gamma_2 y)u_2$$

$$+ \frac{1}{2A}(\alpha_3 + \beta_3 x + \gamma_3 y)u_3 \tag{19.18}$$

where A is the area of the triangle

$$A = \tfrac{1}{2}[x_1(y_2 - y_3) + x_2(y_3 - y_1) + x_3(y_1 - y_2)] \tag{19.19}$$

and

$$\begin{aligned} \alpha_1 &= x_2 y_3 - x_3 y_2, & \beta_1 &= y_2 - y_3, & \gamma_1 &= x_3 - x_2 \\ \alpha_2 &= x_3 y_1 - x_1 y_3, & \beta_2 &= y_3 - y_1, & \gamma_2 &= x_1 - x_3 \\ \alpha_3 &= x_1 y_2 - x_2 y_1, & \beta_3 &= y_1 - y_2, & \gamma_3 &= x_2 - x_1 \end{aligned} \tag{19.20}$$

Similarly, for the v displacement,

$$v(x, y) = \frac{1}{2A}(\alpha_1 + \beta_1 x + \gamma_1 y)v_1$$

$$+ \frac{1}{2A}(\alpha_2 + \beta_2 x + \gamma_2 y)v_2$$

$$+ \frac{1}{2A}(\alpha_3 + \beta_3 x + \gamma_3 y)v_3 \qquad (19.21)$$

The functions that multiply the nodal displacements in Eqs. (19.18) and (19.21) are known as *shape* functions (other common names are *interpolation* and *basis* functions). The shape functions for the CST element are

$$N_1(x, y) = \frac{1}{2A}(\alpha_1 + \beta_1 x + \gamma_1 y)$$

$$N_2(x, y) = \frac{1}{2A}(\alpha_2 + \beta_2 x + \gamma_2 y)$$

$$N_3(x, y) = \frac{1}{2A}(\alpha_3 + \beta_3 x + \gamma_3 y) \qquad (19.22)$$

Then Eqs. (19.18) and (19.21) take the form

$$u(x, y) = \sum_{i=1}^{3} N_i u_i, \qquad v(x, y) = \sum_{i=1}^{3} N_i v_i$$

In matrix notation

$$\{u\} = [N]\{u_i\} \qquad (19.23)$$

where

$$\{u\} = [u(x, y) \quad v(x, y)]^T$$

$$[N] = \begin{bmatrix} N_1 & 0 & N_2 & 0 & N_3 & 0 \\ 0 & N_1 & 0 & N_2 & 0 & N_3 \end{bmatrix} \qquad (19.24)$$

The shape functions for the CST element are illustrated in Fig. 19.3, where $N_1 = 1$ at node 1 and $N_1 = 0$ at nodes 2 and 3. Shape functions N_2 and N_3 behave similarly. Another important characteristic of the shape functions is that

$$\sum_{i=1}^{3} N_i(x, y) = 1.0$$

which is a requirement for shape functions so that the element can represent rigid-body motion.

Element Stiffness Matrix: The Constant Strain Triangle

With the displacement field for the element expressed in terms of the nodal displacements, the remainder of the formulation involves relatively straightforward

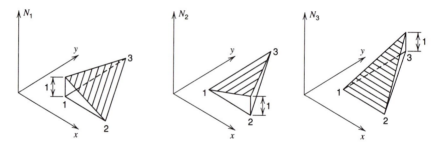

Figure 19.3 Graphical representation of shape functions for the CST element.

manipulation of the virtual work expressions [Eqs. (19.1), (19.6–19.9)]. Consider first the strain-displacement relations. Substitution of Eq. (19.23) into Eq. (19.13) gives the relationship between *continuous* element strains and *nodal* displacements

$$\{\epsilon\} = [L][N]\{u_i\} = [B]\{u_i\} \tag{19.25}$$

where, by Eqs. (19.14) and (19.24),

$$
[B] =
\begin{bmatrix}
\dfrac{\partial N_1}{\partial x} & 0 & \dfrac{\partial N_2}{\partial x} & 0 & \dfrac{\partial N_3}{\partial x} & 0 \\[2mm]
0 & \dfrac{\partial N_1}{\partial y} & 0 & \dfrac{\partial N_2}{\partial y} & 0 & \dfrac{\partial N_3}{\partial y} \\[2mm]
\dfrac{\partial N_1}{\partial y} & \dfrac{\partial N_1}{\partial x} & \dfrac{\partial N_2}{\partial y} & \dfrac{\partial N_2}{\partial x} & \dfrac{\partial N_3}{\partial y} & \dfrac{\partial N_3}{\partial x}
\end{bmatrix}
\tag{19.26}
$$

where $[B]$ is partitioned into nodal submatrices. The matrix $[B]$ is sometimes called the semidiscretized gradient operator. Since the shape functions for the CST element are linear in x and y, $[B]$ contains only constants that depend on the nodal coordinates.

For simplicity, temporarily assume that no body forces or surface tractions are applied to the element. However, concentrated loads at node points are permitted. The virtual work of these loads is [see Eq. (19.8)]

$$\delta W = \delta W_C = \{\delta u_i\}^T\{F_i\} \tag{19.27}$$

Substitution of Eqs. (19.27) and (19.9) into (19.1) leads to

$$\{\delta u_i\}^T\{F_i\} - \int_V \{\delta\epsilon\}^T\{\sigma\}\,dV = 0 \tag{19.28}$$

Note that $\{\sigma\} = [D]\{\epsilon\}$, $\{\epsilon\} = [B]\{u_i\}$, and $\{\delta\epsilon\}^T = \{\delta u_i\}^T[B]^T$. Substitution of these expressions into Eq. (19.28) gives

$$\{\delta u_i\}^T\{F_i\} - \int_V \{\delta u_i\}^T[B]^T[D][B]\{u_i\}\,dV = 0$$

Since $\{u_i\}$ and $\{\delta u_i\}$ are *nodal* quantities, they can be removed from the integral. Thus,

$$\{\delta u_i\}^T\left(\{F_i\} - \left[\int_V [B]^T[D][B]\,dV\right]\{u_i\}\right) = 0 \tag{19.29}$$

Since $\{\delta u_i\}$ is arbitrary, Eq. (19.29) yields the result

$$\{F_i\} = \left[\int_V [B]^T[D][B]\,dV\right]\{u_i\}$$

or

$$\{F_i\} = [K]\{u_i\} \tag{19.30}$$

where

$$[K] = \int_V [B]^T[D][B]\,dV \tag{19.31}$$

The element stiffness matrix $[K]$ relates nodal loads to nodal displacements in a system of linear algebraic equations; see Eq. (19.30). For the CST element, all terms in the integral are constants. Hence, for an element of constant thickness t and area A, the element stiffness matrix is

$$[K] = At[B]^T[D][B] \tag{19.32}$$

The individual terms in $[K]$ are denoted k_{ij}, where $i, j = 1, 2, \ldots, 6$ are the row and column positions, respectively. Since the element has six nodal DOF, $[K]$ has order (6×6). The explicit form of the CST element siffness matrix for a plane stress condition is given in Table 19.1.

Examination of Eq. (19.30) helps to establish a physical interpretation of the stiffness coefficients (the individual terms in $[K]$). Let a unit displacement be assigned to u_1 and take all other DOF to be zero. The resulting displacement vector is

$$\{u_i\} = [1 \quad 0 \quad 0 \quad 0 \quad 0 \quad 0]^T$$

Substitution of this displacement vector into Eq. (19.30) gives the force vector required to maintain the deformed shape.

$$\{F_i\} = [k_{11} \quad k_{21} \quad k_{31} \quad k_{41} \quad k_{51} \quad k_{61}]^T$$

Hence, an individual stiffness coefficient k_{ij} can be interpreted as the nodal force in the direction of DOF i that results from a unit displacement in the direction of DOF j, while all other DOF are set equal to zero. The physical system is illustrated in Fig. 19.4.

Equivalent Nodal Load Vector: The Constant Strain Triangle

Assume that body forces are applied to the CST element (surface tractions will be considered subsequently). The virtual work δW_B of the body forces on the element during an arbitrary virtual displacement $\{\delta u\}$ is given by Eq. (19.7). Substitution

TABLE 19.1

CST Element Stiffness Matrix, Plane Stress Case (Partitioned into 2 × 2 Nodal Submatrices)

$j \rightarrow$	1	2	3	4	5	6	$i \downarrow$
	$y_{23}^2 + \frac{1-\nu}{2}x_{32}^2$	$\frac{1+\nu}{2}x_{32}y_{23}$	$y_{31}y_{23} + \frac{1+\nu}{2}x_{13}x_{32}$	$\nu x_{13}y_{23} + \frac{1-\nu}{2}x_{32}y_{31}$	$y_{12}y_{23} + \frac{1-\nu}{2}x_{21}x_{32}$	$\nu x_{21}y_{23} + \frac{1-\nu}{2}x_{32}y_{12}$	1
	$\frac{1+\nu}{2}x_{32}y_{23}$	$x_{32}^2 + \frac{1-\nu}{2}y_{23}^2$	$\nu x_{32}y_{31} + \frac{1-\nu}{2}x_{13}y_{23}$	$x_{13}x_{32} + \frac{1-\nu}{2}y_{23}y_{31}$	$\nu x_{32}y_{12} + \frac{1-\nu}{2}x_{21}y_{23}$	$x_{21}x_{32} + \frac{1-\nu}{2}y_{12}y_{23}$	2
C	$y_{31}y_{23} + \frac{1+\nu}{2}x_{13}x_{32}$	$\nu x_{32}y_{31} + \frac{1-\nu}{2}x_{13}y_{23}$	$y_{31}^2 + \frac{1-\nu}{2}x_{13}^2$	$\frac{1+\nu}{2}x_{13}y_{31}$	$y_{12}y_{31} + \frac{1-\nu}{2}x_{13}x_{21}$	$\nu x_{21}y_{31} + \frac{1-\nu}{2}x_{13}y_{12}$	3
	$\nu x_{13}y_{23} + \frac{1-\nu}{2}x_{32}y_{31}$	$x_{13}x_{32} + \frac{1-\nu}{2}y_{23}y_{31}$	$\frac{1+\nu}{2}x_{13}y_{31}$	$x_{13}^2 + \frac{1-\nu}{2}y_{31}^2$	$\nu x_{13}y_{12} + \frac{1-\nu}{2}x_{21}y_{31}$	$x_{13}x_{21} + \frac{1-\nu}{2}y_{12}y_{31}$	4
	$y_{12}y_{23} + \frac{1-\nu}{2}x_{21}x_{32}$	$\nu x_{32}y_{12} + \frac{1-\nu}{2}x_{21}y_{23}$	$y_{12}y_{31} + \frac{1-\nu}{2}x_{13}x_{21}$	$\nu x_{13}y_{12} + \frac{1-\nu}{2}x_{21}y_{31}$	$y_{12}^2 + \frac{1-\nu}{2}x_{21}^2$	$\frac{1+\nu}{2}x_{21}y_{12}$	5
	$\nu x_{21}y_{23} + \frac{1-\nu}{2}x_{32}y_{12}$	$x_{21}x_{32} + \frac{1-\nu}{2}y_{12}y_{23}$	$\nu x_{21}y_{31} + \frac{1-\nu}{2}x_{13}y_{12}$	$x_{13}x_{21} + \frac{1-\nu}{2}y_{12}y_{31}$	$\frac{1+\nu}{2}x_{21}y_{12}$	$x_{21}^2 + \frac{1-\nu}{2}y_{12}^2$	6

$$C = \frac{Et}{4A(1-\nu^2)}, \qquad x_{ij} = x_i - x_j, \qquad y_{ij} = y_i - y_j$$

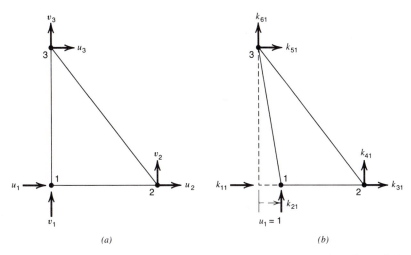

Figure 19.4 Physical interpretation of k_{ij}. (a) Undeformed element. (b) Deformed element, forces k_{i1} required to maintain $u_1 = 1$.

of Eq. (19.23) into Eq. (19.7) gives

$$\delta W_B = \{\delta u_i\}^T \int_V [N]^T \{F_B\} \, dV \tag{19.33}$$

The total external virtual work δW is the sum of the virtual work of the body forces and the virtual work of the concentrated forces so that Eq. (19.29) becomes

$$\{\delta u_i\}^T \left(\{F_i\} + \int_V [N]^T \{F_B\} \, dV - \left[\int_V [B]^T [D][B] \, dV \right] \{u_i\} \right) = 0 \tag{19.34}$$

Comparison of Eq. (19.34) with Eq. (19.29) shows that, with the addition of body forces, the load vector for the element is now

$$\{P_i\} = \{F_i\} + \int_V [N]^T \{F_B\} \, dV \tag{19.35}$$

The vector $\{P_i\}$ is the equivalent nodal load vector for the element. That is, the work of the loads $\{P_i\}$ under the virtual displacement $\{\delta u_i\}$ of the nodes is equivalent to the work of the actual concentrated loads and body forces under the virtual displacement $\{\delta u\}$.

In Eq. (19.35), the body force $\{F_B\}$ is expressed as a continuous function of the spatial coordinates. However, when constructing a finite element model, it is customary for the analyst to define element loads in terms of the intensity of the load at the nodes, rather than in functional form. The nodal force intensity is simply the magnitude of the body force at the node. Thus, for convenience, assume that the body force distribution may be expressed in terms of the force intensities at the nodes according to the relation

$$\{F_B\} = [N]\{f_{Bi}\}$$

where $\{f_{Bi}\}$ is the vector of nodal force intensities. Substitution of this relation into Eq. (19.35) gives

$$\{P_i\} = \{F_i\} + \int_V [N]^T[N]\{f_{Bi}\} \, dV$$

Since $\{f_{Bi}\}$ does not vary over the element, that is, they are nodal quantities that can be removed from the integral,

$$\{P_i\} = \{F_i\} + [Q]\{f_{Bi}\} \tag{19.36}$$

where

$$[Q] = \int_V [N]^T[N] \, dV$$

Thus, for a CST element,

$$[Q] = \frac{At}{9} \begin{bmatrix} 1 & 0 & 1 & 0 & 1 & 0 \\ 0 & 1 & 0 & 1 & 0 & 1 \\ 1 & 0 & 1 & 0 & 1 & 0 \\ 0 & 1 & 0 & 1 & 0 & 1 \\ 1 & 0 & 1 & 0 & 1 & 0 \\ 0 & 1 & 0 & 1 & 0 & 1 \end{bmatrix}$$

Now suppose that, in addition to concentrated nodal loads and body forces, the element is subjected to surface tractions along a single edge and that the continuous load function $\{F_S\}$ is expressed in terms of the nodal force intensities $\{f_{Si}\}$ by use of the shape functions. Since only one edge is loaded, only two of the nodes have nodal intensities and only these two nodes have equivalent nodal load components. Hence, for these two nodes, the interpolation equation is

$$\{\bar{F}_S\} = [\bar{N}]\{\bar{f}_{Si}\}$$

where the overbar indicates that only these two element nodes are included in the equation.

By the same approach as for body forces, the equivalent nodal loads due to surface traction on one edge are

$$\{\bar{P}_i\} = [\bar{Q}]\{\bar{f}_{Si}\} \tag{19.37}$$

where

$$[\bar{Q}] = \int_S [\bar{N}]^T[\bar{N}] \, dS \tag{19.38}$$

and the integral is evaluated over the loaded edge, where $dS = t \, ds$, $t = $ thickness and s is a coordinate along the loaded edge. The equivalent nodal load vector $\{\bar{P}_i\}$ in Eq. (19.37) is then added to $\{P_i\}$ from Eq. (19.36), but first it must be expanded from four to six terms to account for the fact that one node does not participate in the loading.

EXAMPLE 19.1
Equivalent Nodal Loads for Linear Surface Traction

A horizontally directed, linearly varying surface traction is applied to edge 1–3 of the CST element with nodal intensities as shown in Fig. E19.1. Determine the vector of equivalent nodal loads for the element.

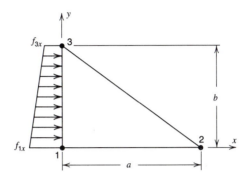

Figure E19.1

SOLUTION

The surface traction function is interpolated from the nodal intensities at nodes 1 and 3 the corresponding shape functions

$$f_x(y) = N_1 f_{1x} + N_3 f_{3x} \tag{a}$$

With the coordinates of the nodes, the shape functions are simplified to

$$N_1 = 1 - \frac{y}{b}, \qquad N_3 = \frac{y}{b} \tag{b}$$

By Eq. (19.38), with $ds = dy$,

$$[\bar{Q}] = t \int_0^b \begin{bmatrix} N_1^2 & 0 & N_1 N_3 & 0 \\ 0 & N_1^2 & 0 & N_1 N_3 \\ N_1 N_3 & 0 & N_3^2 & 0 \\ 0 & N_1 N_3 & 0 & N_3^2 \end{bmatrix} dy \tag{c}$$

By Eqs. (b) and (c),

$$[\bar{Q}] = t \begin{bmatrix} \dfrac{b}{3} & 0 & \dfrac{b}{6} & 0 \\ 0 & \dfrac{b}{3} & 0 & \dfrac{b}{6} \\ \dfrac{b}{6} & 0 & \dfrac{b}{3} & 0 \\ 0 & \dfrac{b}{6} & 0 & \dfrac{b}{3} \end{bmatrix} \tag{d}$$

and by Eq. (a), the vector of nodal intensities $\{\bar{f}_{Si}\}$ is

$$\{\bar{f}_{Si}\} = [f_{1x} \quad 0 \quad f_{3x} \quad 0]^T \tag{e}$$

With Eqs. (d) and (e), the equivalent nodal load vector $\{\bar{P}_i\}$ is obtained from Eq. (19.37) as

$$\{\bar{P}_i\} = \left\{ \begin{array}{c} tb\left(\dfrac{f_{1x}}{3} + \dfrac{f_{3x}}{6}\right) \\ 0 \\ \hline tb\left(\dfrac{f_{1x}}{6} + \dfrac{f_{3x}}{3}\right) \\ 0 \end{array} \right\} \tag{f}$$

Equation (e) is partitioned to identify the equivalent nodal loads associated with nodes 1 and 3.

If the vector $\{\bar{P}_i\}$ is expanded to include positions for node 2, it becomes

$$\{P_i\} = \left\{ \begin{array}{c} tb\left(\dfrac{f_{1x}}{3} + \dfrac{f_{3x}}{6}\right) \\ 0 \\ \hline 0 \\ 0 \\ \hline tb\left(\dfrac{f_{1x}}{6} + \dfrac{f_{3x}}{3}\right) \\ 0 \end{array} \right\}$$

Assembly of the Structure Stiffness Matrix and Load Vector

To solve a plane elasticity problem by the finite element method, it is necessary to combine the individual element stiffness matrices $[K]_j$ and load vectors $\{P\}_j$ to form the *structure stiffness matrix* $[K]$ and *structure load vector* $\{P\}$, respectively. To demonstrate the logic associated with the assembly process, two node numbering systems for the nodes are used. Let numerals in boldface refer to the nodes of the structural system and numerals in lightface the nodes for a particular element. Likewise, lightface $[K]$, $\{u_i\}$, and $\{P_i\}$ refer to element quantities, whereas boldface $[\mathbf{K}]$, $\{\mathbf{u}_i\}$, and $\{\mathbf{P}_i\}$ refer to structure quantities. A specific two-dimensional discretization is shown in Fig. 19.5 to illustrate the node numbering. For this model, there are 6 structure nodes but a total of 12 separate element nodes. The assembly process involves assigning unique identifiers to each of the nodes in the model, using the structure node numbering, and then combining element stiffness matrices and load vectors according to the numbering.

For purposes of demonstration, we consider first a mathematically precise, but computationally inefficient, approach for this assembly. Then, we discuss an approach that is more appropriate for computer implementation.

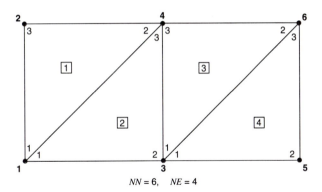

$$NN = 6, \quad NE = 4$$

Figure 19.5   Assembly of CST elements.

For element j, define a matrix $[M]_j$ with order $(6 \times 2NN)$*, where NN is the number of nodes in the structure, to define the mapping from the element DOF vector $\{u_i\}_j$, with order (6×1), to the structure DOF vector $\{u_i\}$, with order $(2NN \times 1)$.

$$\{u_i\}_j = [M]_j\{u_i\} \tag{19.39}$$

By Fig. 19.5, the mapping for element 1 takes the form

$$\{u_i\}_1 = [u_1 \quad v_1 \quad u_2 \quad v_2 \quad u_3 \quad v_3]_1^T$$
$$\{u_i\} = [u_1 \quad v_1 \quad u_2 \quad v_2 \quad u_3 \quad v_3 \quad \cdots \quad v_6]^T$$

and

$$[M]_1 = \begin{bmatrix} 1 & 0 & 0 & 0 & 0 & 0 & 0 & 0 & 0 & 0 & 0 & 0 \\ 0 & 1 & 0 & 0 & 0 & 0 & 0 & 0 & 0 & 0 & 0 & 0 \\ 0 & 0 & 0 & 0 & 0 & 0 & 1 & 0 & 0 & 0 & 0 & 0 \\ 0 & 0 & 0 & 0 & 0 & 0 & 0 & 1 & 0 & 0 & 0 & 0 \\ 0 & 0 & 1 & 0 & 0 & 0 & 0 & 0 & 0 & 0 & 0 & 0 \\ 0 & 0 & 0 & 1 & 0 & 0 & 0 & 0 & 0 & 0 & 0 & 0 \end{bmatrix}$$

By inspection or by Eq. (19.39), the DOF mapping for element 1 is

$$[u_1 \quad v_1 \quad u_2 \quad v_2 \quad u_3 \quad v_3]_1^T \leftrightarrow [u_1 \quad v_1 \quad u_4 \quad v_4 \quad u_2 \quad v_2]^T$$

The double-headed arrow indicates the reversibility of the mapping of the quantities on the left to the quantities on the right. Nodal forces and stiffness coefficients for element 1 follow the same mapping.

Next, the virtual work expressions for the entire structure are written as the sum of the virtual work for all elements

$$\sum_{j=1}^{NE} \{\delta u_i\}_j^T \{F_i\}_j + \sum_{j=1}^{NE} \{\delta u_i\}_j^T \int_S [\bar{N}]_j^T \{F_S\}_j \, dS$$

$$+ \sum_{j=1}^{NE} \{\delta u_i\}_j^T \int_V [N]_j^T \{F_B\}_j \, dV - \sum_{j=1}^{NE} \{\delta u_i\}_j^T [K]_j \{u_i\}_j = 0 \tag{19.40}$$

* The matrix $[M]_j$ in known as a *Boolean connectivity matrix* since it contains only ones and zeros.

where NE is the number of elements in the model. Substitution of Eq. (19.39) into Eq. (19.40) for each element gives

$$\{\delta u_i\}^T \sum_{j=1}^{NE} [M]_j^T \{F_i\}_j + \{\delta u_i\}^T \sum_{j=1}^{NE} [\bar{M}]_j^T \int_S [\bar{N}]_j^T \{F_S\}_j \, dS$$

$$+ \{\delta u_i\}^T \sum_{j=1}^{NE} [M]_j^T \int_V [N]_j^T \{F_B\}_j \, dV - \{\delta u_i\}^T \left[\sum_{j=1}^{NE} [M]_j^T [K]_j [M]_j \right] \{u_i\} = 0$$

(19.41)

Since $\{\delta u_i\}$ is arbitrary, it is eliminated from Eq. (19.41) to obtain

$$[K]\{u_i\} = \{P_i\} \tag{19.42a}$$

where

$$[K] = \left[\sum_{j=1}^{NE} [M]_j^T [K]_j [M]_j \right] \tag{19.42b}$$

and

$$\{P_i\} = \sum_{j=1}^{NE} [M]_j^T \{F_i\}_j + \sum_{j=1}^{NE} [\bar{M}]_j^T \int_S [\bar{N}]_j^T \{F_S\}_j \, dS + \sum_{j=1}^{NE} [M]_j^T \int_V [N]_j^T \{F_B\}_j \, dV$$

(19.42c)

In Eqs. (19.41) and (19.42c), matrix $[\bar{M}]_j$, of order $(4 \times 2NN)$, accounts for the mapping to the structure nodes of the two nodes in element j that participate in the surface tractions. If more than one edge on an element is loaded, then Eqs. (19.41) and (19.42c) are extended accordingly.

The forms of $[K]$ and $\{P_i\}$ in Eq. (19.42) are precise but they are not used in practice. The matrix products involving $[M]_j$, which involve multiplying by 0 or 1, do nothing more than move individual quantities from one position in the element stiffness matrix or load vector to another in the structure stiffness matrix or load vector. Although the above development is not practical, it does demonstrate that the structure stiffness matrix is assembled by successively adding the stiffness terms from each element into appropriate locations of the structure matrix; the same is true for the structure load vector. A more direct approach to assembly is demonstrated in Example 19.2.

EXAMPLE 19.2
Assembly of the Structure Stiffness Matrix

For the model shown in Fig. 19.5, illustrate the assembly of the stiffness matrix for element 1 into the structure stiffness matrix.

SOLUTION

Since the structure has six nodes, each of which has two DOF, the structure stiffness matrix is of order (12×12). The individual stiffness coefficients are designated k_{ij}^e, where now the superscript identifies the element number. With this notation, the stiffness matrix for element 1 is shown in Fig. E19.2. The mapping of element

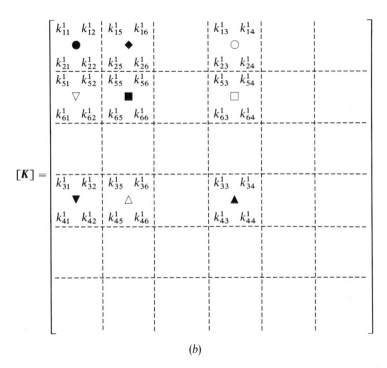

$$(a)$$

$$(b)$$

Figure E19.2   Assembly of element 1 stiffness matrix (with 2 × 2 nodal submatrix partitions). (a) Stiffness matrix for element 1. (b) Structure stiffness matrix with element 1 assembled.

TABLE E19.2
Element to Structure Node Mapping

	Structure Node Numbers			
Element Node No.	**Element 1**	**Element 2**	**Element 3**	**Element 4**
1	1	1	3	3
2	4	3	6	5
3	2	4	4	6

node numbers to structure node numbers is determined by inspection of the model in Fig. 19.5 and is summarized in Table E19.2. The list of structure node numbers that define the nodes for each element is commonly known as the *incidence list*. The incidence list is one of the input requirements for finite element programs. Using the incidence list, one can obtain the mapping of the element 1 nodal submatrices into the structure stiffness matrix (Fig. E19.2). Markers have been added to the nodal submatrices as an aid to visualization of the placement of element stiffness coefficients into the structure stiffness matrix.

As described in the previous example, the incidence list is used to drive the assembly process. Suppose that the node numbers that comprise the incidence list (e.g., Table E19.2) are placed into a matrix $[INCID]$ that contains one column for each element. The i, j term in the matrix is defined as the structure node number that corresponds to element node number i of element j. Then, by using the incidence matrix $[INCID]$, each term from the element stiffness matrix $[K]_j$ is moved into the structure stiffness matrix $[K]$ in a prescribed manner. The method is illustrated by a Fortran subroutine in Table 19.2. The subroutine moves one nodal submatrix at a time. Note that this code is for illustrative purposes only. Because of the symmetry and sparsity of the structure stiffness matrix, it is usually stored in some form other than a square matrix.

Application of Constraints

The model shown in Fig. 19.5 is not fastened to supports. Hence, it represents an unstable structure, a structure that is not capable of resisting external loads. The assembled stiffness matrix for an unstable structure is singular; it has a rank deficiency of 3 due to the three rigid-body modes that the model possesses. Physically, the structure must be supported to prevent rigid-body motion. In a like fashion, if the structure stiffness matrix is modified to reflect the support conditions (commonly known as constraints), it becomes nonsingular. Several methods may be used to apply constraints to the structure stiffness matrix. Only one, the so-called *equation modification method*, will be discussed here.

To demonstrate the equation modification method, consider a model that contains only a single element (Fig. 19.6a). The first step is to switch appropriate rows and columns of the stiffness such that those DOF that are constrained are grouped together. The rearranged stiffness matrix, displacement vector, and load vector for the one-element model are shown in Fig. 19.6b. For simplicity, the rearranged equations are represented in the symbolic form

$$\left[\begin{array}{c|c} K_{cc} & K_{cu} \\ \hline K_{uc} & K_{uu} \end{array} \right] \left\{ \begin{array}{c} u_c \\ u_u \end{array} \right\} = \left\{ \begin{array}{c} P_c \\ P_u \end{array} \right\} \tag{19.43}$$

where the subscript c represents the constrained DOF and the subscript u the unconstrained DOF.* The relationship between the submatrices and subvectors in

---

* The rearrangement of the equations and subsequent partitioning are done for convenience in representing the method. Computer implementation of this approach does not require that the equations be rearranged, nor would such rearrangement be computationally efficient.

TABLE 19.2
FORTRAN Subroutine for Stiffness Assembly

```
      SUBROUTINE ASSMBL(KS,KE,NNE,NDOF,INCID,IE,NRS,NRE,NE)
      DIMENSION KS(NRS, NRS), KE(NRE, NR), INCID(NNE, NE)
      REAL KS, KE
C
C         ASSEMBLE THE STIFFNESS FOR ELEMENT 'IE' INTO THE
C         STRUCTURE STIFFNESS.
C
C         CONTROL VARIABLES:
C
C  KS, KE    = STRUCTURE & ELEMENT STIFFNESS MATRICES.
C  NNE       = NUMBER OF NODES IN AN ELEMENT.
C  NDOF      = NUMBER OF DOF AT EACH NODE.
C  INCID     = INCIDENCE MATRIX.
C  IE        = CURRENT ELEMENT NUMBER.
C  NRS, NRE  = NUMBER OF ROWS IN STRUCTURE & ELEMENT STIFFNESS
C  NE        = NUMBER OF ELEMENTS IN THE MODEL.
C
C         LOCAL VARIABLES:
C
C  INE  = CURRENT ELEMENT SUBMATRIX ROW NUMBER.
C  JNE  = CURRENT ELEMENT SUBMATRIX COLUMN NUMBER.
C  INS  = CURRENT STRUCTURE SUBMATRIX ROW NUMBER.
C  JNS  = CURRENT STRUCTURE SUBMATRIX COLUMN NUMBER.
C  IDOF = CURRENT DOF NUMBER IN SUBMATRIX ROW.
C  JDOF = CURRENT DOF NUMBER IN SUBMATRIX COLUMN.
C  IKE  = ROW ENTRY IN THE ELEMENT STIFFNESS.
C  JKE  = COLUMN ENTRY IN THE ELEMENT STIFFNESS.
C  IKS  = ROW ENTRY IN THE STRUCTURE STIFFNESS.
C  JKS  = COLUMN ENTRY IN THE STRUCTURE STIFFNESS.
C
      DO 10 INE = 1, NNE
        INS = INCID( INE, IE )
      DO 10 JNE = 1, NNE
        JNS = INCID( JNE, IE )
C
C         ASSEMBLE THE ELEMENT SUBMATRIX (INE,JNE) INTO
C         THE STRUCTURE SUBMATRIX (INS, JNS)
C
      DO 10 IDOF = 1, NDOF
        IKE = ( INE - 1 ) * NDOF + IDOF
        IKS = ( INS - 1 ) * NDOF + IDOF
      DO 10 JDOF = 1, NDOF
        JKE = ( JNE - 1 ) * NDOF + JDOF
        JKS = ( JNS - 1 ) * NDOF + JDOF
C
        KS( IKS, JKS ) = KS( IKS, JKS ) + KE( IKE, JKE )
C
   10 CONTINUE
      RETURN
      END
```

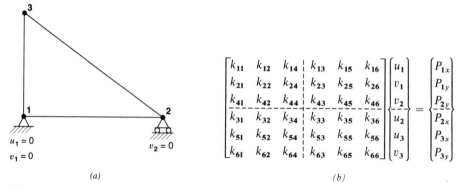

Figure 19.6   Application of constraints by the equation modification method. (a) One-element model with constraints. (b) Rearranged structure equations.

Eq. (19.43) and those in Fig. 19.6b is determined by their respective positions in the equations.

The unknown quantities are the displacements $\{u_u\}$ of the unconstrained DOF and the forces $\{P_c\}$ at the constrained DOF. Rewrite Eq. (19.43) as two separate submatrix/subvector equations.

$$[K_{cc}]\{u_c\} + [K_{cu}]\{u_u\} = \{P_c\} \tag{19.44a}$$
$$[K_{uc}]\{u_c\} + [K_{uu}]\{u_u\} = \{P_u\} \tag{19.44b}$$

Since $\{u_c\}$ is known, it is moved to the load side of Eq. (19.44b) to obtain

$$[K_{uu}]\{u_u\} = \{P_u\} - [K_{uc}]\{u_c\} \tag{19.45}$$

Equation (19.45) is the constrained system of equations. If the imposed constraints $\{u_c\}$ are nonzero, they serve to modify the load vector. If the constraints are all zero, such as in Fig. 19.6a, then the second term on the right side of Eq. (19.45) vanishes. In either case, the system of equations is reduced in order by the number of constrained DOF. If appropriate constraints are applied to render the structure stable, then $[K_{uu}]$ will be nonsingular.

Solution of the System of Equations

After assembly of the stiffness matrix and load vector and application of constraints, the system of linear algebraic equations may be solved. It is common to represent the solution of Eq. (19.45) in the symbolic form

$$\{u_u\} = [K_{uu}]^{-1}(\{P_u\} - [K_{uc}]\{u_c\})$$

However, inversion of the stiffness matrix $[K_{uu}]$ is computationally expensive and can lead to significant numerical error. A more efficient approach, known as *Choleski decomposition*, involves triangular factorization of the stiffness matrix.

$$[K_{uu}] = [U]^T[U]$$

where $[U]$ is an upper triangular matrix; that is, each term in the lower triangle of $[U]$ is zero ($u_{ij} = 0$, $i > j$). Factorization of $[K_{uu}]$ into this form permits direct solution for displacements via two *load-pass* operations. The first of these, known as the *forward load-pass*, yields an intermediate solution vector $\{y\}$.

$$[U]^T\{y\} = \{P_u\} - [K_{uc}]\{u_c\}$$

The second operation, known as the *backward load-pass*, produces the final displacement vector $\{u_u\}$.

$$[U]\{u_u\} = \{y\}$$

Upon solution for $\{u_u\}$, the reactions that result from deformation of the structure can be found from Eq. (19.44a). The total reactions are obtained by subtracting any nodal loads that are applied to the constrained DOF. Such loads frequently exist when element loads, in the form of body forces or surface tractions, are resolved into equivalent nodal loads.

Details of the equation solving methods and discussions of their advantages and disadvantages can be found in books that specialize in the finite element method (see the references at the end of this chapter).

19.3

THE BILINEAR RECTANGLE

The constant strain triangle is the simplest element that can be used for plane elasticity problems. As such, it is an attractive choice for demonstration of the basic formulation of the finite element method. However, because of its simplicity, the CST element exhibits relatively poor performance in a coarse mesh (a few large elements). In order to obtain satisfactory results with the CST element, a very highly refined mesh (many small elements) is generally needed for all but the most trivial problems. Alternatively, one may use a different element that is based on different displacement interpolation functions and that yields better results. The number of alternatives to the CST element is quite large and no attempt is made to discuss all of them here. Instead, we examine two alternatives: the bilinear rectangle and the linear isoparametric quadrilateral. The development of the bilinear rectangle follows. The linear isoparametric quadrilateral is presented in Sec. 19.4.

Consider a rectangular element of width $2a$, height $2b$, and with corner nodes numbered in a counterclockwise order. The (x, y) coordinate axes for the element are parallel to the 1–2 and 1–4 edges of the element, respectively, and the origin of the coordinate system is at the centroid of the element (see Fig. 19.7). As with the CST element, the displacement components (u, v) at any point P are expressed in terms of the nodal displacements. Since there are four nodes in the element, each with two nodal DOF, the displacement functions for $u(x, y)$ and $v(x, y)$ each have four coefficients. Hence, we choose the bilinear functions*

$$u(x, y) = a_1 + a_2 x + a_3 y + a_4 xy$$
$$v(x, y) = a_5 + a_6 x + a_7 y + a_8 xy$$

---

* These functions are said to be bilinear functions of (x, y) because the dependency on x and y comes from the product of two linear expressions, one in x and one in y. The corresponding rectangular element is said to be bilinear. With the given functions (u, v), the straight edges of the bilinear rectangle remain straight under deformation (like the CST element). However, the strain components in the bilinear rectangle element are not constant.

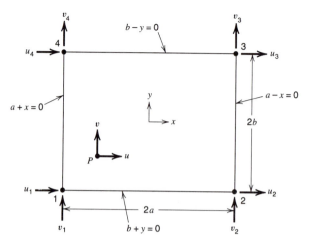

Figure 19.7   Bilinear rectangle element.

Like the CST element, the bilinear rectangle can properly represent rigid-body translation, rigid-body rotation, and constant strain. The bilinear displacement components (a_4xy and a_8xy) result in strain components such that ϵ_{xx} is linear in y, ϵ_{yy} is linear in x, and γ_{xy} is linear in both x and y. This higher-order response, compared to the CST element, results in more efficient and accurate numerical solutions.

Development of the stiffness matrix and load vector proceeds in a manner similar to that for the CST element. Only the stiffness matrix is developed here; development of the load vector is left as an exercise for the reader. The shape functions are expressed as products of one-dimensional Lagrange interpolation functions (Kellison, 1975)

$$N_1(x, y) = \frac{(a - x)(b - y)}{4ab}$$

$$N_2(x, y) = \frac{(a + x)(b - y)}{4ab}$$

$$N_3(x, y) = \frac{(a + x)(b + y)}{4ab}$$

$$N_4(x, y) = \frac{(a - x)(b + y)}{4ab} \tag{19.46}$$

Since the shape function for node i has zero value along any element edge that does not include node i, the shape function can be derived directly as the product of the equations of the lines that define these edges; see Fig. 19.7. The shape functions for the bilinear rectangle are illustrated in Fig. 19.8 where they form straight lines along the element edges. However, over the interior of the element, the functions form curved surfaces, with linearly varying slopes in the x and y directions.

The strain-displacement relations are written in the form of Eq. (19.25), with the nodal displacement vector

$$\{u_i\} = [u_1 \quad v_1 \quad u_2 \quad v_2 \quad u_3 \quad v_3 \quad u_4 \quad v_4]^T$$

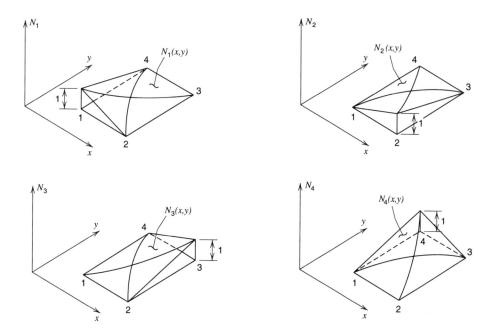

Figure 19.8 Graphical representation of shape functions for the bilinear rectangle element.

and $[B]$ matrix

$$[B] = [B_1 \quad B_2 \quad B_3 \quad B_4] \tag{19.47}$$

where for node i

$$[B_i] = \begin{bmatrix} \dfrac{\partial N_i}{\partial x} & 0 \\[2mm] 0 & \dfrac{\partial N_i}{\partial y} \\[2mm] \dfrac{\partial N_i}{\partial y} & \dfrac{\partial N_i}{\partial x} \end{bmatrix} \tag{19.48}$$

The element stiffness matrix is found from Eq. (19.31). That equation is repeated here with the order of each matrix shown as a subscript.

$$[K]_{8 \times 8} = \int_V [B]_{8 \times 3}^T [D]_{3 \times 3} [B]_{3 \times 8} \, dV$$

The stiffness matrix can be written in terms of (2×2) nodal submatrices as

$$[K_{ij}]_{2 \times 2} = \int_V [B_i]_{2 \times 3}^T [D]_{3 \times 3} [B_j]_{3 \times 2} \, dV$$

where i and j are element node numbers. The explicit form of the bilinear rectangle element stiffness matrix for a plane stress condition is given in Table 19.3.

TABLE 19.3
Bilinear Rectangle Stiffness Matrix, Plane Stress Case (Partitioned into 2 × 2 Nodal Submatrices)

C times:

row index $i\downarrow$ \ column index $j\to$	1	2	3	4	5	6	7	8
1	$4\beta + \dfrac{2(1-v)}{\beta}$	$\dfrac{3}{2}(1+v)$	$-4\beta + \dfrac{(1-v)}{\beta}$	$\dfrac{3}{2}(1-3v)$	$-2\beta - \dfrac{(1-v)}{\beta}$	$-\dfrac{3}{2}(1+v)$	$2\beta - \dfrac{2(1-v)}{\beta}$	$\dfrac{3}{2}(1-3v)$
2	$\dfrac{3}{2}(1+v)$	$\dfrac{4}{\beta} + 2(1-v)\beta$	$\dfrac{3}{2}(1-3v)$	$\dfrac{2}{\beta} - 2(1-v)\beta$	$-\dfrac{3}{2}(1+v)$	$-\dfrac{2}{\beta} - (1-v)\beta$	$-\dfrac{3}{2}(1-3v)$	$-\dfrac{4}{\beta} + (1-v)\beta$
3	$-4\beta + \dfrac{(1-v)}{\beta}$	$\dfrac{3}{2}(1-3v)$	$4\beta + \dfrac{2(1-v)}{\beta}$	$-\dfrac{3}{2}(1+v)$	$2\beta - \dfrac{2(1-v)}{\beta}$	$-\dfrac{3}{2}(1-3v)$	$-2\beta - \dfrac{(1-v)}{\beta}$	$\dfrac{3}{2}(1+v)$
4	$\dfrac{3}{2}(1-3v)$	$\dfrac{2}{\beta} - 2(1-v)\beta$	$-\dfrac{3}{2}(1+v)$	$\dfrac{4}{\beta} + 2(1-v)\beta$	$\dfrac{3}{2}(1-3v)$	$-\dfrac{4}{\beta} + (1-v)\beta$	$\dfrac{3}{2}(1+v)$	$-\dfrac{2}{\beta} - (1-v)\beta$
5	$-2\beta - \dfrac{(1-v)}{\beta}$	$-\dfrac{3}{2}(1+v)$	$2\beta - \dfrac{2(1-v)}{\beta}$	$\dfrac{3}{2}(1-3v)$	$4\beta + \dfrac{2(1-v)}{\beta}$	$\dfrac{3}{2}(1+v)$	$-4\beta + \dfrac{(1-v)}{\beta}$	$-\dfrac{3}{2}(1-3v)$
6	$-\dfrac{3}{2}(1+v)$	$-\dfrac{2}{\beta} - (1-v)\beta$	$-\dfrac{3}{2}(1-3v)$	$-\dfrac{4}{\beta} + (1-v)\beta$	$\dfrac{3}{2}(1+v)$	$\dfrac{4}{\beta} + 2(1-v)\beta$	$\dfrac{3}{2}(1-3v)$	$\dfrac{2}{\beta} - 2(1-v)\beta$
7	$2\beta - \dfrac{2(1-v)}{\beta}$	$-\dfrac{3}{2}(1-3v)$	$-2\beta - \dfrac{(1-v)}{\beta}$	$\dfrac{3}{2}(1+v)$	$-4\beta + \dfrac{(1-v)}{\beta}$	$\dfrac{3}{2}(1-3v)$	$4\beta + \dfrac{2(1-v)}{\beta}$	$-\dfrac{3}{2}(1+v)$
8	$\dfrac{3}{2}(1-3v)$	$-\dfrac{4}{\beta} + (1-v)\beta$	$\dfrac{3}{2}(1+v)$	$-\dfrac{2}{\beta} - (1-v)\beta$	$-\dfrac{3}{2}(1-3v)$	$\dfrac{2}{\beta} - 2(1-v)\beta$	$-\dfrac{3}{2}(1+v)$	$\dfrac{4}{\beta} + 2(1-v)\beta$

$$C = \frac{Et}{12(1-v^2)}, \qquad \beta = \frac{b}{a}$$

By itself, the bilinear rectangle element is limited to rectangular domains. This is potentially a rather severe restriction. However, nonrectangular domains can be modeled with a combination of bilinear rectangle elements and CST elements. Since both elements represent linear displacement variation along their edges, they are compatible; that is, displacements will be continuous across element boundaries.

EXAMPLE 19.3
Performance of the Bilinear Rectangle and CST Elements

Compare the ability of the bilinear rectangle and CST elements to model in-plane bending of a thin, square plate.

SOLUTION

A square plate of width a and thickness t is considered. For simplicity, Poisson's ratio is taken as zero, $v = 0$. To impose a state of pure bending, displacements $u = \pm\delta$ are imposed on the corners of the plate as shown in Fig. E19.3a. From the

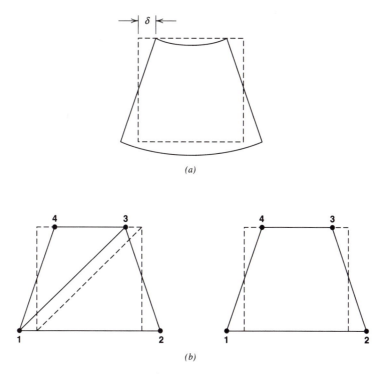

(a)

(b)

Figure E19.3   (a) Deformed shape (elasticity solution). (b) Deformed shape (finite element models).

theory of elasticity, the displacements are

$$u = -\frac{4xy\,\delta}{a^2} \tag{a}$$

$$v = \left(\frac{4x^2}{a} - 1\right)\frac{\delta}{2} \tag{b}$$

Differentiation of Eqs. (a) and (b) gives the strain components

$$\epsilon_{xx} = -\frac{4y\,\delta}{a^2}, \qquad \epsilon_{yy} = 0, \qquad \gamma_{xy} = 0 \tag{c}$$

The strain energy in the plate is

$$U = \int_V U_0\,dV$$

$$= \int_V \frac{E\epsilon_{xx}^2}{2}\,dV$$

$$= \frac{2}{3}Et\,\delta^2 \tag{d}$$

Two finite element models of the square plate are considered. The first uses two CST elements and the second, a single bilinear rectangle. As for the elasticity solution, nodal displacements of $u_i = \pm\delta$ are imposed. The models and their deformed shapes are shown in Fig. E19.3b. The model of two CST elements is considered first. Strains in the CST elements can be determined first since the displacement vector is known.

$$\{u_i\} = [-\delta \quad 0 \mid \delta \quad 0 \mid -\delta \quad 0 \mid \delta \quad 0]^T \tag{e}$$

The $[B]$ matrices for the two CST elements are defined by Eq. (19.26). For the element geometries in Fig. E19.3b, these matrices are

$$[B]_{CST-1} = \frac{1}{a}\begin{bmatrix} 0 & 0 & 1 & 0 & -1 & 0 \\ 0 & -1 & 0 & 0 & 0 & 1 \\ -1 & 0 & 0 & 1 & 1 & -1 \end{bmatrix} \tag{f}$$

$$[B]_{CST-2} = \frac{1}{a}\begin{bmatrix} -1 & 0 & 1 & 0 & 0 & 0 \\ 0 & 0 & 0 & -1 & 0 & 1 \\ 0 & -1 & -1 & 1 & 1 & 0 \end{bmatrix} \tag{g}$$

Thus, the strains in the two elements are obtained by Eq. (19.25) as

$$\{\epsilon\}_{CST-1} = \left[-\frac{2\delta}{a} \quad 0 \quad \frac{2\delta}{a}\right]^T \tag{h}$$

$$\{\epsilon\}_{CST-2} = \left[\frac{2\delta}{a} \quad 0 \quad -\frac{2\delta}{a}\right]^T \tag{i}$$

The structure stiffness for an assembly of two CST elements is

$$[K]_{CST} = Et \begin{bmatrix} 0.75 & 0.0 & -0.5 & 0.0 & 0.0 & -0.25 & -0.25 & 0.25 \\ 0.0 & 0.75 & 0.25 & -0.25 & -0.25 & 0.0 & 0.0 & -0.5 \\ -0.5 & 0.25 & 0.75 & -0.25 & -0.25 & 0.0 & 0.0 & 0.0 \\ 0.0 & -0.25 & -0.25 & 0.75 & 0.25 & -0.5 & 0.0 & 0.0 \\ 0.0 & -0.25 & -0.25 & 0.25 & 0.75 & 0.0 & -0.5 & 0.0 \\ -0.25 & 0.0 & 0.0 & -0.5 & 0.0 & 0.75 & 0.25 & -0.25 \\ -0.25 & 0.0 & 0.0 & 0.0 & -0.5 & 0.25 & 0.75 & -0.25 \\ 0.25 & -0.5 & 0.0 & 0.0 & 0.0 & -0.25 & -0.25 & 0.75 \end{bmatrix} \quad \text{(j)}$$

from which the product $[K]_{CST}\{u_i\}$ gives the nodal forces $\{P_i\}_{CST}$ as

$$\{P_i\}_{CST} = Et\,\delta[-1.5 \quad 0.5 \mid 1.5 \quad -0.5 \mid -1.5 \quad 0.5 \mid 1.5 \quad -0.5]^T \quad \text{(k)}$$

The strain energy in the structure is

$$U_{CST} = \tfrac{1}{2}\{u_i\}^T\{P_i\}_{CST} = 3Et\,\delta^2 \quad \text{(l)}$$

Next, consider the model with only a single bilinear rectangle shown in Fig. E19.3b. The $[B]$ matrix for the element is given by Eqs. (19.47) and (19.48) as

$$[B]_{BR} = \frac{1}{a^2}\begin{bmatrix} -\dfrac{a}{2}+y & 0 & \dfrac{a}{2}-y & 0 & \dfrac{a}{2}+y & 0 & -\dfrac{a}{2}-y & 0 \\[2mm] 0 & -\dfrac{a}{2}+x & 0 & -\dfrac{a}{2}-x & 0 & \dfrac{a}{2}+x & 0 & \dfrac{a}{2}-x \\[2mm] -\dfrac{a}{2}+x & -\dfrac{a}{2}+y & -\dfrac{a}{2}-x & \dfrac{a}{2}-y & \dfrac{a}{2}+x & \dfrac{a}{2}+y & \dfrac{a}{2}-x & -\dfrac{a}{2}-y \end{bmatrix}$$

$$\text{(m)}$$

The strains are obtained by Eq. (19.25) as

$$\{\epsilon\}_{BR} = \begin{bmatrix} -\dfrac{4y\,\delta}{a^2} & 0 & -\dfrac{4x\,\delta}{a^2} \end{bmatrix}^T \quad \text{(n)}$$

The stiffness matrix for the bilinear rectangle is obtained from Table (19.3), which, for this problem, becomes

$$[K]_{BR} = Et \begin{bmatrix} 0.5 & 0.125 & -0.25 & -0.125 & -0.25 & -0.125 & 0.0 & 0.125 \\ 0.125 & 0.5 & 0.125 & 0.0 & -0.125 & -0.25 & -0.125 & -0.25 \\ -0.25 & 0.125 & 0.5 & -0.125 & 0.0 & -0.125 & -0.25 & 0.125 \\ -0.125 & 0.0 & -0.125 & 0.5 & 0.125 & -0.25 & 0.125 & -0.25 \\ -0.25 & -0.125 & 0.0 & 0.125 & 0.5 & 0.125 & -0.25 & -0.125 \\ -0.125 & -0.25 & -0.125 & -0.25 & 0.125 & 0.5 & 0.125 & 0.0 \\ 0.0 & -0.125 & -0.25 & 0.125 & -0.25 & 0.125 & 0.5 & -0.125 \\ 0.125 & -0.25 & 0.125 & -0.25 & -0.125 & 0.0 & -0.125 & 0.5 \end{bmatrix}$$

$$\text{(o)}$$

from which the product $[K]_{BR}\{u_i\}$ gives the nodal forces $\{P_i\}_{BR}$ as

$$\{P_i\}_{BR} = Et\,\delta[-0.5 \quad 0 \mid 0.5 \quad 0 \mid -0.5 \quad 0 \mid 0.5 \quad 0]^T \tag{p}$$

The strain energy in the element is

$$U_{BR} = \tfrac{1}{2}\{u_i\}^T\{P_i\}_{BR} = Et\,\delta^2 \tag{q}$$

This example clearly demonstrates that the bilinear rectangle is superior to the CST element. The bilinear rectangle correctly predicts the normal strains ϵ_{xx} and ϵ_{yy}. In addition, the bilinear rectangle model stores less strain energy than the CST model. If we use the elasticity solution as the *exact* solution, $U_{BR} = 1.5U_{\text{exact}}$, whereas $U_{CST} = 4.5U_{\text{exact}}$. Notice though that both the CST and bilinear rectangle possess nonzero shear stress where none should exist. This defect, known as *parasitic* shear, contributes to excess strain energy in the elements. Although little can be done to improve the performance of the CST element, a more general formulation of the bilinear rectangle, known as the linear isoparametric quadrilateral (Sec. 19.4), can be used to control parasitic shear.

19.4

THE LINEAR ISOPARAMETRIC QUADRILATERAL

Suppose that an analyst wishes to model an irregular domain but wants to avoid the use of CST elements because of their relatively poor performance. Since the domain is irregular, the bilinear rectangle element would be inappropriate. Instead, arbitrarily shaped quadrilateral (four-sided) elements are selected to better fit boundaries. A quadrilateral element may be formulated directly, as was done above for the CST and bilinear rectangle elements. However, the necessary integrations are quite complex. This is due, in part, to the difficulty in defining the limits of integration. Use of isoparametric elements eliminates this difficulty. Isoparametric elements are formulated in *natural* coordinates as square elements and then *mapped* to physical coordinates via coordinate interpolation functions, similar to displacement interpolation functions. Depending on the type of isoparametric element used, the configuration of the element in physical coordinates can be nonrectangular and can have curved sides. If the shape functions used for coordinate interpolation are identical to those used for displacement interpolation, then the element is said to be *isoparametric*. If coordinate interpolation is of higher order than displacement interpolation (i.e., more nodes are used to represent the variation in geometry than the variation in displacements), then the element is called *superparametric*. If coordinate interpolation is of lower order than displacement interpolation (fewer nodes are used to represent the variation in geometry than the variation in displacements), then the element is called *subparametric* (Zienkiewicz and Taylor, 1989, p. 160). Because of their versatility and accuracy, isoparametric elements have become the mainstay of modern finite element programs.

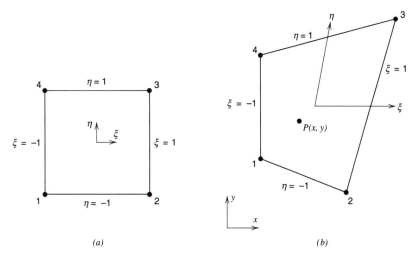

Figure 19.9   Isoparametric coordinate mapping. (*a*) Element in natural coordinates. (*b*) Element in physical coordinates.

Isoparametric Mapping

Consider the mapping of the four-node quadrilateral element from a *natural* (ξ, η) coordinate system (Fig. 19.9*a*) to a physical (x, y) coordinate system (Fig. 19.9*b*). In natural coordinates, the element is a (2×2) square and the origin of the coordinate system is at its center. In physical coordinates, the element is distorted from a rectangular shape. With shape functions in terms of the (ξ, η) coordinate system, the coordinates of any point P can be expressed in terms of the (x, y) coordinates of the nodes.

$$x(\xi, \eta) = \sum_{i=1}^{4} N_i(\xi, \eta) x_i, \qquad y(\xi, \eta) = \sum_{i=1}^{4} N_i(\xi, \eta) y_i \qquad (19.49a)$$

In matrix form, Eq. (19.49a) is

$$\begin{Bmatrix} x(\xi, \eta) \\ y(\xi, \eta) \end{Bmatrix} = [N]\{x_i\} \qquad (19.49b)$$

where $\{x_i\}$ is the vector of nodal coordinates

$$\{x_i\} = [x_1 \quad y_1 \quad x_2 \quad y_2 \quad x_3 \quad y_3 \quad x_4 \quad y_4]^T$$

and $[N]$ is the shape function matrix

$$[N] = \begin{bmatrix} N_1 & 0 & N_2 & 0 & N_3 & 0 & N_4 & 0 \\ 0 & N_1 & 0 & N_2 & 0 & N_3 & 0 & N_4 \end{bmatrix} \qquad (19.50)$$

The shape functions are the Lagrange interpolation functions [refer to Eq. (19.46)] in dimensionless (ξ, η) coordinates.

$$N_1(\xi, \eta) = \frac{(1 - \xi)(1 - \eta)}{4}$$

$$N_2(\xi, \eta) = \frac{(1 + \xi)(1 - \eta)}{4}$$

$$N_3(\xi, \eta) = \frac{(1 + \xi)(1 + \eta)}{4}$$

$$N_4(\xi, \eta) = \frac{(1 - \xi)(1 + \eta)}{4} \tag{19.51}$$

After the element is mapped from natural to physical coordinates, the ξ and η axes need not remain orthogonal.

The principal reason for using isoparametric elements is to avoid integrating in physical coordinates. However, the general expression for the stiffness matrix, Eq. (19.31), is expressed in terms of physical coordinates. Therefore, the differential lengths dx and dy must be expressed in terms of the natural coordinate differentials $d\xi$ and $d\eta$. In addition, strain is defined in terms of the derivatives of the shape functions with respect to physical coordinates. These derivatives are the elements in the $[B]$ matrix, and they must be converted to derivatives with respect to natural coordinates.

The differentials (dx, dy) are related to the differentials $(d\xi, d\eta)$ by means of Eq. (19.49a). Thus,

$$dx = \frac{\partial x}{\partial \xi} d\xi + \frac{\partial x}{\partial \eta} d\eta$$

$$dy = \frac{\partial y}{\partial \xi} d\xi + \frac{\partial y}{\partial \eta} d\eta \tag{19.52}$$

where

$$\frac{\partial x}{\partial \xi} = \sum \frac{\partial N_i}{\partial \xi} x_i, \qquad \frac{\partial y}{\partial \xi} = \sum \frac{\partial N_i}{\partial \xi} y_i$$

$$\frac{\partial x}{\partial \eta} = \sum \frac{\partial N_i}{\partial \eta} x_i, \qquad \frac{\partial y}{\partial \eta} = \sum \frac{\partial N_i}{\partial \eta} y_i$$

The coordinate derivatives are combined in matrix form as

$$[J] = \begin{bmatrix} \dfrac{\partial x}{\partial \xi} & \dfrac{\partial y}{\partial \xi} \\[2mm] \dfrac{\partial x}{\partial \eta} & \dfrac{\partial y}{\partial \eta} \end{bmatrix} \tag{19.53}$$

where $[J]$ is the *Jacobian* of the transformation (Courant, 1950).

Equations (19.52) and (19.53) relate the differentials of the two coordinate systems as

$$\begin{Bmatrix} dx \\ dy \end{Bmatrix} = [J]^T \begin{Bmatrix} d\xi \\ d\eta \end{Bmatrix} \tag{19.54}$$

In a like manner, derivatives of the shape function for node i are related by

$$
\begin{Bmatrix} \dfrac{\partial N_i}{\partial x} \\[2mm] \dfrac{\partial N_i}{\partial y} \end{Bmatrix} = [J]^{-1} \begin{Bmatrix} \dfrac{\partial N_i}{\partial \xi} \\[2mm] \dfrac{\partial N_i}{\partial \eta} \end{Bmatrix}
\tag{19.55}
$$

If $[J]^{-1}$ exists, then the area mapping from (ξ, η) coordinates to (x, y) coordinates is unique and reversible. A physical interpretation of $[J]$ can be obtained by comparing the area of the element in (x, y) coordinates to that in (ξ, η) coordinates. If the determinate $|J| > 0$, then the area of the element is preserved and the mapping is physically meaningful. In precise terms, $|J|$ is the differential area ratio $A_{xy}/A_{\xi\eta}$ at any point in the element.

This physical interpretation of $[J]$ leads to a change in the differential volume for a constant thickness, plane elasticity element from $t\,dx\,dy$ to $t|J|\,d\xi\,d\eta$. The limits of integration are -1 to 1 in ξ and -1 to 1 in η. So, the integral of any function $F(x, y)$ can be transformed to natural coordinates in the manner

$$
\int_A F(x, y)\, dx\, dy = \int_{-1}^{1} \int_{-1}^{1} F(x(\xi, \eta), y(\xi, \eta)) |J|\, d\xi\, d\eta
$$

Element Stiffness Matrix

Equation (19.31) defines the element stiffness matrix for any elasticity element (using displacement DOF), including the isoparametric linear quadrilateral. A change in coordinate system from (x, y) to (ξ, η), with the modified limits of integration, leads to the stiffness matrix

$$
[K] = t \int_{-1}^{1} \int_{-1}^{1} [B]^T [D][B] |J|\, d\xi\, d\eta
\tag{19.56}
$$

where $[B]$ is given by Eq. (19.47) and $[B_i]$ by Eq. (19.48). From Eqs. (19.48) and (19.55), the individual terms in $[B_i]$, in terms of (ξ, η), are

$$
[B_i(\xi, \eta)] = \begin{bmatrix} J_{11}^* \dfrac{\partial N_i}{\partial \xi} + J_{12}^* \dfrac{\partial N_i}{\partial \eta} & 0 \\[4mm] 0 & J_{21}^* \dfrac{\partial N_i}{\partial \xi} + J_{22}^* \dfrac{\partial N_i}{\partial \eta} \\[4mm] J_{21}^* \dfrac{\partial N_i}{\partial \xi} + J_{22}^* \dfrac{\partial N_i}{\partial \eta} & J_{11}^* \dfrac{\partial N_i}{\partial \xi} + J_{12}^* \dfrac{\partial N_i}{\partial \eta} \end{bmatrix}
\tag{19.57}
$$

where J_{ij}^* is the i, j term from $[J]^{-1}$.

It is usually more convenient to work with just a single (2×2) nodal submatrix of $[K]$ at one time. Hence, we write

$$
[K_{ij}] = t \int_{-1}^{1} \int_{-1}^{1} [B_i]^T [D][B_j] |J|\, d\xi\, d\eta
\tag{19.58}
$$

where i and j are node numbers for the element.

Numerical Integration

Although analytical expressions for the individual terms in Eq. (19.58) can be developed, they are quite complex and, thus, prone to errors in algebra or computer programming. As an alternative to direct integration, the required integrals are usually evaluated numerically within the finite element program. The most commonly used numerical integration method is *Gauss quadrature*. The Gauss quadrature method is more efficient than many other methods, such as the Newton–Cotes methods, since fewer sampling points are required to obtain a given level of accuracy. In fact, in one dimension, the use of n sampling points in Gauss quadrature results in exact integration of a polynomial of order $(2n - 1)$. However, the integration of a function that is not a polynomial is approximate.

Consider a function $F(\xi, \eta)$ that is to be integrated over the limits of -1 to 1 in ξ and -1 to 1 in η. The integral is evaluated numerically by the form

$$I = \int_{-1}^{1} \int_{-1}^{1} F(\xi, \eta)\, d\xi\, d\eta = \sum_{k=1}^{m} \sum_{l=1}^{n} w_k w_l F(\xi_k, \eta_l)$$

where m and n are the numbers of sampling points in the ξ and η directions, respectively. Also, ξ_k and η_l are the locations of the kth and lth sampling points and w_k and w_l are weights applied to $F(\xi, \eta)$ after it is evaluated at the sampling points. Usually, m and n are taken as equal, in which case the numerical scheme is symmetric.

If Gauss quadrature is used to evaluate the nodal submatrix $[K_{ij}]$ in Eq. (19.58), the integral becomes

$$[K_{ij}] = t \sum_{k=1}^{m} \sum_{l=1}^{n} w_k w_l [B_i(\xi_k, \eta_l)]^T [D][B_j(\xi_k, \eta_l)] |J(\xi_k, \eta_l)| \qquad (19.59)$$

The accuracy achieved with Gauss quadrature is dependent on the proper selection of sampling point locations and weights. For elements in natural coordinates, the optimal sampling point locations and weights are given in Fig. 19.10. Only symmetric integration and the one-, two-, and three-point rules are considered. Nonsymmetric integration and higher-order integration rules are discussed elsewhere.

The number of integration points used to evaluate Eq. (19.59) influences the ultimate performance of the element. *Full integration* is the integration order needed to *exactly* integrate the stiffness for an undistorted element. For the linear quadrilateral, a two-point rule provides full integration. An integration rule less than that required for full integration is termed *reduced integration*. Reduced integration, although not exactly evaluating Eq. (19.59), can often lead to improved performance of an element, relative to full integration. For instance, reduced integration of the linear quadrilateral can eliminate the parasitic shear that is a common defect in the element (see Example 19.3). A more complete discussion of reduced integration, including justification for its use, can be found in most finite element textbooks.

High-Order Isoparametric Elements

The concept of isoparametric mapping has been applied to a broad list of element geometries. Within the scope of plane elasticity problems, elements with more than four nodes permit greater flexibility in element shape (including curved edges) and

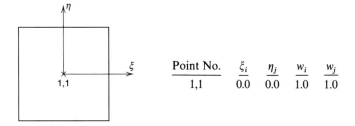

Point No.	ξ_i	η_j	w_i	w_j
1,1	0.0	0.0	1.0	1.0

(a)

Point No.	ξ_i	η_j	w_i	w_j
1,1	$-1/\sqrt{3}$	$-1/\sqrt{3}$	1.0	1.0
2,1	$1/\sqrt{3}$	$-1/\sqrt{3}$	1.0	1.0
1,2	$-1/\sqrt{3}$	$1/\sqrt{3}$	1.0	1.0
2,2	$1/\sqrt{3}$	$1/\sqrt{3}$	1.0	1.0

(b)

Point No.	ξ_i	η_j	w_i	w_j
1,1	$-\sqrt{0.6}$	$-\sqrt{0.6}$	5/9	5/9
2,1	0	$-\sqrt{0.6}$	8/9	5/9
3,1	$\sqrt{0.6}$	$-\sqrt{0.6}$	5/9	5/9
1,2	$-\sqrt{0.6}$	0	5/9	8/9
2,2	0	0	8/9	8/9
3,2	$\sqrt{0.6}$	0	5/9	8/9
1,3	$-\sqrt{0.6}$	$\sqrt{0.6}$	5/9	5/9
2,3	0	$\sqrt{0.6}$	8/9	5/9
3,3	$\sqrt{0.6}$	$\sqrt{0.6}$	5/9	5/9

(c)

Figure 19.10 Optimal sampling point locations and weights for Gauss quadrature. (a) One-point rule. (b) Two-point rule. (c) Three-point rule.

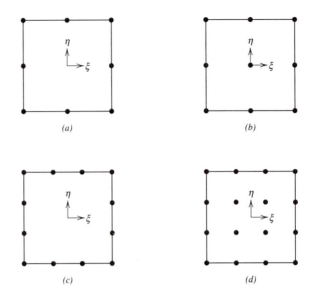

Figure 19.11   Higher-order isoparametric elements. (*a*) Quadratic serendipity element. (*b*) Quadratic Lagrange element. (*c*) Cubic serendipity element. (*d*) Cubic Lagrange element.

are capable of representing greater variation in displacements. Perhaps the most popular of all isoparametric elements is the eight-node quadrilateral. This element has four corner nodes, like the linear quadrilateral, but it also has four midside nodes, one midway along the length of each edge, (Fig. 19.11*a*.). With three nodes along each edge, the element can have curved (parabolic) sides. Another popular high-order isoparametric element is the nine-node quadrilateral (Fig. 19.11*b*.) This element has four corner nodes, four midside nodes, and one interior node. Both the eight- and nine-node elements represent complete quadratic displacement fields. The generalization of these elements to cubic interpolation is straightforward, see Figs. 19.11*c* and *d*.

The eight-node quadrilateral and other high-order elements that contain only boundary nodes are known as *serendipity* elements. The term serendipity is used because shape functions for this family of elements were initially developed by inspection. The nine-node quadrilateral and other high-order elements that contain a regular pattern of nodes are known as Lagrangian elements since their shape functions are based on the Lagrange interpolation functions.

19.5

THE PLANE FRAME ELEMENT

Analysis of framed structures by the *stiffness method* (also known as *matrix analysis*) was fairly well established at the time of the development of the finite element method. The stiffness method for frame analysis can be developed entirely from basic mechanics of material principles, without the need to consider virtual work formulations and interpolation polynomials. As a result, many engineers view the two methods as distinct. However, it is clear that the stiffness method for frames is simply a special case of the finite element method. Hence, in this section, we develop a finite element that represents a plane frame member, using the same approach that was used for plane elasticity problems.

Element Stiffness Matrix

The classical plane frame element has two nodes, it is straight and prismatic, and it has three DOF and three corresponding end actions at each node (see Fig. 19.12a). The element has constant cross-sectional area A, moment of inertia I, and modulus of elasticity E. We assume that the axial response of the member is independent of the bending response. Consequently, the frame element stiffness is formulated as a superposition of the stiffness for an axial rod and that for a beam (Fig. 19.12b). In the following, a *local* $(\bar{x}, \bar{y})$ coordinate system is established for the element. The local $\bar{x}$ axis is aligned with the longitudinal axis of the member, and the $\bar{y}$ axis lies in the plane of the element cross section. The stiffness matrix for the frame element is derived in terms of this local coordinate system. When the element is oriented at some angle ϕ with respect to the *global* (x, y) coordinates for the structure, the nodal DOF of the element must be related to the global coordinate system. Thus, a coordinate rotation from local to global coordinates is required for the displacements, loads, and stiffness. This rotation is discussed following Eq. (19.75).

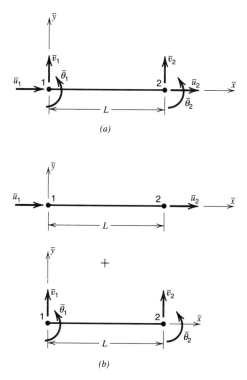

Figure 19.12 Plane frame element. (a) Element with combined axial and bending DOF. (b) Axial and bending DOF treated separately.

Consider first the case of the axial rod. There are two nodal DOF associated with axial response, so the displacement is taken as a linear function

$$\bar{u}(x) = a_0 + a_1 \bar{x}$$

The coefficients a_0 and a_1 are evaluated based on the boundary conditions $\bar{u}(0) = \bar{u}_1$ and $\bar{u}(L) = \bar{u}_2$, where L is the element length. The displacement function, in terms of the nodal displacements, becomes

$$\bar{u}(x) = [N]\{\bar{u}_i\}$$

where $[N] = [1 - \bar{x}/L \quad \bar{x}/L]$ and $\{\bar{u}_i\} = [\bar{u}_1 \quad \bar{u}_2]^T$.

The only nonzero strain component is ϵ_{xx}, which is written in terms of the nodal displacements as

$$\epsilon_{xx} = [B_A]\{\bar{u}_i\}$$

in which the subscript A indicates *axial* response and

$$[B_A] = \left[\frac{\partial N_1}{\partial \bar{x}} \quad \frac{\partial N_2}{\partial \bar{x}} \right] = \left[-\frac{1}{L} \quad \frac{1}{L} \right]$$

The axial stress is written as $\sigma_{xx} = E\epsilon_{xx}$ and the variation of internal energy is

$$\delta U = \int_V \delta\epsilon_{xx}\sigma_{xx}\,dV \tag{19.60}$$

Assume that only concentrated nodal loads are applied. Substitution for σ_{xx} and $\delta\epsilon_{xx}$ in Eq. (19.60), and then substitution of Eq. (19.60) into Eq. (19.1), yield

$$\{\delta\bar{u}_i\}^T\{\bar{F}_i\} - \{\delta\bar{u}_i\}^T\left[A\int_0^L [B_A]^T E[B_A]\,d\bar{x} \right]\{\bar{u}_i\} = 0$$

Since $\{\delta\bar{u}_i\}$ is arbitrary,

$$\{\bar{F}_i\} = \left[A\int_0^L [B_A]^T E[B_A]\,d\bar{x} \right]\{\bar{u}_i\}$$

which leads to the stiffness matrix for the axial rod

$$[\bar{K}_A] = A\int_0^L [B_A]^T E[B_A]\,d\bar{x}$$

For constant E, the integrals are easily evaluated to obtain $[\bar{K}_A]$ in terms of A, E, and L.

$$[\bar{K}_A] = \begin{bmatrix} \dfrac{AE}{L} & -\dfrac{AE}{L} \\ -\dfrac{AE}{L} & \dfrac{AE}{L} \end{bmatrix} \tag{19.61}$$

Next consider the *bending* effect of the frame element. There are four nodal DOF associated with bending (a lateral translation and a rotation at each node), so the displacement is written as a cubic polynomial with four coefficients.

$$\bar{v}(x) = a_0 + a_1\bar{x} + a_2\bar{x}^2 + a_3\bar{x}^3$$

The coefficients a_0 through a_3 are evaluated based on the boundary conditions $\bar{v}(0) = \bar{v}_1$, $\bar{\theta}(0) = \bar{\theta}_1$, $\bar{v}(L) = \bar{v}_2$, and $\bar{\theta}(L) = \bar{\theta}_2$ in which $\bar{\theta} = d\bar{v}/d\bar{x}$. In terms of the of the nodal displacements, the displacement function is

$$\bar{v}(\bar{x}) = [N]\{\bar{v}_i\} \tag{19.62}$$

where $\{\bar{v}_i\} = [\bar{v}_1 \quad \bar{\theta}_1 \quad \bar{v}_2 \quad \bar{\theta}_2]^T$, and the shape function matrix $[N]$ is

$$[N] = [N_1 \quad N_2 \quad N_3 \quad N_4] \tag{19.63a}$$

for which the individual shape functions are

$$N_1 = 1 - 3\frac{\bar{x}^2}{L^2} + 2\frac{\bar{x}^3}{L^3}$$

$$N_2 = \bar{x} - 2\frac{\bar{x}^2}{L} + \frac{\bar{x}^3}{L^2}$$

$$N_3 = 3\frac{\bar{x}^2}{L^2} - 2\frac{\bar{x}^3}{L^3}$$

$$N_4 = -\frac{\bar{x}^2}{L} + \frac{\bar{x}^3}{L^2} \tag{19.63b}$$

These shape functions are illustrated in Fig. 19.13.

The strain energy in a beam subjected to bending is given by Eq. (5.19); that is,

$$U = \int_0^L \frac{M^2}{2EI}\,d\bar{x} \tag{5.19}$$

If the curvature $\bar{v}''$ is taken as a *generalized strain* quantity, the strain-nodal displacement relation is

$$\bar{v}''(\bar{x}) = [B_B]\{\bar{v}_i\} \tag{19.64a}$$

where the subscript B represents *bending* response and

$$[B_B] = \left[\frac{d^2N_1}{d\bar{x}^2} \quad \frac{d^2N_2}{d\bar{x}^2} \quad \frac{d^2N_3}{d\bar{x}^2} \quad \frac{d^2N_4}{d\bar{x}^2}\right] \tag{19.64b}$$

Substitution of $M = EI\bar{v}''$ into Eq. (5.19) gives

$$U = \int_0^L \frac{EI(\bar{v}'')^2}{2}\,d\bar{x} \tag{19.65}$$

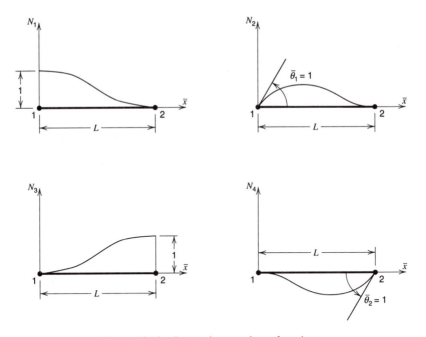

Figure 19.13 Beam element shape functions.

from which the first variation of the strain energy is

$$\delta U = \int_0^L (\delta \bar{v})'' E I \bar{v}'' \, d\bar{x}$$

In terms of nodal DOF, from Eq. (19.64a), δU is

$$\delta U = \int_0^L \{\delta \bar{v}_i\}^T [B_B]^T E I [B_B] \{\bar{v}_i\} \, d\bar{x} \tag{19.66}$$

In the manner followed with other elements, only nodal loads are assumed, Eq. (19.66) is substituted into Eq. (19.1), $\{\delta v_i\}$ is eliminated, and the bending stiffness matrix is found to be

$$[\bar{K}_B] = \int_0^L [B_B]^T E I [B_B] \, d\bar{x}$$

Since EI is constant, integration yields the bending stiffness matrix in terms of E, I, and L as

$$[\bar{K}_B] = \begin{bmatrix} \dfrac{12EI}{L^3} & \dfrac{6EI}{L^2} & -\dfrac{12EI}{L^3} & \dfrac{6EI}{L^2} \\[2mm] \dfrac{6EI}{L^2} & \dfrac{4EI}{L} & -\dfrac{6EI}{L^2} & \dfrac{2EI}{L} \\[2mm] -\dfrac{12EI}{L^3} & -\dfrac{6EI}{L^2} & \dfrac{12EI}{L^3} & -\dfrac{6EI}{L^2} \\[2mm] \dfrac{6EI}{L^2} & \dfrac{2EI}{L} & -\dfrac{6EI}{L^2} & \dfrac{4EI}{L} \end{bmatrix} \tag{19.67}$$

The stiffness matrix for the plane frame element [see Eq. (19.68)] is a combination of the axial stiffness matrix, Eq. (19.61), and the bending stiffness matrix, Eq. (19.67). Note that the ordering of the DOF in the element first lists all three DOF at node 1 and then the three DOF at node 2.

$$[\bar{K}] = \begin{bmatrix} \dfrac{AE}{L} & 0 & 0 & \dfrac{-AE}{L} & 0 & 0 \\[2mm] 0 & \dfrac{12EI}{L^3} & \dfrac{6EI}{L^2} & 0 & \dfrac{-12EI}{L^3} & \dfrac{6EI}{L^2} \\[2mm] 0 & \dfrac{6EI}{L^2} & \dfrac{4EI}{L} & 0 & \dfrac{-6EI}{L^2} & \dfrac{2EI}{L} \\[2mm] \dfrac{-AE}{L} & 0 & 0 & \dfrac{AE}{L} & 0 & 0 \\[2mm] 0 & \dfrac{-12EI}{L^3} & \dfrac{-6EI}{L^2} & 0 & \dfrac{12EI}{L^3} & \dfrac{-6EI}{L^2} \\[2mm] 0 & \dfrac{6EI}{L^2} & \dfrac{2EI}{L} & 0 & \dfrac{-6EI}{L^2} & \dfrac{4EI}{L} \end{bmatrix} \qquad (19.68)$$

The displacement vector $\{\bar{u}_i\}$ for the element is

$$\{\bar{u}_i\} = [\bar{u}_1 \quad \bar{v}_1 \quad \bar{\theta}_1 \quad \bar{u}_2 \quad \bar{v}_2 \quad \bar{\theta}_2]^T \qquad (19.69)$$

and the element end action (load) vector $\{\bar{P}_i\}$ is

$$\{\bar{P}_i\} = [\bar{P}_{x1} \quad \bar{P}_{y1} \quad \bar{M}_1 \quad \bar{P}_{x2} \quad \bar{P}_{y2} \quad \bar{M}_2]^T \qquad (19.70)$$

Finally, the relationship between nodal loads and nodal displacements for an element in local coordinates is given by the familiar form

$$[\bar{K}]\{\bar{u}_i\} = \{\bar{P}_i\} \qquad (19.71)$$

Equivalent Nodal Load Vector

As for most other elements, actual loads that are applied over the element must be converted to equivalent nodal loads. We consider only element loads that affect beam behavior. Two cases are considered: a distributed load over a portion of the element and a transverse concentrated force. Equivalent nodal loads for axial behavior are derived in a like fashion.

For a distributed load along the beam, not necessarily over the full length, the variation of work δW_D of the load is

$$\delta W_D = \int_{L_a}^{L_b} \delta \bar{v} \bar{q}(\bar{x}) \, d\bar{x} \qquad (19.72)$$

where $\bar{q}(\bar{x})$ is the load function that exists over the domain $L_a < \bar{x} < L_b$ (see Fig. 19.14a) and the subscript D denotes a distributed load. Equation (19.62) is substituted into Eq. (19.72), and the equivalent nodal load vector is obtained as

$$\{\bar{P}_{Di}\} = \int_{L_a}^{L_b} [N]^T \bar{q}(\bar{x}) \, d\bar{x} \qquad (19.73)$$

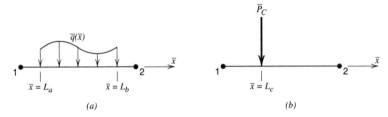

Figure 19.14   Element loads for beam element. (*a*) Distributed load. (*b*) Concentrated load.

For a concentrated load $\bar{P}_C$ located at $\bar{x} = L_c$ along the beam (see Fig. 19.14*b*) the variation of work δW_C of the load is

$$\delta W_C = \delta \bar{v}|_{\bar{x}=L_c} \bar{P}_C \tag{19.74}$$

The variation of displacement $\delta \bar{v}$ at $\bar{x} = L_c$ is written in terms of the variation of nodal displacements by Eq. (19.62) with the shape functions evaluated at $\bar{x} = L_c$. The equivalent nodal load vector is

$$\{\bar{P}_{Ci}\} = [N]|^T_{\bar{x}=L_c} \bar{P}_C \tag{19.75}$$

By Eqs. (19.73) and (19.75), equivalent nodal load vectors for several load patterns on a beam element were determined and are shown in Fig. 19.15.

Coordinate Rotations

Consider an element in a structure oriented at an angle ϕ with respect to the global x axis (Fig. 19.16). To assemble the stiffness matrix and load vector for this element with those of other elements, all nodal DOF must be defined in terms of the global coordinate system. For node i, the displacements in the two coordinate systems are related by

$$\begin{Bmatrix} \bar{u}_i \\ \bar{v}_i \\ \bar{\theta}_i \end{Bmatrix} = [\lambda] \begin{Bmatrix} u_i \\ v_i \\ \theta_i \end{Bmatrix} \tag{19.76}$$

where

$$[\lambda] = \begin{bmatrix} \cos\phi & \sin\phi & 0 \\ -\sin\phi & \cos\phi & 0 \\ 0 & 0 & 1 \end{bmatrix}$$

For a plane frame element, with two nodes, the displacements are related by

$$\{\bar{u}_i\} = [T]\{u_i\} \tag{19.77}$$

where the rotation (transformation) matrix $[T]$ is

$$[T] = \begin{bmatrix} \lambda & 0 \\ \hline 0 & \lambda \end{bmatrix}$$

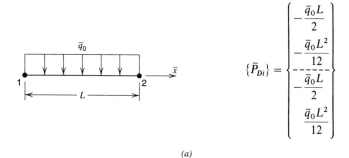

$$\{\bar{P}_{Di}\} = \begin{Bmatrix} -\dfrac{\bar{q}_0 L}{2} \\[2mm] -\dfrac{\bar{q}_0 L^2}{12} \\[1mm] \hdashline \\[-2mm] -\dfrac{\bar{q}_0 L}{2} \\[2mm] \dfrac{\bar{q}_0 L^2}{12} \end{Bmatrix}$$

(a)

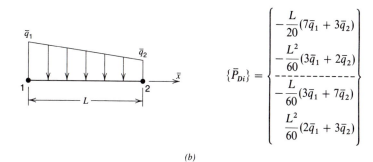

$$\{\bar{P}_{Di}\} = \begin{Bmatrix} -\dfrac{L}{20}(7\bar{q}_1 + 3\bar{q}_2) \\[2mm] -\dfrac{L^2}{60}(3\bar{q}_1 + 2\bar{q}_2) \\[1mm] \hdashline \\[-2mm] -\dfrac{L}{60}(3\bar{q}_1 + 7\bar{q}_2) \\[2mm] \dfrac{L^2}{60}(2\bar{q}_1 + 3\bar{q}_2) \end{Bmatrix}$$

(b)

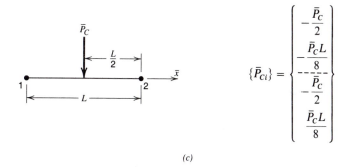

$$\{\bar{P}_{Ci}\} = \begin{Bmatrix} -\dfrac{\bar{P}_C}{2} \\[2mm] -\dfrac{\bar{P}_C L}{8} \\[1mm] \hdashline \\[-2mm] -\dfrac{\bar{P}_C}{2} \\[2mm] \dfrac{\bar{P}_C L}{8} \end{Bmatrix}$$

(c)

Figure 19.15 Equivalent nodal loads for beam element. (a) Uniformly distributed load. (b) Linearly distributed load. (c) Concentrated load.

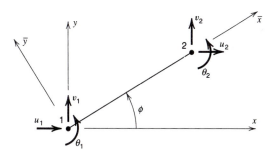

Figure 19.16 Frame element in global coordinates.

TABLE 19.4
Element Stiffness Matrix for Plane Frame Element in Global Coordinates

column index $j \rightarrow$

$$
[K] = \begin{bmatrix}
c^2\dfrac{AE}{L} + s^2\dfrac{12EI}{L^3} & sc\left(\dfrac{AE}{L} - \dfrac{12EI}{L^3}\right) & -s\left(\dfrac{6EI}{L^2}\right) & -c^2\dfrac{AE}{L} - s^2\dfrac{12EI}{L^3} & -sc\left(\dfrac{AE}{L} - \dfrac{12EI}{L^3}\right) & -s\left(\dfrac{6EI}{L^2}\right) \\[2.5ex]
sc\left(\dfrac{AE}{L} - \dfrac{12EI}{L^3}\right) & s^2\dfrac{AE}{L} + c^2\dfrac{12EI}{L^3} & c\left(\dfrac{6EI}{L^2}\right) & -sc\left(\dfrac{AE}{L} - \dfrac{12EI}{L^3}\right) & -s^2\dfrac{AE}{L} - c^2\dfrac{12EI}{L^3} & c\left(\dfrac{6EI}{L^2}\right) \\[2.5ex]
-s\left(\dfrac{6EI}{L^2}\right) & c\left(\dfrac{6EI}{L^2}\right) & \dfrac{4EI}{L} & s\left(\dfrac{6EI}{L^2}\right) & -c\left(\dfrac{6EI}{L^2}\right) & \dfrac{2EI}{L} \\[2.5ex]
-c^2\dfrac{AE}{L} - s^2\dfrac{12EI}{L^3} & -sc\left(\dfrac{AE}{L} - \dfrac{12EI}{L^3}\right) & s\left(\dfrac{6EI}{L^2}\right) & c^2\dfrac{AE}{L} + s^2\dfrac{12EI}{L^3} & sc\left(\dfrac{AE}{L} - \dfrac{12EI}{L^3}\right) & s\left(\dfrac{6EI}{L^2}\right) \\[2.5ex]
-sc\left(\dfrac{AE}{L} - \dfrac{12EI}{L^3}\right) & -s^2\dfrac{AE}{L} - c^2\dfrac{12EI}{L^3} & -c\left(\dfrac{6EI}{L^2}\right) & sc\left(\dfrac{AE}{L} - \dfrac{12EI}{L^3}\right) & s^2\dfrac{AE}{L} + c^2\dfrac{12EI}{L^3} & -c\left(\dfrac{6EI}{L^2}\right) \\[2.5ex]
-s\left(\dfrac{6EI}{L^2}\right) & c\left(\dfrac{6EI}{L^2}\right) & \dfrac{2EI}{L} & s\left(\dfrac{6EI}{L^2}\right) & -c\left(\dfrac{6EI}{L^2}\right) & \dfrac{4EI}{L}
\end{bmatrix}
\begin{matrix}
\text{row index}\ i\downarrow \\
1 \\ 2 \\ 3 \\ 4 \\ 5 \\ 6
\end{matrix}
$$

$$c = \cos\phi, \qquad s = \sin\phi$$

In like manner, element end actions (loads) are rotated by

$$\{\bar{P}_i\} = [T]\{P_i\} \tag{19.78}$$

Substitution of Eqs. (19.77) and (19.78) into Eq. (19.71) yields

$$[\bar{K}][T]\{u_i\} = [T]\{P_i\} \tag{19.79}$$

Premultipling both sides of Eq. (19.79) by $[T]^{-1}$ and observing that $[T]^{-1}=[T]^{T}$, since $[T]$ is an orthogonal matrix, we obtain

$$[T]^{T}[\bar{K}][T]\{u_i\} = \{P_i\}$$

Thus, since $\{u_i\}$ and $\{P_i\}$ are in global coordinates, the stiffness matrix for the plane frame element, in global coordinates, is

$$[K] = [T]^{T}[\bar{K}][T] \tag{19.80}$$

The final form of $[K]$ is given in Table 19.4. The load vector for the element, in global coordinates, is obtained from Eq. (19.78) as

$$\{P_i\} = [T]^{T}\{\bar{P}_i\} \tag{19.81}$$

19.6

CLOSING REMARKS

Requirements for Accuracy

The accuracy of a finite element solution strongly depends on two conditions. First, it is important that the equations of equilibrium be satisfied throughout the model. Second, it is also important that compatibility (continuity of displacements) be maintained. In certain circumstances, these conditions are violated, as noted below.

Equilibrium at the structure nodes is satisfied since the basic system of equations, Eq. (19.45), is fundamentally a system of nodal equilibrium equations. Thus, within the accuracy of the equation-solving process (numerical error), the structure nodes are in equilibrium.

For elements with only displacement DOF, equilibrium along element edges is generally *not* satisfied. This is because although displacements might be continuous across element boundaries, their derivatives are not, and thus, stresses are not continuous. For instance, consider two constant strain triangle elements, such as those shown in Fig. 19.17. Nodes 1, 2, and 3 are fully constrained, whereas node 4 has an imposed displacement in the x direction. Hence, element 1 is unstressed, whereas element 2 has nonzero σ_{xx}. Because of the stress discontinuity, a differential element located at the boundary between the two elements does not satisfy equilibrium in the x direction.

Equilibrium within an element is *not* satisfied, unless body forces are of relatively low order or are entirely absent. For a constant strain triangle, the stress

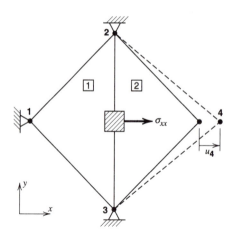

Figure 19.17 Equilibrium along element edges.

state is constant throughout the element. Thus, equilibrium of a differential element is satisfied only when body forces are absent [see Eq. (2.45)]. Similarly, for elements that can represent linear stress variation, body forces must be, at most, constant in magnitude for equilibrium.

Compatibility at the nodes is assured because of the assembly process. That is, the displacements of adjacent elements are the same at their common nodes. However, to assure that compatibility is maintained along the common edge between two adjacent elements, the displacements along that edge, viewed from either element, must be expressed entirely in terms of the displacement of nodes on that edge. Elements that maintain compatibility along common edges are known as *conforming* elements. Generally, this condition is satisfied for elements that possess only translational DOF. However, certain plate-bending and shell elements, for instance, are nonconforming.

Compatibility within an element is assured so long as the displacement interpolation polynomials are continuous.

Requirements for Convergence

As discussed at the beginning of this chapter, a major source of error in a finite element solution is the use of approximation functions to describe element response (formulation error). To reduce formulation error, we successively refine our finite element models with the expectation that the numerical solution will converge to the *exact* solution. Under certain conditions, convergence can be guaranteed. These conditions are the following:

1. The elements must be complete. That is, the shape function must be a complete polynomial. For instance, a complete quadratic contains all possible quadratic terms and omits no linear or constant terms. Inclusion of a few cubic terms, such as for the quadratic serendipity and Lagrange elements, does not destroy completeness of the quadratic polynomial.

2. The elements must be compatible. Hence, continuity of displacements must be assured throughout the entire structural model.

3. The elements must be capable of representing rigid-body motion and constant strain. For two- and three-dimensional elasticity problems, these are assured if the displacement field contains at least a complete linear polynomial. For shell elements, constant strain implies constant curvature and constant twist [see Eq. (13.30)]. Some shell elements cannot represent rigid-body motion.

Generally, a finite element model is too stiff. That is, displacements converge from below. A qualitative explanation is as follows. The elements are *constrained*, by the shape functions, to deform in a specific (unnatural) manner. This constraint adds stiffness, relative to the physical system, that results in smaller displacements when the external influences on the system are loads. If all external loads are zero and the only external influences on the system are imposed (nonzero) displacements, additional energy is required to force the model into the imposed deformed shape.

For isoparametric elements, reduced integration can be used effectively to *soften* the element such that its response improves relative to full integration. Problem 19.1 demonstrates how the use of approximation functions to represent displacements results in a model that is stiff relative to the actual system.

Modeling Recommendations

As an aid to the application of the finite-element method to analysis of practical problems in elasticity, the following recommendations are offered. The list is not exhaustive and the recommendations themselves are not rigid rules that cannot be violated.

1. Avoid abrupt transitions in element size and geometry. Limit the change in *element stiffness* (approximated by E/V_e, where V_e is the volume of the element) from one element to the next to roughly a factor of 3.

2. Avoid unnecessary element irregularity. Keep aspect ratios (the length ratio of the longest side to the shortest side) less than 10:1. Interior angles of quadrilaterals should be as regular as possible. They should not exceed 150° and they should not be less than 30°. Midside nodes on quadratic elements should be within the middle third of the edge.

3. Maintain compatibility between elements. For instance, it is not appropriate to attach one quadratic quadrilateral to two linear quadrilaterals simply because they have three nodes in common. Such an assembly would not maintain compatibility because of the difference in displacement interpolation on the two sides of the boundary; see Fig. 19.18.

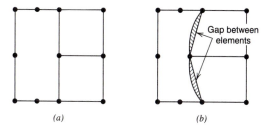

(a) *(b)*

Figure 19.18 Assembly of incompatible elements. (*a*) Undistorted assembly. (*b*) Loss of compatibility under distortion.

4. Use a fine mesh in regions of high stress gradient (stress concentration); use a coarse mesh where gradients are low.

5. When Choleski decomposition, or any other *band* solver, is used, minimize the bandwidth of the assembled structure stiffness matrix by proper node numbering. The nonzero entries in the structure stiffness matrix are clustered about the diagonal in a band. The *bandwidth* is the number of terms across a row (or down a column) of the band. The *half-bandwidth* is the number of terms from the diagonal out to the edge of the band. The nodal half-bandwidth is computed as $(n_{max} - n_{min} + 1)$, where n_{max} and n_{min} are the largest and smallest structure node numbers in the incidence list for an element. Hence, to minimize bandwidth, keep the range of node numbers that define the incidences for a single element as small as possible. Examples of poor and good node numbering schemes are illustrated in Fig. 19.19.

6. Exploit symmetry in the geometry and loads of the physical system to build the smallest reasonable model.

The finite element method and its use in engineering practice are evolving continuously. For instance, not long ago, material and/or geometric nonlinear analyses were rarely attempted. Today, such analyses are not limited to research but are performed by practicing engineers as well. The popularity of the finite element method is due primarily to the greater availability, and affordability, of user-friendly software that integrates sophisticated analysis capabilities with solid modeling and computer-aided design (CAD). Unfortunately, user training and experience are not always equal to the capabilities of the software. Hence, the danger exists that these powerful analytical tools will be used as *black boxes*, without proper understanding of the physical system or algorithms used in the analysis. There is no substitute for common sense and sound judgment, and one should remain skeptical of computer-generated results until they can be verified by some other means.

An effective means for an engineer to gain experience in performing finite element analysis and develop confidence in a finite element program is to solve a series of relatively simple *benchmark* problems. Such problems are specially designed to

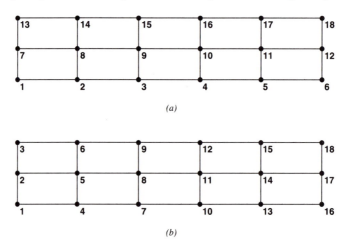

Figure 19.19   Node numbering to minimize bandwidth. (*a*) Poor numbering scheme, half-bandwidth = 8. (*b*) Good numbering scheme, half-bandwidth = 5.

test the accuracy of the individual elements in the program. However, they can also be used as a training device for novice users. A reasonable set of benchmark problems has been proposed by MacNeal and Harder (1984, 1985). Additional problems can be found in (AIAA, 1985).

PROBLEMS
Section 19.2

19.1. A transverse load P is applied to the end of a cantilever beam (Fig. P19.1). The beam has length L, moment of inertia I, and modulus of elasticity E. The displaced shape of the beam is assumed to be of the following forms:

Figure P19.1

   i. $v(x) = a_0 + a_1 x + a_2 x^2$

   ii. $v(x) = b\left(1 - \cos\dfrac{\pi x}{2l}\right)$

   iii. $v(x) = c_0 + c_1 x + c_2 x^2 + c_3 x^3$

Consider only strain energy due to bending as given by Eq. (19.65) and the potential of the load $[\Omega = -Pv(L)]$ with respect to the undeformed beam.

(a) To the extent possible, simplify each of the assumed displaced shapes to account for the boundary conditions.

(b) Calculate the elastic strain energy U and potential Ω of the external load P for each of the assumed displaced shapes.

(c) Solve for the parameters $(a_0, \ldots, c_3)$ using the principle of stationary potential energy, where for equilibrium $\delta\Pi = \delta U + \delta\Omega = 0$. *Hint:* The virtual displacement δv is first written in terms of a variation in the parameters $(\delta a_0, \ldots, \delta c_3)$. Then simultaneous equations are written from $\delta\Pi = (\partial\Pi/\partial a_0)\delta a_0 + \cdots = 0$.

(d) Compute values of Π and $v(L)$ for each of the assumed displaced shapes. Compare the values of $v(L)$ to each other and to the elasticity solution of $v(L) = (PL^3/3EI)$.

(e) Discuss the results.

19.2. For the constant strain triangle element shown in Fig. P19.2

(a) Write the shape function for each node.

(b) Evaluate each shape function at point P.

(c) Show, numerically for each shape function, that the value of the shape function for node i is equal to the ratio A_{Pjk}/A_{ijk}, where A_{Pjk} is the area of triangle Pjk and A_{ijk} the area of the element.

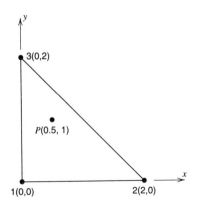

Figure P19.2

19.3. For the mesh shown in Fig. 19.5, construct the boolean connectivity matrix $[M]$ for elements 2, 3, and 4. Refer to Example 19.2.

19.4. For the mesh shown in Fig. 19.5, assemble the complete stiffness matrix for the structure. Use the notation k_{ij}^e to represent each stiffness coefficient, where the superscript identifies the element number. Refer to Example 19.2.

Section 19.4

19.5. A four-node isoparametric element has nodes at the following (x, y) coordinates: 1(0,0), 2(1,0), 3(2,2), 4(0,1).

    (a) Sketch to scale the element and the lines for which $\xi = \pm\frac{1}{2}$, $\xi = \pm\frac{1}{4}$, $\eta = \pm\frac{1}{2}$, and $\eta = \pm\frac{1}{4}$.

    (b) Write the coordinate interpolation functions

$$x(\xi, \eta) = \sum_{i=1}^{4} N_i(\xi, \eta)x_i \quad \text{and} \quad y(\xi, \eta) = \sum_{i=1}^{4} N_i(\xi, \eta)y_i.$$

    (c) Compute the terms in the Jacobian matrix $[J]$ given by Eq. (19.53).

    (d) Evaluate the determinate $|J|$ at $\xi = 0$, $\eta = 0$. Compare this value to the ratio of the area of the element in (x, y) coordinates to that in (ξ, η) coordinates.

19.6. For the linear isoparametric element shown in Fig. P19.6, compute $[B_1]$ at the point $\xi = 0$, $\eta = 0$.

19.7. Using the one-, two-, and three-point Gauss quadrature rules, numerically evaluate the following integrals. Compare the numerical results to the exact solutions.

    (a) $I = \displaystyle\int_{-1}^{1} (6x^3 - 4x^2 + 3x - 2)\, dx$

    (b) $I = \displaystyle\int_{-1}^{1} \cosh \xi \, d\xi$

    (c) $I = \displaystyle\int_{-1}^{1} e^{\xi} \, d\xi$

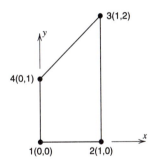

Figure P19.6

19.8. Using the one-, two-, and three-point symmetric Gauss quadrature rules, numerically evaluate the following integrals. Compare the numerical results to the exact solutions.

(a) $I = \int_{-1}^{1} \int_{-1}^{1} \cos \xi \cos \eta \, d\xi \, d\eta$

(b) $I = \int_{-1}^{1} \int_{-1}^{1} \sin^2 \xi \cos \eta \, d\xi \, d\eta$

Section 19.5

19.9. Derive the equivalent nodal load vector for an axial rod element subjected to a concentrated axial force $\bar{P}_C$ acting at L_c from node 1, see Fig. P19.9.

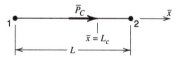

Figure P19.9

19.10. Derive the equivalent nodal load vector for an axial rod subjected to a uniformly distributed axial force of magnitude $\bar{q}_0$ acting over the domain $L_a < \bar{x} < L_b$, see Fig. P19.10

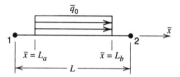

Figure P19.10

19.11. Derive the equivalent nodal load vector for a beam element subjected to a concentrated bending moment $\bar{M}_C$ acting at L_c from node 1; see Fig. P19.11.

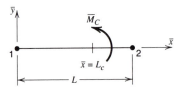

Figure P19.11

19.12.   Derive the stiffness matrix in local coordinates for a *beam on elastic foundation* element. Use the shape functions described in Eq. (19.63b) and a Winkler model for the foundation. With the addition of the elastic foundation, the virtual work of the internal forces δU in the element is the virtual work due to beam bending plus the virtual work due to foundation deformation. That is,

$$\delta U = \int_0^L (\delta \bar{v})'' E I \bar{v}'' \, d\bar{x} + \int_0^L \delta \bar{v} k \bar{v} \, d\bar{x}$$

where k is the foundation modulus [Eq. (10.3)].

REFERENCES

Works Cited in the Chapter

American Institute of Aeronautics and Astronautics (AIAA) (1985). *Proceedings Finite Element Standards Forum.* AIAA/ASME/ASCE/AHS 26th Structural Dynamics and Materials Conference, April 15, Orlando, Florida.

Boresi, A. P. and Chong, K. P. (1987). *Elasticity in Engineering Mechanics.* New York: Elsevier.

Clough, R. W. (1960). The Finite Element Method in Plane Stress Analysis. *Proceedings 2nd ASCE Conference on Electronic Computation.* Pittsburg, Pa. pp. 345–378.

Cook, R. D., Malkus, D. S., and Plesha, M. E. (1989). *Concepts and Applications of Finite Element Analysis,* 3rd ed. New York: Wiley.

Courant, R. (1950). *Differential and Integral Calculus.* New York: Wiley.

Courant, R. (1943). Variational Methods for the Solution of Problems of Equilibrium and Vibrations. *Bull. Amer. Math. Soc.,* **49**: 1–23.

Kellison, S. G. (1975). *Fundamentals of Numerical Analysis,* Homewood, Ill. Richard D. Irwin.

MacNeil, R. H. and Harder, R. L. (1984). A Proposed Standard Set of Problems to Test Finite Element Accuracy. *Proceedings AIAA/ASME/ASCE/AHS 25th Structural Dynamics and Materials Conference,* May 14, Palm Springs, Calif.

MacNeil, R. H. and Harder, R. L. (1985). A Proposed Standard Set of Problems to Test Finite Element Accuracy. *Finite Elements Anal. Des.* **1** (1). 3–20.

Turner, M. J., Clough, R. W., Martin, H. C., and Topp, L. J. (1956). Stiffness and Deflection Analysis of Complex Structures. *J. Aero. Sc.,* **25** (9): 805–823.

Zienkiewicz, O. C. and Taylor, R. L. (1989). *The Finite Element Method,* 4th ed. New York: McGraw-Hill.

Texts on the Finite Element Method

Baker, A. J. and Pepper, D. W. (1991). *Finite Elements 1-2-3*. New York: McGraw-Hill.

Bathe, K.-J. (1982). *Finite Element Procedures in Engineering Analysis*. Englewood Cliffs, N.J.: Prentice-Hall.

Bickford, W. B. (1990). *A First Course in the Finite Element Method*. Homewood, Ill.: Richard D. Irwin.

Burnett, D. S. (1987). *Finite Element Analysis, from Concepts to Applications*. Reading, Mass.: Addison-Wesley.

Cook, R. D., Malkus, D. S., and Plesha, M. E. (1989). *Concepts and Applications of Finite Element Analysis*, 3rd ed. New York: Wiley.

Crisfield, M. A. (1991). *Non-linear Finite Element Analysis of Solids and Structures*. New York: Wiley.

Ghali, A. and Neville, A. M. (1989). *Structural Analysis, A Unified Classical and Matrix Approach*, 3rd ed. London: Chapman and Hall.

Grandin, H., Jr. (1991). *Fundamentals of the Finite Element Method*. Prospect Heights, Ill.: Waveland Press.

Hughes, T. J. R. (1987). *The Finite Element Method, Linear Static and Dynamic Finite Element Analysis*. Englewood Cliffs, N.J.: Prentice-Hall.

Melosh, R. J. (1990). *Structural Engineering Analysis by Finite Elements*. Englewood Cliffs, N.J.: Prentice-Hall.

Potts, J. F. and Oler, J. W. (1989). *Finite Element Applications with Microcomputers*. Englewood Cliffs, N.J.: Prentice-Hall.

Przemieniecki, J.S. (1968). *Theory of Matrix Structural Analysis*. New York: McGraw-Hill.

Rao, S. S. (1989). *The Finite Element Method in Engineering*, 2nd ed. Oxford: Pergamon Press.

Reddy, J. N. (1984). *An Introduction to the Finite Element Method*. New York: McGraw-Hill.

Sack, R. L. (1989). *Matrix Structural Analysis*. Boston: PWS-Kent Publ. Co.

Stasa, F. L. (1985). *Applied Finite Element Analysis for Engineers*. New York: Holt, Rinehart and Winston.

Weaver, W., Jr. and Gere, J. M. (1990). *Matrix Analysis of Framed Structures*, 3rd ed. New York: Van Nostrand Reinhold.

Zienkiewicz, O. C. and Taylor, R. L. (1989). *The Finite Element Method*, 4th ed. New York: McGraw-Hill.

A

AVERAGE MECHANICAL PROPERTIES OF SELECTED MATERIALS

TABLE A.1
Properties in S.I. Units

Material	Density 10^3 kg/m^3	Yield Stress MPa	Ultimate Stress MPa	Poisson's Ratio	Young's Modulus GPa	Percent Elongation at Rupture	Thermal Coefficient (10^{-6})/°C
Steel							
Structural, ASTM A36	7.85	250	400	0.29	200	30	11.7
AISI-C 1030, normalized	7.85	340	530	0.29	200	30	11.7
AISI-C 1040, normalized	7.85	390	590	0.29	200	28	11.7
AISI-C 1080, normalized	7.85	520	1010	0.29	200	11	11.7
AISI-3140, normalized	7.85	620	900	0.29	200	19	11.7
AISI-4340, normalized	7.8	860	1310	0.29	200	12	11.7
AISI 301 Stainless, annealed	7.92	280	760	0.27	193	60	17.3
AISI 301 Stainless, half hard	7.92	760	1030	0.27	193	15	17.3
Cast Iron							
Gray, Class 30 (tension)	7.21		210	0.20	103	nil	12.1
Gray, Class 30 (compression)	7.21		510	0.20	103	nil	12.1
Gray, Class 40 (tension)	7.21		410	0.20	138	nil	12.1
Gray, Class 40 (compression)	7.21		990	0.20	138	nil	12.1
Aluminum Alloys							
1100-H12	2.71	103	110	0.33	70.0	25	23.5
2024 T4	2.77	320	470	0.33	74.5	19	22.5
7075 T6	2.77	500	570	0.33	72.0	11	22.5
Copper Alloys							
Free-Cutting Copper, soft	8.91	62	220	0.35	117	42	17.6
Free-Cutting Copper, hard	8.91	290	320	0.35	117	12	17.6
Yellow Brass, annealed	8.43	117	340	0.35	105	60	20.0
Yellow Brass, half hard	8.43	340	420	0.35	105	23	20.0
Commercial Bronze, annealed	8.84	90	270	0.35	110	45	18.0
Commercial Bronze, half hard	8.84	280	330	0.35	110	25	18.0
Titanium							
Alloy Ti-Al-V	4.54	890	930	0.33	114	12	9.5
Timber							
Douglas Fir	0.45		51		12.5		
Yellow Pine (compression)	0.54		57		13.8		
White Oak (compression)	0.59		47		11.0		
Concrete							
Medium Strength (comp.)	2.32		28	0.15	25.0		9.9

TABLE A.2
Properties in U.S. Customary Units

Material	Density lb/ft^3	Yield Stress 10^3 psi	Ultimate Stress 10^3 psi	Poisson's Ratio	Young's Modulus 10^6 psi	Percent Elongation at Rupture	Thermal Coefficient (10^{-6})/°F
Steel							
Structural, ASTM A36	490	36	58	0.29	29	30	6.5
AISI-C 1030, normalized	490	49	77	0.29	29	30	6.5
AISI-C 1040, normalized	490	56	85	0.29	29	28	6.5
AISI-C 1080, normalized	490	75	145	0.29	29	11	6.5
AISI-3140, normalized	490	90	130	0.29	29	19	6.5
AISI-4340, normalized	490	125	190	0.29	29	12	6.5
AISI 301 Stainless, annealed	495	40	110	0.27	28	60	9.61
AISI 301 Stainless, half hard	495	110	150	0.27	28	15	9.61
Cast Iron							
Gray, Class 30 (tension)	450		30	0.20	15	nil	6.72
Gray, Class 30 (compression)	450		74	0.20	15	nil	6.72
Gray, Class 40 (tension)	450		60	0.20	20	nil	6.72
Gray, Class 40 (compression)	450		144	0.20	20	nil	6.72
Aluminum Alloys							
1100-H12	170	15	16	0.33	10.2	25	13.06
2024 T4	173	46	68	0.33	10.8	19	12.5
7075 T6	173	73	83	0.33	10.4	11	12.5
Copper Alloys							
Free-Cutting Copper, soft	556	9	32	0.35	17	42	9.78
Free-Cutting Copper, hard	556	42	46	0.35	17	12	9.78
Yellow Brass, annealed	526	17	49	0.35	15.2	60	11.11
Yellow Brass, half hard	526	49	61	0.35	15.2	23	11.11
Commercial Bronze, annealed	552	13	39	0.35	16	45	10.0
Commercial Bronze, half hard	552	41	48	0.35	16	25	10.0
Titanium							
Alloy Ti-Al-V	283	130	135	0.33	16.5	12	5.28
Timber							
Douglas Fir	28		7.43		1.8		
Yellow Pine (compression)	344		8.26		2.0		
White Oak (compression)	37		6.80		1.6		
Concrete							
Medium Strength (comp.)	145		4	0.15	3.6		5.5

B

SECOND MOMENT (MOMENT OF INERTIA) OF A PLANE AREA

B.1

MOMENTS OF INERTIA OF A PLANE AREA

The derivation of load-stress formulas for torsion members and beams may require solutions of one or more of the following integrals:

$$I_x = \int y^2 \, dA \tag{B.1}$$

$$I_y = \int x^2 \, dA \tag{B.2}$$

$$J = \int r^2 \, dA \tag{B.3}$$

$$I_{xy} = \int xy \, dA \tag{B.4}$$

where dA is an element of the plane area A lying in the (x, y) plane in Fig. B.1. Area A represents the cross-sectional area of a member subjected to bending and/or torsional loads.

The integrals in Eqs. (B.1), (B.2), and (B.3) are commonly called moments of inertia of the area A because of the similarity with integrals that define the

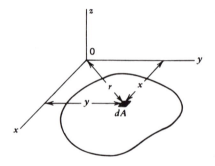

Figure B.1

moment of inertia of bodies in the field of dynamics. Since an area cannot have an inertia, moment of inertia of an area is a misnomer. We use the term because of common usage.

The integral represented by Eq. (B.4) is called the product of inertia. Its sign can be negative. The moment of inertia and product of inertia are given the symbol I if the axes about which the moments are taken lie in the plane of the area [see Eqs. (B.1), (B.2), and (B.4)]. When the axis about which the moment is taken is perpendicular to the area [see Eq. (B.3)], the moment of inertia is given the symbol J and is called the polar moment of inertia of the area.

B.2

PARALLEL AXIS THEOREM

In the application of Eqs. (B.1), (B.2), (B.3), and (B.4) to engineering problems, it is convenient to know these integrals for coordinate axes at the centroid of area A. The values of the integrals for a few cross sections are listed in Table B.1. Often, practical members have cross sections that are composed of two or more simple cross sections (Table B.1). Moments of inertia for composite areas are obtained by application of the parallel axis theorem.

TABLE B.1
Moments of Inertia of Common Plane Areas

Rectangle		$I_x = bh^3/12$ $I_y = hb^3/12$ $J_0 = (bh^3 + hb^3)/12$ $I_{xy} = 0$
Right Triangle		$I_x = bh^3/36$ $I_y = hb^3/36$ $J_0 = (bh^3 + hb^3)/36$ $I_{xy} = -b^2h^2/72$
Circle		$I_x = \pi D^4/64 = \pi R^4/4$ $I_y = \pi D^4/64 = \pi R^4/4$ $J_0 = \pi D^4/32 = \pi R^4/2$ $I_{xy} = 0$
Ellipse		$I_x = \pi bh^3/4$ $I_y = \pi hb^3/4$ $J_0 = \pi bh(h^2 + b^2)/4$ $I_{xy} = 0$

(continues)

TABLE B.1 (*Continued*)

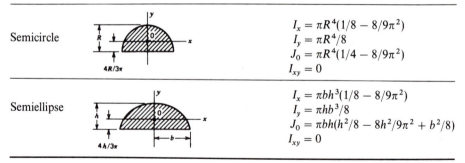

Semicircle		$I_x = \pi R^4(1/8 - 8/9\pi^2)$ $I_y = \pi R^4/8$ $J_0 = \pi R^4(1/4 - 8/9\pi^2)$ $I_{xy} = 0$
Semiellipse		$I_x = \pi bh^3(1/8 - 8/9\pi^2)$ $I_y = \pi hb^3/8$ $J_0 = \pi bh(h^2/8 - 8h^2/9\pi^2 + b^2/8)$ $I_{xy} = 0$

Let it be required to obtain moments of inertia for area A in Fig. B.2 for coordinate axes (x', y', z'). Area A lies in the (x', y') plane. First, locate coordinate axes (x, y, z) with axes parallel, respectively, to the (x', y', z') axes and with the origin 0 at the centroid of A. Let the distances of the centroid 0 from the axes (x', y') be $(\bar{x}, \bar{y})$. Then, $\bar{r} = \sqrt{\bar{x}^2 + \bar{y}^2}$ is the distance between the z' axis and z axis. Using Eqs. (B.1), (B.2), (B.3), and (B.4), we obtain

$$I_{x'} = \int (y + \bar{y})^2 \, dA \qquad = I_x + A\bar{y}^2 \tag{B.5}$$

$$I_{y'} = \int (x + \bar{x})^2 \, dA \qquad = I_y + A\bar{x}^2 \tag{B.6}$$

$$J_{0'} = \int [(x + \bar{x})^2 + (y + \bar{y})^2] \, dA = J_0 + A\bar{r}^2 \tag{B.7}$$

$$I_{x'y'} = \int (x + \bar{x})(y + \bar{y}) \, dA \qquad = I_{xy} + A\bar{x}\bar{y} \tag{B.8}$$

where integrals $\int y \, dA$ and $\int x \, dA$ are zero since the first moment of an area with respect to an axis through the centroid of the area vanishes. Equations (B.5) through (B.8) represent parallel axes formulas for moments of inertia of an area. They may be employed to obtain the moments of inertia of composite areas.

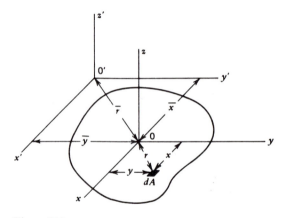

Figure B.2

EXAMPLE B.1
Moments of Inertia for Z-bar

A Z-bar has the cross section shown in Fig. EB.1. Determine I_x, I_y, and I_{xy} for the centroidal axes (x, y) shown.

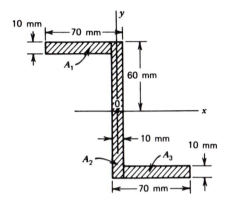

Figure EB.1

SOLUTION

The area is divided into three rectangular areas A_1, A_2, and A_3 (Fig. EB.1). Using Eqs. (B.5), (B.6), and (B.8), and Table B.1, we obtain

$$I_x = \frac{60(10)^3}{12} + 60(10)(55)^2 + \frac{10(120)^3}{12} + 120(10)(0)^2$$

$$+ \frac{60(10)^3}{12} + 60(10)(-55)^2 = 5.08 \times 10^6 \text{ mm}^4$$

$$I_y = \frac{10(60)^3}{12} + 60(10)(-35)^2 + \frac{120(10)^3}{12} + 120(10)(0)^2$$

$$+ \frac{10(60)^3}{12} + 60(10)(35)^2 = 1.84 \times 10^6 \text{ mm}^4$$

$$I_{xy} = 60(10)(-35)(55) + 120(10)(0)(0) + 60(10)(35)(-55)$$
$$= -2.31 \times 10^6 \text{ mm}^2$$

B.3

TRANSFORMATION EQUATIONS FOR MOMENTS AND PRODUCTS OF INERTIA

Let I_x, I_y, and I_{xy} be known moments and product of inertia for area A (Fig. B.3) for (x, y) rectangular axes that lie in the plane of the area. Consider the (X, Y) coordinate axes that have the same origin and same plane as the (x, y) axes. We wish to

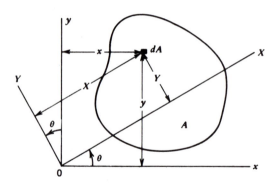

Figure B.3

derive transformation equations by which I_X, I_Y, and I_{XY} are obtained in terms of I_x, I_y, I_{xy}, and θ, the angle through which the (x, y) axes must be rotated to coincide with the (X, Y) axes; θ is positive in the counterclockwise sense. Consider an element of area dA at X and Y coordinates given by the relations

$$X = x \cos \theta + y \sin \theta$$
$$Y = y \cos \theta - x \sin \theta \tag{B.9}$$

Substitution of Eqs. (B.9) into Eqs. (B.1), (B.2), and (B.4) gives

$$I_X = \int (y \cos \theta - x \sin \theta)^2 \, dA = I_x \cos^2 \theta + I_y \sin^2 \theta - 2I_{xy} \sin \theta \cos \theta$$

$$I_Y = \int (x \cos \theta + y \sin \theta)^2 \, dA = I_x \sin^2 \theta + I_y \cos^2 \theta + 2I_{xy} \sin \theta \cos \theta$$

$$I_{XY} = \int (x \cos \theta + y \sin \theta)(y \cos \theta - x \sin \theta) \, dA$$

$$= (I_x - I_y) \sin \theta \cos \theta + I_{xy}(\cos^2 \theta - \sin^2 \theta) \tag{B.10}$$

With double angle identities, Eq. (B.10) can be written in the form

$$I_X = \frac{I_x + I_y}{2} + \frac{I_x - I_y}{2} \cos 2\theta - I_{xy} \sin 2\theta$$

$$I_Y = \frac{I_x + I_y}{2} - \frac{I_x - I_y}{2} \cos 2\theta + I_{xy} \sin 2\theta$$

$$I_{XY} = \frac{I_x - I_y}{2} \sin 2\theta + I_{xy} \cos 2\theta \tag{B.11}$$

Note the similarity between the transformation equations for moments and products of inertia and the transformation equations of stress given by Eq. (2.31). Like stress components and strain components, moments and products of inertia transform according to the rule for second-order symmetric tensors.

Principal Axes of Inertia

There are two values of θ for which $I_{XY} = 0$. To determine these values, let $I_{XY} = 0$. Then, the third of Eqs. (B.11) yields

$$\tan 2\theta = -\frac{2I_{xy}}{I_x - I_y} \tag{B.12}$$

The two values of θ given by Eq. (B.12) locate two positions of axes (X, Y) that represent the principal axes of inertia for a given cross-sectional area. In the discussion that follows, we assume for definitiveness that I_x is greater than I_y. Then, the maximum moment of inertia, which we take to be I_X and is associated with the X axis, will occur for the X axis located at the smallest of the two values of θ from the x axis; the direction is counterclockwise for a positive value of θ and clockwise for a negative value of θ. Since $I_x - I_y > 0$, if we substitute the value of θ given by Eq. (B.12) into the first and second of Eqs. (B.11), we find that

$$I_X = \frac{I_x + I_y}{2} + \sqrt{\left(\frac{I_x - I_y}{2}\right)^2 + I_{xy}^2}$$

$$I_Y = \frac{I_x + I_y}{2} - \sqrt{\left(\frac{I_x - I_y}{2}\right)^2 + I_{xy}^2} \tag{B.13}$$

as the principal moments of inertia for the cross-sectional area A.

EXAMPLE B.2
Principal Axes for Z-Bar

Locate the principal axes and determine the principal moments of inertia I_X and I_Y for the Z-bar whose dimensions are specified in Fig. EB.1.

SOLUTION

Since $I_x = 5.08 \times 10^6$ mm^4, $I_y = 1.84 \times 10^6$ mm^4, and $I_{xy} = -2.31 \times 10^6$ mm^4, the principal values for the moments of inertia are given by Eqs. (B.13).

$$I_X = \frac{5.08 \times 10^6 + 1.84 \times 10^6}{2}$$

$$+ \sqrt{\left(\frac{5.08 \times 10^6 - 1.84 \times 10^6}{2}\right)^2 + (-2.31 \times 10^6)^2}$$

$$= 6.281 \times 10^6 \text{ mm}^4$$

$$I_Y = \frac{5.08 \times 10^6 + 1.84 \times 10^6}{2}$$

$$- \sqrt{\left(\frac{5.08 \times 10^6 - 1.84 \times 10^6}{2}\right) + (-2.31 \times 10^6)^2}$$

$$= 0.639 \times 10^6 \text{ mm}^4$$

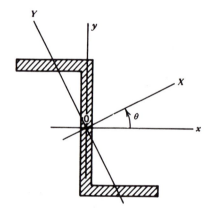

Figure EB.2

The location of the X axis is given by Eq. (B.12). Thus,

$$\tan 2\theta = \frac{-2(-2.31 \times 10^6)}{5.08 \times 10^6 - 1.84 \times 10^6} = 1.4259$$

$$\theta = 0.4796 \text{ rad}$$

Hence, the X axis is located at 0.4796 rad, measured counterclockwise from the x axis, as shown in Fig. EB.2.

PROBLEMS
Section B.2

B.1 Derive the expressions for I_x and I_{xy} for the right triangle in Table B.1.

B.2 Derive the expression for I_x for the semiellipse in Table B.1.

B.3 Determine I_x, I_y, and I_{xy} for the centroidal axes for the cross-sectional area shown in Fig. PB.3.

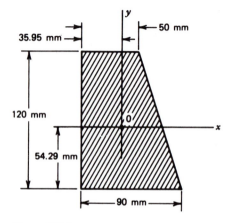

Figure PB.3

Ans. $I_x = 9.806 \times 10^6$ mm^4, $I_y = 3.982 \times 10^6$ mm^4

$I_{xy} = -1.634 \times 10^6$ mm^4

B.4   Determine I_x, I_y, and I_{xy} for the centroidal axes for the cross-sectional area shown in Fig. PB.4.

Ans. $I_x = 333.3 \times 10^3$ mm^4, $I_y = 208.3 \times 10^3$ mm^4, $I_{xy} = 150.0 \times 10^3$ mm^4

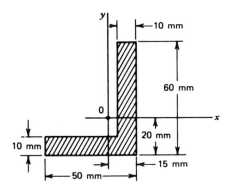

Figure PB.4

Section B.3

B.5   Locate principal axes (X, Y) and determine I_X and I_Y for the cross-sectional area in Problem B.3.

Ans. $I_X = 10.233 \times 10^6$ mm^4, $I_Y = 3.555 \times 10^6$ mm^4, $\theta = 0.2557$ rad

B.6   Locate the principal axes (X, Y) and determine I_X and I_Y for the cross-sectional area in Fig. PB.4.

Ans. $I_X = 433.3 \times 10^3$ mm^4, $I_Y = 108.3 \times 10^3$ mm^4, $\theta = -0.5880$ rad

AUTHOR INDEX

SUBJECT INDEX

Conversion Factors—US Customary Units to SI Units

To Convert From	To	Multiply By
inches (in)	millimeters (mm)	25.400
inches (in)	meters (m)	0.025400
feet (ft)	meters (m)	0.304800
miles (m)	kilometers (km)	1.609344
feet^2 (ft^2)	meters^2 (m^2)	0.09290304
feet^3 (ft^3)	meters^3 (m^3)	0.028316847
feet^3 (ft^3)	liters (l)	28.31605
gallons (gal)	liters (l)	3.785306
pounds (lb)	newtons (N)	4.4482217
pounds/foot^2 (psf)	pascals (Pa = N/m^2)	47.880
pounds/inch^2 (psi)	kilopascals (kPa = kN/m^2)	6.894757
kilopounds/inch^2 (ksi)	megapascals (MPa = MN/m^2 = N/mm^2)	6.894757
kilopounds/inch^2 (ksi)	gigapascals (GPa = GN/m^2 = kN/mm^2)	0.006894757
slugs (lb$\cdot$s^2/ft)	kilograms (kg)	14.59390
pounds/foot^3 (pcf) (unit weight)	kilograms/meter^3 (kg/m^3) (density)	16.018463
inch$\cdot$pounds (in$\cdot$lb)	newton$\cdot$meters (N$\cdot$m)	0.11298483
foot$\cdot$pounds (ft$\cdot$lb)	newton$\cdot$meters (N$\cdot$m)	1.3558179
degrees (angle)	radians (rad)	0.017453293
degrees, Fahrenheit (°F)	degrees, Celsius (°C)	$T_{°C} = (T_{°F} - 32°)/1.8$